NUMERICAL METHODS
FOR ENGINEERS

NUMERICAL METHODS
FOR ENGINEERS
SECOND EDITION

Steven C. Chapra, Ph.D.
Professor of Civil Engineering
The University of Colorado

Raymond P. Canale, Ph.D.
Professor of Civil Engineering
The University of Michigan

McGraw-Hill Publishing Company

New York   St. Louis   San Francisco   Auckland   Bogotá
Caracas   Hamburg   Lisbon   London   Madrid   Mexico   Milan
Montreal   New Delhi   Oklahoma City   Paris   San Juan
São Paulo   Singapore   Sydney   Tokyo   Toronto

To
Margaret and Gabriel Chapra
Helen and Chester Canale

NUMERICAL METHODS FOR ENGINEERS

5 6 7 8 9 10   HDHD   9 9 8 7 6 5 4 3 2 1 0

ISBN 0-07-079984-9

This book was set in Times Roman by Publication Services.
The editors were Anne T. Brown and Steven Tenney;
the production supervisor was Friederich W. Schulte;
the designer was Charles Carson.
Cover photograph by Paul Perry.
Drawings were done by J&R Services, Inc.
Arcata Graphics/Halliday was printer and binder.

Library of Congress Cataloging-in-Publication Data

Chapra, Steven C.
   Numerical methods for engineers.

   Bibliography: p.
   Includes index.
      1. Engineering mathematics—Data processing.
2. Numerical calculations—Data processing. 3. Micro-
computers—Programming. I. Canale, Raymond P.
II. Title.
TA345.C47   1988   511'.02462   88-523
ISBN 0-07-079984-9
ISBN 0-07-010674-5 (solutions manual)
ISBN 0-07-834447-6 (disc)

ABOUT THE AUTHORS

Steven C. Chapra teaches in the Civil, Environmental, and Architectural Engineering Department at the University of Colorado. In addition, he is a visiting lecturer at Duke University's School of Forestry and Environmental Studies. His other books include two graduate texts on mathematical modeling and the undergraduate text *Introduction to Computing for Engineers*.

Dr. Chapra received engineering degrees from Manhattan College and the University of Michigan. Before joining the faculty of the University of Colorado, he worked as an engineer and computer programmer for the Environmental Protection Agency and the National Oceanic and Atmospheric Administration and was an Associate Professor at Texas A&M University. Dr. Chapra has published, in addition to his books, several computer programs and numerous journal articles. All deal with mathematical modeling and the application of computers to engineering problem solving.

Dr. Chapra is active in a number of professional societies and serves on several governmental and professional committees related to applied modeling and computer applications in engineering education. He is the recipient of a number of awards including the Tenneco Award for Teaching Excellence and the Meriam/Wiley Distinguished Author Award.

Raymond P. Canale is a professor of civil and environmental engineering at the University of Michigan, where he teaches courses on computers and numerical methods. He also teaches and performs extensive research in the area of mathematical and computer modeling of environmental systems. He has authored books on mathematical modeling of aquatic ecosystems and is a coauthor of *Introduction to Computing for Engineers*. He has published over 100 scientific papers and reports, and designed and developed personal computer software for engineers. He has been given the Meriam/Wiley Distinguished Author Award by the American Society for Engineering Education for his books and software.

Professor Canale is also a practicing professional engineer. He is an active consultant for engineering firms as well as for industry and governmental agencies. He has served as an expert technical witness on numerous occasions.

Overall, Professor Canale has devoted over twenty years to his profession as a teacher, researcher, author, and practicing engineer.

CONTENTS

PART SIX: ORDINARY DIFFERENTIAL EQUATIONS   565

Chapter 19: One-Step Methods   576

Chapter 20: Adaptive Step Size Control   611

Chapter 21: Boundary-Value and Eigenvalue Problems   641

Chapter 22: Case Studies: Ordinary Differential Equations   671

EPILOGUE: PART SIX   **702**

PREFACE

A lot of water has passed under the bridge in the three years since the publication of the first edition of *Numerical Methods for Engineers*. Our contention that numerical methods and computers would figure more prominently in the engineering curriculum—particularly in the early parts—has been dramatically born out. Many universities now offer freshman, sophomore, and junior courses in both introductory computing and numerical methods. In addition, many of our colleagues are integrating computer-oriented problems into other courses at all levels of the curriculum. Thus, this new edition is still founded on the basic premise that student engineers should be provided with a strong and early introduction to numerical methods. Consequently, although we have expanded our coverage in the new edition, we have tried to maintain many of the features that made the first edition accessible to both lower- and upper-level undergraduates. These include:

1. *Boxed material.* We have endeavored to include important derivations and error analyses in order to enrich the presentation. However, such material sometimes represents a stumbling block for the beginning student. Consequently, we have sequestered the more complicated mathematical material in boxes. Many students will find that they can apply the numerical methods without completely mastering the boxed material.

2. *Introductory material and mathematical background.* Every part of the book includes an introductory section. After a brief statement of the general mathematical problem under study, motivation is provided by describing how the problem would be approached in the absence of computers and how the method would be used in engineering practice. This material is followed by a review of the mathematics required to successfully master the subject at hand. For example, matrix algebra is reviewed prior to introducing linear algebraic equations, and statistics is reviewed prior to regression. Finally, an outline and study objectives are listed to provide some orientation to subsequent materials.

3. *Epilogues.* Just as the introduction is designed to provide motivation and orientation, we include an epilogue at the end of each part of the book to consolidate the newly acquired concepts. An important feature of this epilogue is a section devoted to the

trade-offs involved in choosing the appropriate numerical methods for a particular problem. In addition, important formulas and references for advanced methods are summarized.

4. *Sequential presentations.* Each major part of the book consists of several chapters—the initial ones devoted to theory and the last to case studies. Wherever possible, the theory chapters are structured sequentially; that is, the more elementary and straightforward approaches are presented first. Because many of the more advanced methods build on the simpler ones, this development is intended to provide a sense of the evolution of the techniques. In addition, by first pushing the simpler approaches to their limits, incentive is provided to pursue the higher-order or more complicated methods.

5. *Case studies.* Case studies have been included in each part of the book to demonstrate the practical utility of the numerical methods. A significant effort was made to incorporate examples from the early courses in a typical engineering curriculum. When this was not possible, the theoretical basis and motivation for the problems have been provided.

Although the above features allow the book to be employed at the lower end of the undergraduate engineering curriculum, we have expanded our coverage by adding several advanced topics (Fig. P.1). There are three primary reasons for this expansion. First, although more universities are teaching the subject at lower levels, many institutions still offer numerical methods as an upper undergraduate or graduate course. We wanted to ensure that our book would be flexible enough to be useful at these levels. Second, we wanted to make the book a more comprehensive reference for both students and professionals. Finally, we have found that even when the book is used at the lower levels, introduction to more advanced topics such as eigenvalues and partial differential equations can be presented effectively. This has been facilitated by a number of trends including the growing computer sophistication of our students and the increased credit hours allocated to numerical methods courses at many institutions over the past few years.

The expanded coverage is depicted schematically in Fig. P.1. Notice that most of the new material is added at the end of each part prior to the case studies. In addition, a new part on partial differential equations has been included at the end of the book. The material was incorporated in this way to conform with our sequential approach of presenting numerical methods. Thus, because most of the new topics are of a more advanced nature, they could be very naturally appended to the already existing material. This configuration has the added benefit that those who do not want to explicitly include the new chapters in existing courses can more easily pass over them.

Along with the new material, other modifications have been added to enhance the new edition. Although *Part One* is still intended as an introductory section on modeling, computers, and error analysis, the content has been expanded. *Chapter 1* includes additional material on the major conservation laws employed in engineering. This material was included to broaden the student's conception of modeling and to reinforce the connection with the major case study chapters in the remainder of the book. *Chapter 2* has been completely rewritten to reflect changes in the state-of-computing with particular emphasis on structured programming techniques. *Chapter 3* on errors is essentially intact. However, we have moved some of the formulas on numerical

PART ONE Modeling, Computers and Error Analysis	PART TWO Roots of Equations	PART THREE Systems of Linear Algebraic Equations	PART FOUR Curve Fitting	PART FIVE Numerical Differentiation and Integration	PART SIX Ordinary Differential Equations	PART SEVEN Partial Differential Equations
INTRODUCTION	INTRODUCTION	INTRODUCTION	INTRODUCTION	INTRODUCTION	INTRODUCTION	INTRODUCTION
CHAPTER 1 Modeling and Problem Solving	CHAPTER 4 Bracketing Methods	CHAPTER 7 Gauss Elimination	CHAPTER 11 Regression	CHAPTER 15 Newton-Cotes Integration	CHAPTER 19 One-step Methods	CHAPTER 23 Finite Difference: Elliptic
CHAPTER 2 Computer Programming and Software	CHAPTER 5 Open Methods	CHAPTER 8 Matrix Inversion and Gauss-Seidel	CHAPTER 12 Interpolation	CHAPTER 16 Integration of Equations	CHAPTER 20 Adaptive Step Size Control	CHAPTER 24 Finite Difference: Parabolic
CHAPTER 3 Approximation and Errors	CHAPTER 6 Case Studies	CHAPTER 9 LU Decomposition	CHAPTER 13 Fourier Approximation	CHAPTER 17 Numerical Differentiation	CHAPTER 21 Boundary-value and Eigenvalue Problems	CHAPTER 25 Finite-Element Method
EPILOGUE	EPILOGUE	CHAPTER 10 Case Studies	CHAPTER 14 Case Studies	CHAPTER 18 Case Studies	CHAPTER 22 Case Studies	CHAPTER 26 Case Studies
		EPILOGUE	EPILOGUE	EPILOGUE	EPILOGUE	EPILOGUE

FIGURE P.1
An outline of the present edition. The shaded areas represent new material. In addition, many of the original chapters have been supplemented with new topics, and homework problems and case studies have been revised and new ones added.

differentiation to a new Chapter 17 and have included some new material on computer round-off errors and error propagation.

Part Two on roots of equations includes one major addition. A section on *systems of nonlinear equations* has been appended to *Chapter 5*.

The major addition to *Part Three* is *Chapter 9* on *LU decomposition methods* for solving systems of linear algebraic equations. This chapter begins by illustrating how Gauss elimination can be formulated as an LU decomposition solution. In addition, alternative approaches such as *Crout* and *Cholesky decomposition* are introduced. Aside from these significant additions, former material on banded systems has been expanded into a major section in *Chapter 9*.

Part Four has been augmented in two important ways. First, *Chapter 11* has been supplemented with a section on *nonlinear regression*. Second, material on *Fourier approximation* has been included in a new *Chapter 13*.

As mentioned previously, most of the material on numerical differentiation has been removed from Chapter 3 and expanded into *Chapter 17*. Thus, *Part Five* now treats this topic along with numerical integration. This allows a more satisfying and unified discussion of the effect of data error on these two fundamental mathematical operations. In addition, we have included a new section on improper integrals in *Chapter 16*.

The major addition to *Part Six* is that *boundary-value problems* are treated explicitly in *Chapter 21*. The material also includes an introduction to techniques for determining *eigenvalues*.

Finally, *Part Seven* consists of entirely new material on *partial differential equations*. *Chapter 23 and 24* deal with *finite-difference solutions* for elliptic and parabolic equations. *Chapter 25* provides an introduction to the *finite-element method*.

Aside from the expanded scope, the second edition has been upgraded in a number of ways. For example, many of the homework problems and case studies have been revised or replaced. In addition, our treatment of computer-related material has been significantly modified and improved. This has been necessitated in part by the pronounced changes that have occurred in the computing environments within which both students and faculty work.

Although microcomputers such as the Apple II were beginning to gain acceptance when the first edition was written five years ago, centralized mainframe computing installations represented the major computing environment. Today, although mainframes are still significant, a wide variety of more personalized alternatives are available. These range from local networks of minicomputers or workstations to microcomputer labs. In addition, more of our students are acquiring their own personal computers.

For this reason, and also because we anticipate that students will be using different types of machines to learn numerical methods, we have designed this book so that it can be used with any of these systems. Thus, most of the material can be implemented on machines ranging from personal computers to mainframe systems. At the same time, we believe that the "personal" aspect of the microcomputer revolution lies at the root of the present excitement over computing among students and professionals alike. Therefore, we have tried wherever possible to acknowledge and explore some of the exciting new developments that are directly associated with personal computers. For example, case studies on spreadsheet applications have been added to each part of the book.

Our treatment of computer algorithms and programs has also been upgraded to reflect new developments. Wherever possible, we have expressed algorithms in pseudocode. Additionally, all computer programs have been rewritten in structured format employing Microsoft BASIC, FORTRAN 77, and Turbo Pascal.

The first edition included computer examples and problems using our NUMERI-COMP software package. These features have been retained in the second edition. NUMERICOMP has been replaced by an enhanced software package—the Electronic TOOLKIT.* In addition, we have developed a completely new computer software supplement for the second edition. The supplement, which is available at no cost to students, consists of several example spreadsheets. These are related to the case studies that we have added on numerical applications of spreadsheets. The supplement is designed to allow students to interactively explore the advantages and disadvantages of different numerical methods. Among other things, students can investigate the effects of changing step size, error criteria, and system parameters on numerical efficiency.

Finally, as with the previous edition, we have exerted a conscious effort to make this book as user-friendly as possible. Thus, we have endeavored to keep our explana-

---

*The Electronic TOOLKIT includes the numerical methods programs of NUMERICOMP along with a spreadsheet, statistics package, and graphics generator. The TOOLKIT is included on a diskette in the back of the book.

tions straightforward and oriented practically. Although our primary intent is to provide students with a sound introduction to numerical methods, we have the ancillary objective of making this introduction a pleasurable experience. We believe that students who enjoy numerical methods, computers, and mathematics will, in the end, make better engineers. If our book fosters an enthusiasm for these subjects, we will consider our efforts a success.

ACKNOWLEDGMENTS

We would like to acknowledge reviews by Professors G. Allen, Cleveland State University; Swaminat Balachandran, University of Wisconsin—Platteville; Reinier J.B. Bouwmeester, Michigan State University; Gregory L. Griffin, University of Minnesota; James Hiestand, University of Tennessee; Leendert Kersten, University of Nebraska; Mark Nagurka, Carnegie-Mellon University; James W. Phillips, University of Illinois; Senol Utku, Duke University; R. O. Warrington, Louisiana Technological University; and Theodore L. Weidner, Rensselaer Polytechnic Institute.

   Our gratitude is extended to our friends and colleagues at the University of Colorado, the University of Michigan and Texas A & M University. In particular, Hon Yim Ko, Dave Clough and Ben Wylie, were very supportive of this endeavor. We gained insights from our colleagues, Bill Anderson, Bill Batchelor, Brice Carnahan, Chuck Giammona, Tissa Illangasekare, Ken Strzepek, Walt Weber, and Jim Wilkes. Also, Kathy Van Veen and Florence Petersen provided significant support to our efforts. Our students, particularly Jean Boyer, Jim Waterman, Jim Brannon, and John Lilley, provided suggestions and helped us test our material.

   Thanks are also due to our friends at McGraw-Hill. In particular, Anne Brown, Don Burden, Chuck Carson, B.J. Clark, Jim Dodd, Eric Munson, Fred Schulte, Frank Snyder and Steve Tenney offered us a positive and professional atmosphere for creating this edition.

   Finally, we would like to thank our family, friends, and students for their enduring patience and support. In particular, Kathy Callahan, Christian Chapra, and Carol Humke were always there providing understanding, perspective, and love.

Steven C. Chapra
Raymond P. Canale

PART ONE

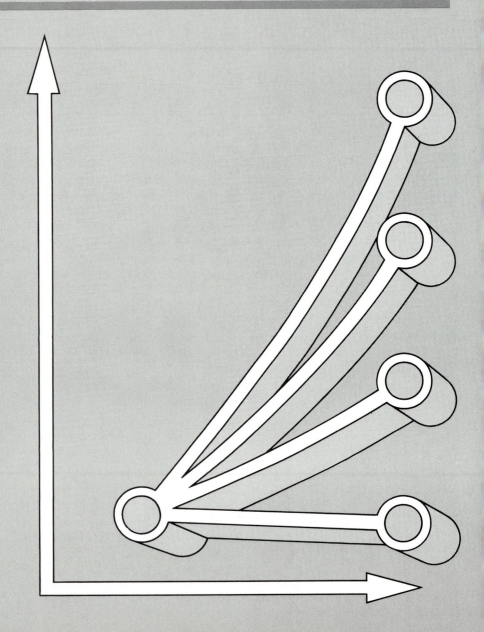

MODELING, COMPUTERS, AND ERROR ANALYSIS

Numerical methods are techniques by which mathematical problems are formulated so that they can be solved with arithmetic operations. Although there are many kinds of numerical methods, they have one common characteristic. Numerical methods invariably involve large numbers of tedious arithmetic calculations. It is little wonder that with the development of fast, efficient digital computers, the role of numerical methods in engineering problem solving has increased dramatically in recent years.

PT1.1.1 Precomputer Methods

Beyond providing increased computational firepower, the widespread availability of computers (especially personal computers) and their partnership with numerical methods has had a significant influence on the actual engineering problem-solving process. In the precomputer era there were generally three different ways engineers approached problem solving:

1. First, solutions were derived for some problems using analytical or exact methods. These solutions were often useful and provided excellent insight into the behavior of some systems. However, analytical solutions can be derived for only a limited class of problems. These problems include those that can be approximated with linear models and those that have simple geometry and low dimensionality. Consequently, analytical solutions are of limited practical value because most real problems are nonlinear and involve complex shapes and processes.

2. Graphical solutions were used to characterize the behavior of systems. These graphical solutions usually took the form of plots or nomographs. Although graphical techniques can often be used to solve complex problems, the results are not very precise. Furthermore, graphical solutions (without the aid of computers) are extremely tedious and awkward to implement. Finally, graphical techniques are often limited to problems that can be described using three or fewer dimensions.

3. Calculators and slide rules were used to implement numerical methods manually. Although in theory such approaches should be perfectly adequate for solving complex problems, in actuality, several difficulties are encountered. Manual calculations are slow and tedious. Furthermore, consistent results are elusive because of simple blunders that arise when numerous manual tasks are performed.

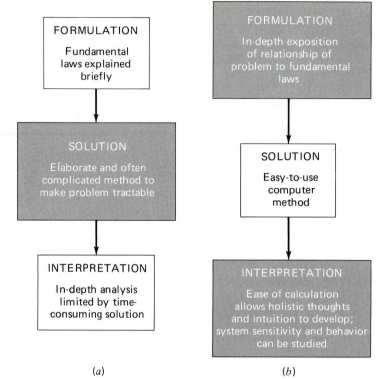

Figure PT1.1
The three phases of engineering problem solving in (a) the precomputer and (b) the computer era. The sizes of the boxes indicate the level of emphasis directed toward each phase. Computers facilitate the implementation of solution techniques and thus allow more emphasis to be placed on the creative aspects of problem formulation and interpretation of results.

During the precomputer era, significant amounts of energy were expended on the solution technique itself, rather than on problem definition and interpretation (Fig. PT1.1a). This unfortunate situation existed because so much time and drudgery were required to obtain numerical answers using precomputer techniques.

Today, computers and numerical methods provide an alternative for such complicated calculations. Using computer power to obtain solutions directly, you can approach these calculations without recourse to simplifying assumptions or time-intensive techniques. Although analytical solutions are still extremely valuable both for problem solving and for providing insight, numerical methods represent alternatives that greatly enlarge your capabilities to confront and solve problems. As a result, more time is available for the use of your creative skills. Thus, more emphasis can be placed on problem formulation and solution interpretation and the incorporation of total system, or "holistic," awareness (Fig. PT1.1b).

PT1.1.2 Numerical Methods and Engineering Practice

Since the late 1940s the widespread availability of digital computers has led to a veritable explosion in the use and development of numerical methods. At first, this growth was somewhat limited by the cost of access to large mainframe computers, and many engineers consequently continued to use simple analytical approaches in a significant portion of their work. Needless to say, the recent evolution of inexpensive personal computers has given most of us ready access to powerful computational capabilities.

There are a number of additional reasons why you should study numerical methods:

1. Numerical methods are extremely powerful problem-solving tools. They are capable of handling large systems of equations, non-linearities, and complicated geometries that are not uncommon in engineering practice and that are often impossible to solve analytically. As such, they greatly enhance your problem-solving skills.

2. During your careers, you may often have occasion to use commercially available prepackaged, or "canned," computer programs that involve numerical methods. The intelligent use of these programs often is predicated on knowledge of the basic theory underlying the methods.

3. Many problems cannot be approached using canned programs. If you are conversant with numerical methods and are adept at computer programming you will have the capability of designing your own programs to solve problems without having to buy or commission expensive software.

4. Numerical methods are an efficient vehicle for learning to use computers. It is well known that an effective way to learn programming is to actually write computer programs. Because numerical methods are for the most part designed for implementation on computers, they are ideal for this purpose. Further, they are especially well-suited to illustrate the power and the limitations of computers. When you successfully implement numerical methods on a computer and then apply them to solve otherwise intractable problems, you will be provided with a dramatic demonstration of how computers can serve your professional development. At the same time, you will also learn to acknowledge and control the errors of approximation that are part and parcel of large-scale numerical calculations.

5. Numerical methods provide a vehicle for you to reinforce your understanding of mathematics. Because one function of numerical methods is to reduce higher mathematics to basic arithmetic operations, they get at the "nuts and bolts" of some otherwise obscure topics. Enhanced understanding and insight can result from this alternative perspective.

PT1.2 MATHEMATICAL BACKGROUND

Every part of this book requires some mathematical background. Consequently, the introductory material for each part includes a section, such as the one you are reading, on mathematical background. Because Part 1 itself is devoted to background material on mathematics and computers, the present section does not involve a review of a specific mathematical topic. Rather, we take this opportunity to introduce you to the types of mathematical subject areas that are covered in this book. As summarized in Fig. PT1.2, these are:

1. Roots of equations (Fig. PT1.2a). These problems are concerned with the value of a variable or a parameter that satisfies a single equation. These

problems are especially valuable in engineering design contexts where it is often impossible to explicitly solve design equations for parameters.

2. **Systems of linear algebraic equations** (Fig. PT1.2b). These problems are similar in spirit to roots of equations in the sense that they are concerned with values that satisfy equations. However, in contrast to satisfying a single equation, a set of values is sought that simultaneously satisfies a set of linear algebraic equations. Such equations arise in a variety of problem contexts and in all disciplines of engineering. In particular, they originate in the mathematical modeling of large systems of interconnected elements such as structures, electric circuits, and fluid networks. However, they are also encountered in other areas of numerical methods such as curve fitting.

3. **Curve fitting** (Fig. PT1.2c). You will often have occasion to fit curves to data points. The techniques developed for this purpose can be divided into two general categories: regression and interpolation. Regression is employed where there is a significant degree of error associated with the data. Experimental results are often of this kind. For these situations, the strategy is to derive a single curve that represents the general trend of the data without necessarily matching any individual points. In contrast, interpolation is used where the objective is to determine intermediate values between relatively error-free data points. Such is usually the case for tabulated information. For these situations, the strategy is to fit a curve directly through the data points and use the curve to predict the intermediate values.

4. **Integration** (Fig. PT1.2d). As depicted, a physical interpretation of numerical integration is the determination of the area under a curve. Integration has many applications in engineering practice, ranging from the determination of the centroids of oddly shaped objects to the calculation of total quantities based on sets of discrete measurements. In addition, numerical integration formulas play an important role in the solution of differential equations.

5. **Ordinary differential equations** (Fig. PT1.2e). Ordinary differential equations are of great significance in engineering practice. This is because many physical laws are couched in terms of the rate of change of a quantity rather than the magnitude of the quantity itself. Examples range from population forecasting models (rate of change of population) to the acceleration of a falling body (rate of change of velocity). Two types of problems are addressed: initial-value and boundary-value problems. In addition, the computation of eigenvalues is covered.

6. **Partial differential equations** (Fig. PT1.2f). Partial differential equations are used to characterize engineering systems where the behavior of a physical quantity is couched in terms of its rate of change with respect to two or more independent variables. Examples include the time-variable vibrations of a string (time and one spatial dimension) or the steady-state distribution of temperature on a heated plate (two spatial dimensions). Two fundamentally different approaches are employed to solve partial differential equations numerically. In the present text, we will emphasize finite-difference methods that approximate the solution in a pointwise fashion (Fig. PT1.2f). However, we will also present an introduction to finite-element methods, which use a piecewise approach.

Figure PT1.2
Summary of the numerical methods covered in this book.

Part 2: Roots of equations

Solve $f(x) = 0$ for x.

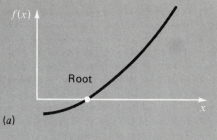

(a)

Part 3: Systems of linear algebraic equations

Given the a's and c's, solve

$$a_{11}x_1 + a_{12}x_2 = c_1$$
$$a_{21}x_1 + a_{22}x_2 = c_2$$

for the x's.

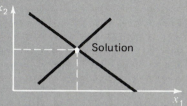

(b)

Part 4: Curve fitting

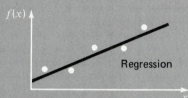

Regression

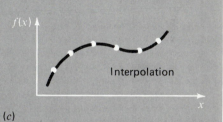

Interpolation

(c)

Part 5: Integration

$$I = \int_a^b f(x)\, dx$$

Find the area under the curve.

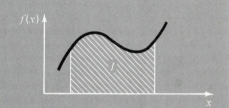

(d)

Part 6: Ordinary differential equations

Given

$$\frac{dy}{dt} \simeq \frac{\Delta y}{\Delta t} = f(t, y)$$

solve for y as a function of t.

$$y_{i+1} = y_i + f(t, y)\, \Delta t$$

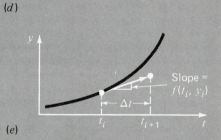

Slope = $f(t_i, y_i)$

(e)

Part 7: Partial differential equations

Given

$$\frac{\partial^2 u}{\partial x^2} + \frac{\partial^2 u}{\partial y^2} = f(x, y)$$

solve for u as a function of x and y

(f)

PT1.3   ORIENTATION

Some orientation might be helpful before proceeding with our introduction to numerical methods. The following is intended as an overview of the material in Part 1. In addition, some objectives have been included to focus your efforts when studying the material.

PT1.3.1   Scope and Preview

Figure PT1.3 is a schematic representation of the material in Part One. We have designed this diagram to provide you with a global overview of this part of the book. We believe that a sense of the "big picture" is critical to developing insight into numerical methods. When reading a text, it is often possible to become lost in technical details. Whenever you feel that you are losing the "big picture," refer back to Fig. PT1.3 to reorient yourself. Every part of this book includes a similar figure.

This figure also serves as a brief preview of the material covered in Part One. *Chapter 1* is designed to orient you to numerical methods and to provide motivation by demonstrating how these techniques can be used in the engineering modeling process. *Chapter 2* is an introduction and review of computer-related aspects of numerical methods and suggests the level of computer communication skills you should acquire to efficiently apply succeeding information. *Chapter 3* deals with the important topic of error analysis, which must be understood for the effective use of numerical methods. In addition, an *epilogue* is included that introduces the trade-offs that have such great significance for the effective implementation of numerical methods.

PT1.3.2   Goals and Objectives

Study Objectives.   Upon completing Part One, you should be adequately prepared to embark on your studies of numerical methods. In general, you should have gained a fundamental understanding of the importance of computers and the role of approximations and errors in the implementation and development of numerical methods. In addition to these general goals, you should have mastered each of the specific study objectives listed in Table PT1.1.

Computer Objectives.   Upon completing Part One you should have mastered sufficient computer skills to develop your own software for the numerical methods in this text. You should be able to develop well-structured and reliable computer programs on the basis of pseudocode, flowcharts, or other forms of algorithms. You should have the capability of saving your software on external storage devices and should have developed the capability to document your programs so that they may be effectively employed by users. Finally, if you have obtained the supplementary software (Electronic TOOLKIT) available with this book, you should know what programs it contains and be familiar with its computer graphics capabilities.

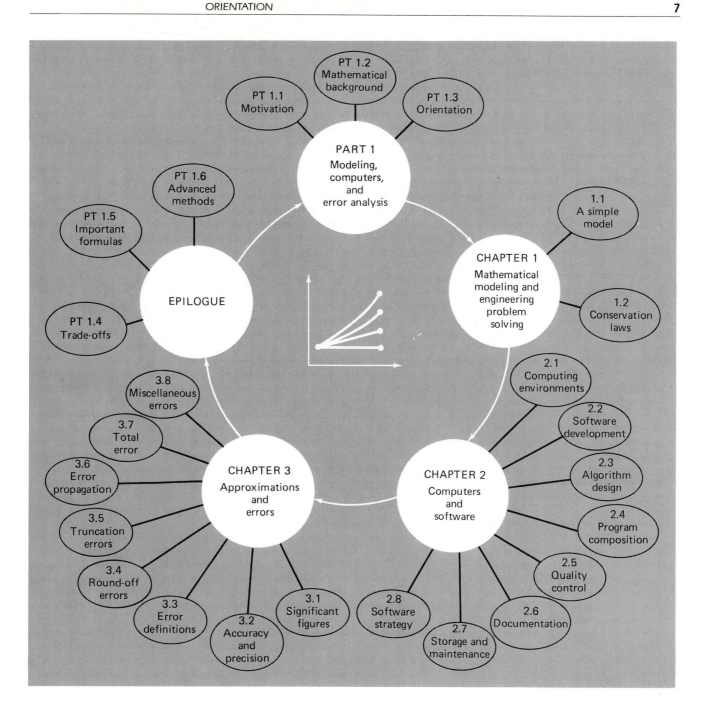

Figure PT1.3
Schematic of the organization of the material in Part One: Modeling, Computers, and Error Analysis.

TABLE PT1.1   Specific Study Objectives for Part One

1. Recognize the difference between analytical and numerical solutions.
2. Understand how conservation laws are employed to develop mathematical models of physical systems.
3. Define top-down and modular design.
4. Delineate the rules that underlie structured programming.
5. Be capable of composing a program in BASIC, FORTRAN, Pascal, or another structured, high-level computer language.
6. Know how to translate structured flowcharts and pseudocode into code in a high-level language.
7. Recognize the distinction between truncation and round-off errors.
8. Understand the concepts of significant figures, accuracy, and precision.
9. Recognize the difference between true relative error ϵ_t, approximate relative error ϵ_a and acceptable error ϵ_s, and understand how ϵ_a and ϵ_s are used to terminate an iterative computation.
10. Understand how numbers are represented in digital computers and how this representation induces round-off error. In particular, know the difference between single and extended precision.
11. Recognize how computer arithmetic can introduce and amplify round-off errors in calculations. In particular appreciate the problem of subtractive cancellation.
12. Understand how the Taylor series and its remainder are employed to represent continuous functions.
13. Know the relationship between finite divided differences and derivatives.
14. Know how to analyze how errors are propagated through functional relationships.
15. Be familiar with the concepts of stability and condition.
16. Familiarize yourself with the trade-offs outlined in the Epilogue of Part One.

CHAPTER 1

Mathematical Modeling and Engineering Problem-Solving

Knowledge and understanding are prerequisites for the effective implementation of any tool. No matter how impressive your tool chest, you will be hard-pressed to repair a car if you do not understand how it works.

This is particularly true when using computers to solve engineering problems. Although they have great potential utility, computers are practically useless without a fundamental understanding of how engineering systems work.

This understanding is initially gained by empirical means—that is, by observation and experiment. However, while such empirically derived information is essential, it is only half the story. Over years and years of observation and experiment, engineers and scientists have noticed that certain aspects of their empirical studies occur repeatedly. Such general behavior can then be expressed as fundamental laws that essentially embody the cumulative wisdom of past experience. Thus, most engineering problem solving employs the two-pronged approach of empiricism and theoretical analysis (Fig. 1.1).

It must be stressed that the two prongs are closely coupled. As new measurements are taken, the generalizations may be modified or new ones developed. Similarly, the generalizations can have a strong influence on the experiments and observations. In particular, generalizations can serve as organizing principles that can be employed to synthesize observations and experimental results into a coherent and comprehensive framework from which conclusions can be drawn. From an engineering problem-solving perspective, such a framework is most useful when it is expressed in the form of a mathematical model.

The primary objective of the present chapter is to introduce you to mathematical modeling and its role in engineering problem solving. We will also illustrate how numerical methods figure in the process.

1.1 A SIMPLE MATHEMATICAL MODEL

A *mathematical model* can be broadly defined as a formulation or equation that expresses the essential features of a physical system or process in mathematical terms. In a very general sense, it can be represented as a functional relationship of the form

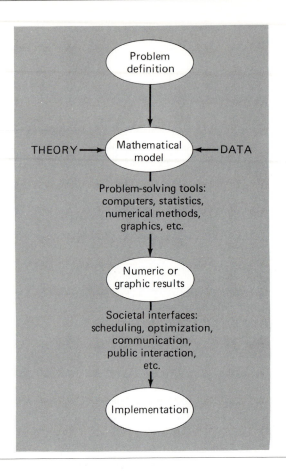

Figure 1.1
The engineering problem-solving process.

$$\begin{array}{c}\text{Dependent}\\\text{variable}\end{array} = f\left(\begin{array}{c}\text{independent}\\\text{variables}\end{array}, \text{parameters}, \begin{array}{c}\text{forcing}\\\text{functions}\end{array}\right) \qquad (1.1)$$

where the *dependent variable* is a characteristic that usually reflects the behavior or state of the system; the *independent variables* are usually dimensions, such as time and space, along which the system's behavior is being determined; the *parameters* are reflective of the system's properties or composition; and the *forcing functions* are external influences acting upon it.

The actual mathematical expression of Eq. (1.1) can range from a simple algebraic relationship to large complicated sets of differential equations. For example, on the basis of his observations, Newton formulated his second law of motion which states that the time rate of change of momentum of a body is equal to the resultant force acting on it. The mathematical expression, or model, of the second law is the well-known equation

$$F = ma \qquad (1.2)$$

where F is the net force acting on the body (in newtons, or kilogram-meters per second squared), m is the mass of the object (in kilograms), and a is its acceleration (in meters per second squared).

The second law can be recast in the format of Eq. (1.1) by merely dividing both sides by m to give

$$a = \frac{F}{m} \tag{1.3}$$

where a is the dependent variable reflecting the system's behavior, F is the forcing function and m is a parameter representing a property of the system. Note that for this simple case there is no independent variable because we are not predicting how acceleration varies in time or space.

Equation (1.2) has a number of characteristics that are typical of mathematical models of the physical world.

1. It describes a natural process or system in mathematical terms.
2. It represents an idealization and simplification of reality. That is, the model ignores negligible details of the natural process and focuses on its essential manifestations. Thus, the second law does not include the effects of relativity that are of minimal importance when applied to objects and forces that interact on or about the earth's surface at velocities and on scales visible to humans.
3. Finally, it yields reproducible results and, consequently, can be used for predictive purposes. For example, if the force on an object and its mass are known, Eq. (1.3) can be used to compute acceleration.

Because of its simple algebraic form, the solution of Eq. (1.2) could be obtained easily. However, other mathematical models of physical phenomena may be much more complex, and either cannot be solved exactly or require more sophisticated mathematical techniques than simple algebra for their solution. To illustrate a more complex model of this kind, Newton's second law can be used to determine the terminal velocity of a free-falling body near the earth's surface. Our falling body will be a parachutist as depicted in Fig. 1.2. A model for this case can be derived by expressing the acceleration as the time rate of change of the velocity (dv/dt) and substituting it into Eq. (1.2) to yield

$$m\frac{dv}{dt} = F \tag{1.4}$$

where v is velocity (in meters per second). Thus, the mass multiplied by the rate of change of the velocity is equal to the net force acting on the body. If the net force is positive, the object will accelerate. If it is negative, the object will decelerate. If the net force is zero, the object's velocity will remain at a constant level.

Next, we will express the net force in terms of measurable variables and parameters. For a body falling within the vicinity of the earth (Fig. 1.2), the net force is composed of two opposing forces: the downward pull of gravity F_D and the upward force of air resistance F_U:

$$F = F_D + F_U \tag{1.5}$$

If the downward force is assigned a positive sign, the second law can be used to formulate the force due to gravity as

$$F_D = mg \tag{1.6}$$

Figure 1.2
Schematic diagram of the forces acting on a falling parachutist. F_D is the downward force due to gravity. F_U is the upward force due to air resistance.

where g is the gravitational constant, or the acceleration due to gravity, which is approximately equal to 9.8 m/s^2.

Air resistance can be formulated in a variety of ways. A simple approach is to assume that it is linearly proportional to velocity and acts in an upward direction, as in

$$F_U = -cv \tag{1.7}$$

where c is a proportionality constant called the *drag coefficient* (in kilograms per second). Thus, the greater the fall velocity, the greater the upward force due to air resistance. The parameter c accounts for properties of the falling object, such as shape or surface roughness, that affect air resistance. For the present case, c might be a function of the type of jumpsuit or the orientation used by the parachutist during free-fall.

The net force is the difference between the downward and upward force. Therefore, Eqs. (1.4) through (1.7) can be combined to yield

$$m\frac{dv}{dt} = mg - cv \tag{1.8}$$

or, dividing each side by m,

$$\frac{dv}{dt} = g - \frac{c}{m}v \tag{1.9}$$

Equation (1.9) is a model that relates the acceleration of a falling object to the forces acting on it. It is a *differential equation* because it is written in terms of the differential rate of change (dv/dt) of the variable that we are interested in predicting. However, in contrast to the solution of Newton's second law in Eq. (1.3), the exact solution of Eq. (1.9) for the velocity of the falling parachutist cannot be obtained using simple algebraic manipulation. Rather, more advanced techniques such as those of calculus must be applied to obtain an exact or analytical solution. For example, if the parachutist is initially at rest ($v = 0$ at $t = 0$), calculus can be used to solve Eq. (1.9) for

$$v(t) = \frac{gm}{c}[1 - e^{-(c/m)t}] \tag{1.10}$$

Note that Eq. (1.10) is cast in the general form of Eq. (1.1) where $v(t)$ is the dependent variable, t is the independent variable, c and m are parameters, and g is the forcing function.

EXAMPLE 1.1:   *Analytical Solution to the Falling Parachutist Problem*

Problem statement: A parachutist with a mass of 68.1 kg jumps out of a stationary hot-air balloon. Use Eq. (1.10) to compute velocity prior to opening the chute. The drag coefficient is equal to 12.5 kg/s.

Solution: Inserting the parameters into Eq. (1.10) yields

$$v = \frac{9.8(68.1)}{12.5}(1 - e^{-(12.5/68.1)t})$$

$$= 53.39(1 - e^{-0.18355t})$$

which can be used to compute

t, s	v, m/s
0	0.00
2	16.40
4	27.77
6	35.64
8	41.10
10	44.87
12	47.49
∞	53.39

According to the model, the parachutist accelerates rapidly (Fig. 1.3). A velocity of 44.87 m/s (100.4 mi/h) is attained after 10 s. Note also that after a sufficiently long time, a constant velocity, called the *terminal velocity,* of 53.39 m/s (119.4 mi/h) is reached. This velocity is constant because, eventually, the force of gravity will be in balance with the air resistance. Thus, the net force is zero and acceleration has ceased.

Figure 1.3
The analytical solution to the falling parachutist problem as computed in Example 1.1. Velocity increases with time and asymptotically approaches a terminal velocity.

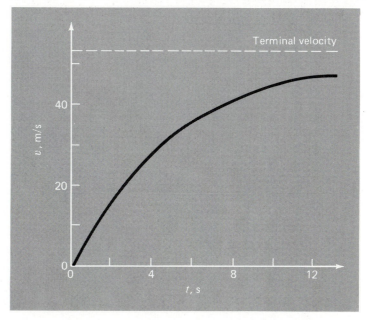

Equation (1.10) is called an *analytical* or *exact solution* because it exactly satisfies the original differential equation. Unfortunately, there are many mathematical models that cannot be solved exactly. In many of these cases, the only alternative is to develop a *numerical solution* that approximates the exact solution.

As mentioned previously, *numerical methods* are those in which the mathematical problem is reformulated so it can be solved by arithmetic operations. This can be illustrated for Newton's second law by realizing that the time rate of change of velocity can be approximated by (Fig. 1.4):

$$\frac{dv}{dt} \simeq \frac{\Delta v}{\Delta t} = \frac{v(t_{i+1}) - v(t_i)}{t_{i+1} - t_i} \tag{1.11}$$

where Δv and Δt are differences in velocity and time computed over finite intervals, $v(t_i)$ is velocity at an initial time t_i, and $v(t_{i+1})$ is velocity at some later time t_{i+1}. Equation (1.11) is called a *finite divided difference* approximation of the derivative at time t_i. It can be substituted into Eq. (1.9) to give

$$\frac{v(t_{i+1}) - v(t_i)}{t_{i+1} - t_i} = g - \frac{c}{m}v(t_i)$$

This equation can then be rearranged to yield

Figure 1.4
The use of a finite difference to approximate the first derivative of v with respect to t.

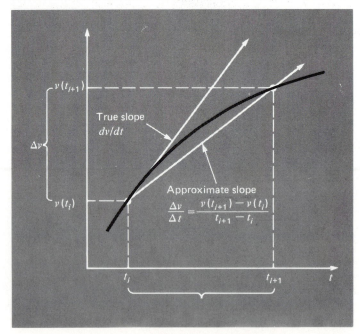

$$v(t_{i+1}) = v(t_i) + [g - \frac{c}{m}v(t_i)](t_{i+1} - t_i) \qquad (1.12)$$

Notice that the term in brackets is the right-hand-side of the differential equation itself [Eq. (1.9)]. That is, it provides a means to compute the rate of change or slope of v. Thus, the differential equation has been transformed into an equation that can be used to determine the velocity algebraically at t_{i+1} using the slope and previous values of v and t. If you are given an initial value for velocity at some time t_i, you can easily compute velocity at a later time t_{i+1}. This new value of velocity at t_{i+1} can in turn be employed to extend the computation to velocity at t_{i+2} and so on. Thus at any time along the way,

New value = old value + slope × step size

EXAMPLE 1.2:   Numerical Solution to the Falling Parachutist Problem

Problem statement: Perform the same computation as in Example 1.1 but use Eq. (1.12) to compute velocity. Employ a step size of 2 s for the calculation.

Solution: At the start of the computation ($t_i = 0$), the velocity of the parachutist is zero. Using this information and the parameter values from Example 1.1, Eq. (1.12) can be used to compute velocity at $t_{i+1} = 2$ s:

$$v = 0 + \left[9.8 - \frac{12.5}{68.1}(0)\right]2 = 19.60 \text{ m/s}$$

For the next interval (from $t = 2$ to 4 s), the computation is repeated, with the result

$$v = 19.60 + \left[9.8 - \frac{12.5}{68.1}(19.60)\right]2 = 32.00 \text{ m/s}$$

The calculation is continued in a similar fashion to obtain additional values:

t, s	v, m/s
0	0.00
2	19.60
4	32.00
6	39.85
8	44.82
10	47.97
12	49.96
∞	53.39

The results are plotted in Fig. 1.5 along with the exact solution. It can be seen that the numerical method captures the essential features of the exact solution. However, because we have employed straight-line segments to approximate a continuously curving function, there is some discrepancy between the two results. One way to minimize such discrepancies is to use a smaller step size. For example, applying Eq. (1.12) at 1-s intervals results in a smaller error, as the straight-line segments track closer to the true

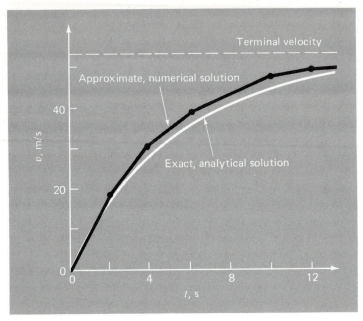

Figure 1.5
Comparison of the numerical and analytical solutions for the falling parachutist problem.

solution. Using hand calculations, the effort associated with using smaller and smaller step sizes would make such numerical solutions impractical. However, with the aid of the computer, large numbers of calculations can be performed easily. Thus you can accurately model the velocity of the falling parachutist without having to solve the differential equation exactly.

As in the previous example, a computational price must be paid for a more accurate numerical result. Each halving of the step size to attain more accuracy leads to a doubling of the number of computations. Thus, we see that there is a trade-off between accuracy and computational effort. Such trade-offs figure prominently in numerical methods and constitute an important theme of this book. Consequently, we have devoted the Epilogue of Part One to an introduction to more of these trade-offs.

1.2 CONSERVATION LAWS AND ENGINEERING

Aside from Newton's second law, there are other major organizing principles in engineering. Among the most important of these are the conservation laws. Although they form the basis for a variety of complicated and powerful mathematical models, the great conservation laws of science and engineering are conceptually easy to understand. They all boil down to

Change = increases − decreases   (1.13)

This is precisely the format that we employed when using Newton's law to develop a force balance for the falling parachutist [Eq. (1.8)].

Although simple, Eq. (1.13) embodies one of the most fundamental ways in which conservation laws are used in engineering—that is, to predict changes with respect to time. We will give it a special name—the *time-variable computation*.

Aside from predicting changes, another way in which conservation laws are applied is for cases where change is nonexistent. If change is zero, Eq. (1.13) becomes

Change = 0 = increases − decreases

or

Increases = decreases $\qquad$ (1.14)

Thus, if no change occurs, the increases and decreases must be in balance. This case, which is also given a special name—the *steady-state computation*—has many applications in engineering. For example, for steady-state incompressible fluid flow in pipes, the flow into a junction must be balanced by flow going out, as in

Flow in = Flow out

For the junction in Fig. 1.6, the balance can be used to compute that the flow out of the fourth pipe must be 60.

For the falling parachutist, the steady state would correspond to the case where the net force was zero or [Eq. (1.8) with $m(dv/dt) = 0$]

$$mg = cv \qquad (1.15)$$

Thus, at steady state, the downward and upward forces are in balance and Eq. (1.15) can be solved for the terminal velocity

$$v = \frac{mg}{c}$$

Although Eqs. (1.13) and (1.14) might appear trivially simple, they embody the two fundamental ways that conservation laws are employed in engineering. As such, they will form an important part of our efforts in subsequent chapters to illustrate the connection between numerical methods and engineering. Our primary vehicles for making this connection are the case studies that appear at the end of each part of the book.

Table 1.1 summarizes some of the simple engineering models and associated conservation laws that will form the basis for many of these case studies. Most of the chemical engineering case studies will focus on mass balances for reactors. The mass balance derives from the conservation of mass. It specifies that the change of mass of a chemical in the reactor depends on the amount of mass flowing in minus the mass flowing out.

Both the civil and mechanical engineering case studies will focus on models developed from the conservation of momentum. For civil engineering, force balances are utilized to analyze structures such as the simple truss in Table 1.1. The same principles are employed for the mechanical engineering case studies to analyze the transient up-and-down motion or vibrations of an automobile.

TABLE 1.1 Devices and types of balances that are commonly used in the four major areas of engineering. For each case, the conservation law upon which the balance is based is specified.

Field	Device	Organizing Principle	Mathematical Expression
Chemical engineering	**Reactors**	Conservation of mass	Mass balance: Over a unit time period Δmass $=$ inputs $-$ outputs
Civil engineering	Structure	Conservation of momentum	Force balance: At each node Σ horizontal forces $(F_H) = 0$ Σ vertical forces $(F_V) = 0$
Mechanical engineering	Machine	Conservation of momentum	Force balance: $$m\frac{d^2 x}{dt^2} = \text{downward force} - \text{upward force}$$
Electrical engineering	Circuit	Conservation of charge	Current balance: For each node Σ current $(i) = 0$
		Conservation of energy	Voltage balance Around each loop Σ emf's $-$ Σ voltage drops for resistors $= 0$ $\Sigma \xi - \Sigma iR = 0$

Finally, the electrical engineering case studies employ both current and energy balances to model electric circuits. The current balance, which results from the conservation of charge, is similar in spirit to the flow balance depicted in Fig. 1.6. Just as flow must balance at the junction of pipes, electric current must balance at the junction of electric

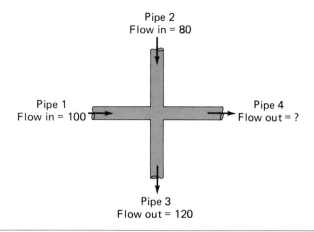

Pipe 2
Flow in = 80

Pipe 1
Flow in = 100

Pipe 4
Flow out = ?

Pipe 3
Flow out = 120

Figure 1.6
A flow balance for incompressible fluid flow at the junction of pipes.

wires. The energy balance specifies that the changes of voltage around any loop of the circuit must add up to zero.

The case studies are designed to illustrate how numerical methods are actually employed in the engineering problem-solving process. As such, they will permit us to explore practical issues (Table 1.2) that arise in real-world applications. Making these connections between mathematical techniques such as numerical methods and engineering practice is a critical step in tapping their true potential. Careful examination of the case studies will help you to take this step.

TABLE 1.2 **Some practical issues that will be explored in the case studies at the end of each part of the book.**

1. **Nonlinear vs. linear** Much of classical engineering depends on linearization to permit analytical solutions. Although this is often appropriate, expanded insight can often be gained if nonlinear problems are examined.
2. **Large vs. small systems** Without a computer, it is often not feasible to examine systems with over three interacting components. With computers and numerical methods, more realistic multicomponent systems can be examined.
3. **Nonideal vs. ideal** Idealized laws abound in engineering. Often there are nonidealized alternatives that are more realistic but more computationally demanding. Approximate numerical approaches can facilitate the application of these nonideal relationships.
4. **Sensitivity analysis** Because they are so involved, many manual calculations require a great deal of time and effort for successful implementation. This sometimes discourages the analyst from implementing the multiple computations that are necessary to examine how a system responds under different conditions. Such sensitivity analyses are facilitated when numerical methods allow the computer to assume the computational burden.
5. **Design** It is often a straightforward proposition to determine the performance of a system as a function of its parameters. It is usually more difficult to solve the inverse problem— that is, determining the parameters when the required performance is specified. Numerical methods and computers often permit this task to be implemented in an efficient manner.

PROBLEMS

1.1 Answer true or false:

(a) Today more attention can be paid to problem formulation and interpretation because the computer and numerical methods facilitate the solution of engineering problems.

(b) The value of a variable that satisfies a single equation is called the root of the equation.

(c) Numerical methods are those in which a mathematical problem is reformulated so that it can be solved by arithmetic operations.

(d) Finite divided differences are used to represent derivatives in approximate terms.

(e) A physical interpretation of integration is the area under a curve.

(f) In the precomputer era, numerical methods were widely employed because they required little computational effort.

(g) Newton's second law is a good example of the fact that most physical laws are based on the rate of change of quantities rather than on their magnitudes.

(h) Interpolation is employed for curve-fitting problems when there is significant error associated with the data points.

(i) The large systems of equations, nonlinearities, and complicated geometries that are common in engineering practice are easy to solve analytically.

(j) Math models should never be used for predictive purposes.

1.2 Read the following problem descriptions and identify which area of numerical methods (as outlined in Fig. PT1.2) relates to their solution.

(a) You are on a survey crew and must determine the area of a field that is bounded by two roads and a meandering stream.

(b) You are responsible for determining the flows in a large interconnected network of pipes to distribute natural gas to a series of communities spread out over a 20-mi^2 area.

(c) For the falling parachutist problem, you must determine the value of the drag coefficient in order that a 200-lb parachutist not exceed 100 mi/h within 10 s of jumping. You must make this evaluation on the basis of the analytical solution [Eq. (1.10)]. The information will be used to design a jumpsuit.

(d) You are performing experiments to determine the voltage drop across a resistor as a function of current. You make measurements of voltage drop for a number of different values of current. Although there is some error associated with your data points, when you plot them they suggest a curvilinear relationship. You are to derive an equation to characterize this relationship.

(e) You must determine the temperature distribution on the two-dimensional surface of a flat gasket as a function of the temperatures of its edges.

(f) You must develop a shock-absorber system for a racing car. Newton's second law can be used to derive an equation to predict the rate of change in position of the front wheel in response to external forces. You must compute the motion of the wheel as a function of time after it hits a 6-in bump at 150 mi/h.

(g) You have to determine the annual revenues required over a 20-year period for an entertainment center to be built for a client. Money can presently be borrowed at an interest rate of 17.6 percent. Although the information to perform this estimate is contained in economics tables, values are listed only for interest rates of 15 and 20 percent.

1.3 Give one example of an engineering problem where each of the following classes of numerical methods can come in handy. If possible, draw from your experience in class and in readings or from any professional experience you have gathered to date.

(a) Roots of equations

(b) Linear algebraic equations

(c) Curve fitting: regression

(d) Curve fitting: interpolation

(e) Integration

(f) Ordinary differential equations

(g) Partial differential equations

1.4 What is the two-pronged approach to engineering problem solving? Into what category should the conservation laws be placed?

1.5 What is the form of the transient conservation law? What is it for steady state?

1.6 The following information is available for a bank account:

Date	Deposits	Withdrawals	Balance
5/1			512.33
	200.13	427.26	
6/1			
	206.80	328.61	
7/1			
	450.25	206.80	
8/1			
	127.31	380.61	
9/1			

Use the conservation of cash to compute the balance on 6/1, 7/1, 8/1, and 9/1. Show each step in the computation. Is this a steady-state or a transient computation?

1.7 Give examples of conservation laws in engineering and in everyday life.

1.8 Repeat Example 1.2 but use a step size of 1 s. Compute the velocity to $t = 12$ s.

CHAPTER 2

Computers and Software

Numerical methods combine two of the most important tools in the engineering repertoire: mathematics and computers. In fact, numerical methods can be loosely defined as computer math. Good computer programming skills will enhance your studies of numerical methods. In particular, the powers and limitations of numerical techniques are best appreciated when you use the methods in tandem with a computer to solve engineering problems.

You will have the opportunity in using this book to develop your own software. Because of the widespread availability of personal computers and magnetic storage devices, this software can be easily retained and applied throughout your career. Therefore, one of the primary objectives of this text is that you come away from it possessing useful, high-quality software.

This text has a number of features that are intended to maximize this possibility. All the numerical techniques are accompanied by material related to effective computer implementation. In addition, supplementary software is available for the application or illustration of several of the most elementary methods discussed in this book. This software, which is compatible with widely used personal computers (IBM-PC and compatibles), can serve as a starting point for your own program library.

The present chapter provides background information that has utility if you plan to use this text as the basis for developing your own programs. It is written under the premise that you have had some prior exposure to computer programming. Because this book is not intended to support a computer programming course per se, the discussion pertains only to those aspects related to developing high-quality software for numerical methods. It is also intended to provide you with specific criteria for evaluating the quality of your efforts.

2.1 COMPUTING ENVIRONMENTS

We believe that an historical perspective is very useful in appreciating recent advances in software design. As in Fig. 2.1, the modern computer age can be conceptualized as a progression of generations. Prior to the early 1970s, computing was monolithic in the sense that large mainframe systems were the only available option. Because they were so expensive to own, operate, and maintain, computers could be acquired only by large

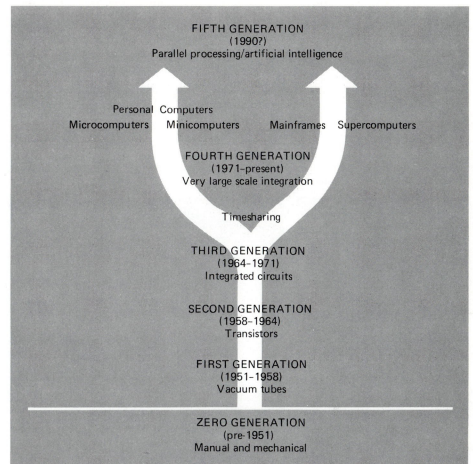

Figure 2.1
The evolution of the modern computer era can be conceptualized as a succession of generations closely linked to technological advances. In the third generation, a divergence emerged with the advent of machines expressly designed for the needs of individuals. Today, a broad range of computers is available. The progression seems to be moving toward a fifth generation in which artificial intelligence will figure prominently. (Redrawn with permission from Chapra and Canale, 1986.)

organizations such as corporations, universities, government agencies, and the military. As might be expected, the centralized way in which computing was configured had a profound impact on the manner in which engineers and scientists interacted with the machines. In particular, a distinct separation existed between the user and the computer.

In the late 1960s, the introduction of integrated circuits resulted in machines that were significantly faster, cheaper, and smaller. This technological breakthrough in turn spawned a number of new developments that have led to a divergence in the previously one-dimensional evolution of the computing industry (Fig. 2.1).

Whereas earlier computers had been developed expressly for large institutions and organizations, the fourth generation resulted in machines that were designed to meet the needs of individuals. Consequently, today's engineer has a broad spectrum of tools with which to implement computations. In terms of size, cost, and number of users, these tools can be broadly divided into three major categories: microcomputers, minicomputers, and mainframes.

Microcomputers (Fig. 2.2*a*) were originally defined as computers whose main functions were contained on a single integrated-circuit chip. They are compact, one-user computers that can fit on a desk top and cost on the order of $1000 to $5000.

Minicomputer (Fig. 2.2*b*) is a somewhat imprecise term referring to computers that are more powerful than micros but yet are still within the means of some individuals and small- to medium-sized firms. They can support multiple users but do not require the level of support needed by larger machines. They cost on the order of $10,000 to $100,000.

Figure 2.2
The three classes of computers commonly used by today's engineer: (*a*) the microcomputer, (*b*) the minicomputer, and (*c*) the mainframe. (Courtesy of Texas Instruments Corporation, Sun Microsystems, Inc., and IBM, respectively.)

(*a*)

(*b*)

(*c*)

Figure 2.3
A supercomputer. These machines will allow engineers and scientists to perform numerical computations that were impossible in the past. (Courtesy of Los Alamos National Laboratory.)

Mainframe computers (Fig. 2.2c) run in the million dollar range and are owned by large organizations. They require a special support staff and typically serve hundreds of users at a time.

Before proceeding, it must be stressed that the foregoing classifications are not hard and fast. For example, a more advanced class of mainframe called the *supercomputer* is emerging as a powerful tool in engineering and scientific research environments (Fig. 2.3). Similarly, the distinction between micros and minis is being blurred by the so-called *supermicros* that can fit on a desk top but can serve multiple users. Additionally, the way in which the computers are deployed or linked tends to blur the distinctions between the categories. For example, an engineer's *workstation* could consist of a terminal tied to a mainframe, a minicomputer with several terminals, or a stand-alone microcomputer.

Whatever definitions are used, the beauty of today's situation is that we have potential access to all types of computers. For example, preliminary work on a large-scale computation might be performed on a microcomputer. Then, a telecommunications link can be used to upload the program to a mainframe for fast implementation. In this way, the strengths of each type of computer are exploited. Engineers with access to a range of machines will have enormous capability at their command.

For this reason, and because we anticipate that you will be using different types of machines to learn numerical methods, we have designed this book so that it can be used with any of these systems. Thus, most of the material can be implemented on systems ranging from personal computers to mainframes.

At the same time, we have tried wherever possible to acknowledge and explore some of the exciting new applications that are directly associated with personal computers. We believe that this "personal" aspect of the microcomputer revolution is a major source of the present excitement over computing among students and professionals alike.

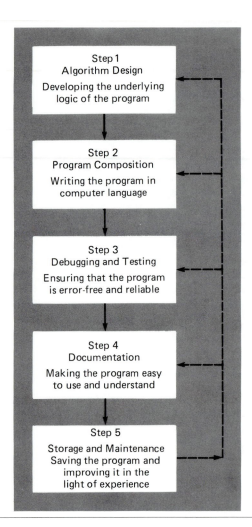

Figure 2.4
The five steps required to
produce high-quality software.
The feedback arrows indicate
that the first four steps can
be improved in the light of
experience.

2.2 THE SOFTWARE DEVELOPMENT PROCESS

No matter what type of computer you use, it has utility only if you provide it with careful instructions. These instructions are the *software*. The next sections deal with the process of composing your own high-quality software to implement numerical methods. Figure 2.4 delineates the five steps that constitute the process. Each of these steps will be discussed in detail in the subsequent sections. However, before embarking on this discussion, we must first mention some qualitative changes that have occurred in software development and which have great relevance to implementing numerical methods on computers.

2.2.1 Programming Style

Just as hardware has evolved dramatically, the software development process has also undergone a radical transformation in recent years. In the early years of computing, programming was somewhat more of an art than a science. Although all languages had very precise vocabularies and grammars, there were no standard definitions of good programming style. Consequently, individuals developed their own criteria for what constituted an excellent piece of software.

In the early years, these criteria were strongly influenced by the available hardware. Early computers were expensive and slow and had small memories. As a consequence, one mark of a good program was that it utilized as little memory as possible. Another might be that it executed rapidly. Still another was how quickly it could be written the first time.

The ultimate result was that programs were highly nonuniform. In particular, because they were strongly influenced by hardware limitations, programmers put little premium on the ease with which programs might be used and maintained. Thus, if you wanted to effectively utilize and maintain a large program over long periods, you usually had to keep the original programmer close by. In other words, the program and the programmer formed a costly package.

Today many of the hardware limitations do not exist or are rapidly disappearing. In addition, as software costs constitute a proportionally larger fraction of the total computing budget (Fig. 2.5), there is a high premium on developing software that is easy to use and modify. In particular, there is a strong emphasis on clarity and readability rather than on terse, noncommunicative code. This emphasis has great significance in engineering where work is often performed by teams, and programs must be shared and modified by several individuals.

Computer scientists have systematically studied the factors and procedures needed to develop high-quality software of this kind. Although the resulting methods are somewhat oriented toward large-scale programming efforts, most of the general techniques are also extremely useful for the types of programs that engineers develop routinely in the

Figure 2.5
Trends in hardware and software costs in the computing industry. (Modified from Branscomb, 1982.)

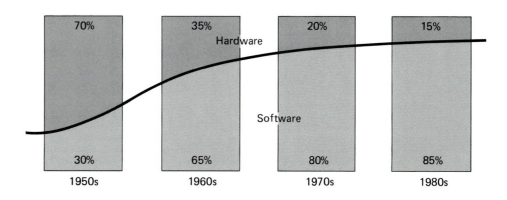

course of their work. Collectively, we will call these techniques *structured design and programming*. In the present section, we discuss three of these approaches—modular design, top-down design, and structured programming.

2.2.2 Modular Design

Imagine how difficult it would be to study a textbook that had no chapters, sections, or paragraphs. Breaking complicated tasks or subjects into more manageable parts is one way to make them easier to handle. In the same spirit, computer programs can be divided into small subprograms, or *modules,* that can be developed and tested separately. This approach is called *modular design*.

The most important attribute of modules is that they be as independent and self-contained as possible. In addition, they are typically designed to perform a specific, well-defined function and have one entry and one exit point. As such, they are usually short (fewer than 50 lines of code) and highly focused.

One of the primary programming elements used to represent each module is the subroutine or procedure. A *subroutine* is a series of computer instructions that together perform a given task. A *calling,* or *main, program* invokes these modules as they are needed. Thus the main program orchestrates each of the parts in a logical fashion.

Modular design has a number of advantages. The use of small, self-contained units makes the underlying logic easier to devise and to understand for both the developer and the user. Development is facilitated because each module can be perfected in isolation. In fact, for large projects, different programmers can work on individual parts. In the same spirit, you can maintain your own library of useful modules for later use in other programs. Modular design also increases the ease with which a program can be debugged and tested because errors can be more easily isolated. Finally, program maintenance and modification are facilitated. This is primarily due to the fact that new modules can be developed to perform additional tasks and then easily incorporated into the already coherent and organized scheme.

2.2.3 Top-down Design

Although the previous section introduced the notion of modular design, it did not specify how the actual modules are identified. The top-down design process is a particularly effective approach for accomplishing this objective.

Top-down design is a systematic development process that begins with the most general statement of a program's objectives and then successively divides it into more detailed segments. Thus the design proceeds from the general to the specific. Most top-down designs start with an English description of the program in its most general terms. This description is then broken down into a few elements that define the program's major functions. Then each major element is divided into subelements. This process is continued until well-defined modules have been identified.

Besides providing a method for dividing the algorithm into well-defined units, top-down design has other benefits. In particular, it ensures that the finished product is comprehensive. By starting with a broad definition and progressively adding detail, the programmer is less likely to overlook important operations.

2.2.4 Structured Programming

The previous two sections have dealt with effective ways to organize a program. In both cases, emphasis was placed on *what* the program is supposed to do. Although this usually guarantees that the program will be coherent and neatly organized, it does not ensure that the actual program instructions or code for each module will manifest an associated clarity and orderliness. Structured programming, on the other hand, deals with *how* the actual program code is developed so that it is easy to understand, correct, and modify.

In essence, structured programming is a set of rules that prescribe good style habits for the programmer. Although structured programming is flexible enough to allow considerable creativity and personal expression, its rules impose enough constraints to render resulting codes far superior to unstructured versions. In particular, the finished product is more elegant and easier to understand.

The rules that constitute structured programming will be described in a subsequent section (2.4.2). Before doing so, we will first turn to the initial step of the program development process—designing the underlying logic or algorithm of the program.

2.3 ALGORITHM DESIGN

The most common problems encountered by inexperienced programmers usually can be traced to the premature preparation of a program that does not encompass an overall strategy or plan. In the early days of computing, this was a particularly thorny problem because the design and programming components were not clearly separated. Programmers would often launch into a computing project without a clearly formulated design. This led to all sorts of problems and inefficiencies. For example, the programmer could get well into a job only to discover that some key preliminary factor had been overlooked. In addition, the logic underlying improvised programs was invariably obscure. This meant that it was usually difficult for someone to modify someone else's program. Today, because of the high costs of software development, much greater emphasis is placed on aspects such as preliminary planning that lead to a more coherent and efficient product. The focus of this planning is algorithm design.

An *algorithm* is the sequence of logical steps required to perform a specific task such as solving a problem. Aside from accomplishing its objectives, a good algorithm must have a number of attributes:

1. Each step must be deterministic; that is, nothing can be left to chance. The final results cannot depend on who is following the algorithm. In this sense, an algorithm is analogous to a recipe. Two chefs working independently from a good recipe should end up with dishes that are essentially identical.
2. The process must always end after a finite number of steps. An algorithm cannot be open-ended.
3. The algorithm must be general enough to deal with any contingency.

Figure 2.6*a* shows an algorithm for the solution of the simple problem of adding a pair of numbers. Two independent programmers working from this algorithm might develop programs exhibiting somewhat different styles. However, given the same data, their programs should yield identical results.

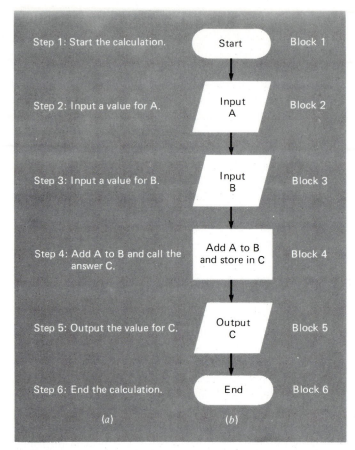

Figure 2.6
(a) Algorithm and (b) flowchart for the solution of a simple addition program.

Step-by-step English descriptions of the sort depicted in Fig. 2.6a are one way to express an algorithm. They are particularly useful for small problems or for specifying the broad tasks of a large programming effort. However, for detailed representations of complicated programs, they rapidly become inadequate. For this reason, more versatile, visual alternatives, called flowcharts, have been developed.

2.3.1 Flowcharting

A *flowchart* is a visual or graphical representation of an algorithm. The flowchart employs a series of blocks and arrows, each of which represents a particular operation or step in the algorithm. The arrows represent the sequence in which the operations are implemented. Figure 2.6b shows a flowchart for the problem of adding two numbers.

Not everyone involved with computer programming agrees that flowcharting is a productive endeavor. In fact, some experienced programmers do not advocate flowcharts.

However, we feel that there are three good reasons for studying them. First, they are still used for expressing and communicating algorithms. Second, even if they are not employed routinely, there will be times when they will prove useful in planning, unraveling, or communicating the logic of your own or someone else's program. Finally, and most important for our purposes, they are excellent pedagogical tools. From a teaching perspective, they are ideal vehicles for visualizing some of the fundamental control structures employed in computer programming.

Flowchart Symbols. Notice that in Fig. 2.6*b* the blocks have different shapes to distinguish different types of operations. In the early years of computing, flowchart symbols were nonstandardized. This led to confusion, because programmers used different symbols to represent the same operation. Today a number of standardized systems have been developed to facilitate communication. In this book, we use the set of symbols shown in Fig. 2.7.

A flowchart is read by beginning at the start terminal and following the flowlines to trace the algorithm's logic. Thus, as shown in Fig. 2.6*b*, we would begin at block 1 and then follow the arrow sequentially until we arrive at the terminal block 6. If the flowchart is written properly there should never be any doubt regarding the correct path.

Although the flowlines in Fig. 2.7 show humps to depict lines that pass over each other without connecting, it is generally best to avoid crossing flowlines. Circular connectors allow the flow to be discontinued at one section of a flowchart and started again at another location. A number is placed within the exit- and entry-point connectors to specify each pair used on a given page (Fig. 2.8).

Figure 2.7
Common ANSI symbols used in flowcharts.

SYMBOL	NAME	FUNCTION
	Terminal	Represents the beginning or end of a program.
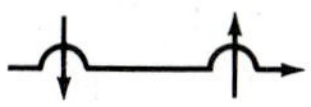	Flowlines	Represents the flow of logic. The humps on the horizontal arrow indicate that it passes over and does not connect with the vertical flowlines.
	Process	Represents calculations or data manipulations.
	Input/output	Represents inputs or outputs of data and information.
	Decision	Represents a comparison, question, or decision that determines alternative paths to be followed.
	On-page connector	Represents a break in the path of a flowchart that is used to continue flow from one point on a page to another. The connector symbol marks where the flow ends at one spot on a page and where it continues at another spot.
	Off-page connector	Similar to the on-page connector but represents a break in the path of a flowchart that is used when it is too large to fit on a single page. The connector symbol marks where the algorithm ends on the first page and where it continues on the second.

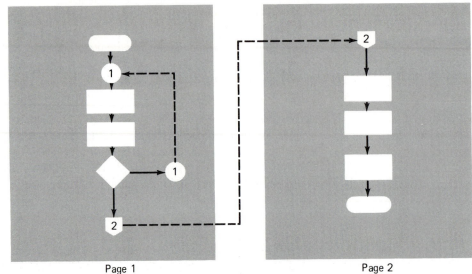

Figure 2.8
Illustration of the use of on-page and off-page connectors. The dashed lines would not actually be a part of the charts but are used here to show how control is transferred between the connectors.

Page 1   Page 2

Some flowcharts are so large that they spread over several pages. The spade-shaped symbols in Figs. 2.7 and 2.8 are used to transfer flow between pages in a fashion similar to the circular on-page connectors.

The construction of flowcharts is facilitated by use of a plastic template (Fig. 2.9). This stencil includes the standard symbols for constructing flowcharts. Templates are available at most office-supply shops and college bookstores.

Levels of Flowchart Complexity. There are no set standards for how detailed a flowchart should be. During the development of a major algorithm, you will often draft flowcharts of various levels of complexity. As an example, Fig. 2.10 depicts a hierarchy of three charts that might be used in the development of an algorithm to determine the grade-point average (GPA) or cumulative index of a student engineer. The system

Figure 2.9
A flowchart template.

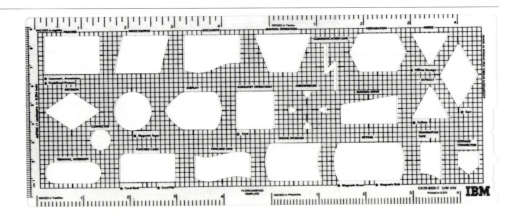

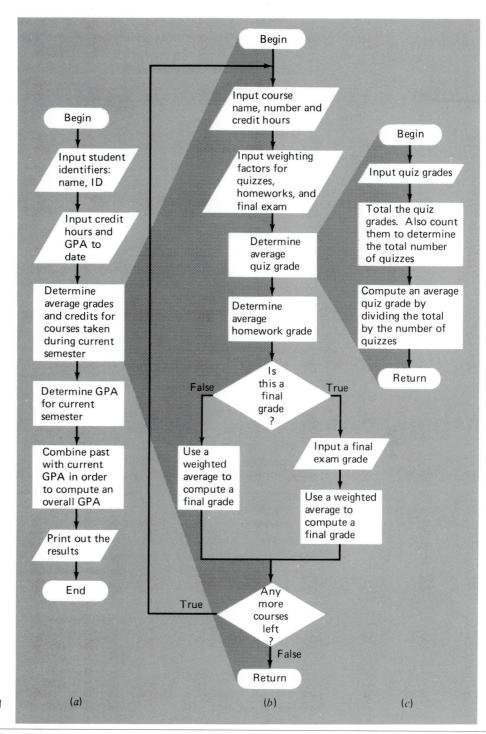

Figure 2.10
A hierarchy of flowcharts dealing with the problem of determining a student's grade-point average. The system flowchart in (a) provides a comprehensive plan. A more detailed chart for a major module is shown in (b). An even more detailed chart for a segment of the major module is seen in (c). This way of approaching a problem is called top-down design.

flowchart (Fig. 2.10*a*) delineates the big picture. It identifies the major tasks, or *modules,* and the sequence that is required to solve the entire problem. Such an overview flowchart is invaluable for ensuring that the total scheme is comprehensive enough to be successful. Next the major modules can be broken off and charted in greater detail (Fig. 2.10*b*). Finally, it is sometimes advantageous to break the major modules down into even more manageable units, as in Fig. 2.10*c*. This is an example of the *top-down design process* that was introduced in Sec. 2.2.3.

In the following sections we will study and improve the flowchart in Fig. 2.10*c*. This will be done to illustrate how the symbols in Fig. 2.7 can be orchestrated to develop an efficient and effective algorithm. In the process we will demonstrate how the power of computers can be reduced to three fundamental types of operations. The first, which should already be apparent, is that computer logic can proceed in a definite *sequence*. As depicted in Fig. 2.10*a* and *c*, this is represented by the flow of logic from box to box. The other two fundamental operations are *selection* and *repetition*.

Selection and Repetition in Flowcharts.   Notice that the operations in Fig. 2.10*a* and *c* specify a simple sequential flow. That is, the operations follow one after the other. However, notice how the first diamond-shaped decision symbol in Fig. 2.10*b* permits the flow to branch and follow one of two possible paths, depending on the answer to a question. For this case, if final grades are being determined, the flow follows the right branch, whereas if intermediate or midterm grades are being determined, the flow follows the left branch. This type of flowchart operation is called *selection*.

Aside from depicting a conditional branch, Fig. 2.10*b* demonstrates another flowchart operation that greatly increases the power of computer programs—*repetition*. Notice that the "true" branch of the second decision symbol allows the flow to return to the head of the module and repeat the calculation process for another course. This ability to perform repetitive tasks is among the most important strengths of computers. The operation of "looping" back to repeat a set of operations is accordingly named a *loop* in computer jargon.

The operations of selection and repetition can be employed to great advantage when deriving algorithms. This can be demonstrated by developing an improved version of Fig. 2.10*c*.

In Fig. 2.10*c*, we specified a process box to total the quiz grades and keep count of the number of quizzes. A more detailed flowchart to accomplish these objectives is presented in Fig. 2.11. Notice that the algorithm consists of a loop and a decision construct. The loop inputs, counts, and sums the grades. These operations are repeated until a negative grade triggers the exit from the loop. At this point, a selection construct is used to compute the average grade. A selection construct is employed for this purpose to avoid division by zero in the event that no positive grade has been entered.

Structured Flowcharts.   Flowcharting symbols can be orchestrated to construct logic patterns of extreme complexity. However, programming experts have recognized that, without rules, flowcharts can become so complicated that they impede rather than facilitate clear thinking.

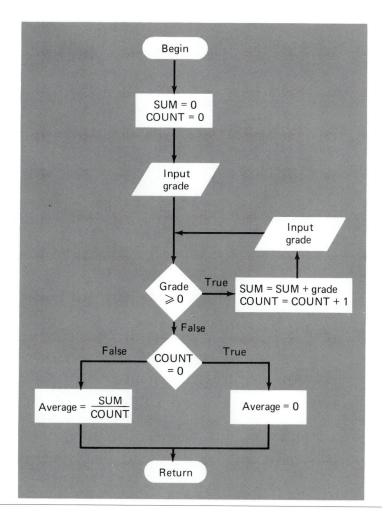

Figure 2.11
A detailed flowchart to compute the average quiz grade. Through the use of an accumulator and a counter, this example improves on the simple version shown in Fig. 2.10c.

One of the main culprits that contributes to algorithm complexity is the unconditional transfer, also known as the GO TO. As the name implies, this control structure allows you to "go to" any other location in the algorithm. For flowcharts, it is simply represented by an arrow. Indiscriminate use of unconditional transfers can lead to algorithms that, at a distance, resemble a plate of spaghetti.

In an effort to avoid this dilemma, advocates of structured programming have demonstrated that any program can be constructed with the three fundamental control structures depicted in Fig. 2.12. As can be seen, they correspond to the three operations—sequence, selection, and repetition—that we have just been discussing.

All flowcharts should be composed of the three fundamental control structures shown in Fig. 2.12. If we limit ourselves to these structures and avoid GO TOs, the resulting algorithm will be much easier to understand.

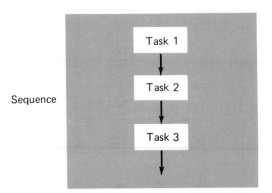

Sequence

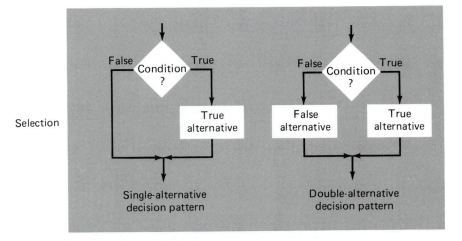

Selection

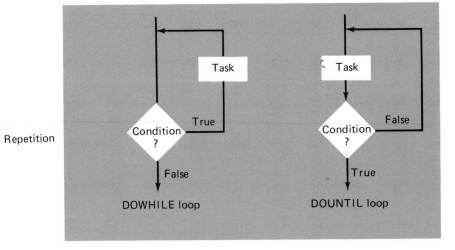

Repetition

Figure 2.12
The three fundamental control structures.

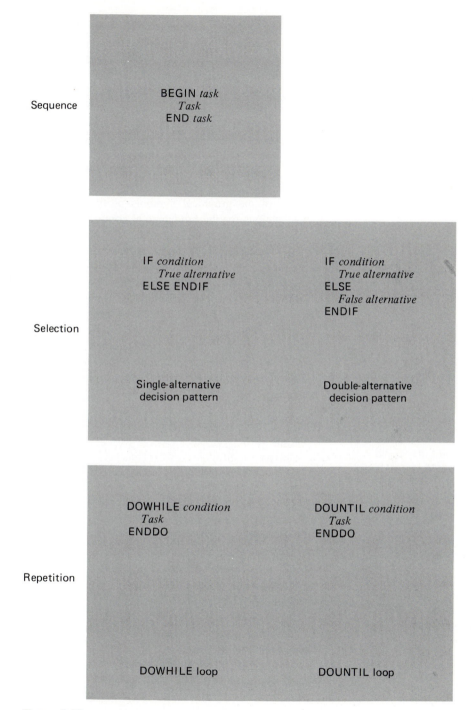

Figure 2.13
Pseudocode representations of the three fundamental control structures.

2.3.2 Pseudocode

The object of flowcharting is obviously to develop a quality computer program consisting of a set of step-by-step instructions called *code*. Inspection of Figs. 2.10 and 2.11 illustrates how top-down flowcharting moves in the direction of a computer program. Recall that in top-down design, more detailed flowcharts are developed as we progress from the general overview (Fig. 2.10*a*) to specific modules (Fig. 2.10*c*). At the highest level of detail, the module flowcharts are almost in the form of a computer program. That is, as seen in Fig. 2.11, each flowchart compartment represents a single well-defined instruction to the computer.

An alternative way to express an algorithm that bridges the gap between flowcharts and computer code is called *pseudocode*. This technique uses Englishlike statements in place of the graphical symbols of the flowchart. Figure 2.13 shows pseudocode representations for the fundamental control structures.

Keywords such as BEGIN, IF, DOWHILE, etc., are capitalized, whereas the conditions, processing steps, and tasks are in lowercase. Additionally, notice that the processing steps are indented. Thus the keywords form a "sandwich" around the steps to visually define the extent of each control structure.

Figure 2.14 shows a pseudocode representation of the flowchart from Fig. 2.11. This version looks much more like a computer program than the flowchart. Thus one advantage of pseudocode is that it is easier to develop a program with it than with a flowchart. The pseudocode is also easier to modify. However, because of their graphic form, flowcharts sometimes are better suited for visualizing complex algorithms. In the present text, pseudocode will be used to communicate algorithms related to numerical methods.

Figure 2.14
Pseudocode to compute the average quiz grade; this algorithm was represented previously by the flowchart in Fig. 2.11.

```
BEGIN  average grade
    Sum = 0
    Count = 0
    INPUT Grade
    DOWHILE (Grade > 0)
        Sum = Sum + Grade
        Count = Count + 1
        INPUT Grade
    ENDDO
    IF (Count = 0)
        Average = 0
    ELSE
        Average = Sum/Count
    ENDIF
END average grade
```

2.4 PROGRAM COMPOSITION

After concocting the algorithm, the next step is to express it as code. Before discussing how this should be done, we will first review the computer languages used in this book.

2.4.1 High-level Languages

Hundreds of computer languages have been developed since the computer age began. The question of which language is "best" has been the subject of heated debate. Unfortunately, many programmers have an almost emotional attachment to "their" language and some go so far as to contend that all others are inferior. We believe that such parochial attitudes can be self-defeating. Although every language has limitations, each can be used to advantage in particular problem contexts.

In the present text, we have chosen to emphasize BASIC, FORTRAN, and Pascal. These languages were chosen because our main audience is not the professional software developer, but rather the engineer who is primarily interested in using programming languages and numerical methods as problem-solving tools. As outlined next, each has advantages in this regard.

FORTRAN.   FORTRAN, which stands for *for*mula *tran*slation was introduced commercially by IBM in 1957. As the name implies, one of its distinctive features is that it uses a notation that facilitates the writing of mathematical formulas. Because of such features, FORTRAN is the native language of most engineers and scientists who attained computer fluency on large mainframe machines in the early 1960s and 1970s. Although most personal computers use BASIC, FORTRAN compilers are available for many of these machines.

Aside from mathematical formulas, FORTRAN has other features that are relevant to numerical methods. For example, it has standard capabilities to handle extended precision and special mathematical functions including complex variables. Additionally, it is highly conducive to modular programming because its subroutines allow local variable scoping and the transfer of values between subprograms and the main program as arguments. Thus, libraries of numerically oriented subprograms can be developed or obtained commercially for use in a FORTRAN code.

Most of FORTRAN's disadvantages are related to the fact that it is commonly implemented as a batch-mode, compiled language. Thus, program development is slower than for interpreted languages such as BASIC that are implemented in interactive environments. FORTRAN also suffers because graphics are not an intrinsic part of the language. Consequently, most FORTRANs require special library functions to implement graphics.

BASIC.   Just as professional engineers and scientists have unique computing needs, so also do students. BASIC, which stands for *b*eginners' *a*ll-purpose *s*ymbolic *i*nstruction *c*ode, was developed expressly as an instructional language at Dartmouth College in the mid-1960s. Its originators, John Kemeny and Thomas Kurtz, recognized that, although FORTRAN was widely used, it had certain characteristics that posed problems for novice programmers. Kemeny and Kurtz came up with BASIC as an alternative that was easier to learn and use. However, because it retained many of the key features of FORTRAN and added some nice ones of its own, it is more than adequate for many engineering and scientific computations. As a consequence, it is presently among the most popular computer languages in the world. In addition, it is unquestionably the most commonly used language on personal computers.

The major advantage of BASIC is the ease with which it can be learned and applied, particularly for the development of short- to medium-sized programs. This is due primarily to the fact that it is an interpreted language. Although this means that programs run slowly, the immediate feedback gained by the user greatly facilitates learning the language and debugging codes. In addition, once a BASIC program has been debugged and tested, compilers are available to speed up execution time by an order of magnitude.

BASIC shares many of the computational advantages of FORTRAN. In particular, current dialects such as Microsoft BASIC allow long variable names, double precision, and adequate libraries of mathematical functions. Also, graphics are a more intrinsic part of the language than for FORTRAN and are, therefore, easier to implement.

Although BASIC is generally well-designed to implement computations, it has several significant disadvantages in this regard. For example, it does not explicitly manipulate or include functions of complex variables. More important, however, is the fact that most of the popular BASICs do not allow arguments and local variables in subroutines (that is, variable scoping). Thus, BASIC is much less suited for a modular approach than languages such as FORTRAN or Pascal. This shortcoming represents the most serious impediment to the use of BASIC for large-scale programming efforts.

Pascal.   As noted in Fig. 2.5, software development constitutes the dominant share of today's computing costs.   Consequently, innovations that increase a programmer's productivity and efficiency can lead to significant savings. Pascal is one of the structured languages that has been created to facilitate large-scale programming. Developed by Nicklaus Wirth and named after Blaise Pascal, the language uses a restricted set of design structures and other organizing principles. By essentially imposing good style habits on the user, Pascal makes for programs that are easier to use and maintain. These advantages become more pronounced for larger programs.

Aside from its structure, most of Pascal's other strong points relate to its capabilities to manipulate nonnumerical information. For example, its superior capacity to handle data structures makes it better suited for capitalizing on some advanced personal computer capabilities such as windowing, mouse input, and menu operation.

Pascal's primary disadvantages relate to computations. For example it lacks double-precision variables and does not have as extensive an arsenal of library functions as FORTRAN or BASIC. In addition, the fact that it is a relatively verbose language makes it inappropriate for smaller programs.

Dialects.   As time passes, new versions or "dialects" of each language are being developed. Not surprisingly, the newer versions of each language often incorporate some of the advantages of the others. For instance, recent dialects of BASIC have improved input–output capabilities that are closer in power to those of FORTRAN. Also, structured versions of both BASIC and FORTRAN are now available that incorporate some of the strengths of Pascal.

The specific dialects used in this book—Microsoft BASIC, FORTRAN 77, and Turbo Pascal—were chosen because of their availability, ease of use, and compatibility with a structured programming approach. Thus, all codes in the book are written in a highly structured style derived from pseudocode. Consequently, all can be easily translated into more advanced languages such as C or Ada. In the following section, we will elaborate on this structured approach.

2.4.2 Structured Programming

Not everyone agrees on a standard definition of structured programming. In general, the major idea is embodied in what can be called the Structure Principle (Dijkstra 1968):

The structure principle: *The static structure (that is, spread out on the page) of the program should correspond in a simple way to the dynamic structure (that is, spread out in time) of the corresponding computation.*

Thus, the "jumping around" due to indiscriminate branching and other poor programming habits will be avoided as a natural consequence of the orderly nature of well-conceived computations.

In addition to the structure principle, there are several other rules that can be considered as part of the structured programming philosophy. The following are among the most generally agreed upon rules:

1. Programs should consist solely of the three fundamental control structures of sequence, selection, and repetition (recall Fig. 2.12). Computer scientists have proved that any program can be constructed from these basic structures.
2. Each of the structures must have only one entrance and one exit.
3. Unconditional transfers (GO TOs) should be avoided.
4. The structures should be clearly identified with comments and visual devices such as indentation, blank lines, and blank spaces.

Although they may appear deceptively simple, adherence to these rules avoids the spaghettilike code that is the hallmark of indiscriminate branching. Because of their benefits, these rules and other structured programming practices will be stressed continually in subsequent parts of this book.

Figures 2.15 and 2.16 show how the selection and repetition constructs are programmed in the three dialects employed in this book. In addition, Fig. 2.17 illustrates an alternative type of repetition construct—the counter loop*—that will also be used in subsequent programs.

Figure 2.18 is a dramatic example of the contrast between unstructured and structured approaches. Both programs are designed to determine an average grade, the algorithm for which was delineated in Fig. 2.14. Although the programs compute identical values, their designs are in marked contrast. The purpose, logic, and organization of the unstructured program in Fig. 2.18*a* is not obvious. The user must delve into the code on a line-by-line basis to decipher the program. In contrast, the major parts of Fig. 2.18*b* are immediately apparent. Title and variable-identification modules are placed at the beginning to orient the user. Program execution progresses in an orderly fashion from the top to the bottom with no major jumps. Every segment has one entrance and one exit, and each is clearly delineated by comments, spaces, and indentation. In essence, the program is distinguished by a visual clarity that is reminiscent of a well-designed flowchart. Additionally, the program is a natural evolution of the pseudocode shown in Fig. 2.14.

On the debit side, there is no question that the program in Fig. 2.18*b* takes longer to develop than the program in Fig. 2.18*a*. Because many pragmatic engineers are

---

*Note that the counter loop is, in fact, a specialized version of the DOWHILE construct.

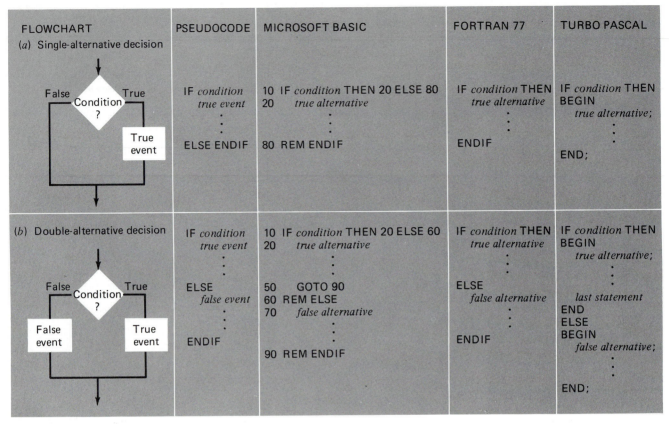

FLOWCHART	PSEUDOCODE	MICROSOFT BASIC	FORTRAN 77	TURBO PASCAL

(a) Single-alternative decision	IF *condition*
	   true event
	.
	.
	.
	ELSE ENDIF
		10 IF *condition* THEN 20 ELSE 80
		20 *true alternative*
		    .
		    .
		    .
		80 REM ENDIF
			IF *condition* THEN
			  true alternative
			   .
			   .
			   .
			ENDIF
				IF *condition* THEN
				BEGIN
				   true alternative;
				    .
				    .
				    .
				END;

(b) Double-alternative decision	IF *condition*
	   true event
	   .
	   .
	ELSE
	   false event
	   .
	   .
	ENDIF
		10 IF *condition* THEN 20 ELSE 60
		20 *true alternative*
		    .
		    .
		    .
		50 GOTO 90
		60 REM ELSE
		70 *false alternative*
		    .
		    .
		90 REM ENDIF
			IF *condition* THEN
			  true alternative
			   .
			   .
			ELSE
			  false alternative
			   .
			   .
			ENDIF
				IF *condition* THEN
				BEGIN
				   true alternative;
				    .
				    .
				   last statement
				END
				ELSE
				BEGIN
				   false alternative;
				    .
				    .
				END;

Figure 2.15
Selection constructs as expressed in Microsoft BASIC, FORTRAN 77, and Turbo Pascal. Note that because of the limitations of its IF/THEN/ELSE statement, the BASIC version requires a GOTO statement. Such is not the case for the more highly-evolved FORTRAN and Pascal versions.

Figure 2.16 (*facing page, top*)
Repetition constructs as expressed in Microsoft BASIC, FORTRAN 77, and Turbo Pascal. Note that FORTRAN has neither a formal DOWHILE or a DOUNTIL construct. Similarly, Microsoft BASIC does not have a DOUNTIL. Thus, these constructs must presently be simulated as shown here.

Figure 2.17 (*facing page, bottom*)
When the number of iterations are known beforehand, counter loops of the types illustrated above can be used to implement repetition in Microsoft BASIC, FORTRAN 77, and Turbo Pascal. Note that Turbo Pascal does not allow steps other than plus or minus one. Therefore, as shown here, a DOWHILE loop is used to simulate other step sizes.

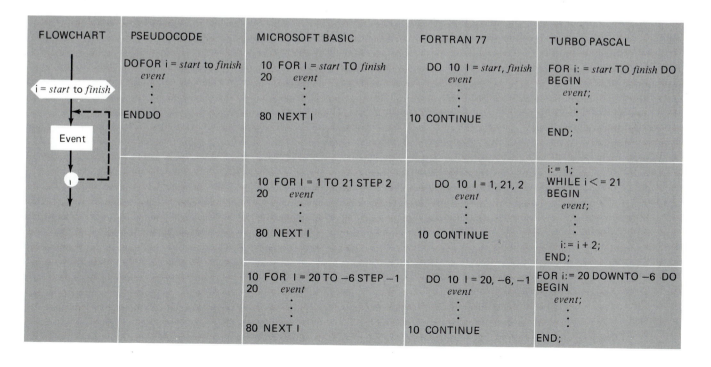

(a) DOWHILE loop

FLOWCHART	PSEUDOCODE	MICROSOFT BASIC	FORTRAN 77	TURBO PASCAL

Event / Condition? / True / False	DOWHILE *condition*
	  event
	  ⋮
	ENDDO
		10 WHILE *condition*
		20 *event*
		   ⋮
		80 WEND
			C WHILE LOOP
			10 IF (*condition*) THEN
			     event
			     ⋮
			     GO TO 10
			   ENDIF
				WHILE *condition* DO
				BEGIN
				  event;
				  ⋮
				END;

(b) DOUNTIL loop

FLOWCHART	PSEUDOCODE	MICROSOFT BASIC	FORTRAN 77	TURBO PASCAL

Event / Condition? / False / True	DOUNTIL *condition*
	  event
	  ⋮
	ENDDO
		10 *set condition false*
		20 WHILE NOT *condition*
		30 *event*
		   ⋮
		80 WEND
			C UNTIL LOOP
			   set condition false
			10 IF (.NOT.*condition*) THEN
			     event
			     ⋮
			     GO TO 10
			   ENDIF
				REPEAT
				BEGIN
				  event;
				  ⋮
				UNTIL *condition*;

DOFOR loop

FLOWCHART	PSEUDOCODE	MICROSOFT BASIC	FORTRAN 77	TURBO PASCAL

i = *start* to *finish* / Event / i	DOFOR i = *start* to *finish*
	  event
	  ⋮
	ENDDO
		10 FOR I = *start* TO *finish*
		20 *event*
		   ⋮
		80 NEXT I
			DO 10 I = *start, finish*
			     event
			     ⋮
			10 CONTINUE
				FOR i: = *start* TO *finish* DO
				BEGIN
				  event;
				  ⋮
				END;

		10 FOR I = 1 TO 21 STEP 2
		20 *event*
		   ⋮
		80 NEXT I
			DO 10 I = 1, 21, 2
			     event
			     ⋮
			10 CONTINUE
				i: = 1;
				WHILE i < = 21
				BEGIN
				  event;
				  ⋮
				  i: = i + 2;
				END;

		10 FOR I = 20 TO −6 STEP −1
		20 *event*
		   ⋮
		80 NEXT I
			DO 10 I = 20, −6, −1
			     event
			     ⋮
			10 CONTINUE
				FOR i: = 20 DOWNTO −6 DO
				BEGIN
				  event;
				  ⋮
				END;

```
100 S=0
105 I=0
110 INPUT G
115 IF G<0 THEN 135
120 S=S+G
125 I=I+1
130 GOTO 110
135 IF I=0 THEN 150
140 A=S/I
145 GOTO 155
150 A=0
155 PRINT A
160 END
```
(a)

```
100 REM       AVERAGE GRADE PROGRAM (BASIC)
105 REM
110 REM ************************************
115 REM *        DEFINITION OF VARIABLES        *
120 REM *                                       *
125 REM * GRADE   = GRADE INPUT BY USER         *
130 REM * TOTAL   = SUMMATION OF GRADES         *
135 REM * COUNT   = NUMBER OF GRADES            *
140 REM * AVERAG  = AVERAGE GRADE               *
145 REM ************************************
150 REM
155 REM *********  MAIN PROGRAM  **********
160 REM
165 GRADE = 0
170 TOTAL = 0
175 COUNT = 0
180 REM            Accumulate Grade Data
185 REM
190 PRINT "GRADE = ";
195 PRINT "(ENTER NEGATIVE TO TERMINATE)";
200 INPUT GRADE
205 WHILE GRADE >= 0
210    TOTAL = TOTAL + GRADE
215    COUNT = COUNT + 1
220    PRINT "GRADE = ";
225    PRINT "(ENTER NEGATIVE TO TERMINATE)";
230    INPUT GRADE
235 WEND
240 REM            Calculate Average Grade
245 REM
250 IF COUNT = 0 THEN 255 ELSE 265
255      AVERAG = 0
260      GOTO 275
265    REM else
270      AVERAG = TOTAL/COUNT
275 REM endif
280 PRINT "AVERAGE GRADE = ";AVERAG
285 END
```
(b)

Figure 2.18
(a) An unstructured and (b) a structured version of the same program. Both versions perform identical computations but differ markedly in terms of clarity. Note that GOTO in line 260 of (b) had to be used in order to program a double-alternative decision in Microsoft BASIC (see Fig. 2.15b). Otherwise, GOTOs should be avoided.

accustomed to "getting the job done" in an efficient and timely manner, the extra effort might not always be justified. Such would be the case for a short program you might develop for a "one-time-only" computation. However, for more important programs that are intended for application by yourself or others, there is no question that a structured approach yields long-term benefits.

2.5 QUALITY CONTROL

With the tools described up to this point, you would be capable of developing a program to implement a calculation on the computer. However, the fact that a program prints out information is no guarantee that these answers are correct. Consequently, we will now turn to some issues related to the reliability of your program. We are emphasizing this subject at this point to stress its importance in the program-development process. Since we are problem-solving engineers, most of our work will directly or indirectly be used by clients. For this reason, it is essential that our programs be reliable—that is, that they do exactly what they are supposed to do. At best, lack of reliability can be embarrassing. At worst, for engineering problems involving public safety, it can be tragic.

2.5.1 Errors or "bugs"

During the course of preparing and executing a program of any size, it is likely that errors will occur. These errors are sometimes called *bugs* in computer jargon. This label was coined by Admiral Grace Hopper, one of the pioneers of computer languages. In 1945, she was working with one of the earliest electromechanical computers when it went dead. She and her colleagues opened the machine and discovered that a moth had become lodged in one of the relays. Thus, the label "bug" came to be synonymous with problems and errors associated with computer operations and the term *debugging* came to be associated with their solution.

When developing and running a program, three types of bugs can occur:

1. *Syntax errors* violate the rules of the language such as spelling, number formation, line numbering, and other conventions. These errors are often the result of mistakes such as typing REED rather than READ. They usually result in the computer printing out an error message. Such messages are called *diagnostics* because the computer is helping you to "diagnose" the problem.
2. *Run-time errors* are those that occur during program execution. An example is when there are an insufficient number of data entries for the number of variables in an input statement. For such situations a diagnostic will usually be printed and you must then reenter the data properly. For other run-time errors the computer may simply terminate the run or print a message providing you with information regarding the error. In any event, you will be cognizant of the fact that an error has occurred.
3. *Logic errors,* as the name implies, are due to faulty program logic. These are the worst type of error because they can occur without diagnostics. Thus, your program will appear to be working properly in the sense that it executes and generates output. However, the output will be incorrect.

All the above types of errors must be removed before a program can be employed for engineering applications. We divide the process of quality control into two categories: debugging and testing. As mentioned above, *debugging* involves correcting known errors. In contrast, *testing* is a broader process that is intended to detect errors that are not known. Additionally, testing should also determine whether the program fulfills the needs for which it was designed.

As will be clear from the following discussion, debugging and testing are interrelated and are often carried out in tandem. For example, we might debug a module to remove the obvious errors. Then, after a test, we might uncover some additional bugs that require correction. This two-pronged process would be repeated until we were convinced that the program was absolutely reliable.

2.5.2 Debugging

As stated above, debugging deals with correcting known errors. There are three ways in which you will become aware of errors. First, an explicit diagnostic can provide you with the exact location and the nature of the error. Second, a diagnostic may occur but the exact location of the error is unclear. For example, the computer may indicate that an error has occurred in a line representing a long equation including some user-defined functions. This error could be due to syntax errors in the equation or it could also stem from errors in the statements defining the functions. Third, no diagnostics occur but the program does not operate properly. These are usually due to logic errors such as infinite loops or mathematical mistakes that do not yield syntax errors. Corrections are easy for the first type of error. However, for the second and third types, some detective work is in order to identify the errors and their exact locations.

Unfortunately, the analogy with detective work is all too apt. That is, ferreting out errors is often difficult and involves a lot of "leg work," collecting clues and running up some blind alleys. In particular, as with detective work, there is no clear-cut methodology involved. However, there are some ways of going about it that are more efficient than others.

First, it should be noted that interpreted BASIC on a personal computer is an ideal tool for debugging. Because of its interactive or "hands-on" mode of operation, the user can efficiently perform the repeated runs that are sometimes needed to zero in on the location of errors. Although it is compiled, Turbo Pascal's efficient run mode offers similar advantages.

One key to identifying bugs is to print out intermediate results. With this approach, it is often possible to determine at what point the computation went awry. Similarly the use of a modular approach will obviously help to localize errors in this way. You could put a different output statement at the start of each module and in this way focus in on the module in which the error occurred. In addition, it is always good practice to debug (and test) each module separately before integrating it with the total package.

A more formal way to determine the location of an error is with a trace. Activating such a command before a run causes the computer to print out every line number as it is executed along with any other input or output for a program. Although for long programs this can result in large quantities of output, it is sometimes the most efficient way to home in on an error.

Although the above procedures can help, there are often times when your only option will be to sit down and read the code line by line following the flow of logic and performing all the computations with a pencil, paper, and a pocket calculator. Just as a detective sometimes has to think like a criminal in order to solve a case, there will be times you will have to think like a computer to debug a program.

2.5.3 Testing

One of the greatest misconceptions of the novice programmer is the belief that, if a program runs and prints out results, it is correct. Of particular danger are those cases where the results appear to be "reasonable" but are in fact wrong. To ensure that such cases do not occur, the program must be subjected to a battery of tests for which the correct answer is known beforehand.

As mentioned in the previous section, it is good practice to debug and test modules prior to integrating them into the total program. Thus, testing usually proceeds in phases. These are module tests, development tests, whole system tests, and operational tests.

Module tests, as the name implies, deal with the reliability of individual modules. Because each is designed to perform a specific, well-defined function, the modules can be run in isolation to determine that they are executing properly. Sample input data can be developed for which the proper output is known. This data can be used to run the module and the outcome compared with the known result to verify successful performance.

Development tests are implemented as the modules are integrated into the total program package. That is, a test is performed after each module is integrated. An effective way to do this is with a top-down approach that starts with the first module and progresses down through the program in execution sequence. By performing a test after each module is incorporated, problems can be isolated more easily because new errors can usually be attributed to the latest module that has been added.

After the total program has been assembled, it can be subjected to a series of *whole-system tests*. These should be designed to expose the program to

1. Typical data
2. Unusual but valid data
3. Incorrect data to check the program's error-handling capabilities

Finally, *operational tests* are designed to check how the program performs in a realistic setting. A standard way to do this is to have a number of independent subjects (including the client) implement the program. Operational tests sometimes turn up bugs. In addition, they provide feedback to the developer regarding ways to improve the program so that it more closely meets the user's needs.

Although the ultimate objective of the testing process is to ensure that programs are error-free, there will be complicated cases where it is impossible to subject a program to every possible contingency. Also, it should be recognized that the level of testing depends on the importance and magnitude of the problem context for which the program is being developed. There will obviously be different levels of rigor applied to testing a program for determining the averages of your intramural softball team as compared with a program to regulate the operation of a nuclear reactor or the space shuttle. In each case, however, adequate testing must be performed that is consistent with the liabilities connected with a particular problem context.

2.6 DOCUMENTATION

After the program has been debugged and tested, it must be documented. Documentation is the addition of English-language descriptions to allow the user to implement the program more easily. Remember, along with other people who might employ your software, you also are a "user." Although a program may seem clear and simple when it is first composed, with time the same code may seem inscrutable. Therefore, sufficient documentation must be included to allow you and other users to immediately understand your program and how it can be implemented. This task has both internal and external aspects.

2.6.1 Internal Documentation

Picture for a moment a text with just words and no paragraphs, subheadings, or chapters. Such a book would be more than a little intimidating to study. The way the pages of a text are structured—that is, all the devices that act to break up and organize the material—makes the text more effective and enjoyable. In the same sense, internal documentation of computer code can greatly enhance the user's understanding of a program and how it works.

The following are some general suggestions for documenting programs internally:

1. Include a module at the head of the program giving the program's title and your name and address. This is your signature and marks the program as your work.
2. Include a second module to define each of the key variables.
3. Select variable names that are reflective of the type of information the variables are to store.
4. Insert spaces within statements to make them easier to read.
5. Use comment statements liberally throughout the program to add explanation and to skip lines for the purposes of labeling, clarity, and separation of modules.
6. In particular, use comment statements to clearly label and set off all the modules.
7. Use indentation to clarify the program structures. In particular, indentation can be used to set off loops and decisions.

2.6.2 External documentation

External documentation refers to instructions in the form of output messages and supplementary printed matter. The messages occur when the program runs and are intended to make the output attractive and "user-friendly." This involves the effective use of spaces, blank lines, or special characters to illuminate the logical sequence and structure of the program output. Attractive output simplifies the detection of errors and enhances the communication of program results.

The supplementary printed matter can range from a single sheet to a comprehensive user's manual. Figure 2.19 is an example of a simple documentation form that we recommend you prepare for every program you develop. These forms can be maintained in a notebook to provide a quick reference for your program library. The user's manual for your computer is an example of comprehensive documentation. This manual tells you how to run your computer system and disk-operating programs. Note that for well-documented programs, the title information in Fig. 2.19 is unnecessary because the title module is included in the program.

2.7 STORAGE AND MAINTENANCE

Remember that when you sign off the computer, your active file will be destroyed. In order to retain a program for later use, you must transfer it to a secondary storage device such as a disk or tape before turning the machine off.

Aside from this physical act of retaining the program, storage and maintenance consists of two major tasks: (1) upgrading the program in light of experience and (2)

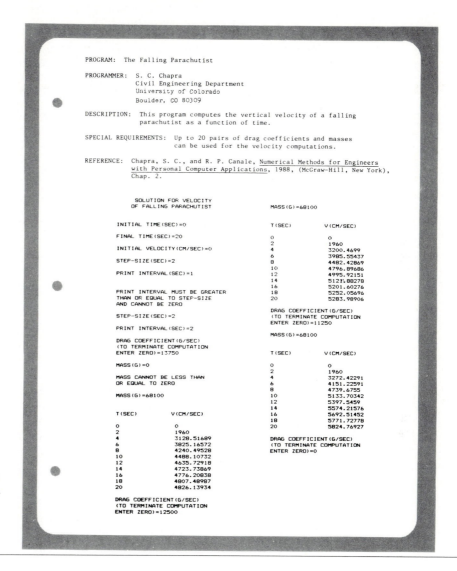

Figure 2.19
An example of external documentation for a program to compute the velocity of the falling parachutist (recall Chap. 1). Note that such a sheet could be incorporated into a notebook along with a listing of the program's source code.

ensuring that the program is safely stored. The former is a matter of communicating with users to gather feedback regarding suggested improvements to your design.

The question of safe storage of programs is especially critical when using soft diskettes or floppy disks. Although these devices offer a handy means for retaining your work and transferring it to others, they can be easily damaged. Table 2.1 outlines a number of tips for effectively maintaining your programs on soft diskettes. Follow the suggestions on this list carefully in order to avoid losing your valuable programs accidentally.

TABLE 2.1   Suggestions and precautions for effective maintenance of software on diskettes.

- Maintain paper printouts of all finished programs in a file or binder.
- Always maintain at least one backup copy of all diskettes.
- Handle a diskette gently. Do not bend it or touch its exposed surfaces.
- Afix a temporary label to each diskette describing its contents. Once the label is attached, only use soft felt-tipped pens to add more information.
- Once a diskette is complete, cover the write-protect notch with foil tape.
- Keep the diskette in its protective envelope when not in use.
- Store diskettes upright in a place that is neither too hot nor too cold.
- Keep diskettes away from water and contaminants such as smoke and dust.
- Store diskettes away from magnetic devices such as televisions, tape recorders, electric motors, etc.
- Remove diskettes from the drives before turning the system off.
- Never remove a diskette when the red light on the drive is lit.

2.8   SOFTWARE STRATEGY

One of your primary objectives in using this book should be to develop your own numerical methods software library. Such a library will be of immense value in your future academic and professional efforts. We have provided a diverse array of computer media to help you accomplish this objective. These include computer programs, pseudocode, and software.

The programs included in the book are summarized in Table 2.2. Versions have been written in Microsoft BASIC, FORTRAN 77, and Turbo Pascal. All BASIC programs are written in an interactive style that is compatible with a personal computer environment. In contrast, the FORTRAN and Pascal versions are presented as subroutines in order to capitalize on the modularity of these languages. Of course, the FORTRAN and Pascal subroutines can be incorporated into interactive programs for implementation on microcomputers.

It should be noted that all the computer code is sparsely documented. This was done intentionally. One of your tasks will be to document and enhance these programs using your own resources and individual style. Among other things, this will encourage you to understand each program's underlying organization and logic.

Aside from these programs, pseudocode and other descriptions of algorithms are provided for the remaining methods in the book. These can form the basis for your own programs. Together with the codes from Table 2.2, these programs will constitute a fairly comprehensive numerical methods capability.

Over and above your own collection of programs, we have also developed a generic software product called the Electronic TOOLKIT that can be used in conjunction with the book. This package includes spreadsheet, statistics, and graphics programs along with six of the more fundamental numerical methods (Table 2.3). Each of the numerical methods programs is illustrated fully by an example problem in the relevant chapter of the book. The examples can provide you with models for the design of input and output screens including the graphical display of results. Homework problems are included

TABLE 2.2 The computer programs included in this book.

Program	Chapter	Purpose
Bisection	4	Determine a single root of an algebraic or transcendental equation.
Gauss elimination	7	Solve a system of linear algebraic equations for a single right-hand-side vector.
LU decomposition	9	Solve a system of linear algebraic equations for multiple right-hand-side vectors.
Linear regression	11	Fit a straight line to a set of paired observations by minimizing the sum of the squares of the residuals.
Newton's polynomial	12	Estimate intermediate values by passing an interpolating polynomial through a set of paired observations.
Trapezoidal rule	15	Estimating the integral of a function defined by a set of paired observations.
Numerical differentiation	17	Determine the derivative of a function defined by a set of equispaced, paired observations.
Euler's method	19	Solve a single, first-order ordinary differential equation.

in the relevant chapters to reinforce your capabilities to use the disk on your own microcomputer. Additionally the software can be employed to check the accuracy of your own programming efforts.

Aside from the numerical methods, the other capabilities on the Electronic TOOLKIT can also be used in conjunction with the text. The statistics program has direct application to the review of that subject in the introduction to Part Four. In addition, each part of the

TABLE 2.3 The programs in the Electronic TOOLKIT.

Program	Capability
Bisection	Determines a real root of a single algebraic or transcendental equation.
Gauss elimination	Solves a system of linear algebraic equations.
Linear regression	Fits a straight line to data points by minimizing the sum of the squares of the residuals.
Lagrange interpolation	Uses interpolating polynomials to estimate intermediate values between data points.
Trapezoidal rule	Integrates a function in equation or tabular form.
Euler's method	Solves a first-order ordinary differential equation.
Spreadsheet	A single-screen spreadsheet that can be employed for "what if?" computations involving interconnected sets of equations.
Statistics	Determines descriptive statistics and generates a histogram for data.
Graph generator	Generates rectilinear, semilog, and log-log plots of data and functions.

book includes a set of homework problems that can be solved directly with the spreadsheet program. Finally, as described in the following example, the graph generator can be employed throughout the text to provide a visual representation of data and functions.

EXAMPLE 2.1   Computer Graphics

Problem Statement: The purpose of this example is to familiarize you with the programs on the optional software (the Electronic TOOLKIT) available with the text and to use its graphics capabilities to plot functions. If you purchased this book without this software, then you should explore ways to perform similar tasks on your own computer. This may be accomplished with your system's software, or it may require that you develop your own programs. The ability to plot functions is important because the problem-solving utility of numerical methods is tremendously enhanced when used in coordination with computer graphics.

Solution: Insert the Electronic TOOLKIT disk into your computer's disk drive and run the program according to the directions in the *User's Manual*.

The screen should produce a pattern similar to that in Fig. 2.20*a*. This is simply a title page. Enter RETURN to continue. The screen should now produce a program selection main menu as shown in Fig. 2.20*b*. The menu contains a list of nine programs along with options to read or save data or to terminate the session. We will use each of these programs at appropriate places throughout the text. For now we will use the GRAPHICS program to plot the velocity of the falling parachutist as a function of time. To do this, simply enter the GRAPHICS program by selecting option 6 whereupon the screen should produce a pattern similar to that of Fig. 2.20*c*. You will need to use only options 1, 4, and 5 to plot functions. Select option 1 to enter the function using Eq. (1.10) with $m = 68.1$ kg, $c = 12.5$ kg/s, and $g = 9.8$ m/s^2 (Fig. 2.20*d*). Return to the GRAPHICS program menu and select option 4 to plot the function. Before the graph can be generated, it is necessary to select the type of plot, its dimensions, and labeling information. Figure 2.20*e* shows parameters that define a rectilinear plot of the falling parachutist model. Try various values for the minimum and maximum of both x (in our case, time) and y (velocity) including negative values to gain familiarity with the plot option design and operation. A well-proportioned plot corresponding to the parameters in Fig. 2.20*e* is shown in Fig. 2.20*f*.

The plot option of your software will have many other uses as you pursue your goal of understanding and applying numerical methods and computers to solve engineering problems. We will explore these uses in appropriate sections of the text.

Figure 2.20 (*facing page*)
(*a*) Title screen from the Electronic TOOLKIT software. (*b*) The main menu of the Electronic TOOLKIT. (*c*) The menu for the GRAPHICS program. (*d*) Screen showing how a function is entered; here the function is Eq. (1.10) which computes the velocity of the falling parachutist. (*e*) Screen used to select plot; (*f*) The resulting plot.

```
┌─────────────────────────────────────┐
│     Electronic Toolkit Main Menu     │
└─────────────────────────────────────┘

┌─────────────────────────────────────┐
│     [1]   Spreadsheet                │
│     [2]   Linear Regression          │
│     [3]   Lagrange Interpolation     │
│     [4]   Trapezoidal Rule           │
│     [5]   Statistics                 │
│     [6]   Graphics                   │
│                                      │
│     [7]   Bisection                  │
│     [8]   Gauss Elimination          │
│     [9]   Euler's Method             │
│                                      │
│    [10]   Read Data                  │
│    [11]   Save Data                  │
│                                      │
│    [12]   Exit Electronic Toolkit    │
└─────────────────────────────────────┘

── Make selection and enter [Return] ──
```

(b)

```
┌─────────────────────────────────────┐
│        Graphical Display of          │
│         Data and Functions           │
└─────────────────────────────────────┘

┌─────────────────────────────────────┐
│   [1]   Enter function               │
│                                      │
│   [2]   Enter new data               │
│                                      │
│   [3]   Modify old data              │
│                                      │
│   [4]   Plot data and function       │
│                                      │
│   [5]   Return to main menu          │
└─────────────────────────────────────┘

── Make selection and enter [Return] ──
```

(c)

```
┌─────────────────────────────────────┐
│            Enter function            │
└─────────────────────────────────────┘


   f(X)  =   53.39*(1-exp(-0.18355*X))




───── Enter f(X) = 0 then [Return] ─────
```

(d)

```
┌─────────────────────────────────────┐
│         Plot data and function       │
└─────────────────────────────────────┘
```

Linear	Semi-log	Log-log
Points	Function	Both

Parameter	Value
Minimum X	0
Maximum X	15
Minimum Y	0
Maximum Y	60
X axis label	TIME
Y axis label	VELOCITY
Plot Title	PARACHUTIST

```
───── Enter parameters then [End] ─────
```

(e)

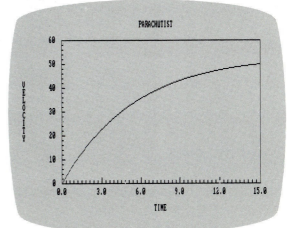

(f)

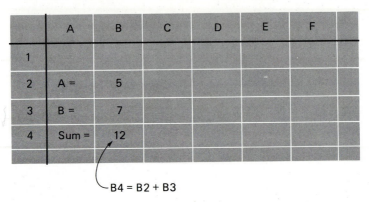

B4 = B2 + B3

Figure 2.21
Three fundamental entries can be made to a spreadsheet cell: a label, a number, or a formula. The latter are themselves functions of the numerical values in other cells. As such, the spreadsheet is a computerized version of a large piece of paper on which an involved computation can be displayed or spread out. Because the entire computation is automatically updated when any number is changed, spreadsheets are ideal for "what if?" sorts of analyses.

2.8.1 Spreadsheets and Engineering

Electronic spreadsheets are a special type of software that allows the user to enter and perform calculations on rows and columns of data displayed on a computer monitor. It is a computerized version of a large piece of paper or worksheet on which an involved calculation can be displayed or spread out.

Spreadsheets have a number of advantages that have contributed to their immense popularity. First, they are extremely easy to use and understand. Second, they provide an organized record of your computation. Third, because the entire calculation may be automatically updated when any number on the sheet is changed, spreadsheets are ideal for "what if?" sorts of analyses (Fig. 2.21). It is this latter aspect that adds a powerful new dimension to engineering problem solving (see Chapra and Canale 1986).

Although they have made their most significant impact in business-related fields, spreadsheets promise to figure more prominently in future engineering work. For this reason, we have included a spreadsheet case study and spreadsheet homework problems in each major part of the book. In addition, we have developed a software supplement, in a diskette at the back of the book, that you can use to implement the case study spreadsheets on IBM PCs and compatibles. The operation of this software supplement will be described in the appropriate case studies.

PROBLEMS

2.1 List and define six steps in the production of high-quality software in engineering.

2.2 Define top-down design, modular design, and structured programming.

2.3 A value for the plasticity of soil samples is recorded on each in a set of index cards. A card marked "end of data" is placed at the end of the set. Write an algorithm to determine the sum and the average of these values.

2.4 Write a structured flowchart for Prob. 2.3.

2.5 Write an algorithm to print out the roots (either real or complex) of a quadratic equation

$$ax^2 + bx + c = 0$$

where a, b, and c are real coefficients. Design your algorithm so that it deals with all possible contingencies that might occur during the computation.

2.6 Develop a structured flowchart for Prob. 2.5.

2.7 Write pseudocode for Prob. 2.5.

2.8 Develop, debug, and document a computer program for Prob. 2.5 using the high-level language of your choice. Perform test runs for the cases
   (a) a = 1   b = 5   c = 2
   (b) a = 0   b = −3   c = 2.8
   (c) a = 1   b = 2.5   c = 7

2.9 The exponential function e^x can be evaluated by the following infinite series:

$$e^x = 1 + x + \frac{x^2}{2} + \frac{x^3}{3!} + \frac{x^4}{4!} + \cdots$$

Write an algorithm to implement this formula so that it computes and prints out the values of e^x as each term in the series is added. In other words, compute and print in sequence the values for

$$e^x \simeq 1$$

$$e^x \simeq 1 + x$$

$$e^x \simeq 1 + x + x^2/2$$

.

.

.

up to the order term of your choosing. For each of the above, compute and print out the percent relative error as

$$\% \text{ error} = \frac{\text{true} - \text{series approximation}}{\text{true}} \times 100\%$$

2.10 Develop a structured flowchart for Prob. 2.9.

2.11 Write pseudocode for Prob. 2.9.

2.12 Develop, debug, and document a computer program for Prob. 2.9 using the high-level language of your choice. Employ the library function for e^x in your computer to determine the true value. Have the program print out the series approximation and the error at each step. As a test case, employ the program to compute $e^{0.5}$ for up to and including the term $x^{20}/20!$. Interpret your results.

2.13 The following algorithm is designed to determine a grade for a course that consists of quizzes, homework, and a final exam:

Step 1: Input course number and name.
Step 2: Input weighting factors for quizzes (WQ), homework (WH), and the final exam (WF).
Step 3: Input quiz grades and determine an average quiz grade (AQ).
Step 4: Input homework grades and determine an average homework grade (AH).
Step 5: If this is a final grade, continue to step 6. If not go to step 9.
Step 6: Input final exam grade (FE).
Step 7: Determine average grade AG according to

$$AG = \frac{WQ*AQ + WH*AH + WF*FE}{WQ + WH + WF}$$

Step 8: Go to step 10.
Step 9: Determine average grade AG according to

$$AG = \frac{WQ*AQ + WH*AH}{WQ + WH}$$

Step 10: Print out course number, name, and average grade.
Step 11: Terminate computation.

Write, debug, and document a structured computer program based on the algorithm. Test it using the following data to calculate a grade without the final exam and a grade with the final exam: WQ = 35; WH = 25; WF = 40; quizzes = 100, 93, 86, 76, 100; homework = 96, 94, 93, 100, 77; and final exam = 89.

2.14 An amount of money P is invested in an account where interest is compounded at the end of the period. The future worth F yielded at an interest rate i after n periods may be determined from the following formulation:

$$F = P(1 + i)^n$$

Write a program which will calculate the future worth of an investment. The input to the program should include the initial investment P, the interest rate i (as a decimal), and the number of years n for which the future worth is to be calculated. The output should include these values

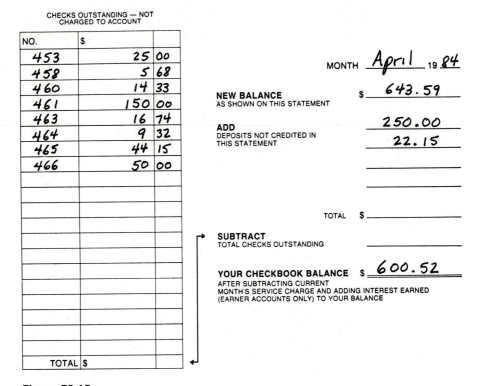

Figure P2.15

also. The output should also include, in a labeled table format, the future worth for each year up to and including the nth year. Run the program for $P = \$1000.00$, $i = 0.1$, and $n = 20$ years.

2.15 Figure P2.15 shows the reverse of a checking account statement. The bank has developed this to help you balance your checkbook. If you look at it closely, you will realize that it is an algorithm. Develop a step-by-step algorithm (in a format similar to Fig. 2.6a) to accomplish the same task.

2.16 Write a structured flowchart for Prob. 2.15.

2.17 Develop, debug, and document a computer program for Prob. 2.15. Test it by verifying that it duplicates the results shown in Fig. P2.15.

2.18 Write, debug, and document a computer program to determine statistics for your favorite sport. Pick anything from football to jogging to bowling. If you play intramural sports, make one up for your team. Design the program

so that it is user-friendly and provides valuable and interesting information to anyone (for example, a coach or a player) who might use it to evaluate athletic performance.

2.19 Economic formulas are available to compute annual payments for loans. Suppose that you borrow an amount of money P and agree to repay it in n annual payments at an interest rate of i. The formula to compute the annual payment A_1 is

$$A_1 = P \frac{i(1 + i)^n}{(1 + i)^n - 1}$$

Write a computer program to compute A_1. Test it with $P = \$10,000$ and an interest rate of 20 percent ($i = 0.20$). Set up the program so that you can evaluate as many values of n as you like. Compute results for $n = 1, 2, 3, 4$, and 5.

2.20 Aside from computing annual payments for a loan, as was done in Prob. 2.19, economic formulas can also be employed to determine annual payments corresponding to other types of cash flow. For example, suppose that you had an expense that increased at a constant rate G as time increased. Such a payment is called an *arithmetic gradient series*. The economic formula to compute an equivalent annual payment for this type of cash flow is

$$A_2 = G\left[\frac{1}{i} - \frac{n}{(1 + i)^n - 1}\right]$$

Now, suppose that you take out a loan of $P = \$10,000$ at an interest rate of $i = 0.20$ and buy a new personal computer system. The maintenance cost for this computer increases according to the arithmetic gradient series at a rate of $G = \$50$/year/year. Aside from these two costs (that is, negative cash flows for the loan repayment and maintenance), you also gain benefits or positive cash flow from your ownership of the system. Your annual consulting profit and enjoyment from the computer can be valued at an annual worth of $A_3 = \$4000$. Therefore, the net worth A_N of owning the machine on an annual basis can be calculated as benefits minus costs, or

$$A_N = A_3 - A_1 - A_2$$

Thus, if A_N is positive, the computer is making you money on an annual basis. If A_N is negative, you are losing money. Develop, debug, test, and document a computer program to calculate A_N. Design the program so that the user can input different values for P, i, G, A_3, and n. Use the program to estimate A_N for your new computer system for values of $n = 1, 2, 3, 4$, and 5. That is, evaluate its worth if it is owned for 1 through 5 years. Plot A_N ver-sus n (use your computer to make the plot if possible). Determine how long you must own it in order to begin to make money. (Note: Additional background information for this problem can be obtained from the first case study in Chap. 6).

2.21 Develop, debug, and test a program to compute the velocity of the falling parachutist as outlined in Example 1.2. Design the program so that it allows the user to input values for the drag coefficient and mass. Test the program by duplicating the results from Example 1.2. Repeat the computation but employ step sizes of 1 and 0.5 s. Compare your results with the analytical solution obtained previously in Example 1.1. Does a smaller step size make the results better or worse? Explain your results.

2.22 Use the GRAPHICS program on the Electronic TOOLKIT to plot several functions of your choice. Try polynomials and transcendental functions (e.g. sines, logs, etc.) whose behavior may be difficult to visualize in advance of plotting. Utilize several choices for the scales of both x and y axes to facilitate your exploration. Make permanent copies of the plots if you have a printer.

2.23 Attain the capability to plot functions in a manner similar to the GRAPHICS program on the Electronic TOOLKIT. Use your own computer and its utility software or library subroutines if appropriate. If your computer does not have programs to help you, then write your own code using the capabilities of your computer. Develop this program in the form of a subroutine and save it on a magnetic storage disk. Be prepared to update this program as you apply it to various problems covered in other parts of the book.

CHAPTER 3
Approximations and Errors

Because so many of the methods in this book are straightforward in description and application, it would be very tempting at this point for us to proceed directly to the main body of the text and teach you how to use these techniques. However, understanding the concept of error is so important to the effective use of numerical methods that we have chosen to devote the present chapter to this topic.

The importance of error was introduced in our discussion of the falling parachutist in Chap. 1. Recall that we determined the velocity of a falling parachutist by both analytical and numerical methods. Although the numerical technique yielded estimates that were close to the exact analytical solution, there was a discrepancy, or *error*, due to the fact that the numerical method involved an approximation. Actually we are fortunate in this case because the availability of the analytical solution allows us to compute the error exactly. For many applied engineering problems we cannot obtain analytical solutions. Therefore we cannot compute the errors associated with our numerical methods exactly. In these cases we must settle for approximations or estimates of the errors.

Such errors are characteristic of most of the techniques described in this book. This statement at first might seem contrary to what one normally conceives of as sound engineering. Students and practicing engineers constantly strive to limit errors in their work. When taking examinations or doing homework problems, you are penalized, not rewarded, for your errors. In professional practice, errors can be costly and sometimes catastrophic. If a structure or device fails, lives can be lost.

Although perfection is a laudable goal, it is rarely, if ever attained. For example, despite the fact that the model developed from Newton's second law is an excellent approximation, it would never in practice exactly predict the parachutist's fall. A variety of factors such as winds and slight variations in air resistance would result in deviations from the prediction. If these deviations are systematically high or low, then we might need to develop a new model. However, if they are randomly distributed and tightly grouped around the prediction, then the deviations might be considered negligible and the model deemed adequate. Numerical approximations also introduce similar discrepancies into the analysis. Again, the question is: How much error is present in our calculations and is it tolerable?

The present chapter covers basic topics related to the identification, quantification, and minimization of these errors. General information concerned with the quantification of error is reviewed in the first sections. This is followed by sections on the two major forms of numerical error: roundoff error and truncation error. *Roundoff error* is due to the fact that computers can only represent quantities with a finite number of digits. *Truncation error* is the discrepancy introduced by the fact that the numerical method employs an approximation to represent exact mathematical operations and quantities. Finally, we briefly discuss errors not directly connected with the numerical methods themselves. These include blunders, formulation or model errors, and data uncertainty.

3.1 SIGNIFICANT FIGURES

This book deals extensively with approximations connected with the manipulation of numbers. Consequently, before discussing the errors associated with numerical methods, it is useful to review basic concepts related to approximate representation of the numbers themselves.

Whenever we employ a number in a computation, we must have assurance that it can be used with confidence. For example, Fig. 3.1 depicts a speedometer and odometer from an automobile. Visual inspection of the speedometer indicates that the car is traveling between 48 and 49 km/h. Because the indicator is higher than the midpoint between the markers on the gauge, we can say with assurance that the car is traveling at approximately 49 km/h. We have confidence in this result because two or more reasonable individuals reading this gauge would arrive at the same conclusion. However, let us say that we insist that the speed be estimated to one decimal place. For this case, one person might say 48.7, whereas another might say 48.8 km/h. Therefore, because of the limits of this instrument, only the first two digits can be used with confidence. Estimates of the third digit (or higher) must be viewed as approximations. It would be ludicrous to claim, on the basis of this speedometer, that the automobile is traveling at 48.7642138 km/h. In contrast, the odometer provides up to six certain digits. From Fig. 3.1, we can conclude that the car has traveled slightly less than 87,324.5 km during its lifetime. In this case, the seventh digit (and higher) is uncertain.

Figure 3.1
An automobile speedometer and odometer illustrating the concept of a significant figure.

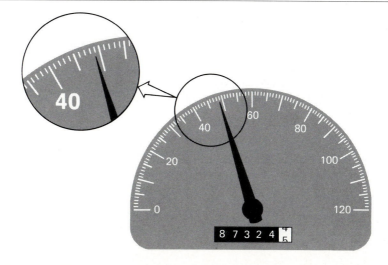

The concept of a significant figure, or digit, has been developed to formally designate the reliability of a numerical value. The *significant digits* of a number are those that can be used with confidence. They correspond to the certain digits plus one estimated digit. For example, the speedometer and the odometer in Fig. 3.1 yield readings of three and seven significant figures, respectively. For the speedometer, the two certain digits are 48. It is conventional to set the estimated digit at one-half of the smallest scale division on the measurement device. Thus the speedometer reading would consist of the three significant figures—48.5. In a similar fashion, the odometer would yield a seven-significant-figure reading of 87,324.45.

Although it is usually a straightforward procedure to ascertain the significant figures of a number, some cases can lead to confusion. For example, zeros are not always significant figures because they may be necessary just to locate a decimal point. The numbers 0.00001845, 0.0001845, and 0.001845 all have four significant figures. Similarly, when trailing zeros are used in large numbers, it is not clear how many, if any, of the zeros are significant. For example, at face value the number 45,300 may have three, four, or five significant digits, depending on whether the zeros are known with confidence. Such uncertainty can be resolved by using scientific notation, where 4.53×10^4, 4.530×10^4, and 4.5300×10^4 designate that the number is known to three, four, and five significant figures, respectively.

The concept of significant figures has two important implications for our study of numerical methods:

1. As introduced in the falling parachutist problem, numerical methods yield approximate results. We must therefore develop criteria to specify how confident we are in our approximate result. One way to do this is in terms of significant figures. For example, we might decide that our approximation is acceptable if it is correct to four significant figures.

2. Although quantities such as π, e, or $\sqrt{7}$ represent specific quantities, they cannot be expressed exactly by a limited number of digits. For example, the quantity π is equal to

3.14159265358979323846264 3. . .

ad infinitum. Because computers only retain a finite number of significant figures, such numbers can never be represented exactly. The omission of the remaining significant figures is called *round-off error*.

Both round-off error and the use of significant figures to express our confidence in a numerical result will be explored in detail in subsequent sections. In addition, the concept of significant figures will have relevance to our definition of accuracy and precision in the next section.

3.2 ACCURACY AND PRECISION

The errors associated with both calculations and measurements can be characterized with regard to their accuracy and precision. *Accuracy* refers to how closely a computed or measured value agrees with the true value. *Precision* refers to how closely individual computed or measured values agree with each other. Thus, precision signifies (1) the number of significant figures representing a quantity or (2) the spread in repeated computations or measurements of a particular value.

These concepts can be illustrated graphically using an analogy from target practice. The bullet holes on each target in Fig. 3.2 can be thought of as the predictions of a numerical technique, whereas the bull's-eye represents the truth. *Inaccuracy* (also called *bias*) is defined as systematic deviation from the truth. Thus, although the shots in Fig. 3.2c are more tightly grouped than in Fig. 3.2a, the two cases are equally biased because they are both centered on the upper left quadrant of the target. *Imprecision* (also called *uncertainty*), on the other hand, refers to the magnitude of the scatter. Therefore, although Fig. 3.2b and d are equally accurate (that is, centered on the bull's-eye), the latter is more precise because the shots are tightly grouped.

Numerical methods should be sufficiently accurate or unbiased to meet the requirements of a particular engineering problem. They also should be precise enough for adequate engineering design. In this book, we will use the collective term *error* to represent both the inaccuracy and imprecision of our predictions. With these concepts as background, we can now discuss the factors that contribute to the error of numerical computations.

Figure 3.2
An example from markmanship illustrating the concepts of accuracy and precision. (a) Inaccurate and imprecise; (b) accurate and imprecise; (c) inaccurate and precise; (d) accurate and precise.

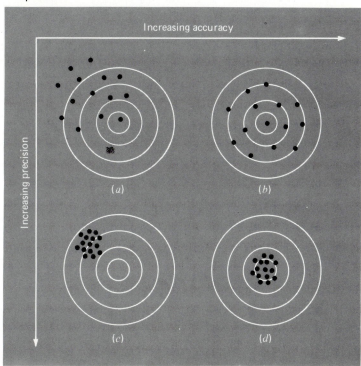

3.3   ERROR DEFINITIONS

Numerical errors arise from the use of approximations to represent exact mathematical operations and quantities. These include *truncation errors,* which result when approximations are used to represent exact mathematical procedures, and *round-off errors,* which result when approximate numbers are used to represent exact numbers. For both types, the relationship between the exact, or true, result and the approximation can be formulated as

$$\text{True value} = \text{approximation} + \text{error} \tag{3.1}$$

By rearranging Eq. (3.1), we find that the numerical error is equal to the discrepancy between the truth and the approximation, as in

$$E_t = \text{true value} - \text{approximation} \tag{3.2}$$

where E_t is used to designate the exact value of the error. The subscript t is included to designate that this is the "true" error. This is in contrast to other cases, as described shortly, where an "approximate" estimate of the error must be employed.

A shortcoming of this definition is that it takes no account of the order of magnitude of the value under examination. For example, an error of a centimeter is much more significant if we are measuring a rivet than a bridge. One way to account for the magnitudes of the quantities being evaluated is to normalize the error to the true value, as in

$$\text{Fractional relative error} = \frac{\text{error}}{\text{true value}}$$

where, as specified by Eq. (3.2), error = true value − approximation. The relative error can also be multiplied by 100 percent in order to express it as

$$\epsilon_t = \frac{\text{true error}}{\text{true value}} 100\% \tag{3.3}$$

where ϵ_t designates the *true percent relative error*.

EXAMPLE 3.1   Calculation of Errors

Problem Statement: Suppose that you have the task of measuring the lengths of a bridge and a rivet and come up with 9999 and 9 cm, respectively. If the true values are 10,000 and 10 cm, respectively, compute (*a*) the error and (*b*) the percent relative error for each case.

Solution: (*a*) The error for measuring the bridge is [Eq. (3.2)]

$$E_t = 10,000 - 9999 = 1 \text{ cm}$$

and for the rivet it is

$$E_t = 10 - 9 = 1 \text{ cm}$$

(*b*) The percent relative error for the bridge is [Eq. (3.3)]

$$\epsilon_t = \frac{1}{10,000} 100\% = 0.01\%$$

and for the rivet it is

$$\epsilon_t = \frac{1}{10} 100\% = 10\%$$

Thus, although both measurements have an error of 1 cm, the relative error for the rivet is much greater. We would conclude that we have done an adequate job of measuring the bridge, whereas our estimate for the rivet leaves something to be desired.

Notice that for Eqs. (3.2) and (3.3), E and ϵ are subscripted with a t to signify that the error is normalized to the true value. In Example 3.1, we were provided with this value. However, in actual situations such information is rarely available. For numerical methods, the true value will only be known when we deal with functions that can be solved analytically. Such will typically be the case when we investigate the theoretical behavior of a particular technique for simple systems. However, in real-world applications, we will obviously not know the true answer a priori. For these situations, an alternative is to normalize the error using the best available estimate of the true value, that is, to the approximation itself, as in

$$\epsilon_a = \frac{\text{approximate error}}{\text{approximation}} 100\% \tag{3.4}$$

where the subscript a signifies that the error is normalized to an approximate value. Note also that for real-world applications, Eq. (3.2) cannot be used to calculate the error term for Eq. (3.4). One of the challenges of numerical methods is to determine error estimates in the absence of knowledge regarding the true value. For example, certain numerical methods use an *iterative approach* to compute answers. In such an approach, a present approximation is made on the basis of a previous approximation. This process is performed repeatedly, or iteratively, in order to successively compute better and better approximations. For such cases, the error is often estimated as the difference between previous and present approximations. Thus, percent relative error is determined according to

$$\epsilon_a = \frac{\text{present approximation} - \text{previous approximation}}{\text{present approximation}} 100\% \tag{3.5}$$

This and other approaches for expressing errors will be elaborated on in subsequent chapters.

The signs of Eqs. (3.2) through (3.5) may be either positive or negative. If the approximation is greater than the true value (or the previous approximation is greater than the present approximation), the error is negative; if the approximation is less than the true value, the error is positive. Also, for Eqs. (3.3) to (3.5), the denominator may be less than zero, which can also lead to a negative error. Often, when performing computations, we may not be concerned with the sign of the error but are interested in whether the absolute value is lower than a prespecified tolerance ϵ_s. Therefore, it is often useful to employ the absolute value of Eqs. (3.2) through (3.5). For such cases, the computation is repeated until

$$|\epsilon_a| < \epsilon_s \tag{3.6}$$

If this relationship holds, our result is assumed to be within the prespecified acceptable level ϵ_s.

It is also convenient to relate these errors to the number of significant figures in the approximation. It can be shown (Scarborough, 1966) that if the following criterion is met, we can be assured that the result is correct to *at least n* significant figures.

$$\epsilon_s = (0.5 \times 10^{2-n})\% \tag{3.7}$$

EXAMPLE 3.2   Error Estimates for Iterative Methods

Problem Statement: In mathematics, functions can often be represented by infinite series. For example, the exponential function can be computed using

$$e^x = 1 + x + \frac{x^2}{2!} + \frac{x^3}{3!} + \cdots + \frac{x^n}{n!} \tag{E3.2.1}$$

Thus, as more terms are added in sequence, the approximation becomes a better and better estimate of the true value of e^x. Equation (E3.2.1) is called a *Maclaurin series expansion*.

Starting with the simplest version, $e^x = 1$, add terms one at a time in order to estimate $e^{0.5}$. After each new term is added, compute the true and approximate percent relative errors with Eqs. (3.3) and (3.5), respectively. Note that the true value is $e^{0.5} = 1.648721271$. Add terms until the absolute value of the approximate error estimate ϵ_a falls below a prespecified error criterion ϵ_s conforming to three significant figures.

Solution: First, Eq. (3.7) can be employed to determine the error criterion that ensures a result that is correct to *at least* three significant figures:

$$\epsilon_s = (0.5 \times 10^{2-3})\% = 0.05\%$$

Thus, we will add terms to the series until ϵ_a falls below this level.

The first estimate is simply equal to Eq. (E3.2.1) with a single term. Thus, the first estimate is equal to 1. The second estimate is then generated by adding the second term, as in

$$e^x \simeq 1 + x$$

or for $x = 0.5$

$$e^{0.5} \simeq 1 + 0.5 = 1.5$$

This represents a true percent relative error of [Eq. (3.3)]

$$\epsilon_t = \frac{1.648721271 - 1.5}{1.648721271}100\% = 9.02\%$$

Equation (3.5) can be used to determine an approximate estimate of the error, as in

$$\epsilon_a = \frac{1.5 - 1}{1.5} 100\% = 33.3\%$$

Because ϵ_a is not less than the required value of ϵ_s, we would continue the computation by adding another term, $x^2/2!$, and repeating the error calculations. The process is continued until $\epsilon_a < \epsilon_s$. The entire computation can be summarized as

Terms	Result	ϵ_t,%	ϵ_a,%
1	1	39.3	
2	1.5	9.02	33.3
3	1.625	1.44	7.69
4	1.645833333	0.175	1.27
5	1.648437500	0.0172	0.158
6	1.648697917	0.00142	0.0158

Thus, after six terms are included, the approximate error falls below $\epsilon_s = 0.05\%$, and the computation is terminated. However, notice that, rather than three significant figures, the result is accurate to five! This is because, for this case, both Eqs. (3.5) and (3.7) are conservative. That is, they ensure that the result is at least as good as they specify. Although as discussed in Chap. 5, this is not always the case for Eq. (3.5), it is true most of the time.

With the preceding definitions as background, we can now proceed to the two types of error connected directly with numerical methods. These are round-off errors and truncation errors.

3.4 ROUND-OFF ERRORS

As mentioned previously, round-off errors originate from the fact that computers retain only a fixed number of significant figures during a calculation. Numbers such as π, e, or $\sqrt{7}$ cannot be expressed by a fixed number of significant figures. Therefore, they cannot be represented exactly by the computer. The discrepancy introduced by this omission of significant figures is called *round-off error*.

3.4.1 Computer Representation of Numbers

Numerical round-off errors are directly related to the manner in which numbers are stored in a computer. The fundamental unit whereby information is represented is called a *word*. This is an entity that consists of a string of *binary digits*, or *bits*. Numbers are typically stored in one or more words. To understand how this is accomplished, we must first review some material related to number systems.

Number Systems. A *number system* is merely a way of representing numbers. Because we have 10 fingers and 10 toes, the number system that we are most familiar with is the *decimal,* or *base-10,* number system. A *base* is the number used as the reference for constructing the system. The base-10 system uses ten digits—0, 1, 2, 3, 4, 5, 6, 7, 8, and 9—to represent numbers. By themselves, these numbers are satisfactory for counting from 0 to 9. For larger quantities, combinations of these basic digits are used, with the position or *place value* specifying the magnitude. The rightmost digit in a whole number represents a number from 0 to 9. The second digit from the right represents a multiple of 10. The third digit from the right represents a multiple of 100 and so on. For example, if we have the number 86,409 then we have eight groups of 10,000, six groups of 1000, four groups of 100, zero groups of 10, and nine more units, or

$$(8 \times 10^4) + (6 \times 10^3) + (4 \times 10^2) + (0 \times 10^1) + (9 \times 10^0) = 86,409$$

Figure 3.3*a* provides a visual representation of how a number is formulated in the base-10 system.

Now, because the base-10 system is so familiar, it is not commonly realized that there are alternatives. For example, if human beings happened to have eight fingers and toes, we would undoubtedly have developed an *octal,* or *base-8,* representation. In the same sense, the computer is like a two-fingered animal that is limited to two states— either 0 or 1. This relates to the fact that the primary logic units of digital computers are electronic components that are either off (corresponding to 0) or on (corresponding to 1). Hence, numbers on the computer are represented with a *binary,* or *base-2,* system. Just

Figure 3.3
How the (*a*) decimal (base 10) and the (*b*) binary (base 2) systems work. In (*b*), the binary number 10101101 is equivalent to the decimal number 173.

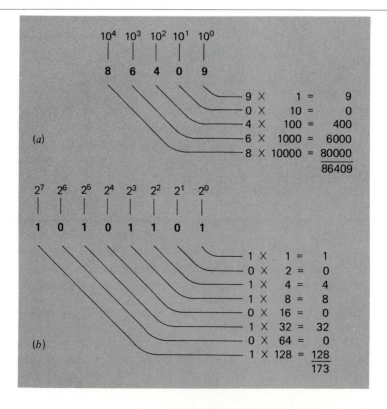

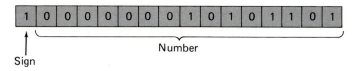

Figure 3.4
The representation of the decimal integer −173 on a 16-bit computer using the signed magnitude method.

as with the base-10 system, each position represents higher powers of the base number. For example, the binary number 11 is equivalent to $(1 \times 2^1) + (1 \times 2^0) = 2 + 1 = 3$ in the base-10 system. Figure 3.3*b* illustrates a more complicated example.

Integer Representation. Now that we have reviewed how base-10 numbers can be represented in binary form, it is simple to conceive how integers are represented on a computer. The most straightforward approach, called the *signed magnitude method,* employs the first bit of a word to indicate the sign, with a 0 for positive and a 1 for negative. The remaining bits are used to store the number. For example, the integer value of −173 would be stored on a 16-bit computer as in Fig. 3.4.

EXAMPLE 3.3   *Range of Integers*

Problem Statement: Determine the range of integers in base 10 that can be represented on a 16-bit computer.

Solution: Of the 16 bits, the first bit holds the sign. The remaining 15 bits can hold binary numbers from 0 to 111111111111111. The upper limit can be converted to a decimal integer as in

$$(1 \times 2^{14}) + (1 \times 2^{13}) + \cdots + (1 \times 2^1) + (1 \times 2^0)$$

which equals 32,767 (note that this expression can be simply evaluated as $2^{15} - 1$). Thus, a 16-bit computer word can store decimal integers ranging from −32,767 to 32,767. In addition, because zero is already defined as 0000000000000000, it is redundant to use the number 1000000000000000 to define a "minus zero," Therefore, it is usually employed to represent an additional negative number: −32,768, and the range is from −32,768 to 32,767.

Note that the signed magnitude method described above is not used to represent integers on conventional computers. A preferred approach called the 2's complement technique directly incorporates the sign into the number's magnitude rather than providing a separate bit to represent plus or minus. However, Example 3.3 still serves to illustrate how all digital computers are limited in their capability to represent integers. That is, numbers above or below the range cannot be represented. A more serious limitation is encountered in the storage and manipulation of fractional quantities as described next.

Figure 3.5
The manner in which a floating-point number is stored in a word.

Floating-Point Representation.   Fractional quantities are typically represented in computers using floating-point form. In this approach, the number is expressed as a fractional part, called a *mantissa* or *significand,* and an integer part, called an *exponent* or *characteristic,* as in

$$m \cdot b^e$$

where m is the mantissa, b is the base of the number system being used, and e is the exponent. For instance, the number 156.78 could be represented as $0.15678 \cdot 10^3$ in a floating-point base-10 system.

Figure 3.5 shows one way that a floating-point number could be stored in a word. The first bit is reserved for the sign, the next series of bits for the signed exponent, and the last bits for the mantissa.

Note that the mantissa is usually *normalized* if its leading digit is zero. For example, suppose the quantity $1/34 = 0.029411765 \ldots$ was stored in a floating-point base-10 system which allowed four decimal places to be stored. Thus, 1/34 would be stored as

$$0.0294 \cdot 10^0$$

However, in the process of doing this, the inclusion of the useless zero to the right of the decimal forces us to drop the digit 1 in the fifth decimal place. The number can be normalized to remove the leading zero by multiplying the mantissa by 10 and lowering the exponent by 1 to give

$$0.2941 \cdot 10^{-1}$$

Thus, we retain an additional significant figure when the number is stored.

The consequence of normalization is that the absolute value of m is limited. That is,

$$\frac{1}{b} \leq m < 1 \tag{3.8}$$

where b is the base. For example, for a base-10 system, m would range between 0.1 and 1, and for a base-2 system, between 0.5 and 1.

Floating-point representation allows both fractions and very large numbers to be expressed on the computer. However, it has some disadvantages. For example, floating-point numbers take up more room and take longer to process than integer numbers. More significantly, however, their use introduces a source of error because the mantissa can hold only a finite number of significant figures. Thus, a round-off error is introduced.

EXAMPLE 3.4   *Hypothetical Set of Floating-Point Numbers*

Problem Statement: Create a hypothetical floating-point number set for a machine that stores information using 7-bit words. Employ the first bit for the sign of the number, the next three for the sign and the magnitude of the exponent, and the last three for the magnitude of the mantissa (Fig. 3.6).

Solution: The smallest possible positive number is depicted in Fig. 3.6. The initial 0 indicates that the quantity is positive. The 1 in the second place designates that the exponent has a negative sign. The 1's in the third and fourth places give a maximum value to the exponent of

$$1 \times 2^1 + 1 \times 2^0 = 3$$

Therefore, the exponent will be -3. Finally, the mantissa is specified by the 100 in the last three places which conforms to

$$1 \times 2^{-1} + 0 \times 2^{-2} + 0 \times 2^{-3} = 0.5$$

Although a smaller mantissa is possible (e.g., 000, 001, 010, and 011), the value of 100 is used because of the limit imposed by normalization [Eq. (3.8)]. Thus, the smallest possible positive number for this system is

$$+0.5 \times 2^{-3}$$

which is equal to 0.0625 in the base-10 system. The next highest numbers are developed by increasing the mantissa as in

$$0111101 = (1 \times 2^{-1} + 0 \times 2^{-2} + 1 \times 2^{-3}) \times 2^{-3} = (0.078125)_{10}$$

$$0111110 = (1 \times 2^{-1} + 1 \times 2^{-2} + 0 \times 2^{-3}) \times 2^{-3} = (0.093750)_{10}$$

$$0111111 = (1 \times 2^{-1} + 1 \times 2^{-2} + 1 \times 2^{-3}) \times 2^{-3} = (0.109375)_{10}$$

Notice that the base-10 equivalents are spaced evenly with an interval of 0.015625.

At this point, to continue increasing, we must decrease the exponent to 10 which gives a value of

$$1 \times 2^1 + 0 \times 2^0 = 2$$

Figure 3.6
The smallest possible positive floating-point number from Example 3.4.

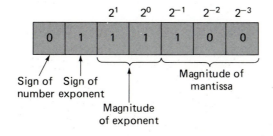

The mantissa is decreased back to its smallest value of 100. Therefore the next number is

$$0110100 = (1 \times 2^{-1} + 0 \times 2^{-2} + 0 \times 2^{-3}) \times 2^{-2} = (0.125000)_{10}$$

This still represents a gap of $0.125000 - 0.109375 = 0.015625$. However, now when higher numbers are generated by increasing the mantissa, the gap is lengthened to 0.03125,

$$0110101 = (1 \times 2^{-1} + 0 \times 2^{-2} + 1 \times 2^{-3}) \times 2^{-2} = (0.156250)_{10}$$

$$0110110 = (1 \times 2^{-1} + 1 \times 2^{-2} + 0 \times 2^{-3}) \times 2^{-2} = (0.187500)_{10}$$

$$0110111 = (1 \times 2^{-1} + 1 \times 2^{-2} + 1 \times 2^{-3}) \times 2^{-2} = (0.218750)_{10}$$

This pattern is repeated as each larger quantity is formulated until a maximum number is reached,

$$0011111 = (1 \times 2^{-1} + 1 \times 2^{-2} + 1 \times 2^{-3}) \times 2^{3} = (7)_{10}$$

The final number set is depicted graphically in Fig. 3.7.

Figure 3.7
The hypothetical number system developed in Example 3.4. Each value is indicated by a tick mark. Only the positive numbers are shown. An identical set would also extend in the negative direction.

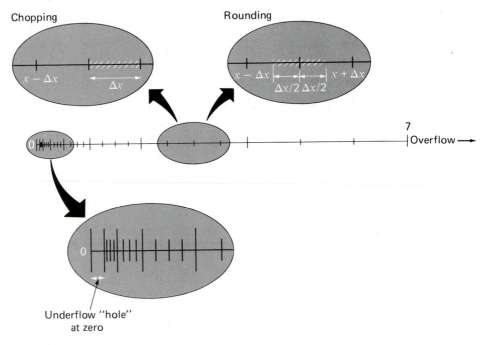

Figure 3.7 manifests a number of aspects of floating-point representation that have significance regarding computer round-off errors:

1. *There Is a Limited Range of Quantities that May Be Represented.* Just as for the integer case, there are large positive and negative numbers that cannot be represented. Attempts to employ these numbers will result in what is called an *overflow* error. However, in addition to large quantities, the floating-point representation has the added limitation that very small numbers cannot be represented. This is illustrated by the *underflow* "hole" between zero and the first positive number in Fig. 3.7. It should be noted that this hole is enlarged because of the normalization constraint of Eq. (3.8).

2. *There Are Only a Finite Number of Quantities that Can Be Represented Within the Range.* Thus, the degree of precision is limited. Obviously, irrational numbers cannot be represented. Furthermore, rational numbers that do not exactly match one of the values in the set also cannot be represented precisely. The errors introduced by approximating both these cases are referred to as *quantizing* errors. The actual approximation is accomplished in either of two ways: chopping or rounding. For example, suppose that the value of $\pi = 3.14159265358\ldots$ is to be stored on a base-10 number system carrying seven significant figures. One method of approximation would be to merely omit or "chop off" the eighth and higher terms as in $\pi = 3.141592$, with the introduction of an associated error of [Eq. (3.2)]

$$E_t = 0.00000065\ldots$$

This technique of retaining only the significant terms was originally dubbed "truncation" in computer jargon. We prefer to call it *chopping* to distinguish it from the truncation errors discussed in Sec. 3.5. Note that for the base-2 number system in Fig. 3.7, chopping means that any quantity falling within an interval of length Δx will be stored as the quantity at the lower end of the interval. Thus, the upper error bound for chopping is Δx. Additionally, a bias is introduced because all errors are positive. The shortcomings of chopping are attributable to the fact that the higher terms in the complete decimal representation have no impact on the shortened version. For example, in our example of π, the first discarded digit is 6. Thus, the last retained digit should be rounded up to yield 3.141593. Such *rounding* reduces the error to

$$E_t = 0.00000035\ldots$$

Consequently, rounding yields a lower error than for chopping. Note that for the base-2 number system in Fig. 3.7, rounding means that any quantity falling within an interval of length Δx will be represented as the nearest allowable number. Thus, the upper error bound for rounding is $\Delta x/2$. Additionally, no bias is introduced because some errors are positive and some are negative. Some computers employ rounding. However, this adds substantially to the computational overhead, and consequently, many machines use simple chopping. This approach is justified under the supposition that the number of significant figures is large enough that resulting round-off error is usually negligible.

3. *The Interval between Numbers,* Δx, *Increases as the Numbers Grow in Magnitude.* It is this characteristic, of course, that allows floating-point repre-

sentation to preserve significant digits. However, it also means that quantizing errors will be proportional to the magnitude of the number being represented. For normalized floating-point numbers, this proportionality can be expressed, for cases where chopping is employed, as

$$\frac{|\Delta x|}{|x|} \le \varepsilon \qquad (3.9)$$

and, for cases where rounding is employed, as

$$\frac{|\Delta x|}{|x|} \le \frac{\varepsilon}{2} \qquad (3.10)$$

where ε is referred to as the *machine epsilon* which can be computed as

$$\varepsilon = b^{1-t} \qquad (3.11)$$

where b is the number base and t is the number of significant digits in the mantissa. Notice that the inequalities in Eqs. (3.9) and (3.10) signify that these are error *bounds*. That is, they specify the worst cases.

EXAMPLE 3.5   Machine Epsilon

Problem Statement: Determine the machine epsilon and verify its effectiveness in characterizing the errors of the number system from Example 3.4. Assume that chopping is used.

Solution: The hypothetical floating-point system from Example 3.4 employed values of the base $b = 2$, and the number of mantissa bits $t = 3$. Therefore, the machine epsilon would be [Eq. (3.11)]

$$\varepsilon = 2^{1-3} = 0.25$$

Consequently, the relative quantizing error should be bounded by 0.25 for chopping. The largest relative errors should occur for those quantities that fall just below the upper bound of the first interval between successive equispaced numbers (Fig. 3.8). Those numbers falling in the succeeding higher intervals would have the same value of Δx but a greater value of x and, hence, would have a lower relative error. An example of a maximum error would be a value falling just below the upper bound of the interval between $(0.125000)_{10}$ and $(0.156250)_{10}$. For this case, the error would be less than

$$\frac{0.03125}{0.125000} = 0.25$$

Thus, the error is as predicted by Eq. (3.9).

Figure 3.8
The largest quantizing error will occur for those values falling just below the upper bound of the first of a series of equispaced intervals.

Largest relative
error

The magnitude dependence of quantizing errors has a number of practical applications in numerical methods. Most of these relate to the commonly employed operation of testing whether two numbers are equal. This occurs when testing convergence of quantities as well as in the stopping mechanism for iterative processes (recall Example 3.2). For these cases, it should be clear that, rather than test whether the two quantities are equal, it is advisable to test whether their difference is less than an acceptably small tolerance. Further, it should also be evident that normalized rather than absolute difference should be compared, particularly when dealing with numbers of large magnitude. In addition, the machine epsilon can be employed in formulating stopping or convergence criteria. This ensures that programs are portable—that is, they are not dependent on the computer on which they are implemented. Figure 3.9 lists pseudocode to automatically determine the machine epsilon of a binary computer.

Figure 3.9
Pseudocode to determine machine epsilon for a binary computer.

```
epsilon = 1
DOWHILE (epsilon+1 > 1)
    epsilon = epsilon/2
ENDDO
epsilon = 2 · epsilon
```

Extended Precision. It should be noted at this point that, although round-off errors can be important in contexts such as testing convergence, the number of significant digits carried on most computers allows most engineering computations to be performed with more than acceptable precision. For example, the hypothetical number system in Fig. 3.7 is a gross exaggeration that was employed for illustrative purposes. Commercial computers use much larger words and, consequently, allow numbers to be expressed with more than adequate precision. For example, the IBM PC family allows 24 bits to be used for the mantissa[*] which translates into about seven significant base-10 digits of precision.

With this acknowledged, there are still cases where round-off error becomes critical. For this reason most computers allow the specification of extended precision. The most common of these is double precision, in which an extra word is employed to store additional significant figures of the mantissa. Note that some computers permit triple and higher precision and others allow more precision in the exponent as well as in the mantissa of the floating-point number.

To illustrate the improvement due to standard double precision, we will use a hypothetical computer which employs 32 bits to store floating-point numbers. For single precision, 1 bit would be used for the sign, 7 for the signed exponent, and 24 for the mantissa. As stated above, this translates into about 7 significant base-10 digits. For double precision, an extra 32 bits would be used to create a 56-bit mantissa which allows about 17 significant base-10 digits. Therefore, despite the use of the word "double," the increase would amount to *more than double* the precision in terms of significant figures.

---

[*]Note that only 23 bits are actually used to store the mantissa. However, because of normalization, the first bit of the mantissa is always one and is, therefore, not stored. Thus, this first bit together with the 23 stored bits gives the IBM PC 24 total bits of precision for its mantissa.

In many cases, the use of double-precision quantities can greatly mitigate the effect of round-off errors. However, a price is paid for such remedies in that they also require more memory and execution time. The latter is illustrated in the following example.

EXAMPLE 3.6   Execution Time for Single- and Double-Precision Computations

Problem Statement: We have suggested that double-precision computations require more computer time than single-precision computations. We can explore this situation experimentally on the IBM PC.

Solution: Figure 3.10 shows a simple BASICA program to add 1 to an accumulator, SUM, for 10,000 iterations in both single and double precision. The single-precision version requires about 35 s of execution time whereas the double-precision version takes 45 s.

```
100 SUM = 0
105 X = 1
110 FOR I = 1 TO 10000
115    SUM = SUM + X
120 NEXT I
125 PRINT SUM
130 END
        (a)
```

```
100 SUM# = 0
105 X# = 1
110 FOR I = 1 TO 10000
115    SUM# = SUM# + X#
120 NEXT I
125 PRINT SUM#
130 END
                (b)
```

Figure 3.10
BASICA programs to sum 10,000 numbers. (a) Single and (b) double precision.

The difference in execution time for the foregoing example might seem insignificant. However, as your programs become larger and more complicated, the added execution time could become considerable and have a negative impact on your effectiveness as a problem solver. Therefore, extended precision should not be used frivolously. Rather, it should be selectively employed where it will yield the maximum benefit at the least cost in terms of execution time. In the following sections, we will look closer at how round-off errors affect computations and in so doing provide a foundation of understanding to guide your use of the double-precision capability.

3.4.2 Arithmetic Manipulations of Computer Numbers

Aside from the limitations of a computer's number system, the actual arithmetic manipulations involving these numbers can also result in round-off error. In the following section, we will first illustrate how common arithmetic operations affect round-off errors. Then we will investigate a number of particular manipulations that are especially prone to round-off errors.

Common Arithmetic Operations.   Because of their familiarity, normalized base 10 numbers will be employed to illustrate the effect of round-off errors on simple addition, subtraction, multiplication, and division. Other number bases would behave in a similar fashion. To simplify the discussion, we will employ a hypothetical decimal computer with a 4-digit mantissa and a 1-digit exponent. In addition, chopping is used. Rounding would lead to similar though less dramatic errors.

When two floating-point numbers are added, the mantissa of the number with the smaller exponent is modified so that the exponents are the same. This has the effect of aligning the decimal points. For example, suppose we want to add $0.1557 \cdot 10^1 + 0.4381 \cdot 10^{-1}$. The decimal of the mantissa of the second number is shifted to the left a number of places equal to the difference of the exponents $[1 - (-1) = 2]$ as in

$$0.4381 \cdot 10^{-1} \rightarrow 0.004381 \cdot 10^1$$

Now the numbers can be added

$$
\begin{array}{r}
0.1557 \quad \cdot 10^1 \\
\underline{0.004381 \cdot 10^1} \\
0.160081 \cdot 10^1
\end{array}
$$

and the result chopped to $0.1600 \cdot 10^1$. Notice how the last two digits of the second number that were shifted to the right have essentially been lost from the computation.

Subtraction is performed identically to addition except that the sign of the subtrahend is reversed. For example, suppose that we are subtracting 26.86 from 36.41. That is,

$$
\begin{array}{r}
0.3641 \cdot 10^2 \\
\underline{-0.2686 \cdot 10^2} \\
0.0955 \cdot 10^2
\end{array}
$$

For this case the result is not normalized, and so we must shift the decimal one place to the right to give $0.9550 \cdot 10^1$. Notice that the zero added to the end of the mantissa is not significant but is merely appended to fill the empty space created by the shift. Even more dramatic results would be obtained when the numbers are very close as in

$$
\begin{array}{r}
0.7642 \cdot 10^3 \\
\underline{-0.7641 \cdot 10^3} \\
0.0001 \cdot 10^3
\end{array}
$$

which would be converted to $0.1000 \cdot 10^0$. Thus, for this case, three nonsignificant zeros are appended. This introduces a substantial computational error because subsequent manipulations would act as if these zeros were significant. As we will see in a later section, the loss of significance during the subtraction of nearly equal numbers is the greatest source of round-off error in numerical methods.

Multiplication and division are somewhat more straightforward than addition or subtraction. The exponents are added and the mantissas multiplied. Because multiplication of two n-digit mantissas will yield a $(2n)$-digit result, most computers hold intermediate results in a double-length register. For example,

$$0.1363 \cdot 10^3 \times 0.6423 \cdot 10^{-1} = 0.08754549 \cdot 10^2$$

If, as in this case, a leading zero is introduced, the result is normalized,

$$0.08754549 \cdot 10^2 \rightarrow 0.8754549 \cdot 10^1$$

and chopped to give

$$0.8754 \cdot 10^1$$

Division is performed in a similar manner, but the mantissas are divided and the exponents are subtracted. Then the results are normalized and chopped.

Large Computations. Certain methods require extremely large numbers of arithmetic manipulations to arrive at their final results. In addition, these computations are often interdependent. That is, the later calculations are dependent on the results of earlier ones. Consequently, even though an individual round-off error could be small, the cumulative effect over the course of a large computation can be significant.

EXAMPLE 3.7   Large Numbers of Interdependent Computations

Problem Statement: Investigate the effect of round-off error on large numbers of interdependent computations. Develop a program to sum a number one hundred thousand times. Sum the number 1 in single precision, and 0.00001 in single and double precision.

Solution: Figure 3.11 shows a BASICA program for the IBM PC that performs the summation. Whereas the single-precision summation of 1 yields the expected result, the single-precision summation of 0.00001 yields a large discrepancy. This error is reduced significantly when 0.00001 is summed in double precision.

Figure 3.11
A BASICA program to sum a number 1 hundred thousand times. The case sums the number 1 in single precision and the number 0.00001 in single and double precision.

```
100 SUM1 = 0
105 SUM2 = 0
110 SUM3# = 0#
115 X1 = 1
120 X2 = .00001
125 X3# = .00001#
130 FOR I = 1 TO 100000!
135     SUM1 = SUM1 + X1
140     SUM2 = SUM2 + X2
145     SUM3# = SUM3# + X3#
150 NEXT I
155 PRINT SUM1
160 PRINT SUM2
165 PRINT SUM3#
170 END
  100000
  1.00099
  .9999999999998739
```

The source of the discrepancies is quantizing errors. Because the integer 1 can be represented precisely within the computer, it can be summed exactly. In contrast, 0.00001 cannot be represented exactly and is quantized by a value that is slightly different than its true value. Whereas this very slight discrepancy would be negligible for a small computation, it accumulates after repeated summations. The problem still occurs in double precision but is greatly mitigated because the quantizing error is much smaller.

Note that the type of error illustrated by the previous example is somewhat atypical in that all the errors in the repeated operation are of the same sign. In most cases the errors of a long computation alternate sign in a random fashion and, thus, often cancel out. However, there are also instances where such errors do not cancel but, in fact, lead to a spurious final result. The following sections are intended to provide insight into ways in which this may occur.

Adding a Large and a Small Number. Suppose we add a small number, 0.0010, to a large number, 4000, using our hypothetical computer with the 4-digit mantissa and the 1-digit exponent. After modifying the smaller number so that its exponent matches the larger,

$$0.4000 \quad \cdot 10^4$$
$$0.0000001 \cdot 10^4$$
$$\overline{0.4000001 \cdot 10^4}$$

which is chopped to $0.4000 \cdot 10^4$. Thus, we might as well have not performed the addition!

This type of error can occur in the computation of an infinite series. The initial terms in such series are often relatively large in comparison with the later terms. Thus, after a few terms have been added, we are in the situation of adding a small quantity to a large quantity.

One way to mitigate this type of error is to sum the series in reverse order—that is, in ascending rather than descending order. In this way, each new term will be of comparable magnitude to the accumulated sum (see Prob. 3.4).

Subtractive Cancellation. This term refers to the round-off induced when subtracting two nearly equal floating-point numbers.

EXAMPLE 3.8   *Subtractive Cancellation*

Problem Statement: The roots of

$$f(x) = 0 = ax^2 + bx + c$$

are given by the quadratic formula

$$\frac{x_1}{x_2} = \frac{-b \pm \sqrt{b^2 - 4ac}}{2a}$$

Compute the values of the roots for $a = 1$, $b = 3000.001$, and $c = 3$. Check the computed values versus the true roots of $x_1 = -0.001$ and $x_2 = -3000$.

Solution: Figure 3.12 shows a BASICA program for the IBM PC that computes the roots R1 and R2 on the basis of the quadratic formula. Note that both single- and double-precision versions are given. Whereas the results for R2 are adequate, the percent relative errors for R1 are much poorer with $\epsilon_t = 1.18\%$ and 2.35% for double and single precision, respectively. These levels would be inadequate for many applied engineering problems. This result is particularly surprising because we are employing an analytical formula to obtain our solution!

```
100 A = 1                        100 A# = 1#
105 B = 3000.001                 105 B# = 3000.001#
110 C = 3                        110 C# = 3#
115 D = SQR(B*B-4*A*C)           115 D# = SQR(B#*B#-4*A#*C#)
120 R1 = (-B+D)/(2*A)            120 R1# = (-B#+D#)/(2*A#)
125 R2 = (-B-D)/(2*A)            125 R2# = (-B#-D#)/(2*A#)
130 PRINT R1                     130 PRINT R1#
135 PRINT R2                     135 PRINT R2#
140 PRINT D                      140 PRINT D#
145 PRINT A*R1*R1+B*R1+C         145 PRINT A#*R1#*R1#+B#*R1#+C#
150 PRINT A*R2*R2+B*R2+C         150 PRINT A#*R2#*R2#+B#*R2#+C#
155 END                          155 END
-9.765625E-04                    -9.882812499881766D-04
-3000                            -3000.00001171875
 2999.999                        2999.9990234375
 .0703125                        3.515623845404942D-02
 0                               3.515623835846782D-02
          (a)                              (b)
```

Figure 3.12
BASICA programs to determine the roots of a quadratic. (a) Single and (b) double precision.

The loss of significance occurs in line 120 of both programs where two relatively large numbers are subtracted. Similar problems do not occur when the same numbers are added in line 130.

On the basis of the above, we can draw the general conclusion that the quadratic formula will be susceptible to subtractive cancellation whenever $b^2 >> 4ac$. One way to circumvent this problem is to recast the quadratic formula. For example, x_1 can be calculated alternatively by

$$x_1 = \frac{-2c}{b + \sqrt{b^2 - 4ac}}$$

This will give a much smaller error because the subtractive cancellation is avoided.

Note that, as in the foregoing example, there are times where subtractive cancellation can be circumvented by using a transformation. However, the only general remedy is to employ extended precision.

Smearing.   Smearing occurs whenever the individual terms in a summation are larger than the summation itself. As in the following example, one case where this occurs is in series of mixed signs.

EXAMPLE 3.9   Evaluation of e^x using Infinite Series

Problem Statement: The exponential function $y = e^x$ is given by the infinite series

$$y = 1 + x + \frac{x^2}{2!} + \frac{x^3}{3!} + \cdots$$

We would like to evaluate this function for $x = 10$ and $x = -10$ and be attentive to the problems of round-off error.

Solution: Figure 3.13a gives a BASICA program to evaluate e^x written for the IBM PC. I is the number of terms in the series, TERM is the value of the current term added to the series, and SUM is the accumulative value of the series. TEST is the preceding accumulative value of the series prior to adding TERM. The series is terminated when the computer cannot detect the difference between TEST and SUM in line 130.

Figure 3.13b shows the results of running the program for $x = 10$. Note that this case is completely satisfactory. The final result is achieved in 31 terms with the series identical to the library function value within seven significant figures.

Figure 3.13

(a) A BASICA program to evaluate e^x using an infinite series. (b) Evaluation of e^{10} and (c) e^{-10}.

```
100 I = 0
105 TERM = 1
110 SUM = 1
115 TEST = 0
120 INPUT "X = ";X
125 PRINT "I","TERM","SUM"
130 WHILE SUM<>TEST
135     PRINT I,TERM,SUM
140     I = I + 1
145     TERM = TERM*X/I
150     TEST = SUM
155     SUM = SUM + TERM
160 WEND
165 PRINT "EXACT VALUE = ",EXP(X)
170 END
```
                    (a)

X = ? 10					
I	TERM	SUM	16	477.9478	21430.84
0	1	1	17	281.1458	21711.98
1	10	11	18	156.1921	21868.18
2	50	61	19	82.20636	21950.38
3	166.6667	227.6667	20	41.10318	21991.49
4	416.6667	644.3334	21	19.57294	22011.06
5	833.3334	1477.667	22	8.896792	22019.95
6	1388.889	2866.556	23	3.868171	22023.82
7	1984.127	4850.683	24	1.611738	22025.43
8	2480.159	7330.842	25	.6446951	22026.08
9	2755.732	10086.57	26	.2479597	22026.33
10	2755.732	12842.31	27	.0918369	22026.42
11	2505.211	15347.52	28	3.279889E-02	22026.45
12	2087.676	17435.19	29	1.130996E-02	22026.46
13	1605.905	19041.1	30	3.769988E-03	22026.47
14	1147.075	20188.17	31	1.216125E-03	22026.47
15	764.7165	20952.89	EXACT VALUE =		22026.47

(b)

Figure 3.13
(cont.)

X = ? -10		
I	TERM	SUM
0	1	1
1	-10	-9
2	50	41
3	-166.6667	-125.6667
4	416.6667	291
5	-833.3334	-542.3334
6	1388.889	846.5557
7	-1984.127	-1137.572
8	2480.159	1342.587
9	-2755.732	-1413.145
10	2755.732	1342.587
11	-2505.211	-1162.624
12	2087.676	925.0522
13	-1605.905	-680.8523
14	1147.075	466.2223
15	-764.7165	-298.4942
16	477.9478	179.4536
17	-281.1458	-101.6921
18	156.1921	54.49994
19	-82.20636	-27.70642
20	41.10318	13.39676
21	-19.57294	-6.176184
22	8.896792	2.720609

23	-3.868171	-1.147562
24	1.611738	.4641758
25	-.6446951	-.1805193
26	.2479597	6.744036E-02
27	-.0918369	-2.439654E-02
28	3.279889E-02	8.402355E-03
29	-1.130996E-02	-2.907608E-03
30	3.769988E-03	8.623798E-04
31	-1.216125E-03	-3.537452E-04
32	3.800391E-04	2.629386E-05
33	-1.151633E-04	-8.886949E-05
34	3.387157E-05	-5.499791E-05
35	-9.677592E-06	-6.467551E-05
36	2.68822E-06	-6.198729E-05
37	-7.265459E-07	-6.271384E-05
38	1.911963E-07	-6.252264E-05
39	-4.902469E-08	-6.257166E-05
40	1.225617E-08	-6.255941E-05
41	-2.989311E-09	-6.25624E-05
42	7.117406E-10	-6.256169E-05
43	-1.655211E-10	-6.256186E-05
44	3.761842E-11	-6.256182E-05
45	-8.35965E-12	-6.256183E-05
EXACT VALUE =		4.539993E-05

(c)

Figure 3.13c shows similar results for $x = -10$. However, for this case, the results of the series calculation are not even the same sign as the true result. As a matter of fact, the negative results are open to serious question because e^x can never be less than zero. The problem here is caused by round-off error. Note that many of the terms that make up the sum are much larger than the final result of the sum. Furthermore, unlike the previous case, the individual terms vary in sign. Thus in effect we are adding and subtracting large numbers (each with some small error) and placing great significance on the differences—that is, subtractive cancellation. Thus, we can see that the culprit behind this example of smearing is, in fact, subtractive cancellation. For such cases it is appropriate to seek some other computational strategy. For example, one might try to compute $y = e^{-10}$ as $y = (e^{-1})^{10}$. Other than such a reformulation, the only general recourse is extended precision.

Inner Products. As should be clear from the last sections, some infinite series are particularly prone to round-off error. Fortunately, the calculation of series is not one of the more common operations in numerical methods. A far more ubiquitous manipulation is the calculation of inner products as in

$$\sum_{i=1}^{n} x_i y_i = x_1 y_1 + x_2 y_2 + \cdots + x_n y_n$$

This operation is very common, particularly in the solution of simultaneous linear algebraic equations. Such summations are especially prone to round-off error. Consequently, it is often desirable to compute such summations in extended precision.

Although the foregoing sections should provide rules of thumb to mitigate round-off error, they do not provide a direct means beyond trial and error to actually determine the effect of such errors on a computation. In the next section, we will introduce the Taylor series, which will provide a mathematical approach for estimating these effects.

3.5 TRUNCATION ERRORS

Truncation errors are those that result from using an approximation in place of an exact mathematical procedure. For example, in Chap. 1 we approximated the derivative of velocity of a falling parachutist by a finite-divided-difference equation of the form [Eq. (1.11)]

$$\frac{dv}{dt} \simeq \frac{\Delta v}{\Delta t} = \frac{v(t_{i+1}) - v(t_i)}{t_{i+1} - t_i} \tag{3.12}$$

A truncation error was introduced into the numerical solution because the difference equation only approximates the true value of the derivative (Fig. 1.4). In order to gain insight into the properties of such errors, we now turn to a mathematical formulation that is used widely in numerical methods to express functions in an approximate fashion—the Taylor series.

Box 3.1   *Taylor's Theorem*

Taylor's theorem: *If the function f and its first n + 1 derivatives are continuous on an interval containing a and x, then the value of the function at x is given by*

$$f(x) = f(a) + f'(a)(x - a) + \frac{f''(a)}{2!}(x - a)^2$$

$$+ \frac{f^{(3)}(a)}{3!}(x - a)^3 + \cdots \tag{B3.1.1}$$

$$+ \frac{f^{(n)}(a)}{n!}(x - a)^n + R_n$$

where the remainder, R_n, is defined as

$$R_n = \int_a^x \frac{(x - t)^n}{n!} f^{(n+1)}(t)\, dt \tag{B3.1.2}$$

where *t* is a dummy variable. Equation (B3.1.1) is called the *Taylor series* or *Taylor's formula*. If the remainder is omitted, the right side of Eq. (B3.1.1) is the Taylor polynomial approxi-

mation to $f(x)$. In essence, the theorem states that smooth functions can be approximated by polynomials.

Equation (B3.1.2) is but one way, called the *integral form*, by which the remainder can be expressed. An alternative formulation can be derived on the basis of the integral mean-value theorem:

First theorem of mean for integrals: *If the function g is continuous and integrable on an interval containing a and x, then there exists a point ξ between a and x such that*

$$\int_a^x g(t)\, dt = g(\xi)(x - a) \tag{B3.1.3}$$

In other words, this theorem states that the integral can be represented by an average value for the function, $g(\xi)$, times the interval length, $x - a$. Because the average must occur between the minimum and maximum values for the interval, there is a point, $x = \xi$, at which the function takes on the average value.

The first theorem is in fact a special case of a second mean-value theorem for integrals:

Second theorem of mean for integrals: *If the functions g and h are continuous and integrable on an interval containing a and x, and h does not change sign in the interval, then there exists a point ξ between a and x such that*

$$\int_a^x g(t)h(t)\,dt = g(\xi)\int_a^x h(t)\,dt \qquad \text{(B3.1.4)}$$

Thus, Eq. (B3.1.3) is equivalent to Eq. (B3.1.4) with $h(t) = 1$.

The second theorem can be applied to Eq. (B3.1.2) with

$$g(t) = f^{(n+1)}(t) \qquad h(t) = \frac{(x-t)^n}{n!}$$

As t varies from a to x, $h(t)$ is continuous and does not change sign. Therefore, if $f^{(n+1)}(t)$ is continuous, then the integral mean-value theorem holds and

$$R_n = \frac{f^{(n+1)}(\xi)}{(n+1)!}(x-a)^{n+1} \qquad \text{(B3.1.5)}$$

This equation is referred to as the *derivative* or *Lagrange form* of the remainder.

3.5.1 The Taylor Series

Taylor's theorem (Box 3.1) and its associated formula, the Taylor series, is of great value in the study of numerical methods. In essence, the Taylor series provides a means to predict a function value at one point in terms of the function value and its derivatives at another point. A useful way to gain insight into the Taylor series is to build it term by term. For example, the first term in the series is

$$f(x_{i+1}) \simeq f(x_i) \tag{3.13}$$

This relationship, which is called the *zero-order approximation*, indicates that the value of f at the new point is the same as the value at the old point. This result makes intuitive sense because if x_i and x_{i+1} are close to each other, it is likely that the new value is probably similar to the old value.

Equation (3.13) provides a perfect estimate if the function being approximated is, in fact, a constant. However, if the function changes at all over the interval, additional terms of the Taylor series are required to provide a better estimate. For example, the *first-order approximation* is developed by adding another term to yield

$$f(x_{i+1}) \simeq f(x_i) + f'(x_i)(x_{i+1} - x_i) \tag{3.14}$$

The additional first-order term consists of a slope $f'(x_i)$ multiplied by the distance between x_i and x_{i+1}. Thus, the expression is now in the form of a straight line and is capable of predicting an increase or decrease of the function between x_i and x_{i+1}.

Although Eq. (3.14) can predict a change, it is only exact for a straight-line, or *linear*, trend. Therefore, a *second-order* term is added to the series in order to capture some of the curvature that the function might exhibit:

$$f(x_{i+1}) \simeq f(x_i) + f'(x_i)(x_{i+1} - x_i) + \frac{f''(x_i)}{2!}(x_{i+1} - x_i)^2 \tag{3.15}$$

In a similar manner, additional terms can be included to develop the complete Taylor series expansion.

$$f(x_{i+1}) = f(x_i) + f'(x_i)(x_{i+1} - x_i) + \frac{f''(x_i)}{2!}(x_{i+1} - x_i)^2$$

$$+ \frac{f'''(x_i)}{3!}(x_{i+1} - x_i)^3 + \cdots + \frac{f^{(n)}(x_i)}{n!}(x_{i+1} - x_i)^n + R_n \qquad (3.16)$$

Note that because Eq. (3.16) is an infinite series, an equal sign replaces the approximate sign that was used in Eqs. (3.13) to (3.15). A remainder term is included to account for all terms from $n + 1$ to infinity:

$$R_n = \frac{f^{(n+1)}(\xi)}{(n+1)!}(x_{i+1} - x_i)^{n+1} \qquad (3.17)$$

where the subscript n connotes that this is the remainder for the nth-order approximation and ξ is a value of x that lies somewhere between x_i and x_{i+1}. The introduction of the ξ is so important that we will devote an entire section (Sec. 3.5.2) to its derivation. For the time being, it is sufficient to recognize that there is such a value that provides an exact estimate of the error.

It is often convenient to simplify the Taylor series by defining a step size $h = x_{i+1} - x_i$ and expressing Eq. (3.16) as

$$f(x_{i+1}) = f(x_i) + f'(x_i)h + \frac{f''(x_i)}{2!}h^2 + \frac{f'''(x_i)}{3!}h^3 + \cdots$$

$$+ \frac{f^{(n)}(x_i)}{n!}h^n + R_n \qquad (3.18)$$

where the remainder term is now

$$R_n = \frac{f^{(n+1)}(\xi)}{(n+1)!}h^{n+1} \qquad (3.19)$$

EXAMPLE 3.10   *Taylor Series Approximation of a Polynomial*

Problem Statement: Use zero- through fourth-order Taylor series expansions to approximate the function

$$f(x) = -0.1x^4 - 0.15x^3 - 0.5x^2 - 0.25x + 1.2$$

from $x_i = 0$ with $h = 1$. That is, predict the function's value at $x_{i+1} = 1$.

Solution: Because we are dealing with a known function, we can compute values for $f(x)$ between 0 and 1. The results (Fig. 3.14) indicate that the function starts at $f(0) = 1.2$ and then curves downward to $f(1) = 0.2$. Thus, the true value that we are trying to predict is 0.2.

The Taylor series approximation with $n = 0$ is [Eq. (3.13)]

$$f(x_{i+1}) \simeq 1.2$$

Figure 3.14
The approximation of $f(x) = -0.1x^4 - 0.15x^3 - 0.5x^2 - 0.25x + 1.2$ at $x = 1$ by zero-order, first-order, and second-order Taylor series expansions.

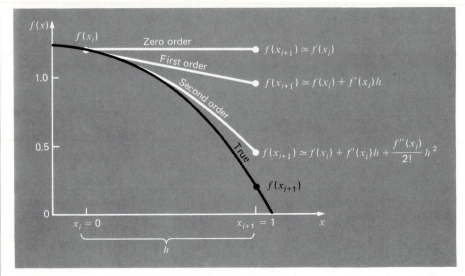

Thus, as in Fig. 3.14, the zero-order approximation is a constant. Using this formulation results in a truncation error [recall Eq. (3.2)] of

$$E_t = 0.2 - 1.2 = -1.0$$

at $x = 1$.

For $n = 1$, the first derivative must be determined and evaluated at $x_i = 0$:

$$f'(0) = -0.4(0.0)^3 - 0.45(0.0)^2 - 1.0(0.0) - 0.25 = -0.25$$

Therefore, the first-order approximation is [Eq. (3.14)]

$$f(x_{i+1}) \simeq 1.2 - 0.25h$$

which can be used to compute $f(1) = 0.95$. Consequently, the approximation begins to capture the downward trajectory of the function in the form of a sloping straight line (Fig. 3.14). This results in a reduction of the truncation error to

$$E_t = 0.2 - 0.95 = -0.75$$

at $x = 1$.

For $n = 2$, the second derivative is evaluated at $x_i = 0$:

$$f''(0) = -1.2(0.0)^2 - 0.9(0.0) - 1.0 = -1.0$$

Therefore, according to Eq. (3.15)

$$f(x_{i+1}) \simeq 1.2 - 0.25h - 0.5h^2$$

and, substituting $h = 1$,

$$f(1) \simeq 0.45$$

The inclusion of the second derivative now adds some downward curvature resulting in an improved estimate, as in Fig. 3.14. The truncation error is reduced further to $0.2 - 0.45 = -0.25$.

Additional terms would improve the approximation even more. In fact, the inclusion of the third and the fourth derivatives results in exactly the same equation as we started with:

$$f(x_{i+1}) \simeq 1.2 - 0.25h - 0.5h^2 - 0.15h^3 - 0.10h^4$$

where the remainder term is

$$R_4 = \frac{f^{(5)}(\xi)}{5!} h^5$$

Thus, because the fifth derivative of a fourth-order polynomial is zero, $R_4 = 0$. Consequently, the Taylor series expansion to the fourth derivative yields an exact estimate at $x_{i+1} = 1$:

$$f(1) \simeq 1.2 - 0.25(1) - 0.5(1)^2 - 0.15(1)^3 - 0.10(1)^4 = 0.2$$

In general, the nth-order Taylor series expansion will be exact for an nth-order polynomial. For other differentiable and continuous functions, such as exponentials and sinusoids, a finite number of terms will probably not yield an exact estimate. Each additional term will contribute some improvement, however slight, to the approximation. This behavior will be demonstrated in Example 3.11. Only if an infinite number of terms are added, will the series yield an exact result.

Although the above is true, the practical value of Taylor series expansions is that, in most cases, the inclusion of only a few of the terms will result in an approximation that is close enough to the true value for practical purposes. The assessment of how many terms are required to get "close enough" is based on the remainder term of the expansion. Recall that the remainder term is of the general form of Eq. (3.19). This relationship has two major drawbacks. First, ξ is not known exactly but merely lies somewhere between x_i and x_{i+1}. Second, in order to evaluate Eq. (3.19), we need to determine the $(n + 1)$th derivative of $f(x)$. To do this, we need to know $f(x)$. However, if we knew $f(x)$, there would be no reason to perform the Taylor series expansion in the first place!

Despite this dilemma, Eq. (3.19) is still useful for gaining insight into truncation errors. This is because we *do* have control over the term h^{n+1} in the equation. In other words, we can choose how far away from x_i we want to evaluate $f(x)$, and we can control the number of terms we include in the expansion. Consequently, Eq. (3.19) is usually expressed as

$$R_n = 0(h^{n+1})$$

where the nomenclature $0(h^{n+1})$ means that the truncation error is of the order of h^{n+1}. That is, the error is proportional to the step size h raised to the $(n + 1)$th power. Although this approximation implies nothing regarding the magnitude of the derivatives that multiply h^{n+1}, it is extremely useful in judging the comparative error of numerical methods based on Taylor series expansions. For example, if the error is $0(h)$, halving the step size will halve the error. On the other hand, if the error is $0(h^2)$, halving the step size will quarter the error.

In general, we can usually assume that the truncation error is decreased by the addition of terms to the Taylor series. In many cases, if h is sufficiently small, the first and other lower-order terms usually account for a disproportionately high percent of the error. Thus, only a few terms are required to obtain an adequate estimate. This property is illustrated by the following example.

EXAMPLE 3.11   Use of Taylor Series Expansion to Approximate a Function with an Infinite Number of Derivatives

Problem Statement: Use Taylor series expansions with $n = 0$ to 6 to approximate

$$f(x) = \cos x$$

at $x_{i+1} = \pi/3$ on the basis of the value of $f(x)$ and its derivatives at $x_i = \pi/4$. Note that this means that $h = \pi/3 - \pi/4 = \pi/12$.

Solution: As with Example 3.10, our knowledge of the true function means that we can determine the correct value of $f(\pi/3) = 0.5$.

The zero-order approximation is [Eq. (3.13)]

$$f\left(\frac{\pi}{3}\right) \simeq \cos\left(\frac{\pi}{4}\right) = 0.707106781$$

which represents a percent relative error of

$$\epsilon_t = \frac{0.5 - 0.707106781}{0.5} 100\% = -41.4\%$$

For the first-order approximation, we add the first derivative term where $f'(x) = -\sin x$:

$$f\left(\frac{\pi}{3}\right) \simeq \cos\left(\frac{\pi}{4}\right) - \sin\left(\frac{\pi}{4}\right)\left(\frac{\pi}{12}\right) = 0.521986659$$

which has $\epsilon_t = -4.40$ percent.

For the second-order approximation, we add the second derivative term where $f''(x) = -\cos x$:

$$f\left(\frac{\pi}{3}\right) \simeq \cos\left(\frac{\pi}{4}\right) - \sin\left(\frac{\pi}{4}\right)\left(\frac{\pi}{12}\right) - \frac{\cos(\pi/4)}{2}\left(\frac{\pi}{12}\right)^2 = 0.497754491$$

with $\epsilon_t = 0.449$ percent. Thus, the inclusion of additional terms results in an improved estimate.

The process can be continued and the results compiled, as in Table 3.1. Notice that the derivatives never go to zero as was the case with the polynomial in Example 3.10. Therefore, each additional term results in some improvement in the estimate. However, also notice how most of the improvement comes with the initial terms. For this case, by the time we have added the third-order term, the error is reduced to 2.62×10^{-2} percent, which means that we have attained 99.9738 percent of the true value. Consequently, although the addition of more terms will reduce the error further, the improvement becomes negligible.

TABLE 3.1 Taylor series approximation of $f(x) = \cos x$ at $x_{i+1} = \pi/3$ using a base point of $x_i = \pi/4$. Values are shown for various orders (n) of approximation.

Order n	$f^{(n)}(x)$	$f(\pi/3)$	ϵ_t
0	$\cos x$	0.707106781	-41.4
1	$-\sin x$	0.521986659	-4.4
2	$-\cos x$	0.497754491	0.449
3	$\sin x$	0.499869147	2.62×10^{-2}
4	$\cos x$	0.500007551	-1.51×10^{-3}
5	$-\sin x$	0.500000304	-6.08×10^{-5}
6	$-\cos x$	0.499999988	2.40×10^{-6}

3.5.2 The Remainder for the Taylor Series Expansion

Before demonstrating how the Taylor series is actually used to estimate numerical errors, we must explain why we included the argument ξ in Eq. (3.19). A mathematical derivation is presented in Box 3.1. We will now develop an alternative exposition based on a somewhat more visual interpretation. Then we can extend this specific case to the more general formulation.

Suppose that we truncated the Taylor series expansion [Eq. (3.18)] after the zero-order term to yield

$$f(x_{i+1}) \simeq f(x_i)$$

A visual depiction of this zero-order prediction is shown in Fig. 3.15. The remainder, or error, of this prediction, which is also shown in the illustration, consists of the infinite series of terms that were truncated:

$$R_0 = f'(x_i)h + \frac{f''(x_i)}{2!}h^2 + \frac{f'''(x_i)}{3!}h^3 + \cdots$$

It is obviously inconvenient to deal with the remainder in this infinite series format. One simplification might be to truncate the remainder itself, as in

$$R_0 \simeq f'(x_i)h \tag{3.20}$$

Although, as stated in the previous section, lower-order derivatives usually account for a greater share of the remainder than the higher-order terms, this result is still inexact because of the neglected second- and higher-order terms. This "inexactness" is implied by the approximate equality symbol ($\simeq$) employed in Eq. (3.20).

An alternative simplification that transforms the approximation into an equivalence is based on a graphical insight. Notice in Fig. 3.15 that a true error for the error R_0 could be determined if we knew the location of the exact value. Obviously, this value is not known because otherwise there would be no need for a Taylor series expansion. However, the derivative mean-value theorem of calculus provides a way to recast the problem to partially circumvent this dilemma.

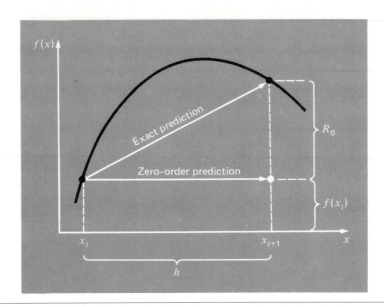

Figure 3.15
Graphical depiction of a zero-order Taylor series prediction and remainder.

The *derivative mean-value theorem* states that if a function $f(x)$ and its first derivative are continuous over an interval from x_i to x_{i+1}, then there exists at least one point on the function that has a slope, designated by $f'(\xi)$, that is parallel to the line joining $f(x_i)$ and $f(x_{i+1})$. The parameter ξ marks the x value where this slope occurs (Fig. 3.16). A physical illustration of this theorem is seen in the fact that if you travel between two points with an average velocity, there will be at least one moment during the course of the trip when you will be moving at that average velocity.

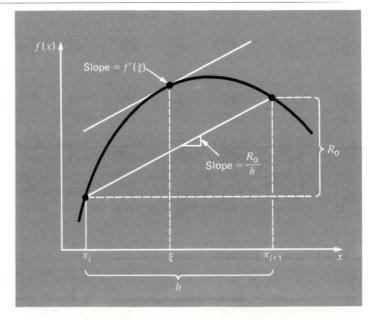

Figure 3.16
Graphical depiction of the derivative mean-value theorem.

By invoking this theorem it is simple to realize that, as illustrated in Fig. 3.16, the slope $f'(\xi)$ is equal to the rise R_0 divided by the run h, or

$$f'(\xi) = \frac{R_0}{h}$$

which can be rearranged to give

$$R_0 = f'(\xi)h \tag{3.21}$$

Thus, we have derived the zero-order version of Eq. (3.19). The higher-order versions are merely a logical extension of the reasoning used to derive Eq. (3.21). Thus, the first-order version is

$$R_1 = \frac{f''(\xi)}{2!}h^2 \tag{3.22}$$

For this case, the value of ξ conforms to the x value corresponding to the second derivative that makes Eq. (3.22) exact. Similar higher-order versions can be developed from Eq. (3.19).

3.5.3 Using the Taylor Series to Estimate Truncation Errors

Although the Taylor series will be extremely useful in estimating truncation errors throughout this book, it may not be clear to you how the expansion can actually be applied to numerical methods. In fact, we have already done so in our example of the falling parachutist. Recall that the objective of both Examples 1.1 and 1.2 was to predict velocity as a function of time. That is, we were interested in determining $v(t)$. As specified by Eq. (3.16), $v(t)$ can be expanded in a Taylor series:

$$v(t_{i+1}) = v(t_i) + v'(t_i)(t_{i+1} - t_i) + \frac{v''(t_i)}{2!}(t_{i+1} - t_i)^2 + \cdots + R_n \tag{3.23}$$

Now let us truncate the series after the first derivative term:

$$v(t_{i+1}) = v(t_i) + v'(t_i)(t_{i+1} - t_i) + R_1 \tag{3.24}$$

Equation (3.24) can be solved for

$$v'(t_i) = \underbrace{\frac{v(t_{i+1}) - v(t_i)}{t_{i+1} - t_i}}_{\substack{\text{First-order} \\ \text{approximation}}} - \underbrace{\frac{R_1}{t_{i+1} - t_i}}_{\substack{\text{Truncation} \\ \text{error}}} \tag{3.25}$$

The first part of Eq. (3.25) is exactly the same relationship that was used to approximate the derivative in Example 1.2 [Eq. (1.11)]. However, because of the Taylor series approach, we have also obtained an estimate of the truncation error associated with this approximation of the derivative. Using Eqs. (3.17) and (3.25) yields

$$\frac{R_1}{t_{i+1} - t_i} = \frac{v''(\xi)}{2!}(t_{i+1} - t_i) \tag{3.26}$$

or

$$\frac{R_1}{t_{i+1} - t_i} = 0(t_{i+1} - t_i) \tag{3.27}$$

Thus, the estimate of the derivative [Eq. (1.11) or the first part of Eq. (3.25)] has a truncation error of order $t_{i+1} - t_i$. In other words, the error of our derivative approximation should be proportional to the step size. Consequently, if we halve the step size, we would expect to halve the error of the derivative.

EXAMPLE 3.12   **The Effect of Nonlinearity and Step Size on the Taylor Series Approximation.**

Problem Statement: Figure 3.17 is a plot of the function

$$f(x) = x^m \tag{E3.12.1}$$

for $m = 1, 2, 3$, and 4 over the range from $x = 1$ to 2. Notice that for $m = 1$ the function is linear and that as m increases more curvature or *nonlinearity* is introduced into the function. Employ the first-order Taylor series to approximate this function for various values of the exponent m and the step size h.

Solution: Equation (E3.12.1) can be approximated by a first-order Taylor series expansion as in

$$f(x_{i+1}) \simeq f(x_i) + mx_i^{m-1}h \tag{E3.12.2}$$

which has a remainder

$$R_1 = \frac{f''(x_i)}{2}h^2 + \frac{f^{(3)}(x_i)}{3!}h^3 + \frac{f^{(4)}(x_i)}{4!}h^4 + \cdots$$

First, we can examine how the approximation performs as m increases—that is, as the function becomes more nonlinear.

For $m = 1$, the actual value of the function at $x = 2$ is 2. The Taylor series yields

$$f(2) = 1 + 1(1) = 2$$

and

$$R_1 = 0$$

The remainder is zero because the second and higher derivatives of a linear function are zero. Thus, as expected, the first-order Taylor series expansion is perfect when the underlying function is linear.

For $m = 2$, the actual value is $f(2) = 2^2 = 4$. The first-order Taylor series approximation is

$$f(2) = 1 + 2(1) = 3$$

and

$$R_1 = \frac{2}{2}(1)^2 + 0 + 0 + \cdots = 1$$

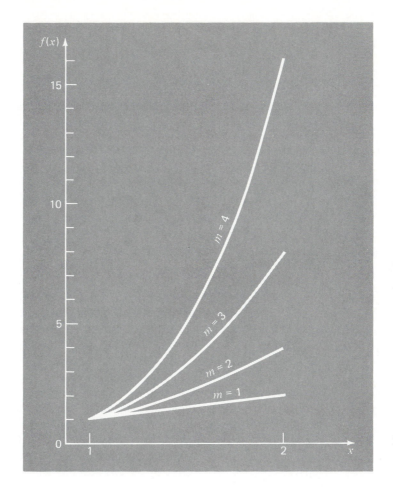

Figure 3.17
Plot of the function $f(x) = x^m$ for $m = 1, 2, 3$, and 4. Notice that the function becomes more nonlinear as m increases.

Thus, because the function is a parabola, the straight-line approximation results in a discrepancy. Note that the remainder is determined exactly.

For $m = 3$, the actual value is $f(2) = 2^3 = 8$. The first-order Taylor series approximation is

$$f(2) = 1 + 3(1)^2(1) = 4$$

and

$$R_1 = \frac{6}{2}(1)^2 + \frac{6}{6}(1)^3 + 0 + 0 + \cdots = 4$$

Again, there is a discrepancy which can be determined exactly from the Taylor series.

For $m = 4$, the actual value is $f(2) = 2^4 = 16$. The Taylor series approximation is

$$f(2) = 1 + 4(1)^3(1) = 5$$

and

$$R_1 = \frac{12}{2}(1)^2 + \frac{24}{6}(1)^3 + \frac{24}{24}(1)^4 + 0 + 0 + \cdots = 11$$

On the basis of these four cases, we observe that R_1 increases as the function becomes more nonlinear. Furthermore, R_1 accounts exactly for the discrepancy. This is due to the fact that Eq. (E3.12.1) is a simple monomial with a finite number of derivatives. This permits a complete determination of the Taylor series remainder.

Next, we will examine Eq. (E3.12.2) for the case where $m = 4$ and observe how R_1 changes as the step size h is varied. For $m = 4$, Eq. (E3.12.2) is

$$f(x_i + h) = f(x_i) + 4x_i^3 h$$

If $x_i = 1$, $f(1) = 1$ and this equation can be expressed as

$$f(1 + h) = 1 + 4h$$

with a remainder of

$$R_1 = 6h^2 + 4h^3 + h^4$$

This leads to the conclusion that the discrepancy will decrease as h is reduced. Also, at sufficiently small values of h, the error should become proportional to h^2. That is, as h is halved, the error will be quartered. This behavior is confirmed by Table 3.2 and Fig. 3.18.

Thus, we conclude that the error of the first-order Taylor series approximation decreases as m approaches 1 and as h decreases. Intuitively, this means that the Taylor series becomes more accurate when the function we are approximating becomes more like a straight line over the interval of interest. This can be accomplished either by reducing the size of the interval or by "straightening" the function by reducing m. Obviously, the latter option is usually not available in the real world because the functions we analyze are typically dictated by the physical problem context. Consequently, we do not

TABLE 3.2 Comparison of the exact value of the function $f(x) = x^4$ with the first-order Taylor series approximation. Both the function and the approximation are evaluated at $x_i + h$, where $x_i = 1$.

h	True	First-order Approximation	R_1
1	16	5	11
0.5	5.0625	3	2.0625
0.25	2.441406	2	0.441406
0.125	1.601807	1.5	0.101807
0.0625	1.274429	1.25	0.024429
0.03125	1.130982	1.125	0.005982
0.015625	1.063980	1.0625	0.001480

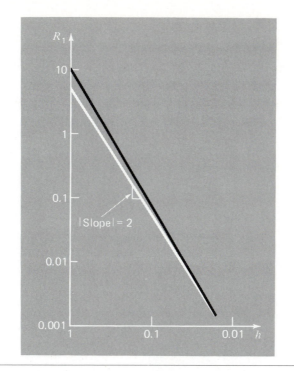

Figure 3.18
Log-log plot of the remainder R_1 of the first-order Taylor series approximation of the function $f(x) = x^4$ versus step size h. A line with a slope of 2 is also shown to indicate that as h decreases, the error becomes proportional to h^2.

have control of their lack of linearity and our only recourse is reducing the step size or including additional terms in the Taylor series expansion.

3.5.4 Numerical Differentiation

Equation (3.25) is given a formal label in numerical methods—it is called a *finite divided difference*. It can be represented generally as

$$f'(x_i) = \frac{f(x_{i+1}) - f(x_i)}{x_{i+1} - x_i} + 0(x_{i+1} - x_i) \tag{3.28}$$

or

$$f'(x_i) = \frac{\Delta f_i}{h} + 0(h) \tag{3.29}$$

where Δf_i is referred to as the *first forward difference* and h is called the *step size*, that is, the length of the interval over which the approximation is made. It is termed a "forward" difference because it utilizes data at i and $i + 1$ to estimate the derivative (Fig. 3.19*a*). The entire term $\Delta f_i/h$ is referred to as a *first finite divided difference*.

Figure 3.19
Graphical depiction of (*a*) forward, (*b*) backward, and (*c*) centered finite-divided-difference approximations of the first derivative.

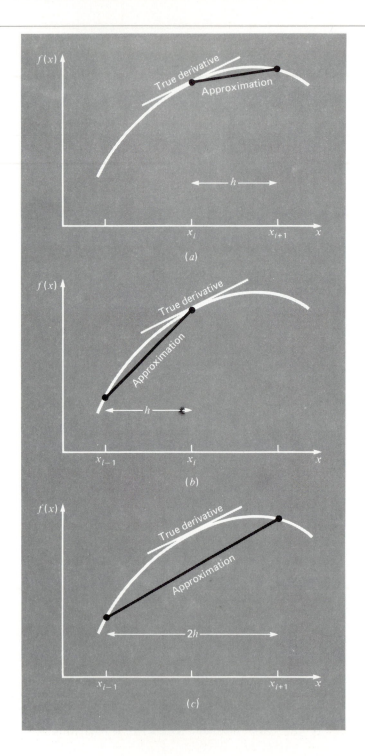

This forward divided difference is but one of many that can be developed from the Taylor series to approximate derivatives numerically. For example, *backward* and *centered difference* approximations of the first derivative can be developed in a fashion similar to the derivation of Eq. (3.25). The former utilizes data at $x_{i=1}$ (Fig. 3.19b), whereas the latter uses information that is equally spaced around the point at which the derivative is estimated (Fig. 3.19c). More accurate approximations of the first derivative can be developed by including higher-order terms of the Taylor series. Finally, all the above versions can also be developed for second, third, and higher derivatives. The following sections provide brief summaries illustrating how some of these cases are derived.

Backward Difference Approximation of the First Derivative. The Taylor series can be expanded backwards to calculate a previous value on the basis of a present value, as in

$$f(x_{i-1}) = f(x_i) - f'(x_i)h + \frac{f''(x_i)}{2}h^2 - \cdots \tag{3.30}$$

Truncating this equation after the first derivative and rearranging yields

$$f'(x_i) \simeq \frac{f(x_i) - f(x_{i-1})}{h} = \frac{\nabla f_i}{h} \tag{3.31}$$

where the error is $0(h)$ and ∇f_i is referred to as the *first backward difference*. See Fig. 3.19b for a graphical representation.

Centered Difference Approximation of the First Derivative. A third way to approximate the first derivative is to subtract Eq. (3.30) from the forward Taylor series expansion:

$$f(x_{i+1}) = f(x_i) + f'(x_i)h + \frac{f''(x_i)}{2}h^2 + \cdots \tag{3.32}$$

to yield

$$f(x_{i+1}) = f(x_{i-1}) + 2f'(x_i)h + \frac{f'''(x_i)}{3}h^3 + \cdots$$

which can be solved for

$$f'(x_i) = \frac{f(x_{i+1}) - f(x_{i-1})}{2h} - \frac{f'''(x_i)}{6}h^2 + \cdots$$

or

$$f'(x_i) = \frac{f(x_{i+1}) - f(x_{i-1})}{2h} - 0(h^2) \tag{3.33}$$

Equation (3.33) is a *centered* (or *central) difference* representation of the first derivative. Notice that the truncation error is of the order of h^2 in contrast to the forward and backward approximations that were of the order of h. Consequently, the Taylor series analysis yields the practical information that the centered difference is a more accurate representation of the derivative (Fig. 3.19c). For example, if we halve the step size using a forward or backward difference, we would approximately halve the truncation error, whereas for the central difference, the error would be quartered.

EXAMPLE 3.13   Finite-Divided-Difference Approximations of Derivatives

Problem Statement: Use forward and backward difference approximations of $0(h)$ and a centered difference approximation of $0(h^2)$ to estimate the first derivative of

$$f(x) = -0.1x^4 - 0.15x^3 - 0.5x^2 - 0.25x + 1.2$$

at $x = 0.5$ using a step size $h = 0.5$. Repeat the computation using $h = 0.25$. Note that the derivative can be calculated directly as

$$f'(x) = -0.4x^3 - 0.45x^2 - 1.0x - 0.25$$

and can be used to compute the true value as $f'(0.5) = -0.9125$.

Solution: For $h = 0.5$, the function can be employed to determine

$$x_{i-1} = 0 \qquad f(x_{i-1}) = 1.2$$
$$x_i = 0.5 \qquad f(x_i) = 0.925$$
$$x_{i+1} = 1.0 \qquad f(x_{i+1}) = 0.2$$

This data can be used to compute the forward divided difference [Eq. (3.28)],

$$f'(0.5) \simeq \frac{0.2 - 0.925}{0.5} = -1.45 \qquad \epsilon_t = -58.9\%$$

the backward divided difference [Eq. (3.31)],

$$f'(0.5) \simeq \frac{0.925 - 1.2}{0.5} = -0.55 \qquad \epsilon_t = 39.7\%$$

and the centered divided difference [Eq. (3.33)],

$$f'(0.5) \simeq \frac{0.2 - 1.2}{1.0} = -1.0 \qquad \epsilon_t = -9.6\%$$

For $h = 0.25$, the data is

$$x_{i-1} = 0.25 \qquad f(x_{i-1}) = 1.10351563$$
$$x_i = 0.50 \qquad f(x_i) = 0.925$$
$$x_{i+1} = 0.75 \qquad f(x_{i+1}) = 0.63632813$$

which can be used to compute the forward divided difference,

$$f'(0.5) \simeq \frac{0.63632813 - 0.925}{0.25} = -1.155 \qquad \epsilon_t = -26.5\%$$

the backward divided difference,

$$f'(0.5) \simeq \frac{0.925 - 1.10351563}{0.25} = -0.714 \qquad \epsilon_t = 21.7\%$$

and the centered divided difference,

$$f'(0.5) \simeq \frac{0.63632813 - 1.10351563}{0.5} = -0.934 \qquad \epsilon_t = -2.4\%$$

For both step sizes, the centered difference approximation is more accurate than forward or backward differences. Also, as predicted by the Taylor series analysis, halving the step size approximately halves the error of the backward and forward differences and quarters the error of the centered difference.

Finite Difference Approximations of Higher Derivatives. Besides first derivatives, the Taylor series expansion can be used to derive numerical estimates of higher derivatives. To do this, we write a forward Taylor series expansion for $f(x_{i+2})$ in terms of $f(x_i)$:

$$f(x_{i+2}) = f(x_i) + f'(x_i)(2h) + \frac{f''(x_i)}{2}(2h)^2 + \cdots \qquad (3.34)$$

Equation (3.32) can be multiplied by 2 and subtracted from Eq. (3.34) to give

$$f(x_{i+2}) - 2f(x_{i+1}) = -f(x_i) + f''(x_i)h^2 + \cdots$$

which can be solved for

$$f''(x_i) = \frac{f(x_{i+2}) - 2f(x_{i+1}) + f(x_i)}{h^2} + 0(h) \qquad (3.35)$$

This relationship is called the *second forward finite divided difference*. Similar manipulations can be employed to derive a backward version

$$f''(x_i) = \frac{f(x_i) - 2f(x_{i-1}) + f(x_{i-2})}{h^2} + 0(h)$$

and a centered version

$$f''(x_i) = \frac{f(x_{i+1}) - 2f(x_i) + f(x_{i-1})}{h^2} + 0(h^2)$$

As was the case with the first-derivative approximations, the centered case is more accurate. Notice also that the centered version can be alternatively expressed as

$$f''(x_i) = \frac{\dfrac{f(x_{i+1}) - f(x_i)}{h} - \dfrac{f(x_i) - f(x_{i-1})}{h}}{h}$$

Thus, just as the second derivative is a derivative of a derivative, the second divided difference approximation is a difference of two first divided differences.

We will return to the topic of numerical differentiation in Chap. 17. We have introduced you to the topic at this point because it is a very good example of how the Taylor series figures in numerical methods. In addition, several of the formulas introduced in this section will be employed prior to Chap. 17.

3.6 ERROR PROPAGATION

The purpose of this section is to study how errors in numbers can propagate through mathematical functions. For example, if we multiply two numbers that have errors, we would like to estimate the error in the product.

3.6.1 Functions of a Single Variable

Suppose that we have a function $f(x)$ that is dependent on a single independent variable x. Assume that $\tilde{x}$ is an approximation of x. We, therefore, would like to assess the effect of the discrepancy between x and $\tilde{x}$ on the value of the function. That is, we would like to estimate

$$\Delta f(\tilde{x}) = |f(x) - f(\tilde{x})|$$

The problem with evaluating $\Delta f(\tilde{x})$ is that $f(x)$ is unknown because x is unknown. We can overcome this difficulty if $\tilde{x}$ is close to x and $f(\tilde{x})$ is continuous and differentiable. If these conditions hold, a Taylor series can be employed to compute $f(x)$ near $f(\tilde{x})$, as in

$$f(x) = f(\tilde{x}) + f'(\tilde{x})(x - \tilde{x}) + \frac{f''(\tilde{x})}{2}(x - \tilde{x})^2 + \cdots$$

Dropping the second- and higher-order terms and rearranging yields

$$f(x) - f(\tilde{x}) \simeq f'(\tilde{x})(x - \tilde{x})$$

or

$$\Delta f(\tilde{x}) = |f'(\tilde{x})|\Delta \tilde{x} \tag{3.36}$$

where $\Delta f(\tilde{x}) = |f(x) - f(\tilde{x})|$ represents an estimate of the error of the function and $\Delta \tilde{x} = |x - \tilde{x}|$ represents an estimate of the error of x. Equation (3.36) provides the capability to approximate the error in $f(x)$ given the derivative of a function and an estimate of the error in the independent variable. Figure 3.20 is a graphical illustration of the operation.

EXAMPLE 3.14   Error Propagation in a Function of a Single Variable

Problem Statement: Given a value of $\tilde{x} = 2.5$ with an error of $\Delta \tilde{x} = 0.01$, estimate the resulting error in the function, $f(x) = x^3$.

Solution: Using Eq. (3.36),

$$\Delta f(\tilde{x}) \simeq 3(2.5)^2(0.01) = 0.1875$$

Because $f(2.5) = 15.625$, we predict that

$$f(2.5) = 15.625 \pm 0.1875$$

or that the true value lies between 15.4375 and 15.8125. In fact, if x were actually 2.49, the function could be evaluated as 15.4382 and if x were 2.51, it would be 15.8132. For this case, the first-order error analysis provides a fairly close estimate of the true error.

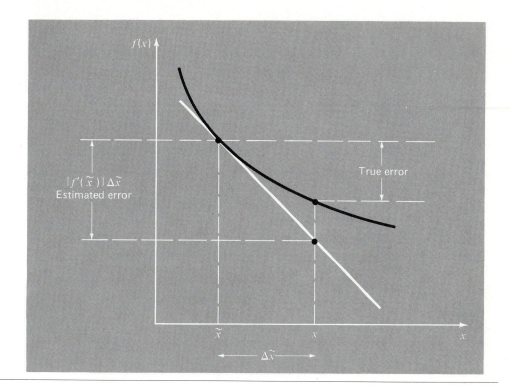

Figure 3.20
Graphical depiction of first-order error propagation.

3.6.2 Functions of More than One Variable

The foregoing approach can be generalized to functions that are dependent on more than one independent variable. This is accomplished with a multivariable version of the Taylor series. For example, if we have a function of two independent variables, u and v, the Taylor series can be written as

$$f(u_{i+1}, v_{i+1}) = f(u_i, v_i) + \frac{\partial f}{\partial u}(u_{i+1} - u_i) + \frac{\partial f}{\partial v}(v_{i+1} - v_i)$$

$$+ \frac{1}{2!}\left[\frac{\partial^2 f}{\partial u^2}(u_{i+1} - u_i)^2 + 2\frac{\partial^2 f}{\partial u \partial v}(u_{i+1} - u_i)(v_{i+1} - v_i)\right. \tag{3.37}$$

$$\left. + \frac{\partial^2 f}{\partial v^2}(v_{i+1} - v_i)^2\right] + \cdots$$

where all partial derivatives are evaluated at the base point i. If all second-order and higher terms are dropped, Eq. (3.37) can be solved for

$$\Delta f(\tilde{u}, \tilde{v}) \simeq \left|\frac{\partial f}{\partial u}\right|\Delta \tilde{u} + \left|\frac{\partial f}{\partial v}\right|\Delta \tilde{v}$$

where $\Delta \tilde{u}$ and $\Delta \tilde{v}$ are estimates of the errors in u and v respectively.

For n independent variables $\tilde{x}_1, \tilde{x}_2, \ldots, \tilde{x}_n$ having errors $\Delta \tilde{x}_1, \Delta \tilde{x}_2, \ldots, \Delta \tilde{x}_n$ the following general relationship holds

$$\Delta f(\tilde{x}_1, \tilde{x}_2, \ldots, \tilde{x}_n) \cong \left|\frac{\partial f}{\partial x_1}\right| \Delta \tilde{x}_1 + \left|\frac{\partial f}{\partial x_2}\right| \Delta \tilde{x}_2 + \cdots + \left|\frac{\partial f}{\partial x_n}\right| \Delta \tilde{x}_n \tag{3.38}$$

EXAMPLE 3.15   Error Propagation in a Multivariable Function

Problem Statement: The deflection y of the top of a sailboat mast is (see Case Study 22.3)

$$y = \frac{FL^4}{8EI}$$

where F is a uniform side loading (lb/ft), L is height (ft), E is the modulus of elasticity (lb/ft^2), and I is the moment of inertia (ft^4). Estimate the error in y given the following data:

$$\tilde{F} = 50 \text{ lb/ft} \qquad\qquad \Delta \tilde{F} = 2 \text{ lb/ft}$$
$$\tilde{L} = 30 \text{ ft} \qquad\qquad\quad \Delta \tilde{L} = 0.1 \text{ ft}$$
$$\tilde{E} = 1.5 \times 10^8 \text{ lb/ft}^2 \quad \Delta \tilde{E} = 0.01 \times 10^8 \text{ lb/ft}^2$$
$$\tilde{I} = 0.06 \text{ ft}^4 \qquad\qquad \Delta \tilde{I} = 0.0006 \text{ ft}^4$$

Solution: Employing Eq. (3.38) gives

$$\Delta y(\tilde{F}, \tilde{L}, \tilde{E}, \tilde{I}) \simeq \left|\frac{\partial y}{\partial F}\right| \Delta \tilde{F} + \left|\frac{\partial y}{\partial L}\right| \Delta \tilde{L} + \left|\frac{\partial y}{\partial E}\right| \Delta \tilde{E} + \left|\frac{\partial y}{\partial I}\right| \Delta \tilde{I}$$

or

$$\Delta y(\tilde{F}, \tilde{L}, \tilde{E}, \tilde{I}) \simeq \frac{\tilde{L}^4}{8\tilde{E}\tilde{I}} \Delta \tilde{F} + \frac{\tilde{F}\tilde{L}^3}{2\tilde{E}\tilde{I}} \Delta \tilde{L} + \frac{\tilde{F}\tilde{L}^4}{8\tilde{E}^2\tilde{I}} \Delta \tilde{E} + \frac{\tilde{F}\tilde{L}^4}{8\tilde{E}\tilde{I}^2} \Delta \tilde{I}$$

Substituting the appropriate values gives

$$\Delta y \simeq 0.0225 + 0.0075 + 0.00375 + 0.005625 = 0.039375$$

Therefore, $y \simeq 0.5625 \pm 0.039375$. In other words, y is between 0.523125 and 0.601875 ft. The validity of these estimates can be verified by substituting the extreme values for the variables into the equation to generate an exact minimum of

$$y_{min} = \frac{48(29.9)^4}{8(1.51 \times 10^8)0.0606} = 0.52407$$

and

$$y_{max} = \frac{52(30.1)^4}{8(1.49 \times 10^8)0.0594} = 0.60285$$

Thus, the first-order estimates are reasonably close to the exact values.

TABLE 3.3 Estimated error bounds associated with common mathematical operations using inexact numbers $\tilde{u}$ and $\tilde{v}$.

Operation		Estimated Error
Addition	$\Delta(\tilde{u} + \tilde{v})$	$\Delta\tilde{u} + \Delta\tilde{v}$
Subtraction	$\Delta(\tilde{u} - \tilde{v})$	$\Delta\tilde{u} + \Delta\tilde{v}$
Multiplication	$\Delta(\tilde{u} \cdot \tilde{v})$	$\lvert\tilde{u}\rvert\,\Delta\tilde{v} + \lvert\tilde{v}\rvert\,\Delta\tilde{u}$
Division	$\Delta(\tilde{u}/\tilde{v})$	$\dfrac{\lvert\tilde{u}\rvert\Delta\tilde{v} + \lvert\tilde{v}\rvert\Delta\tilde{u}}{\lvert\tilde{v}\rvert^2}$

Equation (3.38) can be employed to define error propagation relationships for common mathematical operations. The results are summarized in Table 3.3. We will leave the derivation of these formulas as a homework exercise.

3.6.3 Stability and Condition

The *condition* of a mathematical problem relates to its sensitivity to changes in its data. We say that a computation is *numerically unstable* with respect to round-off error if these uncertainties are grossly magnified by the numerical method.

These ideas can be studied using a first-order Taylor series

$$f(x) \simeq f(\tilde{x}) + f'(\tilde{x})(x - \tilde{x})$$

This relationship can be employed to estimate the *relative error of f(x)* as in

$$\epsilon[f(x)] = \frac{f(x) - f(\tilde{x})}{f(x)} \simeq \frac{f'(\tilde{x})(x - \tilde{x})}{f(\tilde{x})}$$

The *relative error of x* is given by

$$\epsilon(x) = \frac{x - \tilde{x}}{\tilde{x}}$$

A *condition number* can be defined as the ratio of these relative errors

$$\text{Condition number} = \frac{\tilde{x}f'(\tilde{x})}{f(\tilde{x})} \tag{3.39}$$

The condition number provides a measure of the extent to which an uncertainty in x is magnified by $f(x)$. A value of 1 tells us that the function's relative error is identical to the relative error in x. A value greater than 1 tells us that the relative error is amplified, whereas a value less than 1 tells us that it is attenuated. Functions with very large values are said to be *ill-conditioned*. Any combination of factors in Eq. (3.39) that increases the numerical value of the condition number will tend to magnify uncertainties in the computation of $f(x)$.

EXAMPLE 3.16   Condition Number

Problem Statement: Compute and interpret the condition number for

$$f(x) = \tan x \qquad \text{for } \tilde{x} = \pi/2 + 0.1(\pi/2)$$

$$f(x) = \tan x \qquad \text{for } \tilde{x} = \pi/2 + 0.01(\pi/2)$$

Solution: The condition number is computed as

$$\text{Condition number} = \frac{\tilde{x}(1/\cos^2 \tilde{x})}{\tan \tilde{x}}$$

For $\tilde{x} = \pi/2 + 0.1(\pi/2)$

$$\text{Condition number} = \frac{1.7279(40.86)}{-6.314} = -11.2$$

Thus, the function is ill-conditioned. For $x = \pi/2 + 0.01(\pi/2)$, the situation is even worse:

$$\text{Condition number} = \frac{1.5865(4053)}{-63.66} = -101$$

For this case, the major cause of ill conditioning appears to be the derivative. This makes sense because in the vicinity of $\pi/2$, the tangent approaches both positive and negative infinity.

3.7   TOTAL NUMERICAL ERROR

The total numerical error is the summation of the truncation and round-off errors. In general, the only way to minimize round-off errors is to increase the number of significant figures of the computer. Further, we have noted that round-off error will *increase* due to subtractive cancellation or due to an increase in the number of computations in an analysis. In contrast, Example 3.13 demonstrated that the truncation error can be reduced by decreasing the step size. Because a decrease in step size can lead to subtractive cancellation or to an increase in computations, the truncation errors are *decreased* as the round-off errors are *increased*. Therefore, we are faced by the following dilemma: The strategy for decreasing one component of the total error leads to an increase of the other component. In a computation, we could conceivably decrease the step size to minimize truncation errors only to discover that in doing so, the round-off error begins to dominate the solution and the total error grows! Thus, our remedy becomes our problem (Fig. 3.21). One challenge that we face is to determine an appropriate step size for a particular computation. We would like to choose a large step size in order to decrease the amount of calculations and round-off errors without incurring the penalty of a large truncation error. If the total error is as shown in Fig. 3.21, the challenge is to identify the point of diminishing returns where round-off error begins to negate the benefits of step-size reduction.

In actual cases, however, such situations are relatively uncommon because most computers carry enough significant figures that round-off errors do not predominate.

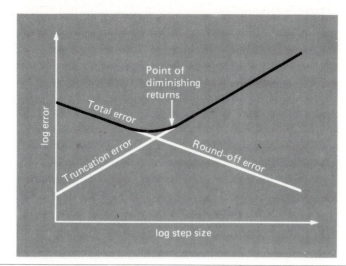

Figure 3.21
A graphical depiction of the trade-off between round-off and truncation error that sometimes comes into play in the course of a numerical method. The point of diminishing returns, where round-off error begins to negate the benefits of step-size reduction, is shown.

Nevertheless, they sometimes do occur and suggest a sort of "numerical uncertainty principle" that places an absolute limit on the accuracy that may be obtained using certain computerized numerical methods.

Because of these shortcomings, we have limitations on our ability to estimate errors. Consequently, error estimation in numerical methods is, to a certain extent, an art that depends in part on trial-and-error solution and the intuition and experience of the analyst.

Although the present chapter has concentrated on one type of numerical problem—numerical differentiation—the foregoing conclusions have general relevance to many of the other techniques in the book. However, it must be stressed that although the subject is, to a certain point, an art, there are a variety of methods that the analyst can use to quantify and control the errors in a computation. The elaboration of these techniques will play a prominent role in the following pages.

EXAMPLE 3.17   Calculation of Total Numerical Error on the Computer

Problem Statement: We would like to compute a numerical estimate of the derivative and the total numerical error of $y = x^3$ using progressively smaller step sizes.

Solution: Figure 3.22 shows a BASICA program to compute the derivative of $y = x^3$ at $x = 10$ and the output obtained from it. The approximate values are given by a forward divided difference

$$\frac{dy}{dx} \simeq \frac{y(x + h) - y(x)}{h}$$

where h varies from 1 to 10^{-10}. The results are compared with the analytical solution given by

$$\frac{dy}{dx} = 3x^2 = 300$$

```
LIST
100 B = 10
105 FOR A = 0 TO 10
110     H = 1/10^A
115     HPLUS = B + H
120     DIFF = HPLUS*HPLUS*HPLUS - B*B*B
125     DER = DIFF/H
130     PRINT H,DIFF,DER,ABS((DER-3*B*B)/(3*B*B))*100
135 NEXT A
140 END
Ok
RUN
 1              331          331          10.33333
 .1             30.30115     303.0115     1.003825
 .01            3.003052     300.3052     .1017253
 .001           .3001709     300.1709     5.696615E-02
 .0001          .0300293     300.293      9.765625E-02
 .00001         2.868652E-03 286.8652     4.378255
 .000001        2.441406E-04 244.1406     18.61979
 .0000001       0            0            100
 1E-08          0            0            100
 1E-09          0            0            100
 1E-10          0            0            100
Ok
```

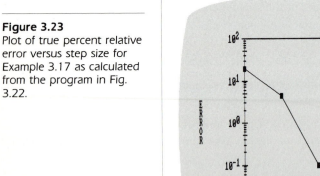

Figure 3.22
BASICA program to compute the derivative of $y = x^3$ at $x = 10$ using a forward divided difference with different values for the step size H.

The program prints out the step size, $\text{DIFF} = y(x + h) - y(x)$, the estimate of the derivative, and the percent relative error. Figure 3.23 shows a plot of the error vs. step size.

Note that a distinct minimum error occurs for a step size of $h = 0.001$. Larger step sizes increase the truncation error. Smaller step sizes increase the round-off error. At a step size between $h = 10^{-6}$ and $h = 10^{-7}$, the numerical estimate of the derivative changes from 244.14 to 0! This is a dramatic example of instability caused by round-

Figure 3.23
Plot of true percent relative error versus step size for Example 3.17 as calculated from the program in Fig. 3.22.

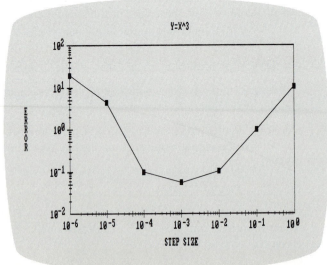

off error. For $h = 10^{-7}$, the program cannot detect any difference between $y(x + h)$ and $y(x)$. Therefore it gives DER $= 0$ and a 100 percent relative error. Clearly, decreasing the step size to decrease truncation error can be too much of a good thing if carried to extremes.

3.7.1 Control of Numerical Errors

For most practical cases, we do not know the exact error associated with numerical methods. The exception, of course, is when we have obtained the exact solution which makes our numerical approximations unnecessary. Therefore for most engineering applications we must settle for some *estimate* of the error in our calculations.

There are no systematic and general approaches to evaluating numerical errors for all problems. In many cases error estimates are based on the experience and judgment of the engineer.

Although error analysis is to a certain extent an art, there are several practical programming guidelines we can suggest. First and foremost, avoid subtracting two nearly equal numbers. Loss of significance almost always occurs when this is done. Sometimes you can rearrange or reformulate the problem to avoid subtractive cancellation. If this is not possible, you may want to use extended-precision arithmetic. Furthermore, when adding and subtracting numbers, it is best to sort the numbers and work with the smallest numbers first. This avoids loss of significance.

Beyond these computational hints, one can attempt to *predict* total numerical errors using theoretical formulations. The Taylor series is our primary tool for analysis of both truncation and round-off errors. Several examples have been presented in this chapter. Prediction of total numerical error is very complicated for even moderately sized problems and tends to be pessimistic. Therefore it is usually attempted for only small-scale tasks.

The tendency is to *push forward* with the numerical computations and try to estimate the accuracy of your results. This can sometimes be done by seeing if the results satisfy some condition or equation as a check. Or it may be possible to substitute the results back into the original equation to check that it is actually satisfied.

Finally you should be prepared to perform numerical *experiments* to increase your awareness of computational errors and possible ill-conditioned problems. Such experiments may involve repeating the computations with a different step size or method and comparing the results. We may employ sensitivity analysis to see how our solution changes when we change model parameters or input data. We may want to try different numerical algorithms that have different theoretical foundations, are based on different computational strategies, or have different convergence properties and stability characteristics.

When the results of numerical computations are extremely critical and may involve loss of human life or have severe economic ramifications, it is appropriate to take special precautions. This may involve the use of two or more *independent* groups to solve the same problem so that their results can be compared.

The roles of errors will be a topic of concern and analysis in all sections of this book. We will leave these investigations to specific sections.

3.8 BLUNDERS, FORMULATION ERRORS, AND DATA UNCERTAINTY

Although the following sources of error are not directly connected with most of the numerical methods in this book, they can sometimes have great impact on the success of a modeling effort. Thus, they must always be kept in mind when applying numerical techniques in the context of real-world problems.

3.8.1 Blunders

We are all familiar with gross errors, or blunders. In the early years of computers, erroneous numerical results could sometimes be attributed to malfunctions of the computer itself. Today, this source of error is highly unlikely, and most blunders must be attributed to human imperfection.

Blunders can occur at any stage of the mathematical modeling process and can contribute to all the other components of error. They can only be avoided by sound knowledge of fundamental principles and by the care with which you approach and design your solution to a problem.

Blunders are usually disregarded in discussions of numerical methods. This is no doubt due to the fact that, try as we may, mistakes are to a certain extent unavoidable. However, we believe that there are a number of ways in which their occurrence can be minimized. In particular, the good programming habits that were outlined in Chap. 2 are extremely useful for mitigating programming blunders. In addition, there are usually simple ways to check whether a particular numerical method is working properly. Throughout this book, we discuss ways to check the results of numerical calculations.

3.8.2 Formulation Errors

Formulation, or model, errors relate to bias that can be ascribed to incomplete mathematical models. An example of a negligible formulation error is the fact that Newton's second law does not account for relativistic effects. This does not detract from the adequacy of the solution in Example 1.1 because these errors are minimal on the time and space scales associated with the falling parachutist problem.

However, suppose that air resistance is not linearly proportional to fall velocity, as in Eq. (1.7), but is a function of the square of velocity. If this were the case, both the analytical and numerical solutions obtained in the first chapter would be erroneous because of formulation error. Further consideration of formulation error is included in some of the case studies in the remainder of the book. You should be cognizant of these problems and realize that, if you are working with a poorly conceived model, no numerical method will provide adequate results.

3.8.3 Data Uncertainty

Errors sometimes enter into an analysis because of uncertainty in the physical data upon which a model is based. For instance, suppose we wanted to test the falling parachutist model by having an individual make repeated jumps and then measuring his or her velocity after a specified time interval. Uncertainty would undoubtedly be associated with these measurements, as the parachutist would fall faster during some jumps than during others. These errors can exhibit both inaccuracy and imprecision. If our instruments consistently underestimate or overestimate the velocity, we are dealing with an inaccurate, or biased, device. On the other hand, if the measurements are randomly high and low, we are dealing with a question of precision.

Measurement errors can be quantified by summarizing the data with one or more well-chosen *statistics* that convey as much information as possible regarding specific characteristics of the data. These descriptive statistics are most often selected to represent (1) the location of the center of the distribution of the data and (2) the degree of spread of the data. As such, they provide a measure of the bias and imprecision, respectively. We will return to the topic of characterizing data uncertainty in Part Four.

Although you must be cognizant of blunders, formulation errors, and uncertain data, the numerical methods used for building models can be studied, for the most part, independently of these errors. Therefore, for most of this book, we will assume that we have not made gross errors, we have a sound model, and we are dealing with errorless measurements. Under these conditions, we can study numerical errors without complicating factors.

PROBLEMS

3.1 Compose your own program based on Fig. 3.9 and use it to determine your computer's machine epsilon.

3.2 In a fashion similar to that in Fig. 3.9, write a short program to determine the smallest number, x_{min}, used on the computer you will be employing along with this book. Note that your computer will be unable to reliably distinguish between zero and a quantity that is smaller than this number.

3.3 Determine a theoretical relationship to predict the smallest floating point number for a digital computer on the basis of parameters such as its word size, number of mantissa bits, number of exponent bits, etc.

3.4 The infinite series

$$f(N) = \sum_{n=1}^{N} \frac{1}{n^2}$$

converges on a value of $f(N) = \pi^2/6$ as N approaches infinity. Write a program to compute $f(N)$ for $N = 10,000$ by computing the sum from $n = 1$ to 10,000. Then repeat the computation but in reverse order—that is, from $n = 10,000$ to 1 using increments of -1. Explain the results.

3.5 In Example 3.2, we used the infinite series

$$f(x) = 1 + x + \frac{x^2}{2!} + \frac{x^3}{3!} + \cdots$$

to approximate e^x.

(a) Prove that this Maclaurin series expansion is a special case of the Taylor series expansion [Eq. (3.18)] with $x_i = 0$ and $h = x$.

(b) Use the Taylor series to estimate $f(x) = e^{-x}$ at $x_{i+1} = 2$ for three separate cases: $x_i = 0.5$, 1.0, and 1.5. Employ the zero-, first-, second-, and third-order versions and compute the $|\epsilon_t|$ for each case.

3.6 The Maclaurin series expansion for $\cos x$ is

$$\cos x = 1 - \frac{x^2}{2!} + \frac{x^4}{4!} - \frac{x^6}{6!} + \frac{x^8}{8!} - \cdots$$

Starting with the simplest version, $\cos x \cong 1$, add terms one at a time in order to estimate $\cos (\pi/3)$. After each new term is added, compute the true and approximate percent relative errors. Use your pocket calculator to determine the true value. Add terms until the absolute value of the approximate error estimate falls below an error criterion conforming to two significant figures.

3.7 Perform the same computation as in Prob. 3.6, but use the Maclaurin series expansion for $\sin x$

$$\sin x = x - \frac{x^3}{3!} + \frac{x^5}{5!} - \frac{x^7}{7!} + \cdots$$

to estimate $\sin (\pi/2)$.

3.8 Use zero- through third-order Taylor series expansions to predict $f(3)$ for

$$f(x) = 25x^3 - 6x^2 + 7x - 88$$

using a base point at $x = 2$. Compute the true percent relative error ϵ_t for each approximation.

3.9 Use zero- through fourth-order Taylor series expansions to predict $f(4)$ for $f(x) = \ln x$ using a base point at $x = 2$. Compute the true percent relative error ϵ_t for each approximation.

3.10 Use zero- through fourth-order Taylor series expansions to predict $f(3)$ for $f(x) = e^{-x}$ using a base point at $x = 1$. Compute the percent relative error ϵ_t for each approximation.

3.11 Use forward and backward difference approximations of $0(h)$ and a centered difference approximation of $0(h^2)$ to estimate the first derivative of the function examined in Prob. 3.8. Evaluate the derivative at $x = 2.5$ using a step size of $h = 0.25$. Compare your results with the true value of the derivative at $x = 2.5$. Interpret your results on the basis of the remainder term of the Taylor series expansion.

3.12 Use a centered difference approximation of $0(h^2)$ to estimate the second derivative of the function examined in Prob. 3.8. Perform the evaluation at $x = 2.6$ using step sizes of $h = 0.2$ and 0.1. Compare your estimates with the true value of the second derivative at $x = 2.6$. Interpret your results on the basis of the remainder term of the Taylor series expansion.

3.13 Recall that the velocity of the falling parachutist can be computed by [Eq. (1.10)],

$$v = \frac{gm}{c}\left[1 - e^{-(c/m)t}\right]$$

Use a first-order error analysis to estimate the error of v at $t = 7$, if $g = 9.8$ and $m = 68.1$ but $c = 12.5 \pm 2$.

3.14 Repeat Prob. 3.13 with $g = 9.8$, $t = 7$, $c = 12.5 \pm 2$, and $m = 68.1 \pm 0.5$.

3.15 The Stefan-Boltzmann law can be employed to estimate the rate of radiation of energy H, from a surface, as in

$$H = Ae\sigma T^4$$

where H is in watts, A is the surface area (m^2), e is the emissivity which characterizes the emitting properties of the surface (dimensionless), σ is a universal constant called the Stefan-Boltzmann constant ($=5.67 \times 10^{-8} W \cdot m^{-2} \cdot K^{-4}$), and T is absolute temperature (K). Determine the error of H for a steel plate with $A = 0.1m^2$, $e = 1.0$, and $T = 600 \pm 20$. Compare your results with the exact error. Repeat the computation but with $T = 600 \pm 40$. Interpret your results.

3.16 Repeat Prob. 3.15 but for a copper sphere with radius $=0.1 \pm 0.01$ m, $e = 0.5 \pm 0.05$, and $T = 500 \pm 20$.

3.17 Evaluate and interpret the condition numbers for

(a) $f(x) = \sqrt{|x - 1|} + 1 \qquad$ for $x = 1.001$

(b) $f(x) = e^{-x} \qquad$ for $x = 10$

(c) $f(x) = \sqrt{x^2 + 1} - x \qquad$ for $x = 10^3$

(d) $f(x) = \dfrac{e^x - 1}{x} \qquad$ for $x = 0.001$

(e) $f(x) = \dfrac{\sin x}{1 + \cos x} \qquad$ for $x = 1.0001\pi$

3.18 Employing ideas from Sec. 3.6, derive the relationships from Table 3.3.

3.19 How can the machine epsilon be employed to formulate a stopping criterion, ϵ_s, for your programs? Provide an example.

EPILOGUE: PART ONE

PT1.4 TRADE-OFFS

Numerical methods are scientific in the sense that they represent systematic techniques for solving mathematical problems. However, there is a certain degree of art, subjective judgment, and compromise associated with their effective use in engineering practice. For each problem, you may be confronted with several alternative numerical methods and many different types of computers. Thus, the elegance and efficiency of different approaches to problems is highly individualistic and correlated with your ability to choose wisely among options. Unfortunately, as with any intuitive process, the factors influencing this choice are difficult to communicate. Only by experience can these skills be fully comprehended and honed. However, because these skills play such a prominent role in the effective implementation of the methods, we have included this section as an introduction to some of the trade-offs that you must consider when selecting a numerical method and the tools for implementing the method. It is hoped that the discussion that follows will influence your orientation when approaching subsequent material. Also, it is hoped that you will refer back to this material when you are confronted with choices and trade-offs in the remainder of the book.

Figure PT1.4 illustrates seven different factors, or trade-offs, that should be considered when selecting a numerical method for a particular problem.

1. Type of Mathematical Problem. As delineated previously in Fig. PT1.2, several types of mathematical problems are discussed in this book:
   a. Roots of equations
   b. Systems of simultaneous linear algebraic equations
   c. Curve fitting
   d. Numerical integration
   e. Ordinary differential equations
   f. Partial differential equations
   You will probably be introduced to the applied aspects of numerical methods by confronting a problem in one of the above areas. Numerical methods will be required because the problem cannot be solved efficiently using analytical techniques. You should be cognizant of the fact that your professional activities will eventually involve problems in all the above areas. Thus,

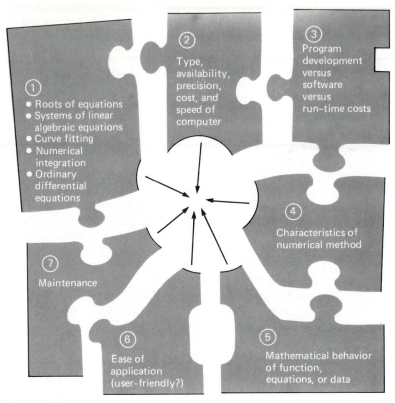

Figure PT1.4
Seven considerations in choosing a numerical method for the solution of engineering problems.

the study of numerical methods and the selection of automatic computation equipment should, at the minimum, consider these basic types of problems. More advanced problems may require capabilities of handling parameter optimization, linear programming, integral equations, etc. These areas typically demand greater computation power and advanced methods not covered in this text. Other references such as Carnahan, Luther, and Wilkes (1969); Hamming (1973); and Ralston and Rabinowitz (1978) should be consulted for problems beyond the scope of this book. In addition, at the end of each part of this text, we include a brief summary and references for advanced methods to provide you with avenues for pursuing further studies of numerical methods.

2. Type, Availability, Precision, Cost, and Speed of Computer. You may have the option of working with a variety of computation tools. These range from pocket calculators to large mainframe computers. Of course, any of the tools can be used to implement any numerical method (including simple paper and pencil). It is usually not a question of ultimate capability but rather of cost, convenience, speed, dependability, repeatability, and precision. Although each of the tools will continue to have utility, the recent rapid advances in the

performance of personal computers have already had a major impact on the engineering profession. We expect this revolution will spread as technological improvements continue, because personal computers offer an excellent compromise in convenience, cost, precision, speed, and storage capacity. Furthermore, they can be readily applied to most practical engineering problems.

3. Program Development Cost versus Software Cost versus Run-Time Cost. Once the types of mathematical problems to be solved have been identified and the computer system has been selected, it is appropriate to consider software and run-time costs. Software development may represent a substantial effort in many engineering projects and may therefore be a significant cost. In this regard, it is particularly important that you be very well acquainted with the theoretical and practical aspects of the relevant numerical methods. Professionally developed software may be available for a limited number of engineering problems at a reasonable cost. However, these programs should be used with great care because you will not ordinarily be intimately familiar with the program logic. Alternatively, general utility software of low cost (such as that associated with this book) is available to implement numerical methods that may be readily adapted to a wide variety of problems. Program development costs and software costs may be recovered during execution if the programs are efficiently written and well-tested.

4. Characteristics of the Numerical Method. When computer hardware and software costs are high, or if computer availability is limited (for example, on some timeshare systems), it pays to choose carefully the numerical method to suit the situation. On the other hand, if the problem is still at the exploratory stage and computer access and cost are not problems, it may be appropriate for you to select a numerical method that always works but may not be the most computationally efficient. The numerical methods available to solve any particular type of problem involve the types of trade-offs just discussed and others:

   a. Number of Initial Guesses or Starting Points. Some of the numerical methods for finding roots of equations or solving differential equations require the user to specify initial guesses or starting points. Simple methods usually require one value, whereas complicated methods may require more than one value. The advantages of complicated methods that are computationally efficient may be offset by the requirement for multiple starting points. You must use your experience and judgment to assess the trade-offs for each particular problem.

   b. Rate of Convergence. Certain numerical methods converge more rapidly than others. However, this rapid convergence may require more initial guesses and more complex programming than a method with slower convergence. Again, you must use your judgment in selecting a method. Faster is not always better!

   c. Stability. Some numerical methods for finding roots of equations or solutions for systems of linear equations may diverge rather than converge on the correct answer for certain problems. Why would you tolerate this possibility when confronted with design or planning problems? The answer is that these methods may be highly efficient when they work. Thus, trade-offs again emerge. You must decide if your problem requirements justify the effort needed to apply a method that may not always converge.

d. Accuracy and Precision. Some numerical methods are simply more accurate or precise than others. Good examples are the various equations available for numerical integration. Usually, the performance of low-accuracy methods can be improved by decreasing the step size or increasing the number of applications over a given interval. Is it better to use a low-accuracy method with small step sizes or a high-accuracy method with large step sizes? This question must be addressed on a case-by-case basis taking into consideration the additional factors such as cost and ease of programming. In addition, you must also be concerned with round-off errors when you are using multiple applications of low-accuracy methods and when the number of computations becomes large. Here the number of significant figures handled by the computer may be the deciding factor.

e. Breadth of Application. Some numerical methods can only be applied to a limited class of problems or to problems that satisfy certain mathematical restrictions. Other methods are not affected by such limitations. You must evaluate whether it is worth your effort to develop programs that employ techniques that are appropriate for only a limited number of problems. The fact that such techniques may be widely used suggests that they have advantages that will often outweigh their disadvantages. Obviously, trade-offs are occurring.

f. Special Requirements. Some numerical techniques attempt to increase accuracy and rate of convergence using additional or special information. An example would be to use estimated or theoretical values of errors to improve accuracy. However, these improvements are generally not achieved without some inconvenience in terms of added computing costs or increased program complexity.

g. Programming Effort Required. Efforts to improve rates of convergence, stability, and accuracy can be creative and ingenious. When improvements can be made without increasing the programming complexity, they may be considered elegant and will probably find immediate use in the engineering profession. However, if they require more complicated programs, you are again faced with a trade-off situation that may or may not favor the new method.

It is clear that the above discussion concerning a choice of numerical methods reduces to one of cost and accuracy. The costs are those involved with computer time and program development. Appropriate accuracy is a question of professional judgment and ethics.

5. Mathematical Behavior of the Function, Equation, or Data. In selecting a particular numerical method, type of computer, and type of software, you must consider the complexity of your functions, equations, or data. Simple equations and smooth data may be appropriately handled by simple numerical algorithms and inexpensive computers. The opposite is true for complicated equations and data exhibiting discontinuities.

6. Ease of Application (User-Friendly?). Some numerical methods are easy to apply, others are difficult. This may be a consideration when choosing one method over another. This same idea applies to decisions regarding program development costs versus professionally developed software. It may take considerable effort to convert a difficult program to one that is user-friendly. Ways to do this were introduced in Chap. 2 and are elaborated throughout the book. In addition, the Electronic TOOLKIT software supplement is an example of user-friendly programming.

7. Maintenance. Programs for solving engineering problems require maintenance because during application, difficulties invariably occur. Maintenance may require changing the program code or expanding the documentation. Simple programs and numerical algorithms are simpler to maintain.

The chapters that follow involve the development of various types of numerical methods for various types of mathematical problems. Several alternative methods will be given in each chapter. These various methods (rather than a single method chosen by the authors) are presented because there is no single "best" method. There is no "best" method because there are many trade-offs that must be considered when applying the methods to practical problems. A table that highlights the trade-offs involved in each method will be found at the end of each part of the book. This table should assist you in selecting the appropriate numerical procedure for your particular problem context.

PT1.5 IMPORTANT RELATIONSHIPS AND FORMULAS

Table PT1.2 summarizes important information that was presented in Part One. The table can be consulted to quickly access important relationships and formulas. The epilogue of each part of the book will contain such a summary.

PT1.6 ADVANCED METHODS AND ADDITIONAL REFERENCES

The epilogue of each part of the book will also include a section designed to facilitate and encourage further studies of numerical methods. This section will reference other books on the subject as well as material related to more advanced methods.*

To extend the background provided in Part One, numerous manuals on computer programming are available. It would be difficult to reference all the excellent books and manuals pertaining to specific languages and computers. In addition, you probably already have material from your previous exposure to programming. However, if this is your first experience with computers, Chapra and Canale (1986) provide a general introduction to BASIC and FORTRAN. This book also includes a supplement on Turbo Pascal that is available from the publisher. Your instructor

---

*Books are referenced only by author here; a complete bibliography will be found at the back of this text.

TABLE PT1.2   Summary of important information presented in Part One.

The Structure Principle

The static structure of the program should correspond in a simple way to the dynamic structure of the corresponding computation.

Error Definitions

True error

$$E_t = \text{true value} - \text{approximation}$$

True percent
   relative error

$$\epsilon_t = \frac{\text{true value} - \text{approximation}}{\text{true value}} 100\%$$

Approximate
   percent relative error

$$\epsilon_a = \frac{\text{present approximation} - \text{previous approximation}}{\text{present approximation}} 100\%$$

Stopping criterion

Terminate computation when

$$\epsilon_a < \epsilon_s$$

where ϵ_s is the desired percent relative error

Taylor Series

Taylor series expansion

$$f(x_{i+1}) = f(x_i) + f'(x_i)h + \frac{f''(x_i)}{2!}h^2$$

$$+ \frac{f'''(x_i)}{3!}h^3 + \cdots + \frac{f^{(n)}(x_i)}{n!}h^n + R_n$$

where

Remainder

$$R_n = \frac{f^{(n+1)}(\xi)}{(n+1)!}h^{n+1}$$

or

$$R_n = O(h^{n+1})$$

Numerical Differentiation

First forward
   finite divided difference

$$f'(x_i) = \frac{f(x_{i+1}) - f(x_i)}{h} + O(h)$$

(Other divided differences are summarized in Chaps. 3 and 17.)

Error Propagation

For n independent variables $x_1, x_2, \ldots, x_n$ having errors $\Delta \tilde{x}_1, \Delta \tilde{x}_2, \ldots, \Delta \tilde{x}_n$, the error in the function f can be estimated via

$$\Delta f = \left| \frac{\partial f}{\partial x_1} \right| \Delta \tilde{x}_1 + \left| \frac{\partial f}{\partial x_2} \right| \Delta \tilde{x}_2 + \cdots + \left| \frac{\partial f}{\partial x_n} \right| \Delta \tilde{x}_n$$

and fellow students should also be able to advise you regarding good reference books for the machines and languages available at your school.

As for error analysis, any good introductory calculus book will include supplementary material related to subjects such as the Taylor series expansion. Texts by Swokowski (1979), Thomas and Finney (1979), and Simmons (1985) provide very readable discussions of these subjects. In addition, Taylor (1982) presents a nice introduction to error analysis.

Finally, although we hope that our book serves you well, it is always good to consult other sources when trying to master a new subject. Ralston and Rabinowitz (1978) and Carnahan, Luther, and Wilkes (1969) provide comprehensive discussions of most numerical methods, including many advanced methods that are beyond our scope. Other enjoyable books on the subject are Gerald and Wheatley (1984), James, Smith, and Wolford (1985), Rice (1983), Cheney and Kincaid (1985), Yakowitz and Szidarovszky (1986), and Al-Khafaji and Tooley (1986). In addition, Press et al. (1986) and Atkinson and Harley (1983) include computer codes to implement a variety of methods.

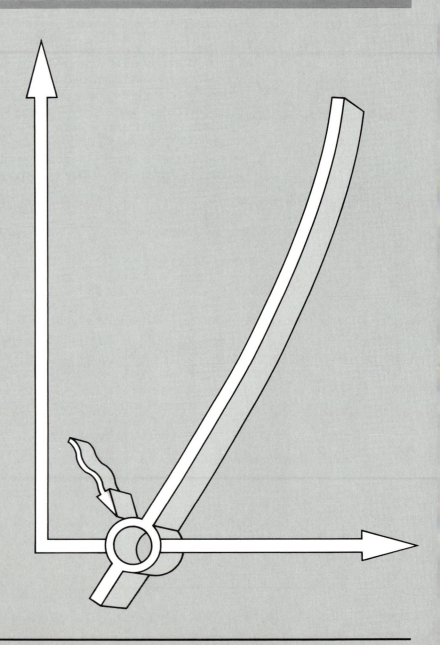

ROOTS OF EQUATIONS

PT2.1 MOTIVATION

Years ago, you learned to use the *quadratic formula*

$$x = \frac{-b \pm \sqrt{b^2 - 4ac}}{2a} \tag{PT2.1}$$

to solve

$$f(x) = ax^2 + bx + c = 0 \tag{PT2.2}$$

The values calculated with Eq. (PT2.1) are called the "roots" of Eq. (PT2.2). They represent the values of x that make Eq. (PT2.2) equal to zero. Thus, we can define the *root* of an equation as the value of x that makes $f(x) = 0$. For this reason, roots are sometimes called the *zeros* of the equation.

Although the quadratic formula is handy for solving Eq. (PT2.2), there are many other functions for which the root cannot be determined so easily. For these cases, the numerical methods described in Chaps. 4 and 5 provide efficient means to obtain the answer.

PT2.1.1 Precomputer Methods for Determining Roots

Before the advent of digital computers, there were a number of ways to solve for roots of algebraic and transcendental equations. For some cases, the roots could be obtained by direct methods, as was done with Eq. (PT2.1). Although there were equations like this that could be solved directly, there were many more that could not. For example, even an apparently simple function such as $f(x) = e^{-x} - x$ cannot be solved analytically. In such instances, the only alternative is an approximate solution technique.

One method to obtain an approximate solution is to plot the function and determine where it crosses the x axis. This point, which represents the x value for which $f(x) = 0$, is the root. Graphical techniques are discussed at the beginning of Chaps. 4 and 5.

Although graphical methods are useful for obtaining rough estimates of roots, they are limited because of their lack of precision. An alternative approach is to use trial and error. This "technique" consists of guessing a value of x and evaluating whether $f(x)$ is zero. If not (as is almost always the case), another guess is made, and $f(x)$ is again evaluated to determine whether the new value provides a better estimate of the root. The process is repeated until a guess is obtained that results in an $f(x)$ that is close to zero.

Such haphazard methods are obviously inefficient and inadequate for the requirements of engineering practice. The techniques described in Part Two represent alternatives that are also approximate but employ systematic strategies to home in on the true root. As elaborated in the following pages, the combination of these systematic methods and computers makes the solution of most applied roots-of-equations problems a simple and efficient task.

PT2.1.2   Roots of Equations and Engineering Practice

Although they arise in other problem contexts, roots of equations frequently occur in the area of engineering design. Table PT2.1 lists a number of fundamental principles that are routinely used in design work. As introduced in Chap. 1, mathematical equations or models derived from these principles are employed to predict dependent variables as a function of independent variables, forcing functions, and parameters. Note that in each case, the dependent variables reflect the state or performance of the system, whereas the parameters represent its properties or composition.

An example of such a model is the equation, derived from Newton's second law, used in Chap. 1 for the parachutist's velocity:

$$v = \frac{gm}{c}\left[1 - e^{-(c/m)t}\right]$$ (PT2.3)

where velocity v is the dependent variable; time t is the independent variable; the gravitational constant g is the forcing function; and the drag coefficient c and the mass m are parameters. If the parameters are known, Eq. (PT2.3) can be used to predict the parachutist's velocity as a function of time. Such computations can be performed directly because v is expressed *explicitly* as a function of time. That is, it is isolated on one side of the equal sign.

However, suppose that we had to determine the drag coefficient for a parachutist of a given mass to attain a prescribed velocity in a set time period. Although Eq. (PT2.3) provides a mathematical representation of the interrelationship among the model variables and parameters, it cannot be solved explicitly for the drag coefficient. Try it. There is no way to rearrange the equation so that c is isolated on one side of the equal sign. In such cases, c is said to be *implicit*.

This represents a real dilemma, because many engineering design problems involve specifying the properties or composition of a system (as represented by its parameters) in order to ensure that it performs in a desired manner (as represented

TABLE PT2.1 Fundamental principles used in engineering design problems

Fundamental Principle	Dependent Variable	Independent Variable	Parameters
Heat balance	Temperature	Time and position	Thermal properties of material and geometry of system
Mass balance	Concentration or quantity of mass	Time and position	Chemical behavior of material, mass transfer coefficients, and geometry of system
Force balance	Magnitude and direction of forces	Time and position	Strength of material, structural properties, and geometry of system
Energy balance	Changes in the kinetic- and potential-energy states of the system	Time and position	Thermal properties, mass of material, and system geometry
Newton's laws of motion	Acceleration, velocity, or location	Time and position	Mass of material, system geometry, and dissipative parameters such as friction or drag
Kirchhoff's laws	Currents and voltages in electric circuits	Time	Electrical properties of systems such as resistance, capacitance, and inductance

by its variables). Thus, these problems often require the determination of implicit parameters.

The solution to the dilemma is provided by numerical methods for roots of equations. To solve the problem using numerical methods, it is conventional to reexpress Eq. (PT2.3). This is done by subtracting the dependent variable v from both sides of the equation to give

$$f(c) = \frac{gm}{c}\left[1 - e^{-(c/m)t}\right] - v \qquad \text{(PT2.4)}$$

The value of c that makes $f(c) = 0$ is, therefore, the root of the equation. This value also represents the drag coefficient that solves the design problem.

Part Two of this book deals with a variety of numerical and graphical methods for determining roots of relationships such as Eq. (PT2.4). These techniques can be applied to engineering design problems that are based on the fundamental principles outlined in Table PT2.1 as well as to many other problems confronted routinely in engineering practice.

PT2.2   MATHEMATICAL BACKGROUND

For most of the subject areas in this book, there is usually some prerequisite mathematical background needed to successfully master the topic. For example, the concepts of error estimation and the Taylor series expansion discussed in Chap. 3 have direct relevance to our discussion of roots of equations. Additionally, prior to this point we have mentioned the terms "algebraic" and "transcendental" equations. It might be helpful to formally define these terms and discuss how they relate to the scope of this part of the book.

By definition, a function given by $y = f(x)$ is *algebraic* if it can be expressed in the form

$$f_n y^n + f_{n-1} y^{n-1} + \cdots + f_1 y + f_0 = 0 \qquad \text{(PT2.5)}$$

where f_i is an ith-order polynomial in x. *Polynomials* are a simple class of algebraic functions that are represented generally by

$$f_n(x) = a_0 + a_1 x + \cdots + a_n x^n \qquad \text{(PT2.6)}$$

where the a's are constants. Some specific examples are

$$f(x) = 1 - 2.37x + 7.5x^2 \qquad \text{(PT2.7)}$$

and

$$f(x) = 5x^2 - x^3 + 7x^6 \qquad \text{(PT2.8)}$$

A *transcendental* function is one that is nonalgebraic. These include trigonometric, exponential, logarithmic, and other, less familiar, functions. Examples are

$$f(x) = e^{-x} - x \qquad \text{(PT2.9)}$$

$$f(x) = \sin x \qquad \text{(PT2.10)}$$

$$f(x) = \ln x^2 - 1 \qquad \text{(PT2.11)}$$

Roots of equations may be either real or complex. Although there are cases where complex roots of nonpolynomials are of interest, such situations are less common than for polynomials. As a consequence, the standard methods for locating roots typically fall into two somewhat related but primarily distinct problem areas:

1. *The determination of the real roots of algebraic and transcendental equations.* These techniques are designed to determine the value of a single real root on the basis of foreknowledge of its approximate location.
2. *The determination of all real and complex roots of polynomials.* These methods are specifically designed for polynomials. They systematically determine all the roots of the polynomial rather than determining a single real root given an approximate location.

In the present book, we focus on the first problem area. Because they are somewhat beyond the scope of this part of the text, we will not discuss methods for locating all the roots of polynomials here. However, we will describe one such approach, Bairstow's method, when we cover eigenvalues in Chap. 21.

PT2.3 ORIENTATION

Some orientation is helpful before proceeding to the numerical methods for determining roots of equations. The following is intended to give you an overview of the material in Part Two. In addition, some objectives have been included to help you focus your efforts when studying the material.

PT2.3.1 Scope and Preview

Figure PT2.1 is a schematic representation of the organization of Part Two. Examine this figure carefully, starting at the top and working clockwise.

After the present introduction, *Chap. 4* is devoted to *bracketing methods* for finding roots. These methods start with guesses that bracket, or contain, the root and then systematically reduce the width of the bracket. Two specific methods are covered: *bisection* and *false position.* Graphical methods are used to provide visual insight into the techniques. Error formulations are developed to help you determine how much computational effort is required to estimate the root to a prespecified level of precision.

Chapter 5 covers *open methods.* These methods also involve systematic trial-and-error iterations but do not require that the initial guesses bracket the root. We will discover that these methods are usually more computationally efficient than bracketing methods, but that they do not always work. *One-point iteration, Newton-Raphson,* and *secant* methods are described. Graphical methods are used to provide geometric insight into cases where the open methods do not work. Formulas are developed that provide an idea of how fast open methods home in on the root. In addition, an approach to extend the Newton-Raphson method to *systems of nonlinear equations* is explained.

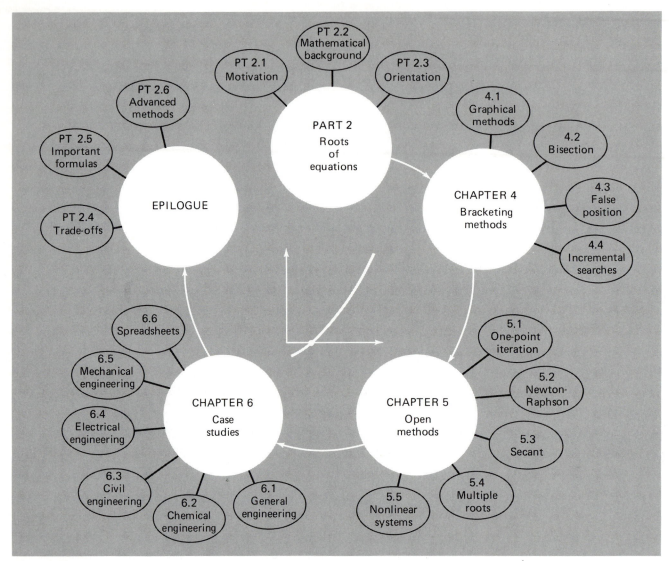

Figure PT2.1
Schematic of the organization of the material in Part Two: Roots of Equations.

Chapter 6 extends the above concepts to actual engineering problems. Case studies are used to illustrate the strengths and weaknesses of each method and provide insight into the application of the techniques in professional practice. The case studies in Chap. 6 also highlight the trade-offs (as discussed in Part One) associated with the various methods.

An epilogue is included at the end of Part Two. It contains a detailed comparison of the methods discussed in Chaps. 4 and 5. This comparison includes a description of trade-offs related to the proper use of each technique. This section also provides a summary of important formulas, along with references for some numerical methods that are beyond the scope of this text.

PT2.3.2   Goals and Objectives

Study Objectives.   After completing Part Two, you should have sufficient information to successfully approach a wide variety of engineering problems dealing with roots of equations. In general, you should have mastered the techniques, have learned to assess their reliability, and be capable of choosing the best method (or methods) for any particular problem. In addition to these general goals, the specific concepts in Table PT2.2 should be assimilated for a comprehensive understanding of the material in Part Two.

Computer Objectives.   The book provides you with software, simple computer programs, and pseudocode to implement the techniques discussed in Part Two. All have utility as learning tools.

The optional software, the Electronic TOOLKIT, is user-friendly. It contains the bisection method to determine the real roots of algebraic and transcendental equations. The graphics associated with this software will enable you to easily visualize the behavior of the function being analyzed. The software can be used

TABLE PT2.2   Specific study objectives for Part Two

1. Understand the graphical interpretation of a root
2. Know the graphical interpretation of the false-position method and why it is usually superior to the bisection method
3. Understand the difference between bracketing and open methods for root location
4. Understand the concepts of convergence and divergence. Use the two-curve graphical method to provide a visual manifestation of the concepts
5. Know why bracketing methods always converge, whereas open methods may sometimes diverge
6. Realize that convergence of open methods is more likely if the initial guess is close to the true root
7. Understand the concepts of linear and quadratic convergence and their implications for the efficiencies of the one-point iteration and Newton-Raphson methods
8. Know the fundamental difference between the false-position and secant methods and how it relates to convergence
9. Understand the problems posed by multiple roots and the modifications available to mitigate them
10. Know how to extend the single-equation Newton-Raphson approach to solve systems of nonlinear equations

to conveniently determine roots of equations to a desired degree of precision. The Electronic TOOLKIT is easy to apply in solving many practical problems and can be used to check the results of any computer programs you may develop yourself.

BASIC, FORTRAN, and Pascal programs for the bisection method are also supplied directly in the text. In addition, general algorithms are provided for most of the other methods in Part Two. This information will allow you to expand your software library to include programs that are more efficient than the bisection method. For example, you may also want to have your own software for the false-position, Newton-Raphson, and secant techniques, which are usually more efficient than the bisection method.

CHAPTER 4

Bracketing Methods

This chapter on roots of equations deals with methods that exploit the fact that a function typically changes sign in the vicinity of a root. These techniques are called *bracketing methods* because two initial guesses for the root are required. As the name implies, these guesses must "bracket," or be on either side of, the root. The particular methods described herein employ different strategies to systematically reduce the width of the bracket and, hence, home in on the correct answer.

As a prelude to these techniques, we will briefly discuss graphical methods for depicting functions and their roots. Beyond their utility for providing rough guesses, graphical techniques are also useful for visualizing the properties of the functions and the behavior of the various numerical methods.

4.1 GRAPHICAL METHODS

A simple method for obtaining an estimate of the root of the equation $f(x) = 0$ is to make a plot of the function and observe where it crosses the x axis. This point, which represents the x value for which $f(x) = 0$, provides a rough approximation of the root.

EXAMPLE 4.1   The Graphical Approach

Problem Statement: Use the graphical approach to determine the drag coefficient c needed for a parachutist of mass $m = 68.1$ kg to have a velocity of 40 m/s after free-falling for time $t = 10$ s. *Note:* The acceleration due to gravity is 9.8 m/s^2.

Solution: This problem can be solved by determining the root of Eq. (PT2.4) using the parameters $t = 10, g = 9.8, v = 40$, and $m = 68.1$:

$$f(c) = \frac{9.8(68.1)}{c}(1 - e^{-(c/68.1)10}) - 40$$

or

$$f(c) = \frac{667.38}{c}(1 - e^{-0.146843c}) - 40 \qquad \text{(E4.1.1)}$$

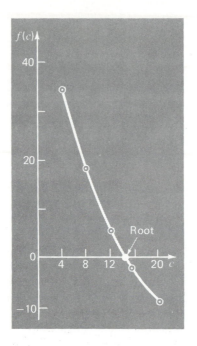

Figure 4.1
The graphical approach for determining the roots of an equation.

Various values of c can be substituted into the right-hand side of this equation to compute

c	f(c)
4	34.115
8	17.653
12	6.067
16	−2.269
20	−8.401

These points are plotted in Fig. 4.1. The resulting curve crosses the c axis between 12 and 16. Visual inspection of the plot provides a rough estimate of the root of 14.75. The validity of the graphical estimate can be checked by substituting it into Eq. (E4.1.1) to yield

$$f(14.75) = \frac{667.38}{14.75}(1 - e^{-0.146843(14.75)}) - 40$$

$$= 0.059$$

which is close to zero. It can also be checked by substituting it into Eq. (PT2.4) along with the parameter values from this example to give

$$v = \frac{9.8(68.1)}{14.75}(1 - e^{-(14.75/68.1)10}) = 40.059$$

which is very close to the desired fall velocity of 40 m/s.

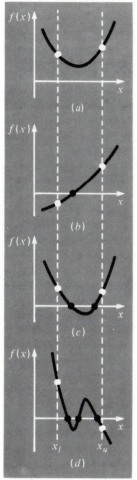

Graphical techniques are of limited practical value because they are not precise. However, graphical methods can be utilized to obtain rough estimates of roots. These estimates can be employed as starting guesses for numerical methods discussed in this and the next chapter. For example, the Electronic TOOLKIT software that supplements this text allows you to plot the function over a specified range. This plot can be used to select guesses that bracket the root prior to implementing the numerical method. The plotting option greatly enhances the utility of the software.

Aside from providing rough estimates of the root, graphical interpretations are important tools for understanding the properties of the functions and anticipating the pitfalls of the numerical methods. For example, Fig. 4.2 shows a number of ways in which roots can occur (or be absent) in an interval prescribed by a lower bound x_l and an upper bound x_u. Figure 4.2b depicts the case where a single root is bracketed by negative and positive values of $f(x)$. However, Fig. 4.2d, where $f(x_l)$ and $f(x_u)$ are also on opposite sides of the x axis, shows three roots occurring within the interval. In general, if $f(x_l)$ and $f(x_u)$ have opposite signs, there are an odd number of roots in the interval. As indicated by Fig. 4.2a and c, if $f(x_l)$ and $f(x_u)$ have the same sign, there are either no roots or an even number of roots between the values.

Although these generalizations are usually true, there are cases where they do not hold. For example, functions that are tangential to the x axis (Fig. 4.3a) and discontinuous functions (Fig. 4.3b) can violate these principles. An example of a function that is tangential to the axis is the cubic equation $f(x) = (x - 2)(x - 2)(x - 4)$. Notice that $x = 2$ makes two terms in this polynomial equal to zero. Mathematically, $x = 2$ is called a *multiple root*. At the end of Chap. 5, we will present techniques that are expressly designed to locate multiple roots.

Figure 4.2
Illustration of a number of general ways that a root may occur in an interval prescribed by a lower bound x_l and an upper bound x_u. Parts (a) and (c) indicate that if both $f(x_l)$ and $f(x_u)$ have the same sign, either there will be no roots or there will be an even number of roots within the interval. Parts (b) and (d) indicate that if the function has different signs at the end points, there will be an odd number of roots in the interval.

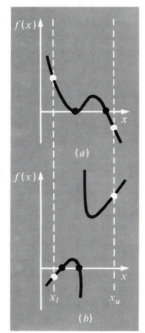

Figure 4.3
Illustration of some exceptions to the general cases depicted in Fig. 4.2. (a) Multiple root that occurs when the function is tangential to the x axis. For this case, although the end points are of opposite signs, there are an even number of axis intersections for the interval. (b) Discontinuous function where end points of opposite sign bracket an even number of roots. Special strategies are required for determining the roots for these cases.

The existence of cases of the type depicted in Fig. 4.3 makes it difficult to develop general computer algorithms guaranteed to locate all the roots in an interval. However, when used in conjunction with graphical approaches, the methods described in the following sections are extremely useful for solving many roots of equations problems confronted routinely by engineers and applied mathematicians.

EXAMPLE 4.2   Use of Computer Graphics to Locate Roots

Problem Statement: Computer graphics can expedite and inform your efforts to locate roots of equations. The present example was developed using the Electronic TOOLKIT software supplement. However, the insights and issues raised are relevant to computer graphics in general.

The function

$$f(x) = \sin 10x + \cos 3x$$

has several roots over the range from $x = -5$ to $x = 5$. Use computer graphics to gain insight into the behavior of this function.

Solution: As previously illustrated in Example 2.1, the Electronic TOOLKIT can be used to generate plots. Figure 4.4a is a plot of $f(x)$ from $x = -5$ to $x = 5$. This plot suggests the presence of several roots, including a possible double root at about $x = 4.2$ where $f(x)$ appears to be tangent to the x axis. A more detailed picture of the behavior of $f(x)$ is obtained by changing the plotting range from $x = 3$ to $x = 5$, as shown in Fig. 4.4b. Finally, in Fig. 4.4c, the vertical scale is narrowed further to $f(x) = -0.15$ to $f(x) = 0.15$ and the horizontal scale narrowed to $x = 4.2$ to $x = 4.3$. This plot shows clearly that a double root does not exist in this region and that in fact there are two distinct roots at about $x = 4.229$ and $x = 4.264$.

Computer graphics will have great utility in your studies of numerical methods. This capability will also find many other applications in your other classes and professional activities as well.

4.2   THE BISECTION METHOD

When applying the graphical technique in Example 4.1, you have observed (Fig. 4.1) that $f(x)$ changed sign on opposite sides of the root. In general, if $f(x)$ is real and continuous in the interval from x_l to x_u and $f(x_l)$ and $f(x_u)$ have opposite signs, that is,

$$f(x_l) f(x_u) < 0 \tag{4.1}$$

then there is at least one real root between x_l and x_u.

Incremental search methods capitalize on this observation by locating an interval where the function changes sign. Then the location of the sign change (and consequently, the root) is identified more precisely by dividing the interval into a number of subintervals. Each of these subintervals is searched to locate the sign change. The process is repeated and the root estimate refined by dividing the subintervals into finer increments. We will return to the general topic of incremental searches in Sec. 4.4.

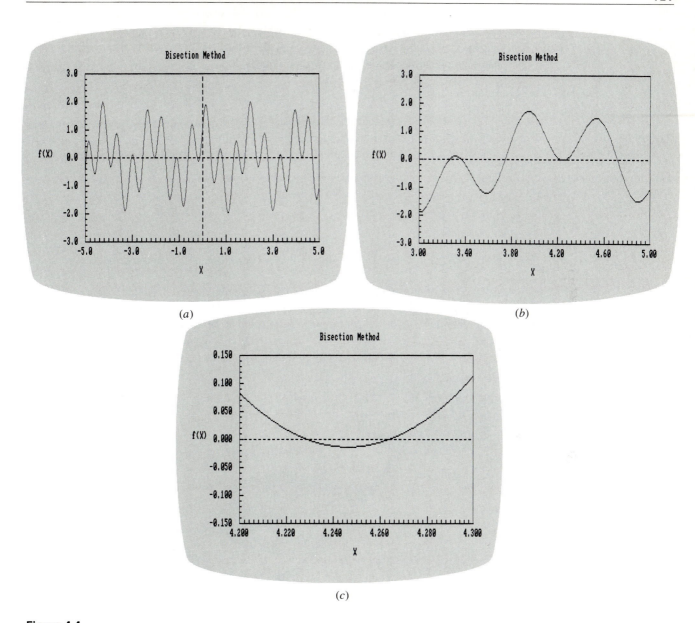

Figure 4.4
The progressive enlargement of $f(x) = \sin 10x + \cos 3x$ by the computer. Such interactive graphics permits the analyst to determine that two roots exist between $x = 4.2$ and $x = 4.3$.

The *bisection method,* which is alternatively called binary chopping, interval halving, or Bolzano's method, is one type of incremental search method in which the interval is always divided in half. If a function changes sign over an interval, the function value at the midpoint is evaluated. The location of the root is then determined as lying at the midpoint of the subinterval within which the sign change occurs. The process is repeated to obtain refined estimates. An algorithm for bisection is listed in Fig. 4.5, and a graphical depiction of the method is provided in Fig. 4.6. The following example goes through the actual computations involved in the method.

Figure 4.5
An algorithm for bisection. This procedure is continued until the root estimate is accurate enough to meet your requirements.

Step 1: Choose lower x_l and upper x_u guesses for the root, so that the function changes sign over the interval. This can be checked by ensuring that $f(x_l)f(x_u) < 0$.

Step 2: An estimate of the root x_r is determined by

$$x_r = \frac{x_l + x_u}{2}$$

Step 3: Make the following evaluations to determine in which subinterval the root lies:
(a) If $f(x_l)f(x_r) < 0$, the root lies in the lower subinterval. Therefore set $x_u = x_r$ and return to step 2.
(b) If $f(x_l)\,f(x_r) > 0$, the root lies in the upper subinterval. Therefore set $x_l = x_r$ and return to step 2.
(c) If $f(x_l)f(x_r) = 0$, the root equals x_r; terminate the computation.

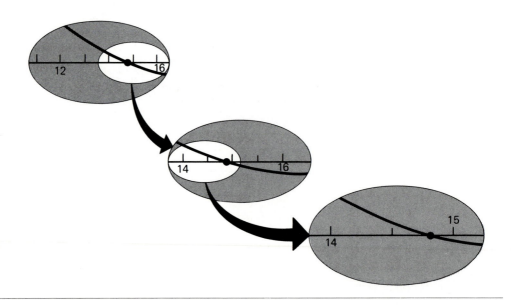

Figure 4.6
A graphical depiction of the bisection method. This plot conforms to the first three iterations from Example 4.3.

EXAMPLE 4.3   Bisection

Problem Statement: Use bisection to solve the same problem approached graphically in Example 4.1.

Solution: The first step in bisection is to guess two values of the unknown (in the present problem, c) that give values for $f(c)$ with different signs. From Fig. 4.1, we can see that the function changes sign between values of 12 and 16. Therefore, the initial estimate of the root x_r lies at the midpoint of the interval

$$x_r = \frac{12 + 16}{2} = 14$$

This estimate represents a true relative error of $\epsilon_t = 5.3\%$ (note that the true value of the root is 14.7802). Next we compute the product of the function value at the lower bound and at the midpoint:

$$f(12)f(14) = 6.067(1.569) = 9.517$$

which is greater than zero, and hence no sign change occurs between the lower bound and the midpoint. Consequently, the root must be located between 14 and 16. Therefore we create a new interval by redefining the lower bound as 14 and determining a revised root estimate as

$$x_r = \frac{14 + 16}{2} = 15$$

which represents a true error of $\epsilon_t = -1.5\%$. The process can be repeated to obtain refined estimates. For example,

$$f(14)f(15) = 1.569(-0.425) = -0.666$$

Therefore the root is between 14 and 15. The upper bound is redefined as 15, and the root estimate for the third iteration is calculated as

$$x_r = \frac{14 + 15}{2} = 14.5$$

which represents an error of $\epsilon_t = 1.9\%$. The method can be repeated until the result is accurate enough to satisfy your needs.

In the previous example, you may have noticed that the true error does not decrease with each iteration. However, the interval within which the root is located is halved with each step in the process. As discussed in the next section, the interval width provides an exact estimate of the upper bound of the error for the bisection method.

4.2.1 Termination Criteria and Error Estimates

We ended Example 4.3 with the statement that the method could be continued in order to obtain a refined estimate of the root. We must now develop an objective criterion for deciding when to terminate the method.

An initial suggestion might be to end the calculation when the error falls below some prespecified level. For instance, in Example 4.3, the relative error dropped from 5.3 to 1.9 percent during the course of the computation. We might decide that we should terminate when the error drops below, say, 0.1 percent. This strategy is flawed because the error estimates in the example were based on knowledge of the true root of the function. Such would not be the case in an actual situation because there would be no point in using the method if we already knew the root.

Therefore, we require an error estimate that is not contingent on foreknowledge of the root. As developed previously in Sec. 3.3, an approximate relative error ϵ_a can be calculated as in [recall Eq. (3.5)]

$$|\epsilon_a| = \left| \frac{x_r^{new} - x_r^{old}}{x_r^{new}} \right| 100\% \tag{4.2}$$

where x_r^{new} is the root for the present iteration and x_r^{old} is the root from the previous iteration. The absolute value is used because we are usually concerned with the magnitude of ϵ_a rather than with its sign. When $|\epsilon_a|$ becomes less than a prespecified stopping criterion ϵ_s, the computation is terminated.

EXAMPLE 4.4 Error Estimates for Bisection

Problem Statement: Continue Example 4.3 until the approximate error falls below a stopping criterion of $\epsilon_s = 0.5\%$. Use Eq. (4.2) to compute the errors.

Solution: The results of the first two iterations for Example 4.3 were 14 and 15. Substituting these values into Eq. (4.2) yields

$$|\epsilon_a| = \left| \frac{15 - 14}{15} \right| 100\% = 6.667\%$$

Recall that the true error for the root estimate of 15 was 1.5%. Therefore, ϵ_a is greater than ϵ_t. This behavior is manifested for the other iterations:

Iteration	x_1	x_u	x_r	$\|\epsilon_a\|$, %	$\|\epsilon_t\|$, %
1	12	16	14		5.279
2	14	16	15	6.667	1.487
3	14	15	14.5	3.448	1.896
4	14.5	15	14.75	1.695	0.204
5	14.75	15	14.875	0.840	0.641
6	14.75	14.875	14.8125	0.422	0.219

Thus after six iterations $|\epsilon_a|$ finally falls below $\epsilon_s = 0.5\%$, and the computation can be terminated.

These results are summarized in Fig. 4.7. The "ragged" nature of the true error is due to the fact that, for bisection, the true root can lie anywhere within the bracketing interval. The true and approximate errors are far apart when the interval happens to be centered on the true root. They are close when the true root falls at either end of the interval.

Although the approximate error does not provide an exact estimate of the true error, Fig. 4.7 suggests that ϵ_a captures the general downward trend of ϵ_t. In addition, the plot exhibits the extremely attractive characteristic that ϵ_a is always greater than ϵ_t. Thus, when ϵ_a falls below ϵ_s, the computation could be terminated with confidence that the root is known to be *at least* as accurate as the prespecified acceptable level.

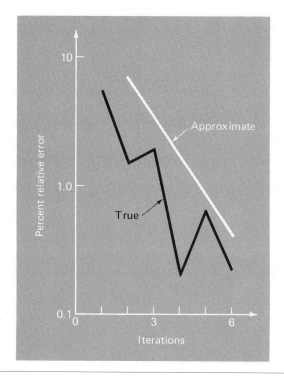

Figure 4.7
Errors for the bisection method. True and estimated errors are plotted versus the number of iterations.

Although it is always dangerous to draw general conclusions from a single example, it can be demonstrated that ϵ_a will always be greater than ϵ_t for the bisection method. This is due to the fact that each time an approximate root is located using bisection as $x_r = (x_l + x_u)/2$, we know that the true root lies somewhere within an interval of $(x_u - x_l)/2 = \Delta x/2$. Therefore, the root must lie within $\pm \Delta x/2$ of our estimate (Fig.

Figure 4.8
Three ways in which the interval may bracket the root. In (a) the true value lies at the center of the interval, whereas in (b) and (c) the true value lies near the extreme. Notice that the discrepancy between the true value and the midpoint of the interval never exceeds half the interval length, or $\Delta x/2$.

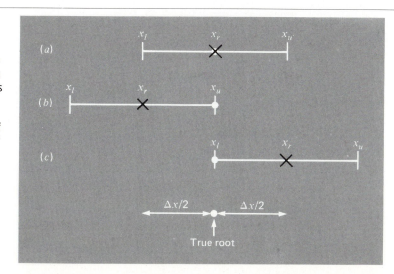

4.8). For instance, when Example 4.3 was terminated, we could make the definitive statement that

$$x_r = 14.5 \pm 0.5$$

Because $\Delta x/2 = x_r^{new} - x_r^{old}$ (Fig. 4.9), Eq. (4.2) provides an exact upper bound on the true error. For this bound to be exceeded, the true root would have to fall outside the bracketing interval, which, by definition, could never occur for the bisection method. As illustrated in a subsequent example (Example 4.7), other root-locating techniques do not always behave as nicely. Although bisection is generally slower than other methods, the neatness of its error analysis is certainly a positive aspect that could make it attractive for certain engineering applications.

Before proceeding to the computer program for bisection, we should note that the relationships (Fig. 4.9)

$$x_r^{new} - x_r^{old} = \frac{x_u - x_l}{2}$$

and (Fig. 4.2)

$$x_r^{new} = \frac{x_l + x_u}{2}$$

can be substituted into Eq. (4.5) to develop an alternative formulation for the approximate error

$$|\epsilon_a| = \left| \frac{x_u - x_l}{x_u + x_l} \right| 100\% \qquad\qquad (4.3)$$

This equation yields identical results to Eq. (4.2) for bisection. In addition, it allows us to calculate an error estimate on the basis of our initial guesses—that is, on our first iteration. For instance, on the first iteration of Example 4.2, an approximate error can be computed as

$$|\epsilon_a| = \left| \frac{16 - 12}{16 + 12} \right| 100\% = 14.29\%$$

Figure 4.9

Graphical depiction of why the error estimate for bisection ($\Delta x/2$) is equivalent to the root estimate for the present iteration (x_r^{new}) minus the root estimate for the previous iteration (x_r^{old}).

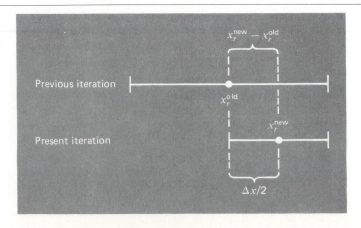

4.2.2 Computer Programs for the Bisection Method

The algorithm in Fig. 4.5 can now be expanded into computer code. Figure 4.10 shows a BASIC program that has been designed for an interactive mode of implementation. Figure 4.11 lists FORTRAN and Pascal codes that are written as subroutines. The programs employ user-defined functions to make root location and function evaluation more efficient. In addition, an upper limit is placed on the number of iterations. Finally, error checks are included to ensure that the initial guesses bracket a root and to avoid division by zero during the error evaluation. Such would be the case when the bracketing interval is centered on zero. For this situation Eq. (4.2) becomes infinite. If this occurs, the program skips over the error evaluation for that iteration.

The programs in Figs. 4.10 and 4.11 are not user-friendly; they are designed strictly to come up with the answer. In Prob. 4.13 at the end of this chapter, you will have the task of making these skeletal computer codes easier to use and understand. An example of a user-friendly program for implementing bisection is included in the Electronic TOOLKIT software associated with this text. The following example demonstrates the use of this

Figure 4.10
An interactive computer program for bisection written in Microsoft BASIC.

```
100 REM          BISECTION (BASIC VERSION)        290 RETURN
105 REM                                           300 REM ******* SUBROUTINE BISECT *******
110 REM ********************************           305 REM
115 REM *       DEFINITION OF VARIABLES    *      310 ITER=0
120 REM *                                  *      315 EA=1.1*ES
125 REM * XL    = LOWER GUESS              *      320 WHILE EA>ES AND ITER<MAXIT
130 REM * XU    = UPPER GUESS              *      325    XR=(XL+XU)/2
135 REM * XR    = ROOT ESTIMATE            *      330    ITER=ITER+1
140 REM * ES    = STOPPING CRITERION(%)    *      335    IF XL+XU<>0 THEN 340 ELSE 345
145 REM * EA    = APPROXIMATE ERROR(%)     *      340      EA=ABS((XU-XL)/(XL+XU))*100
150 REM * MAXIT = MAXIMUM ITERATIONS       *      345    REM endif
155 REM * ITER  = NUMBER OF ITERATIONS     *      350    TEST=FNF(XL)*FNF(XR)
160 REM ********************************           355    IF TEST=0 THEN 360 ELSE 370
165 REM                                           360       EA=0
170 DEF FNF(X)=9.8*68.1/X*                        365       GOTO 405
                (1-EXP(-X/68.1*10))-40            370     REM else
175 CLS                                           375        IF TEST<0 THEN 380 ELSE 390
180 REM ********* MAIN PROGRAM *********           380            XU=XR
185 REM                                           385            GOTO 400
190 GOSUB 250 'input data                         390         REM else
195 GOSUB 300 'perform bisection                  395            XL=XR
200 GOSUB 450 'output results                     400       REM endif
205 END                                           405    REM endif
250 REM ******** SUBROUTINE INPUT ********         410 WEND
255 CLS                                           415 RETURN
260 INPUT "Lower,upper guess= ";XL,XU             450 REM ******* SUBROUTINE OUTPUT *******
265 WHILE FNF(XL)*FNF(XU)>=0                       455 REM
270    INPUT "Lower,upper guess= ";XL,XU           460 CLS
275 WEND                                           465 PRINT:PRINT "root,error,iterations:"
280 INPUT "maximum error(%) = ";ES                470 PRINT:PRINT XR,EA,ITER
285 INPUT "maximum iterations = ";MAXIT           475 RETURN
```

```
        SUBROUTINE BISECT(XL,XU,ES,XR,EA,
      *                          MAXIT,ITER)
**************************************************
*           DEFINITION OF VARIABLES            *
*     XL     = LOWER GUESS                      *
*     XU     = UPPER GUESS                      *
*     XR     = ROOT ESTIMATE                    *
*     ES     = STOPPING CRITERION(%)            *
*     EA     = APPROXIMATE ERROR(%)             *
*     MAXIT  = MAXIMUM ITERATIONS               *
*     ITER   = NUMBER OF ITERATIONS             *
**************************************************
        ITER = 0
        EA=1.1*ES
   10 IF (EA.GT.ES.AND.ITER.LT.MAXIT) THEN
        XR = (XL + XU)/2.
        ITER = ITER + 1
        IF (XL+XU.NE.0.) THEN
          EA=ABS((XU-XL)/(XL+XU))*100.
        ENDIF
        TEST=F(XL)*F(XR)
        IF (TEST.EQ.0.) THEN
          EA = 0.
        ELSE
          IF (TEST.LT.0.) THEN
            XU = XR
          ELSE
            XL = XR
          ENDIF
        ENDIF
        GO TO 10
      ENDIF
      RETURN
      END
```
(a)

```
PROCEDURE Bisection(VAR Xl,Xu,Es,Xr,Ea:real;
                       Maxit,Iter:integer);
{
          Definition of variables
     Xl    = lower guess
     Xu    = upper guess
     Xr    = root estimate
     Es    = stopping criterion(%)
     Ea    = approximate error(%)
     Maxit = maximum iterations
     Iter  = number of iterations          }
VAR
   Test: Real;
Begin                    { procedure Bisection}
Iter := 0;
Ea := 1.1*Es;
While (Ea>Es) and (Iter<Maxit)
   Begin
   Xr := (Xl + Xu)/2;
   Iter := Iter + 1;
   If Xl + Xu <> 0 Then
        Ea := Abs((Xu-Xl)/(Xl+Xu))*100.;
   Test := F(Xl)*F(Xr);
   If (Test = 0.0) Then
        Ea := 0
   Else if (Test < 0.0) Then
        Xu := Xr
   Else { (Test > 0.0) }
        Xl := Xr;
   End;
End;                  {of procedure Bisection}
```
(b)

Figure 4.11
Subroutines for bisection written in (a) FORTRAN 77 and (b) Turbo Pascal. Note: user must define functions.

software for root location. It also provides a good reference for assessing and testing your own software.

EXAMPLE 4.5   Root Location Using the Computer

Problem Statement: A user-friendly computer program to implement the bisection method is contained in the Electronic TOOLKIT software associated with the text.

We can use this software to solve a design problem associated with the falling parachutist example discussed in Examples 4.1, 4.3, and 4.4. Assume that you would like the parachutist to attain a velocity of 40 m/s after a period of 7 s. Thus, you must determine a value of c such that

$$0 = f(c) = \frac{gm}{c}[1 - e^{-(c/m)t}] - v \tag{E4.5.1}$$

with $t = 7$ s and $v = 40$ m/s.

Solution: In order to implement the bisection method, it is required that we obtain an initial interval that brackets the value of c that satisfies Eq. (E4.5.1). It is convenient to select this interval in association with the BISECTION plot option on the disk (option 3). The program prompts you for minimum and maximum values for both x and $f(x)$ and produces the plot shown in Fig. 4.12a after you have entered the plot dimensions. It is seen that a root exists between 10 and 15 kg/s.

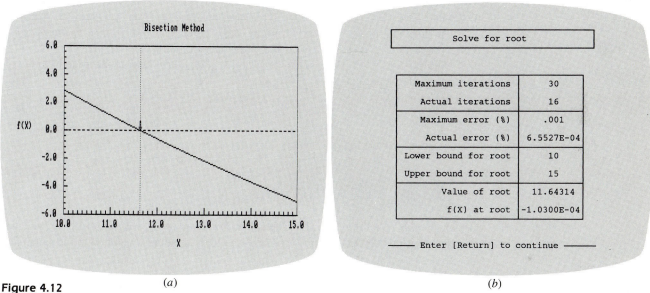

Figure 4.12
(a) Plot of Eq. (E4.5.1) and (b) results using BISECTION to determine the drag coefficient for the falling parachutist.

The BISECTION program prompts you for a limit on the maximum number of iterations, a convergence error ϵ_s, and upper and lower bounds for the root. These inputs, along with a calculated root of 11.64314 kg/s, are shown in Fig. 4.12b. Note that an estimated value of the root with an error of less than ϵ_s was obtained in 16 iterations. Furthermore, the computer displays an error check of

$$f(11.64314) = -1.0300 \times 10^{-4}$$

to confirm the results. If the required accuracy was not achieved within the specified number of iterations, the solution algorithm would have terminated after 30 iterations.

These results are based on a simple algorithm for the bisection method with user-friendly input and output routines. The algorithm employed is similar to those shown in Figs. 4.10 and 4.11. You should be able to write your own program for the bisection method. If you have obtained our software, you can use it as a model and to check the adequacy of your own program.

4.3 THE FALSE-POSITION METHOD

Although bisection is a perfectly valid technique for determining roots, its "brute-force" approach is relatively inefficient. False position is an improved alternative based on a graphical insight.

A shortcoming of the bisection method is that in dividing the interval from x_l to x_u into equal halves, no account is taken of the magnitudes of $f(x_l)$ and $f(x_u)$. For example, if $f(x_l)$ is much closer to zero than $f(x_u)$, it is likely that the root is closer to x_l than to x_u (Fig. 4.13). An alternative method that exploits this graphical insight is to join the points by a straight line. The intersection of this line with the x axis represents an improved estimate of the root. The fact that the replacement of the curve by a straight line gives a "false position" of the root is the origin of the name, *method of false position*, or, in Latin, *regula falsi*. It is also called the *linear interpolation method*.

With the use of similar triangles (Fig. 4.13), the intersection of the straight line with the x axis can be estimated as

$$\frac{f(x_l)}{x_r - x_l} = \frac{f(x_u)}{x_r - x_u} \tag{4.4}$$

which can be solved for (see Box 4.1 for details).

$$x_r = x_u - \frac{f(x_u)(x_l - x_u)}{f(x_l) - f(x_u)} \tag{4.5}$$

This is the *false-position formula*. The value of x_r computed with Eq. (4.5) then replaces whichever of the two initial guesses, x_l or x_u, yields a function value with the same sign

Figure 4.13
A graphical depiction of the method of false position. Similar triangles used to derive the formula for the method are shaded.

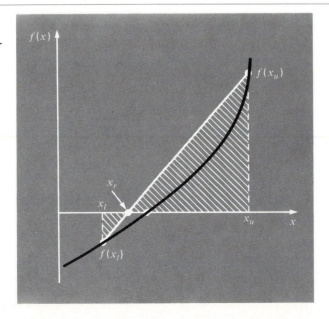

as $f(x_r)$. In this way the values of x_l and x_u always bracket the true root. The process is repeated until the root is estimated adequately. The algorithm is identical to the one for bisection (Fig. 4.5) with the exception that Eq. (4.5) is used for step 2. In addition, the same stopping criterion [Eq. (4.2)] is used to terminate the computation.

EXAMPLE 4.6   *False Position*

Problem Statement: Use the false-position method to determine the root of the same equation investigated in Example 4.1 [Eq. (E4.1.1)].

Solution: As in Example 4.3, initiate the computation with guesses of $x_l = 12$ and $x_u = 16$.

First iteration:

$$x_l = 12 \qquad f(x_l) = 6.0669$$

$$x_u = 16 \qquad f(x_u) = -2.2688$$

$$x_r = 16 - \frac{-2.2688(12 - 16)}{6.0669 - (-2.2688)} = 14.9113$$

which has a true relative error of 0.89 percent.

Second iteration:

$$f(x_l)f(x_r) = -1.5426$$

Box 4.1   *Derivation of the Method of False Postition*

Cross-multiply Eq. (4.4) to yield

$$f(x_l)(x_r - x_u) = f(x_u)(x_r - x_l)$$

Collect terms and rearrange:

$$x_r[f(x_l) - f(x_u)] = x_u f(x_l) - x_l f(x_u)$$

Divide by $f(x_l) - f(x_u)$:

$$x_r = \frac{x_u f(x_l) - x_l f(x_u)}{f(x_l) - f(x_u)} \qquad \text{(B4.1.1)}$$

This is one form of the method of false position. Note that it allows the computation of the root x_r as a function of the lower and upper guesses x_l and x_u. It can be put in an alternative form by expanding it:

$$x_r = \frac{f(x_l)x_u}{f(x_l) - f(x_u)} - \frac{f(x_u)x_l}{f(x_l) - f(x_u)}$$

then adding and subtracting x_u on the right hand side:

$$x_r = x_u + \frac{f(x_l)x_u}{f(x_l) - f(x_u)} - x_u - \frac{f(x_u)x_l}{f(x_l) - f(x_u)}$$

Collecting terms yields

$$x_r = x_u + \frac{f(x_u)x_u}{f(x_l) - f(x_u)} - \frac{f(x_u)x_l}{f(x_l) - f(x_u)}$$

or

$$x_r = x_u - \frac{f(x_u)(x_l - x_u)}{f(x_l) - f(x_u)} \qquad \text{(B4.1.2)}$$

which is the same as Eq. (4.5). We use this form because it involves one less function evaluation and one less multiplication than Eq. (B4.1.1). In addition, it is directly comparable with the secant method which will be discussed in Chap. 5.

Therefore, the root lies in the first subinterval, and x_r becomes the upper limit for the next iteration, $x_u = 14.9113$.

$$x_l = 12 \qquad f(x_l) = 6.0669$$

$$x_u = 14.9113 \qquad f(x_u) = -0.2543$$

$$x_r = 14.9113 - \frac{-0.2543(12 - 14.9113)}{6.0669 - (-0.2543)} = 14.7942$$

which has true and approximate relative errors of 0.09 and 0.79 percent. Additional iterations can be performed to refine the estimate of the roots.

A feeling for the relative efficiency of the bisection and false-position methods can be appreciated by referring to Fig. 4.14, where we have plotted the true percent relative errors for Examples 4.4 and 4.6. Note how the error for false position decreases much faster than for bisection because of the more efficient scheme for root location in the false-position method.

Figure 4.14
Comparison of the relative errors of the bisection and the false-position methods.

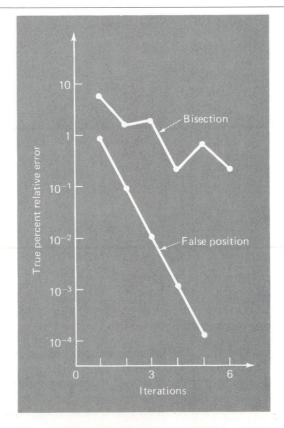

Recall in the bisection method that the interval between x_l and x_u grew smaller during the course of a computation. The interval, as defined by $\Delta x/2 = |x_u - x_l|/2$, therefore provided a measure of the error for this approach. This is not the case for the method of false position because one of the initial guesses may stay fixed throughout the computation as the other guess converges on the root. For instance, in Example 4.6 the lower guess x_l remained at 12 while x_u converged on the root. For such cases, the interval does not shrink but rather approaches a constant value.

Example 4.6 suggests that Eq. (4.2) represents a very conservative error criterion. In fact, Eq. (4.2) actually constitutes an approximation of the discrepancy of the *previous* iteration. This is due to the fact that for a case such as Example 4.6, where the method is converging quickly (for example, the error is being reduced nearly an order of magnitude per iteration), the root for the present iteration x_r^{new} is a much better estimate of the true value than the result of the previous iteration x_r^{old}. Thus, the quantity in the numerator of Eq. (4.2) actually represents the discrepancy of the previous iteration. Consequently, we are assured that satisfaction of Eq. (4.2) ensures that the root will be known with greater accuracy than the prescribed tolerance. However, as described in the next section, there are cases where false position converges slowly. For these cases, Eq. (4.2) becomes unreliable, and an alternative stopping criterion must be developed.

4.3.1 Pitfalls of the False-Position Method

Although the false-position method would seem to always be the bracketing method of preference, there are cases where it performs poorly. In fact, as in the following example, there are certain cases where bisection yields superior results.

EXAMPLE 4.7   A Case Where Bisection Is Preferable to False Position

Problem Statement: Use bisection and false position to locate the root of

$$f(x) = x^{10} - 1$$

between $x = 0$ and 1.3.

Solution: Using bisection, the results can be summarized as

Iteration	x_l	x_u	x_r	$|\epsilon_t|$, %	$|\epsilon_a|$, %
1	0	1.3	0.65	35	100.0
2	0.65	1.3	0.975	2.5	33.3
3	0.975	1.3	1.1375	13.8	14.3
4	0.975	1.1375	1.05625	5.6	7.7
5	0.975	1.05625	1.015625	1.6	4.0

Thus, after five iterations, the true error is reduced to less than 2 percent. For false position, a very different outcome is obtained:

Iteration	x_1	x_u	x_r	$\left| \epsilon_t \right|$, %	$\left| \epsilon_a \right|$, %
1	0	1.3	0.09430	90.6	
2	0.09430	1.3	0.18176	81.8	48.1
3	0.18176	1.3	0.26287	73.7	30.9
4	0.26287	1.3	0.33811	66.2	22.3
5	0.33811	1.3	0.40788	59.2	17.1

After five iterations, the true error has only been reduced to about 59 percent. In addition, note that $\left| \epsilon_a \right| < \left| \epsilon_t \right|$. Thus, the approximate error is misleading. Insight into these results can be gained by examining a plot of the function. As in Fig. 4.15, the curve violates the premise upon which false position was based—that is, if $f(x_l)$ is much closer to zero than $f(x_u)$ is, then the root is closer to x_l than to x_u (recall Fig. 4.13). Because of the shape of the present function, the opposite is true.

Figure 4.15
Plot of $f(x) = x^{10} - 1$, illustrating slow convergence of the false-position method.

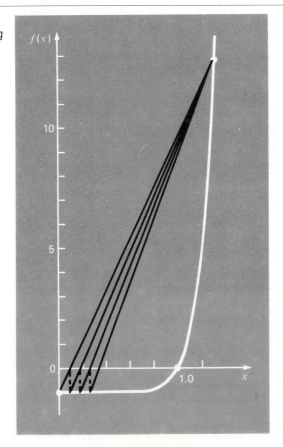

The foregoing example illustrates that blanket generalizations regarding root-location methods are usually not possible. Although a method such as false position is usually superior to bisection, there are invariably special cases that violate this general conclusion. Therefore, in addition to using Eq. (4.2), the results can be checked by substituting the root estimate into the original equation and determining whether the result is close to zero. Such a check should be incorporated into all computer programs for root location.

4.3.2 Computer Program for the False-Position Method

A computer program for the false-position method can be developed directly from the bisection codes in Figs. 4.10 and 4.11. The only modifications are to substitute Eq. (4.5) and to use Eq. (4.2) to compute the approximate error. In addition, the zero check suggested in the last section should also be incorporated into the code.

4.4 INCREMENTAL SEARCHES AND DETERMINING INITIAL GUESSES

Besides checking an individual answer, you must determine whether all possible roots have been located. As mentioned previously, a plot of the function is usually very useful in guiding you in this task. Another option is to incorporate an *incremental search* at the beginning of the computer program. This consists of starting at one end of the region of interest and then making function evaluations at small increments across the region. When the function changes sign, it is assumed that a root falls within the increment. The x values at the beginning and the end of the increment can then serve as the initial guesses for one of the bracketing techniques described in this chapter.

A potential problem with an incremental search is the choice of the increment length. If the length is too small, the search can be very time-consuming. On the other hand, if the length is too great, there is a possibility that closely spaced roots might be missed

Figure 4.16
Cases where roots could be missed because the increment length of the search procedure is too large. Note that the last root on the right is multiple and would be missed regardless of increment length.

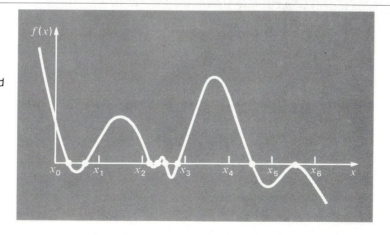

(Fig. 4.16). The problem is compounded by the possible existence of multiple roots. A partial remedy for such cases is to compute the first derivative of the function $f'(x)$ at the beginning and the end of each interval. If the derivative changes sign, it suggests that a minimum or maximum may have occurred and that the interval should be examined more closely for the existence of a possible root.

Although such modifications or the employment of a very fine increment can alleviate the problem, it should be clear that brute-force methods such as incremental search are not foolproof. You would be wise to supplement such automatic techniques with any other information that provides insight into the location of the roots. Such information can be found in plotting and in understanding the physical problem from which the equation originated.

PROBLEMS

Hand Calculations

4.1 Determine the real roots of

$$f(x) = -0.9x^2 + 1.70x + 2.5$$

(a) Graphically.
(b) Using the quadratic formula.
(c) Using three iterations of the bisection method to determine the highest root. Employ initial guesses of $x_l = 2.8$ and $x_u = 3.0$. Compute the estimated error ϵ_a and the true error ϵ_t after each iteration.

4.2 Determine the real roots of

$$f(x) = -2.0 + 6.2x - 4.0x^2 + 0.70x^3$$

(a) Graphically.
(b) Using bisection to locate the lowest root. Employ initial guesses of $x_l = 0.4$ and $x_u = 0.6$ and iterate until the estimated error ϵ_a falls below a level of $\epsilon_s = 10\%$.

4.3 Determine the real roots of

$$f(x) = -24 + 80x - 90x^2 + 42x^3 - 8.7x^4 + 0.66x^5$$

(a) Graphically.
(b) Using bisection to determine the highest root to $\epsilon_s = 1\%$. Employ initial guesses of $x_l = 4.5$ and $x_u = 5$.
(c) Perform the same computation as in (b) but use the false-position method.

4.4 Determine the real roots of

$$f(x) = 9.34 - 21.97x + 16.3x^2 - 3.704x^3$$

(a) Graphically.
(b) Using the false-position method with a value of ϵ_s corresponding to two significant figures to determine the lowest root.

4.5 Locate the first nontrivial root of $\tan x = 1.2x$ where x is in radians. Use a graphical technique and bisection with the initial interval from 0.4 to 0.8. Perform the computation until ϵ_a is less than $\epsilon_s = 10\%$. Also perform an error check by substituting your final answer into the original equation.

4.6 Determine the real root of $\ln x = 0.6$.
(a) Graphically.
(b) Using three iterations of the bisection method, with initial guesses of $x_l = 1$ and $x_u = 2$.
(c) Using three iterations of the false-position method, with the same initial guesses as in (b).

4.7 Determine the real root of

$$f(x) = \frac{1 - 0.61x}{x}$$

(a) Analytically.
(b) Graphically.
(c) Using three iterations of the false-position method and initial guesses of 1.5 and 2.0. Compute the approximate error ϵ_a and the true error ϵ_t after each iteration.

4.8 Find the positive square root of 11 using the false-position method to within $\epsilon_s = 0.5\%$. Employ initial guesses of $x_l = 3$ and $x_u = 3.4$.

4.9 Find the smallest positive root of the function (x is in radians)

$$x^2|\sin x| = 4.1$$

using the false-position method. To locate the region in which the root lies, first plot this function for values of x between 0 and 4. Perform the computation until ϵ_a falls below $\epsilon_s = 1\%$. Check your final answer by substituting it into the original function.

4.10 Find the positive real root of

$$f(x) = x^4 - 8.5x^3 - 35.50x^2 + 465x - 1000$$

using the false-position method. Use a plot to make your initial guesses, and perform the computation to within $\epsilon_s = 0.1\%$.

4.11 Determine the real root of

$$f(x) = x^3 - 98$$

(a) Analytically.
(b) With the false-position method to within $\epsilon_s = 0.1\%$.

4.12 The velocity of a falling parachutist is given by

$$v = \frac{gm}{c}[1 - e^{-(c/m)t}]$$

where $g = 9.8$. For a parachutist with a drag coefficient $c = 13.5$ kg/s, compute the mass m so that velocity is $v = 36$ m/s at $t = 6$ s. Use the false-position method to determine m to a level of $\epsilon_s = 0.1\%$.

Computer-Related Problems

4.13 Reprogram one of the codes in Figs. 4.10 or 4.11 so that it is user-friendly. Among other things:
(a) Place documentation statements throughout the program to identify what each section is intended to accomplish.
(b) Label the input and output.
(c) Add an answer check that substitutes the root estimate into the original function to verify whether the final result is close to zero.

4.14 Test the program you developed in Prob. 4.13 by duplicating the computations from Example 4.3.

4.15 Use the program you developed in Prob. 4.13 to repeat Probs. 4.1 through 4.6.

4.16 Repeat Probs. 4.14 and 4.15, except use the Electronic TOOLKIT software available with the text. Use the plotting capabilities of this program to verify your results.

4.17 Use the Electronic TOOLKIT software to find the real roots of two polynomial functions of your choice. Plot each function over a range you specify to obtain upper and lower bounds on the roots.

4.18 Repeat Prob. 4.17 except use two transcendental functions of your choice.

4.19 This problem uses only the graphics capability of the Electronic TOOLKIT software available with the text. The software plots the function over smaller and smaller intervals to increase the number of significant figures to which a root can be estimated. Start with $f(x) = e^{-x} \sin (10x)$. Plot the function with a full-scale range of $x = 0$ to $x = 2.5$. Estimate the root. Plot the function again with $x = 0.5$ to $x = 1.0$. Estimate the root. Finally, plot the function over a range of 0.6 to 0.7. This permits you to estimate the root to two significant figures.

4.20 Develop a user-friendly program for the false-position method based on Sec. 4.3.2. Test the program by duplicating Example 4.6.

4.21 Use the program you developed in Prob. 4.20 to duplicate the computation from Example 4.7. Perform a number of runs of 5, 10, 15, and more iterations until the true percent relative error falls below 0.1%. Plot the true and approximate percent relative errors versus number of iterations on semilog paper. Interpret your results.

CHAPTER 5
Open Methods

For the bracketing methods in the previous chapter, the root is located within an interval prescribed by a lower and an upper bound. Repeated application of these methods always results in closer estimates of the true value of the root. Such methods are said to be *convergent* because they move closer to the truth as the computation progresses (Fig. 5.1*a*).

In contrast, the *open methods* described in the present chapter are based on formulas that require a single starting value of x or two starting values that do not necessarily bracket the root. As such, they sometimes *diverge* or move away from the true root as the computation progresses (Fig. 5.1*b*). However, when the open methods converge (Fig. 5.1*c*) they usually do so much more quickly than the bracketing methods. We will begin our discussion of open techniques with a simple version that is useful for illustrating their general form and also for demonstrating the concept of convergence.

5.1 SIMPLE ONE-POINT ITERATION

As mentioned above, open methods employ a formula to predict the root. Such a formula can be developed for simple one-point iteration by rearranging the function $f(x) = 0$ so that x is on the left-hand side of the equation:

$$x = g(x) \tag{5.1}$$

This transformation can be accomplished either by algebraic manipulation or by simply adding x to both sides of the original equation. For example,

$$x^2 - 2x + 3 = 0$$

can be simply manipulated to yield

$$x = \frac{x^2 + 3}{2}$$

whereas $\sin x = 0$ would be put into the form of Eq. (5.1) by adding x to both sides to yield

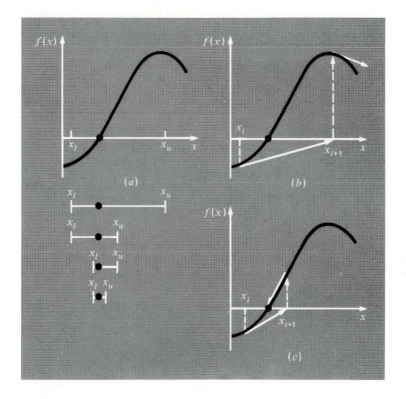

Figure 5.1
Graphical depiction of the fundamental difference between the (a) bracketing and (b) and (c) open methods for root location. In (a), which is the bisection method, the root is constrained within the interval prescribed by x_1 and x_u. In contrast, for the open method depicted in (b) and (c), a formula is used to project from x_i to x_{i+1} in an iterative fashion. Thus, the method can either (b) diverge or (c) converge rapidly, depending on the value of the initial guess.

$$x = \sin x + x$$

The utility of Eq. (5.1) is that it provides a formula to predict a value of x as a function of x. Thus, given an initial guess at the root x_i, Eq. (5.1) can be used to compute a new estimate x_{i+1}, as expressed by the iterative formula

$$x_{i+1} = g(x_i) \tag{5.2}$$

As with other iterative formulas in this book, the approximate error for this equation can be determined using the error estimator [Eq. (3.5)]:

$$|\epsilon_a| = \left| \frac{x_{i+1} - x_i}{x_{i+1}} \right| 100\%$$

EXAMPLE 5.1   Simple One-Point Iteration

Problem Statement: Use simple one-point iteration to locate the root of $f(x) = e^{-x} - x$.

Solution: The function can be separated directly and expressed in the form of Eq. (5.2) as $x_{i+1} = e^{-x_i}$. Starting with an initial guess of $x_0 = 0$, this iterative equation can be applied to compute:

i	x_i	$|\epsilon_t|$, %	$|\epsilon_a|$, %
0	0	100	
1	1.000000	76.3	100.0
2	0.367879	35.1	171.8
3	0.692201	22.1	46.9
4	0.500473	11.8	38.3
5	0.606244	6.89	17.4
6	0.545396	3.83	11.2
7	0.579612	2.20	5.90
8	0.560115	1.24	3.48
9	0.571143	0.705	1.93
10	0.564879	0.399	1.11

Thus, each iteration brings the estimate closer to the true value of the root, or 0.56714329.

5.1.1 Convergence

Notice that the true relative error for each iteration of Example 5.1 is roughly proportional (by a factor of about 0.5 to 0.6) to the error from the previous iteration. This property, called *linear convergence,* is characteristic of one-point iteration.

Aside from the "rate" of convergence, we must comment at this point about the "possibility" of convergence. The concepts of convergence and divergence can be depicted graphically. Recall that in Sec. 4.1, we graphed a function in order to visualize its structure and behavior (Example 4.1). Such an approach is employed in Fig. 5.2a for the function $f(x) = e^{-x} - x$. An alternative graphical approach is to separate the equation into two component parts, as in

$$f_1(x) = f_2(x)$$

Then the two equations

$$y_1 = f_1(x) \tag{5.3}$$

and

$$y_2 = f_2(x) \tag{5.4}$$

can be plotted separately (Fig. 5.2b). The x values corresponding to the intersections of these functions represent the roots of $f(x) = 0$.

EXAMPLE 5.2   The Two-Curve Graphical Method

Problem Statement: Separate the equation $e^{-x} - x = 0$ into two parts and determine its root graphically.

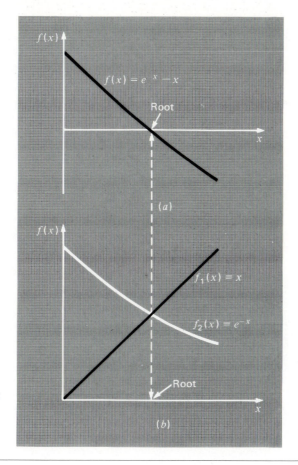

Figure 5.2
Two alternative graphical methods for determining the root of $f(x) = e^{-x} - x$. (a) Root at the point where it crosses the x axis; (b) root at the intersection of the component functions.

Solution: Reformulate the equation as $y_1 = x$ and $y_2 = e^{-x}$. The following values can be computed:

x	y_1	y_2
0.0	0.0	1.000
0.2	0.2	0.819
0.4	0.4	0.670
0.6	0.6	0.549
0.8	0.8	0.449
1.0	1.0	0.368

These points are plotted in Fig. 5.2b. The intersection of the two curves indicates a root estimate of $x \cong 0.57$, which corresponds to the point where the single curve in Fig. 5.2a crosses the x axis.

The two-curve method can now be used to illustrate the convergence and divergence of one-point iteration. First, Eq. (5.1) can be reexpressed as a pair of equations: $y_1 = x$ and $y_2 = g(x)$. These two equations can then be plotted separately. As was the case with Eqs. (5.3) and (5.4), the roots of $f(x) = 0$ correspond to the abscissa value at the intersection of the two curves. The function $y_1 = x$ and four different shapes for $y_2 = g(x)$ are plotted in Fig. 5.3.

For the first case (Fig. 5.3a), the initial guess of x_0 is used to determine the corresponding point on the y_2 curve $[x_0, g(x_0)]$. The point $[x_1, x_1]$ is located by moving left horizontally to the y_1 curve. These movements are equivalent to the first iteration in the one-step method:

$$x_1 = g(x_0)$$

Figure 5.3
Graphical depiction of (a) and (b) convergence and (c) and (d) divergence of simple one-point iteration. Graphs (a) and (c) are called monotone patterns, whereas (b) and (d) are called oscillating or spiral patterns. Note that convergence occurs when $|g'(x)| < 1$.

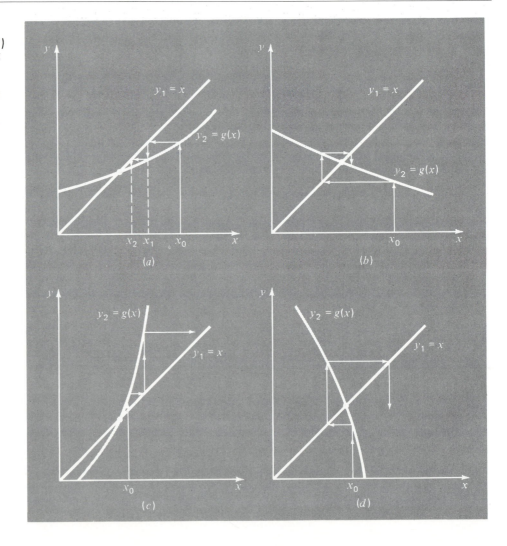

Thus, in both the equation and in the plot, a starting value of x_0 is used to obtain an estimate of x_1. The next iteration consists of moving to $[x_1, g(x_1)]$ and then to $[x_2, x_2]$. This iteration is equivalent to the equation

$$x_2 = g(x_1)$$

The solution in Fig. 5.3a is *convergent* because the estimates of x move closer to the root with each iteration. The same is true for Fig. 5.3b. However, this is not the case for Figs. 5.3c and d, where the iterations diverge from the root. Notice that convergence occurs only when the absolute value of the slope of $y_2 = g(x)$ is less than the slope of $y_1 = x$, that is, when $|g'(x)| < 1$. Box 5.1 provides a theoretical derivation of this result.

Box 5.1   Convergence of One-Point Iteration

From studying Fig. 5.3, it should be clear that one-point iteration converges if, in the region of interest $|g'(x)| < 1$. In other words, convergence occurs if the magnitude of the slope of $g(x)$ is less than the slope of the line $f(x) = x$. This observation can be demonstrated theoretically. Recall that the approximate equation is

$$x_{i+1} = g(x_i)$$

Suppose that the true solution is

$$x_r = g(x_r)$$

Subtracting these equations yields

$$x_r - x_{i+1} = g(x_r) - g(x_i) \qquad \text{(B5.1.1)}$$

In calculus, there is a principle called the *derivative mean-value theorem* (Sec. 3.5.2). It states that if a function $g(x)$ and its first derivative are continuous over an interval $a \le x \le b$, then there exists at least one value of $x = \xi$ within the interval such that

$$g'(\xi) = \frac{g(b) - g(a)}{b - a} \qquad \text{(B5.1.2)}$$

The right-hand side of this equation is the slope of the line joining $g(a)$ and $g(b)$. Thus, the mean-value theorem states that there is at least one point between a and b that has a slope, designated by $g'(\xi)$, which is parallel to the line joining $g(a)$ and $g(b)$ (Fig. 3.16).

Now, if we let $a = x_i$ and $b = x_r$, the right-hand side of Eq. (B5.1.1) can be expressed as

$$g(x_r) - g(x_i) = (x_r - x_i)g'(\xi)$$

where ξ is somewhere between x_i and x_r. This result can then be substituted into Eq. (B5.1.1) to yield

$$x_r - x_{i+1} = (x_r - x_i)g'(\xi) \qquad \text{(B5.1.3)}$$

If the true error for iteration i is defined as

$$E_{t,i} = x_r - x_i$$

then Eq. (B5.1.3) becomes

$$E_{t,i+1} = g'(\xi)E_{t,i}$$

Consequently, if $|g'(\xi)| < 1$, the errors decrease with each iteration. For $|g'(\xi)| > 1$ the errors grow. Notice also that if the derivative is positive, the errors will be positive, and hence, the iterative solution will be monotonic (Fig. 5.3a and c). If the derivative is negative, the errors will oscillate (Fig. 5.3b and d).

An offshoot of the analysis is that it also demonstrates that when the method converges, the error is roughly proportional to and less than the error of the previous step. For this reason, simple one-point iteration is said to be *linearly convergent*.

5.1.2   Computer Algorithm for One-Point Iteration

The computer algorithm for one-point iteration is extremely simple. It consists of a loop to iteratively compute new estimates until the termination criterion has been met. Figure

5.4 presents pseudocode for the algorithm. Other open methods can be programmed in a similar way, the major modification being to change the iterative formula that is used to compute the new root estimate.

Figure 5.4

Pseudocode for one-point iteration. Note that other open methods can be cast in this general format.

```
INPUT x₀ (initial guess)
INPUT εₛ (stopping criterion)
INPUT maxit (maximum iterations)
iter = 0
εₐ = 1.1 εₛ
DOWHILE (εₐ > εₛ) and (iter < maxit)
   iter = iter + 1
   xᵢₜₑᵣ = g (xᵢₜₑᵣ₋₁)
   IF (xᵢₜₑᵣ ≠ 0)
```

$$|\epsilon_a| = \left| \frac{x_{iter} - x_{iter-1}}{x_{iter}} \right| \cdot 100\ \%$$

```
   ELSE ENDIF
ENDDO
OUTPUT xᵢₜₑᵣ, εₐ and iter
```

Figure 5.4

5.2   THE NEWTON-RAPHSON METHOD

Perhaps the most widely used of all root-locating formulas is the Newton-Raphson equation (Fig. 5.5). If the initial guess at the root is x_i, a tangent can be extended from

Figure 5.5

Graphical depiction of the Newton-Raphson method. A tangent to the function of x_i [that is, $f'(x_i)$] is extrapolated down to the x axis to provide an estimate of the root at x_{i+1}.

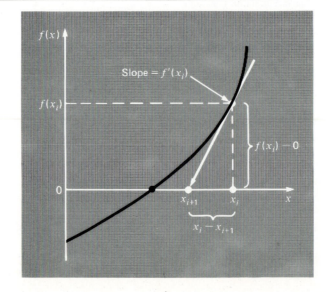

the point $[x_i, f(x_i)]$. The point where this tangent crosses the x axis usually represents an improved estimate of the root.

The Newton-Raphson method can be derived on the basis of this geometrical interpretation (an alternative method based on the Taylor series is described in Box 5.2). As in Fig. 5.5, the first derivative at x_i is equivalent to the slope:

$$f'(x_i) = \frac{f(x_i) - 0}{x_i - x_{i+1}} \tag{5.5}$$

which can be rearranged to yield

$$x_{i+1} = x_i - \frac{f(x_i)}{f'(x_i)} \tag{5.6}$$

which is called the *Newton-Raphson formula*.

Box 5.2   *Derivation and Error Analysis of the Newton-Raphson Method from a Taylor Series Expansion*

Aside from the geometric derivation [Eqs. (5.5) and (5.6)], the Newton-Raphson method may also be developed from the Taylor series expansion. This alternative derivation is useful in that it provides insight into the rate of convergence of the method.

Recall from Chap. 3 that the Taylor series expansion can be represented as

$$f(x_{i+1}) = f(x_i) + f'(x_i)(x_{i+1} - x_i) + \frac{f''(\xi)}{2}(x_{i+1} - x_i)^2 \tag{B5.2.1}$$

where ξ lies somewhere in the interval from x_i to x_{i+1}. An approximate version is obtainable by truncating the series after the first derivative term:

$$f(x_{i+1}) \simeq f(x_i) + f'(x_i)(x_{i+1} - x_i)$$

At the intersection with the x axis, $f(x_{i+1})$ would be equal to zero, or

$$0 \simeq f(x_i) + f'(x_i)(x_{i+1} - x_i) \tag{B5.2.2}$$

which can be solved for

$$x_{i+1} = x_i - \frac{f(x_i)}{f'(x_i)}$$

which is identical to Eq. (5.6). Thus, we have derived the Newton-Raphson formula using a Taylor series.

Aside from the derivation, the Taylor series can also be used to estimate the error of the formula. This can be done by realizing that if the complete Taylor series were employed, an exact result would be obtained. For this situation $x_{i+1} = x_r$,

where x_r is the true value of the root. Substituting this value along with $f(x_r) = 0$ into Eq. (B5.2.1) yields

$$0 = f(x_i) + f'(x_i)(x_r - x_i) + \frac{f''(\xi)}{2}(x_r - x_i)^2 \tag{B5.2.3}$$

where the third- and higher-order terms have been dropped. Equation (B5.2.2) can be subtracted from Eq. (B5.2.3) to give

$$0 = f'(x_i)(x_r - x_{i+1}) + \frac{f''(\xi)}{2}(x_r - x_i)^2 \tag{B5.2.4}$$

Now, realize that the error is equal to the discrepancy between x_{i+1} and the true value x_r, as in

$$E_{t,i+1} = x_r - x_{i+1}$$

and Eq. (B5.2.4) can be expressed as

$$0 = f'(x_i)E_{t,i+1} + \frac{f''(\xi)}{2}E_{t,i}^2 \tag{B5.2.5}$$

If we assume convergence, both x_i and ξ should be eventually be approximated by the root x_r, and Eq. (B5.2.5) can be rearranged to yield

$$E_{t,i+1} \simeq \frac{-f''(x_r)}{2f'(x_r)}E_{t,i}^2 \tag{B5.2.6}$$

According to Eq. (B5.2.6), the error is roughly proportional to the square of the previous error. This means that the number of correct decimal places approximately doubles with each iteration. Such behavior is referred to as *quadratic convergence*. Example 5.4 manifests this property.

EXAMPLE 5.3   Newton-Raphson Method

Problem Statement: Use the Newton-Raphson method to estimate the root of $e^{-x} - x$ employing an initial guess of $x_0 = 0$.

Solution: The first derivative of the function can be evaluated as

$$f'(x) = -e^{-x} - 1$$

which can be substituted along with the original function into Eq. (5.6) to give

$$x_{i+1} = x_i - \frac{e^{-x_i} - x_i}{-e^{-x_i} - 1}$$

Starting with an initial guess of $x_0 = 0$, this iterative equation can be applied to compute:

i	x_i	$|\epsilon_t|\%$
0	0	100
1	0.500000000	11.8
2	0.566311003	0.147
3	0.567143165	0.0000220
4	0.567143290	$< 10^{-8}$

Thus, the approach rapidly converges on the true root. Notice that the relative error at each iteration decreases much faster than it does in simple one-point iteration (compare with Example 5.1).

5.2.1 Termination Criteria and Error Estimates

As with other root-location methods, Eq. (3.5) can be used as a termination criterion. In addition, however, the Taylor series derivation of the method (Box 5.2) provides theoretical insight regarding the rate of convergence as expressed by $E_{i+1} = 0(E_i^2)$. Thus the error should be roughly proportional to the square of the previous error. In other words, the number of significant figures of accuracy approximately doubles with each iteration. This behavior is examined in the following example.

EXAMPLE 5.4   Error Analysis of Newton-Raphson Method

Problem Statement: As derived in Box 5.2, the Newton-Raphson method is quadratically convergent. That is, the error is roughly proportional to the square of the previous error, as in

$$E_{t,i+1} \simeq -\frac{f''(x_r)}{2f'(x_r)} E_{t,i}^2 \qquad \text{(E5.4.1)}$$

Examine this formula and see if it applies to the results of Example 5.3.

Solution: The first derivative of $f(x) = e^{-x} - x$ is

$$f'(x) = -e^{-x} - 1$$

which can be evaluated at $x_r = 0.56714329$ as

$$f'(0.56714329) = -1.56714329$$

The second derivative is

$$f''(x) = e^{-x}$$

which can be evaluated as

$$f''(0.56714329) = 0.56714329$$

These results can be substituted into Eq. (E5.4.1) to yield

$$E_{t,i+1} \simeq -\frac{0.56714329}{2(-1.56714329)} E_{t,i}{}^2$$

or

$$E_{t,i+1} \simeq 0.18095 E_{t,i}{}^2$$

From Example 5.3, the initial error was $E_{t,0} = 0.56714329$, which can be substituted into the error equation to predict

$$E_{t,1} \simeq 0.18095(0.56714329)^2 = 0.0582$$

which is close to the true error of 0.06714329. For the next iteration,

$$E_{t,2} \simeq 0.18095(0.06714329)^2 = 0.0008158$$

which also compares favorably with the true error of 0.0008323. For the third iteration,

$$E_{t,3} \simeq 0.18095(0.0008323)^2 = 0.000000125$$

which is exactly the error obtained in Example 5.3. The error estimate improves in this manner because as we come closer to the root, x_i and ξ are better approximated by x_r [recall our assumption in going from Eq. (B5.2.5) to Eq. (B5.2.6) in Box 5.2]. Finally,

$$E_{t,4} \simeq 0.18095(0.000000125)^2 = 2.83 \times 10^{-15}$$

Thus, this example illustrates that the error of the Newton-Raphson method for this case is, in fact, roughly proportional (by a factor of 0.18095) to the square of the error of the previous iteration.

5.2.2 Pitfalls of the Newton-Raphson Method

Although the Newton-Raphson method is often very efficient, there are situations where it performs poorly. A special case—multiple roots—will be addressed later in this chapter. However, even when dealing with simple roots, difficulties sometimes arise, as in the following example.

EXAMPLE 5.5   Example of a Slowly Converging Function Using the Newton-Raphson Method

Problem Statement: Determine the positive root of $f(x) = x^{10} - 1$ using the Newton-Raphson method and an initial guess of $x = 0.5$.

Solution: The Newton-Raphson formula for this case is

$$x_{i+1} = x_i - \frac{x_i{}^{10} - 1}{10x_i{}^9}$$

which can be used to compute

Iteration	x_i
0	0.5
1	51.65
2	46.485
3	41.8365
4	37.65285
5	33.887565
.	.
.	.
.	.
∞	1.00000

Thus, after the first poor prediction, the technique is converging on the true root of 1, but at a very slow rate.

Aside from slow convergence due to the nature of the function, other difficulties can arise, as illustrated in Fig. 5.6. For example, Fig. 5.6a depicts the case where an inflection point—that is, $f''(x) = 0$—occurs in the vicinity of a root. Notice that iterations beginning at x_0 progressively diverge from the root. Figure 5.6b illustrates the tendency of the Newton-Raphson technique to oscillate around a local maximum or minimum. Such oscillations may persist, or, as in Fig. 5.6b, a near-zero slope is reached whereupon the solution is sent far from the area of interest. Figure 5.6c shows how an initial guess that is close to one root can jump to a location several roots away. This tendency to move away from the area of interest is due to the fact that near-zero slopes are encountered. Obviously, a zero slope [$f'(x) = 0$] is a real disaster as it causes division by zero in the Newton-Raphson formula [Eq. (5.6)]. Graphically (Fig. 5.6d), it means that the solution shoots off horizontally and *never* hits the x axis.

Thus, there is no general convergence criterion for Newton-Raphson. Its convergence depends on the nature of the function and on the accuracy of the initial guess. The only remedy is to have an initial guess that is "sufficiently" close to the root. This knowledge is, of course, predicated on knowledge of the physical problem setting or on devices such as graphs that provide insight into the behavior of the solution. It also suggests that good computer software should be designed to recognize slow convergence or divergence. The next section addresses some of these issues.

5.2.3   Computer Program for the Newton-Raphson Method

A computer program for the Newton-Raphson method is readily obtained by substituting Eq. (5.6) for the predictive formula [Eq. (5.2)] in Fig. 5.4. Note, however, that the

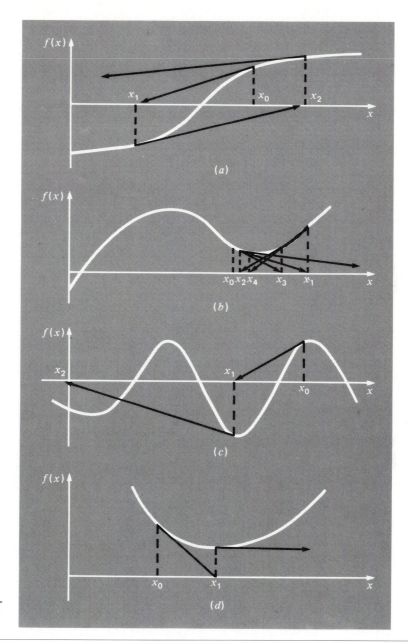

Figure 5.6
Four cases where the Newton-Raphson method exhibits poor convergence.

program must also be modified to compute the first derivative. This can be simply accomplished by the inclusion of a user-defined function.

Additionally, in light of the foregoing discussion of potential problems of the Newton-Raphson method, the program would be improved by incorporating a number of additional features:

1. If possible, a plotting routine should be included in the program.

2. At the end of the computation, the final root estimate should always be substituted into the original function to compute whether the result is close to zero. This check partially guards against those cases where slow or oscillating convergence may lead to a small value of ϵ_a, while the solution is still far from a root.
3. The program should always include an upper limit on the number of iterations to guard against oscillating, slowly convergent, or divergent solutions that could persist interminably.

5.3   THE SECANT METHOD

A potential problem in implementing the Newton-Raphson method is the evaluation of the derivative. Although this is not inconvenient for polynomials and many other functions, there are certain functions whose derivatives may be extremely difficult to evaluate. For these cases, the derivative can be approximated by a finite divided difference, as in (Fig. 5.7)

$$f'(x_i) \simeq \frac{f(x_{i-1}) - f(x_i)}{x_{i-1} - x_i}$$

This approximation can be substituted into Eq. (5.6) to yield the following iterative equation:

$$x_{i+1} = x_i - \frac{f(x_i)(x_{i-1} - x_i)}{f(x_{i-1}) - f(x_i)} \tag{5.7}$$

Equation (5.7) is the formula for the *secant method*. Notice that the approach requires two initial estimates of x. However, because $f(x)$ is not required to change signs between the estimates, it is not classified as a bracketing method.

Figure 5.7
Graphical depiction of the secant method. This technique is similar to the Newton-Raphson technique (Fig. 5.5) in the sense that an estimate of the root is predicted by extrapolating a tangent of the function to the x axis. However, the secant method uses a difference rather than a derivative to estimate the slope.

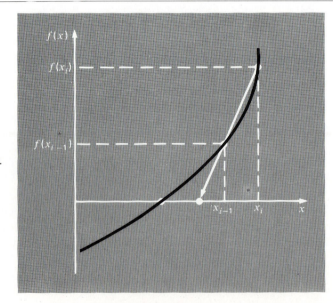

EXAMPLE 5.6   The Secant Method

Problem Statement: Use the secant method to estimate the root of $f(x) = e^{-x} - x$. Start with initial estimates of $x_{-1} = 0$ and $x_0 = 1.0$.

Solution: Recall that the true root is 0.56714329 . . .
First iteration:

$$x_{-1} = 0 \quad f(x_{-1}) = 1.00000$$

$$x_0 = 1 \quad f(x_0) = -0.63212$$

$$x_1 = 1 - \frac{-0.63212(0 - 1)}{1 - (-0.63212)} = 0.61270 \quad |\epsilon_t| = 8.0\%$$

Second iteration:

$$x_0 = 1 \qquad f(x_0) = -0.63212$$

$$x_1 = 0.61270 \qquad f(x_1) = -0.07081$$

(Note that both estimates are now on the same side of the root.)

$$x_2 = 0.61270 - \frac{-0.07081(1 - 0.61270)}{-0.63212 - (-0.07081)} = 0.56384$$

$$|\epsilon_t| = 0.58\%$$

Third iteration:

$$x_1 = 0.61270 \quad f(x_1) = -0.07081$$

$$x_2 = 0.56384 \quad f(x_2) = 0.00518$$

$$x_3 = 0.56384 - \frac{0.00518(0.61270 - 0.56384)}{-0.07081 - (0.00518)} = 0.56717$$

$$|\epsilon_t| = 0.0048\%$$

5.3.1 The Difference Between the Secant and False-Position Methods

Note the similarity between the secant method and the false-position method. For example, Eqs. (5.7) and (4.5) are identical on a term-by-term basis. Both use two initial estimates to compute an approximation of the slope of the function that is used to project to the x axis for a new estimate of the root. However, a critical difference between the methods is how one of the initial values is replaced by the new estimate. Recall that in the false-position method the latest estimate of the root replaces whichever of the original values yielded a function value with the same sign as $f(x_r)$. Consequently, the two estimates always bracket the root. Therefore, for all practical purposes, the method always converges because the root is kept within the bracket. In contrast, the secant method replaces the values in strict sequence, with the new value x_{i+1} replacing x_i and

x_i replacing x_{i-1}. As a result, the two values can sometimes lie on the same side of the root. For certain cases, this can lead to divergence.

EXAMPLE 5.7   Comparison of Convergence of the Secant and False-Position Techniques

Problem Statement: Use the false-position and secant methods to estimate the root of $f(x) = \ln x$. Start the computation with values of $x_l = x_{i-1} = 0.5$ and $x_u = x_i = 5.0$.

Solution: For the false-position method the use of Eq. (4.5) and the bracketing criterion for replacing estimates results in the following iterations:

Iteration	x_l	x_u	x_r
1	0.5	5.0	1.8546
2	0.5	1.8546	1.2163
3	0.5	1.2163	1.0585

As can be seen (Fig. 5.8a and c), the estimates are converging on the true root which is equal to 1.

Figure 5.8
Comparison of the false-position and the secant methods. The first iterations (a) and (b) for both techniques are identical. However, for the second iterations (c) and (d), the points used differ. As a consequence, the secant method can diverge, as indicated in (d).

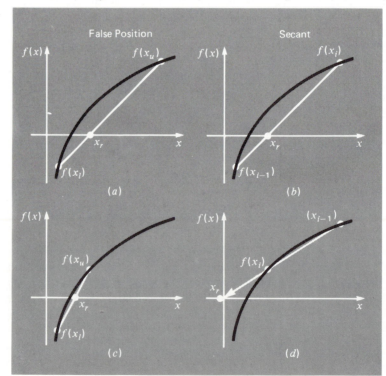

For the secant method, using Eq. (5.7) and the sequential criterion for replacing estimates results in:

Iteration	x_i	x_u	x_r
1	0.5	5.0	1.8546
2	5.0	1.8546	−0.10438

As in Fig. 5.8d, the approach is divergent.

Although the secant method may be divergent, when it converges it usually does so at a quicker rate than the false-position method does. For instance, Fig. 5.9 demonstrates the superiority of the secant method in this regard. The inferiority of the false-position method is due to the fact that one end stays fixed in order to maintain the bracketing of the root. This property, which is an advantage in that it prevents divergence, is a shortcoming with regard to the rate of convergence; it makes the finite-difference estimate a less-accurate approximation of the derivative.

Figure 5.9
Comparison of the true percent relative errors ϵ_t for the methods to determine the roots of $f(x) = e^{-x} - x$.

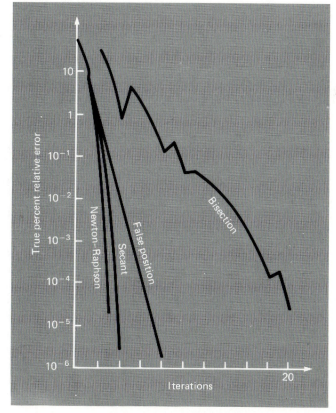

5.3.2 Computer Program for the Secant Method

As with the other open methods, a computer program for the secant method is simply obtained by modifying Fig. 5.4 so that two initial guesses are input and by using Eq. (5.7) to calculate the root. In addition, the options suggested in Sec. 5.2.3 for the Newton-Raphson method can also be applied to good advantage for the secant program.

5.4 MULTIPLE ROOTS

A *multiple root* corresponds to a point where a function is tangential to the x axis. For example, a *double root* results from

$$f(x) = (x - 3)(x - 1)(x - 1) \tag{5.8}$$

or, multiplying terms,

$$f(x) = x^3 - 5x^2 + 7x - 3 \tag{5.9}$$

The equation has a double root because one value of x makes two terms in Eq. (5.8) equal to zero. Graphically, this corresponds to the curve touching the x axis tangentially at the double root. Examine Fig. 5.10a at $x = 1$. Notice that the function touches the axis but does not cross it at the root.

A *triple root* corresponds to the case where one x value makes three terms in an equation equal to zero, as in

$$f(x) = (x - 3)(x - 1)(x - 1)(x - 1)$$

or, multiplying terms,

$$f(x) = x^4 - 6x^3 + 12x^2 - 10x + 3$$

Notice that the graphical depiction (Fig. 5.10b) again indicates that the function is tangential to the axis at the root but that for this case the axis is crossed. In general, odd multiple roots cross the axis, whereas even ones do not. For example, the quadruple root in Fig. 5.10c does not cross the axis.

Multiple roots pose a number of difficulties for many of the numerical methods described in Part Two:

1. The fact that the function does not change sign at an even multiple root precludes the use of the reliable bracketing methods that were discussed in Chap. 4. Thus, of the methods covered in this book, you are limited to the open methods that may diverge.
2. Another possible problem is related to the fact that not only $f(x)$ but also $f'(x)$ goes to zero at the root. This poses problems for both the Newton-Raphson and secant methods, which both contain the derivative (or its estimate) in the denominator of their respective formulas. This could result in division by zero when the solution converges very close to the root. A simple way to circumvent these problems is based on the fact that it can be demonstrated theoretically (Ralston and Rabinowitz, 1978) that $f(x)$ will always reach zero before $f'(x)$. Therefore, if a zero check for $f(x)$

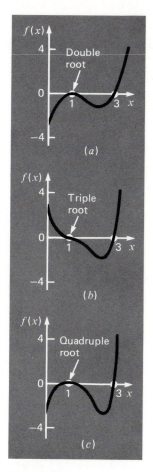

Figure 5.10
Examples of multiple roots that are tangential to the x axis. Notice that the function does not cross the axis on either side of even multiple roots (a) and (c), whereas it crosses the axis for odd cases (b).

is incorporated into the computer program, the computation can be terminated before $f'(x)$ reaches zero.

3. It can be demonstrated that the Newton-Raphson and secant methods are linearly, rather than quadratically, convergent for multiple roots (Ralston and Rabinowitz, 1978). Modifications have been proposed to alleviate this problem. Ralston and Rabinowitz (1978) have indicated that a slight change in the formulation returns it to quadratic convergence, as in

$$x_{i+1} = x_i - m\frac{f(x_i)}{f'(x_i)}$$

where m is the *multiplicity* of the root (that is, $m = 2$ for a double root, $m = 3$ for a triple root, etc.). Of course, this may be an unsatisfactory alternative because it hinges on foreknowledge of the multiplicity of the root.

Another alternative, also suggested by Ralston and Rabinowitz (1978), is to define a new function $u(x)$, that is, the ratio of the function to its derivative, as in

$$u(x) = \frac{f(x)}{f'(x)} \tag{5.10}$$

It can be shown that this function has roots at all the same locations as the original function. Therefore, Eq. (5.10) can be substituted into Eq. (5.6) in order to develop an alternative form of the Newton-Raphson method:

$$x_{i+1} = x_i - \frac{u(x_i)}{u'(x_i)} \tag{5.11}$$

Equation (5.10) can be differentiated to give

$$u'(x) = \frac{f'(x)f'(x) - f(x)f''(x)}{[f'(x)]^2} \tag{5.12}$$

Equations (5.10) and (5.12) can be substituted into Eq. (5.11) and the result simplified to yield

$$x_{i+1} = x_i - \frac{f(x_i)f'(x_i)}{[f'(x_i)]^2 - f(x_i)f''(x_i)} \tag{5.13}$$

EXAMPLE 5.8   Modified Newton-Raphson Method for Multiple Roots

Problem Statement: Use both the standard and modified Newton-Raphson methods to evaluate the multiple root of Eq. (5.9), with an initial guess of $x_0 = 0$.

Solution: The first derivative of Eq. (5.9) is $f'(x) = 3x^2 - 10x + 7$, and, therefore, the standard Newton-Raphson method for this problem is [Eq. (5.6)]

$$x_{i+1} = x_i - \frac{x_i^3 - 5x_i^2 + 7x_i - 3}{3x_i^2 - 10x_i + 7}$$

which can be solved iteratively for

i	x_i	$\|\epsilon_t\|$, %
0	0	100
1	0.428571429	57
2	0.685714286	31
3	0.832865400	17
4	0.913328983	8.7
5	0.955783293	4.4
6	0.977655101	2.2

As anticipated, the method is linearly convergent toward the true value of 1.0.

For the modified method, the second derivative is $f''(x) = 6x - 10$, and the iterative relationship is [Eq. (5.13)]

$$x_{i+1} = x_i - \frac{(x_i^3 - 5x_i^2 + 7x_i - 3)(3x_i^2 - 10x_i + 7)}{(3x_i^2 - 10x_i + 7)^2 - (x_i^3 - 5x_i^2 + 7x_i - 3)(6x_i - 10)}$$

which can be solved for

i	x_i	$\|\epsilon_t\|$, %
0	0	100
1	1.105263158	11
2	1.003081664	0.31
3	1.000002382	0.00024

Thus, the modified formula is quadratically convergent. We can also use both methods to search for the single root at $x = 3$. Using an initial guess of $x_0 = 4$ gives the following results:

i	Standard $(\|\epsilon_t\|)$	Modified $(\|\epsilon_t\|)$
0	4 (33%)	4 (33%)
1	3.4 (13%)	2.636363637 (12%)
2	3.1 (3.3%)	2.820224720 (6.0%)
3	3.008695652 (0.29%)	2.961728211 (1.3%)
4	3.000074641 (2.5×10^{-3}%)	2.998478719 (0.051%)
5	3.000000006 (2×10^{-7}%)	2.999997682 (7.7×10^{-5}%)

Thus, both methods converge quickly, with the standard method being somewhat more efficient.

The above example illustrates the trade-offs involved in opting for the modified Newton-Raphson method. Although it is much preferable for multiple roots, it is somewhat less efficient and requires much more computational effort than the standard method for simple roots.

It should be noted that a modified version of the secant method suited for multiple roots can also be developed by substituting Eq. (5.10) into Eq. (5.7). The resulting formula is (Ralston and Rabinowitz, 1978)

$$x_{i+1} = x_i - \frac{u(x_i)(x_{i-1} - x_i)}{u(x_{i-1}) - u(x_i)}$$

5.5 SYSTEMS OF NONLINEAR EQUATIONS

To this point, we have focused on the determination of the roots of a single equation. A related problem is to locate the roots of a set of simultaneous equations,

$$f_1(x_1, x_2, \ldots, x_n) = 0$$

$$f_2(x_1, x_2, \ldots, x_n) = 0$$

$$\cdot \qquad \cdot$$

$$\cdot \qquad \cdot \tag{5.14}$$

$$\cdot \qquad \cdot$$

$$f_n(x_1, x_2, \ldots, x_n) = 0$$

The solution of this system consists of a set of x's that simultaneously result in all the equations equaling zero.

In Part Three, we will present methods for the case where the simultaneous equations are linear—that is, they can be expressed in the general form

$$f(x) = a_1 x_1 + a_2 x_2 + \cdots + a_n x_n - c = 0 \tag{5.15}$$

where the c and the a's are constants. Algebraic and transcendental equations that do not fit this format are called *nonlinear equations*. For example,

$$x^2 + xy = 10$$

and

$$y + 3xy^2 = 57$$

are two simultaneous nonlinear equations with two unknowns, x and y. They can be expressed in the form of Eq. (5.14) as

$$u(x, y) = x^2 + xy - 10 = 0 \tag{5.16a}$$

$$v(x, y) = y + 3xy^2 - 57 = 0 \tag{5.16b}$$

Thus, the solution would be the values of x and y that make the functions $u(x, y)$ and $v(x, y)$ equal to zero. Most approaches for determining such solutions are extensions of the open methods for solving single equations. In this section, we will investigate two of these: one-point iteration and Newton-Raphson.

5.5.1 One-Point Iteration

The one-point iteration approach (Sec. 5.1) can be modified to solve two simultaneous, nonlinear equations. This approach will be illustrated in the following example.

EXAMPLE 5.9   One-Point Iteration for Nonlinear System

Problem Statement: Use one-point iteration to determine the roots of Eq. (5.16). Note that a correct pair of roots is $x = 2$ and $y = 3$. Initiate the computation with guesses of $x = 1.5$ and $y = 3.5$.

Solution: Equation (5.16a) can be solved for

$$x_{i+1} = \frac{10 - x_i^2}{y_i} \tag{E5.9.1}$$

and Eq. (5.16b) can be solved for

$$y_{i+1} = 57 - 3x_i y_i^2 \tag{E5.9.2}$$

Note that we will drop the subscripts for the remainder of the example.

On the basis of the initial guesses, Eq. (E5.9.1) can be used to determine a new value of x:

$$x = \frac{10 - (1.5)^2}{3.5} = 2.21429$$

This result and the value of $y = 3.5$ can be substituted into Eq. (E5.9.2) to determine a new value of y:

$$y = 57 - 3(2.21429)(3.5)^2 = -24.37516$$

Thus, the approach seems to be diverging. This behavior is even more pronounced on the second iteration

$$x = \frac{10 - (2.21429)^2}{-24.37516} = -0.20910$$

$$y = 57 - 3(-0.20910)(-24.37516)^2 = 429.709$$

Obviously, the approach is deteriorating.

Now we will repeat the computation but with the original equations set up in a different format. For example, an alternative formulation of Eq. (5.16a) is

$$x = \sqrt{10 - xy}$$

and of Eq. (5.16b) is

$$y = \sqrt{\frac{57 - y}{3x}}$$

Now the results are more satisfactory:

$$x = \sqrt{10 - 1.5(3.5)} = 2.17945$$

$$y = \sqrt{\frac{57 - 3.5}{3(2.17945)}} = 2.86051$$

$$x = \sqrt{10 - 2.17945(2.86051)} = 1.94053$$

$$y = \sqrt{\frac{57 - 2.86051}{3(1.94053)}} = 3.04955$$

$$x = \sqrt{10 - 1.94053(3.04955)} = 2.02046$$

$$y = \sqrt{\frac{57 - 3.04955}{3(2.0246)}} = 2.98340$$

Thus, the approach is converging on the true values of $x = 2$ and $y = 3$.

The previous example illustrates the most serious shortcoming of simple one-point iteration—that is, that its convergence often depends on the manner in which the equations are formulated. Additionally, even in those instances where convergence is possible, divergence can occur if the initial guesses are insufficiently close to the true solution. Using reasoning similar to that in Box 5.1, it can be demonstrated that sufficient conditions for convergence for the two-equation case are

$$\left| \frac{\partial u}{\partial x} \right| + \left| \frac{\partial v}{\partial x} \right| < 1$$

and

$$\left| \frac{\partial u}{\partial y} \right| + \left| \frac{\partial v}{\partial y} \right| < 1$$

These criteria are so restrictive that one-point iteration is rarely used in practice.

5.5.2 Newton-Raphson

Recall that the Newton-Raphson method was predicated on employing the derivative (that is, the slope) of a function to estimate its intercept with the axis of the independent variable—that is, the root (Fig. 5.5). This estimate was based on a first-order Taylor series expansion (recall Box 5.2),

$$f(x_{i+1}) = f(x_i) + (x_{i+1} - x_i)f'(x_i) \tag{5.17}$$

where x_i is the initial guess at the root and x_{i+1} is the point at which the slope intercepts the x axis. At this intercept, $f(x_{i+1})$ by definition equals zero and Eq. (5.17) can be rearranged to yield

$$x_{i+1} = x_i - \frac{f(x_i)}{f'(x_i)} \tag{5.18}$$

which is the single-equation form of the Newton-Raphson method.

The multiequation form is derived in an identical fashion. However, a multivariable Taylor series must be used in order to account for the fact that more than one independent variable contributes to the determination of the root. For the two-variable case, a first-order Taylor series can be written [recall Eq. (3.37)] for each nonlinear equation as

$$u_{i+1} = u_i + (x_{i+1} - x_i)\frac{\partial u_i}{\partial x} + (y_{i+1} - y_i)\frac{\partial u_i}{\partial y} \qquad (5.19a)$$

and

$$v_{i+1} = v_i + (x_{i+1} - x_i)\frac{\partial v_i}{\partial x} + (y_{i+1} - y_i)\frac{\partial v_i}{\partial y} \qquad (5.19b)$$

Just as for the single-equation version, the root estimate corresponds to the points at which u_{i+1} and v_{i+1} equal zero. For this situation, Eq. (5.19) can be rearranged to give

$$\frac{\partial u_i}{\partial x}x_{i+1} + \frac{\partial u_i}{\partial y}y_{i+1} = -u_i + x_i\frac{\partial u_i}{\partial x} + y_i\frac{\partial u_i}{\partial y} \qquad (5.20a)$$

$$\frac{\partial v_i}{\partial x}x_{i+1} + \frac{\partial v_i}{\partial y}y_{i+1} = -v_i + x_i\frac{\partial v_i}{\partial x} + y_i\frac{\partial v_i}{\partial y} \qquad (5.20b)$$

Because all values subscripted with i's are known (they correspond to the latest guess or approximation), the only unknowns are x_{i+1} and y_{i+1}. Thus, Eq. (5.20) is a set of two *linear* equations with two unknowns [compare with Eq. (5.15)]. Consequently, algebraic manipulations (for example, Cramer's rule) can be employed to solve for

$$x_{i+1} = x_i - \frac{u_i\dfrac{\partial v_i}{\partial y} - v_i\dfrac{\partial u_i}{\partial y}}{\dfrac{\partial u_i}{\partial x}\dfrac{\partial v_i}{\partial y} - \dfrac{\partial u_i}{\partial y}\dfrac{\partial v_i}{\partial x}} \qquad (5.21a)$$

and

$$y_{i+1} = y_i + \frac{u_i\dfrac{\partial v_i}{\partial x} - v_i\dfrac{\partial u_i}{\partial x}}{\dfrac{\partial u_i}{\partial x}\dfrac{\partial v_i}{\partial y} - \dfrac{\partial u_i}{\partial y}\dfrac{\partial v_i}{\partial x}} \qquad (5.21b)$$

The denominator of each of these equations is formally referred to as the determinant of the *Jacobian* of the system.

Equation (5.21) is the two-equation version of the Newton-Raphson method. As in the following example, it can be employed iteratively to simultaneously home in on the roots of two simultaneous equations. In addition, Case Study 6.6 also deals with nonlinear systems.

EXAMPLE 5.10   Roots of Simultaneous Nonlinear Equations

Problem Statement: Use the multiple-equation Newton-Raphson method to determine roots of Eq. (5.16). Note that a correct pair of roots is $x = 2$ and $y = 3$. Initiate the computation with guesses of $x = 1.5$ and $y = 3.5$.

Solution: First compute the partial derivatives and evaluate them at the initial guesses:

$$\frac{\partial u_0}{\partial x} = 2x + y = 2(1.5) + 3.5 = 6.5$$

$$\frac{\partial u_0}{\partial y} = x = 1.5$$

$$\frac{\partial v_0}{\partial x} = 3y^2 = 3(3.5)^2 = 36.75$$

$$\frac{\partial v_0}{\partial y} = 1 + 6xy = 1 + 6(1.5)(3.5) = 32.5$$

Thus, the determinant of the Jacobian for the first iteration is

$$6.5(32.5) - 1.5(36.75) = 156.125$$

The values of the functions can be evaluated at the initial guesses as

$$u_0 = (1.5)^2 + 1.5(3.5) - 10 = -2.5$$
$$v_0 = 3.5 + 3(1.5)(3.5)^2 - 57 = 1.625$$

These values can be substituted into Eq. (5.21) to give

$$x_1 = 1.5 - \frac{-2.5(32.5) - 1.625(1.5)}{156.125} = 2.03603$$

and

$$x_2 = 3.5 + \frac{-2.5(36.75) - 1.625(6.5)}{156.125} = 2.84388$$

Thus, the results are converging on the true values of $x_1 = 2$ and $x_2 = 3$. The computation can be repeated until an acceptable accuracy is obtained.

Just as with one-point iteration, the Newton-Raphson approach will often diverge if the initial guesses are not sufficiently close to the true roots. Whereas graphical methods could be employed to derive good guesses for the single-equation case, no such simple procedure is available for the multiequation version. Although there are some advanced approaches for obtaining acceptable first estimates, often the initial guesses must be obtained on the basis of trial and error and knowledge of the physical system being modeled.

The two-equation Newton-Raphson approach can be generalized to solve n simultaneous equations. Because the most efficient way to do this involves matrix algebra and the solution of simultaneous linear equations, we will defer discussion of the general approach to Part Three.

PROBLEMS

Hand Calculations

5.1 Use the Newton-Raphson method to determine the highest root of

$$f(x) = -0.9x^2 + 1.7x + 2.5$$

Employ an initial guess of $x_i = 3.1$. Perform the computation until ϵ_a is less than $\epsilon_s = 0.01\%$. Also perform an error check of your final answer.

5.2 Determine the real roots of

$$f(x) = -2.0 + 6.2x - 4.0x^2 + 0.70x^3$$

(a) Graphically.
(b) Using the Newton-Raphson method to within $\epsilon_s = 0.01\%$.

5.3 Employ the Newton-Raphson method to determine a real root for

$$f(x) = -24 + 80x - 90x^2 + 42x^3$$
$$-8.7x^4 + 0.66x^5$$

using initial guesses of (a) $x_i = 3.5$; (b) $x_i = 4.0$; and (c) $x_i = 4.5$. Discuss and use graphical methods to explain any peculiarities in your results.

5.4 Determine the lowest real root of

$$f(x) = 9.34 - 21.97x + 16.3x^2 - 3.704x^3$$

(a) Graphically.
(b) Using the secant method, to a value of ϵ_s corresponding to three significant figures.

5.5 Locate the first positive root of

$$f(x) = 0.51x - \sin x$$

where x is in radians. Use the graphical method and then use three iterations of the Newton-Raphson method with an initial guess of $x_i = 2.0$ to locate the root. Repeat the computation but with an initial guess of $x_i = 1.0$. Use the graphical method to explain your results.

5.6 Find the positive real root of

$$f(x) = x^4 - 8.5x^3 - 35.50x^2 + 465x - 1000$$

using the secant method. Employ initial guesses of $x_{i-1} = 7$ and $x_i = 9$ and perform four iterations. Compute ϵ_a and interpret your results.

5.7 Perform the same computation as in Prob. 5.6 but use the Newton-Raphson method, with an initial guess of $x_i = 7$.

5.8 Find the positive square root of 11 using three iterations of
(a) The Newton-Raphson method, with an initial guess of $x_i = 3$.
(b) The secant method, with an initial guess of $x_{i-1} = 3$ and $x_i = 4$.

5.9 Determine the real root of

$$f(x) = \frac{1 - 0.61x}{x}$$

using three iterations of the secant method and initial guesses of $x_{i-1} = 1.5$ and $x_i = 2.0$. Compute the approximate error ϵ_a after the second and the third iterations.

5.10 Determine the real root of

$$f(x) = x^3 - 98$$

with the secant method, to within $\epsilon_s = 0.1\%$.

5.11 Determine the highest real root of

$$f(x) = x^3 - 6x^2 + 11x - 5.9$$

(a) Graphically.
(b) Using the bisection method (two iterations, $x_l = 2.5$ and $x_u = 3.5$).
(c) Using the false-position method (two iterations, $x_l = 2.5$ and $x_u = 3.5$).
(d) Using the Newton-Raphson method (two iterations, $x_i = 3.5$).
(e) Using the secant method (two iterations, $x_{i-1} = 2.5$ and $x_i = 3.5$).

5.12 Use the Newton-Raphson method to determine all the roots of $f(x) = x^2 + 5.8x - 11.45$ to within an $\epsilon_s = 0.001\%$.

5.13 Determine the lowest real root of

$$f(x) = 9.34 - 21.97x + 16.3x^2 - 3.704x^3$$

(a) Graphically.
(b) Using the bisection method (two iterations, $x_l = 0.5$ and $x_u = 1.05$).
(c) Using the false-position method (two iterations, $x_l = 0.5$ and $x_u = 1.05$).
(d) Using the Newton-Raphson method (two iterations, $x_i = 0.5$).

(**e**) Using the secant method (two iterations, $x_{i-1} = 0.5$ and $x_i = 1.05$).

5.14 Determine the smallest positive real root of

$$f(x) = x^3 - 4.8x^2 + 7.56x - 3.92$$

(**a**) Graphically.

(**b**) Using the most efficient available method, employ initial guesses of $x_l = x_{i-1} = 0.5$ and $x_u = x_i = 1.5$ and perform the computation to within $\epsilon_s = 15\%$.

5.15 Determine the roots of

$$f(x) = x^3 - 7x^2 - 3.75x + 12.5$$

(**a**) Graphically.

(**b**) Using the most efficient available method to within $\epsilon_s = 0.1\%$.

5.16 Repeat Prob. 4.12, but use the Newton-Raphson method.

5.17 Repeat Prob. 4.12, but use the secant method.

5.18 Determine the roots of the simultaneous nonlinear equations

$$y = -x^2 + x + 0.5$$

$$y + 5xy = x^3$$

Employ initial guesses of $x = y = 1.2$ and perform two iterations using the Newton-Raphson approach from Sec. 5.5.2. Compute the estimated error.

5.19 Determine the roots of the simultaneous nonlinear equations

$$y + 1 = x^2$$

$$x^2 = 5 - y^2$$

Use a graphical approach to obtain your initial guesses. Determine refined estimates with the two-equation Newton-Raphson method described in Sec. 5.5.2.

Computer-Related Problems

5.20 Develop a user-friendly program for the Newton-Raphson method based on Fig. 5.4 and Sec. 5.2.3. Test it by duplicating the computation from Example 5.3.

5.21 Use the program you developed in Prob. 5.20 to duplicate the computation from Example 5.5. Determine the root using the initial guess of $x_i = 0.5$. Perform a number of runs of 5, 10, 15, and more iterations until the true percent relative error falls below 0.1%. Plot true and approximate percent relative errors versus number of iterations on semilog paper. Interpret your results.

5.22 Use the program developed in Prob. 5.20 to solve Probs. 5.1 through 5.5. For all cases, perform the computations to within a tolerance of $\epsilon_s = 0.001\%$.

5.23 Develop a user-friendly program for the secant method based on Fig. 5.4 and Sec. 5.3.2. Test it by duplicating the computation from Example 5.6.

5.24 Use the program you developed in Prob. 5.23 to solve Probs. 5.6, 5.9, and 5.10. For all cases, perform the computations to within a tolerance of $\epsilon_s = 0.001\%$.

5.25 Develop a user-friendly program for the two-equation Newton-Raphson method based on Sec. 5.5. Test it by solving Example 5.10.

5.26 Use the program you developed in Prob. 5.25 to solve Probs. 5.18 and 5.19 to within a tolerance of $\epsilon_s = 0.01\%$.

CHAPTER 6
Case Studies: Roots of Equations

The purpose of this chapter is to use the numerical procedures discussed in Chaps. 4 and 5 to solve actual engineering problems. Numerical techniques are important for practical applications because engineers frequently encounter problems that cannot be approached using analytical techniques. For example, simple mathematical models that can be solved analytically may not be applicable when real problems are involved. Thus, more complicated models must be employed. For these cases, it is appropriate to implement a numerical solution on a computer. In other situations, engineering design problems may require solutions for implicit variables in complicated equations (recall Eq. (PT2.3) and Example 4.5).

The following case studies are typical of those that are routinely encountered during upper-class and graduate studies. Furthermore, they are representative of problems you will address professionally. The problems are drawn from the general area of engineering economics as well as from the four major disciplines of engineering: chemical, civil, electrical, and mechanical. These case studies also serve to illustrate the trade-offs among the various numerical techniques.

For example, *Case Study 6.1* uses all the methods with the exception of Newton-Raphson to perform an economic break-even analysis. The Newton-Raphson method is not employed because the function in the case study is inconvenient to differentiate. Among other things, the example demonstrates how the secant method may diverge if the initial guess is not close enough to the root.

Case Study 6.2, which is taken from chemical engineering, provides an excellent example of how root-location methods allow you to use realistic formulas in engineering practice. In addition, it also demonstrates how the efficiency of the Newton-Raphson technique is used to advantage when a large number of root-location computations is required.

Case Studies 6.3, 6.4, and 6.5 are engineering design problems taken from civil, electrical, and mechanical engineering. *Case Study 6.3* uses bisection to design a catenary cable. *Case Study 6.4* performs a similar analysis for an electric circuit. *Case Study 6.5* deals with a vibration analysis of an automobile. Aside from investigating the efficiency of the various methods, this example has the added feature of illustrating how graphical methods provide insight into the root-location process.

Finally, *Case Study 6.6* uses a spreadsheet to solve a system of nonlinear equations. This exercise utilizes the supplementary software that has been included on the diskette at the back of this book.

CASE STUDY 6.1   BREAK-EVEN ANALYSIS (GENERAL ENGINEERING)

Background: Good engineering practice requires that projects, products, and planning be approached in a cost-effective manner. A well-trained engineer must therefore be familiar with cost analysis. The present problem is called a "break-even problem." It is used to determine the point at which two alternative options are of equivalent value. Such choices are confronted in all fields of engineering. Although the case study is couched in personal terms, it is a prototype of other break-even problems that you may be called upon to address professionally.

TABLE 6.1   **Cost and benefits for two personal computers. The negative signs indicate a cost or loss, whereas positive signs indicate a benefit.**

	COMPUTER	
	Lean Machine	Ultimate
Purchase cost, $	−3000	−10,000
Increase in maintenance cost per year, $/yr/yr	−200	−50
Annual profit and enjoyment, $/yr	1000	4000

You are considering the purchase of one of two personal computers: the "Lean Machine" and the "Ultimate." The estimated expenses and benefits for each computer are summarized in Table 6.1. If money can currently be borrowed at 20 percent interest ($i = 0.20$), how long must the machines be owned so that they would have equivalent worth? In other words, what is the break-even point in years?

Solution: As is common in economic problems, we have a mixture of present and future costs. For example, as depicted in Fig. 6.1, buying the Lean Machine involves an initial outlay of $3000. In addition to this one-time cost, money must be spent every year to

Figure 6.1
Cash-flow diagram for the costs and benefits of the Lean Machine computer. The abscissa is the number of years that you own this device. Cash flow is measured on the ordinate, with benefits positive and costs negative.

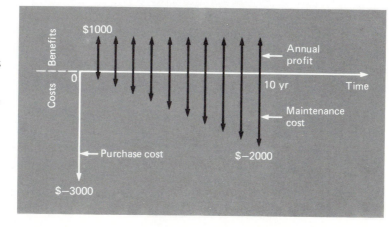

maintain the machine. Because such costs tend to increase as the computer gets older, the maintenance costs are assumed to increase linearly with time. For example, $2000 would be required annually, by year 10, to keep the machine in working condition (Fig. 6.1). Finally, aside from these costs, you will also derive benefit from owning the computer. The annual profit and enjoyment derived from the Lean Machine are characterized by a constant annual income of $1000 each year.

In order to assess the two options, we must convert these costs into comparable measures. One way to do this is to express all the individual costs as equivalent annual payments, that is, the equivalent dollar value per year over the life span of the computer. The annual profit and enjoyment income is already in this form. Economic formulas are available to express the purchase and maintenance costs in the same manner. For example, the initial purchase cost can be transformed into a series of uniform annual payments by the formula (Fig. 6.2*a*)

$$A_p = P\frac{i(1 + i)^n}{(1 + i)^n - 1} \tag{6.1}$$

where A_p is the amount of the annual payment, P is the purchase cost, i is the interest rate, and n is the number of years. For example, the initial payment for the Lean Machine is -3000, where the negative sign indicates a loss to you. If the interest rate is 20 percent ($i = 0.2$),

$$A_p = -3000\frac{0.2(1.2)^n}{(1.2)^n - 1}$$

For example, if the payments are to be spread over 10 years ($n = 10$), this formula can be used to compute that the equivalent annual payment would be -715.57 per year.

Figure 6.2
Graphical depiction of the use of an economic formula (*a*) to transform an initial payment to a series of equivalent annual payments using Eq. (6.1) and (*b*) to transform an arithmetic gradient series into a series of equivalent annual payments using Eq. (6.2).

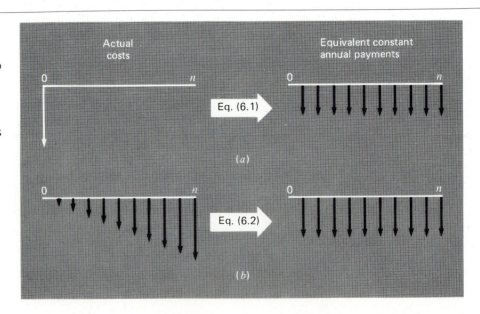

The maintenance cost is called an *arithmetic gradient series* because it increases at a constant rate. Conversion of such a series to an annual rate A_m can be accomplished by the formula

$$A_m = G\left[\frac{1}{i} - \frac{n}{(1 + i)^n - 1}\right] \tag{6.2}$$

where G is the arithmetic rate of increase of maintenance. As depicted in Fig. 6.2*b*, this formula transforms the increasing maintenance cost into an equivalent series of constant annual payments.

These equations can be combined to express the value of each computer in terms of a uniform series of payments. For example, for the Lean Machine,

$$A_t = -3000\frac{0.2(1.2)^n}{(1.2)^n - 1} - 200\left[\frac{1}{0.2} - \frac{n}{(1.2)^n - 1}\right] + 1000$$

Total worth $= -$ purchase cost $-$ maintenance cost $+$ profit

where A_t designates the total annual worth. This equation can be simplified by collecting terms:

$$A_t = \frac{-600(1.2)^n}{(1.2)^n - 1} + \frac{200n}{(1.2)^n - 1} \tag{6.3}$$

Substituting $n = 2$ into Eq. (6.3) yields the result that if you decide to discard the Lean Machine after owning it for only 2 years it will have cost \$1055 per year. If the computer is discarded after 10 years ($n = 10$), Eq. (6.3) indicates that it will have cost \$330 per year.

Similarly for the Ultimate, an equation for annual worth can be developed, as in

$$A_t = \frac{-2000(1.2)^n}{(1.2)^n - 1} + \frac{50n}{(1.2)^n - 1} + 3750 \tag{6.4}$$

The values for Eq. (6.4) for $n = 2$ and $n = 10$ are \$$-2568$ and \$$+1461$ per year. Thus, although the Ultimate is more costly on a short-term basis, if owned long enough, it will not only be more cost-effective but will actually earn money for you. Equations (6.3) and (6.4) are plotted for various values of n in Fig. 6.3*a*.

The identification of the point at which the two machines have equivalent value designates when the Ultimate becomes the better buy. Graphically, it corresponds to the intersection of the two curves in Fig. 6.3*a*. From a mathematical perspective, the break-even point is the value of n for which Eqs. (6.3) and (6.4) are equivalent, that is,

$$\frac{-600(1.2)^n}{(1.2)^n - 1} + \frac{200n}{(1.2)^n - 1} = \frac{-2000(1.2)^n}{(1.2)^n - 1} + \frac{50n}{(1.2)^n - 1} + 3750$$

By bringing all the terms of this equation to one side, the problem reduces to finding the root of

$$f(n) = \frac{-1400(1.2)^n}{(1.2)^n - 1} - \frac{150n}{(1.2)^n - 1} + 3750 = 0 \tag{6.5}$$

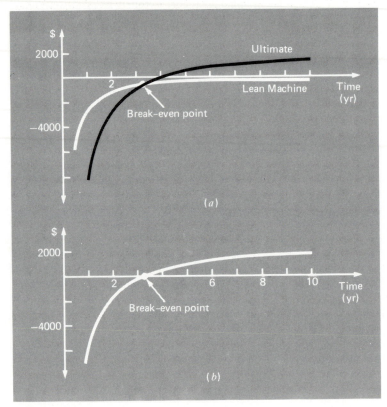

Figure 6.3
(a) Net cost curves for the Lean Machine [Eq. (6.3)] and the Ultimate [Eq. (6.4)] computers.
(b) The break-even function [Eq. (6.5)].

Note that because of the way in which we have derived this equation, the Lean Machine is more cost-effective when $f(n) < 0$, and the Ultimate is more cost-effective when $f(n) > 0$ (Fig. 6.3b). The roots of Eq. (6.5) cannot be determined analytically. On the other hand, the equivalent annual payments are easy to calculate for a given n. Thus, as in our discussions of Sec. PT2.1.2, the design aspects of this problem create the need for a numerical approach.

The roots of Eq. (6.5) can be computed using some of the numerical methods described in Chaps. 4 and 5. The bracketing approaches and the secant method can be applied with minimal effort, whereas the Newton-Raphson method is awkward to use because it is time-consuming to determine df/dn from Eq. (6.5).

On the basis of Fig. 6.3, we know that the root is between $n = 2$ and 10. These values provide starting values for the bisection method. Interval halving can be repeated for 18 iterations to yield a result with an ϵ_a less than 0.001 percent. The break-even point occurs at $n = 3.23$ years. This result can be checked by substituting it back into Eq.(6.5) to verify that $f(3.23) \simeq 0$.

Substituting $n = 3.23$ into either Eq. (6.3) or Eq. (6.4) yields the result that at the break-even point both machines cost about \$542 per year. Beyond this point, the Ultimate becomes more cost-effective. Consequently, if you intend to own your machine for more than 3.23 years, the Ultimate is the better buy.

The method of false position can also be easily applied to this problem. A similar root and accuracy are attained after 12 iterations for the same bracketing interval of 2 to 10. On the other hand, the secant method converges to a root of -24.83 for this same bracketing interval. However, if the bracketing interval is reduced to 3 to 4, the secant method converges on $n = 3.23$ in only five iterations. Interestingly, the secant method also converges rapidly to the proper root when the initial interval is 2 to 3 and does not bracket the root. These results are typical of the trade-offs discussed subsequently in the epilogue. The best numerical method for this problem then depends on your judgment concerning the trade-offs among factors such as numerical efficiency, computer costs, and the dependability of the method.

CASE STUDY 6.2   IDEAL AND NONIDEAL GAS LAWS (CHEMICAL ENGINEERING)

Background: The *ideal gas law* is given by

$$pV = nRT \tag{6.6}$$

where p is the absolute pressure, V is the volume, n is the number of moles, R is the universal gas constant, and T is the absolute temperature. Although this equation is widely used by engineers and scientists, it is only accurate over a limited range of pressure and temperature. Furthermore, Eq. (6.6) is more appropriate for some gases than for others.

An alternative equation of state for gases is given by

$$\left(p + \frac{a}{v^2}\right)(v - b) = RT \tag{6.7}$$

which is known as the *van der Waals equation*, where $v = V/n$ is the molal volume, and a and b are empirical constants that depend on the particular gas.

A chemical engineering design project requires that you accurately estimate the molal volume (v) of both carbon dioxide and oxygen for a number of different temperature and pressure combinations so that appropriate containment vessels can be selected. It is also of interest to examine how well each gas conforms to the ideal gas law by comparing the molal volume as calculated by Eqs. (6.6) and (6.7). The following data is provided:

$$R = 0.082054 \text{ L} \cdot \text{atm/(mol} \cdot \text{K)}$$

$$\left.\begin{array}{l} a = 3.592 \\ b = 0.04267 \end{array}\right\} \text{carbon dioxide}$$

$$\left.\begin{array}{l} a = 1.360 \\ b = 0.03183 \end{array}\right\} \text{oxygen}$$

The design pressures of interest are 1, 10, and 100 atm for temperature combinations of 300, 500, and 700 K.

Solution: Molal volumes for both gases are calculated using the ideal gas law, with $n = 1$. For example, if $p = 1$ atm and $T = 300$ K,

$$v = \frac{V}{n} = \frac{RT}{p} = 0.082054 \frac{\text{L} \cdot \text{atm}}{\text{mol} \cdot \text{K}} \frac{300 \text{ K}}{1 \text{ atm}}$$

$$v = 24.6162 \text{ L/mol}$$

These calculations are repeated for all temperature and pressure combinations and presented in Table 6.2.

TABLE 6.2 **Computations of molal volume for Case Study 6.2**

Temperature, K	Pressure, atm	Molal Volume (Ideal Gas Law), L /mol	Molal Volume (van der Waals) Carbon Dioxide, L /mol	Molal Volume (van der Waals) Oxygen, L /mol
300	1	24.6162	24.5126	24.5928
	10	2.4616	2.3545	2.4384
	100	0.2462	0.0795	0.2264
500	1	41.0270	40.9821	41.0259
	10	4.1027	4.0578	4.1016
	100	0.4103	0.3663	0.4116
700	1	57.4378	57.4179	57.4460
	10	5.7438	5.7242	5.7521
	100	0.5744	0.5575	0.5842

The computation of molal volume from the van der Waals equation can be accomplished using any of the numerical methods for finding roots of equations discussed in Chaps. 4 and 5, with

$$f(v) = \left(p + \frac{a}{v^2} \right)(v - b) - RT \tag{6.8}$$

In this case, the derivative of $f(v)$ is easy to determine and the Newton-Raphson method is convenient and efficient to implement. The derivative of f with respect to v is given by

$$f'(v) = p - \frac{a}{v^2} + \frac{2ab}{v^3}$$

The Newton-Raphson method is described by Eq. (5.6):

$$v_{i+1} = v_i - \frac{f(v_i)}{f'(v_i)}$$

which can be used to estimate the root. For example, using the initial guess of 24.6162, the molal volume of carbon dioxide at 300 K and 1 atm is computed as 24.5126 L/mol. This result was obtained after two iterations and has an ϵ_a of less than 0.001 percent.

Similar computations for all combinations of pressure and temperature for both gases are presented in Table 6.2. It is seen that the results for the ideal gas law differ from those for van der Waals equation for both gases, depending on specific values for p and T. Furthermore, because some of these results are significantly different, your design of the containment vessels would be quite different, depending on which equation of state was used.

In this case, a complicated equation of state was examined using the Newton-Raphson method. The results varied significantly from the ideal gas law for several cases. From a practical standpoint, the Newton-Raphson method was appropriate for this application because $f'(v)$ was easy to calculate. Thus, the rapid convergence properties of the Newton-Raphson method could be exploited.

In addition to demonstrating its power for a single computation, the present case study also illustrates how the Newton-Raphson method is especially attractive when numerous computations are required. Because of the speed of digital computers, the efficiency of various numerical methods for most roots of equations are indistinguishable for a single computation. Even a 10-s difference between the crude bisection approach and the efficient Newton-Raphson does not amount to a significant time loss when only one computation is performed. However, suppose that millions of root evaluations are required to solve a problem. In this case, the efficiency of the method could be a deciding factor in the choice of a technique.

For example, suppose that you are called upon to design an automatic computerized control system for a chemical production process. This system requires accurate estimates of molal volumes on an essentially continuous basis in order to properly manufacture the final product. Gauges are installed that provide instantaneous readings of pressure and temperature. Evaluations of v must be obtained for a variety of gases that are used in the process.

For such an application, bracketing methods such as bisection or false position would probably be too time-consuming. In addition, the two initial guesses that are required for these approaches would also interject a critical delay in the procedure. This shortcoming is relevant to the secant method, which also needs two initial estimates.

In contrast, the Newton-Raphson method requires only one guess for the root. The ideal gas law could be employed to obtain this guess at the initiation of the process. Then, assuming that the time frame is short enough so that pressure and temperature do not vary wildly between computations, the previous root solution would provide a good guess for the next application. Thus, the close guess that is often a prerequisite for convergence of the Newton-Raphson method would automatically be available. All of the above considerations would greatly favor the Newton-Raphson technique for such problems.

CASE STUDY 6.3   CATENARY CABLE (CIVIL ENGINEERING)

Background: In Chap. 1, we stated that force balances were fundamental to the development of mathematical models in civil engineering. In the present case study, a force balance is employed to derive a model for a catenary cable.

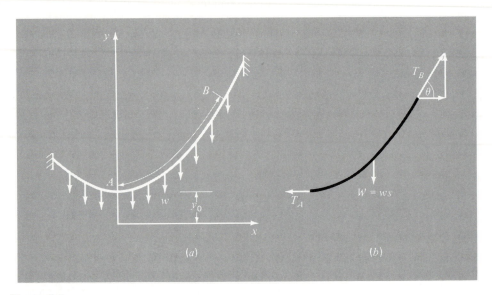

Figure 6.4
(a) Forces acting on a section *AB* of a flexible hanging cable. The load is uniform along the cable (but not uniform per the horizontal distance *x*). (b) A free-body diagram of section *AB*.

A *catenary cable* is one which is hung between two points not in the same vertical line. As depicted in Fig. 6.4a, it is subject to no loads other than its own weight. Thus, its weight acts as a uniform load per unit length along the cable, *w* (N/m). A free-body diagram of a section *AB* is depicted in Fig. 6.4b. Notice that the section is subject to three forces: a horizontal tension acting at A, T_A, a tangential tension acting at B, T_B, and the weight of the cable, $W = ws$. If the system is at rest, the horizontal and vertical components of the forces must be in balance; that is,

$$T_B \cos \theta = T_A \qquad T_B \sin \theta = W$$

Dividing these equations gives

$$\frac{T_B \sin \theta}{T_B \cos \theta} = \tan \theta = \frac{W}{T_A}$$

or because $\tan \theta = dy/dx$

$$\frac{dy}{dx} = \frac{ws}{T_A} \tag{6.9}$$

The arc length *s* can be computed by (Thomas and Finney, 1979)

$$s = \int \sqrt{1 + \left(\frac{dy}{dx}\right)^2} \, dx \tag{6.10}$$

Equations (6.9) and (6.10) can be differentiated and combined to give

$$\frac{d^2y}{dx^2} = \frac{w}{T_A}\sqrt{1 + \left(\frac{dy}{dx}\right)^2}$$

(6.11)

Thus, Eq. (6.11) is a mathematical model of section AB of the catenary cable. Because it is a second-order ordinary differential equation, two conditions are required for its solution. For the present case, the first condition is

$$\frac{dy}{dx} = 0 \quad \text{at } x = 0$$

That is, the cable is horizontal at A. The second condition is that the height equals y_0 at A,

$$y = y_0 \quad \text{at } x = 0$$

Under these conditions, calculus can be employed to obtain the solution

$$y = \frac{T_A}{w} \cosh\left(\frac{w}{T_A}x\right) + y_0 - \frac{T_A}{w}$$

(6.12)

where the hyperbolic cosine can be computed by

$$\cosh x = \frac{1}{2}(e^x + e^{-x})$$

(6.13)

The model now provides a simple means to predict the value of the dependent variable, the cable's height y, given values for the independent variable x and the parameters T_A, w, and y_0. For example, the values $w = 10$, $T_A = 2000$, and $y_0 = 1$ were used to compute the heights for the cable shown in Fig. 6.4a.

However, suppose that it was required to calculate a value for the parameter, T_A, given values for the parameters $w = 10$ and $y_0 = 5$, such that the cable has a height of $y = 12$ at $x = 50$. For this case, the unknown, T_A, is implicit and a numerical method is required to obtain a solution.

Solution: Although FORTRAN has an intrinsic function to compute the hyperbolic cosine, Microsoft BASIC and Turbo Pascal do not. Therefore, to keep this example as general as possible, we will substitute Eq. (6.13) into Eq. (6.12) along with the given values to yield

$$12 = \frac{T_A}{10}\left\{\frac{1}{2}\left[\exp\left(\frac{10}{T_A}50\right) + \exp\left(-\frac{10}{T_A}50\right)\right]\right\} + 5 - \frac{T_A}{10}$$

Therefore, a root must be determined for

$$f(T_A) = \frac{T_A}{20}\left[\exp\left(\frac{500}{T_A}\right) + \exp\left(\frac{-500}{T_A}\right)\right] - 7 - \frac{T_A}{10}$$

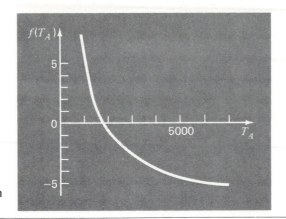

Figure 6.5
Plot of function from Case
Study 6.3 used to determine an
initial guess.

This function can be plotted (Fig. 6.5) in order to determine initial guesses of 1000 and 5000. Using bisection with a stopping criterion of $\epsilon_s = 0.1\%$ gives a result of $T_A = 1797.852$ in 12 iterations.

CASE STUDY 6.4   DESIGN OF AN ELECTRIC CIRCUIT (ELECTRICAL ENGINEERING)

Background: Electrical engineers often use Kirchhoff's laws to study the steady-state (not time-varying) behavior of electric circuits. Such steady-state behavior will be examined in Case Study 10.4. Another important problem involves circuits that are transient in nature where sudden temporal changes take place. Such a situation occurs following the closing of the switch in Fig. 6.6. In this case, there will be a period of adjustment following the closing of the switch as a new steady state is reached. The length of this adjustment period is closely related to the storage properties of the capacitor and the inductor. Energy storage may oscillate between these two elements during a transient period. However, resistance in the circuit will dissipate the magnitude of the oscillations.

Figure 6.6
An electric circuit. When the
switch is closed, the current will
undergo a series of oscillations
until a new steady state is
reached.

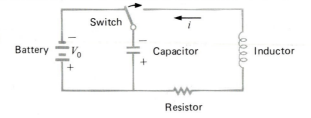

The flow of current through the resistor causes a voltage drop (V_R) given by

$$V_R = iR$$

where i is the current and R is the resistance of the resistor. When R and i have units of ohms and amperes, then V_R has units of volts.

Similarly, an inductor resists changes in current, such that the voltage drop V_L across it is

$$V_L = L\frac{di}{dt}$$

where L is the inductance. When L and i have units of henrys and amperes, V_L has units of volts and t has units of seconds.

The voltage drop across the capacitor (V_C) depends on the charge (q) on it:

$$V_C = \frac{q}{C}$$

where C is the capacitance. When the charge is expressed in units of coulombs, the unit of C is the farad.

Kirchhoff's second law states that the algebraic sum of voltage drops around a closed circuit is zero. After the switch is closed we have

$$L\frac{di}{dt} + Ri + \frac{q}{C} = 0$$

However, the current is related to the charge according to

$$i = \frac{dq}{dt}$$

Therefore,

$$L\frac{d^2q}{dt^2} + R\frac{dq}{dt} + \frac{q}{C} = 0$$

This is a second-order linear ordinary differential equation that can be solved using the methods of calculus. This solution is given by

$$q(t) = q_0 e^{-Rt/2L}\cos\left(\sqrt{\frac{1}{LC} - \left(\frac{R}{2L}\right)^2}\, t\right) \tag{6.14}$$

where at $t = 0$, $q = q_0 = V_0 C$, and V_0 is the voltage from the charging battery. Equation (6.14) describes the time variation of the charge on the capacitor. The solution $q(t)$ is plotted in Fig. 6.7.

Figure 6.7
The charge on a capacitor as a function of time following the closing of the switch in Fig. 6.6

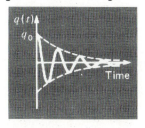

A typical electrical-engineering design problem might involve determining the proper resistor to dissipate energy at a specified rate, with known values for L and C. For the present case study, assume the charge must be dissipated to 1 percent of its original value ($q/q_0 = 0.01$) in $t = 0.05$ s, with $L = 5$ H and $C = 10^{-4}$ F.

Solution: It is necessary to solve Eq. (6.14) for R, with known values of q, q_0, L, and C. However, a numerical approximation technique must be employed because R is an implicit variable in Eq. (6.14). The bisection method will be used for this purpose. The other methods discussed in Chaps. 4 and 5 are also appropriate, although the Newton-

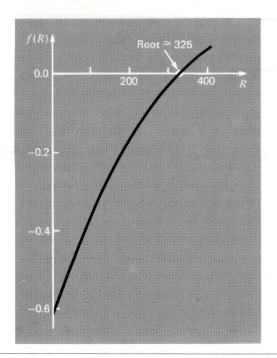

Figure 6.8
Plot of Eq. (6.15) used to obtain initial guesses for R that bracket the root.

Raphson method would be cumbersome to use because the derivative of Eq. (6.14) is somewhat complicated. Rearranging Eq. (6.14),

$$f(R) = e^{-Rt/2L} \cos\left(\sqrt{\frac{1}{LC} - \left(\frac{R}{2L}\right)^2}\, t\right) - \frac{q}{q_0}$$

or, using the numerical values given,

$$f(R) = e^{-0.005R} \cos\left[\sqrt{2000 - 0.01R^2}(0.05)\right] - 0.01 \tag{6.15}$$

Examination of this equation suggests that a reasonable initial range for R is 0 to 400 Ω (because $2000 - 0.01R^2$ must be greater than zero). Figure 6.8, a plot of Eq. (6.15), confirms this. Twenty-one iterations of the bisection method give $R = 328.1515 \ \Omega$, with an error of less than 0.0001 percent.

Thus, you can specify a resistor with this rating for the circuit shown in Fig. 6.6 and expect to achieve a dissipation performance that is consistent with the requirements of the problem. This design problem could not be solved efficiently without using the numerical methods in Chaps. 4 and 5.

CASE STUDY 6.5   VIBRATION ANALYSIS (MECHANICAL ENGINEERING)

Background: Differential equations are often used to model the behavior of engineering systems. A class of such models broadly applicable to most fields of engineering is harmonic oscillators. Some basic examples of harmonic oscillators are a simple pendulum, a mass on a spring, and an inductance-capacitance electric circuit (Fig. 6.9). Although these are very different physical systems, their oscillations can all be described by the same mathematical model. Thus, although the present problem deals with the design of an automobile shock absorber, the general approach is applicable to a variety of other problems in all fields of engineering.

As depicted in Fig. 6.10, a car of mass m is supported by springs. Shock absorbers offer a resistance to the motion of the car that is proportional to the vertical speed (up-and-down motion) of the car. Disturbance of the car from equilibrium causes the system to move with an oscillating motion $x(t)$. At any instant the net forces acting on m are the resistance of the spring and the damping force of the shock absorber. According to Hooke's law, the resistance of the spring is proportional to a spring constant (k) and the distance from equilibrium (x):

$$\text{Spring force} = -kx \tag{6.16}$$

where the negative sign indicates that the restoring force acts to return the car toward the position of equilibrium. The damping force of the shock absorber is given by

$$\text{Damping force} = -c\frac{dx}{dt}$$

where c is a damping coefficient and dx/dt is the vertical velocity. The negative sign indicates that the damping force acts in the opposite direction against the velocity.

The equations of motion for the system are given by Newton's second law ($F = ma$), which for the present problem is expressed as

$$\underbrace{m\ \frac{d^2x}{dt^2}}_{\text{Mass}\ \times\ \text{acceleration}} = \underbrace{-c\frac{dx}{dt}}_{\text{damping force}} + \underbrace{(-kx)}_{\text{spring force}}$$

or

$$\frac{d^2x}{dt^2} + \frac{c}{m}\frac{dx}{dt} + \frac{k}{m}x = 0$$

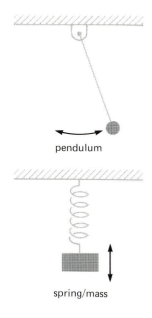

pendulum

spring/mass

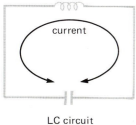

current

LC circuit

Figure 6.9
Three examples of simple harmonic oscillators. The two-way arrows illustrate the oscillations for each system.

Figure 6.10
A car of mass m.

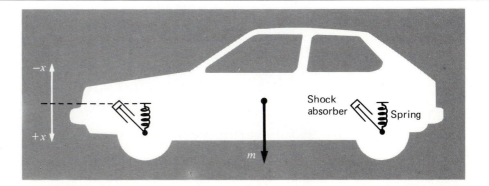

This is a second-order linear differential equation that can be solved using the methods of calculus. For example, if the car hits a hole in the road at $t = 0$, such that it is displaced from equilibrium with $x = x_0$ and $dx/dt = 0$, then

$$x(t) = e^{-nt}(x_0 \cos pt + x_0 \frac{n}{p} \sin pt) \tag{6.17}$$

where $n = c/(2m)$, $p = \sqrt{k/m - c^2/(4m^2)}$ and $k/m > c^2/(4m^2)$. Equation (6.17) gives the vertical position of the car as a function of time. The parameter values are $c = 1.4 \times 10^7$ g/s, $m = 1.2 \times 10^6$ g, and $k = 1.25 \times 10^9$ g/s^2. If $x_0 = 0.3$ m, mechanical engineering design considerations require that estimates be provided for the first three times the car passes through the equilibrium point.

Solution: This design problem can be solved using numerical methods of Chaps. 4 and 5. The bracketing or secant methods are preferable because the derivative of Eq. (6.17) is complicated.

Estimates of the initial guesses are easily obtained by reference to Fig. 6.11. This case study illustrates how graphical methods often provide information that is essential for the successful application of the numerical techniques. The plot indicates that this problem is complicated by the existence of several roots. Thus, in this case, rather narrow bracketing intervals must be used to avoid overlap.

Table 6.3 lists the results of using the bisection, false-position, and secant methods, with a stopping criterion of 0.1 percent. All the techniques converge quickly. As expected, the false-position and secant methods are more efficient than bisection.

Figure 6.11
Plot of the position versus time for a shock absorber after the wheel of a car hits a hole in the road.

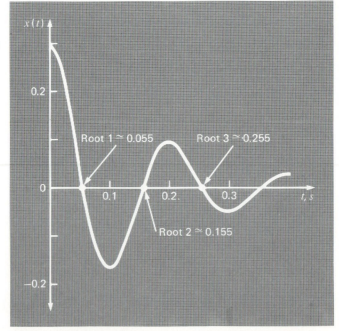

TABLE 6.3   Results of using the bisection, false-position, and secant methods to locate the first
three roots for vibrations of a shock absorber[*].

| Method | Lower Guess | Upper Guess | Root Estimate | Number of Iterations | Percent Relative Error | |
					Approximate	True
Bisection	0.0	0.1	0.0552246	11	0.088	0.027
	0.1	0.2	0.1541922	10	0.063	0.014
	0.2	0.3	0.2533203	9	0.077	0.069
False position	0.0	0.1	0.0552095	5	0.002	0.0001
	0.1	0.2	0.1541790	4	0.069	0.0006
	0.2	0.3	0.2531475	4	0.043	0.0003
Secant	0.0	0.1	0.0552095	5	0.038	0.0001
	0.1	0.2	0.1541780	5	0.020	0.0001
	0.2	0.3	0.2531465	5	0.017	0.0001

[*] A stopping criterion of 0.1 percent was used to obtain these results. Note that the exact values of the roots
are 0.0552095329, 0.15417813, and 0.253146726.

Notice how, for all methods, the approximate percent relative errors are greater
than the true errors. Thus, the results are *at least* as accurate as the stopping criterion of
0.1 percent. However, also observe that the false-position and secant methods are very
conservative in this regard. Recall from our discussion in Sec. 4.3 that the termination
criterion usually constitutes an approximation of the discrepancy of the previous iteration.
Thus, for rapidly convergent approaches such as the false-position and secant methods,
the improvement in accuracy between successive iterations is so great that ϵ_t will usually
be much less than ϵ_a. The practical significance of this behavior is of little importance
when determining a single root. However, if numerous root locations are involved, rapid
convergence becomes a valuable property to be considered when choosing a particular
technique.

CASE STUDY 6.6   SPREADSHEET SOLUTION OF ROOTS OF NONLINEAR SYSTEMS

Background: The spreadsheet can be used to solve systems of nonlinear algebraic
equations. This case study employs the two-variable Newton-Raphson method (Sec.
5.5.2) to solve for values of x and y that satisfy the following equations,

$$f(x, y) = 4 - y - 2x^n$$

$$g(x, y) = 8 - y^m - 4x$$

where n and m are parameters. When n and m are either 0 or 1, the above system is
linear and can be solved analytically or using methods described in Part Three of the
text. Other values of n and m result in a nonlinear system of equations that must be
solved numerically.

Solution: Insert the disk that accompanies the text into an IBM-PC or compatible computer and type NUMMET following the A prompt. A title screen appears on the monitor
as shown in Fig. 6.12a. The Return key advances the screen to the main menu as shown

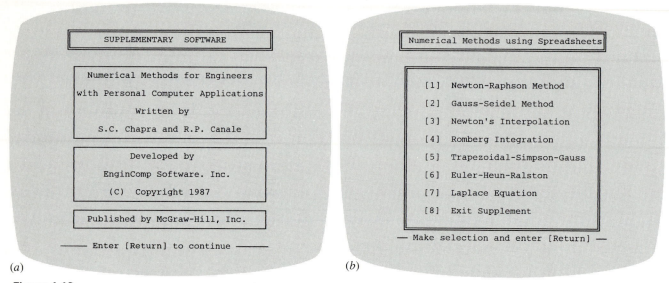

Figure 6.12
(a) Title screen and (b) main menu of the spreadsheet software supplement.

in Fig. 6.12b. The cursor can be moved using the arrow keys or the numeric keys to select any case study on the disk. Move the cursor to selection 1 and enter Return. This results in the appearance of a spreadsheet as shown in Fig. 6.13a. This spreadsheet contains the solution of the above algebraic equations using the Newton-Raphson method. The elements of the spreadsheet can be viewed on the bottom line of the screen by moving the cursor using the arrow keys. Study the contents of the cells of the B column carefully. First, the initial values of x and y are required. Next the function and partial derivatives are computed. Finally elements B19 and B20 contain equations that represent the solution of the equivalent linear system as required by the method. These new values of x and y are subsequently transferred to rows 8 and 9 of column C to perform the next iteration. Note that seven iterations are shown on the screen with convergence to the exact values after five iterations with initial values of $x = 0$ and $y = 0$.

There are only 5 cells on the spreadsheet that you can change. The values of elements B3 and B4 can be modified to enable you to change the nature of the function. Changing the values in elements B8 and B9 allows you to change the initial values of x and y. The value of element B6 allows you to change the number of the iteration.

Move the cursor and change the first values to $x = 10$ and $y = 10$. Now strike the key [F3] to update all the other elements of the spreadsheet. The results of the computation are shown in Fig. 6.13b where convergence is attained in six iterations. Now change the values to $x = 50$ and $y = 50$ and note incomplete convergence in seven iterations with $x = 1.01794$ and $y = 1.98197$. Transfer these values to cells B8 and B9, change cell B6 to 8, and strike [F3]. Figure 6.13c results showing convergence in nine iterations.

Next, we can keep $n = 2$ and $m = 2$ but use $x = -5$ and $y = 5$. Note that a new solution is determined at $x = -0.618034$ and $y = 3.23606$. This can be confirmed by verifying that cells H11 and H12 are close to zero (see Fig. 6.13d).

(a)

Cell	A	B	C	D	E	F	G	H
1	NEWTON RAPHSON METHOD FOR SYSTEMS OF NONLINEAR EQUATIONS							
2	-------							
3	N=	2		0=F(X,Y)=4-Y-2X^N				
4	M=	2		0=G(X,Y)=8-Y^M-4X				
5	-------							
6	ITER=	1	2	3	4	5	6	7
7								
8	OLD X=	0	2	1.2	1.01176	1.00004	1	1
9	OLD Y=	0	4	2.4	2.02353	2.00009	2	2
10								
11	F(X,Y)=	4	-8	-1.28	-.709E-01	-.275E-03	0	0
12	G(X,Y)=	8	-16	-2.56	-.141730	-.549E-03	0	0
13								
14	DF/DX=	0	-8	-4.8	-4.04705	-4.00018	-4	-4
15	DF/DY=	-1	-1	-1	-1	-1	-1	-1
16	DG/DX=	-4	-4	-4	-4	-4	-4	-4
17	DG/DY=	0	-8	-4.8	-4.04705	-4.00018	-4	-4
18								
19	NEW X=	2	1.2	1.01176	1.00004	1	1	1
20	NEW Y=	4	2.4	2.02353	2.00009	2	2	2

[A1] NEWTON RA

(b)

Cell	A	B	C	D	E	F	G	H
1	NEWTON RAPHSON METHOD FOR SYSTEMS OF NONLINEAR EQUATIONS							
2	-------							
3	N=	2		0=F(X,Y)=4-Y-2X^N				
4	M=	2		0=G(X,Y)=8-Y^M-4X				
5	-------							
6	ITER=	1	2	3	4	5	6	7
7								
8	OLD X=	10	4.98995	2.59873	1.54611	1.10129	1.00598	1.00002
9	OLD Y=	10	4.40201	1.92897	1.43508	1.97001	1.99415	1.99998
10								
11	F(X,Y)=	-206	-50.2012	-11.4358	-2.21602	-.395724	-.182E-01	-.710E-04
12	G(X,Y)=	-132	-31.3374	-6.11589	-.243928	-.286149	-.583E-03	-.339E-04
13								
14	DF/DX=	-40	-19.9598	-10.3949	-6.18445	-4.40518	-4.02394	-4.00008
15	DF/DY=	-1	-1	-1	-1	-1	-1	-1
16	DG/DX=	-4	-4	-4	-4	-4	-4	-4
17	DG/DY=	-20	-8.80402	-3.85795	-2.87017	-3.94003	-3.98830	-3.99997
18								
19	NEW X=	4.98995	2.59873	1.54611	1.10129	1.00598	1.00002	1
20	NEW Y=	4.40201	1.92897	1.43508	1.97001	1.99415	1.99998	2

[B10]

(c)

Cell	A	B	C	D	E	F	G	H
1	NEWTON RAPHSON METHOD FOR SYSTEMS OF NONLINEAR EQUATIONS							
2	-------							
3	N=	2		0=F(X,Y)=4-Y-2X^N				
4	M=	2		0=G(X,Y)=8-Y^M-4X				
5	-------							
6	ITER=	8	9	10	11	12	13	14
7								
8	OLD X=	1.01794	1.00018	1	1	1	1	1
9	OLD Y=	1.98197	1.99989	2	2	2	2	2
10								
11	F(X,Y)=	-.544E-01	-.630E-03	0	0	0	0	0
12	G(X,Y)=	+.353E-04	-.321E-03	0	0	0	0	0
13								
14	DF/DX=	-4.07176	-4.00073	-4	-4	-4	-4	-4
15	DF/DY=	-1	-1	-1	-1	-1	-1	-1
16	DG/DX=	-4	-4	-4	-4	-4	-4	-4
17	DG/DY=	-3.96394	-3.99979	-4	-4	-4	-4	-4
18								
19	NEW X=	1.00018	1	1	1	1	1	1
20	NEW Y=	1.99989	2	2	2	2	2	2

[B6] 8

(d)

Cell	A	B	C	D	E	F	G	H
1	NEWTON RAPHSON METHOD FOR SYSTEMS OF NONLINEAR EQUATIONS							
2	-------							
3	N=	2		0=F(X,Y)=4-Y-2X^N				
4	M=	2		0=G(X,Y)=8-Y^M-4X				
5	-------							
6	ITER=	8	9	10	11	12	13	14
7								
8	OLD X=	-5	-2.48529	-1.27557	-.766570	-.629609	-.618115	-.618034
9	OLD Y=	5	4.29411	3.67266	3.34290	3.24470	3.23613	3.23606
10								
11	F(X,Y)=	-51	-12.6474	-2.92684	-.518168	-.375E-01	-.264E-03	-.596E-07
12	G(X,Y)=	3	-.498269	-.386198	-.108742	-.964E-02	-.727E-04	-.477E-06
13								
14	DF/DX=	20	9.94117	5.10229	3.06628	2.51843	2.47246	2.47213
15	DF/DY=	-1	-1	-1	-1	-1	-1	-1
16	DG/DX=	-4	-4	-4	-4	-4	-4	-4
17	DG/DY=	-10	-8.58823	-7.34533	-6.68581	-6.48940	-6.47226	-6.47213
18								
19	NEW X=	-2.48529	-1.27557	-.766570	-.629609	-.618115	-.618034	-.618034
20	NEW Y=	4.29411	3.67266	3.34290	3.24470	3.23613	3.23606	3.23606

[B9] 5

Figure 6.13
Applications of the spreadsheet software supplement to solve two simultaneous nonlinear equations. The screens are described in the text.

The above examples show how the spreadsheet can be used to solve systems of nonlinear algebraic equations. One advantage of the spreadsheet is that the values of all intermediate computations are readily available to you. This helps you understand the method more completely. No programming is required, thus implementation is easy. Finally, it is convenient to interactively change values of some parameters. This encourages numerical experimentation. On the other hand, the spreadsheet computations are slow and inefficient compared to algorithms specifically designed to implement a given method. Finally, the spreadsheet has a limited number of cells that can be viewed at one time, which restricts the size of the system. Thus the spreadsheet is best used as a learning tool and as a way to solve practical problems of limited size where computational efficiency is not an important issue.

PROBLEMS

General Engineering

6.1 Using your own software, reproduce the computation performed in Case Study 6.1.

6.2 Perform the same computation as in Case Study 6.1, but use an interest rate of 17% ($i = 0.17$). If possible, use your computer software to determine the break-even point. Otherwise, use any of the numerical methods discussed in Chaps. 4 and 5 to perform the computation. Justify your choice of technique.

6.3 For Case Study 6.1, determine the number of years the Ultimate computer must be owned for it to earn money for you. That is, compute the value of n at which A_t for Eq. (6.4) becomes positive.

6.4 Using an approach similar to Case Study 6.1, the following equation can be developed for determining the total annual worth of a personal computer:

$$A_t = \frac{-1800(1.18)^n}{(1.18)^n - 1} + \frac{45n}{(1.18)^n - 1} + 3000$$

Find the value of n such that A_t is zero.

6.5 You are interested in buying an automobile and have narrowed the choices down to two options. Just as in Case Study 6.1, the net annual worth of owning either car is a composite of purchase cost, maintenance cost, and profit:

	Luxury Model	Economy Model
Purchase cost, $	−15,000	−5000
Increase in maintenance, $/yr/yr	−400	−200
Annual profit and enjoyment, $	7500	3000

If the interest rate is 12.5% ($i = 0.125$), compute the break-even point (n) for the cars.

6.6 You buy a $20,000 piece of equipment for nothing down and $5000 per year for 5 years. What interest rate are you paying? The formula relating present worth (P), annual payments (A), number of years (n), and interest rate (i) is

$$A = P\frac{i(1 + i)^n}{(1 + i)^n - 1}$$

6.7 Because many engineering-economics tables were developed years ago, they are not designed to handle the large interest rates that are sometimes prevalent today. In addition, they are often not designed to handle fractional interest rates. As in the following problem, numerical methods can be used to determine economic estimates for such situations.

A new entertainment complex is estimated to cost $10 million and to produce a net annual revenue of $2 million. If the debt is to be paid off in 10 years, at what interest rate must the funds be borrowed? Present cost (P), annual payments (A), and interest rate (i) are related to each other by the following economic formula:

$$\frac{P}{A} = \frac{(1 + i)^n - 1}{i(1 + i)^n}$$

where n is the number of annual payments. For the present problem,

$$\frac{P}{A} = \frac{10,000,000}{2,000,000} = 5$$

Therefore, the equation becomes

$$5 = \frac{(1 + i)^{10} - 1}{i(1 + i)^{10}}$$

The interest rate that satisfies this equation can be determined by finding the root of

$$f(i) = \frac{(1 + i)^{10} - 1}{i(1 + i)^{10}} - 5$$

(a) Sketch $f(i)$ versus i to make an initial graphical guess at the root.

(b) Solve for i using the bisection method (count iterations).

(c) Solve for i using the false-position method (count iterations). In both (b) and (c), use initial guesses of $i = 0.1$ and 0.2. Attain an error level of 2% for both cases.

6.8 The "rate of return" is a concept that is familiar to most people. A return of $100 interest per year on a deposit of $1000 is easily understood as a 10% rate of return. In more complex cases, the rate of return is the interest rate for which the benefits of an activity are equivalent to its costs. Among other things, the rate of return provides a handy measure for comparing alternative activities. For example, an engineering project with a return of 20% is superior to one with a return of 8%.

The determination of a project's rate of return is often complicated by the fact that the benefits and costs occur at different times. For example, the initial investment is a present cost whereas profits and expenses are accrued annually. Economic formulas are available to express these values in common terms (Case Study 6.1). When these formulas are combined to set up a balance between expenditures and income, the resulting equation usually has interest rate as an implicit variable. Thus, the problem must be solved by trial and error or, preferably, with a numerical method for determining the root of the equation.

A fleet of company vehicles can be purchased for $210,000. The anticipated income from the vehicles is $100,000 per year with direct expenses of $55,000 per year. After 5 years, the market value of the autos will be $80,000. What is the rate of return?

6.9 The upward velocity of a rocket can be computed by the following formula:

$$v = u \ln \frac{m_0}{m_0 - qt} - gt$$

where v is upward velocity, u is the velocity at which fuel is expelled relative to the rocket, m_0 is the initial mass of the rocket at time $t = 0$, q is the fuel consumption rate, and g is the downward acceleration of gravity (assumed constant = 9.8 m/s²). If $u = 2200$ m/s, $m_0 = 160,000$ kg, and $q = 2680$ kg/s, compute the time at which $v = 1000$ m/s. *Hint:* t is somewhere between 10 and 50 s. Determine your result so that it is within 1% of the true value. Check you answer.

Chemical Engineering

6.10 Using your own software, reproduce the computation performed in Case Study 6.2.

6.11 Perform the same computation as in Case Study 6.2, but for ethyl alcohol ($a = 12.02$ and $b = 0.08407$) at a temperature of 350 K and p of 1.7 atm. Compare your results with the ideal gas law. If possible, use your computer software to determine the molal volume. Otherwise, use any of the numerical methods discussed in Chaps. 4 and 5 to perform the computation. Justify your choice of technique.

6.12 Repeat Prob. 6.11, but use nitrous oxide ($a = 3.782$ and $b = 0.04415$) at a temperature of 450 K; $p = 1.75$ atm.

6.13 The temperature (in kelvin) of a system varies over the course of a day according to

$$T = 400 + 225 \cos \frac{2\pi t}{1440}$$

where t is expressed in minutes. Pressure is lost from the system according to $p = p_o e^{-t/1440}$. Develop a computer program to calculate the molal volume of oxygen at minute

intervals over the course of the day. Note that $p_o = 1$ atm. Plot the results. If you have computer graphics capabilities, plot all the data. If not, plot the results at 60-min intervals. Background for this problem can be found in Case Study 6.2.

6.14 In chemical engineering, plug flow reactors (that is, those in which fluid flows from one end to the other with minimal mixing along the longitudinal axis) are often used to convert reactants into products. It has been determined that the efficiency of the conversion can sometimes be improved by recycling a portion of the product stream so that it returns to the entrance for an additional pass through the reactor (Fig. P6.14). The recycle rate is defined as

$$R = \frac{\text{volume of fluid returned to entrance}}{\text{volume leaving the system}}$$

Suppose that we are processing a chemical A in order to generate a product B. For the case where A forms B according to an autocatalytic reaction (that is, in which one of the products acts as a catalyst or stimulus for the reaction), or

$$A + B \rightarrow B + B$$

it can be shown that an optimal recycle rate must satisfy

$$\ln \frac{1 + R(1 - X_{Af})}{R(1 - X_{Af})} = \frac{R + 1}{R[1 + R(1 - X_{Af})]}$$

where X_{Af} is the fraction of the reactant A that is converted to the product B. The optimal recycle rate corresponds to the minimum-sized reactor needed to attain the desired level of conversion.

Use the bisection method to determine the recycle ratios needed to minimize reactor size for fractional conversions of

(**a**) $X_{Af} = 0.99$
(**b**) $X_{Af} = 0.995$
(**c**) $X_{Af} = 0.999$

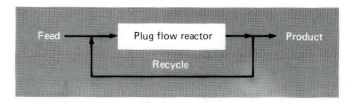

Figure P6.14
Schematic representation of a plug flow reactor with recycle.

6.15 In a chemical engineering process, water vapor (H_2O) is heated to sufficiently high temperatures that a significant portion of the water dissociates, or splits apart, to form oxygen (O_2) and hydrogen (H_2):

$$H_2O \rightleftharpoons H_2 + 1/2\, O_2$$

If it is assumed that this is the only reaction involved, the mole fraction (x) of H_2O that dissociates can be represented by

$$k_p = \frac{x}{1-x}\sqrt{\frac{2p_t}{2+x}} \qquad \text{(P6.15)}$$

where k_p is the reaction's equilibrium constant and p_t is the total pressure of the mixture. If $p_t = 2$ atm and $k_p = 0.04568$, determine the value of x that satisfies Eq. (P6.15).

6.16 The following equation pertains to the concentration of a chemical in a completely mixed reactor:

$$c = c_{in}(1 - e^{-0.05t}) + c_0 e^{-0.05t}$$

If the initial concentration $c_0 = 5$ and the inflow concentration $c_{in} = 20$, compute the time required for c to be 95 percent of c_{in}. Employ bisection.

6.17 A reversible chemical reaction

$$2A + B \rightleftharpoons C$$

can be characterized by the equilibrium relationship

$$K = \frac{C_C}{C_A^2 C_B}$$

Suppose that we define a variable x as representing the number of moles of C that are produced. Conservation of mass can be used to reformulate the equilibrium relationship as

$$K = \frac{(C_{C,0} + x)}{(C_{A,0} - 2x)^2 (C_{B,0} - x)}$$

where the subscript 0 designates the initial concentration of each constituent. If $K = 1.25 \times 10^{-2}$, $C_{A,0} = 50$, $C_{B,0} = 40$, and $C_{C,0} = 5$, compute x.

6.18 The following chemical reactions take place in a closed system

$$2A + B \rightleftharpoons C$$

$$A + D \rightleftharpoons C$$

At equilibrium, they can be characterized by

$$K_1 = \frac{C_C}{C_A^2 C_B}$$

$$K_2 = \frac{C_C}{C_A C_D}$$

If x_1 and x_2 are the rates of formation of C due to the first and second reactions, respectively, use an approach similar to that of Prob. 6.17 to reformulate the equilibrium relationships in terms of the initial concentrations of the constituents. Then, use the Newton-Raphson method to solve the pair of simultaneous nonlinear equations for x_1 and x_2 if $K_1 = 4 \times 10^{-4}$, $K_2 = 3.7 \times 10^{-2}$, $C_{A,0} = 50$, $C_{B,0} = 20$, $C_{C,0} = 5$, and $C_{D,0} = 10$. Use a graphical approach to develop your initial guesses.

Civil Engineering

6.19 Using you own software, reproduce the computation of Case Study 6.3.

6.20 Perform the same computation as in Case Study 6.3, but use a value of $w = 12$ N/m.

6.21 Perform the same computation as in Case Study 6.3 but solve for w given $y_0 = 6$, $T_A = 2000$, and $y = 14$ at $x = 40$.

6.22 The concentration of pollutant bacteria C in a lake decreases according to

$$C = 80 e^{-2t} + 20 e^{-0.1t}$$

Determine the time required for the bacteria to be reduced to 10 using (a) the graphical method and (b) the Newton-Raphson method.

6.23 Many fields of engineering require accurate population estimates. For example, transportation engineers might find it necessary to determine separately the population growth trends of a city and adjacent suburb. The population of the urban area is declining with time according to

$$P_u(t) = P_{u,max} e^{-k_u t} + P_{u,min}$$

while the suburban population is growing, as in

$$P_s(t) = \frac{P_{s,max}}{1 + \left(\frac{P_{s,max}}{P_o} - 1\right) e^{-k_s t}}$$

where $P_{u,max}$, k_u, $P_{u,min}$, $P_{s,max}$, P_o, and k_s are empirically derived parameters.

Determine the time and corresponding values of $P_u(t)$ and $P_s(t)$ when the populations are equal. The parameter values are $P_{u,max} = 60{,}000$; $k_u = 0.04$ yr^{-1}; $P_{u,min} = 120{,}000$; $P_{s,max} = 300{,}000$; $P_o = 5000$; and $k_s = 0.06$ yr^{-1}. To obtain your solutions, use (a) the graphical and (b) the false-position methods.

6.24 In environmental engineering (a specialty area in civil engineering), the following equation can be used to compute the oxygen level in a river downstream from a sewage discharge:

$$c = 10 - 15(e^{-0.1x} - e^{-0.5x})$$

where x is the distance downstream in miles. Determine the distance downstream where the oxygen level first falls to a reading of 4. (*Hint:* It is within 5 mi of the discharge.) Determine you answer to a 1% error.

6.25 The displacement of a structure is defined by the following equation for a damped oscillation:

$$y = 10e^{-kt} \cos wt$$

where $k = 0.5$ and $w = 2$.
 (a) Use the graphical method to make an initial estimate of the time required for the displacement to decrease to 4.
 (b) Use the Newton-Raphson method to determine the root to $\epsilon_s = 0.01\%$.
 (c) Use the secant method to determine the root to $\epsilon_s = 0.01\%$.

6.26 Figure P6.26 shows an open channel of constant dimensions with a rectangular cross-sectional area A. Under uniform flow conditions, the following relationship, based on Manning's equation, holds

$$Q = \frac{y_n B}{n} \left(\frac{y_n B}{B + 2y_n} \right)^{2/3} S^{1/2} \qquad \text{(P6.26)}$$

where Q is flow, y_n is normal depth, B is the width of the channel, n is a roughness coefficient used to parameterize the frictional effects of the channel material, and S is the slope of the channel. This equation is used by fluid and water-resource engineers to determine the normal depth. If this value is less than the critical depth,

$$y_c = \left(\frac{Q^2}{B^2 g} \right)^{1/3}$$

where g is acceleration due to gravity (980 cm/s^2), then flow will be subcritical.

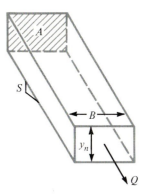

Figure P6.26

Use the graphical and bisection methods to determine y_n if $Q = 14.15$ m^3/s; $B = 4.572$ m; $n = 0.017$; and $S = 0.0015$. Compute whether flow is sub- or super-critical.

6.27 Figure P6.27a shows a uniform beam subject to a linearly increasing distributed load. The equation for the resulting elastic curve is (see Fig. P6.27b)

$$y = \frac{w_0}{120EIL}(-x^5 + 2L^2x^3 - L^4x) \qquad \text{(P6.27)}$$

If the derivative of the elastic curve is

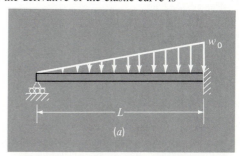

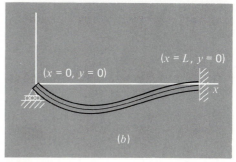

Figure P6.27

$$\frac{dy}{dx} = \frac{w_0}{120EIL}(-5x^4 + 6L^2x^2 - L^4)$$

use bisection to determine the point of maximum deflection (that is, the value of x where $dy/dx = 0$). Then substitute this value into Eq. (P6.27) to determine the value of the maximum deflection. Use the following parameter values in your computation: $L = 180$ in, $E = 29 \times 10^6$ lb/in^2, $I = 723$ in^4, and $w_0 = 12$ kips/ft. Express your results in inches.

6.28 In ocean engineering, the equation for a reflected standing wave in a harbor is given by $\lambda = 20$, $t = 10$, $v = 50$

$$h = h_0 \left[\sin\left(\frac{2\pi x}{\lambda}\right) \cos\left(\frac{2\pi t v}{\lambda}\right) + e^{-x} \right]$$

Solve for the lowest positive value of x if $h = 0.5h_0$.

6.29 The secant formula defines the force per unit area, P/A, that causes a maximum stress σ_m in a column of given slenderness ratio L_e/r:

$$\frac{P}{A} = \frac{\sigma_m}{1 + (ec/r^2)\sec[1/2(\sqrt{P/EA})(L_e/r)]}$$

If $E = 29 \times 10^3$ ksi, $ec/r^2 = 0.2$, and $\sigma_m = 36$ ksi, compute P/A for $L_e/r = 100$. [Hint: Recall that sec $x = 1/\cos x$.]

Electrical Engineering

6.30 Using your own software, reproduce the computation of Case Study 6.4.

6.31 Perform the same computation as in Case Study 6.4, but assume that the charge must be dissipated to 3% of its original value in 0.04 s.

6.32 Perform the same computation as in Case Study 6.4, but determine the time required for the circuit to dissipate to 15% of its original value, given $R = 300\,\Omega$, $C = 10^{-4}$ F, and $L = 4$ H.

6.33 Perform the same computation as in Case Study 6.4, but determine the value of L required for the circuit to dissipate to 1% of its original value in $t = 0.05$ s, given $R = 280\,\Omega$ and $C = 10^{-4}$ F.

6.34 An oscillating current in an electric circuit is described by

$$I = 10e^{-t}\sin(2\pi t)$$

where t is in seconds. Determine all values of t such that $I = 2$.

Mechanical Engineering

6.35 Using your own software, reproduce the computation performed in Case Study 6.5.

6.36 Perform the same computation as in Case Study 6.5, but use $c = 1.7 \times 10^7$ g/s, $k = 1.5 \times 10^9$ g/s^2, and $m = 2 \times 10^6$ g.

6.37 Perform the same computation as in Case Study 6.5, but determine the value of k so that the first root occurs at $t = 0.075$ s.

6.38 Perform the same computation as in Case Study 6.5, but determine the value of m so that the first root occurs at $t = 0.03$ s.

6.39 Perform the same computation as in Case Study 6.5, but determine the value of c so that the second root occurs at $t = 0.22$ s.

Spreadsheet

6.40 Use a fully active spreadsheet program such as the Electronic TOOLKIT or Lotus 123 to solve the following equations using the Newton-Raphson method. Explore the behavior of the system for various initial values of x, y, z.

$$f(x, y, z) = 4x + 5\sin y + 0.1z - 5 = 0$$

$$g(x, y, z) = x^2 + 2y + \exp(-0.5z) - 5 = 0$$

$$h(x, y, z) = x + y + z^2 - 12 = 0$$

6.41 Repeat Prob. 6.40 with

$$f(x, y) = 1 - x^2 - y^2 = 0$$

$$g(x, y) = 0.1 + x - y = 0$$

6.42 Change the function on the disk enclosed with the text by using $n = 1.5$ and $m = 2.5$. Examine the efficiency of the method for the following cases:
  (a) $x_{old} = 1$ and $y_{old} = 1$.
  (b) $x_{old} = 2$ and $y_{old} = 2$.
  (c) $x_{old} = 4$ and $y_{old} = 4$.
  (d) $x_{old} = 1.2$ and $y_{old} = 1.2$.
  (e) Explore other values of x_{old} and y_{old}. Discuss the convergence of the method based on the results of your observations.

6.43 Repeat Prob. 6.42 with $n = 0.5$ and $m = 0.5$ with your own initial values of x_{old} and y_{old}.

6.44 Repeat Prob. 6.43 with $n = 3$ and $m = 3$.

6.45 Repeat Prob. 6.43 with $n = 0.5$ and $m = 4$.

Miscellaneous

6.46 Read all the case studies in Chap. 6. On the basis of your reading and experience, make up your own case study for any one of the fields of engineering. This may involve modifying or reexpressing one of our case studies. However, it can also be totally original. As with our examples, it must be drawn from an engineering problem context and must demonstrate the use of the numerical methods for solving roots of equations. Write up your results using our case studies as models.

EPILOGUE: PART TWO

PT2.4 TRADE-OFFS

Table PT2.3 provides a summary of the trade-offs involved in solving for roots of algebraic and transcendental equations. Although graphical methods are time-consuming, they provide insight into the behavior of the function and are useful in identifying initial guesses and potential problems such as multiple roots. Therefore, if time permits, a quick sketch (or better yet, a computerized graph) can yield valuable information regarding the behavior of the function.

The numerical methods themselves are divided into two general categories: bracketing and open methods. The former require two initial guesses that are on either side of a root. This "bracketing" is maintained as the solution proceeds, and thus, these techniques are always convergent. However, a price is paid for this property in that the rate of convergence is relatively slow. Of the bracketing techniques, the false-position method is usually the method of preference because for most problems it converges much faster than bisection.

Open techniques differ from bracketing methods in that they use information at a single point (or two values that need not bracket the root) to extrapolate to a new root estimate. This property is a double-edged sword. Although it leads to quicker convergence, it also allows the possibility that the solution may diverge. In general, the convergence of open techniques is partially dependent on the quality of the initial guess and the nature of the function. The closer it is to the true root, the more likely the methods will converge.

Of the open techniques, the standard Newton-Raphson method is often used because of its property of quadratic convergence. However, its major shortcoming is that it requires that the derivative of the function be obtained analytically. For some functions this is impractical. In these cases, the secant method, which employs a finite-difference representation of the derivative, provides a viable alternative. Because of the finite-difference approximation, the rate of convergence of the secant method is initially slower than for the Newton-Raphson method. However, as the root estimate is refined, the difference approximation becomes a better representation of the true derivative, and convergence accelerates rapidly. The modified Newton-Raphson technique can be used to attain rapid convergence for multiple roots. However, this technique requires an analytical expression for both the first and second derivative.

All the numerical methods are easy to program on computers and require minimal

TABLE PT2.3   Comparison of the characteristics of alternative methods for finding roots of algebraic and transcendental equations. The comparisions are based on general experience and do not account for the behavior of specific functions.

Method	Initial Guesses	Relative rate of Convergence	Stability	Accuracy	Breadth of Application	Programming Effort	Comments
Direct	—	—	—	—	Very limited		
Graphical	—	—	—	Poor	General	—	May take more time than the numerical method
Bisection	2	Slow	Always converges	Good	General	Easy	
False position	2	Medium	Always converges	Good	General	Easy	
One–point iteration	1	Slow	May not converge	Good	General	Easy	
Newton-Raphson	1	Fast	May not converge	Good	Limited if f′(x) = 0	Easy	Requires evaluation of f′(x)
Modified Newton-Raphson	1	Fast for multiple roots; medium for single roots	May not converge	Good	Specifically designed for multiple roots	Easy	Requires evaluation of f″(x) and f′(x)
Secant	2	Medium to fast	May not converge	Good	General	Easy	Initial guesses do not have to bracket the root

time to determine a single root. On this basis, you might conclude that simple methods such as bisection would be good enough for practical purposes. This would be true if you were exclusively interested in determining the root of an equation once. However, there are many cases in engineering where numerous root locations are required and where speed becomes important. For these cases, slow methods are very time-consuming and, hence, costly. On the other hand, the fast open methods may diverge, and the accompanying delays can also be costly. Some computer algorithms attempt to capitalize on the strong points of both classes of techniques by initially employing a bracketing method to approach the root, then switching to an open method to rapidly refine the estimate. Whether a single approach or a combination is used, the trade-offs between convergence and speed are at the heart of the choice of a root-location technique.

PT2.5 IMPORTANT RELATIONSHIPS AND FORMULAS

Table PT2.4 summarizes important information that was presented in Part Two. This table can be consulted to quickly access important relationships and formulas.

PT2.6 ADVANCED METHODS AND ADDITIONAL REFERENCES

The methods in this text have focused on determining a single real root of an algebraic or transcendental equation based on foreknowledge of its approximate location. Additional references on the subject are Ralston and Rabinowitz (1978) and Carnahan, Luther, and Wilkes (1969).

As for specific techniques, the *Newton-Raphson method* can be employed in certain cases to locate complex roots on the basis of an initial complex guess. Because languages such as BASIC and Pascal cannot perform complex arithmetic, the technique is somewhat limited. However, Stark (1970) illustrates a way to circumvent this dilemma.

Muller's method is similar to the false-position method but uses quadratic rather than linear interpolation to locate the root. The approach can be employed to determine complex as well as real roots (Muller, 1956; Gerald and Wheatley, 1984; and Rice, 1983).

Several techniques are available to determine all the roots of polynomials. *Bairstow's method* will be reviewed in Chap. 21. In addition to Bairstow's method, several techniques are available to determine all the roots of polynomials. In particular, the *quotient-difference (QD) algorithm* (Henrici, 1964, and Gerald and Wheatley, 1984) determines all roots without initial guesses. Ralston and Rabinowitz (1978) and Carnahan, Luther, and Wilkes (1969) contain discussions of this method as well as of other techniques for locating roots of polynomials.

In summary, the foregoing is intended to provide you with avenues for deeper exploration of the subject. Additionally, all of the above references provide descriptions of the basic techniques covered in Part Two. We urge you to consult these alternative sources to broaden your understanding of numerical methods for root location.[*]

---

[*]Books are referenced only by author here; a complete bibliography will be found at the back of this text.

TABLE PT2.4 Summary of important information presented in Part Two.

Method	Formulation	Graphical Interpretation	Errors and Stopping Criteria
		Bracketing methods:	
Bisection	$$x_r = \frac{x_l + x_u}{2}$$ If $f(x_l)f(x_r) < 0$, $x_u = x_r$ $f(x_l)f(x_r) > 0$, $x_l = x_r$		Stopping criterion: $$\left| \frac{x_r^{new} - x_r^{old}}{x_r^{new}} \right| 100\% \leq \epsilon_s$$
False Position	$$x_r = x_u - \frac{f(x_u)(x_l - x_u)}{f(x_l) - f(x_u)}$$ If $f(x_l)f(x_r) < 0$, $x_u = x_r$ $f(x_l)f(x_r) > 0$, $x_l = x_r$		Stopping criterion: $$\left| \frac{x_r^{new} - x_r^{old}}{x_r^{new}} \right| 100\% \leq \epsilon_s$$
		Open methods:	
Newton-Raphson	$$x_{i+1} = x_i - \frac{f(x_i)}{f'(x_i)}$$		Stopping criterion: $$\left| \frac{x_{i+1} - x_i}{x_{i+1}} \right| 100\% \leq \epsilon_s$$ Error: $E_{i+1} = 0(E_i^2)$
Secant	$$x_{i+1} = x_i - \frac{f(x_i)(x_{i-1} - x_i)}{f(x_{i-1}) - f(x_i)}$$		Stopping criterion: $$\left| \frac{x_{i+1} - x_i}{x_{i+1}} \right| 100\% \leq \epsilon_s$$

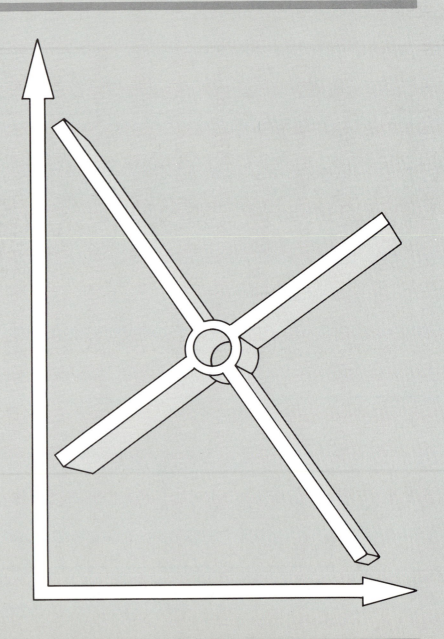

PART THREE

SYSTEMS OF LINEAR
ALGEBRAIC EQUATIONS

PT3.1 MOTIVATION

In Part Two, we determined the value x that satisfied a single equation, $f(x) = 0$. Now, we deal with the case of determining the values $x_1, x_2, \ldots, x_n$, that simultaneously satisfy a set of equations:

$$f_1(x_1, x_2, \ldots, x_n) = 0$$

$$f_2(x_1, x_2, \ldots, x_n) = 0$$

$$\vdots$$

$$f_n(x_1, x_2, \ldots, x_n) = 0$$

Such systems can be either linear or nonlinear. In Part Three, we deal with *linear algebraic equations* that are of the general form:

$$a_{11}x_1 + a_{12}x_2 + \cdots + a_{1n}x_n = c_1$$

$$a_{21}x_1 + a_{22}x_2 + \cdots + a_{2n}x_n = c_2$$

$$\vdots \qquad \qquad \qquad \qquad \vdots \qquad \qquad \text{(PT3.1)}$$

$$a_{n1}x_1 + a_{n2}x_2 + \cdots + a_{nn}x_n = c_n$$

where the a's are constant coefficients, the c's are constants, and n is the number of equations. All other equations are nonlinear. Nonlinear systems were discussed in Chap. 5 and will be covered briefly again in Chap. 7.

PT3.1.1 Precomputer Methods for Solving Systems of Equations

For small numbers of equations ($n \leq 3$), linear (and sometimes nonlinear) equations can be solved readily by simple techniques. Some of these methods will be reviewed at the beginning of Chap. 7. However, for four or more equations, solutions become arduous and computers must be utilized. Historically, the inability to solve all but the smallest sets of equations by hand has limited the scope of problems addressed in many engineering applications.

Before computers, techniques to solve systems of linear algebraic equations were time-consuming and awkward. These approaches placed a constraint on creativity because the methods were often difficult to implement and understand. Consequently, the techniques were sometimes overemphasized at the expense of other aspects of the problem-solving process such as formulation and interpretation (recall Fig. PT1.1 and accompanying discussion).

The advent of easily accessible computers makes it possible and practical for you to solve large sets of simultaneous linear algebraic equations. Thus, you can approach more complex and realistic examples and problems. Furthermore, you will have more time to test your creative skills because you will be able to place more emphasis on problem formulation and solution interpretation.

PT3.1.2 Linear Algebraic Equations and Engineering Practice

Many of the fundamental equations of engineering are based on conservation laws (recall Table 1.1). Some familiar quantities that conform to such laws are mass, energy, and momentum. In mathematical terms, these principles lead to balance or continuity equations that relate system *behavior* as represented by the *levels* or *response* of the quantity being modeled to the *properties* or *characteristics* of the system and the external *stimuli* or *forcing functions* acting on the system.

As an example, the conservation of mass can be used to formulate a mass balance for a series of chemical reactors (Fig. PT3.1*a*). For this case, the quantity being modeled is the mass of the chemical in each reactor. The system properties are the reaction characteristics of the chemical and the reactors' sizes and flow rates. The forcing functions are the feed rates of the chemical into the system.

Figure PT3.1
Two types of system that can be modeled using linear algebraic equations: (*a*) lumped-variable system that involves coupled finite components and (*b*) distributed-variable system that involves a continuum.

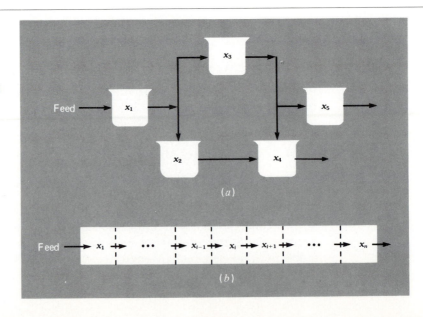

In the previous part of the book, you saw how single-component systems result in a single equation that can be solved using root-location techniques. Multicomponent systems result in a coupled set of mathematical equations that must be solved simultaneously. The equations are coupled because the individual parts of the system are influenced by other parts. For example, in Fig. PT3.1a, reactor 4 receives chemical inputs from reactors 2 and 3. Consequently, its response is dependent on the quantity of chemical in these other reactors.

When these dependencies are expressed mathematically, the resulting equations are often of the linear algebraic form of Eq. (PT3.1). The x's are usually measures of the magnitudes and responses of the individual components. Using Fig. PT3.1a as an example, x_1 might quantify the amount of mass in the first reactor, x_2 might quantify the amount in the second, and so forth. The a's typically represent the properties and characteristics that bear on the interactions between components. For instance, the a's for Fig. PT3.1a might be reflective of the flow rates of mass between the reactors. Finally, the c's usually represent the forcing functions acting on the system, such as the feed rate in Fig. PT3.1a. The case studies in Chap. 10 provide other examples of such equations derived from engineering practice.

Multicomponent problems of the above types arise from both *lumped* (macro-) or *distributed* (micro-) variable mathematical models (Fig. PT3.1). Lumped-variable problems involve coupled finite components. Examples include trusses (Case Study 10.3), reactors (Fig. PT3.1a and Case Study 10.2), and electric circuits (Case Study 10.4). These types of problems use models that provide little or no spatial detail.

Conversely, distributed-variable problems attempt to describe spatial detail of systems on a continuous or semicontinuous basis. The distribution of chemicals along the length of an elongated, rectangular reactor (Fig. PT3.1b) is an example of a continuous-variable model. Differential equations derived from the conservation laws specify the distribution of the dependent variable for such systems. These differential equations can be solved numerically by converting them to an equivalent system of simultaneous algebraic equations. The solution of such sets of equations represents a major engineering application area for the methods in the following chapters. These equations are coupled because the variables at one location are dependent on the variables in adjoining regions. For example, the concentration at the middle of the reactor is a function of the concentration in adjoining regions. Similar examples could be developed for the spatial distribution of temperature or momentum. We will address such problems when we discuss differential equations later in the book.

Aside from physical systems, simultaneous linear algebraic equations also arise in a variety of mathematical-problem contexts. These result when mathematical functions are required to satisfy several conditions simultaneously. Each condition results in an equation that contains known coefficients and unknown variables. The techniques discussed in this part can be used to solve for the unknowns when the equations are linear and algebraic. Some widely used numerical techniques that employ simultaneous equations are regression analysis (Chap. 11) and spline interpolation (Chap. 12).

PT3.2 MATHEMATICAL BACKGROUND

All parts of this book require some mathematical background. For Part Three, matrix notation and algebra are useful because they provide a concise way to represent and manipulate systems of linear algebraic equations. If you are already familiar with matrices, feel free to skip to Sec. PT3.3. For those who are unfamiliar or require a review, the following material provides a brief introduction to the subject.

PT3.2.1 Matrix Notation

A *matrix* consists of a rectangular array of elements represented by a single symbol. As depicted in Fig. PT3.2, [A] is the shorthand notation for the matrix and a_{ij} designates an individual *element* of the matrix.

Figure PT3.2
A matrix.

A horizontal set of elements is called a *row* and a vertical set is called a *column*. The first subscript i always designates the number of the row in which the element lies. The second subscript j designates the column. For example, element a_{23} is in row 2 and column 3.

The matrix in Fig. PT3.2 has m rows and n columns and is said to have a *dimension* of m by n (or $m \times n$). It is referred to as an *m-by-n* matrix.

Matrices with row dimension $m = 1$, such as

$$[B] = [b_1 \quad b_2 \cdots b_n]$$

are called *row vectors*. Note that for simplicity, the first subscript of each element is dropped. Also, it should be mentioned that there are times when it is desirable to employ a special shorthand notation to distinguish a row matrix from other types of matrices. One way to accomplish this is to employ special open-topped brackets as in $\lfloor B \rfloor$.

Matrices with column dimension $n = 1$, such as

$$[C] = \begin{bmatrix} c_1 \\ c_2 \\ \cdot \\ \cdot \\ \cdot \\ c_m \end{bmatrix}$$

are referred to as *column vectors.* For simplicity, the second subscript is dropped. As with the row vector, there are occasions when it is desirable to employ a special shorthand notation to distinguish a column matrix from other types of matrices. One way to accomplish this is to employ special brackets as in $\{B\}$.

Matrices where $m = n$ are called *square matrices.* For example, a 4-by-4 matrix is

$$[A] = \begin{bmatrix} a_{11} & a_{12} & a_{13} & a_{14} \\ a_{21} & a_{22} & a_{23} & a_{24} \\ a_{31} & a_{32} & a_{33} & a_{34} \\ a_{41} & a_{42} & a_{43} & a_{44} \end{bmatrix}$$

The diagonal consisting of the elements a_{11}, a_{22}, a_{33}, and a_{44} is termed the *principal* or *main diagonal* of the matrix.

Square matrices are particularly important when solving sets of simultaneous linear equations. For such systems, the number of equations (corresponding to rows) and the number of unknowns (corresponding to columns) must be equal in order for a unique solution to be possible. Consequently, square matrices of coefficients are encountered when dealing with such systems. Some special types of square matrices are described in Box PT3.1.

Box PT3.1 Special Types of Square Matrices

There are a number of special forms of square matrices that are important and should be noted:

A *symmetric matrix* is one where $a_{ij} = a_{ji}$ for all i's and j's. For example,

$$[A] = \begin{bmatrix} 5 & 1 & 2 \\ 1 & 3 & 7 \\ 2 & 7 & 8 \end{bmatrix}$$

is a 3-by-3 symmetric matrix.

A *diagonal matrix* is a square matrix where all elements off the main diagonal are equal to zero, as in

$$[A] = \begin{bmatrix} a_{11} & & & \\ & a_{22} & & \\ & & a_{33} & \\ & & & a_{44} \end{bmatrix}$$

Note that where large blocks of elements are zero, they are left blank.

An *identity matrix* is a diagonal matrix where all elements on the main diagonal are equal to 1, as in

$$[I] = \begin{bmatrix} 1 & & & \\ & 1 & & \\ & & 1 & \\ & & & 1 \end{bmatrix}$$

The symbol $[I]$ is used to denote the identity matrix. The identity matrix has properties similar to unity.

An *upper triangular matrix* is one where all the elements below the main diagonal are zero, as in

$$[A] = \begin{bmatrix} a_{11} & a_{12} & a_{13} & a_{14} \\ & a_{22} & a_{23} & a_{24} \\ & & a_{33} & a_{34} \\ & & & a_{44} \end{bmatrix}$$

A *lower triangular matrix* is one where all elements above the main diagonal are zero, as in

$$[A] = \begin{bmatrix} a_{11} & & & \\ a_{21} & a_{22} & & \\ a_{31} & a_{32} & a_{33} & \\ a_{41} & a_{42} & a_{43} & a_{44} \end{bmatrix}$$

A *banded matrix* has all elements equal to zero, with the exception of a band centered on the main diagonal:

$$[A] = \begin{bmatrix} a_{11} & a_{12} & & \\ a_{21} & a_{22} & a_{23} & \\ & a_{32} & a_{33} & a_{34} \\ & & a_{43} & a_{44} \end{bmatrix}$$

The above matrix has a bandwidth of 3 and is given a special name—the *tridiagonal matrix.*

PT3.2.2 Matrix Operating Rules

Now that we have specified what we mean by a matrix, we can define some operating rules that govern its use. Two *m-by-n* matrices are equal if, and only if, every element in the first is equal to every element in the second; that is $[A] = [B]$ if $a_{ij} = b_{ij}$ for all i and j.

Addition of two matrices, say $[A]$ and $[B]$, is accomplished by adding corresponding terms in each matrix. The elements of the resulting matrix $[C]$ are computed as

$$c_{ij} = a_{ij} + b_{ij}$$

for $i = 1, 2, \ldots, m$ and $j = 1, 2, \ldots, n$. Similarly, the *subtraction* of two matrices, say $[E]$ minus $[F]$, is obtained by subtracting corresponding terms, as in

$$d_{ij} = e_{ij} - f_{ij}$$

for $i = 1, 2, \ldots, m$ and $j = 1, 2, \ldots, n$. It follows directly from the above definitions that addition and subtraction can only be performed between matrices having the same dimensions.

Both addition and subtraction are *commutative:*

$$[A] + [B] = [B] + [A]$$

and

$$[E] - [F] = -[F] + [E]$$

Addition and subtraction are also *associative,* that is,

$$[A] + ([B] + [C]) = ([A] + [B]) + [C]$$

The *multiplication* of a matrix $[A]$ by a *scalar g* is obtained by multiplying every element of $[A]$ by g, as in

$$[B] = g[A] = \begin{bmatrix} ga_{11} & ga_{12} & \cdot & \cdot & ga_{1n} \\ ga_{21} & ga_{22} & \cdot & \cdot & ga_{2n} \\ \cdot & & \cdot & & \cdot \\ \cdot & & \cdot & & \cdot \\ \cdot & & \cdot & & \cdot \\ ga_{m1} & ga_{m2} & \cdot & \cdot & ga_{mn} \end{bmatrix}$$

The *product* of two matrices is represented as $[C] = [A][B]$, where the elements of $[C]$ are defined as (see Box PT3.2 on p. 210 for a simple way to conceptualize matrix multiplication)

$$c_{ij} = \sum_{k=1}^{n} a_{ik}b_{kj} \tag{PT3.2}$$

where n = the column dimension of $[A]$ and the row dimension of $[B]$. That is, the c_{ij} element is obtained by adding the product of individual elements from the ith row of the first matrix, in this case $[A]$, by the jth column of the second matrix $[B]$.

According to this definition, multiplication of two matrices can only be performed if the first matrix has as many columns as the number of rows in the second matrix. Thus, if [A] is an m-by-n matrix, [B] could be an n-by-l matrix. For this case, the resulting [C] matrix would have dimension of m by l. However, if [B] were a l-by-n matrix, the multiplication could not be performed. Figure PT3.3 provides an easy way to check whether two matrices can be multiplied.

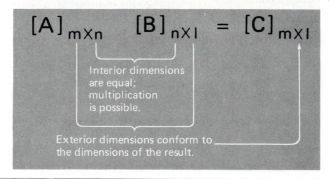

Figure PT3.3
A simple way to check whether matrix multiplication is possible.

If the dimensions of the matrices are suitable, matrix multiplication is *associative,*

$$([A][B])[C] = [A]([B][C])$$

and *distributive,*

$$[A]([B] + [C]) = [A][B] + [A][C]$$

or

$$([A] + [B])[C] = [A][C] + [B][C]$$

However, multiplication is not generally *commutative:*

$$[A][B] \neq [B][A]$$

That is, *the order of multiplication is important.*

Figure PT3.4 shows pseudocode to multiply an m-by-n matrix, [A], by an n-by-l matrix, [B], and store the result in an m-by-l matrix, [C]. Notice that, instead of the inner product being directly accumulated in [C], it is collected in a temporary variable, sum. This is done for two reasons. First, it is a bit more efficient,

Figure PT3.4
Pseudocode to multiply an m-by-n matrix [A] by an n-by-l matrix [B] and store the result in an m-by-l matrix [C].

```
DOFOR i = 1 to m
    DOFOR j = 1 to l
        sum = 0
        DOFOR k = 1 to n
            sum = sum + a  · b
                         i,k   k,j
        ENDDO
        c    = sum
         i,j
    ENDDO
ENDDO
```

because the computer need determine the location of $c_{i,j}$ only $m \times l$ times rather than $m \times l \times n$ times. Second, the precision of the multiplication can be greatly improved by declaring sum as a double-precision variable (recall the discussion of inner products in Sec. 3.4.2).

Although multiplication is possible, matrix division is not a defined operation. However, if a matrix $[A]$ is square and nonsingular, there is another matrix $[A]^{-1}$, called the *inverse* of $[A]$, for which

$$[A][A]^{-1} = [A]^{-1}[A] = [I] \tag{PT3.3}$$

Thus, the multiplication of a matrix by the inverse is analogous to division, in the sense that a number divided by itself is equal to 1. That is, multiplication of a matrix by its inverse leads to the identity matrix (recall Box PT3.1).

The inverse of a two-dimensional square matrix can be represented simply by

$$[A]^{-1} = \frac{1}{a_{11}a_{22} - a_{12}a_{21}} \begin{bmatrix} a_{22} & -a_{12} \\ -a_{21} & a_{11} \end{bmatrix} \tag{PT3.4}$$

Similar formulas for higher-dimensional matrices are much more involved. Sections in Chaps. 8 and 9 will be devoted to techniques for using numerical methods and the computer to calculate the inverse for such systems.

Two other matrix manipulations that will have utility in our discussion are the transpose and the trace of a matrix. The *transpose* of a matrix involves transforming its rows into columns and its columns into rows. For the matrix

$$[A] = \begin{bmatrix} a_{11} & a_{12} & \cdot & \cdot & \cdot & a_{1n} \\ a_{21} & a_{22} & \cdot & \cdot & \cdot & a_{2n} \\ \cdot & \cdot & & & & \cdot \\ \cdot & \cdot & & & & \cdot \\ \cdot & \cdot & & & & \cdot \\ a_{m1} & a_{m2} & \cdot & \cdot & \cdot & a_{mn} \end{bmatrix}$$

the transpose, designated $[A]^T$, is defined as

$$[A]^T = \begin{bmatrix} a_{11} & a_{21} & \cdot & \cdot & \cdot & a_{m1} \\ a_{12} & a_{22} & \cdot & \cdot & \cdot & a_{m2} \\ \cdot & \cdot & & & & \cdot \\ \cdot & \cdot & & & & \cdot \\ \cdot & \cdot & & & & \cdot \\ a_{1n} & a_{2n} & \cdot & \cdot & \cdot & a_{mn} \end{bmatrix}$$

In other words, the element a_{ij} of the transpose is equal to the a_{ji} element of the original matrix.

The transpose has a variety of functions in matrix algebra. One simple advantage is that it allows a column vector to be written as a row. For example, if

$$\{C\} = \begin{bmatrix} c_1 \\ c_2 \\ c_3 \\ c_4 \end{bmatrix}$$

then

$$\{C\}^T = [c_1 \quad c_2 \quad c_3 \quad c_4]$$

where the superscript T designates the transpose. For example, this can save space when writing a column vector in a manuscript. In addition, the transpose has numerous mathematical applications.

The *trace* of a matrix is the sum of the elements on its principal diagonal. It is designated as tr[A] and is computed as

$$\text{tr}[A] = \sum_{i=1}^{n} a_{ii}$$

The trace will figure prominently in our discussion of eigenvalues in Chap. 21.

The final matrix manipulation that will have utility in our discussion is augmentation. A matrix is *augmented* by the addition of a column (or columns) to the original matrix. For example, suppose we have a matrix of coefficients:

$$[A] = \begin{bmatrix} a_{11} & a_{12} & a_{13} \\ a_{21} & a_{22} & a_{23} \\ a_{31} & a_{32} & a_{33} \end{bmatrix}$$

We might wish to augment this matrix [A] with an identity matrix (recall Box PT3.1) to yield a 3-by-6 dimensional matrix:

$$\begin{bmatrix} a_{11} & a_{12} & a_{13} & | & 1 & 0 & 0 \\ a_{21} & a_{22} & a_{23} & | & 0 & 1 & 0 \\ a_{31} & a_{32} & a_{33} & | & 0 & 0 & 1 \end{bmatrix}$$

Such an expression has utility when we must perform a set of identical operations on two matrices. Thus, we can perform the operations on the single augmented matrix rather than on the two individual matrices.

PT3.2.3 Representing Simultaneous Linear Algebraic Equations in Matrix Form

It should be clear that matrices provide a concise notation for representing simultaneous linear equations. For example, Eq. (PT3.1) can be expressed as

$$[A]\{X\} = \{C\} \tag{PT3.5}$$

where [A] is the n-by-n square matrix of coefficients,

$$[A] = \begin{bmatrix} a_{11} & a_{12} & \cdot & \cdot & \cdot & a_{1n} \\ a_{21} & a_{22} & \cdot & \cdot & \cdot & a_{2n} \\ \cdot & \cdot & & & & \cdot \\ \cdot & \cdot & & & & \cdot \\ \cdot & \cdot & & & & \cdot \\ a_{n1} & a_{n2} & \cdot & \cdot & \cdot & a_{nn} \end{bmatrix}$$

$\{C\}$ is the n-by-1 column vector of constants,

$$\{C\}^T = [c_1 \quad c_2 \cdots c_n]$$

and $\{X\}$ is the n-by-1 column vector of unknowns:

$$\{X\}^T = [x_1 \quad x_2 \cdots x_n]$$

Recall the definition of matrix multiplication [Eq. (PT3.2) or Box PT3.2] to convince yourself that Eqs. (PT3.1) and (PT3.5) are equivalent. Also, realize that Eq. (PT3.5) is a valid matrix multiplication because the number of columns (n) of the first matrix ([A]) is equal to the number of rows (n) of the second matrix ($\{X\}$).

Box PT3.2   *A Simple Method for Multiplying Two Matrices*

Although Eq. (PT3.2) is well-suited for implementation on a computer, it is not the simplest means for visualizing the mechanics of multiplying two matrices. What follows gives more tangible expression to the operation. Suppose that we want to multiply [A] by [B] to yield [C]:

$$[C] = [A][B] = \begin{bmatrix} 3 & 1 \\ 8 & 6 \\ 0 & 4 \end{bmatrix} \begin{bmatrix} 5 & 9 \\ 7 & 2 \end{bmatrix}$$

A simple way to represent the computation of [C] is to raise [B], as in

$$\begin{bmatrix} 5 & 9 \\ 7 & 2 \end{bmatrix} \quad \leftarrow [B]$$

$$[A] \rightarrow \begin{bmatrix} 3 & 1 \\ 8 & 6 \\ 0 & 4 \end{bmatrix} \begin{bmatrix} & ? & \end{bmatrix} \quad \leftarrow [C]$$

Now the answer [C] can be computed in the space vacated by [B]. This format has utility because it aligns the appropriate rows and columns that are to be multiplied. For example, according to Eq. (PT3.2) the element c_{11} is obtained by multiplying the first row of [A] by the first column of [B]. This amounts to adding the product of a_{11} and b_{11} to the product of a_{12} and b_{21}, as in

Thus, c_{11} is equal to 22. Element c_{21} can be computed in a similar fashion, as in

The computation can be continued in this way, following the alignment of the rows and columns, to yield the result:

$$[C] = \begin{bmatrix} 22 & 29 \\ 82 & 84 \\ 28 & 8 \end{bmatrix}$$

Note how this simple method makes it clear why it is impossible to multiply two matrices if the number of columns of the first matrix does not equal the number of rows in the second matrix. Also, note how it demonstrates that the order of multiplication matters. Problem 7.3 also illustrates these points.

This part of the book is devoted to solving Eq. (PT3.5) for $\{X\}$. A formal way to obtain a solution using matrix algebra is to multiply each side of the equation by the inverse of [A] to yield

$$[A]^{-1}[A]\{X\} = [A]^{-1}\{C\}$$

Because $[A]^{-1}[A]$ equals the identity matrix, the equation becomes

$$\{X\} = [A]^{-1}\{C\} \tag{PT3.6}$$

Therefore, the equation has been solved for $\{X\}$. This is another example of how the inverse plays a role in matrix algebra that is similar to division. (It should be noted that this is not a very efficient way to solve a system of equations. Thus, other approaches are employed in numerical algorithms. However, as discussed in Chap. 8, the matrix inverse itself has great value in the engineering analyses of such systems.)

Finally, we will sometimes find it useful to augment $[A]$ with $\{C\}$. For example, if $n = 3$, this results in a 3-by-4 dimensional matrix:

$$[A] = \begin{bmatrix} a_{11} & a_{12} & a_{13} & | & c_1 \\ a_{21} & a_{22} & a_{23} & | & c_2 \\ a_{31} & a_{32} & a_{33} & | & c_3 \end{bmatrix} \tag{PT3.7}$$

Expressing the equations in this form has utility, because several of the techniques for solving linear systems perform identical operations on a row of coefficients and the corresponding right-hand-side constant. As expressed in Eq. (PT3.7), we can perform the manipulation once on an individual row of the augmented matrix rather than separately on the coefficient matrix and the right-hand-side vector.

PT3.3 ORIENTATION

Before proceeding to the numerical methods, some further orientation might be helpful. The following is intended as an overview of the material discussed in Part Three. In addition, we have formulated some objectives to help focus your efforts when studying the material.

PT3.3.1 Scope and Preview

Figure PT3.5 provides an overview for Part Three. *Chapter 7* is devoted to the most fundamental technique for solving linear algebraic systems: Gauss elimination. Before launching into a detailed discussion of this technique, a preliminary section deals with simple methods for solving small systems. These approaches are presented to provide you with visual insight and because one of the methods—the elimination of unknowns—represents the basis for Gauss elimination.

After the preliminary material, "naive" Gauss elimination is discussed. We start with this "stripped-down" version because it allows the fundamental technique to be elaborated without complicating details. Then, in subsequent sections, we discuss potential problems of the naive approach and present a number of modifications to minimize and circumvent these problems. The focus of this discussion will be the process of switching rows or "partial pivoting."

Chapter 8 begins with a discussion of the Gauss-Jordan method. Although this technique is quite similar to that of Gauss elimination, it is presented because

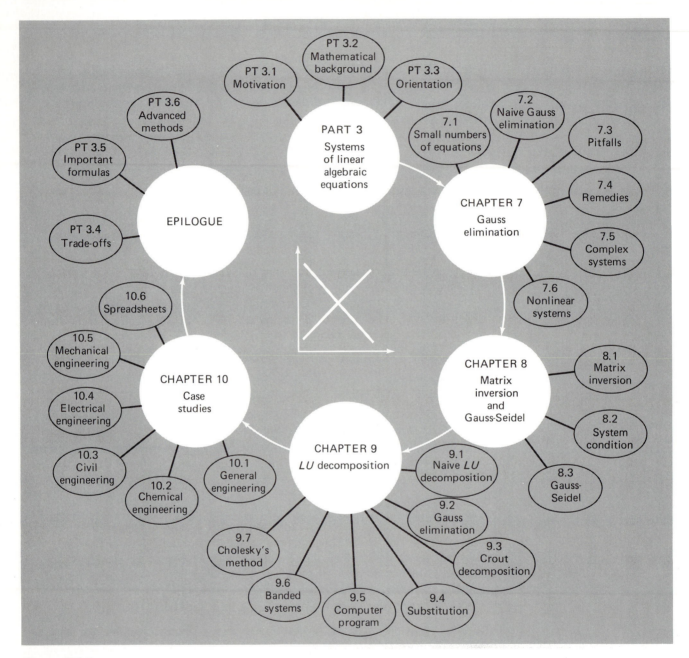

Figure PT3.5
Schematic of the organization of the material in Part Three: Systems of Linear Algebraic Equations.

it allows the computation of the matrix inverse which has tremendous utility in engineering practice.

Both the Gauss elimination and the Gauss-Jordan methods are exact techniques. Chapter 8 concludes with an alternative type of approach called the Gauss-Seidel method. This technique is similar in spirit to the approximate methods for roots of equations that were discussed in Chap. 5. That is, the technique involves guessing a solution and then iterating to obtain a refined estimate.

Chapter 9 is devoted to *LU* decomposition methods. This chapter begins by illustrating how Gauss elimination can be formulated as an *LU* decomposition solution. In addition, other popular *LU* decomposition approaches such as Crout and Cholesky decomposition are introduced. Finally, a short section on *LU* decomposition schemes for banded systems is included.

Chapter 10 demonstrates how the methods can actually be applied for problem solving. As with other parts of the book, case studies are drawn from all fields of engineering. In addition, we include a case study on the use of spreadsheets to solve linear algebraic equations.

Finally, an epilogue is included at the end of Part Three. This review includes discussion of trade-offs that are relevant to implementation of the methods in engineering practice. This section also summarizes the important formulas and advanced methods related to linear algebraic equations. As such, it can be used before exams or as a refresher after you have graduated and must return to linear algebraic equations as a professional.

PT3.3.2   Goals and Objectives

Study objectives.   After completing Part Three, you should be able to solve problems involving linear algebraic equations and appreciate the application of these equations in many fields of engineering. You should strive to master several techniques and assess their reliability. You should understand the trade-offs involved in selecting the "best" method (or methods) for any particular problem. In addition to these general objectives, the specific concepts listed in Table PT3.1 should be assimilated and mastered.

Computer objectives.   Your most fundamental computer objective is to be able to use a program to successfully solve systems of linear algebraic equations. You should understand how to use the Electronic TOOLKIT software or, alternatively, how to use the computer program for naive Gauss elimination given in this book. These programs will allow you to adequately handle many practical problems that involve several simultaneous linear algebraic equations. As you progress to problems that contain more equations, you can utilize the other programs and algorithms provided in Part Three. Eventually you may want to incorporate partial pivoting, determinant calculation, and condition evaluation into your programs. You may also want to have your own software for the Gauss-Seidel and *LU* decomposition methods. Additionally, you may develop the capability to calculate the matrix inverse.

TABLE PT3.1   Specific study objectives for Part Three

1. Understand the graphical interpretation of ill-conditioned systems and how it relates to the determinant
2. Be familiar with terminology: forward elimination, back substitution, pivot equation, and pivot coefficient
3. Understand the problems of division by zero, round-off error, and ill conditioning
4. Know how to compute the determinant using Gauss elimination
5. Understand the advantages of pivoting; realize the difference between partial and complete pivoting
6. Know the fundamental difference between Gauss elimination and the Gauss-Jordan method and which is more efficient
7. Understand how the Gauss-Jordan method is used to compute the matrix inverse
8. Know how to interpret the elements of the matrix inverse in evaluating stimulus-response computations in engineering
9. Realize how to use the inverse and matrix norms to evaluate system condition
10. Understand why the Gauss-Seidel method is particularly well-suited for large, sparse systems of equations
11. Know how to assess diagonal dominance of a system of equations and how it relates to whether the system can be solved with the Gauss-Seidel method
12. Understand the rationale behind relaxation; know where underrelaxation and overrelaxation are appropriate
13. Recognize how Gauss elimination can be formulated as an *LU* decomposition
14. Understand the difference between Crout and Doolittle decomposition.
15. Know how to incorporate pivoting and matrix inversion into an *LU* decomposition algorithm.
16. Understand how banded and symmetric systems can be decomposed and solved efficiently.

Finally, spreadsheet software to implement Gauss-Seidel is contained on the diskette that accompanies the book. This software is designed to be used in conjunction with Case Study 10.6 which illustrates how systems of linear algebraic equations can be implemented on a spreadsheet.

CHAPTER 7
Gauss Elimination

The present chapter deals with simultaneous linear algebraic equations which can be represented generally as

$$a_{11}x_1 + a_{12}x_2 + \cdots + a_{1n}x_n = c_1$$

$$a_{21}x_1 + a_{22}x_2 + \cdots + a_{2n}x_n = c_2$$

$$\cdot \qquad \cdot \qquad\qquad \cdot \qquad \cdot$$

$$\cdot \qquad \cdot \qquad\qquad\quad \cdot \qquad \cdot \qquad\qquad (7.1)$$

$$\cdot \qquad \cdot \qquad\qquad\quad \cdot \qquad \cdot$$

$$a_{n1}x_1 + a_{n2}x_2 + \cdots + a_{nn}x_n = c_n$$

where the a's are constant coefficients and the c's are constants.

The technique described in this chapter is called *Gauss elimination* because it involves combining equations in order to eliminate unknowns. Although it is one of the earliest methods for solving simultaneous equations, it remains among the most important algorithms in use today. In particular, it is easy to program and apply using personal computers.

7.1 SOLVING SMALL NUMBERS OF EQUATIONS

Before proceeding to the computer methods, we will describe several methods that are appropriate for solving small ($n \leq 3$) sets of simultaneous equations and that do not require a computer. These are the graphical method, Cramer's rule, and the elimination of unknowns.

7.1.1 The Graphical Method

A graphical solution is obtainable for two equations by plotting them on cartesian coordinates with one axis corresponding to x_1 and the other to x_2. Because we are dealing with linear systems, each equation is a straight line. This can be easily illustrated for the general equations

$$a_{11}x_1 + a_{12}x_2 = c_1$$

$$a_{21}x_1 + a_{22}x_2 = c_2$$

215

Both equations can be solved for x_2:

$$x_2 = -\left(\frac{a_{11}}{a_{12}}\right)x_1 + \frac{c_1}{a_{12}}$$

$$x_2 = -\left(\frac{a_{21}}{a_{22}}\right)x_1 + \frac{c_2}{a_{22}}$$

Thus, the equations are now in the form of straight lines; that is, $x_2 = (\text{slope})\, x_1 +$ intercept. These lines can be graphed on cartesian coordinates with x_2 as the ordinate and x_1 as the abscissa. The values of x_1 and x_2 at the intersection of the lines represent the solution.

EXAMPLE 7.1   The Graphical Method for Two Equations

Problem Statement: Use the graphical method to solve

$$3x_1 + 2x_2 = 18 \tag{E7.1.1}$$

$$-x_1 + 2x_2 = 2 \tag{E7.1.2}$$

Solution: Let x_1 be the abscissa. Solve Eq. (E7.1.1) for x_2:

$$x_2 = (-3/2)x_1 + 9$$

which, when plotted on Fig. 7.1, is a straight line with an intercept of 9 and a slope of $-3/2$.

Equation (E7.1.2) can also be solved for x_2:

$$x_2 = (1/2)x_1 + 1$$

which is also plotted on Fig. 7.1. The solution is the intersection of the two lines at $x_1 = 4$ and $x_2 = 3$. This result can be checked by substituting these values into the original equations to yield

$$3(4) + 2(3) = 18$$

$$-4 + 2(3) = 2$$

Thus, the results are equivalent to the right-hand sides of the original equations.

For three simultaneous equations, each equation would be represented by a plane in a three-dimensional coordinate system. The point where the three planes intersect would represent the solution. Beyond three equations, graphical methods break down and, consequently, have little practical value for solving simultaneous equations. However, they sometimes prove useful in visualizing properties of the solutions. For example, Fig. 7.2 depicts three cases which can pose problems when solving sets of linear equations. Figure 7.2a shows the case where the two equations represent parallel lines. For such situations, there is no solution because the lines never cross. Figure 7.2b depicts the case where the two lines are coincident. For such situations there is an infinite number of solutions. Both types of systems are said to be *singular*. In addition, systems that are

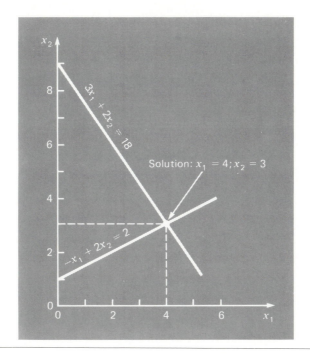

Figure 7.1
Graphical solution of a set of two simultaneous linear algebraic equations. The intersection of the lines represents the solution.

very close to being singular (Fig. 7.2c) can also cause problems. These systems are said to be *ill-conditioned*. Graphically, this corresponds to the fact that it is difficult to identify the exact point at which the lines intersect. Ill-conditioned systems will also pose problems when they are encountered during the numerical solution of linear equations. This is due to the fact that they will be extremely sensitive to round-off error (recall Sec. 3.6.3).

Figure 7.2
Graphical depiction of singular and ill-conditioned systems: (a) no solution, (b) infinite solutions, and (c) ill-conditioned system where the slopes are so close that the point of intersection is difficult to detect visually.

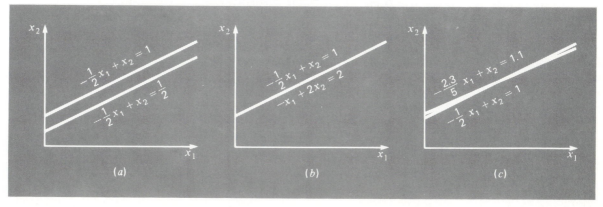

7.1.2 Determinants and Cramer's Rule

Cramer's rule is another solution technique that is best suited to small numbers of equations. Before describing this method, we will briefly introduce the concept of the determinant, which is used to implement Cramer's rule. In addition, the determinant has relevance to the evaluation of ill conditioning of a matrix.

Determinants. The determinant can be illustrated for a set of three equations:

$$a_{11}x_1 + a_{12}x_2 + a_{13}x_3 = c_1$$

$$a_{21}x_1 + a_{22}x_2 + a_{23}x_3 = c_2$$

$$a_{31}x_1 + a_{32}x_2 + a_{33}x_3 = c_3$$

or, in matrix form,

$$[A]\{X\} = \{C\}$$

where $[A]$ is the coefficient matrix:

$$[A] = \begin{bmatrix} a_{11} & a_{12} & a_{13} \\ a_{21} & a_{22} & a_{23} \\ a_{31} & a_{32} & a_{33} \end{bmatrix}$$

The *determinant* D of this system is formed from the coefficients of the equation, as in

$$D = \begin{vmatrix} a_{11} & a_{12} & a_{13} \\ a_{21} & a_{22} & a_{23} \\ a_{31} & a_{32} & a_{33} \end{vmatrix} \tag{7.2}$$

Although the determinant D and the coefficient matrix $[A]$ are composed of the same elements, they are completely different mathematical concepts. That is why they are distinguished visually by using brackets to enclose the matrix and straight lines to enclose the determinant. In contrast to a matrix, the determinant is a single number. For example, the value of the second-order determinant

$$D = \begin{vmatrix} a_{11} & a_{12} \\ a_{21} & a_{22} \end{vmatrix}$$

is calculated by

$$D = a_{11}a_{22} - a_{12}a_{21} \tag{7.3}$$

For the third-order case [Eq. (7.2)], a single numerical value for the determinant can be computed as

$$D = a_{11} \begin{vmatrix} a_{22} & a_{23} \\ a_{32} & a_{33} \end{vmatrix} - a_{12} \begin{vmatrix} a_{21} & a_{23} \\ a_{31} & a_{33} \end{vmatrix} + a_{13} \begin{vmatrix} a_{21} & a_{22} \\ a_{31} & a_{32} \end{vmatrix} \tag{7.4}$$

where the 2-by-2 determinants are called *minors*.

EXAMPLE 7.2   Determinants

Problem Statement: Compute values for the determinants of the systems represented in Figs. 7.1 and 7.2.

Solution: For Fig. 7.1:

$$D = \begin{vmatrix} 3 & 2 \\ -1 & 2 \end{vmatrix} = 3(2) - 2(-1) = 8$$

For Fig. 7.2a:

$$D = \begin{vmatrix} -1/2 & 1 \\ -1/2 & 1 \end{vmatrix} = \frac{-1}{2}(1) - 1\left(\frac{-1}{2}\right) = 0$$

For Fig. 7.2b:

$$D = \begin{vmatrix} -1/2 & 1 \\ -1 & 2 \end{vmatrix} = \frac{-1}{2}(2) - 1(-1) = 0$$

For Fig. 7.2c:

$$D = \begin{vmatrix} -1/2 & 1 \\ -2.3/5 & 1 \end{vmatrix} = \frac{-1}{2}(1) - 1\left(\frac{-2.3}{5}\right) = -0.04$$

In the foregoing example, the singular systems had zero determinants. Additionally, the results suggest that the system which is almost singular (Fig. 7.2c) has a determinant that is close to zero. These ideas will be pursued further in our subsequent discussion of ill conditioning (Sec. 7.3.3).

Cramer's Rule.   This rule states that each unknown in a system of linear algebraic equations may be expressed as a fraction of two determinants with denominator D and with the numerator obtained from D by replacing the column of coefficients of the unknown in question by the constants $c_1, c_2, \ldots, c_n$. For example, x_1 would be computed as

$$x_1 = \frac{\begin{vmatrix} c_1 & a_{12} & a_{13} \\ c_2 & a_{22} & a_{23} \\ c_3 & a_{32} & a_{33} \end{vmatrix}}{D} \tag{7.5}$$

EXAMPLE 7.3   Cramer's Rule

Problem Statement: Use Cramer's rule to solve

$$0.3x_1 + 0.52x_2 + x_3 = -0.01$$
$$0.5x_1 + x_2 + 1.9x_3 = 0.67$$
$$0.1x_1 + 0.3x_2 + 0.5x_3 = -0.44$$

Solution: The determinant D can be written as [Eq. (7.2)]

$$D = \begin{vmatrix} 0.3 & 0.52 & 1 \\ 0.5 & 1 & 1.9 \\ 0.1 & 0.3 & 0.5 \end{vmatrix}$$

The minors are [Eq. (7.3)]

$$A_1 = \begin{vmatrix} 1 & 1.9 \\ 0.3 & 0.5 \end{vmatrix} = 1(0.5) - 1.9(0.3) = -0.07$$

$$A_2 = \begin{vmatrix} 0.5 & 1.9 \\ 0.1 & 0.5 \end{vmatrix} = 0.5(0.5) - 1.9(0.1) = 0.06$$

$$A_3 = \begin{vmatrix} 0.5 & 1 \\ 0.1 & 0.3 \end{vmatrix} = 0.5(0.3) - 1(0.1) = 0.05$$

These can be used to evaluate the determinant, as in [Eq. (7.4)]

$$D = 0.3(-0.07) - 0.52(0.06) + 1(0.05) = -0.0022$$

Applying Eq. (7.5), the solution is

$$x_1 = \frac{\begin{vmatrix} -0.01 & 0.52 & 1 \\ 0.67 & 1 & 1.9 \\ -0.44 & 0.3 & 0.5 \end{vmatrix}}{-0.0022} = \frac{0.03278}{-0.0022} = -14.9$$

$$x_2 = \frac{\begin{vmatrix} 0.3 & -0.01 & 1 \\ 0.5 & 0.67 & 1.9 \\ 0.1 & -0.44 & 0.5 \end{vmatrix}}{-0.0022} = \frac{0.0649}{-0.0022} = -29.5$$

$$x_3 = \frac{\begin{vmatrix} 0.3 & 0.52 & -0.01 \\ 0.5 & 1 & 0.67 \\ 0.1 & 0.3 & -0.44 \end{vmatrix}}{-0.0022} = \frac{-0.04356}{-0.0022} = 19.8$$

For more than three equations, Cramer's rule becomes impractical because, as the number of equations increases, the determinants are time-consuming to evaluate by hand (or by computer). Consequently, more efficient alternatives are used. Some of these alternatives are based on the last noncomputer solution technique covered in this book—the elimination of unknowns.

7.1.3 The Elimination of Unknowns

The elimination of unknowns by combining equations is an algebraic approach that can be illustrated for a set of two equations:

$$a_{11}x_1 + a_{12}x_2 = c_1 \tag{7.6}$$

$$a_{21}x_1 + a_{22}x_2 = c_2 \tag{7.7}$$

The basic strategy is to multiply the equations by constants in order that one of the unknowns will be eliminated when the two equations are combined. The result is a single equation that can be solved for the remaining unknown. This value can then be substituted into either of the original equations to compute the other variable.

For example, Eq. (7.6) might be multiplied by a_{21} and Eq. (7.7) by a_{11} to give

$$a_{11}a_{21}x_1 + a_{12}a_{21}x_2 = c_1a_{21} \tag{7.8}$$

$$a_{21}a_{11}x_1 + a_{22}a_{11}x_2 = c_2a_{11} \tag{7.9}$$

Subtracting Eq. (7.8) from Eq. (7.9) will, therefore, eliminate the x_1 term from the equations to yield

$$a_{22}a_{11}x_2 - a_{12}a_{21}x_2 = c_2a_{11} - c_1a_{21}$$

which can be solved for

$$x_2 = \frac{a_{11}c_2 - a_{21}c_1}{a_{11}a_{22} - a_{12}a_{21}} \tag{7.10}$$

Equation (7.10) can then be substituted into Eq. (7.6), which can be solved for

$$x_1 = \frac{a_{22}c_1 - a_{12}c_2}{a_{11}a_{22} - a_{12}a_{21}} \tag{7.11}$$

Notice that Eqs. (7.10) and (7.11) follow directly from Cramer's rule, which states

$$x_1 = \frac{\begin{vmatrix} c_1 & a_{12} \\ c_2 & a_{22} \end{vmatrix}}{\begin{vmatrix} a_{11} & a_{12} \\ a_{21} & a_{22} \end{vmatrix}} = \frac{c_1a_{22} - a_{12}c_2}{a_{11}a_{22} - a_{12}a_{21}}$$

and

$$x_2 = \frac{\begin{vmatrix} a_{11} & c_1 \\ a_{21} & c_2 \end{vmatrix}}{\begin{vmatrix} a_{11} & a_{12} \\ a_{21} & a_{22} \end{vmatrix}} = \frac{a_{11}c_2 - c_1a_{21}}{a_{11}a_{22} - a_{12}a_{21}}$$

EXAMPLE 7.4   Elimination of Unknowns

Problem Statement: Use the elimination of unknowns to solve (recall Example 7.1),

$$3x_1 + 2x_2 = 18$$

$$-x_1 + 2x_2 = 2$$

Solution: Using Eqs. (7.11) and (7.10)

$$x_1 = \frac{2(18) - 2(2)}{3(2) - 2(-1)} = 4$$

$$x_2 = \frac{3(2) - (-1)18}{3(2) - 2(-1)} = 3$$

which is consistent with our graphical solution (Fig. 7.1).

The elimination of unknowns can be extended to systems with more than two or three equations. However, the numerous calculations that are required for larger systems make the method extremely tedious to implement by hand. However, as described in the next section, the technique can be formalized and readily programmed for the computer.

7.2 NAIVE GAUSS ELIMINATION

In the previous section, the elimination of unknowns was used to solve a pair of simultaneous equations. The procedure consisted of two steps:

1. The equations were manipulated in order to eliminate one of the unknowns from the equations. The result of this *elimination* step was that we had one equation with one unknown.
2. Consequently, this equation could be solved directly and the result *back-substituted* into one of the original equations in order to solve for the remaining unknown.

This basic approach can be extended to large sets of equations by developing a systematic scheme or algorithm to eliminate unknowns and to back-substitute. *Gauss elimination* is the most common of these schemes.

The present section includes the systematic techniques for forward elimination and back-substitution that comprise Gauss elimination. Although these techniques are ideally suited for implementation on computers, some modifications will be required in order to obtain a reliable algorithm. In particular, the computer program must avoid division by zero. The following method is called *"naive" Gauss elimination* because it does not avoid this problem. Subsequent sections will deal with the additional features required for an effective computer program.

The approach is designed to solve a general set of n equations:

$$a_{11}x_1 + a_{12}x_2 + a_{13}x_3 + \cdots + a_{1n}x_n = c_1 \tag{7.12a}$$

$$a_{21}x_1 + a_{22}x_2 + a_{23}x_3 + \cdots + a_{2n}x_n = c_2 \tag{7.12b}$$

$$a_{n1}x_1 + a_{n2}x_2 + a_{n3}x_3 + \cdots + a_{nn}x_n = c_n \tag{7.12c}$$

As was the case with the solution of two equations, the technique for n equations consists of two phases: elimination of unknowns and solution through back-substitution.

Forward Elimination of Unknowns. The first phase is designed to reduce the set of equations to an upper triangular system (Fig. 7.3). The initial step will be to eliminate the first unknown, x_1, from the second through the nth equations. To do this, multiply Eq. (7.12a) by a_{21}/a_{11} to give

$$a_{21}x_1 + \frac{a_{21}}{a_{11}}a_{12}x_2 + \cdots + \frac{a_{21}}{a_{11}}a_{1n}x_n = \frac{a_{21}}{a_{11}}c_1 \tag{7.13}$$

Now this equation can be subtracted from Eq. (7.12b) to give

$$\left(a_{22} - \frac{a_{21}}{a_{11}}a_{12}\right)x_2 + \cdots + \left(a_{2n} - \frac{a_{21}}{a_{11}}a_{1n}\right)x_n = c_2 - \frac{a_{21}}{a_{11}}c_1$$

or

$$a'_{22}x_2 + \cdots + a'_{2n}x_n = c'_2$$

where the prime indicates that the elements have been changed from their original values.

Figure 7.3
The two phases of Gauss elimination: forward elimination and back substitution. The primes indicate the number of times that the coefficients and constants have been modified.

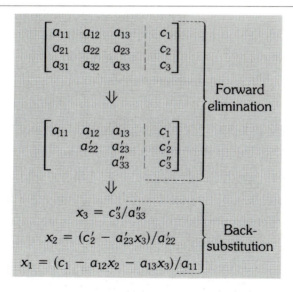

The procedure is then repeated for the remaining equations. For instance, Eq. (7.12a) can be multiplied by a_{31}/a_{11} and the result subtracted from the third equation. Repeating the procedure for the remaining equations results in the following modified system:

$$a_{11}x_1 + a_{12}x_2 + a_{13}x_3 + \cdots + a_{1n}x_n = c_1 \tag{7.14a}$$

$$a'_{22}x_2 + a'_{23}x_3 + \cdots + a'_{2n}x_n = c'_2 \tag{7.14b}$$

$$a'_{32}x_2 + a'_{33}x_3 + \cdots + a'_{3n}x_n = c'_3 \tag{7.14c}$$

$$\vdots \quad \vdots \qquad \vdots \quad \vdots$$

$$\vdots \quad \vdots \qquad \vdots \quad \vdots$$

$$\vdots \quad \vdots \qquad \vdots \quad \vdots$$

$$a'_{n2}x_2 + a'_{n3}x_3 + \cdots + a'_{nn}x_n = c'_n \qquad\qquad (7.14d)$$

For the foregoing steps, Eq. $(7.12a)$ is called the *pivot equation*, and a_{11} is called the *pivot coefficient*. Note that the process of multiplying the first row by a_{21}/a_{11} is equivalent to dividing it by a_{11} and multiplying it by a_{21}. Sometimes the division operation is referred to as *normalization*. We make this distinction because a zero pivot element can interfere with normalization by causing a division by zero. We will return to this important issue after we complete our description of naive Gauss elimination.

Now repeat the above in order to eliminate the second unknown from Eq. $(7.14c)$ through $(7.14d)$. To do this multiply Eq. $(7.14b)$ by a'_{32}/a'_{22} and subtract the result from Eq. $(7.14c)$. Perform a similar elimination for the remaining equations to yield

$$a_{11}x_1 + a_{12}x_2 + a_{13}x_3 + \cdots + a_{1n}x_n = c_1$$

$$a'_{22}x_2 + a'_{23}x_3 + \cdots + a'_{2n}x_n = c'_2$$

$$a''_{33}x_3 + \cdots + a''_{3n}x_n = c''_3$$

$$\vdots \qquad\qquad \vdots \quad \vdots$$

$$\vdots \qquad\qquad \vdots \quad \vdots$$

$$\vdots \qquad\qquad \vdots \quad \vdots$$

$$a''_{n3}x_3 + \cdots + a''_{nn}x_n = c''_n$$

where the double prime indicates that the elements have been modified twice.

The procedure can be continued using the remaining pivot equations. The final manipulation in the sequence is to use the $(n-1)$th equation to eliminate the x_{n-1} term from the nth equation. At this point, the system will have been transformed to an upper triangular system (recall Box PT3.1).

$$a_{11}x_1 + a_{12}x_2 + a_{13}x_3 + \cdots + a_{1n}x_n = c_1 \qquad\qquad (7.15a)$$

$$a'_{22}x_2 + a'_{23}x_3 + \cdots + a'_{2n}x_n = c'_2 \qquad\qquad (7.15b)$$

$$a''_{33}x_3 + \cdots + a''_{3n}x_n = c''_3 \qquad\qquad (7.15c)$$

$$\vdots \qquad\qquad \vdots$$

$$\vdots \qquad\qquad \vdots$$

$$\vdots \qquad\qquad \vdots$$

$$a_{nn}^{(n-1)}x_n = c_n^{(n-1)} \qquad\qquad (7.15d)$$

Pseudocode to implement forward elimination is presented in Fig. 7.4a. Notice that three nested loops provide a concise representation of the process. The outer loop moves down the matrix from one pivot row to the next. The middle loop moves below the pivot row to each of the subsequent rows where elimination is to take place. Finally, the

(a)
```
DOFOR k = 1 to n-1
    DOFOR i = k+1 to n
        factor = a_{i,k}/a_{k,k}
        DOFOR j = k+1 to n
            a_{i,j} = a_{i,j} - factor·a_{k,j}
        ENDDO
        c_{i} = c_{i} - factor·c_{k}
    ENDDO
ENDDO
```

(b)
```
x_{n} = c_{n}/a_{n,n}
DOFOR i = n-1 to 1 step -1
    sum = 0
    DOFOR j = i+1 to n
        sum = sum + a_{i,j}·x_{j}
    ENDDO
    x_{i} = (c_{i} - sum)/a_{i,i}
ENDDO
```

Figure 7.4
Pseudocode to perform (a) forward elimination and (b) back substitution.

innermost loop progresses across the columns to eliminate or transform the elements of a particular row.

Backward Substitution. Equation (7.15d) can now be solved for x_n:

$$x_n = \frac{c_n^{(n-1)}}{a_{nn}^{(n-1)}} \tag{7.16}$$

This result can be back-substituted into the $(n-1)$th equation to solve for x_{n-1}. The procedure, which is repeated to evaluate the remaining x's, can be represented by the following formula:

$$x_i = \frac{c_i^{(i-1)} - \sum\limits_{j=i+1}^{n} a_{ij}^{(i-1)}x_j}{a_{ii}^{(i-1)}} \tag{7.17}$$

for $i = n-1, n-2, \ldots, 1$.

Pseudocode to implement Eqs. (7.16) and (7.17) is presented in Fig. 7.4b. Notice the similarity between this pseudocode and that in Fig. PT3.4 for matrix multiplication. As with Fig. PT3.4, a temporary variable, *sum*, is used to accumulate the summation from Eq. (7.17). This results in a somewhat faster execution time than if the summation were accumulated in c_i. More importantly, it allows efficient improvement in precision if the variable, *sum*, is declared in double precision.

EXAMPLE 7.5   Naive Gauss Elimination

Problem Statement: Use Gauss elimination to solve

$$3x_1 - 0.1x_2 - 0.2x_3 = 7.85 \tag{E7.5.1}$$

$$0.1x_1 + 7x_2 - 0.3x_3 = -19.3 \tag{E7.5.2}$$

$$0.3x_1 - 0.2x_2 + 10x_3 = 71.4 \tag{E7.5.3}$$

Carry six significant figures during the computation.

Solution: The first part of the procedure is forward elimination. Multiply Eq. (E7.5.1) by $(0.1)/3$ and subtract the result from Eq. (E7.5.2) to give

$$7.00333x_2 - 0.293333x_3 = -19.5617$$

Then multiply Eq. (E7.5.1) by $(0.3)/3$ and subtract it from Eq. (E7.5.3) to eliminate x_1. After these operations, the set of equations is

$$3x_1 - 0.1x_2 - 0.2x_3 = 7.85 \tag{E7.5.4}$$

$$7.00333x_2 - 0.293333x_3 = -19.5617 \tag{E7.5.5}$$

$$-0.190000x_2 + 10.0200x_3 = 70.6150 \tag{E7.5.6}$$

To complete the forward elimination, x_2 must be removed from Eq. (E7.5.6). To accomplish this, multiply Eq. (E7.5.5) by $-0.190000/7.00333$ and subtract the result from Eq. (E7.5.6). This eliminates x_2 from the third equation and reduces the system to an upper triangular form, as in

$$3x_1 - 0.1x_2 - 0.2x_3 = 7.85$$

$$7.00333x_2 - 0.293333x_3 = -19.5617 \tag{E7.5.7}$$

$$10.0120x_3 = 70.0843 \tag{E7.5.8}$$

We can now solve these equations by back substitution. First, Eq. (E7.5.8) can be solved for

$$x_3 = 7.00003 \tag{E7.5.9}$$

This result can be back-substituted into Eq. (E7.5.7):

$$7.00333x_2 - 0.293333(7.00003) = -19.5617$$

which can be solved for

$$x_2 = -2.50000 \tag{E7.5.10}$$

Finally, Eqs. (E7.5.9) and (E7.5.10) can be substituted into Eq. (E7.5.4):

$$3x_1 - 0.1(-2.50000) - 0.2(7.00003) = 7.85$$

which can be solved for

$$x_1 = 3.00000$$

Although there is a slight round-off error in Eq. (E7.5.9), the results are very close to the exact solution of $x_1 = 3, x_2 = -2.5$, and $x_3 = 7$. This can be verified by substituting the results into the original equation set

$$3(3) - 0.1(-2.5) - 0.2(7.00003) = \quad 7.84999 \simeq \quad 7.85$$

$$0.1(3) + \quad 7(-2.5) - 0.3(7.00003) = -19.3000 = -19.3$$

$$0.3(3) - 0.2(-2.5) + \quad 10(7.00003) = \quad 71.4003 \simeq \quad 71.4$$

7.3 PITFALLS OF ELIMINATION METHODS

Whereas there are many systems of equations that can be solved with naive Gauss elimination, there are some pitfalls that must be explored before writing a general computer program to implement the method. Although the following material relates directly to naive Gauss elimination, the information has relevance to other elimination techniques.

7.3.1 Division by Zero

The primary reason that the foregoing technique is called "naive" is that during both the elimination and the back-substitution phases, it is possible that a division by zero could occur. For example, if we used naive Gauss elimination to solve

$$2x_2 + 3x_3 = \quad 8$$
$$4x_1 + 6x_2 + 7x_3 = -3$$
$$2x_1 + \quad x_2 + 6x_3 = \quad 5$$

the normalization of the first row would involve division by $a_{11} = 0$. Problems also can arise when a coefficient is very close to zero. The technique of pivoting has been developed to partially avoid these problems. It will be described in Sec. 7.4.2.

7.3.2 Round-Off Errors

Even though the solution in Example 7.5 was close to the true answer, there was a slight discrepancy in the result for x_3 [Eq. (E7.5.9)]. This discrepancy, which amounted to a relative error of -0.00043 percent, was due to our use of six significant figures during the computation. If we had used more significant figures, the error in the results would be reduced further. If we had used fractions instead of decimals (and consequently avoided round-off altogether), the answers would have been exact. However, because computers carry only a limited number of significant figures (recall Sec. 3.4.1), round-off errors can occur and must be considered when evaluating the results.

The problem of round-off error can become particularly important when large numbers of equations are to be solved. This is due to the fact that every result is dependent on previous results. Consequently, an error in the early steps will tend to propagate — that is, it will cause errors in subsequent steps.

Specifying the system size where round-off error becomes significant is complicated by the fact that the type of computer and the properties of the equations are determining factors. A rough rule of thumb is that round-off error may be important when dealing with 100 or more equations. In any event, you should always substitute your answers back into the original equations to check whether a substantial error has occurred. However, as discussed below, the magnitudes of the coefficients themselves can influence whether such an error check ensures a reliable result.

7.3.3   Ill-Conditioned Systems

The adequacy of the solution depends on the condition of the system. In Sec. 7.1.1, a graphical depiction of system condition was developed. As discussed in Sec. 3.6.3, *well-conditioned systems* are those where a small change in one or more of the coefficients results in a similar small change in the solution. *Ill-conditioned systems* are those where small changes in coefficients result in large changes in the solution. An alternative interpretation of ill conditioning is that a wide range of answers can approximately satisfy the equations. Because round-off errors can induce small changes in the coefficients, these artificial changes can lead to large solution errors for ill-conditioned systems, as illustrated in the following example.

EXAMPLE 7.6   Ill-Conditioned Systems

Problem Statement: Solve the following system:

$$x_1 + 2x_2 = 10 \tag{E7.6.1}$$

$$1.1x_1 + 2x_2 = 10.4 \tag{E7.6.2}$$

Then, solve it again, but with the coefficient of x_1 in the second equation modified slightly to 1.05.

Solution: Using Eqs. (7.10) and (7.11), the solution is

$$x_1 = \frac{2(10) - 2(10.4)}{1(2) - 2(1.1)} = 4$$

$$x_2 = \frac{1(10.4) - 1.1(10)}{1(2) - 2(1.1)} = 3$$

However, with the slight change of the coefficient a_{21} from 1.1 to 1.05, the result is changed dramatically to

$$x_1 = \frac{2(10) - 2(10.4)}{1(2) - 2(1.05)} = 8$$

$$x_2 = \frac{1(10.4) - 1.05(10)}{1(2) - 2(1.05)} = 1$$

Notice that the primary reason for the discrepancy between the two results is that the denominator represents the difference of two almost-equal numbers. As illustrated previously in Sec. 3.4.2, such differences are highly sensitive to slight variations in the numbers being manipulated.

At this point, you might suggest that substitution of the results into the original equations would alert you to the problem. Unfortunately, for ill-conditioned systems this is often not the case. Substitution of the erroneous values of $x_1 = 8$ and $x_2 = 1$ into Eqs. (E7.6.1) and (E7.6.2) yields

$$8 + 2(1) = 10 \quad = 10$$

$$1.1(8) + 2(1) = 10.8 \simeq 10.4$$

Therefore, although $x_1 = 8$ and $x_2 = 1$ is not the true solution to the original problem, the error check is close enough to possibly mislead you into believing that your solutions are adequate.

As was done previously in our section on graphical methods, a visual representative of ill conditioning can be developed by plotting Eqs. (E7.6.1) and (E7.6.2) (recall Fig. 7.2). Because the slopes of the lines are almost equal, it is visually difficult to see exactly where they intersect. This visual difficulty is reflected quantitatively in the nebulous results of Example 7.6. We can mathematically characterize this situation by writing the two equations in general form:

$$a_{11}x_1 + a_{12}x_2 = c_1 \tag{7.18}$$

$$a_{21}x_1 + a_{22}x_2 = c_2 \tag{7.19}$$

Dividing Eq. (7.18) by a_{12} and Eq. (7.19) by a_{22} and rearranging yields alternative versions that are in the format of straight lines [$x_2 = $ (slope) $x_1 + $ intercept]:

$$x_2 = -\frac{a_{11}}{a_{12}}x_1 + \frac{c_1}{a_{12}}$$

$$x_2 = -\frac{a_{21}}{a_{22}}x_1 + \frac{c_2}{a_{22}}$$

Consequently, if the slopes are nearly equal,

$$\frac{a_{11}}{a_{12}} \simeq \frac{a_{21}}{a_{22}}$$

or, cross-multiplying,

$$a_{11}a_{22} \simeq a_{21}a_{12}$$

which can be also expressed as

$$a_{11}a_{22} - a_{12}a_{21} \simeq 0 \tag{7.20}$$

Now, recalling that $a_{11}a_{22} - a_{12}a_{21}$ is the determinant of a two-dimensional system [Eq. (7.3)], we arrive at the general conclusion that an ill-conditioned system is one with a determinant close to zero. In fact, if the determinant is exactly zero, the two slopes are identical, which connotes either no solution or an infinite number of solutions, as is the case for the singular systems depicted in Fig. 7.2a and b.

It is difficult to specify *how close* to zero the determinant must be to indicate ill conditioning. This is complicated by the fact that the determinant can be changed by multiplying one or more of the equations by a scale factor without changing the solution. Consequently, the determinant is a relative value that is influenced by the magnitude of the coefficients.

EXAMPLE 7.7   Effect of Scale on the Determinant

Problem Statement: Evaluate the determinant of the following systems:
(*a*) From Example 7.1:

$$3x_1 + 2x_2 = 18 \tag{E7.7.1}$$

$$-x_1 + 2x_2 = \ 2 \tag{E7.7.2}$$

(*b*) From Example 7.6:

$$x_1 + 2x_2 = 10 \tag{E7.7.3}$$

$$1.1x_1 + 2x_2 = 10.4 \tag{E7.7.4}$$

(*c*) Repeat (*b*) but with the equations multiplied by 10.

Solution:
(*a*) The determinant of Eqs. (E7.7.1) and (E7.7.2), which are well-conditioned, is

$$D = 3(2) - 2(-1) = 8$$

(*b*) The determinant of Eqs. (E7.7.3) and (E7.7.4), which are ill-conditioned, is

$$D = 1(2) - 2(1.1) = -0.2$$

(*c*) The results of (*a*) and (*b*) seem to bear out the contention that ill-conditioned systems have near-zero determinants. However, suppose that the ill-conditioned system in (*b*) is multiplied by 10 to give

$$10x_1 + 20x_2 = 100$$

$$11x_1 + 20x_2 = 104$$

The multiplication of an equation by a constant has no effect on its solution. In addition, it is still ill-conditioned. This can be verified by the fact that multiplying by a constant has no effect on the graphical solution. However, the determinant is dramatically affected:

$$D = 10(20) - 20(11) = -20$$

Not only has it been raised two orders of magnitude, but it is now over twice as large as the determinant of the well-conditioned system in (*a*).

As illustrated by the previous example, the magnitude of the coefficients interjects a scale effect that complicates the relationship between system condition and determinant size. One way to partially circumvent this difficulty is to scale the equations so that the maximum element in any row is equal to 1.

EXAMPLE 7.8   Scaling

Problem Statement: Scale the systems of equations in Example 7.7 to a maximum value of 1 and recompute their determinants.

Solution:

(*a*) For the well-conditioned system, scaling results in

$$x_1 + 0.667x_2 = 6$$

$$-0.5x_1 + \quad x_2 = 1$$

for which the determinant is

$$1(1) - 0.667(-0.5) = 1.333$$

(*b*) For the ill-conditioned system, scaling gives

$$0.5x_1 + x_2 = 5$$

$$0.55x_1 + x_2 = 5.2$$

for which the determinant is

$$0.5(1) - 1(0.55) = -0.05$$

(*c*) For the last case, scaling changes the system to the same form as in (*b*) and the determinant is also -0.05. Thus, the scale effect is removed.

In a previous section (Sec. 7.1.2), we suggested that the determinant is difficult to compute for more than three simultaneous equations. Therefore, it might seem that it does not provide a practical means for evaluating system condition. However, as described in Box 7.1, there is a simple algorithm that results from Gauss elimination that can be used to evaluate the determinant.

Box 7.1   *Determinant Evaluation Using Gauss Elimination*

In Sec. 7.1.2, we stated that determinant evaluation by expansion of minors was impractical for large sets of equations. Thus, we concluded that Cramer's rule would only be applicable to small systems. However, as mentioned in Sec. 7.3.3, the determinant has value in assessing system condition. It would, therefore, be useful to have a practical method for computing this quantity.

Fortunately, Gauss elimination provides a simple way to do this. The method is based on the fact that the determinant of a triangular matrix can be simply computed as the product of its diagonal elements:

$$D = a_{11}a_{22}a_{33} \cdots a_{nn} \tag{B7.1.1}$$

The validity of this formulation can be illustrated for a 3-by-3 system:

$$D = \begin{vmatrix} a_{11} & a_{12} & a_{13} \\ 0 & a_{22} & a_{23} \\ 0 & 0 & a_{33} \end{vmatrix}$$

where the determinant can be evaluated as [recall Eq. (7.4)]

$$D = a_{11}\begin{vmatrix} a_{22} & a_{23} \\ 0 & a_{33} \end{vmatrix} - a_{12}\begin{vmatrix} 0 & a_{23} \\ 0 & a_{33} \end{vmatrix} + a_{13}\begin{vmatrix} 0 & a_{22} \\ 0 & 0 \end{vmatrix}$$

or, by evaluating the minors (that is, the 2-by-2 determinants),

$$D = a_{11}a_{22}a_{33} - a_{12}(0) + a_{13}(0) = a_{11}a_{22}a_{33}$$

Recall that the forward-elimination step of Gauss elimination results in an upper triangular system. Because the value of the determinant is not changed by the forward-elimination process, the determinant can be simply evaluated at the end of this step via

$$D = a_{11}a'_{22}a''_{33} \cdots a_{nn}^{(n-1)} \qquad (B7.1.2)$$

where the superscripts signify the number of times that the elements have been modified by the elimination process. Thus, we can capitalize on the effort that has already been expended in reducing the system to triangular form and, in the bargain, come up with a simple estimate of the determinant.

There is a slight modification to the above approach when the program employs partial pivoting (Sec. 7.4.2). For this case, the determinant changes sign every time a row is pivoted. One way to represent this is to modify Eq. (B7.1.2):

$$D = a_{11}a'_{22}a''_{33} \cdots a_{nn}^{(n-1)}(-1)^p \qquad (B7.1.3)$$

where p represents the number of times that rows are pivoted. This modification can be incorporated simply into a program; merely keep track of the number of pivots that take place during the course of the computation and then use Eq. (B7.1.3) to evaluate the determinant.

Aside from the approach used in the previous example, there are a variety of other ways to evaluate system condition. For example, there are alternative methods for normalizing the elements (see Stark, 1970). In addition, as described in the next chapter (Sec. 8.2.2), the matrix inverse and matrix norms can be employed to evaluate system condition. Finally, a simple (but time-consuming) test is to modify the coefficients slightly and repeat the solution. If such modifications lead to drastically different results, the system is likely to be ill-conditioned.

As you might gather from the foregoing discussion, there is no general and definitive test for ill conditioning. Fortunately, most linear algebraic equations derived from engineering-problem settings are naturally well-conditioned. In addition, some of the techniques outlined in the following section help to alleviate the problem.

7.4   TECHNIQUES FOR IMPROVING SOLUTIONS

The following techniques can be incorporated into the naive Gauss elimination algorithm to circumvent some of the pitfalls discussed in the previous section.

7.4.1   Use of More Significant Figures

The simplest remedy for ill conditioning is to use more significant figures in the computation. If your computer has the capability of being extended to handle larger word size, such a feature will greatly reduce the problem. However, a price must be paid in the form of the computational and memory overhead connected with using extended precision (recall Sec. 3.4.1).

7.4.2   Pivoting

As mentioned at the beginning of Sec. 7.3, obvious problems occur when a pivot element is zero because the normalization step leads to division by zero. Problems may also arise when the pivot element is close to, rather than exactly equal to, zero because if the magnitude of the pivot element is small compared to the other elements, then round-off errors can be introduced.

Therefore, before each row is normalized, it is advantageous to determine the largest available coefficient. The rows can then be switched so that the largest element is the pivot element. This is called *partial pivoting*. If columns as well as rows are searched for the largest element and then switched, the procedure is called *complete pivoting*. Complete pivoting is rarely used because switching columns changes the order of the x's and, consequently, adds significant and usually unjustified complexity to the computer program. The following example illustrates the advantages of partial pivoting. Aside from avoiding division by zero, pivoting also minimizes round-off error. As such, it also serves as a partial remedy for ill conditioning.

EXAMPLE 7.9   Partial Pivoting

Problem Statement: Use Gauss elimination to solve

$$0.0003x_1 + 3.0000x_2 = 2.0001$$

$$1.0000x_1 + 1.0000x_2 = 1.0000$$

Note that in this form the first pivot element, $a_{11} = 0.0003$, is very close to zero. Then repeat the computation, but partial pivot by reversing the order of the equations. The exact solution is $x_1 = 1/3$ and $x_2 = 2/3$.

Solution: Multiplying the first equation by $1/(0.0003)$ yields

$$x_1 + 10,000x_2 = 6667$$

which can be used to eliminate x_1 from the second equation:

$$-9999x_2 = -6666$$

which can be solved for

$$x_2 = 2/3$$

This result can be substituted back into the first equation to evaluate x_1:

$$x_1 = \frac{2.0001 - 3(2/3)}{0.0003} \tag{E7.9.1}$$

However, due to subtractive cancellation, the result is very sensitive to the number of significant figures carried in the computation:

Significant Figures	x_2	x_1	Absolute Value of Percent Relative Error for x_1
3	0.667	−3.33	1099
4	0.6667	0.0000	100
5	0.66667	0.30000	10
6	0.666667	0.330000	1
7	0.6666667	0.3330000	0.1

Note how the solution for x_1 is highly dependent on the number of significant figures. This is due to the fact that in Eq. (E7.9.1), we are subtracting two almost-equal numbers. On the other hand, if the equations are solved in reverse order, the row with the larger pivot element is normalized. The equations are

$$1.0000x_1 + 1.0000x_2 = 1.0000$$

$$0.0003x_1 + 3.0000x_2 = 2.0001$$

Elimination and substitution yield $x_2 = 2/3$. For different numbers of significant figures, x_1 can be computed from the first equation, as in

$$x_1 = \frac{1 - (2/3)}{1} \qquad\qquad (E7.9.2)$$

This case is much less sensitive to the number of significant figures in the computation:

Significant Figures	x_2	x_1	Absolute Value of Percent Relative Error for x_1
3	0.667	0.333	0.1
4	0.6667	0.3333	0.01
5	0.66667	0.33333	0.001
6	0.666667	0.333333	0.0001
7	0.6666667	0.3333333	0.00001

Thus, a pivot strategy is much more satisfactory.

General-purpose computer programs must include a pivot strategy. Figure 7.5 provides a simple algorithm to implement such a strategy. Notice that the algorithm consists of two major loops. After storing the current pivot element and its row number as the variables, *big* and *pivot*, the first loop compares the pivot element with the elements below it to check whether any of these is larger than the pivot element. If so, the new largest element and its row number are stored in *big* and *pivot*. Then, the second loop switches the original pivot row with the one with the largest element so that the latter becomes the new pivot row. This pseudocode can be integrated into a program based on the other elements of Gauss elimination outlined in Fig. 7.4. The best way to do this is to employ a modular approach and write Fig. 7.5 as a subroutine (or procedure) which would be called directly after the beginning of the first loop in Fig. 7.4*a*.

Note that the second IF/THEN construct in Fig. 7.5 physically interchanges the rows. For large matrices, this can become quite time-consuming. Consequently, most codes do not actually exchange rows but rather keep track of the pivot rows by storing the appropriate subscripts in a vector. This vector then provides a basis for specifying the proper row ordering during the forward elimination and back-substitution operations. Thus, the operations are said to be implemented *in place*. The programs introduced later in this chapter employ such an approach.

```
pivot = k
big = |a_{k,k}|
DOFOR ii = k+1 to n
    dummy = |a_{ii,k}|
    IF (dummy > big)
        big = dummy
        pivot = ii
    ENDIF
ENDDO
IF (pivot ≠ k)
    DOFOR jj = k to n
        dummy = a_{pivot,jj}
        a_{pivot,jj} = a_{k,jj}
        a_{k,jj} = dummy
    ENDDO
    dummy = c_{pivot}
    c_{pivot} = c_{k}
    c_{k} = dummy
ENDIF
```

Figure 7.5
Pseudocode to implement
partial pivoting.

7.4.3 Scaling

In Sec. 7.3.3, we proposed that scaling had value in standardizing the size of the determinant. Beyond this application, it has utility in minimizing round-off errors for those cases where some of the equations in a system have much larger coefficients than others. Such situations are frequently encountered in engineering practice when widely different units are used in the development of simultaneous equations. For instance, in electric-circuit problems, the unknown voltages can be expressed in units ranging from microvolts to kilovolts. Similar examples can arise in all fields of engineering. As long as each equation is consistent, the system will be technically correct and solvable. However, the use of widely differing units can lead to coefficients of widely differing magnitudes. This, in turn, can have an impact on round-off error as it affects pivoting, as illustrated by the following example.

EXAMPLE 7.10   *Effect of Scaling on Pivoting and Round Off*

Problem Statement:
(*a*) Solve the following set of equations using Gauss elimination and a pivoting strategy:

$$2x_1 + 100,000x_2 = 100,000$$

$$x_1 + x_2 = 2$$

(*b*) Repeat the solution after scaling the equations so that the maximum coefficient in each row is 1.

(*c*) Finally, use the scaled coefficients to determine whether pivoting is necessary. However, actually solve the equations with the original coefficient values. For all cases, retain only three significant figures. Note that the correct answers are $x_1 = 1.00002$ and

$x_2 = 0.99998$ or, for three significant figures, $x_1 = x_2 = 1.00$.

Solution:

(a) Without scaling, forward elimination is applied to give

$$2x_1 + 100,000x_2 = 100,000$$
$$-50,000x_2 = -50,000$$

which can be solved by back substitution for

$$x_2 = 1.00$$
$$x_1 = 0.00$$

Although x_2 is correct, x_1 is 100 percent in error because of round off.

(b) Scaling transforms the original equations to

$$0.00002x_1 + x_2 = 1$$
$$x_1 + x_2 = 2$$

Therefore, the rows should be pivoted to put the greatest value on the diagonal.

$$x_1 + x_2 = 2$$
$$0.00002x_1 + x_2 = 1$$

Forward elimination yields

$$x_1 + x_2 = 2$$
$$x_2 = 1.00$$

which can be solved for

$$x_1 = x_2 = 1$$

Thus, scaling leads to the correct answer.

(c) The scaled coefficients indicate that pivoting is necessary. We therefore pivot but retain the original coefficients to give

$$x_1 + \qquad x_2 = \qquad 2$$
$$2x_1 + 100,000x_2 = 100,000$$

Forward elimination yields

$$x_1 + \qquad x_2 = \qquad 2$$
$$100,000x_2 = 100,000$$

which can be solved for the correct answer: $x_1 = x_2 = 1$. Thus, scaling was useful in determining whether pivoting was necessary, but the equations themselves did not require scaling to arrive at a correct result.

As in the previous example, scaling has utility in minimizing round off. However, it should be noted that scaling itself also leads to round off. For example, given the equation

$$2x_1 + 300,000x_2 = 1$$

and using three significant figures, scaling leads to

$$0.00000667x_1 + x_2 = 0.00000333$$

Thus, scaling introduces a round-off error to the first coefficient and the right-hand-side constant. For this reason, it is sometimes suggested that scaling should only be employed as in part (c) of the preceding example. That is, it is used to calculate scaled values for the coefficients solely as a criterion for pivoting, but the original coefficient values are retained for the actual elimination and substitution computations. This involves a trade-off if the determinant is being calculated as part of the program. That is, the resulting determinant will be unscaled. However, because many applications of Gauss elimination do not require determinant evaluation, it is the most common approach and will be used in the computer codes in the next section.

7.4.4 Computer Programs for Gauss Elimination

The algorithms from Figs. 7.4 and 7.5 can now be expressed as computer code. Figure 7.6 shows a BASIC program that was designed for an interactive mode of implementation. Figure 7.7 lists FORTRAN and Pascal codes that are written as subroutines.

Note that all the programs include modules for the three primary operations of the Gauss elimination algorithm: forward elimination, back substitution, and pivoting. In addition, there are several aspects of the code that differ from the pseudocode of Figs. 7.4 and 7.5. These are:

1. Pivoting is performed in place. That is, the rows are not actually interchanged, but their correct ordering is kept track of by an ordering vector.
2. The equations are not scaled, but scaled values of the elements are used to determine whether pivoting is to be implemented.

EXAMPLE 7.11 Solution of Linear Algebraic Equations Using the Computer

Problem Statement: A user-friendly computer program to implement Gauss elimination is contained on the Electronic TOOLKIT software. We can use this software to solve a problem associated with the falling parachutist example discussed in Chap. 1. Suppose that a team of three parachutists is connected by a weightless cord while free-falling at a velocity of 5 m/s (Fig. 7.8). Calculate the tension in each section of cord and the acceleration of the team, given the following:

Parachutist	Mass, kg	Drag Coefficient, kg/s
1	70	10
2	60	14
3	40	17

Solution: Free-body diagrams for each of the parachutists are depicted in Fig. 7.9. Summing the forces in the vertical direction and using Newton's second law gives a set of three simultaneous linear equations:

```
100 REM    GAUSS ELIMINATION (BASIC VERSION)
110 REM
120 REM **********************************
130 REM *       DEFINITION OF VARIABLES      *
140 REM *                                     *
150 REM * N   = NUMBER OF EQUATIONS          *
160 REM * A[] = MATRIX OF COEFFICIENTS       *
170 REM * C() = RIGHT-HAND-SIDE VECTOR       *
180 REM * X() = UNKNOWNS                     *
190 REM * O() = ORDER VECTOR                 *
200 REM * S() = SCALE VECTOR                 *
210 REM **********************************
220 REM
230 DIM A(15,15),C(15),X(15),O(15),S(15)
240 REM
250 REM *********   MAIN PROGRAM   *********
260 REM
270 GOSUB 350 'input data
280 GOSUB 450 'scale and order vectors
290 GOSUB 500 'forward elimination
300 GOSUB 600 'back substitution
310 GOSUB 700 'output results
320 END
350 REM ********  SUBROUTINE INPUT *********
355 CLS
360 INPUT "NUMBER OF EQUATIONS? ", N
365 PRINT
370 FOR I = 1 TO N
375    FOR J = 1 TO N
380       PRINT "A(";I;",";J;") = ";
385       INPUT A(I,J)
390    NEXT J
395    PRINT "C(";I;") = ";: INPUT C(I)
400    PRINT
405 NEXT I
410 RETURN
450 REM ********  SUBROUTINE ORDER  ********
455 REM
460 FOR I = 1 TO N
465    O(I) = I
470    S(I)=ABS(A(I,1))
475    FOR J = 2 TO N
480       IF ABS(A(I,J))>S(I) THEN
                 S(I)=ABS(A(I,J))
485    NEXT J
490 NEXT I
495 RETURN
500 REM ******  SUBROUTINE ELIMINATE  ******
505 REM
510 FOR K = 1 TO N-1
515    GOSUB 750      'partial pivot
520    FOR I = K+1 TO N
525       FACTOR=A(O(I),K)/A(O(K),K)
530       FOR J = K+1 TO N
535          A(O(I),J)=A(O(I),J)
                        -FACTOR*A(O(K),J)
540       NEXT J
545       C(O(I))=C(O(I))-FACTOR*C(O(K))
550    NEXT I
555 NEXT K
560 RETURN
600 REM ******   SUBROUTINE SUBSTITUTE ******
605 REM
610 X(N) = C(O(N))/A(O(N),N)
615 FOR I = N-1 TO 1 STEP -1
620    SUM = 0
625    FOR J = I+1 TO N
630       SUM = SUM + A(O(I),J)*X(J)
635    NEXT J
640    X(I) = (C(O(I))-SUM)/A(O(I),I)
645 NEXT I
650 RETURN
700 REM  *******  SUBROUTINE OUTPUT  *******
705 REM
710 CLS
715 PRINT:PRINT "SOLUTIONS TO EQUATIONS"
720 FOR I = 1 TO N
725    PRINT "X(" I ") ="; X(I)
730 NEXT I
735 RETURN
750 REM *******  SUBROUTINE PIVOT  ********
755 REM
760 PIVOT=K
765 BIG = ABS(A(O(K),K)/S(O(K)))
770 FOR II = K+1 TO N
775    DUMMY=ABS(A(O(II),K)/S(O(II)))
780    IF DUMMY>BIG THEN 785 ELSE 795
785       BIG = DUMMY
790       PIVOT=II
795    REM endif
800 NEXT II
805 DUMMY=O(PIVOT)
810 O(PIVOT)=O(K)
815 O(K)=DUMMY
835 RETURN
```

Figure 7.6
An interactive computer program for Gauss elimination written in Microsoft BASIC.

```fortran
      SUBROUTINE GAUSS(A,C,X,N)
******************************************
*           DEFINITION OF VARIABLES      *
*                                        *
*     N   = NUMBER OF EQUATIONS          *
*     A[] = MATRIX OF COEFFICIENTS       *
*     C() = RIGHT-HAND-SIDE VECTOR       *
*     X() = UNKNOWNS                      *
*     O() = ORDER VECTOR                 *
*     S() = SCALE VECTOR                 *
******************************************
      DIMENSION A(50,50),C(50),X(50)
      INTEGER O(50)
      CALL ORDER(S,O,A,N)
      CALL ELIM(S,C,A,O,N)
      CALL SUBST(C,X,A,O,N)
      RETURN
      END
******************************************
      SUBROUTINE ORDER(S,O,A,N)
      DIMENSION A(50,50),S(50)
      INTEGER O(50)
      DO 10 I=1,N
         O(I)=I
         S(I)=ABS(A(I,1))

         DO 20 J=2,N
            IF (ABS(A(I,J)).GT.S(I)) THEN
               S(I)=ABS(A(I,J))
            ENDIF
20       CONTINUE
10    CONTINUE
      RETURN
      END
```

(a)

```pascal
PROCEDURE GausElim (VAR A: Matrix;
               VAR C,X: Vector; N: integer);
{       Driver program type definitions
   Matrix = a 2 dimensional real array
   Vector = a 1 dimensional real array
   Row    = a 1 dimensional integer array  }
{       Definition of variables
   N    = number of equations
   A[]  = matrix of coefficients
   C()  = right-hand-side vector
   X()  = unknowns
   O()  = order vector
   S()  = scale vector                      }
PROCEDURE Order(VAR S: Vector;
                VAR O: Row;
                    N: integer);
VAR
   i,j: integer;
Begin                        {procedure Order}
For i := 1 to N do
   Begin
     O[i] := i;
     S[i] := Abs(A[i,1]);

     For j := 2 to N do
        Begin
        If Abs(A[i,j]) > S[i] Then
           S[i] := Abs(A[i,j]);
        End;
   End;
End;                      { of procedure Order}
PROCEDURE Pivot(VAR S: Vector;
                A : Matrix; VAR O:Row;
                    k: integer);
VAR
   i,j: integer;
   pivot,idum: integer;
   big,dummy: real;
Begin                        {procedure Pivot}
pivot := k;
big := Abs(A[O[k],k]/S[O[k]]);
For i := k+1 to N do
   Begin
   dummy := Abs(A[O[i],k]/S[O[i]]);
   If dummy > big Then
      Begin
      big := dummy;
      pivot := i;
      End;
   End;
```

(b)

Figure 7.7
Subroutines for Gauss elimination written in (*a*) FORTRAN 77 and (*b*) Turbo Pascal.

(*Continues* **Figure 7.7**)

```
*********************************************
      SUBROUTINE ELIM(S,C,A,O,N)
      DIMENSION A(50,50),C(50),S(50)
      INTEGER O(50)
      INTEGER*2 I,J,K
      DO 10 K=1,N-1
         CALL PIVOT(S,A,O,N,K)
         DO 20 I=K+1,N
            FACTOR = A(O(I),K)/A(O(K),K)
            DO 30 J=K+1,N
               A(O(I),J)=A(O(I),J)
     *                     -FACTOR*A(O(K),J)
30          CONTINUE
            C(O(I))=C(O(I))-FACTOR*C(O(K))
20       CONTINUE
10    CONTINUE
      RETURN
      END
*********************************************
      SUBROUTINE SUBST(C,X,A,O,N)
      DIMENSION A(50,50),C(50),X(50)
      INTEGER O(50)
      X(N) = C(O(N))/A(O(N),N)
      DO 10 I=N-1,1,-1
         SUM = 0.0
         DO 20 J=I+1,N
            SUM=SUM+A(O(I),J)*X(J)
20       CONTINUE
         X(I) = (C(O(I))-SUM)/A(O(I),I)
10    CONTINUE
      RETURN
      END
*********************************************
      SUBROUTINE PIVOT(S,A,O,N,K)
      DIMENSION A(50,50),S(50)
      INTEGER O(50)
      INTEGER*2 K,PIVIT,II,IDUM
      PIVIT = K
      BIG=ABS(A(O(K),K)/S(O(K)))
      DO 10 II=K+1,N
         DUMMY=ABS(A(O(II),K)/S(O(II)))
         IF (DUMMY.GT.BIG) THEN
            BIG=DUMMY
            PIVIT=II
         ENDIF
10    CONTINUE
      IDUM=O(PIVIT)
      O(PIVIT)=O(K)
      O(K)=IDUM
      RETURN
      END
```

(*a*)

```
  idum := O[pivot];
  O[pivot] := O[k];
  O[k] := idum;
End;                        {of procedure Pivot}
  PROCEDURE Eliminate(VAR C: Vector;
                          VAR A: Matrix;
                          VAR O:Row);

  VAR
     i,j,k: integer;
     factor: real;
  Begin                   {procedure Eliminate}
  For k := 1 to N-1 do
     Begin
     Pivot (S,A,O,k);
     For i := k+1 to N do
        Begin
        factor := A[O[i],k]/A[O[k],k];
        For j := k+1 to N do
           Begin
           A[O[i],j] := A[O[i],j]
                      - factor*A[O[k],j];
           End;
        C[O[i]] := C[O[i]]
              - (A[O[i],k]/A[O[k],k])*C[O[k]];
        End;
     End;
End;                  {of procedure Eliminate}
PROCEDURE Substitute(VAR C,X: Vector;
                         A: Matrix;
                         VAR O:Row);

  VAR
     i,j: integer;
     Sum: real;
  Begin                  {procedure Substitute}
  X[N] := C[O[N]]/A[O[N],N];
  For i := N-1 downto 1 do
     Begin
     Sum := 0.0;
     For j := i+1 to N do
        Begin
        Sum := Sum + A[O[i],j]*X[j];
        End;
     X[i] := (C[O[i]] - Sum)/A[O[i],i];
     End;
End;                  {of procedure Substitute}
{of Procedure definitions
                 within procedure GaussElim}
VAR
   i,j,k: Integer;
Begin                      {procedure GuassElim}
Order(S,O,N);
Eliminate(C,A,O);
Substitute(C,X,A,O);
End;                     {of procedure GaussElim}
```

(*b*)

Figure 7.8
Three parachutists free-falling while connected by weightless cords.

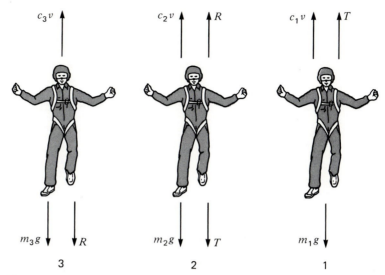

Figure 7.9
Free-body diagrams for each of the three falling parachutists.

$$m_1 g \; - \; T \; - \; c_1 v \qquad\quad = \; m_1 a$$

$$m_2 g \; + \; T \; - \; c_2 v \; - \; R \; = \; m_2 a$$

$$m_3 g \qquad\quad - \; c_3 v \; + \; R \; = \; m_3 a$$

These equations have three unknowns: a, T, and R. After substituting the known values, the equations can be expressed in matrix form (as $g = 9.8$ m/s^2)

$$\begin{bmatrix} 70 & 1 & 0 \\ 60 & -1 & 1 \\ 40 & 0 & -1 \end{bmatrix} \begin{Bmatrix} a \\ T \\ R \end{Bmatrix} = \begin{Bmatrix} 636 \\ 518 \\ 307 \end{Bmatrix}$$

This system can be solved using your own software or the Gauss elimination option on the Electronic TOOLKIT. Figure 7.10a shows the Electronic TOOLKIT solution of $a = 8.5941$ m/s^2; $T = 34.4118$ N; and $R = 36.7647$ N. As depicted in Fig. 7.10b, the software also includes an option to check the accuracy of the computation.

The foregoing results are based on a simple algorithm for the Gauss elimination method with user-friendly input and output routines. The algorithm employed is similar to those shown in Figs. 7.6 and 7.7. You should be able to write your own program for the Gauss elimination method. If you have our software, use it to check the adequacy of your own program.

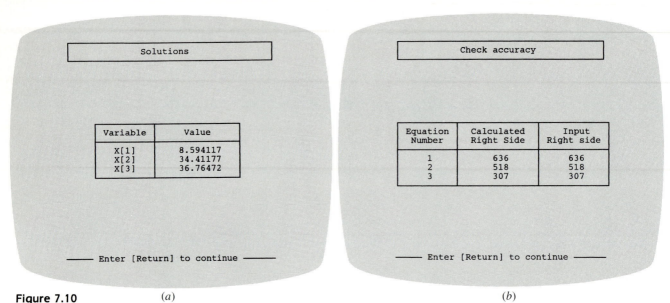

Figure 7.10   (a)   (b)

Computer screens showing (a) the Electronic TOOLKIT solution for the three falling parachutists and (b) accuracy check obtained by substituting the solution into the original equations to verify that the results equal the original right-hand-side constants.

7.5   COMPLEX SYSTEMS

In some problems, it is possible to obtain a complex system of equations

$$[C]\{Z\} = \{W\} \tag{7.21}$$

where

$$[C] = [A] + i[B] \quad \{Z\} = \{X\} + i\{Y\} \quad \{W\} = \{U\} + i\{V\} \tag{7.22}$$

where $i = \sqrt{-1}$.

The most straightforward way to solve such a system is to employ one of the algorithms described in this part of the book, but replace all real operations with complex ones. Of course, this is only possible for those languages, such as FORTRAN, that allow complex variables.

For languages such as BASIC and Pascal that do not permit the declaration of complex variables, it is possible to write code to convert real to complex operations. However, this is not a trivial task. An alternative is to convert the complex system into an equivalent one dealing with real variables. This can be done by substituting Eq. (7.22) into Eq. (7.21) and equating real and complex parts of the resulting equation to yield

$$[A]\{X\} - [B]\{Y\} = \{U\} \tag{7.23}$$

and

$$[B]\{X\} + [A]\{Y\} = \{V\} \tag{7.24}$$

Thus, the system of n complex equations is converted to a set of $2n$ real ones. This means that storage and execution time will be increased significantly. Consequently, a trade-off exists regarding this option. If you evaluate complex systems infrequently, it is preferable to use Eqs. (7.23) and (7.24) because of their convenience. However, if you use them often and desire to employ BASIC or Pascal, it may be worth the up-front programming effort to write a customized equation solver that converts real to complex operations.

7.6 NONLINEAR SYSTEMS OF EQUATIONS

Recall that at the end of Chap. 5 we presented an approach to solve two simultaneous equations with two unknowns. This approach can be extended to the general case of solving n simultaneous nonlinear equations.

$$f_1(x_1, x_2, \ldots, x_n) = 0$$

$$f_2(x_1, x_2, \ldots, x_n) = 0$$

$$\cdot$$

$$\cdot \tag{7.25}$$

$$\cdot$$

$$f_n(x_1, x_2, \ldots, x_n) = 0$$

The solution of this system consists of the set of x values that simultaneously result in all the equations equaling zero.

As described in Sec. 5.5, one approach to solving such systems is based on a multidimensional version of the Newton-Raphson method. Thus, a Taylor series expansion is written for each equation. For example, for the lth equation

$$f_{l,i+1} = f_{l,i} + (x_{1,i+1} - x_{1,i})\frac{\partial f_{l,i}}{\partial x_1} + (x_{2,i+1} - x_{2,i})\frac{\partial f_{l,i}}{\partial x_2}$$

$$+ \cdots + (x_{n,i+1} - x_{n,i})\frac{\partial f_{l,i}}{\partial x_n} \tag{7.26}$$

where the first subscript, l, represents the equation or unknown and the second subscript denotes whether the value or function in question is at the present value (i) or at the next value ($i + 1$).

Equations of the form of (7.26) are written for each of the original nonlinear equations. Then, as was done in deriving Eq. (5.20) from (5.19), all $f_{l,i+1}$ terms are set to zero as would be the case at the root and Eq. (7.26) can be written as

$$-f_{l,i} + x_{1,i}\frac{\partial f_{l,i}}{\partial x_1} + x_{2,i}\frac{\partial f_{l,i}}{\partial x_2} + \cdots + x_{n,i}\frac{\partial f_{l,i}}{\partial x_n}$$

$$= x_{1,i+1}\frac{\partial f_{l,i}}{\partial x_1} + x_{2,i+1}\frac{\partial f_{l,i}}{\partial x_2} + \cdots + x_{n,i+1}\frac{\partial f_{l,i}}{\partial x_n} \tag{7.27}$$

Notice that the only unknowns in Eq. (7.27) are the $x_{l,i+1}$ terms on the right-hand side. All other quantities are located at the present value (i) and, thus, are givens at any iteration. Consequently, the set of equations generally represented by Eq. (7.27) (that is, with $l = 1, 2, \ldots, n$) constitutes a set of linear simultaneous equations that can be solved by methods elaborated in this part of the book.

Matrix notation can be employed to express Eq. (7.27) concisely. The partial derivatives can be expressed as

$$
[Z] = \begin{bmatrix}
\dfrac{\partial f_{1,i}}{\partial x_1} & \dfrac{\partial f_{1,i}}{\partial x_2} & \cdots & \dfrac{\partial f_{1,i}}{\partial x_n} \\[2ex]
\dfrac{\partial f_{2,i}}{\partial x_1} & \dfrac{\partial f_{2,i}}{\partial x_2} & \cdots & \dfrac{\partial f_{2,i}}{\partial x_n} \\[2ex]
\cdot & \cdot & & \cdot \\
\cdot & \cdot & & \cdot \\
\cdot & \cdot & & \cdot \\[1ex]
\dfrac{\partial f_{n,i}}{\partial x_1} & \dfrac{\partial f_{n,i}}{\partial x_2} & \cdots & \dfrac{\partial f_{n,i}}{\partial x_n}
\end{bmatrix}
\tag{7.28}
$$

The initial and final values can be expressed in vector form as

$$
\{X_i\}^T = [x_{1,i} \ \ x_{2,i} \ \ \cdots \ \ x_{n,i}]
$$

and

$$
\{X_{i+1}\}^T = [x_{1,i+1} \ \ x_{2,i+1} \ \ \cdots \ \ x_{n,i+1}]
$$

Finally, the function values at i can be expressed as

$$
\{F_i\}^T = [f_{1,i} \ \ f_{2,i} \ \ \cdots \ \ f_{n,i}]
$$

Using these relationships, Eq. (7.27) can be represented concisely as

$$
[Z]\{X_{i+1}\} = \{W_i\}
\tag{7.29}
$$

where

$$
\{W_i\} = -\{F_i\} + [Z]\{X_i\}
$$

Equation (7.29) can be solved using a technique such as Gauss elimination. This process can be repeated iteratively to obtain refined estimates in a fashion similar to the two-equation case in Sec. 5.5.2.

It should be noted that there are two major shortcomings to the foregoing approach. First, Eq. (7.28) is often inconvenient to evaluate. Therefore, variations of the Newton-Raphson approach have been developed to circumvent this dilemma. As might be expected, most are based on using finite-difference approximations for the partial derivatives that comprise $[Z]$.

The second shortcoming of the multiequation Newton-Raphson method is that excellent initial guesses are usually required to ensure convergence. Because these are often difficult to obtain, alternative approaches that are slower than Newton-Raphson but which have better convergence behavior have been developed. One common approach is to reformulate the nonlinear system as a single function

$$F(x) = \sum_{i=1}^{n} [f_i(x_1, x_2, \ldots, x_n)]^2$$

where $f_i(x_1, x_2, \ldots, x_n)$ is the ith member of the original system of Eq. (7.25). The values of x that minimize this function also represent the solution of the nonlinear system. As we will see in Chap. 11, this reformulation belongs to a class of problems called *nonlinear regression*. As such it can be approached with a number of optimization techniques such as the steepest-descent method and the Levenberg-Marquardt algorithm (Ralston and Rabinowitz, 1978).

7.7 SUMMARY

In summary, we have devoted this chapter to Gauss elimination, the most fundamental method for solving simultaneous linear algebraic equations. Although it is one of the earliest techniques developed for this purpose, it is nevertheless an extremely effective algorithm for obtaining solutions for many engineering problems. Aside from this practical utility, this chapter also provided a context for our discussion of general issues such as round off, scaling, and conditioning.

Answers obtained using Gauss elimination may be checked by substituting them into the original equations. However, this does not always represent a reliable check for ill-conditioned systems. Therefore, some measure of condition, such as the determinant of the scaled system, should be computed if round-off error is suspected. Using partial pivoting and more significant figures in the computation are two options for mitigating round-off error. In the next chapter, we will return to the topic of system condition when we discuss the matrix inverse.

PROBLEMS

Hand Calculations

7.1 Write the following set of equations in matrix form:

$$45 = 7x_1 + 6x_3$$

$$35 = 3x_2 + 8x_1$$

$$22 = 9x_3 + 4x_2$$

Write the transpose of the matrix of coefficients.

7.2 A number of matrices are defined as

$$[A] = \begin{bmatrix} 1 & 5 & 6 \\ 7 & 1 & 3 \\ 4 & 0 & 5 \end{bmatrix} \quad [B] = \begin{bmatrix} 4 & 3 & 7 \\ 1 & 2 & 6 \\ 1 & 0 & 4 \end{bmatrix}$$

$$[C] = \begin{bmatrix} 2 \\ 6 \\ 1 \end{bmatrix} \quad [D] = \begin{bmatrix} 5 & 4 & 3 & 6 \\ 2 & 1 & 7 & 5 \end{bmatrix} \quad [E] = \begin{bmatrix} 4 & 5 \\ 1 & 2 \\ 5 & 6 \end{bmatrix}$$

$$[F] = \begin{bmatrix} 2 & 0 & 1 \\ 1 & 6 & 3 \end{bmatrix} \quad [G] = [8 \quad 6 \quad 4]$$

Answer the following questions regarding these matrices:
(a) What are the dimensions of the matrices?
(b) Identify the square, column, and row matrices.
(c) What are the values of the elements

$$a_{12} \quad b_{23} \quad d_{32} \quad e_{22} \quad f_{12} \quad g_{12}$$

(d) Perform the operations

    (1) $[A] + [B]$ (6) $[B] \times [A]$

    (2) $[B] - [A]$ (7) $[A] \times [C]$

    (3) $[A] + [F]$ (8) $[C]^T$

    (4) $5 \times [B]$ (9) $[D]^T$

    (5) $[A] \times [B]$ (10) $[I] \times [B]$

7.3 Three matrices are defined as

$$[X] = \begin{bmatrix} 2 & 6 \\ 3 & 10 \\ 7 & 4 \end{bmatrix} \quad [Y] = \begin{bmatrix} 6 & 0 \\ 1 & 4 \end{bmatrix} \quad [Z] = \begin{bmatrix} 2 & 1 \\ 6 & 8 \end{bmatrix}$$

(a) Perform all possible multiplications that can be computed between pairs of these matrices.
(b) Use the method in Box PT3.2 to justify why the remaining pairs cannot be multiplied.
(c) Use the results of (a) to illustrate why the order of multiplication is important.

7.4 Use the graphical method to solve

$$4x_1 - 6x_2 = -20$$

$$-x_1 + 12x_2 = 60$$

Check your results by substituting them back into the equations.

7.5 Given the system of equations

$$0.75x_1 + x_2 = 14.25$$

$$1.2x_1 + 1.7x_2 = 20$$

(a) Solve graphically.
(b) On the basis of the graphical solution, what do you expect regarding the condition of the system?
(c) Solve by the elimination of unknowns.
(d) Check your answers by substituting them back into the original equations.

7.6 For the set of equations

$$7x_1 - 3x_2 + 3x_3 = -49$$

$$x_1 - 2x_2 - 5x_3 = 5$$

$$3x_1 - 6x_2 + 10x_3 = -84$$

(a) Compute the determinant.
(b) Use Cramer's rule to solve for the x's.
(c) Substitute your results back into the original equation to check your results.

7.7 Given the equations

$$0.5x_1 - x_2 = -9.5$$

$$0.28x_1 - 0.5x_2 = -4.8$$

(a) Solve graphically.
(b) After scaling, compute the determinant.
(c) On the basis of (a) and (b) what would you expect regarding the system's condition?
(d) Solve by the elimination of unknowns.
(e) Solve again, but with a_{11} modified slightly to 0.55. Interpret your results in light of the discussion of ill conditioning in Sec. 7.3.3.

7.8 Given the system

$$-12x_1 + x_2 - 8x_3 = -80$$

$$x_1 - 6x_2 + 4x_3 = 13$$

$$-2x_1 - x_2 + 10x_3 = 90$$

(a) Solve by naive Gauss elimination. Show all steps of the computation.
(b) Substitute your results into the original equation to check your answers.

7.9 Use Gauss elimination to solve:

$$4x_1 + 5x_2 - 6x_3 = 28$$

$$2x_1 - 7x_3 = 29$$

$$-5x_1 - 5x_2 = -65$$

Employ partial pivoting and check your answers by substituting them into the original equations.

7.10 Use Gauss elimination to solve:

$$3x_2 - 13x_3 = -50$$

$$2x_1 - 6x_2 + x_3 = 45$$

$$4x_1 + 8x_3 = 4$$

Employ partial pivoting. Check your answers by substituting them into the original equations.

7.11 Repeat Example 7.11 but halve the drag coefficients.

7.12 Perform the same computation as in Example 7.11, but use five parachutists with the following characteristics.

Parachutist	Mass, kg	Drag Coefficient, kg/s
1	60	15
2	80	14
3	70	16
4	75	12
5	90	10

The parachutists have a velocity of 8 m/s.

7.13 Three parachutists are connected by a weightless cord while free-falling. The upward force due to drag is represented by

$$F_U = -c'v^2$$

where c' is a drag coefficient (in kilograms per meter). If the team has a velocity of 10 m/s, calculate the tension in each rope and the acceleration, given the following:

Parachutist	Mass, kg	Drag Coefficient, kg/m
1 (top)	50	0.6
2 (middle)	30	0.5
3 (bottom)	70	0.7

Computer-Related Problems

7.14 Write a general computer program to multiply two matrices, that is, $[X] = [Y][Z]$ where $[Y]$ is m by n and $[Z]$ is n by p. Test the program using

$$[Y] = \begin{bmatrix} 3 & 6 \\ 2 & 0 \\ 8 & 7 \end{bmatrix} \quad [Z] = \begin{bmatrix} 5 & 6 & 7 & 4 \\ 3 & 0 & 1 & 2 \end{bmatrix}$$

7.15 Write a computer program to generate the transpose of a matrix. Test it on

$$[A] = \begin{bmatrix} 5 & 0 & 7 \\ 4 & 1 & 5 \\ 2 & 6 & 1 \end{bmatrix}$$

7.16 Reprogram one of the codes from Fig. 7.6 or 7.7 so that it is user-friendly. Among other things provide documentation statements throughout the program to identify what each section is intended to accomplish.

7.17 Test the program you developed in Prob. 7.16 by duplicating the computations from Examples 7.5 and 7.11.

7.18 Use the program you developed in Prob. 7.16 to repeat Probs. 7.8 through 7.11.

7.19 Repeat Probs. 7.17 and 7.18, but use the Electronic TOOLKIT software. Also use the program to perform an error check for each problem.

7.20 Solve Probs. 7.12 and 7.13 using the software developed in Prob. 7.16.

7.21 Develop a user-friendly program for nonlinear systems based on Sec. 7.6.

7.22 Test the program developed in Prob. 7.21 by using it to duplicate the computations in Example 5.10 and Prob. 5.19.

CHAPTER 8
Matrix Inversion and Gauss-Seidel

In the present chapter, we describe two additional methods for solving simultaneous linear equations. The first, *Gauss-Jordan,* is very similar to Gauss elimination. Our primary motive for introducing this method to you is that the Gauss-Jordan technique provides a simple and convenient numerical method for computing the *matrix inverse*. The inverse has a number of valuable applications in engineering practice. It also provides a means for evaluating system condition.

The second method, *Gauss-Seidel,* is fundamentally different from Gauss elimination and the Gauss-Jordan method in that it is an *approximate, iterative* method. That is, it employs initial guesses and then iterates to obtain refined estimates of the solution. The Gauss-Seidel method is particularly well-suited for large numbers of equations. In these cases, elimination methods can be subject to round-off errors. Because the error of the Gauss-Seidel method is controlled by the number of iterations, round-off error is not an issue of concern with this method. However, there are certain instances where the Gauss-Seidel technique will not converge on the correct answer. These and other trade-offs between elimination and iterative methods will be discussed in subsequent pages.

8.1 THE MATRIX INVERSE

8.1.1 Gauss-Jordan

The Gauss-Jordan method is a variation of Gauss elimination. The major difference is that when an unknown is eliminated in the Gauss-Jordan method, it is eliminated from all other equations rather than just the subsequent ones. In addition, all rows are normalized by dividing them by their pivot elements. Thus, the elimination step results in an identity matrix rather than a triangular matrix (Fig. 8.1). Consequently, it is not necessary to employ back-substitution to obtain the solution. The method is best illustrated by an example.

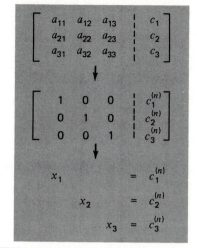

Figure 8.1
Graphical depiction of the Gauss-Jordan method. Compare with Fig. 7.3 to elucidate the differences between this technique and Gauss elimination. The superscript (n)'s mean that the elements of the right-hand-side vector have been modified n times (for this case, $n = 3$).

EXAMPLE 8.1   Gauss-Jordan Method

Problem Statement: Use the Gauss-Jordan technique to solve the same system as in Example 7.5:

$$3x_1 - 0.1x_2 - 0.2x_3 = 7.85$$
$$0.1x_1 + 7x_2 - 0.3x_3 = -19.3$$
$$0.3x_1 - 0.2x_2 + 10x_3 = 71.4$$

Solution: First, reexpress the coefficients and the right-hand side as an augmented matrix:

$$\begin{bmatrix} 3 & -0.1 & -0.2 & | & 7.85 \\ 0.1 & 7 & -0.3 & | & -19.3 \\ 0.3 & -0.2 & 10 & | & 71.4 \end{bmatrix}$$

Then normalize the first row by dividing it by the pivot element, 3, to yield

$$\begin{bmatrix} 1 & -0.0333333 & -0.0666667 & | & 2.61667 \\ 0.1 & 7 & -0.3 & | & -19.3 \\ 0.3 & -0.2 & 10 & | & 71.4 \end{bmatrix}$$

The x_1 term can be eliminated from the second row by subtracting 0.1 times the first row from the second row. Similarly, subtracting 0.3 times the first row from the third row will eliminate the x_1 term from the third row:

$$\begin{bmatrix} 1 & -0.0333333 & -0.0666667 & | & 2.61667 \\ 0 & 7.00333 & -0.293333 & | & -19.5617 \\ 0 & -0.190000 & 10.0200 & | & 70.6150 \end{bmatrix}$$

Next, normalize the second row by dividing it by 7.00333:

$$\begin{bmatrix} 1 & -0.0333333 & -0.0666667 & | & 2.61667 \\ 0 & 1 & -0.0418848 & | & -2.79320 \\ 0 & -0.190000 & 10.0200 & | & 70.6150 \end{bmatrix}$$

Reduction of the x_2 terms from the first and third equations gives

$$\begin{bmatrix} 1 & 0 & -0.0680629 & | & 2.52356 \\ 0 & 1 & -0.0418848 & | & -2.79320 \\ 0 & 0 & 10.0120 & | & 70.0843 \end{bmatrix}$$

The third row is then normalized by dividing it by 10.0120:

$$\begin{bmatrix} 1 & 0 & -0.0680629 & | & 2.52356 \\ 0 & 1 & -0.0418848 & | & -2.79320 \\ 0 & 0 & 1 & | & 7.00003 \end{bmatrix}$$

Finally, the x_3 terms can be reduced from the first and the second equations to give

$$\begin{bmatrix} 1 & 0 & 0 & | & 3.00000 \\ 0 & 1 & 0 & | & -2.50001 \\ 0 & 0 & 1 & | & 7.00003 \end{bmatrix}$$

Thus, as depicted in Fig. 8.1, the coefficient matrix has been transformed to the identity matrix, and the solution is obtained in the right-hand-side vector. Notice that no back-substitution was required to obtain the solution.

All the material in Chap. 7 regarding the pitfalls and improvements in Gauss elimination also applies to the Gauss-Jordan method. For example, a similar pivoting strategy can be used to avoid division by zero and to reduce round-off error.

Although the Gauss-Jordan technique and Gauss elimination might appear almost identical, the former requires approximately 50 percent more operations. Therefore, Gauss elimination is the simple method of preference for obtaining exact solutions of simultaneous linear equations. One of the primary reasons that we have introduced the Gauss-Jordan, however, is that it provides a straightforward method for obtaining the matrix inverse, as will be described in Sec. 8.1.2.

Computer Algorithm for Gauss-Jordan. Pseudocode for the Gauss-Jordan method without partial pivoting is shown in Fig. 8.2. A pivoting scheme similar to the one for Gauss elimination (recall Fig. 7.5) can be incorporated into this algorithm.

Notice that the equations solved by Fig. 8.2 are stored as an augmented matrix of coefficients and the right-hand-side constants. In addition, notice that dummy variables are used to store the multiplying factors prior to each of the j loops. This was done for efficiency and because the values would otherwise be replaced in the course of implementing the loops.

```
DOFOR k = 1 to n
   dummy = a_{k,k}
   DOFOR j = 1 to n+1
      a_{k,j} = a_{k,j}/dummy
   ENDDO
   DOFOR i = 1 to n
      IF (i ≠ k)
         dummy = a_{i,k}
         DOFOR j = 1 to n+1
            a_{i,j} = a_{i,j} - dummy·a_{k,j}
         ENDDO
      ENDIF
   ENDDO
ENDDO
```

Figure 8.2
Pseudocode to implement the Gauss-Jordan method without partial pivoting.

8.1.2 Matrix Inversion

In our discussion of matrix operations (Sec. PT3.2.2), we introduced the notion that if a matrix $[A]$ is square, there is another matrix, $[A]^{-1}$, called the inverse of $[A]$, for which [Eq. (PT3.3)]

$$[A][A]^{-1} = [A]^{-1}[A] = [I]$$

We also demonstrated how the inverse can be used to solve a set of simultaneous equations, as in [Eq. (PT3.6)]

$$\{X\} = [A]^{-1}\{C\} \tag{8.1}$$

One application of the inverse occurs when it is necessary to solve several systems of equations of the form

$$[A]\{X\} = \{C\}$$

differing only by the right-hand-side vector $\{C\}$. Rather than solve each system individually, an alternative approach would be to determine the inverse of the matrix of coefficients. Then, Eq. (8.1) can be used to obtain each solution by simply multiplying $[A]^{-1}$ by the particular right-hand-side vector $\{C\}$. Because matrix multiplication is much quicker than inversion, we only perform the time-consuming step once and then obtain the additional solutions in an efficient manner.

One straightforward way to compute the inverse is using the Gauss-Jordan method. To do this, the coefficient matrix is augmented with an identity matrix (Fig. 8.3). Then the Gauss-Jordan method is applied in order to reduce the coefficient matrix to an identity matrix. When this is accomplished, the right-hand side of the augmented matrix will contain the inverse. The technique is illustrated in the following example.

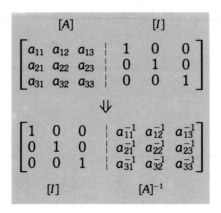

Figure 8.3
Graphical depiction of the Gauss-Jordan method, with matrix inversion. Note that the superscript -1s denote that the original values have been converted to the matrix inverse. They do not represent the value $1/a_{ij}$.

EXAMPLE 8.2   Use of the Gauss-Jordan Method to Compute the Matrix Inverse

Problem Statement: Determine the matrix inverse of the system solved previously in Example 7.5. Obtain the solution by multiplying $[A]^{-1}$ by the right-hand-side vector: $\{C\}^T = [7.85 \quad -19.3 \quad 71.4]$. In addition, obtain the solution for a different right-hand-side vector: $\{C\}^T = [20 \quad 50 \quad 15]$.

Solution: Augment the coefficient matrix with an identity matrix:

$$[A] = \begin{bmatrix} 3 & -0.1 & -0.2 & | & 1 & 0 & 0 \\ 0.1 & 7 & -0.3 & | & 0 & 1 & 0 \\ 0.3 & -0.2 & 10 & | & 0 & 0 & 1 \end{bmatrix}$$

Using a_{11} as the pivot element, normalize row 1 and use it to eliminate x_1 from the other rows:

$$\begin{bmatrix} 1 & -0.0333333 & -0.0666667 & | & 0.333333 & 0 & 0 \\ 0 & 7.00333 & -0.293333 & | & -0.0333333 & 1 & 0 \\ 0 & -0.190000 & 10.0200 & | & -0.0999999 & 0 & 1 \end{bmatrix}$$

Next, a_{22} can be used as the pivot element and x_2 is eliminated from the other rows:

$$\begin{bmatrix} 1 & 0 & -0.068057 & | & 0.333175 & 0.004739329 & 0 \\ 0 & 1 & -0.0417061 & | & -0.00473933 & 0.142180 & 0 \\ 0 & 0 & 10.0121 & | & -0.10090 & 0.0270142 & 1 \end{bmatrix}$$

Finally, a_{33} is used as the pivot element and x_3 is eliminated from the other rows:

$$\begin{bmatrix} 1 & 0 & 0 & | & 0.332489 & 0.00492297 & 0.00679813 \\ 0 & 1 & 0 & | & -0.0051644 & 0.142293 & 0.00418346 \\ 0 & 0 & 1 & | & -0.0100779 & 0.00269816 & 0.0998801 \end{bmatrix}$$

Therefore, the inverse is

$$[A]^{-1} = \begin{bmatrix} 0.332489 & 0.00492297 & 0.00679813 \\ -0.0051644 & 0.142293 & 0.00418346 \\ -0.0100779 & 0.00269816 & 0.0998801 \end{bmatrix}$$

Now, the inverse can be multiplied by the first right-hand-side vector to determine the solution:

$$x_1 = 7.85(0.332489) - 19.3(0.00492297) + 71.4(0.00679813)$$

$$= 3.00041181$$

$$x_2 = 7.85(-0.0051644) - 19.3(0.142293) + 71.4(0.00418346)$$

$$= -2.48809640$$

$$x_3 = 7.85(-0.0100779) - 19.3(0.00269816) , + 71.4(0.0998801)$$

$$= 7.00025314$$

The second solution is simply obtained by performing another multiplication, as in

$$x_1 = 20(0.332489) + 50(0.00492297) + 15(0.00679813)$$

$$= 6.99790045$$

$$x_2 = 20(-0.0051644) + 50(0.142293) + 15(0.00418346)$$

$$= 7.0741139$$

$$x_3 = 20(-0.0100779) + 50(0.00269816) + 15(0.0998801)$$

$$= 1.43155150$$

Although the matrix inverse provides a handy and straightforward way to evaluate multiple right-hand-side vectors, it is not the most efficient algorithm for making such evaluations. In the next chapter, we will present *LU* decomposition methods that employ fewer operations to perform the same task. However, as will be elaborated in Sec. 8.1.3, the elements of the inverse are extremely useful in their own right.

Computer Algorithm for Matrix Inversion. The computer algorithm from Fig. 8.2 can be modified to calculate the matrix inverse. This involves augmenting the coefficient matrix with an identity matrix at the beginning of the program. In addition, some of the loop indices must be increased in order that the computations are performed for all the columns of the augmented coefficient matrix.

Note that an alternative method for determining the inverse will be presented in Chap. 9. This approach, which is based on *LU* decomposition, will be described in Sec. 9.5.1.

8.1.3 Stimulus-Response Computations

As discussed in Sec. PT3.1.2, many of the linear systems of equations confronted in engineering practice are derived from conservation laws. The mathematical expression of these laws is some form of balance equation to ensure that a particular property—mass, force, heat, momentum, or other—is conserved. For a force balance on a structure, the properties might be horizontal or vertical components of the forces acting on each node of the structure (see Case Study 10.3). For a mass balance, the properties might be the mass in each reactor of a chemical process (see Case Study 10.2). Other fields of engineering would yield similar examples.

A single balance equation can be written for each part of the system, resulting in a set of equations defining the behavior of the property for the entire system. These equations are interrelated, or coupled, in that each equation includes one or more of the variables from the other equations. For many cases, these systems are linear and, therefore, of the exact form dealt with in the present chapter.

$$[A]\{X\} = \{C\} \tag{8.2}$$

Now, for balance equations, the terms of Eq. (8.2) have a definite physical interpretation. For example, the elements of $\{X\}$ are the levels of the property being balanced for each part of the system. In a force balance of a structure, they represent the horizontal and vertical forces in each member. For the mass balance, they are the mass of chemical in each reactor. In either case, they represent the system's state or response, which we are trying to determine.

The right-hand-side vector $\{C\}$ contains those elements of the balance that are independent of behavior of the system—that is, they are constants. As such, they often represent the external forces or stimuli that drive the system.

Finally, the matrix of coefficients $[A]$ usually contains the parameters that express how the parts of the system interact or are coupled. Consequently, Eq. (8.2) might be reexpressed as

$$[\text{Interactions}]\{\text{response}\} = \{\text{stimuli}\}$$

Thus, Eq. (8.2) can be seen as an expression of the fundamental mathematical model that we formulated previously as a single equation in Chap. 1 [recall Eq. (1.1)]. We can now see that Eq. (8.2) represents a version that is designed for coupled systems involving several dependent variables, $\{X\}$.

Now, as we know from the present chapter, there are a variety of ways to solve Eq. (8.2). However, using the matrix inverse yields a particularly interesting result. The formal solution [Eq. (8.1)] can be expressed as

$$\{X\} = [A]^{-1}\{C\}$$

or (recalling our definition of matrix multiplication from Box PT3.2)

$$x_1 = a_{11}^{-1}c_1 + a_{12}^{-1}c_2 + a_{13}^{-1}c_3$$

$$x_2 = a_{21}^{-1}c_1 + a_{22}^{-1}c_2 + a_{23}^{-1}c_3$$

$$x_3 = a_{31}^{-1}c_1 + a_{32}^{-1}c_2 + a_{33}^{-1}c_3$$

Thus, we find that the inverted matrix itself, aside from providing a solution, has extremely useful properties. That is, each of its elements represents the response of a single part of the system to a unit stimulus of any other part of the system.

Notice that these formulations are linear and, therefore, superposition and proportionality hold. *Superposition* means that if a system is subject to several different stimuli (the c's), the responses can be computed individually and the results summed to obtain a total response. *Proportionality* means that multiplying the stimuli by a quantity results in the response to those stimuli being multiplied by the same quantity. Thus, the coefficient a_{11}^{-1} is a proportionality constant that gives the value of x_1 due to a unit level of c_1. This result is independent of the effects of c_2 and c_3 on x_1, which are reflected in the coefficients a_{12}^{-1} and a_{13}^{-1}, respectively. Therefore, we can draw the general conclusion that the element a_{ij}^{-1} of the inverted matrix represents the value of x_i due to a unit quantity of c_j. Using the example of the structure, element a_{ij}^{-1} of the matrix inverse would represent the force in member i due to a unit external force at node j. Even for small systems, such behavior of individual stimulus-response interactions would not be intuitively obvious. As such, the matrix inverse provides a powerful technique for understanding the interrelationships of component parts of complicated systems. This power will be demonstrated in Case Studies 10.2, 10.3, and 10.5.

8.2 ERROR ANALYSIS AND SYSTEM CONDITION

Aside from its engineering applications, the inverse also provides a means to discern whether systems are ill-conditioned. Three methods are available for this purpose:

1. Scale the matrix of coefficients, $[A]$, so that the largest element in each row is 1. If there are elements of $[A]^{-1}$ that are several orders of magnitude greater than one, it is likely that the system is ill-conditioned (see Box 8.1).
2. Multiply the inverse by the original coefficient matrix and assess whether the result is close to the identity matrix. If not, it indicates ill conditioning.
3. Invert the inverted matrix and assess whether the result is sufficiently close to the original coefficient matrix. If not, it again indicates that the system is ill-conditioned.

Although these methods can indicate ill conditioning, it would be preferable to obtain a single number (such as the condition number from Sec. 3.6.3) that could serve as an indicator of the problem. Attempts to formulate such a matrix condition number are based on the mathematical concept of the norm.

8.2.1 Vector and Matrix Norms

A *norm* is a real-valued function that provides a measure of the size or "length" of multicomponent mathematical entities such as vectors and matrices (see Box 8.2).

A simple example is a vector in three-dimensional Euclidean space (Fig. 8.4) which can be represented as

$$[F] = [a \quad b \quad c]$$

where a, b, and c are the distances along the x, y, and z axes, respectively. The length of this vector—that is, the distance from the coordinate $(0, 0, 0)$ to (a, b, c)—can be simply computed as

$$\|F\|_e = \sqrt{a^2 + b^2 + c^2}$$

where the nomenclature $\|F\|_e$ indicates that this length is referred to as the *Euclidean norm* of $[F]$.

Similarly for an n-dimensional vector, $[X] = [x_1 \; x_2 \cdots x_n]$, a Euclidean norm would be computed as

$$\|X\|_e = \sqrt{\sum_{i=1}^{n} x_i^2}$$

The concept can be extended further to a matrix $[A]$ as in

$$\|A\|_e = \sqrt{\sum_{i=1}^{n}\sum_{j=1}^{n} a_{ij}^2} \tag{8.3}$$

Box 8.1   Interpreting the Elements of the Matrix Inverse as a Measure of Ill Conditioning

---

One method for assessing a system's condition is to scale $[A]$ so that the largest element in each row is 1 and then compute $[A]^{-1}$. If elements of $[A]^{-1}$ are several orders of magnitude greater than the elements of the original scaled matrix, it is likely that the system is ill-conditioned.

Insight into this approach can be gained by recalling that a way to check whether an approximate solution, $\{\tilde{X}\}$, is acceptable is to substitute it into the original equations and see whether the original right-hand-side constants result. This is equivalent to

$$\{R\} = \{C\} - [A]\{\tilde{X}\} \tag{B8.1.1}$$

where $\{R\}$ is the residual between the right-hand-side constants and the values computed with the solution $\{\tilde{X}\}$. If $\{R\}$ is small, we might conclude that the $\{\tilde{X}\}$ values are adequate. However, suppose that $\{X\}$ is the exact solution which yields a zero residual as in

$$\{0\} = \{C\} - [A]\{X\} \tag{B8.1.2}$$

Subtracting Eq. (B8.1.2) from (B8.1.1) yields

$$\{R\} = [A](\{X\} - \{\tilde{X}\})$$

Multiplying both sides of this equation by $[A]^{-1}$ gives

$$\{X\} - \{\tilde{X}\} = [A]^{-1}\{R\}$$

This result indicates why checking a solution by substitution can be misleading. For cases where elements of $[A]^{-1}$ are large, a small discrepancy in the right-hand-side residual $\{R\}$ could correspond to a large error, $\{X\} - \{\tilde{X}\}$, in the calculated value of the unknowns. In other words, a small residual does not guarantee an accurate solution. Thus, we can conclude that if the largest element of $[A]^{-1}$ is on the order of magnitude of unity, the system can be considered to be well-conditioned. Conversely, if $[A]^{-1}$ includes elements much larger than unity, we conclude that the system is ill-conditioned.

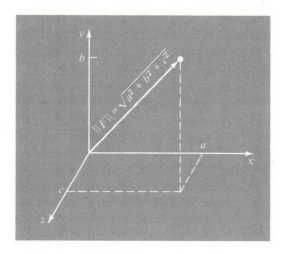

Figure 8.4
Graphical depiction of a vector $[F] = [a\ b\ c]$ in Euclidean space.

which is given a special name—the *Frobenius norm*. However, just as with the vector norms, it provides a single value to quantify the "size" of $[A]$.

It should be noted that there are alternatives to the Euclidean and Frobenius norms (see Box 8.2). For example, a *uniform vector norm* is defined as

$$\|X\|_\infty = \max_{1 \le i \le n} |x_i|$$

That is, the element with the largest absolute value is taken as the measure of the vector's size. Similarly, a *uniform matrix norm* or *row-sum norm* is defined as

$$\|A\|_\infty = \max_{1 \le i \le n} \sum_{j=1}^{n} |a_{ij}| \qquad (8.4)$$

In this case, the sum of the absolute value of the elements is computed for each row and the largest of these is taken as the norm.

Although there are theoretical benefits for using certain of the norms, often the choice is influenced by practical considerations. For example, the uniform-row norm is widely used because of the ease with which it can be calculated and the fact that it usually provides an adequate measure of matrix size.

8.2.2 Matrix Condition Number

Now that we have introduced the concept of the norm, we can use it to define

$$\text{Cond } [A] = \|A\| \cdot \|A^{-1}\| \qquad (8.5)$$

where Cond $[A]$ is called the *matrix condition number*. Note that for a matrix, $[A]$,

Box 8.2   *Matrix Norms*

As developed in this section, Euclidean norms can be employed to quantify the size of a vector,

$$||X||_e = \sqrt{\sum_{i=1}^{n} x_i^2}$$

or matrix,

$$||A||_e = \sqrt{\sum_{i=1}^{n} \sum_{j=1}^{n} a_{ij}^2}$$

For vectors, there are alternatives called *p* norms which can be represented generally by

$$||X||_p = \left(\sum_{i=1}^{n} |x_i|^p \right)^{1/p}$$

Two important examples are

$$||X||_1 = \sum_{i=1}^{n} |x_i|$$

which represents the norm as the sum of the absolute values of the elements. Another is the *maximum-magnitude* or *uniform-vector norm*

$$||X||_\infty = \max_{1 \le i \le n} |x_i|$$

which defines the norm as the element with the largest absolute value. We can also see that the Euclidean norm and the 2 norm, $||X||_2$, are identical for vectors.

Using a similar approach, norms can be developed for matrices. For example,

$$||A||_1 = \max_{1 \le j \le n} \sum_{i=1}^{n} |a_{ij}|$$

That is, a summation of the absolute values of the coefficients is performed for each column and the largest of these summations is taken as the norm. This is called the *column-sum norm*.

A similar determination can be made for the rows, resulting in a *uniform-matrix* or *row-sum norm*.

$$||A||_\infty = \max_{1 \le i \le n} \sum_{j=1}^{n} |a_{ij}|$$

It should be noted that, in contrast to vectors, the 2 norm and the Euclidean norm for a matrix are not the same. Whereas the Euclidean norm $||A||_e$ can be easily determined by Eq. (8.3), the matrix 2 norm, $||A||_2$, is related to its eigenvalues and is not readily computed. It is of special importance, though, because it is the minimum norm and, therefore, provides the tightest measure of size.

this number will be greater than or equal to 1. It can be shown (Ralston and Rabinowitz, 1978; Gerald and Wheatley, 1984), that

$$\frac{||\Delta X||}{||X||} \le \text{Cond } [A] \frac{||\Delta A||}{||A||}$$

That is, the relative error of the norm of the computed solution can be as large as the relative error of the norm of the coefficients of [A] multiplied by the condition number. For example, if the coefficients of [A] are known to *t*-digit precision (that is, rounding errors are on the order of 10^{-t}) and Cond [A] = 10^c, the solution [X] may be valid to only $t - c$ digits (rounding errors $\simeq 10^{c-t}$).

EXAMPLE 8.3   Matrix Condition Evaluation

Problem Statement: The Hilbert matrix, which is notoriously ill-conditioned, can be represented generally as

$$
\begin{bmatrix}
1 & 1/2 & 1/3 & \cdots & 1/n \\
1/2 & 1/3 & 1/4 & \cdots & 1/(n+1) \\
\cdot & & & & \cdot \\
\cdot & & & & \cdot \\
\cdot & & & & \cdot \\
1/n & 1/(n+1) & 1/(n+2) & \cdots & 1/(2n)
\end{bmatrix}
$$

An example of a system of linear algebraic equations based on the Hilbert matrix is ($n = 3$)

$$
\begin{bmatrix}
1 & 1/2 & 1/3 \\
1/2 & 1/3 & 1/4 \\
1/3 & 1/4 & 1/5
\end{bmatrix}
\begin{Bmatrix}
x_1 \\
x_2 \\
x_3
\end{Bmatrix}
=
\begin{Bmatrix}
1 + 1/2 + 1/3 \\
1/2 + 1/3 + 1/4 \\
1/3 + 1/4 + 1/5
\end{Bmatrix}
$$

Note that the exact solution for this system is $\{X\}^T = [1 \quad 1 \quad 1]$. Use the row-sum norm to estimate the matrix condition number for this system.

Solution: The row-sum norm for the normalized matrix can be computed as

$$\|A\|_\infty = 1 + 3/4 + 3/5 = 2.35$$

The matrix inverse can be calculated as

$$
[A]^{-1} =
\begin{bmatrix}
9 & -18 & 10 \\
-36 & 96 & -60 \\
30 & -90 & 60
\end{bmatrix}
$$

Note that the elements of this matrix are much larger than the original matrix. This is also reflected in its row-sum norm which is computed as

$$\|A^{-1}\|_\infty = 36 + 96 + 60 = 192$$

Thus, the condition number can be calculated as

$$\text{Cond}[A] = 2.35(192) = 451.2$$

The fact that the condition number is considerably greater than unity suggests that the system is ill-conditioned. The extent of the ill conditioning can be quantified by calculating $c = \log 451.2 = 2.65$. A computer such as the IBM PC has approximately $t = \log 2^{-24} = 7.2$ significant base-10 digits (recall Sec. 3.4.1). Therefore, the solution could exhibit rounding errors of up to $10^{(2.65-7.2)} = 3 \times 10^{-5}$. Note that such estimates almost always overpredict the actual error. However, they are useful in alerting you to the possibility that round-off errors may be significant.

Practically speaking, the problem with implementing Eq. (8.5) is the computational price required to obtain $\|A^{-1}\|$. Rice (1983) outlines some possible strategies to mitigate this problem. Further, he suggests an alternative way to assess system condition: run the same solution on two different compilers. Because the resulting codes will likely implement the arithmetic differently, the effect of ill conditioning should be evident from such an experiment.

8.2.3 Iterative Refinement

In some cases, round-off errors can be reduced by the following procedure. Consider a system of linear equations of the form

$$
\begin{aligned}
a_{11}x_1 + a_{12}x_2 + &\cdots + a_{1n}x_n = c_1 \\
a_{21}x_1 + a_{22}x_2 + &\cdots + a_{2n}x_n = c_2 \\
&\vdots \\
a_{n1}x_1 + a_{n2}x_2 + &\cdots + a_{nn}x_n = c_n
\end{aligned}
\tag{8.6}
$$

Suppose an approximate solution vector is given by $\tilde{x}_1, \tilde{x}_2, \ldots, \tilde{x}_n$. These results are substituted into Eq. (8.6), giving

$$
\begin{aligned}
a_{11}\tilde{x}_1 + a_{12}\tilde{x}_2 + &\cdots + a_{1n}\tilde{x}_n = \tilde{c}_1 \\
a_{21}\tilde{x}_1 + a_{22}\tilde{x}_2 + &\cdots + a_{2n}\tilde{x}_n = \tilde{c}_2 \\
&\vdots \\
a_{n1}\tilde{x}_1 + a_{n2}\tilde{x}_2 + &\cdots + a_{nn}\tilde{x}_n = \tilde{c}_n
\end{aligned}
\tag{8.7}
$$

Now suppose that the exact solutions $x_1, x_2, \ldots, x_n$ are expressed as a function of $\tilde{x}_1, \tilde{x}_2, \ldots, \tilde{x}_n$ and correction factors $\Delta x_1, \Delta x_2, \ldots, \Delta x_n$, where

$$
\begin{aligned}
x_1 &= \tilde{x}_1 + \Delta x_1 \\
x_2 &= \tilde{x}_2 + \Delta x_2 \\
&\vdots \\
x_n &= \tilde{x}_n + \Delta x_n
\end{aligned}
\tag{8.8}
$$

If these results are substituted into Eq. (8.6) the following system results:

$$a_{11}(\tilde{x}_1 + \Delta x_1) + a_{12}(\tilde{x}_2 + \Delta x_2) + \cdots + a_{1n}(\tilde{x}_n + \Delta x_n) = c_1$$

$$a_{21}(\tilde{x}_1 + \Delta x_1) + a_{22}(\tilde{x}_2 + \Delta x_2) + \cdots + a_{2n}(\tilde{x}_n + \Delta x_n) = c_2$$

$$\vdots \qquad\qquad \vdots \qquad\qquad \vdots \qquad\qquad \vdots$$

$$a_{n1}(\tilde{x}_1 + \Delta x_1) + a_{n2}(\tilde{x}_2 + \Delta x_2) + \cdots + a_{nn}(\tilde{x}_n + \Delta x_n) = c_n$$

(8.9)

Now Eq. (8.7) can be subtracted from Eq. (8.9) to yield

$$a_{11}\Delta x_1 + a_{12}\Delta x_2 + \cdots + a_{1n}\Delta x_n = c_1 - \tilde{c}_1 = E_1$$

$$a_{21}\Delta x_1 + a_{22}\Delta x_2 + \cdots + a_{2n}\Delta x_n = c_2 - \tilde{c}_2 = E_2$$

$$\vdots \qquad\qquad \vdots \qquad\qquad \vdots \qquad\qquad \vdots$$

$$a_{n1}\Delta x_1 + a_{n2}\Delta x_2 + \cdots + a_{nn}\Delta x_n = c_n - \tilde{c}_n = E_n$$

(8.10)

This system itself is a set of simultaneous linear equations that can be solved to obtain the correction factors. The factors can then be applied to improve the solution as specified by Eq. (8.8).

It is relatively straightforward to integrate an iterative refinement procedure into computer programs for elimination methods. It is especially effective for those methods such as Gauss-Jordan with matrix inversion or, better yet, the *LU* decomposition approaches of the next chapter, which are designed to evaluate different right-hand-side vectors efficiently. Note that in order to be effective for correcting ill-conditioned systems, the E's in Eq. (8.10) must be expressed in extended precision.

8.3   THE GAUSS-SEIDEL METHOD

The exact elimination methods discussed previously can be used to solve approximately 100 simultaneous linear equations. This number can often be expanded if the system is well-conditioned, pivot strategy is employed, extended precision is used, or the matrix is sparse. However, because of round-off error, elimination methods sometimes prove inadequate for larger systems. For these types of problems, iterative or approximate methods can often be used to advantage.

We used similar types of techniques to obtain the roots of a single equation in Chap. 5. Those approaches consisted of guessing a value and then using a systematic method to obtain a refined estimate of the root. Because the present part of the book deals with a similar problem—obtaining the values that simultaneously satisfy a set of equations—we might suspect that such approximate methods could be useful in this context.

The reason that iterative methods are useful for round-off-prone systems is that an approximate method can be continued until it has converged within some prespecified tolerance of error. Thus, round off is no longer an issue, because you control the level of error that is acceptable.

The *Gauss-Seidel method* is the most commonly-used iterative method. Assume that we are given a set of n equations:

$$[A]\{X\} = \{C\}$$

If the diagonal elements are all nonzero, the first equation can be solved for x_1, the second for x_2, and so on to yield

$$x_1 = \frac{c_1 - a_{12}x_2 - a_{13}x_3 - \cdots - a_{1n}x_n}{a_{11}} \tag{8.11a}$$

$$x_2 = \frac{c_2 - a_{21}x_1 - a_{23}x_3 - \cdots - a_{2n}x_n}{a_{22}} \tag{8.11b}$$

$$x_3 = \frac{c_3 - a_{31}x_1 - a_{32}x_2 - \cdots - a_{3n}x_n}{a_{33}} \tag{8.11c}$$

$$x_n = \frac{c_n - a_{n1}x_1 - a_{n2}x_2 - \cdots - a_{n,n-1}x_{n-1}}{a_{nn}} \tag{8.11d}$$

Now, we can start the solution process by using guesses for the x's. A simple way to obtain initial guesses is to assume that they are all zero. These zeros can be substituted into Eq. (8.11a), which can be used to calculate a new value for $x_1 = c_1/a_{11}$. Then, we substitute this new value of x_1, along with the previous guesses of zero for $x_3, \ldots, x_n$, into Eq. (8.11b) to compute a new value for x_2. The process is repeated for each of the equations until we use Eq. (8.11d) to calculate a new estimate for x_n. Then we return to the first equation and repeat the entire procedure until our solution converges closely enough to the true values. Convergence can be checked using the criterion [recall Eq. (3.5)]

$$|\epsilon_{a,i}| = \left| \frac{x_i^j - x_i^{j-1}}{x_i^j} \right| 100\% < \epsilon_s \tag{8.12}$$

for all i, where j and $j - 1$ are the present and previous iterations.

EXAMPLE 8.4   Gauss-Seidel Method

Problem Statement: Use the Gauss-Seidel method to obtain the solution of the same system used in Example 8.1:

$$3x_1 - 0.1x_2 - 0.2x_3 = 7.85$$
$$0.1x_1 + 7x_2 - 0.3x_3 = -19.3$$
$$0.3x_1 - 0.2x_2 + 10x_3 = 71.4$$

Recall that the true solution is $x_1 = 3$, $x_2 = -2.5$, and $x_3 = 7$.

Solution: First, solve each of the equations for its unknown on the diagonal.

$$x_1 = \frac{7.85 + 0.1x_2 + 0.2x_3}{3} \qquad (E8.4.1)$$

$$x_2 = \frac{-19.3 - 0.1x_1 + 0.3x_3}{7} \qquad (E8.4.2)$$

$$x_3 = \frac{71.4 - 0.3x_1 + 0.2x_2}{10} \qquad (E8.4.3)$$

By assuming that x_2 and x_3 are zero, Eq. (E8.4.1) can be used to compute

$$x_1 = \frac{7.85}{3} = 2.616666667$$

This value, along with the assumed value of $x_3 = 0$, can be substituted into Eq. (E8.4.2) to calculate

$$x_2 = \frac{-19.3 - 0.1(2.616666667) + 0}{7} = -2.794523810$$

The first iteration is completed by substituting the calculated values for x_1 and x_2 into Eq. (E8.4.3) to yield

$$x_3 = \frac{71.4 - 0.3(2.616666667) + 0.2(-2.794523810)}{10}$$

$$= 7.005609524$$

For the second iteration, the same process is repeated to compute

$$x_1 = \frac{7.85 + 0.1(-2.794523810) + 0.2(7.005609524)}{3}$$

$$= 2.990556508 \qquad |\epsilon_t| = 0.31\%$$

$$x_2 = \frac{-19.3 - 0.1(2.990556508) + 0.3(7.005609524)}{7}$$

$$= -2.499624684 \qquad |\epsilon_t| = 0.015\%$$

$$x_3 = \frac{71.4 - 0.3(2.990556508) + 0.2(-2.499624684)}{10}$$

$$= 7.000290811 \qquad |\epsilon_t| = 0.0042\%$$

The method is, therefore, converging on the true solution. Additional iterations could be applied to improve the answers. However, in an actual problem, we would not know the true answer a priori. Consequently, Eq. (8.12) provides a means to estimate the error.

$$\epsilon_{a1} = \left| \frac{2.990556508 - 2.616666667}{2.990556508} \right| 100 = 12.5\%$$

$$\epsilon_{a2} = \left| \frac{-2.499624684 - (-2.794523810)}{-2.499624684} \right| 100 = 11.8\%$$

$$\epsilon_{a3} = \left| \frac{7.000290811 - 7.005609524}{7.000290811} \right| 100 = 0.076\%$$

Note that, as was the case when determining roots of a single equation, formulations such as Eq. (8.12) usually provide a conservative appraisal of convergence. Thus, when they are met, they ensure that the result is known to *at least* the tolerance specified by ϵ_s.

As each new x value is computed for the Gauss-Seidel method, it is immediately used in the next equation to determine another x value. Thus, if the solution is converging, the best available estimates will be employed. An alternative approach, called *Jacobi iteration,* utilizes a somewhat different tactic. Rather than using the latest available x's, this technique uses Eq. (8.11) to compute a set of new x's on the basis of a set of old x's. Thus, as new values are generated, they are not immediately used but rather are retained for the next iteration.

The difference between the Gauss-Seidel method and Jacobi iteration is depicted in Fig. 8.5. Although there are certain cases where the Jacobi method converges faster, Gauss-Seidel's utilization of the best available estimates usually makes it the method of preference.

8.3.1 Convergence Criterion for the Gauss-Seidel Method

Note that the Gauss-Seidel method is similar in spirit to the technique of simple one-point iteration that was used in Sec. 5.1 to solve for the roots of a single equation. Recall that simple one-point iteration had two fundamental problems: (1) It was sometimes nonconvergent and (2) when it converged it often did so very slowly. The Gauss-Seidel method can also exhibit these shortcomings.

Convergence criteria can be developed by recalling from Sec. 5.5.1 that sufficient conditions for convergence of two nonlinear equations, $u(x, y)$ and $v(x, y)$ are

$$\left| \frac{\partial u}{\partial x} \right| + \left| \frac{\partial v}{\partial x} \right| < 1 \tag{8.13a}$$

and

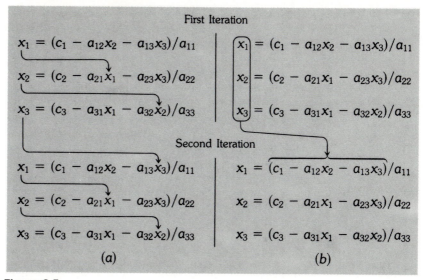

First Iteration

$$x_1 = (c_1 - a_{12}x_2 - a_{13}x_3)/a_{11}$$

$$x_2 = (c_2 - a_{21}x_1 - a_{23}x_3)/a_{22}$$

$$x_3 = (c_3 - a_{31}x_1 - a_{32}x_2)/a_{33}$$

$$x_1 = (c_1 - a_{12}x_2 - a_{13}x_3)/a_{11}$$

$$x_2 = (c_2 - a_{21}x_1 - a_{23}x_3)/a_{22}$$

$$x_3 = (c_3 - a_{31}x_1 - a_{32}x_2)/a_{33}$$

Second Iteration

$$x_1 = (c_1 - a_{12}x_2 - a_{13}x_3)/a_{11}$$

$$x_2 = (c_2 - a_{21}x_1 - a_{23}x_3)/a_{22}$$

$$x_3 = (c_3 - a_{31}x_1 - a_{32}x_2)/a_{33}$$

$$x_1 = (c_1 - a_{12}x_2 - a_{13}x_3)/a_{11}$$

$$x_2 = (c_2 - a_{21}x_1 - a_{23}x_3)/a_{22}$$

$$x_3 = (c_3 - a_{31}x_1 - a_{32}x_2)/a_{33}$$

(a)   (b)

Figure 8.5
Graphical depiction of the difference between (a) the Gauss-Seidel and (b) the Jacobi iterative methods for solving simultaneous linear algebraic equations.

$$\left| \frac{\partial u}{\partial y} \right| + \left| \frac{\partial v}{\partial y} \right| < 1 \tag{8.13b}$$

These criteria also apply to linear equations of the sort we are solving with the Gauss-Seidel method. For example, in the case of two simultaneous equations, the Gauss-Seidel algorithm [Eq. (8.11)] can be expressed as

$$u(x_1, x_2) = \frac{c_1}{a_{11}} - \frac{a_{12}}{a_{11}}x_2 \tag{8.14a}$$

and

$$v(x_1, x_2) = \frac{c_2}{a_{22}} - \frac{a_{21}}{a_{22}}x_1 \tag{8.14b}$$

The partial derivatives of these equations can be evaluated with respect to each of the unknowns as

$$\frac{\partial u}{\partial x_1} = 0 \qquad \frac{\partial v}{\partial x_1} = -\frac{a_{21}}{a_{22}}$$

and

$$\frac{\partial u}{\partial x_2} = -\frac{a_{12}}{a_{11}} \qquad \frac{\partial v}{\partial x_2} = 0$$

which can be substituted into Eq. (8.13) to give

$$\left| \frac{a_{21}}{a_{22}} \right| < 1 \tag{8.15a}$$

and

$$\left| \frac{a_{12}}{a_{11}} \right| < 1 \tag{8.15b}$$

In other words, the absolute values of the slopes of Eq. (8.14) must be less than unity to ensure convergence. This is displayed graphically in Fig. 8.6. Equation (8.15) can also be reformulated as

$$|a_{22}| > |a_{21}|$$

and

$$|a_{11}| > |a_{12}|$$

That is, the diagonal element must be greater than the off-diagonal element for each row.

The extension of the above to n equations is straightforward and can be expressed as

$$|a_{ii}| > \sum |a_{ij}| \tag{8.16}$$

where the summation is taken from $j = 1$ to n, excluding $j = i$. That is, the diagonal coefficient in each of the equations must be larger than the sum of the absolute values of the other coefficients in the equation. This criterion is sufficient but not necessary for convergence. That is, although the method may sometimes work if Eq. (8.16) is not met, convergence is guaranteed if the condition is satisfied. Systems where Eq. (8.16) holds are called *diagonally dominant*. Fortunately, many engineering problems of practical importance fulfill this requirement.

8.3.2 Improvement of Convergence Using Relaxation

Relaxation represents a slight modification of the Gauss-Seidel method and is designed to enhance convergence. After each new value of x is computed using Eq. (8.11), that value is modified by a weighted average of the results of the previous and the present iterations:

$$x_i^{\text{new}} = \lambda x_i^{\text{new}} + (1 - \lambda)x_i^{\text{old}} \tag{8.17}$$

where λ is a weighting factor that is assigned a value between 0 and 2.

If $\lambda = 1, (1 - \lambda)$ is equal to 0 and the result is unmodified. However, if λ is set at a value between 0 and 1, the result is a weighted average of the present and the previous results. This type of modification is called *underrelaxation*. It is typically employed to make a nonconvergent system converge or to hasten convergence by dampening out oscillations.

Figure 8.6
Illustration of (a) convergence and (b) divergence of the Gauss-Seidel method. Notice that the same functions are plotted in both cases (u: $11x_1 + 13x_2 = 286$; v: $11x_1 - 9x_2 = 99$). Thus, the order in which the equations are implemented (as depicted by the direction of the first arrow from the origin) dictates whether the computation converges.

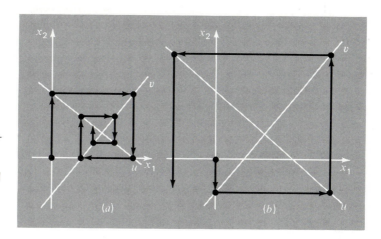

For values of λ from 1 to 2, extra weight is placed on the present value. In this instance, there is an implicit assumption that the new value is moving in the correct direction toward the true solution but at too slow a rate. Thus, the added weight of λ is intended to improve the estimate by pushing it closer to the truth. Hence, this type of modification, which is called *overrelaxation,* is designed to accelerate the convergence of an already convergent system. The approach is also called *successive* or *simultaneous overrelaxation* or *SOR.*

The choice of a proper value for λ is highly problem-specific and is often determined empirically. For a single solution of a set of equations it is often unnecessary. However, if the system under study is to be solved repeatedly, the efficiency introduced by a wise choice of λ can be extremely important. Good examples are the very large systems of partial differential equations that often arise when modeling continuous variations of variables (recall the distributed system depicted in Fig. PT3.1b). We will return to this topic in Part Seven.

8.3.3 Algorithm for Gauss-Seidel

An algorithm for the Gauss-Seidel method, with relaxation, is depicted in Fig. 8.7. Note that this algorithm is not guaranteed to converge if the equations are not input in a diagonally dominant form.

The pseudocode has two features that bear mentioning. First, there is an initial set of nested loops to divide each equation by its diagonal element. This reduces the total number of operations in the algorithm. Second, notice that the error check is designated by a variable called *sentinel.* If any of the equations has an approximate error greater than the stopping criterion (ϵ_s), then the iterations are allowed to continue. The use of the *sentinel* allows us to circumvent unnecessary error estimates once one of the equations exceeds the criterion.

8.3.4 Problem Contexts for the Gauss-Seidel Method

Aside from circumventing the round-off dilemma, the Gauss-Seidel technique has a number of other advantages that make it particularly attractive in the context of certain engineering problems. For example, when the matrix in question is very large and very sparse (that is, most of the elements are zero), elimination methods waste large amounts of computer memory storing zeros.

At the end of Chap. 10, we will see how this shortcoming could be circumvented if the coefficient matrix is banded. For nonbanded systems, there is usually no simple way to avoid large memory requirements when using elimination methods. Because all computers have a finite amount of memory, this inefficiency can place a constraint on the size of systems for which elimination methods are practical.

Although a general algorithm such as the one in Fig. 8.7 is prone to the same constraint, the structure of the Gauss-Seidel equations [Eq. (8.11)] permits concise

Figure 8.7
Pseudocode for Gauss-Seidel with relaxation.

```
DOFOR i = 1 to n
    dummy = a i,i
    DOFOR j = 1 to n
        a i,j = a i,j /dummy
    ENDDO
    c i = c i /dummy
ENDDO
sentinel = 0
iter = 0
DOWHILE (iter < maxit) and (sentinel = 0)
    sentinel = 1
    iter = iter + 1
    DOFOR i = 1 to n
        old = x i
        sum = c i
        DOFOR j = 1 to n
            IF i ≠ j
                sum = sum - a i,j ·x j
            ENDIF
        ENDDO
        x i = λ·sum + (1 - λ) old
        IF (sentinel = 1) and (x i ≠ 0)
```

$$\epsilon_a = \left| \frac{x_i - old}{x_i} \right| \cdot 100$$

```
            IF (ε a > ε s ) THEN sentinel = 0
        ENDIF
    ENDDO
ENDDO
```

programs to be developed for specific systems. Because only nonzero coefficients need be included in Eq. (8.11), large savings of computer memory are possible. Although this entails more up-front investment in software development, the long-term advantages are substantial when dealing with large systems for which many simulations are to be performed. Both lumped- and distributed-variable systems can result in large, sparse matrices for which the Gauss-Seidel method has utility.

PROBLEMS

Hand Calculations

8.1 Use the Gauss-Jordan method to solve Prob. 7.6.

8.2 Determine the matrix inverse for Prob. 7.6. Check your results by multiplying $[A]$ by $[A]^{-1}$ to give the identity matrix.

8.3 Using the Gauss-Jordan method, repeat Prob. 7.9.

8.4 Determine the matrix inverse for Prob. 7.9. Check your results by verifying that $[A][A]^{-1} = [I]$. Do not use a pivoting strategy.

8.5 Using the Gauss-Jordan method, with partial pivoting, compute the matrix inverse for Prob. 7.10.

8.6 Use the Gauss-Jordan method to solve:

$$10x_1 - 3x_2 + 6x_3 = 24.5$$
$$1x_1 + 8x_2 - 2x_3 = -10$$
$$-2x_1 + 4x_2 - 9x_3 = -50$$

8.7 Determine the matrix inverse for Prob. 8.6. Use the inverse to solve the original problem as well as to solve the additional case where the right-hand-side vector is $\{C\}^T = [110 \quad 55 \quad -105]$.

8.8 Solve Prob. 8.6 using the Gauss-Seidel method, with a stopping criterion of $\epsilon_s = 10\%$.

8.9 Solve Prob. 7.8 using the Gauss-Seidel method, with a stopping criterion of $\epsilon_s = 10\%$.

8.10 Use the Gauss-Seidel method to solve ($\epsilon_s = 5\%$):

$$x_1 + 7x_2 - 4x_3 = -51$$
$$4x_1 - 4x_2 + 9x_3 = 62$$
$$12x_1 - x_2 + 3x_3 = 8$$

8.11 Use the Gauss-Seidel method to solve ($\lambda = 0.90$ and $\epsilon_s = 5\%$):

$$-5x_1 \qquad + 12x_3 = 60$$
$$4x_1 - x_2 - x_3 = -2$$
$$6x_1 + 8x_2 \qquad = 45$$

8.12 Given the following set of equations:

$$4x_1 - 2x_2 - x_3 = 40$$
$$x_1 - 6x_2 + 2x_3 = -28$$
$$x_1 - 2x_2 + 12x_3 = -86$$

solve using (**a**) Gauss elimination, (**b**) the Gauss-Jordan method, and (**c**) the Gauss-Seidel method ($\epsilon_s = 5\%$).

8.13 Given the following set of equations:

$$x_1 - 3x_2 + 12x_3 = 10$$
$$5x_1 - 13x_2 + 2x_3 = -34$$
$$x_1 - 14x_2 \qquad = -103$$

solve using (**a**) Gauss elimination, (**b**) the Gauss-Jordan method, and (**c**) the Gauss-Seidel method ($\epsilon_s = 5\%$).

8.14 Determine $\|A\|_e$, $\|A\|_1$, and $\|A\|_\infty$ for

$$[A] = \begin{bmatrix} 6 & -2 & 5 \\ 8.5 & 1.1 & -3.5 \\ -1.6 & -1 & 10.3 \end{bmatrix}$$

8.15 Determine the Euclidean and the row-sum norms for the systems in Probs. 8.10 and 8.11.

8.16 Determine the condition number for the system in Prob. 8.14 using the row-sum norm.

8.17 Repeat Prob. 8.14 for the matrix

$$[A] = \begin{bmatrix} 1 & 0.5 & 0.3 \\ 0.5 & 0.3 & 0.25 \\ 0.3 & 0.25 & 0.2 \end{bmatrix}$$

8.18 Repeat Prob. 8.16 for the matrix in Prob. 8.17.

8.19 Redraw Fig. 8.6 for the case where the slopes of the equations are plus and minus unity. What is the result of applying Gauss-Seidel to such a system?

Computer-Related Problems

8.20 Develop a user-friendly program for the Gauss-Jordan method, based on Fig. 8.2. Incorporate a scheme similar to the one depicted in Fig. 7.5 so that partial pivoting is employed.

8.21 Test the program developed in Prob. 8.20 by duplicating the computation from Example 8.1.

8.22 Repeat Probs. 7.6 and 7.8 through 7.11 using the software developed in Prob. 8.20.

8.23 Develop a user-friendly program for the Gauss-Jordan method with matrix inversion and partial pivoting. Include the features suggested in Sec. 8.1.2 into your program.

8.24 Repeat Probs. 8.5 and 8.7 using the software developed in Prob. 8.23.

8.25 Develop a user-friendly computer program for the Gauss-Seidel method based on Fig. 8.7.

8.26 Test the program developed in Prob. 8.25 by using it to duplicate Example 8.4.

8.27 Using the software developed in Prob. 8.25, repeat Probs. 8.8 through 8.11.

CHAPTER 9
LU Decomposition Methods

Recall that there are two fundamentally different approaches for solving systems of linear algebraic equations: elimination methods and iterative approaches. To this point, we have presented two examples of the former—Gauss elimination and Gauss-Jordan—and two of the latter—Gauss-Seidel and Jacobi. The present chapter deals with a class of elimination methods called *LU* decomposition techniques.

The primary appeal of *LU* decomposition is that the time-consuming elimination step can be formulated so that it only involves operations on the matrix of coefficients, [*A*]. Thus, as with the Gauss-Jordan/matrix inverse approach of Sec. 8.1, it is well-suited for those situations where many right-hand-side vectors {*C*} must be evaluated for a single value of [*A*]. It is actually preferable for such applications because it attains these solutions much more efficiently than the Gauss-Jordan/matrix inverse approach. Furthermore, it can be modified to determine the inverse. It will also be demonstrated that Gauss elimination can, in fact, be expressed as an *LU* decomposition. Consequently, *LU* decomposition is presently the most popular class of techniques for solving systems of linear algebraic equations.

9.1 NAIVE *LU* DECOMPOSITION

Just as was the case with Gauss elimination, *LU* decomposition requires pivoting to avoid division by zero. However, to simplify the following description, we will defer the issue of pivoting until after the fundamental approach is elaborated. In addition, the following explanation is limited to a set of four simultaneous equations. At the end of the description, the results will be generalized to *n*-dimensional systems.

The equations to be solved can be represented in matrix notation as

$$[A]\{X\} = \{C\} \tag{9.1}$$

which can be rearranged to give

$$[A]\{X\} - \{C\} = 0 \tag{9.2}$$

Suppose that Eq. (9.1) could be reexpressed as an upper triangular system with 1's on the diagonal:

$$
\begin{bmatrix}
1 & u_{12} & u_{13} & u_{14} \\
0 & 1 & u_{23} & u_{24} \\
0 & 0 & 1 & u_{34} \\
0 & 0 & 0 & 1
\end{bmatrix}
\begin{Bmatrix}
x_1 \\ x_2 \\ x_3 \\ x_4
\end{Bmatrix}
=
\begin{Bmatrix}
d_1 \\ d_2 \\ d_3 \\ d_4
\end{Bmatrix}
\tag{9.3}
$$

This is similar to the manipulation that occurs in the first step of Gauss elimination. It can also be expressed in matrix notation and rearranged to give

$$
[U]\{X\} - \{D\} = 0
\tag{9.4}
$$

Now, assume that there is a lower diagonal matrix,

$$
[L] =
\begin{bmatrix}
l_{11} & 0 & 0 & 0 \\
l_{21} & l_{22} & 0 & 0 \\
l_{31} & l_{32} & l_{33} & 0 \\
l_{41} & l_{42} & l_{43} & l_{44}
\end{bmatrix}
\tag{9.5}
$$

that has the property that when Eq. (9.4) is premultiplied by it, Eq. (9.2) is the result. That is,

$$
[L]\{[U]\{X\} - \{D\}\} = [A]\{X\} - \{C\}
\tag{9.6}
$$

If this equation holds, it follows from the rules for matrix multiplication that

$$
[L][U] = [A]
\tag{9.7}
$$

and

$$
[L]\{D\} = \{C\}
\tag{9.8}
$$

Equation (9.7) is referred to as the *LU* decomposition of $[A]$. After it is accomplished, solutions can be obtained very efficiently by a two-step substitution procedure employing $[L]$ and $[U]$ and based on Eqs. (9.8) and (9.4). The entire process is illustrated schematically in Fig. 9.1. Before illustrating how this process is implemented, we will first show that Gauss elimination can be formulated as an *LU* decomposition.

9.2   GAUSS ELIMINATION AND *LU* DECOMPOSITION

Although it might appear at face value to be unrelated to *LU* decomposition, Gauss elimination can be used to decompose $[A]$ into $[L]$ and $[U]$. This can be easily seen for $[U]$ which is a direct product of the forward elimination. Recall that the forward elimination step is intended to reduce the original coefficient matrix $[A]$ to the form

$$
[U] =
\begin{bmatrix}
a_{11} & a_{12} & a_{13} \\
0 & a'_{22} & a'_{23} \\
0 & 0 & a''_{33}
\end{bmatrix}
\tag{9.9}
$$

which is in the upper triangular format.

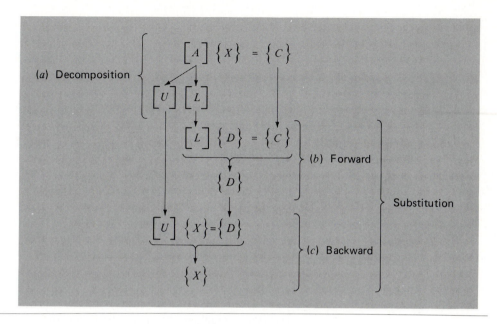

Figure 9.1
The steps in *LU* decomposition.

Though it might not be as apparent, the matrix $[L]$ is also produced during the step. This can be readily illustrated for a three-equation system,

$$\begin{bmatrix} a_{11} & a_{12} & a_{13} \\ a_{21} & a_{22} & a_{23} \\ a_{31} & a_{32} & a_{33} \end{bmatrix} \begin{Bmatrix} x_1 \\ x_2 \\ x_3 \end{Bmatrix} = \begin{Bmatrix} c_1 \\ c_2 \\ c_3 \end{Bmatrix}$$

The first step in Gauss elimination is to multiply row 1 by the factor [recall Eq. (7.13)]

$$f_{21} = \frac{a_{21}}{a_{11}}$$

and subtract the result from the second row to eliminate a_{21}. Similarly, row 1 is multiplied by

$$f_{31} = \frac{a_{31}}{a_{11}}$$

and the result subtracted from the third row to eliminate a_{31}. The final step for a 3×3 system is to multiply the modified second row by

$$f_{32} = \frac{a'_{32}}{a'_{22}}$$

and subtract the result from the third row to eliminate a_{32}.

The values f_{21}, f_{31}, and f_{32} are in fact the elements of an $[L]$ matrix:

$$[L] = \begin{bmatrix} 1 & 0 & 0 \\ f_{21} & 1 & 0 \\ f_{31} & f_{32} & 1 \end{bmatrix} \tag{9.10}$$

which yields the original matrix $[A]$ when multiplied by the upper triangular matrix resulting from forward elimination. Thus, if the factors are retained (rather than discarded as is usually the case), Gauss elimination represents an LU decomposition of $[A]$.

EXAMPLE 9.1   Gauss Elimination and LU Decomposition

Problem Statement: Derive an LU decomposition based on the Gauss elimination performed in Example 7.5.

Solution: In Example 7.5, we solved the matrix

$$[A] = \begin{bmatrix} 3 & -0.1 & -0.2 \\ 0.1 & 7 & -0.3 \\ 0.3 & -0.2 & 10 \end{bmatrix}$$

After forward elimination, the following upper triangular matrix was obtained:

$$[U] = \begin{bmatrix} 3 & -0.1 & -0.2 \\ 0 & 7.00333 & -0.293333 \\ 0 & 0 & 10.0120 \end{bmatrix}$$

The factors employed to obtain the upper triangular matrix can be assembled into a lower triangular matrix. The elements a_{21} and a_{31} were eliminated by using the factors

$$f_{21} = \frac{0.1}{3} = 0.0333333 \qquad f_{31} = \frac{0.3}{3} = 0.1000000$$

Finally, the element f_{32} was eliminated by using the factor

$$f_{32} = \frac{-0.19}{7.00333} = -0.0271300$$

Thus, the lower triangular matrix is

$$[L] = \begin{bmatrix} 1 & 0 & 0 \\ 0.0333333 & 1 & 0 \\ 0.100000 & -0.0271300 & 1 \end{bmatrix}$$

Consequently, the LU decomposition is

$$[A] = [L][U] = \begin{bmatrix} 1 & 0 & 0 \\ 0.0333333 & 1 & 0 \\ 0.100000 & -0.0271300 & 1 \end{bmatrix} \begin{bmatrix} 3 & -0.1 & -0.2 \\ 0 & 7.00333 & -0.293333 \\ 0 & 0 & 10.0120 \end{bmatrix}$$

This result can be verified by performing the multiplication of $[L][U]$ to give

$$[L][U] = \begin{bmatrix} 3 & -0.1 & -0.2 \\ 0.0999999 & 7 & -0.3 \\ 0.3 & -0.2 & 9.99996 \end{bmatrix}$$

where the minor discrepancies are due to round off.

Notice that the $[L]$ and $[U]$ matrices of Eqs. (9.5) and (9.3) are slightly different from those of Eqs. (9.10) and (9.9). For Gauss elimination, the $[L]$ matrix has 1's on the diagonal whereas for LU decomposition the $[U]$ matrix has 1's on the diagonal. The former approach is called a *Doolittle decomposition* or *factorization*, whereas the latter is attributed to *Crout*. Although there are some differences between the approaches (Atkinson, 1978; Ralston and Rabinowitz, 1979), their performance is comparable. In the present book, we will focus on Crout decomposition.

9.3 CROUT DECOMPOSITION

Recall that the Gauss elimination method involves two major steps: forward elimination and back substitution (Fig. 7.3). Of these, the forward-elimination step comprises the bulk of the computational effort. As in Fig. 9.2, this is particularly true for large systems of equations. Consequently, most efforts to develop improved elimination methods have focused on economizing this step and separating it from the computations involving the right-hand-side vector.

One of the most effective of these improved methods is called *Crout decomposition* which represents an efficient algorithm for decomposing $[A]$ into $[L]$ and $[U]$. As such, it lies at the heart of the LU decomposition approach.

For $n = 4$, Eq. (9.7) can be written as

$$\begin{bmatrix} l_{11} & 0 & 0 & 0 \\ l_{21} & l_{22} & 0 & 0 \\ l_{31} & l_{32} & l_{33} & 0 \\ l_{41} & l_{42} & l_{43} & l_{44} \end{bmatrix} \begin{bmatrix} 1 & u_{12} & u_{13} & u_{14} \\ 0 & 1 & u_{23} & u_{24} \\ 0 & 0 & 1 & u_{34} \\ 0 & 0 & 0 & 1 \end{bmatrix} = \begin{bmatrix} a_{11} & a_{12} & a_{13} & a_{14} \\ a_{21} & a_{22} & a_{23} & a_{24} \\ a_{31} & a_{32} & a_{33} & a_{34} \\ a_{41} & a_{42} & a_{43} & a_{44} \end{bmatrix} \qquad (9.11)$$

Crout's method can be derived by using matrix multiplication to evaluate the left side of Eq. (9.11) and then equating the results to the right side. Recalling the rules for

Figure 9.2
A plot showing the percentage of the total arithmetic operations of naive Gauss elimination that are devoted to the forward-elimination step versus the number of equations that are solved. For large numbers of equations, forward elimination will clearly represent the predominant time-consuming step in the solution procedure.

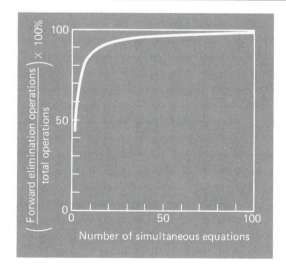

matrix multiplication (Box PT3.2), the first step is to multiply the rows of $[L]$ by the first column of $[U]$. This yields

$$l_{11} = a_{11} \qquad l_{21} = a_{21} \qquad l_{31} = a_{31} \qquad l_{41} = a_{41}$$

Thus, the first column of $[L]$ is merely the first column of $[A]$. This fact can be represented in general terms by

$$l_{i1} = a_{i1} \qquad \text{for } i = 1, 2, \ldots, n \tag{9.12}$$

Next, the first row of $[L]$ can be multiplied by the columns of $[U]$ to give

$$l_{11} = a_{11} \qquad l_{11}u_{12} = a_{12} \qquad l_{11}u_{13} = a_{13} \qquad l_{11}u_{14} = a_{14}$$

The first result has already been established, but the remaining relationships can be solved for

$$u_{12} = \frac{a_{12}}{l_{11}} \qquad u_{13} = \frac{a_{13}}{l_{11}} \qquad u_{14} = \frac{a_{14}}{l_{11}}$$

Thus, the first row of $[U]$ has been determined. This operation can be represented concisely as

$$u_{1j} = \frac{a_{1j}}{l_{11}} \qquad \text{for } j = 2, 3, \ldots, n \tag{9.13}$$

The progress to this point is illustrated by Fig. 9.3*a*. Similar operations are repeated to evaluate the remaining columns of $[L]$ and the rows of $[U]$ (Fig. 9.3*b* through *d*). For example, the second through fourth rows of $[L]$ can be multiplied by the second column of $[U]$ to yield

$$l_{21}u_{12} + l_{22} = a_{22} \qquad l_{31}u_{12} + l_{32} = a_{32} \qquad l_{41}u_{12} + l_{42} = a_{42}$$

which can be solved for l_{22}, l_{32}, and l_{42}, respectively. These operations can be represented generally by

$$l_{i2} = a_{i2} - l_{i1}u_{12} \qquad \text{for } i = 2, 3, \ldots, n \tag{9.14}$$

Next, the remaining unknowns on the second row of $[U]$ can be evaluated by multiplying the second row of $[L]$ by the third and fourth columns of $[U]$ to give

$$l_{21}u_{13} + l_{22}u_{23} = a_{23} \qquad l_{21}u_{14} + l_{22}u_{24} = a_{24}$$

which can be solved for

$$u_{23} = \frac{a_{23} - l_{21}u_{13}}{l_{22}} \qquad u_{24} = \frac{a_{24} - l_{21}u_{14}}{l_{22}}$$

which can be expressed generally as

$$u_{2j} = \frac{a_{2j} - l_{21}u_{1j}}{l_{22}} \qquad \text{for } j = 3, 4, \ldots, n \tag{9.15}$$

The process can be repeated to evaluate the other elements. General formulas that result are

$$l_{i3} = a_{i3} - l_{i1}u_{13} - l_{i2}u_{23} \qquad \text{for } i = 3, 4, \ldots, n \tag{9.16}$$

Figure 9.3
A schematic depicting the evaluations involved in Crout *LU* decomposition.

(a)

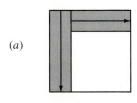

(b)

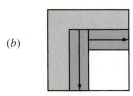

(c)

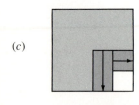

(d)

$$u_{3j} = \frac{a_{3j} - l_{31}u_{1j} - l_{32}u_{2j}}{l_{33}} \qquad \text{for } j = 4, 5, \ldots, n \qquad (9.17)$$

$$l_{i4} = a_{i4} - l_{i1}u_{14} - l_{i2}u_{24} - l_{i3}u_{34} \qquad \text{for } i = 4, 5, \ldots, n \qquad (9.18)$$

and so forth.

Inspection of Eqs. (9.12) through (9.18) leads to the following concise formulas for implementing the method.

$$l_{i1} = a_{i1} \qquad \text{for } i = 1, 2, \ldots, n \qquad (9.19)$$

$$u_{1j} = \frac{a_{1j}}{l_{11}} \qquad \text{for } j = 2, 3, \ldots, n \qquad (9.20)$$

For $j = 2, 3, \ldots, n - 1$,

$$l_{ij} = a_{ij} - \sum_{k=1}^{j-1} l_{ik}u_{kj} \qquad \text{for } i = j, j + 1, \ldots, n \qquad (9.21)$$

$$u_{jk} = \frac{a_{jk} - \sum_{i=1}^{j-1} l_{ji}u_{ik}}{l_{jj}} \qquad \text{for } k = j + 1, j + 2, \ldots, n \qquad (9.22)$$

and

$$l_{nn} = a_{nn} - \sum_{k=1}^{n-1} l_{nk}u_{kn} \qquad (9.23)$$

EXAMPLE 9.2 Crout Decomposition

Problem Statement: Perform the LU decomposition of the system of equations:

$$2x_1 - 5x_2 + x_3 = 12$$

$$-x_1 + 3x_2 - x_3 = -8$$

$$3x_1 - 4x_2 + 2x_3 = 16$$

Solution: According to Eq. (9.19), the first column of $[L]$ is identical to the first column of $[A]$:

$$l_{11} = 2 \qquad l_{21} = -1 \qquad l_{31} = 3$$

Equation (9.20) can be used to compute the first row of $[U]$:

$$u_{12} = \frac{a_{12}}{l_{11}} = \frac{-5}{2} = -2.5$$

$$u_{13} = \frac{a_{13}}{l_{11}} = \frac{1}{2} = 0.5$$

The second column of [L] is computed with Eq. (9.21):

$$l_{22} = a_{22} - l_{21}u_{12} = 3 - (-1)(-2.5) = 0.5$$

$$l_{32} = a_{32} - l_{31}u_{12} = -4 - (3)(-2.5) = 3.5$$

Equation (9.22) is used to compute the last element of [U],

$$u_{23} = \frac{a_{23} - l_{21}u_{13}}{l_{22}} = \frac{-1 - (-1)(0.5)}{0.5} = -1$$

and Eq. (9.23) can be employed to determine the last element of [L],

$$l_{33} = a_{33} - l_{31}u_{13} - l_{32}u_{23} = 2 - 3(0.5) - 3.5(-1) = 4$$

Thus, the *LU* decomposition is

$$[L] = \begin{bmatrix} 2 & 0 & 0 \\ -1 & 0.5 & 0 \\ 3 & 3.5 & 4 \end{bmatrix} \qquad [U] = \begin{bmatrix} 1 & -2.5 & 0.5 \\ 0 & 1 & -1 \\ 0 & 0 & 1 \end{bmatrix}$$

It can be easily verified that the product of these two matrices is equal to the original matrix [A].

Aside from the fact that it consists of a few concise loops, the foregoing approach also has the benefit that storage space can be economized. There is no need to store the 1's on the diagonal of [U] or the 0's for [L] or [U] because they are givens in the method. Consequently, the values of [U] can be stored in the zero space of [L]. Further, close examination of the foregoing derivation makes it clear that after each element of [A] is employed once, it is never used again. Therefore, as each element [L] and [U] is computed, it can be substituted for the corresponding element (as designated by its subscripts) of [A]. Pseudocode to accomplish this is presented in Fig. 9.4.

Notice that Eq. (9.19) is not included in the pseudocode because the first column of [L] is already stored in [A]. Otherwise, the algorithm directly follows from Eqs. (9.20) through (9.23).

9.4   THE SUBSTITUTION STEP

Recall that in Gauss elimination, the transformations involved in forward elimination are simultaneously performed on the coefficient matrix [A] and the augmented right-hand-side vector {C}. This means that the entire Gauss elimination procedure must be implemented for each particular {C}. If several {C}'s are to be evaluated, one alternative is to augment all of them to [A] and then, after forward elimination, perform back-substitution individually on each of the transformed right-hand-side vectors.

Aside from the complexities this adds to the algorithm, there are many problems where the {C}'s to be evaluated are not all known prior to the analysis. Such would be the case in the "what if?" computations that are among the major strengths of personal computers. Another example is the iterative refinement procedure of Sec. 8.2.3. In these situations, it would be an advantage to perform the time-consuming computations on [A]

```
DOFOR j = 2 to n
   a      = a    /a
    1,j      1,j   1,1
ENDDO
DOFOR j = 2 to n-1
   DOFOR i = j to n
      sum = 0
      DOFOR k = 1 to j-1
         sum = sum + a   ·a
                      i,k  k,j
      ENDDO
      a    = a    - sum
       i,j    i,j
   ENDDO
   DOFOR k = j+1 to n
      sum = 0
      DOFOR i = 1 to j-1
         sum = sum + a   ·a
                      j,i  i,k
      ENDDO
      a    = (a    - sum)/a
       j,k     j,k          j,j
   ENDDO
ENDDO
sum = 0
DOFOR k = 1 to n-1
   sum = sum + a   ·a
                n,k  k,n
ENDDO
a    = a    - sum
 n,n    n,n
```

Figure 9.4
Pseudocode for Crout's *LU*
decomposition algorithm.

independently from those on $\{C\}$. This was the approach employed with Gauss-Jordan and matrix inversion in Chap. 8. *LU* decomposition can be utilized in the same fashion but is much more efficient.

Once the original matrix is decomposed, $[L]$ and $[U]$ can be employed to solve for $\{X\}$. This is accomplished in a two-step process (recall Fig. 9.1). First, Eq. (9.8) can be employed to determine $\{D\}$ for a particular $\{C\}$ by forward substitution. This procedure can be represented concisely as

$$d_1 = \frac{c_1}{l_{11}} \tag{9.24}$$

$$d_i = \frac{c_i - \sum_{j=1}^{i-1} l_{ij}d_j}{l_{ii}} \qquad \text{for } i = 2, 3, \ldots, n \tag{9.25}$$

Then, Eq. (9.4) can be employed to compute x by back-substitution

$$x_n = d_n \tag{9.26}$$

$$x_i = d_i - \sum_{j=i+1}^{n} u_{ij}x_j \qquad \text{for } i = n-1, n-2, \ldots, 1 \tag{9.27}$$

The substitution phase is illustrated in the following example.

EXAMPLE 9.3   Solution by Substitution

Problem Statement: Use the matrices $[L]$ and $[U]$ to obtain a solution for the system from Example 9.2.

Solution: The first step is to solve Eq. (9.8), which for this problem has the value

$$\begin{bmatrix} 2 & 0 & 0 \\ -1 & 0.5 & 0 \\ 3 & 3.5 & 4 \end{bmatrix} \begin{Bmatrix} d_1 \\ d_2 \\ d_3 \end{Bmatrix} = \begin{Bmatrix} 12 \\ -8 \\ 16 \end{Bmatrix}$$

The forward substitution procedure of Eqs. (9.24) and (9.25) can be used to solve for

$$d_1 = \frac{12}{2} = 6$$

$$d_2 = \frac{c_2 - l_{21}d_1}{l_{22}} = \frac{-8 - (-1)6}{0.5} = -4$$

$$d_3 = \frac{c_3 - l_{31}d_1 - l_{32}d_2}{l_{33}} = \frac{16 - 3(6) - 3.5(-4)}{4} = 3$$

The second step is to solve Eq. (9.4), which for the present problem has the value

$$\begin{bmatrix} 1 & -2.5 & 0.5 \\ 0 & 1 & -1 \\ 0 & 0 & 1 \end{bmatrix} \begin{Bmatrix} x_1 \\ x_2 \\ x_3 \end{Bmatrix} = \begin{Bmatrix} 6 \\ -4 \\ 3 \end{Bmatrix}$$

The back-substitution procedure of Eqs. (9.26) and (9.27) can be used to solve for

$$x_3 = d_3 = 3$$
$$x_2 = d_2 - u_{23}x_3 = -4 - (-1)3 = -1$$
$$x_1 = d_1 - u_{12}x_2 - u_{13}x_3 = 6 - (-2.5)(-1) - 0.5(3) = 2$$

These values can be verified by substituting them into the original equations.

9.5   COMPUTER APPLICATION OF *LU* DECOMPOSITION

Just as for Gauss elimination, *LU* decomposition algorithms must employ partial pivoting to avoid division by zero and to minimize round-off error. The pivoting is implemented immediately after computing each column of $[L]$.

In contrast to Gauss elimination, the process is complicated by the fact that the right-hand-side vector $\{C\}$ is not operated on simultaneously with $[A]$. This means that the computer algorithm must keep track of any row switches that occur during the decomposition step. Such a scheme is employed in the programs outlined in Figs. 9.5 and 9.6. Notice that the program in Fig. 9.5 is set up to evaluate as many right-hand-

Figure 9.5
An interactive computer program for *LU* decomposition written in Microsoft BASIC.

```
100 REM    LU DECOMPOSITION (BASIC VERSION)
105 REM *******************************
110 REM *      DEFINITION OF VARIABLES      *
120 REM * N   = NUMBER OF EQUATIONS         *
125 REM * A[] = MATRIX OF COEFFICIENTS      *
130 REM * C() = RIGHT-HAND-SIDE VECTOR      *
135 REM * X() = UNKNOWNS                    *
140 REM * O() = ORDER VECTOR                *
145 REM * S() = SCALE VECTOR                *
150 REM *******************************
160 DIM A(15,15),C(15),X(15),O(15),S(15)
170 REM ********** MAIN PROGRAM **********
180 GOSUB 300 'input matrix [A]
185 GOSUB 400 'scale and order vectors
190 GOSUB 450 'decompose matrix [A]
195 PRINT "Solution? (y or n)";
200 INPUT CONTIN$
205 WHILE CONTIN$ = "y" OR CONTIN$ = "Y"
210    CLS
215    FOR NC = 1 TO N
220       PRINT "C(";NC;") = ";
225       INPUT C(NC)
230    NEXT NC
235    GOSUB 600 'solve by substitution
240    GOSUB 700 'output results
245    PRINT
250    PRINT "Solution? (y or n)";
255    INPUT CONTIN$
260 WEND
265 END
300 REM ******** SUBROUTINE INPUT ********
310 CLS: INPUT "NUMBER OF EQUATIONS? ", N
320 PRINT
325 FOR I = 1 TO N
330    FOR J = 1 TO N
335       PRINT "A(";I;",";J;") = ";
340       INPUT A(I,J)
345    NEXT J
350    PRINT
355 NEXT I
360 RETURN
400 REM ******** SUBROUTINE ORDER ********
410 FOR I = 1 TO N
415    O(I) = I
420    S(I)=ABS(A(I,1))
425    FOR J = 2 TO N
430       IF ABS(A(I,J))>S(I)
               THEN S(I)=ABS(A(I,J))
435    NEXT J
440 NEXT I
445 RETURN
450 REM ****** SUBROUTINE DECOMPOSE ******
460 J = 1
465 GOSUB 750 'pivot
470 FOR J = 2 TO N
475    A(O(1),J) = A(O(1),J)/A(O(1),1)
480 NEXT J
485 FOR J = 2 TO N-1
490    FOR I = J TO N
```

```
495       SUM = 0
500       FOR K = 1 TO J-1
505          SUM=SUM+A(O(I),K)*A(O(K),J)
510       NEXT K
515       A(O(I),J) = A(O(I),J) - SUM
520    NEXT I
525    GOSUB 750 'pivot
530    FOR K = J+1 TO N
535       SUM = 0
540       FOR I = 1 TO J-1
545          SUM=SUM+A(O(J),I)*A(O(I),K)
550       NEXT I
555       A(O(J),K)=
                  (A(O(J),K)-SUM)/A(O(J),J)
560    NEXT K
565 NEXT J
570 SUM = 0
575 FOR K = 1 TO N-1
580    SUM = SUM + A(O(N),K)*A(O(K),N)
585 NEXT K
590 A(O(N),N) = A(O(N),N) - SUM
595 RETURN
600 REM ******* SUBROUTINE SOLVE ********
610 X(1) = C(O(1))/A(O(1),1)
615 FOR I = 2 TO N
620    SUM = 0
625    FOR J = 1 TO I-1
630       SUM = SUM + A(O(I),J)*X(J)
635    NEXT J
640    X(I)=(C(O(I))-SUM)/A(O(I),I)
645 NEXT I
650 FOR I = N-1 TO 1 STEP -1
655    SUM = 0
660    FOR J = I+1 TO N
665       SUM = SUM + A(O(I),J)*X(J)
670    NEXT J
675    X(I) = X(I) - SUM
680 NEXT I
685 RETURN
700 REM ******* SUBROUTINE OUTPUT *******
710 PRINT: PRINT "SOLUTIONS TO THE EQUATIONS"
720 FOR I = 1 TO N
725    PRINT "X(" I ") ="; X(I)
730 NEXT I
735 RETURN
750 REM ******* SUBROUTINE PIVOT ********
760 PIVOT = J
765 BIG = ABS(A(O(J),J)/S(O(J)))
770 FOR II = J+1 TO N
775    DUMMY=ABS(A(O(II),J)/S(O(II)))
780    IF DUMMY>BIG THEN 785 ELSE 795
785       BIG = DUMMY
790       PIVOT = II
795    REM endif
800 NEXT II
805 IDUM = O(PIVOT)
810 O(PIVOT) = O(J)
815 O(J) = IDUM
835 RETURN
```

Figure 9.6
Subroutines for *LU* decomposition written in (*a*) FORTRAN 77 and (*b*) Turbo Pascal.

```fortran
      SUBROUTINE LUD(A,C,X,N)
*********************************************
*          DEFINITION OF VARIABLES          *
*                                           *
*    N   = NUMBER OF EQUATIONS               *
*    A[] = MATRIX OF COEFFICIENTS            *
*    C() = RIGHT-HAND-SIDE VECTOR            *
*    X() = UNKNOWNS                          *
*    O() = ORDER VECTOR                      *
*    S() = SCALE VECTOR                      *
*********************************************
      DIMENSION A(50,50),C(50),X(50),S(50)
      INTEGER O(50)
      CALL ORDER(S,A,O,N)
      CALL DECOMP(A,O,S,N)
      CALL SOLVE(A,C,X,O,N)
      RETURN
      END
*********************************************
      SUBROUTINE ORDER(S,A,O,N)
      DIMENSION A(50,50),S(50)
      INTEGER O(50)
      DO 10 I=1,N
         O(I)=I
         S(I)=ABS(A(I,1))
         DO 20 J=2,N
            IF (ABS(A(I,J)).GT.S(I)) THEN
               S(I)=ABS(A(I,J))
            ENDIF
20       CONTINUE
10    CONTINUE
      RETURN
      END
*********************************************
      SUBROUTINE DECOMP(A,O,S,N)
      DIMENSION A(50,50),S(50)
      INTEGER O(50)
      INTEGER*2 I,J,K,N
      J=1
      CALL PIVOT(A,S,O,N,J)
      DO 10 J=2,N
         A(O(1),J) = A(O(1),J)/A(O(1),1)
10    CONTINUE
      DO 20 J=2,N-1
         DO 30 I=J,N
            SUM = 0.0
            DO 40 K=1,J-1
               SUM=SUM+A(O(I),K)*A(O(K),J)
40          CONTINUE
            A(O(I),J)=A(O(I),J) - SUM
30       CONTINUE
```

```pascal
PROCEDURE Lud (VAR A: Matrix; VAR C,X: Vector
                                ;N:integer);

{          Driver program type definitions
   Matrix = a 2 dimensional real array
   Vector = a 1 dimensional real array
   Row    = a 1 dimensional integer array  }

{          Definition of variables
   N    = number of equations
   A[]  = matrix of coefficients
   C()  = right-hand-side vector
   X()  = unknowns
   O()  = order vector
   S()  = scale vector                      }
PROCEDURE Order(VAR S: Vector; A: Matrix;
                VAR O: Row;  N: integer);
VAR
   i,j: integer;
Begin                     {procedure Order}
For i := 1 to N do
   Begin
   O[i] := i;
   S[i] := Abs(A[i,1]);
   For j := 2 to N do
      Begin
      If Abs(A[i,j]) > S[i] Then
         S[i] := Abs(A[i,j]);
      End;
   End;
End;                   {of procedure Order}
PROCEDURE Pivot(VAR S: Vector; A: Matrix;
                VAR O:Row; j,N: integer);
VAR
   ii: integer;
   pivot,idum, dum: integer;
   big,dummy: real;
Begin
pivot := j;
big := Abs(A[O[j],j]/S[O(j)]);
For ii := j+1 to N do
   Begin
   dummy := Abs(A[O[ii],j]/S[O(ii)]);
   If (dummy > big) Then
      Begin
      big := dummy;
      pivot := ii;
      End;
   End;
idum := O[pivot];
O[pivot] := O[j];
```

(*Continues* **Figure 9.6**)

```
        CALL PIVOT(A,S,O,N,J)
        DO 50 K=J+1,N
            SUM = 0.0
            DO 60 I=1,J-1
                SUM=SUM+A(O(J),I)*A(O(I),K)
60          CONTINUE
            A(O(J),K)=
    *                  (A(O(J),K)-SUM)/A(O(J),J)
50      CONTINUE
20  CONTINUE
    SUM = 0.0
    DO 70 K=1,N-1
        SUM=SUM+A(O(N),K)*A(O(K),N)
70  CONTINUE
    A(O(N),N)=A(O(N),N)-SUM
    RETURN
    END
************************************************
    SUBROUTINE SOLVE(A,C,X,O,N)
    DIMENSION A(50,50),C(50),X(50)
    INTEGER O(50)
    X(1)=C(O(1))/A(O(1),1)
    DO 10 I=2,N
        SUM = 0.0
        DO 20 J=1,I-1
            SUM=SUM+A(O(I),J)*X(J)
20      CONTINUE
        X(I) = (C(O(I))-SUM)/A(O(I),I)
10  CONTINUE
    DO 30 I=N-1,1,-1
        SUM = 0.0
        DO 40 J=I+1,N
            SUM=SUM+A(O(I),J)*X(J)
40      CONTINUE
        X(I)=X(I)-SUM
30  CONTINUE
    RETURN
    END
************************************************
    SUBROUTINE PIVOT(A,S,O,N,J)
    DIMENSION A(50,50),S(50)
    INTEGER O(50)
    INTEGER*2 J,PIVIT,II,IDUM
    PIVIT = J
    BIG=ABS(A(O(J),J)/S(O(J)))
    DO 10 II=J+1,N
        DUMMY=ABS(A(O(II),J)/S(O(II)))
        IF (DUMMY.GT.BIG) THEN
            BIG=DUMMY
            PIVIT=II
        ENDIF
10  CONTINUE
    IDUM=O(PIVIT)
```

```
O[j] := idum;
End;                    {of procedure Pivot}
Procedure Decmp(S: Vector; VAR A: Matrix;
                VAR O:Row; N:integer);
VAR
    r,i,j,k: integer;
    Sum,dummy: real;
Begin                   {procedure Decomp}
j := 1;
Pivot (S,A,O,j,N);
For j := 2 to N do
    Begin
    A[O[1],j] := A[O[1],j]/A[O[1],1];
    End;
For j := 2 to N-1 do
    Begin
    For i := j to N do
        Begin
        Sum := 0.0;
        For k := 1 to j-1 do
            Begin
            Sum := Sum+A[O[i],k]*A[O[k],j];
            End;
        A[O[i],j] := A[O[i],j] - Sum;
        End;
    Pivot(S,A,O,j,N);
    For k := j+1 to N do
        Begin
        Sum := 0.0;
        For i := 1 to j-1 do
            Begin
            Sum := Sum +
                A[O[j],i]*A[O[i],k];
            End;
        A[O[j],k] :=
                (A[O[j],k]-Sum)/A[O[j],j];
        End;
    End;
    Sum := 0.0;
    For k := 1 to N-1 do
        Begin
        Sum := A[O[N],k]*A[O[k],N]+Sum;
        End;
    A[O[N],N] := A[O[N],N] - Sum;
End;                    {of procedure Decomp}
PROCEDURE Solve(A: Matrix; C: Vector;
                VAR X: Vector; O: Row;
                        N: integer);
VAR
    Sum: real;
    i, j : integer;
Begin                   {procedure Solve}
X[1] := C[O[1]]/A[O[1],1];
For i := 2 to N do
    Begin
```

(*Continues* **Figure 9.6**)

```
O(PIVIT)=O(J)
O(J)=IDUM
RETURN
END
```

```
                              Sum := 0.0;
                              For j := 1 to i-1 do
                                  Begin
                                      Sum := Sum + A[O[i],j]*X[j];
                                  End;
                              X[i] := (C[O[i]]-Sum)/A[O[i],i];
                              End;
                          For i := N-1 downto 1 do
                              Begin
                              Sum := 0.0;
                              For j := i+1 to N do
                                  Begin
                                      Sum := Sum+A[O[i],j]*X[j];
                                  End;
                              X[i] := X[i] - Sum;
                              End;
                          End;                    {of procedure Solve}
                      {of Procedure definitions
                                          within procedure Lud}
                      Begin                            {procedure Lud}
                      Order(S,A,O,N);
                      Decmp(S,A,O,N);
                      Solve(A,C,X,O,N);
                      End;                     {of procedure Lud}
```

side vectors as the user desires. Also, note that *in-place* partial pivoting is employed and that a vector, *O*, is used to keep track of the row interchanges. The codes in Fig. 9.6 also employ an order vector and in-place pivoting but are designed to obtain a solution for a single right-hand-side vector. This design is more compatible with their subroutine format.

9.5.1   Matrix Inverse

Aside from obtaining a solution, *LU* decomposition can be employed to evaluate the inverse of the original coefficient matrix. Each column j, of the inverse, can be determined by using a unit vector (with the 1 in the jth row) as the right-hand-side vector in the substitution solution algorithms described in Sec. 9.4. This procedure is illustrated by the following example.

EXAMPLE 9.4   Matrix Inversion

Problem Statement: Employ *LU* decomposition to determine the matrix inverse for the system from Example 9.2.

Solution: The first column of the matrix inverse can be determined by performing the forward- and back-substitution solution procedures with a unit vector (with 1 in the first row) as the right-hand-side vector. Thus, Eq. (9.8) is formulated as

$$
\begin{bmatrix} 2 & 0 & 0 \\ -1 & 0.5 & 0 \\ 3 & 3.5 & 4 \end{bmatrix} \begin{Bmatrix} d_1 \\ d_2 \\ d_3 \end{Bmatrix} = \begin{Bmatrix} 1 \\ 0 \\ 0 \end{Bmatrix}
$$

which can be solved with Eqs. (9.24) and (9.25) for

$$\{D\}^T = [0.5 \quad 1 \quad -1.25]$$

This result, in turn, can be used in conjunction with Eqs. (9.26) and (9.27) to determine

$$\{X\}^T = [0.5 \quad -0.25 \quad -1.25]$$

This is the first column of the matrix inverse.

To determine the second column, Eq. (9.8) is formulated as

$$
\begin{bmatrix} 2 & 0 & 0 \\ -1 & 0.5 & 0 \\ 3 & 3.5 & 4 \end{bmatrix} \begin{Bmatrix} d_1 \\ d_2 \\ d_3 \end{Bmatrix} = \begin{Bmatrix} 0 \\ 1 \\ 0 \end{Bmatrix}
$$

This can be solved for $\{D\}$ and the results used with Eqs. (9.26) and (9.27) to determine

$$\{X\}^T = [1.5 \quad 0.25 \quad -1.75]$$

which is the second column of the inverse.

Finally, the forward- and back-substitution procedures can be implemented with $\{C\}^T = [0 \quad 0 \quad 1]$ to give

$$\{X\}^T = [0.5 \quad 0.25 \quad 0.25]$$

which is the third column of the inverse.

Therefore, the matrix inverse is

$$
[A]^{-1} = \begin{bmatrix} 0.5 & 1.5 & 0.5 \\ -0.25 & 0.25 & 0.25 \\ -1.25 & -1.75 & 0.25 \end{bmatrix}
$$

The validity of this result can be checked by verifying that $[A][A]^{-1} = [I]$.

9.6 BANDED SYSTEMS

As mentioned in Box PT3.1, a banded matrix is a square matrix that has all elements equal to zero, with the exception of a band centered on the main diagonal. Banded systems are frequently encountered in engineering and scientific practice. For example, they typically occur in the solution of differential equations. In addition, other numerical methods such as cubic splines (Sec. 12.5) involve the solution of banded systems.

The dimensions of a banded system can be quantified by two parameters: the bandwidth BW and the half-bandwidth HBW (Fig. 9.7). These two values are related by BW = 2HBW + 1. In general, then, a banded system is one for which $a_{ij} = 0$ if $|i - j| > $ HBW.

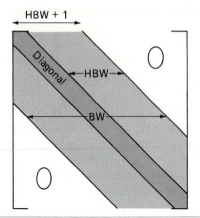

Figure 9.7
Parameters used to quantify the dimensions of a banded system. BW and HBW designate the bandwidth and the half-bandwidth, respectively.

Although Gauss elimination or conventional *LU* decomposition can be employed to solve banded equations, they are inefficient because if pivoting is unnecessary none of the elements outside the band would change from their original values of zero. Thus, unnecessary space and time would be expended on the storage and manipulation of these useless zeros. If it is known beforehand that pivoting is unnecessary, very efficient algorithms can be developed that do not involve the zero elements outside the band. Because many problems involving banded systems do not require pivoting, these alternative algorithms, as described next, are the methods of choice.

9.6.1 Tridiagonal Systems

A tridiagonal system—that is, one with a bandwidth of 3—can be expressed generally as

$$
\begin{bmatrix}
f_1 & g_1 & & & & \\
e_2 & f_2 & g_2 & & & \\
& e_3 & f_3 & g_3 & & \\
& & \cdot & \cdot & \cdot & \\
& & & \cdot & \cdot & \cdot \\
& & & \cdot & \cdot & \cdot \\
& & & & e_{n-1} & f_{n-1} & g_{n-1} \\
& & & & & e_n & f_n
\end{bmatrix}
\begin{Bmatrix}
x_1 \\ x_2 \\ x_3 \\ \cdot \\ \cdot \\ \cdot \\ x_{n-1} \\ x_n
\end{Bmatrix}
=
\begin{Bmatrix}
r_1 \\ r_2 \\ r_3 \\ \cdot \\ \cdot \\ \cdot \\ r_{n-1} \\ r_n
\end{Bmatrix}
\tag{9.28}
$$

Notice that we have changed our notation for the coefficients from a's and c's to e's, f's, g's, and r's. This was done to avoid storing large numbers of useless zeros in the square matrix of a's. This space-saving modification is advantageous because the resulting algorithm requires less computer memory.

Figure 9.8 shows pseudocode for an efficient method, called the *Thomas algorithm*, to solve Eq. (9.28). As with conventional *LU* decomposition, the algorithm consists of three steps: decomposition and forward and back substitution.

Thus, all the advantages of *LU* decomposition, such as convenient evaluation of multiple right-hand-side vectors and the matrix inverse, can be accomplished by proper application of this algorithm.

(a) decomposition

```
DOFOR k = 2 to n
    e  = e  /f
     k    k    k-1
    f  = f  - e ·g
     k    k    k   k-1
ENDDO
```

(b) forward substitution

```
DOFOR k = 2 to n
    r  = r  - e ·r
     k    k    k  k-1
ENDDO
```

(c) back substitution

```
x  = r /f
 n    n   n
DOFOR k = n-1 to 1 step -1
    x  = (r  - g ·x   )/f
     k     k    k  k+1    k
ENDDO
```

Figure 9.8
Pseudocode to implement the Thomas algorithm, an *LU* decomposition method for tridiagonal systems.

Aside from storage in vectors, the coefficients of banded systems can also be stored in a compact matrix $[B]$ as depicted in Fig. 9.9. Note that if a_{ij} is defined as an element of the full matrix $[A]$, then the corresponding element in compressed storage will be located at $b_{i,HBW+1+j-i}$ of the compressed matrix. Pseudocode for a compressed matrix version of the Thomas algorithm is presented in Fig. 9.10.

Note that the foregoing economies are even greater when the matrix is banded and *symmetric*. For these cases, only the elements on the diagonal and in the upper half need be stored. Thus, the compressed matrix could be stored in a matrix of dimension $HBW+1$. For these cases, if a_{ij} is desired, it will be located at $b_{i,1+j-i}$.

We have introduced the notion of compressed storage at this point because many problem contexts in engineering involve systems that are banded and, often, symmetric. Thus, many algorithms, of which the ones in this section are simple examples, are available to analyze them efficiently. In the next section we will review a final method that can be employed to solve symmetric systems.

Figure 9.9
Compact banded storage of a tridiagonal system in a matrix of dimension $n \times$ BW. Note that the right-hand side can be augmented to give a matrix of dimension $n \times$ (BW + 1).

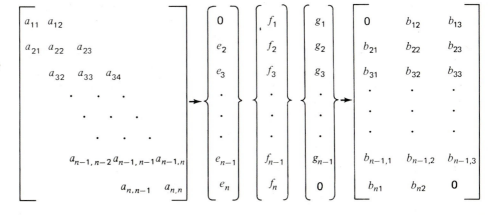

(*a*)   decomposition

```
DOFOR k = 2 to n
    b     = b     /b
     k,1     k,1    k-1,2
    b     = b     - b     ·b
     k,2     k,2     k,1   k-1,3
ENDDO
```

(*b*)   forward substitution

```
DOFOR k = 2 to n
    b     = b     - b     ·b
     k,4     k,4     k,1   k-1,4
ENDDO
```

(*c*)   back substitution

```
b     = b     /b
 n,4     n,4    n,2
DOFOR k = n-1 to 1 step -1
    b     = (b     - b     ·b       )/b
     k,4      k,4     k,3   k+1,4      k,2
ENDDO
```

Figure 9.10
Reformulation of the Thomas algorithm for the compressed matrix representation of the triangular system from Fig. 9.9.

9.7   CHOLESKY DECOMPOSITION

Recall from Box PT3.1 that a symmetric matrix is one where $a_{ij} = a_{ji}$ for all i and j. In other words, $[A] = [A]^T$. Such systems occur commonly in both mathematical and engineering problem contexts. They offer computational advantages because only half the storage is needed and, in most cases, only half the computation time is required for their solution.

One of the most popular approaches involves *Cholesky decomposition*. This algorithm is based on the fact that a symmetric matrix can be decomposed as in

$$[A] = [L][L]^T \tag{9.29}$$

That is, the resulting triangular factors are the transpose of each other.

Just as for Crout decomposition, the terms of Eq. (9.29) can be multiplied out and set equal to each other (see Prob. 9.4 at the end of the chapter). The result can be expressed simply by recurrence relations. For the kth row,

$$l_{ki} = \frac{a_{ki} - \sum_{j=1}^{i-1} l_{ij} l_{kj}}{l_{ii}} \qquad \text{for } i = 1, 2, \ldots, k-1 \tag{9.30}$$

and

$$l_{kk} = \sqrt{a_{kk} - \sum_{j=1}^{k-1} l_{kj}^2} \tag{9.31}$$

EXAMPLE 9.5   Cholesky Decomposition

Problem Statement: Apply Cholesky decomposition to the symmetric matrix

$$[A] = \begin{bmatrix} 6 & 15 & 55 \\ 15 & 55 & 225 \\ 55 & 225 & 979 \end{bmatrix}$$

Solution: For the first row ($k = 1$), Eq. (9.30) is skipped and Eq. (9.31) is employed to compute

$$l_{11} = \sqrt{a_{11}} = \sqrt{6} = 2.4495$$

For the second row ($k = 2$), Eq. (9.30) gives

$$l_{21} = \frac{a_{21}}{l_{11}} = \frac{15}{2.4495} = 6.1237$$

and Eq. (9.31) yields

$$l_{22} = \sqrt{a_{22} - l_{21}^2} = \sqrt{55 - (6.1237)^2} = 4.1833$$

For the third row ($k = 3$), Eq. (9.30) gives ($i = 1$)

$$l_{31} = \frac{a_{31}}{l_{11}} = \frac{55}{2.4495} = 22.454$$

and ($i = 2$)

$$l_{32} = \frac{a_{32} - l_{21}l_{31}}{l_{22}} = \frac{225 - 6.1237(22.454)}{4.1833} = 20.916$$

and Eq. (9.31) yields

$$l_{33} = \sqrt{a_{33} - l_{31}^2 - l_{32}^2} = \sqrt{979 - (22.454)^2 - (20.916)^2} = 6.1106$$

Thus, the Cholesky decomposition yields

$$[L] = \begin{bmatrix} 2.4495 & 0 & 0 \\ 6.1237 & 4.1833 & 0 \\ 22.454 & 20.916 & 6.1106 \end{bmatrix}$$

The validity of this decomposition can be verified by substituting it and its transpose into Eq. (9.29) to see if their product yields the original matrix $[A]$. This will be left for an exercise.

Figure 9.11 presents pseudocode for implementing the Cholesky decomposition algorithm. It should be noted that the algorithm in Fig. 9.11 could result in an execution error if the evaluation of a_{kk} involves taking the square root of a negative number. However, for cases where the matrix is positive definite,[*] this will never occur. Because

[*] A *positive definite matrix* is one for which the product $\{X\}^T[A]\{X\}$ is greater than zero for all nonzero vectors $\{X\}$.

```
DOFOR k = 1 to n
    DOFOR i = 1 to k-1
        sum = 0
        DOFOR j = 1 to i-1
            sum = sum + a_ij·a_kj
        ENDDO
        a_ki = (a_ki - sum)/a_ii
    ENDDO
    sum = 0
    DOFOR j = 1 to k-1
        sum = sum + a²_kj
    ENDDO
    a_kk = √(a_kk - sum)
ENDDO
```

Figure 9.11
Pseudocode for Cholesky's *LU* decomposition algorithm.

many symmetric matrices dealt with in engineering are, in fact, positive definite, the Cholesky algorithm has wide application. Another benefit of dealing with positive definite symmetric matrices is that pivoting is not required to avoid division by zero. Thus, we can implement the algorithm in Fig. 9.11 without the complication of pivoting.

PROBLEMS

Hand Calculations

9.1 Use the rules of matrix multiplication to prove that Eqs. (9.7) and (9.8) follow from Eq. (9.6).

9.2 Use naive Gauss elimination to decompose the following system according to the description in Sec. 9.2:

$$7x_1 + 2x_2 - 5x_3 = -18$$

$$x_1 + 5x_2 - 3x_3 = -40$$

$$2x_1 - x_2 - 9x_3 = -26$$

Then, multiply the resulting $[L]$ and $[U]$ matrices to determine that $[A]$ is produced.

9.3 Use *LU* decomposition to solve the system of equations in Prob. 9.2. Show all the steps in the computation. Also solve the system for an alternative right-hand-side vector

$$\{C\}^T = [-6 \quad 5 \quad 0.5]$$

9.4 Derive Eqs. (9.30) and (9.31) by multiplying out and setting equal the terms of Eq. (9.29). Employ a 4×4 system for your derivation.

9.5 Confirm the validity of the Cholesky decomposition of Example 9.5 by substituting the results into Eq. (9.29) to see if the product of $[L]$ and $[L]^T$ yields $[A]$.

9.6 Solve Prob. 7.9 using *LU* decomposition with partial pivoting.

9.7 Solve Prob. 8.6 using *LU* decomposition with partial pivoting.

9.8 Solve Prob. 8.7 using *LU* decomposition with partial pivoting.

Computer-Related Problems

9.9 Develop a user-friendly program for *LU* decomposition based on one of the codes from Fig. 9.5 or 9.6.

9.10 Develop a user-friendly program for *LU* decomposition, including the capability to evaluate the matrix inverse. Base the program on Fig. 9.5 or 9.6 and the discussion in Sec. 9.5.1.

9.11 Develop a user-friendly program for the Thomas algorithm based on Sec. 9.6.

9.12 Develop a user-friendly program for Cholesky decomposition based on Sec. 9.7.

CHAPTER 10

Case Studies: Systems of Linear Algebraic Equations

The purpose of this chapter is to use the numerical procedures discussed in Chaps. 7, 8, and 9 to solve systems of linear algebraic equations for some engineering applications. These systematic numerical techniques have practical significance because engineers frequently encounter problems involving systems of equations that are too large to solve by hand. The numerical algorithms in these applications are particularly convenient to implement on personal computers.

Among other things, we have designed the case studies to provide realistic illustrations of the characteristics and trade-offs mentioned in the theory chapters. For example, *Case Study 10.1* is a very simple illustration of how linear algebraic equations are used to simultaneously satisfy a number of independent conditions. In addition, the study is also used to illustrate the utility of the matrix inverse as an analytical tool for such problem contexts. Although the example is taken from engineering management, the basic idea has relevance to a wide variety of other technical and analytical contexts.

In contrast, Case Studies 10.2, 10.3, 10.4, and 10.5 deal with lumped (or macro-) variable systems. *Case Study 10.2* shows how a mass balance can be employed to model a series of reactors. *Case Study 10.3* places special emphasis on the use of the matrix inverse to determine the complex cause-effect interactions between forces in the members of a truss. *Case Study 10.4* is an example of the use of Kirchhoff's laws to compute the currents and voltages in a resistor circuit. Finally, *Case Study 10.5* is an illustration of how linear equations are employed to determine the steady-state configuration of a mass-spring system.

In addition to engineering examples, we have also included *Case Study 10.6* which deals with spreadsheet applications. This example illustrates how a spreadsheet is used to solve equations with the Gauss-Seidel method.

CASE STUDY 10.1   INPUT-OUTPUT MODELING (GENERAL ENGINEERING)

Background: Interestingly, a significant fraction of the world's top business leaders started their careers as engineers. This is partially due to our ability to formulate and solve problems quantitatively. Thus, whether heading up a small consulting firm or a large corporation, engineers continually face management decisions where their quantitative training serves them well.

Many of these situations involve the proper allocation of resources to solve a particular problem. Such situations arise when organizing construction schedules, manufacturing and distributing products, and managing large-scale engineering research projects. Although the following technique originated in the field of macroeconomics, it has relevance to a wide variety of other problems involving coupled systems. Beyond its potential application to engineering management, it is instructive because it represents a way of formulating a linear system that is strikingly different than the approaches employed by engineers. In particular, whereas our models start with conservation laws, economists often generate their models directly from measurements.

Suppose that an economy is composed of three production sectors: agriculture, manufacturing, and services. Each of these sectors will produce a total output over a finite time period that can be quantified by some metric of value such as dollars, yen, or marks. Part of this total output will be fed back to each of the sectors. For example, some of the manufacturing products will be used by the agricultural sector. The remainder, or excess, production will then be delivered to consumers. Thus, a web of interconnecting value flow will link the sectors as depicted in Fig. 10.1 and listed in Table 10.1. This is the kind of data that economists routinely measure to characterize such systems. For example, the total production by agriculture is valued at $115 million of which $60 million is delivered to consumers and $10, $30, and $15 million are fed back to agriculture, manufacturing, and service, respectively.

Figure 10.1

A schematic showing the value flows over a year among the sectors of an economy consisting of three production sectors and a consumer sector. The numbers indicate the magnitude of the flows in millions of dollars. Notice that value flow does not balance for any of the production sectors due to factors not included in this model such as the value added by labor.

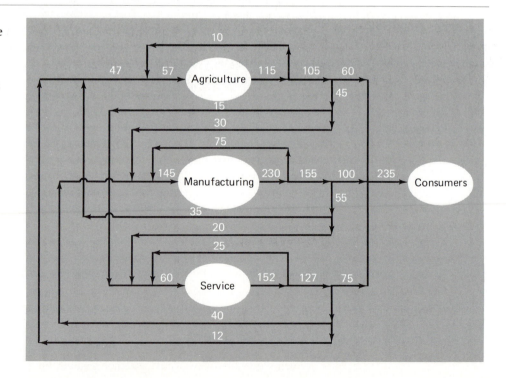

TABLE 10.1 An input-output table for a hypothetical economy composed of three producing sectors and a consumer sector. All numbers in millons of dollars.

| Production by | Consumption by | | | | Total Production |
	Agriculture (1)	Manufacture (2)	Services (3)	Consumers	
Agriculture (1)	10	30	15	60	115
Manufacture (2)	35	75	20	100	230
Services (3)	12	40	25	75	152
Total consumption	57	145	60	235	

Use linear algebraic equations to analyze this system. In particular, use the matrix inverse to determine the interrelationships between production and consumption by the various sectors of the economy.

Solution: As engineers our first inclination might be to apply the "conservation of cash" to develop a "cash balance" for the system. However, before doing this, a procedure must be developed to parameterize the transfers of value. Economists do this in the form of input-output coefficients. These are ratios of numbers from Table 10.1 which represent the amount of production from one sector (input) used to produce a unit amount from another sector (output). For example, \$0.0870 ($= 10/115$), \$0.3043 ($= 35/115$), and \$0.1043 ($= 12/115$) are required from agriculture, manufacturing, and services, respectively, to produce \$1 of output of agriculture. These coefficients can be compiled in a matrix

$$[A] = \begin{bmatrix} \frac{10}{115} & \frac{30}{230} & \frac{15}{152} \\ \frac{35}{115} & \frac{75}{230} & \frac{20}{152} \\ \frac{12}{115} & \frac{40}{230} & \frac{25}{152} \end{bmatrix} = \begin{bmatrix} 0.0870 & 0.1304 & 0.0987 \\ 0.3043 & 0.3261 & 0.1316 \\ 0.1043 & 0.1739 & 0.1645 \end{bmatrix}$$

where a_{ij} = the input from i to produce a unit amount of j.

Now we are ready to develop our "cash balance" which can be formulated in words as

$$\frac{\text{Production}}{\text{of } i} = \frac{\text{consumption}}{\text{of } i} + \sum_{j=1}^{3} \frac{\text{use of } i}{\text{by } j} \tag{10.1}$$

That is, all our production is either consumed or fed back for use by the sectors. This can be expressed mathematically as

$$\begin{Bmatrix} x_1 \\ x_2 \\ x_3 \end{Bmatrix} = \begin{Bmatrix} c_1 \\ c_2 \\ c_3 \end{Bmatrix} + \begin{Bmatrix} a_{11}x_1 + a_{12}x_2 + a_{13}x_3 \\ a_{21}x_1 + a_{22}x_2 + a_{23}x_3 \\ a_{31}x_1 + a_{32}x_2 + a_{33}x_3 \end{Bmatrix} \tag{10.2}$$

where x_i = the total production of sector i and c_i = the amount of x_i that is delivered to consumers. All the x's can be moved to the left side of the equation.

$$
\begin{array}{rcrcrcl}
(1 - a_{11})x_1 & & -a_{12}x_2 & & -a_{13}x_3 & = & c_1 \\
-a_{21}x_1 & + & (1 - a_{22})x_2 & & -a_{23}x_3 & = & c_2 \\
-a_{31}x_1 & & -a_{32}x_2 & + & (1 - a_{33})x_3 & = & c_3
\end{array}
$$

or

$$[B]\{X\} = \{C\} \tag{10.3}$$

where

$$[B] = [I] - [A] \tag{10.4}$$

Equation (10.3) is now in a format that we can use to analyze the system—that is, the stimulus-response format of Sec. 8.1.3. Thus, $\{X\}$ are the dependent variables (the amount of production of each sector), $\{C\}$ are the forcing functions (the amount of consumption), and $[B]$ are the system parameters. Equation (10.3) can be solved by

$$\{X\} = [B]^{-1}\{C\}$$

Thus, the model now provides a way to determine how much the production of each sector must be changed to accommodate new patterns of consumption. For the present case, the $[B]$ matrix is

$$
[B] = \begin{bmatrix}
0.91304 & -0.13043 & -0.09868 \\
-0.30435 & 0.67391 & -0.13158 \\
-0.10435 & -0.17391 & 0.83553
\end{bmatrix}
$$

Gauss-Jordan or LU decomposition can be used to compute

$$
[B]^{-1} = \begin{bmatrix}
1.2109 & 0.2828 & 0.1875 \\
0.6008 & 1.6870 & 0.3366 \\
0.2763 & 0.3865 & 1.2903
\end{bmatrix} \tag{10.5}
$$

This matrix can be employed to generate new production levels. However, more importantly, the coefficients provide insight themselves. For example, the numbers in column 1 of $[B]^{-1}$ tell us that the production of the agricultural, manufacturing, and service should be increased by \$1.2109, \$0.6008, \$0.2763 million, respectively, if the agricultural consumption is increased by \$1 million. Thus, the increase in agricultural consumption actually leads to increases in production in all sectors.

The foregoing is formally called an *input-output analysis* and was the primary contribution for which W. W. Leontief won the 1973 Nobel Prize for Economics. Aside from its possible applications to engineering management, we have included it as an example of how another qualitatively-oriented field formulates and solves a coupled linear system. In particular, you should recognize the fundamental differences between the engineering and economic approaches. In the former, we primarily use conservation laws and mechanisms to construct our models. In contrast, the economists work from a more empirical perspective because the mechanisms that underlie their systems are not as well-established and predictable as for engineering and the sciences.

CASE STUDY 10.2   STEADY-STATE ANALYSIS OF A SERIES OF REACTORS
                    (CHEMICAL ENGINEERING)

Background: One of the most important organizing principles in chemical engineering is the *conservation of mass* (recall Table 1.1). In quantitative terms, the principle is expressed as a mass balance that accounts for all sources and sinks of a material that pass in and out of a unit volume (Fig. 10.2). Over a finite period of time, this can be expressed as

$$\text{Accumulation} = \text{inputs} - \text{outputs} \tag{10.6}$$

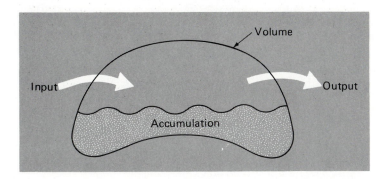

Figure 10.2
A schematic representation of mass balance.

The mass balance represents a bookkeeping exercise for the particular substance being modeled. For the period of the computation, if the inputs are greater than the outputs, the mass of the substance within the volume increases. If the outputs are greater than the inputs, the mass decreases. If inputs are equal to the outputs, accumulation is zero and mass remains constant. For this stable condition, or *steady-state,* Eq. (10.6) can be expressed as

$$\text{Inputs} = \text{outputs} \tag{10.7}$$

Employ the conservation of mass to determine the steady-state concentrations of a system of coupled reactors.

Solution: The mass balance can be used for engineering problem solving by expressing the inputs and outputs in terms of measurable variables and parameters. For example, if we were performing a mass balance for a conservative substance (that is, one that does not increase or decrease due to chemical transformations) in a reactor (Fig. 10.3), we would have to quantify the rate at which mass flows into the reactor through the two inflow pipes and out of the reactor through the outflow pipe. This can be done by taking the product of the flow rate, Q (in cubic meters per minute), and the concentration c (in milligrams per cubic meter) for each pipe. For example, for pipe 1 in Fig. 10.3, $Q_1 = 2$ m^3/min and $c_1 = 25$ mg/m^3; therefore the rate at which mass flows into the reactor through pipe 1 is $Q_1 c_1 = (2$ m^3/min$)(25$ mg/m$^3) = 50$ mg/min. Thus, 50 mg of chemical flows into the reactor through this pipe each minute. Similarly, for pipe 2 the mass inflow rate can be calculated as $Q_2 c_2 = (1.5$ m^3/min$)(10$ mg/m$^3) = 15$ mg/min.

Notice that the concentration out of the reactor through pipe 3 is not specified by Fig. 10.3. This is because we already have sufficient information to calculate it on the

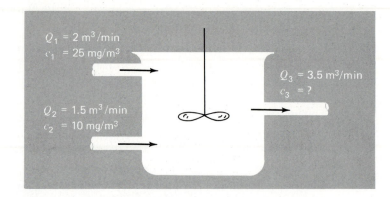

Figure 10.3
A steady-state, completely mixed reactor with two inflow pipes and one outflow pipe. The flows Q are in cubic meters per minute, and the concentrations c are in milligrams per cubic meter.

basis of the conservation of mass. Because the reactor is at steady state, Eq. (10.7) holds and the inputs should be in balance with the outputs, as in

$$Q_1c_1 + Q_2c_2 = Q_3c_3$$

Substituting known values into this equation yields

$$50 + 15 = 3.5c_3$$

which can be solved for $c_3 = 18.6$ mg/m^3. Thus we have determined the concentration in the third pipe. However, the computation yields an additional bonus. Because the reactor is well mixed (as represented by the propeller in Fig. 10.3), the concentration will be uniform, or homogeneous, throughout the tank. Therefore the concentration in pipe 3 should be identical to the concentration throughout the reactor. Consequently, the mass balance has allowed us to compute both the concentration in the reactor and in the outflow pipe. Such information is of great utility to chemical engineers who must design reactors to yield mixtures of a specified concentration.

Because simple algebra was used to determine the concentration for the single reactor in Fig. 10.3, it might not be obvious how computers figure in mass-balance calculations. Figure 10.4 shows a problem setting where computers are not only useful but are a practical necessity. Because there are five interconnected, or coupled, reactors, five simultaneous mass-balance equations are needed to characterize the system. For reactor 1, the rate of mass flow in is

$$5(10) + Q_{31}c_3$$

and the rate of mass flow out is

$$Q_{12}c_1 + Q_{15}c_1$$

Because the system is at steady state, the inflows and outflows must be equal:

$$5(10) + Q_{31}c_3 = Q_{12}c_1 + Q_{15}c_1$$

or, substituting the values for flow from Fig. 10.4,

$$6c_1 - c_3 = 50$$

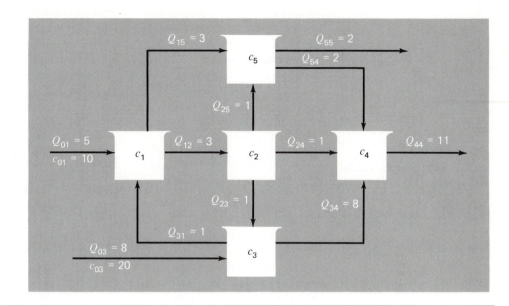

Figure 10.4
Five reactors linked by pipes.

Similar equations can be developed for the other reactors:

$$-3c_1 + 3c_2 = 0$$

$$-c_2 + 9c_3 = 160$$

$$-c_2 - 8c_3 + 11c_4 - 2c_5 = 0$$

$$-3c_1 - c_2 + 4c_5 = 0$$

A numerical method can be used to solve these five equations for the five unknown concentrations:

$$c_1 = 11.51$$

$$c_2 = 11.51$$

$$c_3 = 19.06$$

$$c_4 = 17.00$$

$$c_5 = 11.51$$

In addition, if Gauss-Jordan or *LU* decomposition with matrix inversion is employed, $[A]^{-1}$ can be computed as

$$[A]^{-1} = \begin{bmatrix} 0.1698 & 0.006289 & 0.01887 & 0 & 0 \\ 0.1698 & 0.3369 & 0.01887 & 0 & 0 \\ 0.01887 & 0.03774 & 0.1132 & 0 & 0 \\ 0.06003 & 0.07461 & 0.08748 & 0.09091 & 0.04545 \\ 0.1698 & 0.08962 & 0.01887 & 0 & 0.25 \end{bmatrix}$$

Each of the elements a_{ij} signifies the change in concentration of reactor i due to a unit change in loading to reactor j. Thus, the zeros in column 4 indicate that a loading to reactor 4 will have no impact on reactors 1, 2, 3, and 5. This is consistent with the system configuration (Fig. 10.4) which indicates that flow out of reactor 4 does not feed back into any of the other reactors. In contrast, loadings to any of the first three reactors will affect the entire system as indicated by the lack of zeros in the first three columns. Such information is of great utility to chemical engineers who design and manage such systems.

CASE STUDY 10.3   ANALYSIS OF A STATICALLY DETERMINANT TRUSS (CIVIL ENGINEERING)

Background: An important problem in structural engineering is that of finding the forces and reactions associated with a statically determinant truss. Figure 10.5 shows an example of such a truss.

The forces (F) represent either tension or compression on the members of the truss. External reactions (H_2, V_2, and V_3) are forces which characterize how the truss interacts with the supporting surface. The hinge at node 2 can transmit both horizontal and vertical forces to the surface, whereas the roller at node 3 only transmits vertical forces. It is observed that the effect of the external loading of 1000 lb is distributed among the various members of the truss.

Solution: This type of structure can be described as a system of coupled linear algebraic equations. Free-body force diagrams are shown for each node in Fig. 10.6. The sum of the forces in both horizontal and vertical directions must be zero at each node, because the system is at rest. Therefore, for node 1,

$$\Sigma F_H = 0 = -F_1 \cos 30° + F_3 \cos 60° + F_{1,h} \tag{10.8}$$

$$\Sigma F_V = 0 = -F_1 \sin 30° - F_3 \sin 60° + F_{1,v} \tag{10.9}$$

Figure 10.5
Forces on a statically determinant truss.

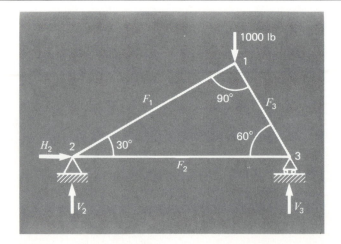

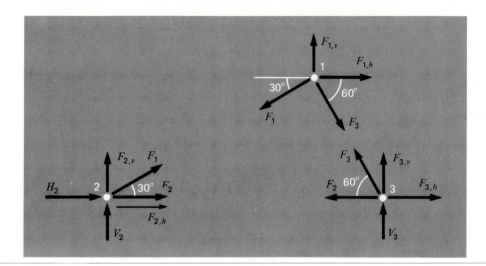

Figure 10.6
Free-body force diagrams for
the nodes of a statically-
determinant truss.

for node 2,

$$\Sigma F_H = 0 = F_2 + F_1 \cos 30° + F_{2,h} + H_2 \tag{10.10}$$

$$\Sigma F_V = 0 = F_1 \sin 30° + F_{2,v} + V_2 \tag{10.11}$$

for node 3,

$$\Sigma F_H = 0 = -F_2 - F_3 \cos 60° + F_{3,h} \tag{10.12}$$

$$\Sigma F_V = 0 = F_3 \sin 60° + F_{3,v} + V_3 \tag{10.13}$$

where $F_{i,h}$ is the external horizontal force applied to node i (where a positive force is from left to right) and $F_{i,v}$ is the external vertical force applied to node i (where a positive force is upward). Thus, in this problem, the 1000-lb downward force on node 1 corresponds to $F_{1,v} = -1000$. For this case all other $F_{i,v}$'s and $F_{i,h}$'s are zero. Note that the directions of the internal forces and reactions are unknown. Proper application of Newton's laws only requires consistent assumptions regarding direction. Solutions are negative if the directions are assumed incorrectly. Also note that in this problem, the forces in all members are assumed to be in tension and act to pull adjoining nodes together. A negative solution therefore corresponds to compression. This problem can be written as the following system of six equations and six unknowns:

$$\begin{bmatrix} 0.866 & 0 & -0.5 & 0 & 0 & 0 \\ 0.5 & 0 & 0.866 & 0 & 0 & 0 \\ -0.866 & -1 & 0 & -1 & 0 & 0 \\ -0.5 & 0 & 0 & 0 & -1 & 0 \\ 0 & 1 & 0.5 & 0 & 0 & 0 \\ 0 & 0 & -0.866 & 0 & 0 & -1 \end{bmatrix} \begin{Bmatrix} F_1 \\ F_2 \\ F_3 \\ H_2 \\ V_2 \\ V_3 \end{Bmatrix} = \begin{Bmatrix} 0 \\ -1000 \\ 0 \\ 0 \\ 0 \\ 0 \end{Bmatrix} \tag{10.14}$$

Notice that as formulated in Eq. (10.14), partial pivoting is required to avoid division by zero diagonal elements. Employing a pivot strategy, the system can be solved using any of the elimination techniques discussed in Chaps. 7, 8, and 9. However, because this problem is an ideal case study for demonstrating the utility of the matrix inverse, the *LU* decomposition can be used to compute:

$$F_1 = -500 \quad F_2 = 433 \quad F_3 = -866$$
$$H_2 = 0 \quad V_2 = 250 \quad V_3 = 750$$

and the matrix inverse is

$$
[A]^{-1} = \begin{bmatrix}
0.866 & 0.5 & 0 & 0 & 0 & 0 \\
0.25 & -0.433 & 0 & 0 & 1 & 0 \\
-0.5 & 0.866 & 0 & 0 & 0 & 0 \\
-1 & 0 & -1 & 0 & -1 & 0 \\
-0.433 & -0.25 & 0 & -1 & 0 & 0 \\
0.433 & -0.75 & 0 & 0 & 0 & -1
\end{bmatrix}
$$

Now, realize that the right-hand-side vector represents the externally applied horizontal and vertical forces on each node, as in

$$\{\text{Right-hand-side vector}\}^T = [F_{1,h} \quad F_{1,v} \quad F_{2,h} \quad F_{2,v} \quad F_{3,h} \quad F_{3,v}] \tag{10.15}$$

Because the external forces have no effect on the *LU* decomposition, the method need not be implemented over and over again to study the effect of different external forces on the truss. Rather, all that we have to do is perform the forward and the backward substitution steps for each right-hand-side vector to efficiently obtain alternative solutions. For example, we might want to study the effect of horizontal forces induced by a wind blowing from left to right. If the wind force can be idealized as two point forces of 1000 lb on nodes 1 and 2 (Fig. 10.7), the right-hand-side vector is

$$\{\text{Right-hand-side vector}\}^T = [1000 \quad 0 \quad 1000 \quad 0 \quad 0 \quad 0]$$

Figure 10.7
Two test cases showing (a) winds from the left and (b) winds from the right.

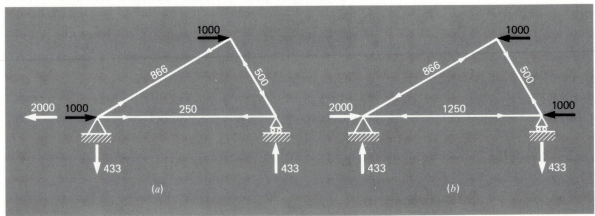

which can be used to compute

$$
\begin{array}{lll}
F_1 = 866 & F_2 = 250 & F_3 = -500 \\
H_2 = -2000 & V_2 = -433 & V_3 = 433
\end{array}
$$

For a wind from the right, $F_{1,h} = -1000$, $F_{3,h} = -1000$, and all other external forces are zero, with the result that

$$
\begin{array}{lll}
F_1 = -866 & F_2 = -1250 & F_3 = 500 \\
H_2 = 2000 & V_2 = 433 & V_3 = -433
\end{array}
$$

The results indicate that the winds have markedly different effects on the structure. Both cases are depicted in Fig. 10.7.

The individual elements of the inverted matrix also have direct utility in elucidating stimulus-response interactions for the structure. Each element represents the change of one of the unknown variables to a unit change of one of the external stimuli. For example, element a_{32}^{-1} indicates that the third unknown (F_3) will change 0.866 due to a unit change of the second external stimulus ($F_{1,v}$). Thus, if the vertical load at the first node were increased by 1, F_3 would increase by 0.866. The fact that elements are 0 indicates that certain unknowns are unaffected by some of the external stimuli. For instance $a_{13}^{-1} = 0$ means that F_1 is unaffected by changes in $F_{2,h}$. This ability to isolate interactions has a number of engineering applications including the identification of those components that are most sensitive to external stimuli and, as a consequence, most prone to failure. In addition, it can be used to determine components that may be unnecessary (see Prob. 10.21).

The foregoing approach becomes particularly useful when applied to large complex structures. In engineering practice it may be necessary to solve trusses with hundreds or even thousands of structural members. Linear equations provide one powerful approach for gaining insight into the behavior of these structures.

CASE STUDY 10.4   CURRENTS AND VOLTAGES IN RESISTOR CIRCUITS (ELECTRICAL ENGINEERING)

Background: A common problem in electrical engineering involves determining the currents and voltages at various locations in resistor circuits. These problems are solved using Kirchhoff's current and voltage rules. The current (or point) rule states that the algebraic sum of all currents entering a node must be zero (see Fig. 10.8a), or

$$
\Sigma i = 0 \tag{10.16}
$$

where all current entering the node is considered positive in sign. The current rule is an application of the principle of conservation of charge (recall Table 1.1).

The voltage (or loop) rule specifies that the algebraic sum of the potential differences (that is, voltage changes) in any loop must equal zero. For a resistor circuit, this is expressed as

$$
\Sigma \xi - \Sigma iR = 0 \tag{10.17}
$$

where ξ is the emf (electromotive force) of the voltage sources and R is the resistance of

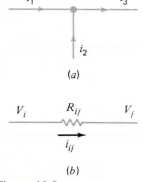

Figure 10.8
Schematic representations of (a) Kirchhoff's current rule and (b) Ohm's law.

any resistors on the loop. Note that the second term derives from Ohm's law (Fig. 10.8b), which states that the voltage drop across an ideal resistor is equal to the product of the current and the resistance. Kirchhoff's voltage rule is an expression of the conservation of energy.

Solution: Application of these rules results in systems of simultaneous linear algebraic equations because the various loops within a circuit are coupled. For example, consider the circuit shown in Fig. 10.9. The currents associated with this circuit are unknown both in magnitude and direction. This presents no great difficulty because one simply assumes a direction for each current. If the resultant solution from Kirchhoff's laws is negative, then the assumed direction was incorrect. For example, Fig. 10.10 shows some assumed currents.

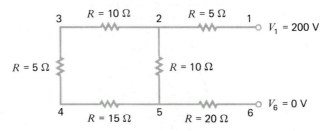

Figure 10.9
A resistor circuit to be solved using simultaneous linear algebraic equations.

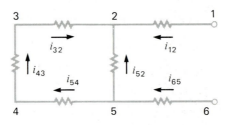

Figure 10.10
Assumed currents.

Given these assumptions, Kirchhoff's current rule is applied at each node to yield

$$i_{12} + i_{52} + i_{32} = 0$$

$$i_{65} - i_{52} - i_{54} = 0$$

$$i_{43} - i_{32} = 0$$

$$i_{54} - i_{43} = 0$$

Application of the voltage rule to each of the two loops gives

$$-i_{54}R_{54} - i_{43}R_{43} - i_{32}R_{32} + i_{52}R_{52} = 0$$

$$-i_{65}R_{65} - i_{52}R_{52} + i_{12}R_{12} - 200 = 0$$

or, substituting the resistances from Fig. 10.9 and bringing constants to the right-hand side,

$$-15i_{54} - 5i_{43} - 10i_{32} + 10i_{52} = 0$$

$$-20i_{65} - 10i_{52} + 5i_{12} = 200$$

Therefore the problem amounts to solving the following set of six equations with six unknown currents:

$$
\begin{array}{cccccccccccc}
i_{12} & + & i_{52} & + & i_{32} & & & & & & & = & 0 \\
& - & i_{52} & & & + & i_{65} & - & i_{54} & & & = & 0 \\
& & & - & i_{32} & & & & & + & i_{43} & = & 0 \\
& & & & & & & i_{54} & - & i_{43} & & = & 0 \\
& & 10i_{52} & - & 10i_{32} & & & - & 15i_{54} & - & 5i_{43} & = & 0 \\
5i_{12} & - & 10i_{52} & & & - & 20i_{65} & & & & & = & 200 \\
\end{array}
$$

Although impractical to solve by hand, this system is easily handled using an elimination method. Proceeding in this manner, the solution is

$$i_{12} = 6.1538$$
$$i_{52} = -4.6154$$
$$i_{32} = -1.5385$$
$$i_{65} = -6.1538$$
$$i_{54} = -1.5385$$
$$i_{43} = -1.5385$$

Thus, with proper interpretation of the signs of the result, the circuit currents and voltages are as shown in Fig. 10.11. The advantages of using numerical algorithms and personal computers for problems of this type should be evident.

Figure 10.11
The solution for currents and voltages obtained using an elimination method.

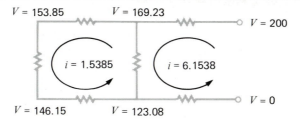

$V = 153.85$ $V = 169.23$ $V = 200$
$i = 1.5385$ $i = 6.1538$
$V = 146.15$ $V = 123.08$ $V = 0$

CASE STUDY 10.5   SPRING-MASS SYSTEMS (MECHANICAL ENGINEERING)

Background: Idealized spring-mass systems play an important role in mechanical and other engineering problems. Figure 10.12 shows such a system. After they are released,

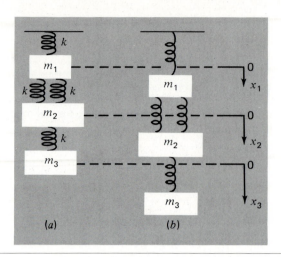

Figure 10.12
A system composed of three masses suspended vertically by a series of springs. (*a*) The system before release, that is, prior to extension or compression of the springs. (*b*) The system after release. Note that the positions of the masses are referenced to local coordinates with origins at their position before release.

the masses are pulled downward by the force of gravity. Notice that the resulting displacement of each spring in Fig. 10.12*b* is measured along local coordinates referenced to its initial position in Fig. 10.12*a*.

As introduced in Chap. 1, Newton's second law can be employed in conjunction with force balances to develop a mathematical model of the system. For each mass, the second law can be expressed as

$$m\frac{d^2x}{dt^2} = F_D - F_U \tag{10.18}$$

To simplify the analysis we will assume that all the springs are identical and follow Hooke's law [recall Eq. (6.16)]. A free-body diagram for the first mass is depicted in Fig. 10.13*a*. The upward force is merely a direct expression of Hooke's law

$$F_U = kx_1 \tag{10.19}$$

Figure 10.13
Free-body diagrams for the three masses from Fig. 10.12.

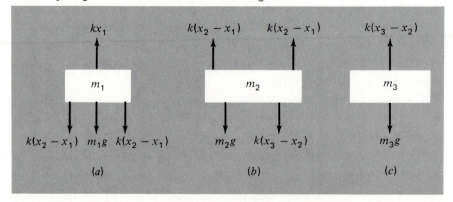

The downward component consists of the two spring forces along with the action of gravity on the mass,

$$F_D = k(x_2 - x_1) + k(x_2 - x_1) + m_1 g \qquad (10.20)$$

Note how the force component of the two springs is proportional to the displacement of the second mass, x_2, corrected for the displacement of the first mass, x_1.

Equations (10.19) and (10.20) can be substituted into Eq. (10.18) to give

$$m_1 \frac{d^2 x_1}{dt^2} = 2k(x_2 - x_1) + m_1 g - kx_1 \qquad (10.21)$$

Thus, we have derived a second-order ordinary differential equation to describe the displacement of the first mass with respect to time. However, notice that the solution cannot be obtained because the model includes a second dependent variable x_2. Consequently, free-body diagrams must be developed for the second and the third masses (Fig. 10.13b and c) which can be employed to derive

$$m_2 \frac{d^2 x_2}{dt^2} = k(x_3 - x_2) + m_2 g - 2k(x_2 - x_1) \qquad (10.22)$$

and

$$m_3 \frac{d^2 x_3}{dt^2} = m_3 g - k(x_3 - x_2) \qquad (10.23)$$

Solution: Now, Eqs. (10.21), (10.22), and (10.23) form a system of three differential equations with three unknowns. With the appropriate initial conditions, they could be used to solve for the displacements of the masses as a function of time (that is, their oscillations). We will discuss numerical methods for obtaining such solutions in Part Six. For the present, we can obtain the displacements that occur when the system eventually comes to rest, that is, to the steady state. To do this, the derivatives in Eqs. (10.21), (10.22), and (10.23) are set to zero to give

$$
\begin{aligned}
3kx_1 \quad - \quad 2kx_2 \qquad\qquad &= \quad m_1 g \\
-2kx_1 \quad + \quad 3kx_2 \quad - \quad kx_3 &= \quad m_2 g \\
- \quad kx_2 \quad + \quad kx_3 &= \quad m_3 g
\end{aligned}
$$

or, in matrix form,

$$[K]\{X\} = \{W\}$$

where $[K]$, called the stiffness matrix, is

$$[K] = \begin{bmatrix} 3k & -2k & \\ -2k & 3k & -k \\ & -k & k \end{bmatrix}$$

and $\{X\}$ and $\{W\}$ are the column vectors of the unknowns X and the weights mg, respectively.

At this point, numerical methods can be employed to obtain a solution. If $m_1 = 2$ kg, $m_2 = 3$ kg, $m_3 = 2.5$ kg, and the k's $= 10$ kg/s^2, use *LU* decomposition to solve for the displacements and generate the inverse of $[K]$.

Substituting the model parameters gives

$$[K] = \begin{bmatrix} 30 & -20 & \\ -20 & 30 & -10 \\ & -10 & 10 \end{bmatrix} \qquad \{W\} = \begin{Bmatrix} 19.6 \\ 29.4 \\ 24.5 \end{Bmatrix}$$

LU decomposition can be employed to solve for $x_1 = 7.35$, $x_2 = 10.045$, and $x_3 = 12.495$. These displacements were used to construct Fig. 10.12*b*.

The inverse of the stiffness matrix is computed as

$$[K]^{-1} = \begin{bmatrix} 0.1 & 0.1 & 0.1 \\ 0.1 & 0.15 & 0.15 \\ 0.1 & 0.15 & 0.25 \end{bmatrix}$$

Each element of this matrix k_{ij}^{-1} tells us the displacement of mass i due to a unit force imposed on mass j. Thus, the values of 0.1 in column 1 tell us that a downward unit load to the first mass will displace all of the masses 0.1 m downward. The other elements can be interpreted in a similar fashion. Therefore, the inverse of the stiffness matrix provides a fundamental summary of how the system's components respond to externally applied forces.

CASE STUDY 10.6   SPREADSHEET SOLUTION OF SIMULTANEOUS EQUATIONS WITH GAUSS-SEIDEL

This case study uses the spreadsheet to solve systems of linear algebraic equations with the Gauss-Seidel technique. A spreadsheet to implement the method is contained on the supplementary disk associated with the text. Insert the disk into the computer and type NUMMET following the A prompt. Advance to the main menu and activate the second selection. This causes Fig. 10.14 to appear on the screen.

The spreadsheet is designed to solve three simultaneous linear algebraic equations. Input the first values of X, Y, and Z to start the solution. Note that the parameter M allows you to change the value of the diagonal term of the system of equations. When M = 1, the diagonal term equals the sum of the absolute value of the off-diagonal terms. When M > 1, then the system becomes diagonally dominant, and when M < 1, the system is dominated by the off-diagonal terms. When M is greater than 1, the Gauss-Seidel method is guaranteed to converge. You can also select a relaxation factor (lambda) to adjust the computed values of X, Y, and Z. Move the cursor around the screen and carefully study the equations in all the cells of the spreadsheet. Note that only cells B3 through B7 are active. This allows you to vary the initial guesses, the nature of the equations, and the relaxation factor.

Observe how the values of X, Y, and Z vary with each iteration. The final values (ITER = 7) are close to the true solution of X = 1.3076, Y = 5.6154, and Z = 2.4615 obtained using an elimination method. Change the initial values of X = 10,000,

```
Cell   A         B        C        D        E        F        G        H
 1  GAUSS SEIDEL METHOD FOR SYSTEMS OF LINEAR EQUATIONS
 2  ------------------------------------------------------------------------
 3   FIRST X=       0
 4   FIRST Y=       0               8*M*X  +   3*Y  -   5*Z  =  15
 5   FIRST Z=       0                 2*X  + 4*M*Y  +   2*Z  =  30
 6     LAMDA=       1                 5*X  +   2*Y  + 7*M*Z  =  35
 7         M=       1
 8  ------------------------------------------------------------------------
 9     ITER=        1        2        3        4        5        6        7
10  ------------------------------------------------------------------------
11
12    OLD X=    1.875  .530134  1.25254  1.41330  1.31225  1.29342  1.30746
13    NEW X=    1.875  .530134  1.25254  1.41330  1.31225  1.29342  1.30746
14
15    OLD Y=   6.5625  6.34207  5.46907  5.52197  5.63748  5.62730  5.61210
16    NEW Y=   6.5625  6.34207  5.46907  5.52197  5.63748  5.62730  5.61210
17
18    OLD Z= 1.78571   2.80931  2.54273  2.41278  2.45197  2.46832  2.46263
19    NEW Z= 1.78571   2.80931  2.54273  2.41278  2.45197  2.46832  2.46263
20
[A1] GAUSS SEI
```

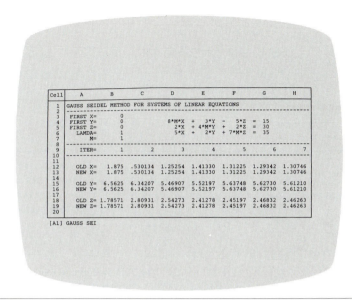

Figure 10.14
An implementation of the Gauss-Seidel technique on a spreadsheet.

Y = 10,000, and Z = 10,000, and strike [F3]. Note that convergence is incomplete. Therefore, transfer the last values of X, Y, and Z (located in cells H13, H16, and H19) to cells B3, B4, and B5 and strike [F3] again (see Fig. 10.15). Note that convergence is more complete. Transfer the last values of X, Y, and Z, strike [F3] again, and note that convergence is now almost complete.

Now try to speed the convergence by changing the value of the relaxation factor. Change the initial values of X, Y, and Z back to 10,000 and try lambda = 1.1 and 0.9.

Figure 10.15
A spreadsheet implementation of Gauss-Seidel showing slow convergence due to poor initial guesses.

```
Cell   A         B        C        D        E        F        G        H
 1  GAUSS SEIDEL METHOD FOR SYSTEMS OF LINEAR EQUATIONS
 2  ------------------------------------------------------------------------
 3   FIRST X=-7.95591
 4   FIRST Y= 22.0623               8*M*X  +   3*Y  -   5*Z  =  15
 5   FIRST Z= 4.37925                 2*X  + 4*M*Y  +   2*Z  =  30
 6     LAMDA=       1                 5*X  +   2*Y  + 7*M*Z  =  35
 7         M=       1
 8  ------------------------------------------------------------------------
 9     ITER=        1        2        3        4        5        6        7
10  ------------------------------------------------------------------------
11
12    OLD X=-3.66133   2.68144  1.93639  1.10686  1.22887   1.3367  1.31747
13    NEW X=-3.66133   2.68144  1.93639  1.10686  1.22887   1.3367  1.31747
14
15    OLD Y= 7.14104   3.37180  5.47115  5.91972  5.62655  5.57432  5.61498
16    NEW Y= 7.14104   3.37180  5.47115  5.91972  5.62655  5.57432  5.61498
17
18    OLD Z= 5.57493   2.12130  2.05367  2.51803  2.51465  2.45255  2.45466
19    NEW Z= 5.57493   2.12130  2.05367  2.51803  2.51465  2.45255  2.45466
20
[A1] GAUSS SEI
```

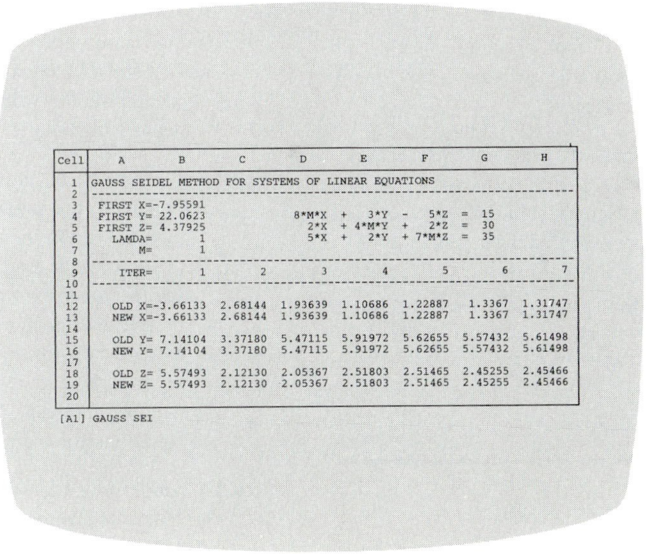

Note that the rate of convergence is increased when lambda = 0.9 and decreased when lambda = 1.1.

This example shows how the spreadsheet can be used to solve systems of linear algebraic equations. It is easy to observe how changes in the first guesses for X, Y, and Z, and the nature of the system, and the relaxation factor affect the efficiency of the Gauss-Seidel technique.

PROBLEMS

General Engineering

10.1 Reproduce the computation performed in Case Study 10.1 using your own software.

10.2 Perform the same computation as in Case Study 10.1, but change the consumption by agriculture of manufacturing output to 50.

10.3 An engineer supervises the production of three types of automobiles. Three kinds of material—metal, plastic, and rubber—are required for production. The amounts needed to produce each car are

Car	Metal, lb/car	Plastic, lb/car	Rubber, lb/car
1	1500	25	100
2	1700	33	120
3	1900	42	160

If totals of 106, 2.17, and 8.2 tons of metal, plastic, and rubber, respectively, are available each day, how many automobiles can be produced per day?

10.4 An engineer requires 4800, 5810, and 5690 yd^3 of sand, fine gravel, and coarse gravel, respectively, for a building project. There are three pits from which these materials can be obtained. The composition of these pits is

	Sand, %	Fine Gravel, %	Coarse Gravel, %
Pit 1	52	30	18
Pit 2	20	50	30
Pit 3	25	20	55

How many cubic yards must be hauled from each pit in order to meet the engineer's needs?

Chemical Engineering

10.5 Reproduce the computation performed in Case Study 10.2 using your own software.

10.6 Perform the same computation as in Case Study 10.2, but change c_{01} to 15 and c_{03} to 4.

10.7 If the input to reactor 1 in Case Study 10.2 is decreased 20 percent, what is the percent change in the concentration of reactors 2 and 3.

10.8 Because the system shown in Fig. 10.4 is at steady state, what can be said regarding the four flows: Q_{01}, Q_{03}, Q_{44}, and Q_{55}?

10.9 Recompute the concentrations for the five reactors shown in Fig. 10.3, if the flows are changed to:

$$Q_{01} = 5 \quad Q_{31} = 2 \quad Q_{25} = 3 \quad Q_{23} = 1$$
$$Q_{15} = 3 \quad Q_{55} = 4 \quad Q_{54} = 2 \quad Q_{34} = 5$$
$$Q_{12} = 4 \quad Q_{03} = 6 \quad Q_{24} = 0$$

10.10 Solve the same system as specified in Prob. 10.9, but set Q_{12} and Q_{54} equal to zero. Use conservation of flow to recompute the values for the other flows. What does the answer indicate to you regarding the physical system?

10.11 Figure P10.11 shows three reactors linked by pipes. As indicated, the rate of transfer of chemicals through each pipe is equal to a flow rate (Q, with units of cubic meters per second) multiplied by the concentration of the reactor from which the flow originates (c, with units of milligrams per cubic meter). If the system is at a steady state, the transfer into each reactor will balance the transfer out. Develop mass balance equations for the reactors and solve the three simultaneous linear algebraic equations for their concentrations.

10.12 Employing the same basic approach as in Case Study 10.2, determine the concentration of chloride in each of the Great Lakes using the information shown in Fig. P10.12.

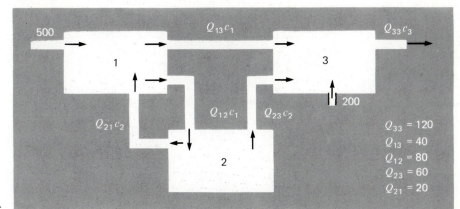

Figure P10.11
Three reactors linked by pipes.
The rate of mass transfer
through each pipe is equal to
the product of flow Q and
concentration c of the reactor
from which the flow originates.

$Q_{33} = 120$
$Q_{13} = 40$
$Q_{12} = 80$
$Q_{23} = 60$
$Q_{21} = 20$

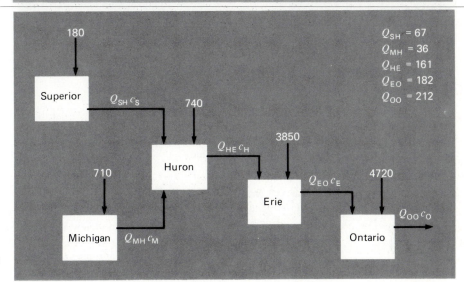

Figure P10.12
A chloride balance for the
Great Lakes. Numbered arrows
are direct inputs.

$Q_{SH} = 67$
$Q_{MH} = 36$
$Q_{HE} = 161$
$Q_{EO} = 182$
$Q_{OO} = 212$

10.13 A stage extraction process is depicted in Fig. P10.13. In such systems, a stream containing a weight fraction Y_{in} of a chemical enters from the left at a mass flow rate of F_1. Simultaneously, a solvent carrying a weight fraction X_{in} of the same chemical enters from the right at a flow rate of F_2. Thus, for stage i, a mass balance can be represented as

$$F_1 Y_{i-1} + F_2 X_{i+1} = F_1 Y_i + F_2 X_i \qquad (P10.13a)$$

At each stage, an equilibrium is assumed to be established between Y_i and X_i as in

$$K = \frac{X_i}{Y_i} \qquad (P10.13b)$$

where K is called a distribution coefficient. Equation (P10.13b) can be solved for X_i and substituted into Eq. (P10.13a) to yield

$$Y_{i-1} - \left(1 + \frac{F_2}{F_1}K\right)Y_i + \left(\frac{F_2}{F_1}K\right)Y_{i+1} = 0 \quad (P10.13c)$$

If $F_1 = 500$ kg/h, $Y_{in} = 0.1$, $F_2 = 750$ kg/h, $X_{in} = 0$, and $K = 5$, determine the values of Y_{out} and X_{out} if a five-stage reactor is used. Note that Eq. (P10.13c) must be modified to account for the inflow weight fractions when applied to the first and last stages.

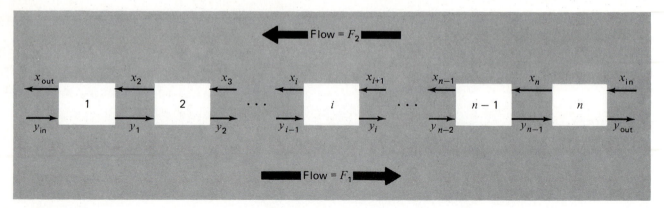

Figure P10.13
A state extraction process.

Civil Engineering

10.14 Reproduce the computation performed in Case Study 10.3 using your own software.

10.15 Perform the same computation as in Case Study 10.3, but change the angle at node 2 to 40° and at node 3 to 55°.

10.16 Perform the same computation as in Case Study 10.3, but for the truss depicted in Fig. P10.16.

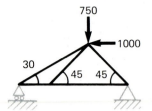

Figure P10.16

10.17 Perform the same computation as in Case Study 10.3, but for the truss depicted in Fig. P10.17.

10.18 Calculate the forces and reactions for the truss in Fig. 10.5 if a downward force of 2000 lb and a horizontal force to the right of 1500 lb are applied at node 1.

10.19 In the example for Fig. 10.5, where a 1000-lb downward force is applied at node 1, the external reactions V_2 and V_3 were calculated. But if the lengths of the truss members had been given, we could have calculated V_2 and V_3 by utilizing the fact that $V_2 + V_3$ must equal 1000 and by summing moments around node 2. However, because we do know V_2 and V_3, we can work backward to solve for the lengths of the truss members. Note that because there are three unknown lengths and only two equations, we can only solve for the relationship between lengths. Solve for this relationship.

10.20 Employing the same methods as used to analyze Fig. 10.5, determine the forces and reactions for the truss shown in Fig. P10.20.

10.21 Solve for the forces and reaction for the truss in Fig. P10.21. Determine the matrix inverse for the system. Does the vertical-member force in the middle member seem reasonable? Why?

Electrical Engineering

10.22 Reproduce the computation performed in Case Study 10.4 using your own software.

10.23 Perform the same computation as in Case Study 10.4, but change the resistance between nodes 3 and 4 to 15 Ω and change V_6 to 50 V.

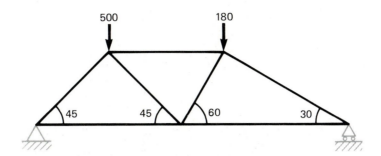

Figure P10.17

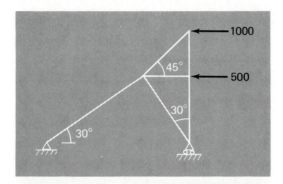

Figure P10.20

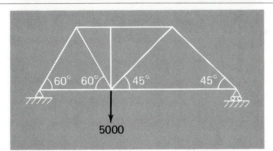

Figure P10.21

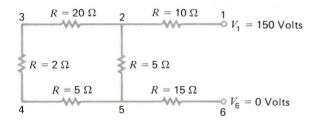

Figure P10.24

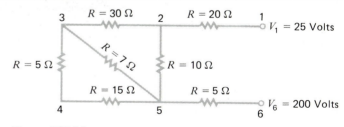

Figure P10.25

10.24 Perform the same computation as in Case Study 10.4, but for the circuit depicted in Fig. P10.24.

10.25 Perform the same computation as in Case Study 10.4, but for the circuit depicted in Fig. P10.25.

10.26 Solve the resistor circuit in Fig. 10.9, using Gauss elimination, if $V_1 = 110$ V and $V_6 = -110$ V. Notice that the voltages do not cancel each other out.

10.27 Solve the circuit in Fig. P10.27 for the currents in each wire. Use Gauss elimination with pivoting.

Mechanical Engineering

10.28 Reproduce the computation performed in Case Study 10.5 using your own software.

10.29 Perform the same computation as in Case Study 10.5, but double the spring constants.

10.30 Perform the same computation as in Case Study 10.5, but add a third spring between masses 1 and 2.

10.31 Perform the same computation as in Case Study 10.5, but change the masses from 2, 3, and 2.5 kg to 10, 2, and 2 kg.

10.32 Idealized spring-mass systems have numerous applications throughout engineering. Figure P10.32 shows an arrangement of four springs in series being depressed with a force of 2000 lb. At equilibrium, force-balance equations can be developed defining the interrelationships between the springs:

$$k_2(x_2 - x_1) = k_1 x_1$$
$$k_3(x_3 - x_2) = k_2(x_2 - x_1)$$
$$k_4(x_4 - x_3) = k_3(x_3 - x_2)$$
$$F = k_4(x_4 - x_3)$$

where the k's are spring constants. If k_1 through k_4 are 100, 50, 75, and 200 lb/in, respectively, compute the x's.

10.33 Three blocks are connected by a weightless cord and rest on an inclined plane (Fig. P10.33a). Employing a procedure similar to the one used in the analysis of the falling parachutists in Example 7.11 yields the following set of simultaneous equations (free-body diagrams are shown in Fig. P10.33b):

$$
\begin{aligned}
100a &+ T & &= 519.72 \\
50a &- T &+ R &= 216.55 \\
20a & &- R &= 86.62
\end{aligned}
$$

Solve for acceleration a and the tensions T and R in the two ropes.

10.34 Perform a computation similar to that called for in Prob. 10.33, but for the system shown in Fig. P10.34.

10.35 Perform the same computation as in Prob. 10.33, but for the system depicted in Fig. P10.35 (angles are $45°$).

Spreadsheet

10.36 Use an active spreadsheet that may be available to you (like Lotus 123 or the one on the Electronic TOOLKIT) to solve the following system of equations using the Gauss-Seidel method.

$$
\begin{aligned}
7x_1 &+ x_2 &+ x_3 &+ x_4 &= 19 \\
2x_1 &+ 11x_2 &+ x_3 &- x_4 &= -5 \\
x_1 &+ 2x_2 &+ 7x_3 &+ 2x_4 &= 23 \\
x_1 &- x_2 &- x_3 &+ 8x_4 &= -10
\end{aligned}
$$

Try various initial guesses and discuss the results.

10.37 Repeat Prob. 10.36 except for the system from Prob. 8.10.

10.38 Activate the spreadsheet associated with the text and set lambda and M equal to 1. Now try various weights, M, on the diagonal so that the system ranges from strongly diagonally-dominant to non-diagonally dominant. Discuss the results.

10.39 Activate the spreadsheet associated with the text, and set M $=0.8$ with first values of X, Y, and Z $= 0$. Study the convergence of this system for lambda $= 1.5$ and 0.8. Discuss your observations.

10.40 Repeat Prob. 10.39 with M $=0.5$ and lambda $= 0.3$ and 1.0.

10.41 Repeat Prob. 10.39 with M $=5$ and 10 and lambda $= 0.9, 1.0$, and 1.1.

10.42 Repeat Prob. 10.39 with M $=0.1$ and lambda $= 0.1, 0.5$, and 1.0.

Miscellaneous

10.43 Read all the case studies in Chap. 10. On the basis of your reading and experience make up your own case study for any one of the fields of engineering. This may involve modifying or reexpressing one of our case studies. However, it can also be totally original. As with our examples, it must be drawn from an engineering problem context and must demonstrate the use of numerical methods for solving linear algebraic equations. Write up your results using our case studies as models.

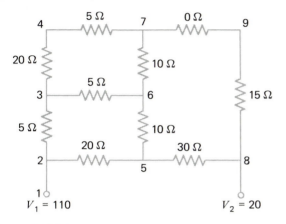

Figure P10.27

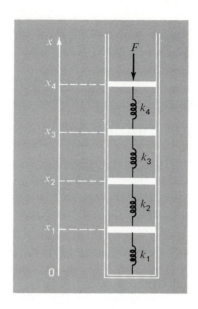

Figure P10.32

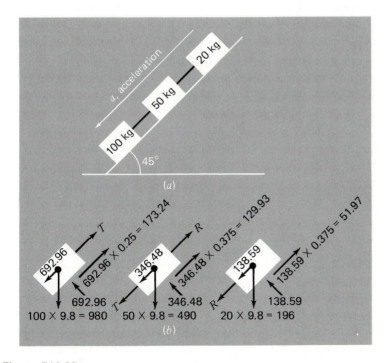

Figure P10.33

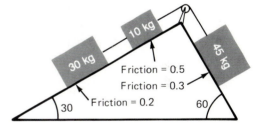

Figure P10.34

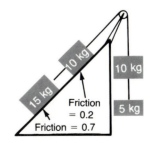

Figure P10.35

EPILOGUE: PART THREE

PT3.4   TRADE-OFFS

Table PT3.2 provides a summary of the trade-offs involved in solving simultaneous linear algebraic equations. Three methods—graphical, Cramer's rule, and algebraic manipulation—are limited to small (≤ 3) numbers of equations and thus have little utility for practical problem solving. However, these techniques are useful didactic tools for understanding the behavior of linear systems in general.

The numerical methods themselves are divided into two general categories: exact and approximate methods. As the name implies, the former are intended to yield exact answers. However, because they are affected by round-off errors they sometimes yield imprecise results. The magnitude of the round-off error varies from system to system and is dependent on a number of factors. These include the system's dimensions, its condition, and whether the matrix of coefficients is sparse or full. In addition, computer precision will affect round-off error. In general, exact methods are usually the method of choice for small equation sets (that is, those less than 100).

It is recommended that a pivoting strategy be employed in any computer program implementing exact elimination methods. The inclusion of such a strategy minimizes round-off error and avoids problems such as division by zero. All other things being equal, *LU*-decomposition-based algorithms are the methods of choice because of their efficiency and flexibility.

Although elimination methods have great utility, their use of the entire matrix of coefficients can be somewhat limiting when dealing with very large, sparse systems. This is due to the fact that large portions of computer memory would be devoted to storage of meaningless zeros. For banded systems, techniques are available to implement elimination methods without having to store the entire coefficient matrix.

The approximate technique described in this book is called the Gauss-Seidel method. It differs from the exact techniques in that it employs an iterative scheme to obtain progressively closer estimates of the solution. Thus, the effect of round off is a moot point with the Gauss-Seidel method, because the iterations can be continued as long as is necessary to obtain the desired precision. In addition, versions of the Gauss-Seidel method can be developed to efficiently utilize computer storage requirements for sparse systems. Consequently, the Gauss-Seidel technique has utility for large systems of equations (> 100), where round-off errors and storage requirements would pose significant problems for the exact techniques.

TABLE PT3.2 Comparison of the characteristics of alternative methods for finding solutions of simultaneous linear algebraic equations.

Method	Approximate Maximum Number of Equations*	Stability	Precision	Breadth of Application	Programming Effort	Comments
Graphical	2	—	Poor	Limited	—	May take more time than the numerical method
Cramer's rule	3	—	Affected by round-off error	Limited	—	Excessive computational effort required for more than three equations
Algebraic manipulation (elimination of unknowns)	3	—	Affected by round-off error	Limited		
Gauss elimination (with partial pivoting)	100†	—	Affected by round-off error	General	Moderate	
Gauss-Jordon (with partial pivoting)	100†	—	Affected by round-off error	General	Moderate	Allows computation of matrix inverse
LU decomposition	100†	—	Affected by round-off error	General	Moderate	Preferred elimination method
Gauss-Seidel	1000†	May not converge if not diagonally dominant	Excellent	Appropriate only for diagonally dominant systems	Easy	

*Does not account for conditioning or sparseness or computer hardware.
†Upper limit depends on computer system and degree of sparseness of the equation. These are very general limits.

The disadvantage of the Gauss-Seidel method is that it does not always converge or sometimes converges slowly on the true solution. It is only strictly reliable for those systems which are diagonally dominant. However, relaxation methods are available that sometimes offset these disadvantages. In addition, because many sets of linear algebraic equations originating from physical systems exhibit diagonal dominance, the Gauss-Seidel method has great utility for engineering problem solving.

In summary, a variety of factors will bear on your choice of a technique for a particular problem involving linear algebraic equations. However, as outlined above, the size and sparseness of the system are particularly important factors in determining your choice.

PT3.5   IMPORTANT RELATIONSHIPS AND FORMULAS

Every part of this book includes a section that summarizes important formulas. Although Part Three does not really deal with single formulas, we have used Table PT3.3 to summarize the algorithms that were covered. The table provides an overview that should be helpful for review and in elucidating the major differences between the methods.

PT3.6   ADVANCED METHODS AND ADDITIONAL REFERENCES

General references on the solution of simultaneous linear equations can be found in Faddeev and Faddeeva (1963), Stewart (1973), Varga (1962), and Young (1971). Ralson and Rabinowitz (1978) provide a general summary.

Many advanced techniques are available to increase time and/or space savings of linear algebraic equations. Most of these focus on exploiting properties of the equations such as symmetry and bandedness. In particular, algorithms are available to operate on sparse matrices in order to convert them to a minimum banded format. Tewerson (1973) and Jacobs (1977) include information on this area. Once they are in a minimum banded format, there are a variety of efficient solution strategies that are employed such as the active column storage approach of Bathe and Wilson (1976).

Aside from $n \times n$ sets of equations, there are other systems where the number of equations, m, and number of unknowns, n, are not equal. Systems where $m < n$ are called *underdetermined*. In such cases there can be either no solution or else more than one. Systems where $m > n$ are called *overdetermined*. For such situations, there is in general no exact solution. However, it is often possible to develop a compromise solution that attempts to determine answers that come "closest" to satisfying all the equations simultaneously. A common approach is to solve the equation in a "least squares" sense (Lawson and Hanson, 1974, and Wilkinson and Reinsch, 1971). Alternatively, *linear programming* methods can be used where the equations are solved in an "optimal" sense by minimizing some objective function (Rabinowitz, 1968, Dantzig, 1963, and Luenberger, 1973).

TABLE PT3.3 Summary of important information presented in Part Three.

Method	Procedure	Potential Problems and Remedies
Gauss elimination	$$\left[\begin{array}{ccc|c} a_{11} & a_{12} & a_{13} & c_1 \\ a_{21} & a_{22} & a_{23} & c_2 \\ a_{31} & a_{32} & a_{33} & c_3 \end{array}\right] \Rightarrow \left[\begin{array}{ccc|c} a_{11} & a_{12} & a_{13} & c_1 \\ & a'_{22} & a'_{23} & c'_2 \\ & & a''_{33} & c''_3 \end{array}\right]$$ $$\Rightarrow \quad \begin{array}{l} x_3 = c''_3/a''_{33} \\ x_2 = (c'_2 - a'_{23}x_3)/a'_{22} \\ x_1 = (c_1 - a_{12}x_2 - a_{13}x_3)/a_{11} \end{array}$$	Problems: Ill conditioning Round-off Division by zero Remedies: Higher precision Partial pivoting
Matrix inversion (Gauss-Jordan)	$$\left[\begin{array}{ccc|ccc} a_{11} & a_{12} & a_{13} & 1 & 0 & 0 \\ a_{21} & a_{22} & a_{23} & 0 & 1 & 0 \\ a_{31} & a_{32} & a_{33} & 0 & 0 & 1 \end{array}\right] \Rightarrow \left[\begin{array}{ccc|ccc} 1 & 0 & 0 & a^{-1}_{11} & a^{-1}_{12} & a^{-1}_{13} \\ 0 & 1 & 0 & a^{-1}_{21} & a^{-1}_{22} & a^{-1}_{23} \\ 0 & 0 & 1 & a^{-1}_{31} & a^{-1}_{32} & a^{-1}_{33} \end{array}\right]$$	Problems: Ill conditioning Round-off Division by zero Remedies: Higher precision Partial pivoting
LU decomposition	$$\left[\begin{array}{ccc} a_{11} & a_{12} & a_{13} \\ a_{21} & a_{22} & a_{23} \\ a_{31} & a_{32} & a_{33} \end{array}\right] \Rightarrow \left[\begin{array}{ccc} l_{11} & 0 & 0 \\ l_{21} & l_{22} & 0 \\ l_{31} & l_{32} & l_{33} \end{array}\right]$$ Decomposition $$\left\{\begin{array}{c} d_1 \\ d_2 \\ d_3 \end{array}\right\} = \left\{\begin{array}{c} c_1 \\ c_2 \\ c_3 \end{array}\right\} \Rightarrow \left[\begin{array}{ccc} 1 & u_{12} & u_{13} \\ 0 & 1 & u_{23} \\ 0 & 0 & 1 \end{array}\right]\left\{\begin{array}{c} x_1 \\ x_2 \\ x_3 \end{array}\right\} = \left\{\begin{array}{c} d_1 \\ d_2 \\ d_3 \end{array}\right\} \Rightarrow \left\{\begin{array}{c} x_1 \\ x_2 \\ x_3 \end{array}\right\}$$ Forward Substitution Back Substitution	Problems: Ill conditioning Round-off Division by zero Remedies: Higher precision Partial pivoting
Gauss-Seidel method	$$\begin{array}{l} x_1^j = (c_1 - a_{12}x_2^{j-1} - a_{13}x_3^{j-1})/a_{11} \\ x_2^j = (c_2 - a_{21}x_1^j - a_{23}x_3^{j-1})/a_{22} \\ x_3^j = (c_3 - a_{31}x_1^j - a_{32}x_2^j)/a_{33} \end{array}$$ continue iteratively until $$\left|\frac{x_i^j - x_i^{j-1}}{x_i^j}\right| 100\% < \epsilon_s$$ for all x_i's	Problems: Divergent or converges slowly Remedies: Diagonal dominance Relaxation

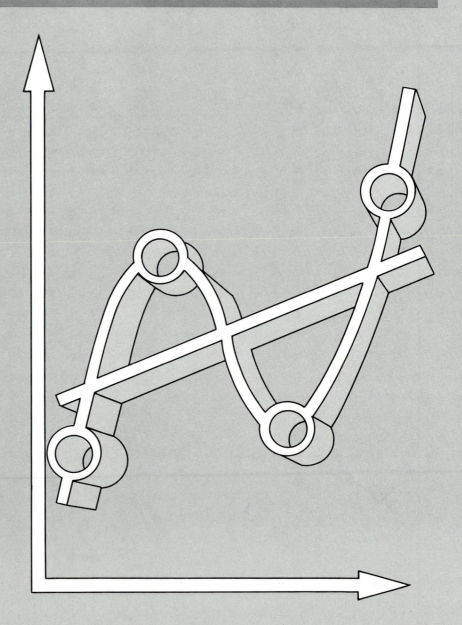

PART FOUR

CURVE FITTING

PT4.1 MOTIVATION

Data is often given for discrete values along a continuum. However, you may require estimates at points between the discrete values. The present part of this book describes techniques to fit curves to such data in order to obtain intermediate estimates. In addition, you may require a simplified version of a complicated function. One way to do this is to compute values of the function at a number of discrete values along the range of interest. Then a simpler function may be derived to fit these values. Both of these applications are known as *curve fitting*.

There are two general approaches for curve fitting that are distinguished from each other on the basis of the amount of error associated with the data. First, where the data exhibits a significant degree of error or "noise," the strategy is to derive a single curve that represents the general trend of the data. Because any individual data point may be incorrect, we make no effort to intersect every point. Rather, the curve is designed to follow the pattern of the points taken as a group. One approach of this nature is called *least-squares regression* (Fig. PT4.1*a*).

Second, where the data is known to be very precise, the basic approach is to fit a curve or a series of curves that pass directly through each of the points. Such data usually originates from tables. Examples are values for the density of water or for the heat capacity of gases as a function of temperature. The estimation of values between well-known discrete points is called *interpolation* (Fig. PT4.1*b* and *c*).

PT4.1.1 Precomputer Methods for Curve Fitting

The simplest method for fitting a curve to data is to plot the points and then sketch a line that visually conforms to the data. Although this is a valid option when quick estimates are required, the results are dependent on the subjective viewpoint of the person sketching the curve.

For example, Fig. PT4.1 shows sketches developed from the same set of data by three students. The first did not attempt to connect the points but rather characterized the general upward trend of the data with a straight line (Fig. PT4.1*a*). The second student used straight-line segments or linear interpolation to connect the points (Fig. PT4.1*b*). This is a very common practice in engineering. If the values are truly close to being linear or are spaced closely, such an approximation provides estimates that are adequate for many engineering calculations. However, where

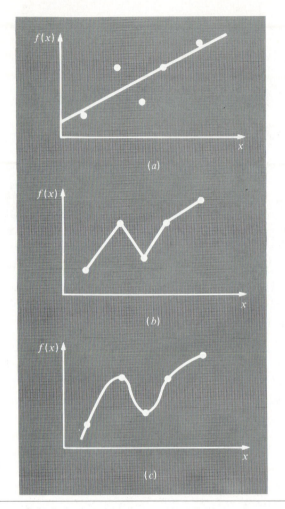

Figure PT4.1
Three attempts to fit a "best" curve through five data points. (*a*) Least-squares regression, (*b*) linear interpolation, and (*c*) curvilinear interpolation.

the underlying relationship is highly curvilinear or the data is widely spaced, significant errors can be introduced by such linear interpolation. The third student used curves to try to capture the meanderings suggested by the data (Fig. PT4.1*c*). A fourth or fifth student would likely develop alternative fits. Obviously, our goal here is to develop systematic and objective methods for the purpose of deriving such curves.

PT4.1.2 Curve Fitting and Engineering Practice

Your first exposure to curve fitting may have been to determine intermediate values from tabulated data—for instance, from interest tables for engineering economics or from steam tables for thermodynamics. Throughout the remainder of your career, you will have frequent occasion to estimate intermediate values from such tables.

Although many of the widely used engineering properties have been tabulated, there are a great many more that are not available in this convenient form. Special cases and new problem contexts often require that you measure your own data and develop your own predictive relationships. Two types of applications are generally encountered when fitting experimental data: trend analysis and hypothesis testing.

Trend analysis represents the process of using the pattern of the data to make predictions. For cases where the data is measured with high precision, you might utilize interpolating polynomials. Imprecise data is usually analyzed with least-squares regression.

Trend analysis may be used to predict or forecast values of the dependent variable. This can involve *extrapolation* beyond the limits of the observed data or *interpolation* within the range of the data. All fields of engineering commonly involve problems of this type.

A second engineering application of experimental curve fitting is *hypothesis testing*. Here an existing mathematical model is compared with measured data. If the model coefficients are unknown, it may be necessary to determine values that best fit the observed data. On the other hand, if estimates of the model coefficients are already available, it may be appropriate to compare predicted values of the model with observed values to test the adequacy of the model. Often, alternative models are compared and the "best" selected on the basis of empirical observations.

In addition to the above engineering applications, curve fitting is important in other numerical methods such as integration and the approximate solution of differential equations. Finally, curve-fitting techniques can be used to derive simple functions to approximate complicated functions.

PT4.2   MATHEMATICAL BACKGROUND

The prerequisite mathematical background for interpolation is found in the material on Taylor series expansions and finite divided differences introduced in Chap. 3. Least-squares regression requires additional information from the field of statistics. If you are familiar with the concepts of the mean, standard deviation, residual sum of the squares, and normal distribution, feel free to skip the following pages and proceed directly to PT4.3. If you are unfamiliar with these concepts or are in need of a review, the following material is designed as a brief introduction to these topics.

PT4.2.1   Simple Statistics

Suppose that in the course of an engineering study, several measurements were made of a particular quantity. For example, Table PT4.1 contains 24 readings of the coefficient of thermal expansion of a structural steel. Taken at face value, the data provides a limited amount of information—that is, that the values range from a minimum of 6.395 to a maximum of 6.775. Additional insight can be gained by summarizing the data in one or more well-chosen statistics that convey as much information as possible about specific characteristics of the data set. These descriptive statistics are most often selected to represent (1) the location of the center of the distribution of the data and (2) the degree of spread of the data set.

TABLE PT4.1 **Measurements of the coefficient of thermal expansion of structural steel [$\times 10^{-6}$ in / (in · °F)].**

6.495	6.625	6.635	6.655
6.665	6.515	6.625	6.775
6.755	6.615	6.575	6.555
6.565	6.435	6.395	6.655
6.595	6.715	6.485	6.605
6.505	6.555	6.715	6.685

The most common location statistic is the arithmetic mean. The *arithmetic mean* ($\bar{y}$) of a sample is defined as the sum of the individual data points (y_i) divided by the number of points (n), or

$$\bar{y} = \frac{\sum y_i}{n}$$

(PT4.1)

where the summation is from $i = 1$ through n.

The most common measure of spread for a sample is the *standard deviation* (s_y) about the mean:

$$s_y = \sqrt{\frac{S_t}{n-1}}$$

(PT4.2)

where S_t is the total sum of the squares of the residuals between the data points and the mean, or

$$S_t = \sum (y_i - \bar{y})^2$$

(PT4.3)

Thus, if the individual measurements are spread out widely around the mean, S_t (and, consequently, s_y) will be large. If they are grouped tightly, the standard deviation will be small. The spread can also be represented by the square of the standard deviation, which is called the *variance*:

$$s_y^2 = \frac{S_t}{n-1}$$

(PT4.4)

Note that the denominator in both Eqs. (PT4.2) and (PT4.4) is $n - 1$. The quantity $n - 1$ is referred to as the *degrees of freedom*. Hence s_y and s_y^2 are said to be based on $n - 1$ degrees of freedom. This nomenclature derives from the fact that the sum of the quantities upon which S_t is based (that is, $\bar{y} - y_1, \bar{y} - y_2, \ldots, \bar{y} - y_n$) is zero. Consequently, if $n - 1$ of the values are specified, the remaining value is fixed. Thus only $n - 1$ of the values are said to be freely determined. Another justification for dividing by $n - 1$ is the fact that there is no such thing as the spread of a single data point. For the case where $n = 1$, Eq. (PT4.4) yields a meaningless result of infinity.

A final statistic that has utility in quantifying the spread of data is the *coefficient of variation* (*c.v.*). This statistic is the ratio of the standard deviation to the mean. As such it provides a normalized measure of the spread. It is often multiplied by 100 so that it can be expressed in the form of a percent:

$$\text{c.v.} = \frac{s_y}{\bar{y}} 100\%$$

(PT4.5)

Notice that the coefficient of variation is similar in spirit to the percent relative error (ϵ_t) discussed in Sec. 3.3. That is, it is the ratio of a measure of error (s_y) to an estimate of the true value ($\bar{y}$).

EXAMPLE PT4.1   Simple Statistics of a Sample

Problem Statement: Compute the mean, variance, standard deviation, and coefficient of variation for the data in Table PT4.1.

Solution: The data is added (Table PT4.2) and the results are used to compute [Eq. (PT4.1)]:

$$\bar{y} = \frac{158.400}{24} = 6.6$$

As in Table PT4.2, the sum of the squares of the residuals is 0.21700, which can be used to compute the standard deviation [Eq. (PT4.2)]:

$$s_y = \sqrt{\frac{0.217000}{24 - 1}} = 0.097133$$

and the variance [Eq. (PT4.4)]:

$$s_y^2 = 0.009435$$

and the coefficient of variation [Eq. (PT4.5)]:

$$\text{c.v.} = \frac{0.097133}{6.6} 100\% = 1.47\%$$

TABLE PT4.2   Computations for statistics and histogram for the readings of the coefficient of thermal expansion.

				Interval	
i	y_i	$(y_i - \bar{y})^2$	Frequency	Lower Bound	Upper Bound
1	6.395	0.042025	1	6.36	6.40
2	6.435	0.027225	1	6.40	6.44
3	6.485	0.013225			
4	6.495	0.011025	4	6.48	6.52
5	6.505	0.009025			
6	6.515	0.007225			
7	6.555	0.002025	2	6.52	6.56
8	6.555	0.002025			
9	6.565	0.001225			
10	6.575	0.000625	3	6.56	6.60
11	6.595	0.000025			
12	6.605	0.000025			
13	6.615	0.000225			
14	6.625	0.000625	5	6.60	6.64
15	6.625	0.000625			
16	6.635	0.001225			
17	6.655	0.003025			
18	6.655	0.003025	3	6.64	6.68
19	6.665	0.004225			
20	6.685	0.007225			
21	6.715	0.013225	3	6.68	6.72
22	6.715	0.013225			
23	6.755	0.024025	1	6.72	6.76
24	6.775	0.030625	1	6.76	6.80
Σ	158.400	0.217000			

PT4.2.2   The Normal Distribution

The final characteristic that bears on the present discussion is the *data distribution*—that is, the shape with which the data is spread around the mean. A *histogram* provides a simple visual representation of the distribution. As in Table PT4.2, the histogram is constructed by sorting the measurements into intervals. The units of measurement are plotted on the abscissa and the frequency of occurrence of each interval is plotted on the ordinate. Thus, five of the measurements fall between 6.60 and 6.64. As in Fig. PT4.2, the histogram suggests that most of the data is grouped close to the mean value of 6.6.

If we have a very large set of data, the histogram often can be approximated by a smooth curve. The symmetric, bell-shaped curve superimposed on Fig. PT4.2 is one such characteristic shape—the *normal distribution*. Given enough additional measurements, the histogram for this particular case could eventually approach the normal distribution.

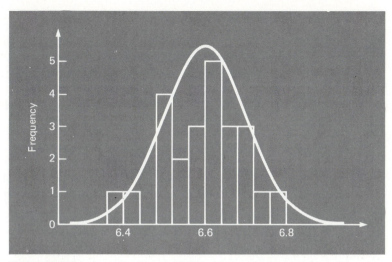

Figure PT4.2
A histogram used to depict the distribution of data. As the number of data points increases, the histogram could approach the smooth, bell-shaped curve called the normal distribution.

The concepts of the mean, standard deviation, residual sum of the squares, and normal distribution all have great relevance to engineering practice. A very simple example is their use to quantify the confidence that can be ascribed to a particular measurement. If a quantity is normally distributed, the range defined by $\bar{y} - s_y$ to $\bar{y} + s_y$ will encompass approximately 68 percent of the total number of measurements. Similarly, the range defined by $\bar{y} - 2s_y$ to $\bar{y} + 2s_y$ will encompass approximately 95 percent.

For example, for the coefficients of thermal expansion in Table PT4.1 ($\bar{y} = 6.6$ and $s_y = 0.097133$), we can make the statement that approximately 95 percent of the readings should fall between 6.405734 and 6.794266. If someone told us that they had measured a value of 7.35, we would suspect that the measurement might be erroneous.

The above is just one simple example of how statistics can be used to make judgments regarding uncertain data. These concepts will also have direct relevance to our discussion of regression models. You can consult any basic statistics book (for example, Ang and Tang, 1975, or Lapin, 1983) to obtain additional information on the subject.

PT4.3 ORIENTATION

Before proceeding to numerical methods for curve fitting, some orientation might be helpful. The following is intended as an overview of the material discussed in Part Four. In addition, we have formulated some objectives to help focus your efforts when studying the material.

PT4.3.1 Scope and Preview

Figure PT4.3 provides a visual overview of the material to be covered in Part Four. *Chapter 11* is devoted to *least-squares regression*. We will first learn how to fit the "best" straight line through a set of uncertain data points. This technique is called *linear regression*. Besides discussing how to calculate the slope and intercept of this straight line, we also present quantitative and visual methods for evaluating the validity of the results.

In addition to fitting a straight line, we also present a general technique for fitting a "best" polynomial. Thus, you will learn to derive a parabolic, cubic, or higher-order polynomial that optimally fits uncertain data. Linear regression is a subset of this more general approach, which is called *polynomial regression*.

The next topic covered in Chap. 11 is *multiple linear regression*. It is designed for the case where the dependent variable y is a linear function of two or more independent variables $x_1, x_2, \ldots, x_n$. This approach has special utility for evaluating experimental data where the variable of interest is dependent on a number of different factors.

After multiple regression, we illustrate how polynomial and multiple regression are both subsets of a *general linear least-squares model*. Among other things, this will allow us to introduce a concise matrix representation of regression and discuss its general statistical properties.

Finally, the last sections of Chap. 11 are devoted to *nonlinear regression*. This approach is designed to compute a least-squares fit of a nonlinear equation to data.

In *Chapter 12,* the alternative curve-fitting technique called *interpolation* is described. As discussed previously, interpolation is used for estimating interme-diate values between precise data points. In Chap. 12, polynomials are derived for this purpose. We introduce the basic concept of polynomial interpolation by using straight lines and parabolas to connect points. Then, we develop a general-ized procedure for fitting an nth-order polynomial. Two formats are presented for expressing these polynomials in equation form. The first, called *Newton's interpolating polynomial,* is preferable when the appropriate order of the poly-nomial is unknown. The second, called the *Lagrange interpolating polynomial,* has advantages when the proper order is known beforehand. In addition, a sec-tion is devoted to a method that can be used to generate the coefficients of the interpolating polynomial.

The last section of Chap. 12 presents an alternative technique for fitting precise data points. This technique, called *spline interpolation,* fits polynomials to data but in a piecewise fashion. As such, it is particularly well-suited for fitting data that is generally smooth but exhibits abrupt local changes.

Chapter 13 deals with the Fourier transform approach to curve fitting where periodic functions are fit to data. Our emphasis in this section will be on the *fast Fourier transform.*

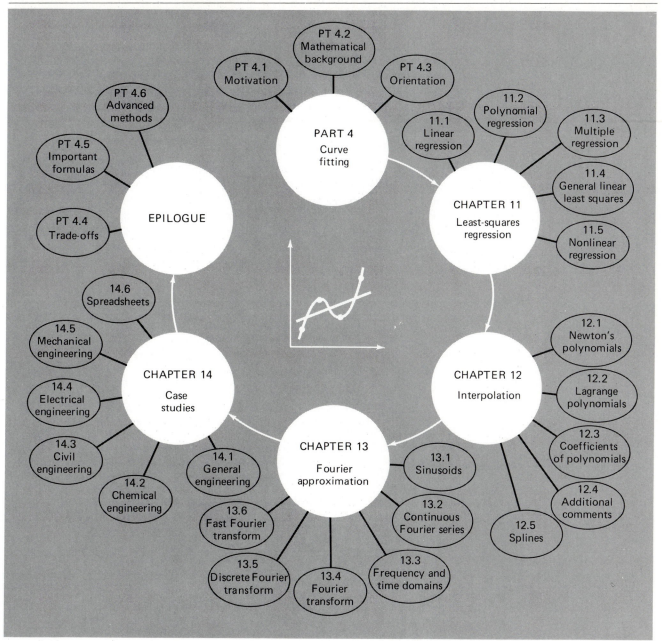

Figure PT4.3
Schematic of the organization of the material in Part Four: Curve Fitting.

Chapter 14 is devoted to case studies that illustrate the utility of the numerical methods in engineering problem contexts. Examples are drawn from general engineering as well as from the four major specialty areas of chemical, civil, electrical, and mechanical engineering. In addition, a spreadsheet case study is included.

Finally, an epilogue is included at the end of Part Four. It contains a summary of the important formulas and concepts related to curve fitting as well as a discussion of trade-offs among the techniques and suggestions for future study.

PT4.3.2 Goals and Objectives

Study Objectives.   After completing Part Four, you should have greatly enhanced your capability to fit curves to data. In general, you should have mastered the techniques, have learned to assess the reliability of the answers, and be capable of choosing the preferred method (or methods) for any particular problem. In addition to these general goals, the specific concepts in Table PT4.3 should be assimilated and mastered.

TABLE PT4.3   **Specific study objectives for Part Four.**

1. Understand the fundamental difference between regression and interpolation and realize why confusing the two could lead to serious problems
2. Understand the derivation of linear least-squares regression and be able to assess the reliability of the fit using graphical and quantitative assessments
3. Know how to linearize data by transformation
4. Understand situations where polynomial, multiple, and nonlinear regression are appropriate
5. Understand the general matrix formulation of linear least squares
6. Understand that there is one and only one polynomial of degree n or less that passes exactly through $n + 1$ points
7. Know how to derive the first-order Newton's interpolating polynomial
8. Understand the analogy between Newton's polynomial and the Taylor series expansion and how it relates to the truncation error
9. Recognize that the Newton and Lagrange equations are merely different formulations of the same interpolating polynomial and understand their respective advantages and disadvantages
10. Realize that more accurate results are obtained if data used for interpolation is centered around and close to the unknown point
11. Realize that data points do not have to be equally spaced nor in any particular order for either the Newton or Lagrange polynomials
12. Know why equispaced interpolation formulas have utility
13. Recognize the liabilities and risks associated with extrapolation
14. Understand why spline functions have utility for data with local areas of abrupt change
15. Recognize how the Fourier series is used to fit data with periodic functions
16. Understand the difference between the frequency and the time domains

Computer Objectives.   You have been provided with software, simple computer programs and algorithms to implement the techniques discussed in Part Four. All have utility as learning tools.

The optional Electronic TOOLKIT software includes linear regression and Lagrange interpolation programs. The graphics associated with this software will enable you to easily visualize your problem and the associated mathematical operations. The graphics are a critical part of your assessment of the validity of a regression fit. They also provide guidance regarding the proper order of polynomial interpolation and the potential dangers of extrapolation. The software is very easy to apply to solve practical problems and can be used to check the results of any computer programs you may develop yourself.

In addition, computer codes and algorithms are provided for most of the other methods in Part Four. This information will allow you to expand your software library to include techniques beyond linear regression and Lagrange interpolation. For example, you may find it useful from a professional viewpoint to have software to implement polynomial and multiple regression, Newton's interpolating polynomial, cubic spline interpolation, and the fast Fourier transform.

Finally, spreadsheet software to determine Newton's polynomial is contained on the diskette that accompanies the book. This software is designed to accompany Case Study 14.6 which illustrates how interpolation can be implemented on a spreadsheet.

CHAPTER 11
Least-Squares Regression

Where substantial error is associated with data, polynomial interpolation is inappropriate and may yield unsatisfactory results when used to predict intermediate values. Experimental data is often of this type. For example, Fig. 11.1a shows seven experimentally derived data points exhibiting significant variability. Visual inspection of the data suggests a positive relationship between y and x. That is, the overall trend indicates that higher values of y are associated with higher values of x. Now, if a sixth-order interpolating polynomial is fitted to this data (Fig. 11.1b), it will pass exactly through all of the points. However, because of the variability in the data, the curve oscillates widely in the interval between the points. In particular, the interpolated values at $x = 1.5$ and $x = 6.5$ appear to be well beyond the range suggested by the data.

A more appropriate strategy for such cases is to derive an approximating function that fits the shape or general trend of the data without necessarily matching the individual points. Figure 11.1c illustrates how a straight line can be used to generally characterize the trend of the data without passing through any particular point.

One way to determine the line in Fig. 11.1c is to visually inspect the plotted data and then sketch a "best" line through the points. Although such "eyeball" approaches have commonsense appeal and are valid for "back-of-the-envelope" calculations, they are deficient because they are arbitrary. That is, unless the points define a perfect straight line (in which case, interpolation would be appropriate), different analysts would draw different lines.

In order to remove this subjectivity, some criterion must be devised to establish a basis for the fit. One way to do this is to derive a curve that minimizes the discrepancy between the data points and the curve. A technique for accomplishing this objective, called *least-squares regression,* will be discussed in the present chapter.

11.1 LINEAR REGRESSION

The simplest example of a least-squares approximation is fitting a straight line to a set of paired observations: $(x_1, y_1), (x_2, y_2), \ldots, (x_n, y_n)$. The mathematical expression for the straight line is

$$y = a_0 + a_1 x + e \tag{11.1}$$

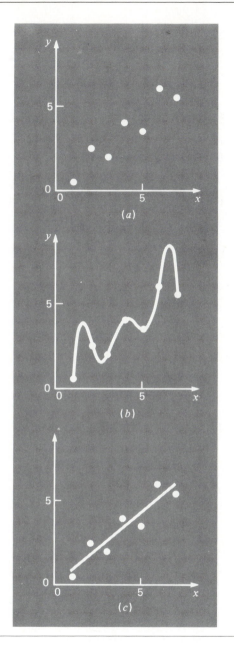

Figure 11.1
(a) Data exhibiting significant error. (b) Polynomial fit oscillating beyond the range of the data. (c) More satisfactory result using the least-squares fit.

where a_0 and a_1 are coefficients representing the intercept and the slope, respectively, and e is the error, or residual, between the model and the observations, which can be represented by rearranging Eq. (11.1) as

$$e = y - a_0 - a_1 x$$

Thus, the error, or residual, is the discrepancy between the true value of y and the approximate value, $a_0 + a_1 x$, predicted by the linear equation.

11.1.1   Criteria for a "Best" Fit

One strategy for fitting a "best" line through the data would be to minimize the sum of the residual errors for all the available data, as in

$$\sum_{i=1}^{n} e_i = \sum_{i=1}^{n} (y_i - a_0 - a_1 x_i) \tag{11.2}$$

where n is the total number of points. However, this is an inadequate criterion, as illustrated by Fig. 11.2*a* which depicts the fit of a straight line to two points. Obviously, the best fit is the line connecting the points. However, any straight line passing through the midpoint of the connecting line (except a perfectly vertical line) results in a minimum value of Eq. (11.2) equal to zero because the errors cancel.

Another criterion would be to minimize the sum of the absolute values of the discrepancies, as in

$$\sum_{i=1}^{n} |e_i| = \sum_{i=1}^{n} |y_i - a_0 - a_1 x_i|$$

Figure 11.2*b* demonstrates why this criterion is also inadequate. For the four points shown, any straight line falling within the dashed lines will minimize the absolute value of the sum. Thus, this criterion also does not yield a unique best fit.

A third strategy for fitting a best line is the *minimax* criterion. In this technique, the line is chosen that minimizes the maximum distance that an individual point falls from the line. As depicted in Fig. 11.2*c*, this strategy is ill-suited for regression because it gives undue influence to an outlier, that is, a single point with a large error. It should be noted that the minimax principle is sometimes well-suited for fitting a simple function to a complicated function (Carnahan, Luther, and Wilkes, 1969).

A strategy that overcomes the shortcomings of the aforementioned approaches is to minimize the sum of the squares of the residuals, S_r, as in

$$S_r = \sum_{i=1}^{n} e_i^2 = \sum_{i=1}^{n} (y_i - a_0 - a_1 x_i)^2 \tag{11.3}$$

This criterion has a number of advantages, including the fact that it yields a unique line for a given set of data. Before discussing these properties, we will present a technique for determining the values of a_0 and a_1 that minimize Eq. (11.3).

11.1.2   Least-Squares Fit of a Straight Line

In order to determine values for a_0 and a_1, Eq. (11.3) is differentiated with respect to each coefficient:

$$\frac{\partial S_r}{\partial a_0} = -2 \sum (y_i - a_0 - a_1 x_i)$$

$$\frac{\partial S_r}{\partial a_1} = -2 \sum [(y_i - a_0 - a_1 x_i) x_i]$$

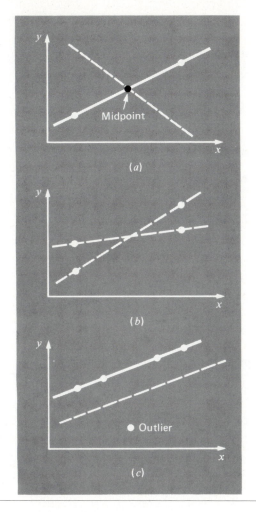

Figure 11.2
Examples of some criteria for "best fit" that are inadequate for regression: (a) minimizes the sum of the residuals, (b) minimizes the sum of the absolute values of the residuals, and (c) minimizes the maximum error of any individual point.

Note that we have simplified the summation symbols; unless otherwise indicated, all summations are from $i = 1$ to n. Setting these derivatives equal to zero will result in a minimum S_r. If this is done, the equations can be expressed as

$$0 = \sum y_i - \sum a_0 - \sum a_1 x_i$$

$$0 = \sum y_i x_i - \sum a_0 x_i - \sum a_1 x_i^2$$

Now, realizing that $\sum a_0 = n a_0$, the equations can be expressed as a set of two simultaneous linear equations with two unknowns (a_0 and a_1):

$$n a_0 + \sum x_i a_1 = \sum y_i \tag{11.4}$$

$$\sum x_i a_0 + \sum x_i^2 a_1 = \sum x_i y_i \qquad (11.5)$$

These are called the *normal equations*. They can be solved simultaneously for [recall Eq. (7.10)]

$$a_1 = \frac{n \sum x_i y_i - \sum x_i \sum y_i}{n \sum x_i^2 - (\sum x_i)^2} \qquad (11.6)$$

This result can then be used in conjunction with Eq. (11.4) to solve for

$$a_0 = \bar{y} - a_1 \bar{x} \qquad (11.7)$$

where $\bar{y}$ and $\bar{x}$ are the means of y and x, respectively.

EXAMPLE 11.1   Linear Regression

Problem Statement: Fit a straight line to the x and y values in the first two columns of Table 11.1.

TABLE 11.1
Computations for an error analysis of the linear fit.

x_i	y_i	$(y_i - \bar{y})^2$	$(y_i - a_0 - a_1 x_i)^2$
1	0.5	8.5765	0.1687
2	2.5	0.8622	0.5625
3	2.0	2.0408	0.3473
4	4.0	0.3265	0.3265
5	3.5	0.0051	0.5896
6	6.0	6.6122	0.7972
7	5.5	4.2908	0.1993
Σ	24	22.7143	2.9911

Solution: The following quantities can be computed:

$$n = 7 \qquad \sum x_i y_i = 119.5 \qquad \sum x_i^2 = 140$$

$$\sum x_i = 28 \qquad \bar{x} = \frac{28}{7} = 4$$

$$\sum y_i = 24 \qquad \bar{y} = \frac{24}{7} = 3.428571429$$

Using Eqs. (11.6) and (11.7),

$$a_1 = \frac{7(119.5) - 28(24)}{7(140) - (28)^2} = 0.839285714$$

$$a_0 = 3.428571429 - 0.839285714(4) = 0.07142857$$

Therefore, the least-squares fit is

$$y = 0.07142857 + 0.839285714x$$

The line, along with the data, is shown in Fig. 11.1c.

11.1.3 Quantification of Error of Linear Regression

Any line other than the one computed in Example 11.1 results in a larger sum of the squares of the residuals. Thus, the line is unique and in terms of our chosen criterion is a "best" line through the points. A number of additional properties of this fit can be elucidated by examining more closely the way in which residuals were computed. Recall that the sum of the squares is defined as [Eq. (11.3)]

$$S_r = \sum_{i=1}^{n} (y_i - a_0 - a_1 x_i)^2 \tag{11.8}$$

Notice the similarity between Eqs. (PT4.3) and (11.8). In the former case, the square of the residual represented the square of the discrepancy between the data and a single estimate of the measure of central tendency—the mean. In Eq. (11.8), the square of the residual represents the square of the vertical distance between the data and another measure of central tendency—the straight line (Fig. 11.3).

The analogy can be extended further for cases where (1) the spread of the points around the line is of similar magnitude along the entire range of the data and (2) the distribution of these points about the line is normal. It can be demonstrated that if these criteria are met, least-squares regression will provide the best (that is, the most likely) estimates of a_0 and a_1 (Draper and Smith, 1981). This is called the *maximum likelihood*

Figure 11.3

The residual in linear regression represents the vertical distance between a data point and the straight line.

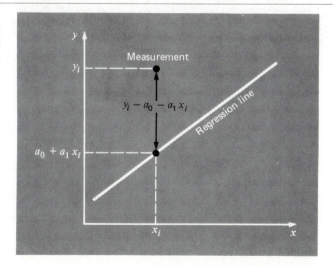

principle in statistics. In addition, if these criteria are met, a "standard deviation" for the regression line can be determined as [compare with Eq. (PT4.2)]

$$s_{y/x} = \sqrt{\frac{S_r}{n - 2}}$$

(11.9)

where $s_{y/x}$ is called the *standard error of the estimate*. The subscript notation "*y/x*" designates that the error is for a predicted value of *y* corresponding to a particular value of *x*. Also, notice that we now divide by $n - 2$ because two data-derived estimates—a_0 and a_1—were used to compute S_r; thus, we have lost two degrees of freedom. As with our discussion of the standard deviation in PT4.2.1, another justification for dividing by $n - 2$ is that there is no such thing as the "spread of data" around a straight line connecting two points. Thus, for the case where $n = 2$, Eq. (11.9) yields a meaningless result of infinity.

Just as was the case with the standard deviation, the standard error of the estimate quantifies the spread of the data. However, $s_{y/x}$ quantifies the spread *around the regression line* as shown in Fig. 11.4*b* in contrast to the original standard deviation s_y that quantified the spread *around the mean* (Fig. 11.4*a*).

The above concepts can be used to quantify the "goodness" of our fit. This is particularly useful for comparison of several regressions (see Fig. 11.5). To do this, we return to the original data and determine the *total sum of the squares* around the mean for the dependent variable (in our case, *y*). As was the case for Eq. (PT4.3), this quantity is designated S_t. This is the uncertainty associated with the dependent variable prior to regression. After performing the regression, we can compute S_r, the sum of the squares of the residuals around the regression line. This represents the uncertainty that remains after the regression. It is, therefore, sometimes called the *unexplained sum of the squares*. The difference between the two quantities, $S_t - S_r$, quantifies the improvement

Figure 11.4
Regression data showing (*a*) the spread of the data around the mean of the dependent variable and (*b*) the spread of the data around the best-fit line. The reduction in the spread in going from (*a*) to (*b*), as indicated by the bell-shaped curves at the right, represents the improvement due to linear regression.

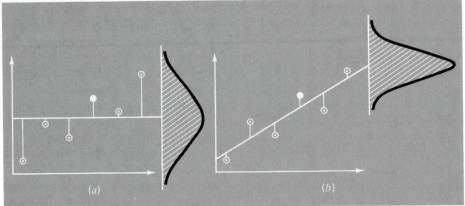

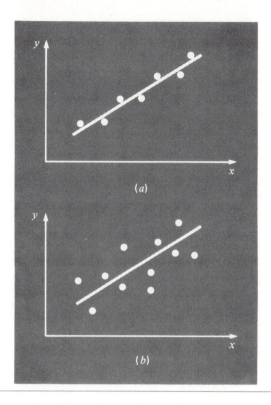

Figure 11.5
Examples of linear regression
with (a) small and (b) large
residual errors.

or error reduction due to describing the data in terms of a straight line rather than as an average value. Because the magnitude of this quantity is scale-dependent, the difference is normalized to the total error to yield

$$r^2 = \frac{S_t - S_r}{S_t} \tag{11.10}$$

where r^2 is called the *coefficient of determination* and r is the *correlation coefficient* ($= \sqrt{r^2}$). For a perfect fit, $S_r = 0$ and $r = r^2 = 1$, signifying that the line explains 100 percent of the variability of the data. For $r = r^2 = 0$, $S_r = S_t$ and the fit represents no improvement. An alternative formulation for r that is more convenient for computer implementation is

$$r = \frac{n \sum x_i y_i - (\sum x_i)(\sum y_i)}{\sqrt{n \sum x_i^2 - (\sum x_i)^2} \sqrt{n \sum y_i^2 - (\sum y_i)^2}} \tag{11.11}$$

EXAMPLE 11.2   Estimation of Errors for the Linear Least-Squares Fit

Problem Statement: Compute the total standard deviation, the standard error of the estimate, and the correlation coefficient for the data in Example 11.1.

Solution: The summations are performed and presented in Table 11.1. The total standard deviation is [Eq. (PT4.2)]

$$s_y = \sqrt{\frac{22.7143}{7-1}} = 1.9457$$

and the standard error of the estimate is [Eq. (11.9)]

$$s_{y/x} = \sqrt{\frac{2.9911}{7-2}} = 0.7735$$

Thus, because $s_{y/x} < s_y$, the linear regression model has merit. The extent of the improvement is quantified by [(Eq. 11.10)]

$$r^2 = \frac{22.7143 - 2.9911}{22.7143} = 0.868$$

or

$$r = \sqrt{0.868} = 0.932$$

These results indicate that 86.8 percent of the original uncertainty has been explained by the linear model.

Before proceeding to the computer program for linear regression, a word of caution is in order. Although the correlation coefficient provides a handy measure of goodness-of-fit, you should be careful not to ascribe more meaning to it than is warranted. Just because r is "close" to 1 does not mean that the fit is necessarily "good." For example, it is possible to obtain a relatively high value of r when the underlying relationship between y and x is not even linear. Draper and Smith (1981) provide guidance and additional material regarding assessment of results for linear regression. In addition, at the minimum, you should *always* inspect a plot of the data along with your regression line whenever you fit regression curves. As described in the next section, the Electronic TOOLKIT software includes such a capability.

11.1.4 Computer Program for Linear Regression

It is a relatively trivial matter to develop a program for linear regression. A BASIC version is contained in Fig. 11.6. In addition, FORTRAN and Pascal subroutines are listed in Fig. 11.7. Because the graphical capabilities of personal computers are so varied, we have not included plot routines in these programs. However, as mentioned above, such an option is critical to the effective use and interpretation of regression and is included in the supplementary Electronic TOOLKIT software. If your computer system has plotting capabilities, we recommend that you expand your program to include a plot of y versus x showing both the data and the regression line. The inclusion of the capability will greatly enhance the utility of the program in problem-solving contexts.

```
100 REM    LINEAR REGRESSION (BASIC VERSION)
110 REM
120 REM ***********************************
130 REM *      DEFINITION OF VARIABLES     *
140 REM *                                  *
150 REM * N   = NUMBER OF DATA POINTS      *
160 REM * X() = INDEPENDENT VARIABLE       *
170 REM * Y() = DEPENDENT VARIABLE         *
180 REM ***********************************
190 REM
200 DIM X(100),Y(100)
210 REM
220 REM ********* MAIN PROGRAM *********
230 REM
240 GOSUB 300 'input data
250 GOSUB 400 'perform regression
260 GOSUB 700 'output results
270 END
300 REM ******* SUBROUTINE INPUT *********
310 CLS
320 INPUT "NUMBER OF DATA POINTS? ", N
330 PRINT
340 FOR I = 1 TO N
350    INPUT "X,Y = ";X(I),Y(I)
360 NEXT I
370 RETURN
400 REM ***** SUBROUTINE REGRESSION ******
410 REM
420 SUMX = 0:  SUMXY = 0:  ST = 0
430 SUMY = 0:  SUMX2 = 0:  SR = 0
440 FOR I = 1 TO N
450    SUMX = SUMX + X(I)
460    SUMY = SUMY + Y(I)
470    SUMXY = SUMXY + X(I)*Y(I)
480    SUMX2 = SUMX2 + X(I)*X(I)
490 NEXT I
500 XMEAN = SUMX/N
510 YMEAN = SUMY/N
520 A1 = (N*SUMXY-SUMX*SUMY)/
                   (N*SUMX2-SUMX*SUMX)
530 A0 = YMEAN - A1*XMEAN
540 FOR I = 1 TO N
550    ST = ST + (Y(I)-YMEAN)^2
560    SR = SR + (Y(I)-A1*X(I)-A0)^2
570 NEXT I
580 SYX = SQR(SR/(N-2))
590 R2 = (ST-SR)/ST
600 R = SQR(R2)
610 RETURN
700 REM ****** SUBROUTINE OUTPUT  ******
710 REM
720 CLS
730 PRINT:PRINT
740 PRINT "SLOPE = ";A1
750 PRINT "INTERCEPT = ";A0
760 PRINT "STANDARD ERROR = ";SYX
770 PRINT "CORRELATION COEFFICIENT(R) = ";R
780 PRINT "COEFFICIENT DETERMINATION(R2) =";
790 PRINT R2
800 RETURN
```

Figure 11.6
An interactive computer program for linear regression written in Microsoft BASIC.

Figure 11.7
Subroutines for linear regression written in (a) FORTRAN 77 and (b) Turbo Pascal.

```
      SUBROUTINE LINREG(X,Y,N,A0,A1,R2,SYX)
*********************************************
*         DEFINITION OF VARIABLES          *
*                                          *
*    N   = NUMBER OF DATA POINTS           *
*    X() = INDEPENDENT VARIABLE            *
*    Y() = DEPENDENT VARIABLE              *
*********************************************
      DIMENSION X(100),Y(100)
      SUMX = 0.0
      SUMY = 0.0
      SUMXY = 0.0
      SUMX2 = 0.0
```

```
PROCEDURE LinReg(X,Y: Vector;
                  N: integer;
                  VAR A1,A0: real;
                  VAR R2,Syx: real);
{    Driver program type definitions
   Vector = a 1 dimensional real array    }
{      Definition of variables
   N   = number of data points
   X() = independent variable
   Y() = dependent variable              }
VAR
   Sumx, Sumy, Sumxy, Sumx2 : real;
   Xmean, Ymean: real;
```

```
      ST = 0.0
      SR = 0.0
      DO 10 I=1,N
         SUMX = SUMX + X(I)
         SUMY = SUMY + Y(I)
         SUMXY = SUMXY + X(I)*Y(I)
         SUMX2 = SUMX2 + X(I)*X(I)
   10 CONTINUE
      XMEAN = SUMX/N
      YMEAN = SUMY/N
      A1=(N*SUMXY-SUMX*SUMY)/
     *                     (N*SUMX2-SUMX*SUMX)
      A0 = YMEAN - A1*XMEAN
      DO 20 I=1,N
         ST = ST + (Y(I)-YMEAN)**2
         SR = SR + (Y(I)-A1*X(I)+A0)**2
   20 CONTINUE
      SYX = SQRT(SR/(N-2))
      R2 = (ST-SR)/ST
      RETURN
      END
```

```
St, Sr: real;
Begin                        {procedure LinRegr}
Sumx := 0.0; Sumy := 0.0;
Sumx2 := 0.0; Sumxy := 0.0;
St := 0.0; Sr := 0.0;
For i := 1 to N do
   Begin
   Sumx := Sumx  + X[i];
   Sumy := Sumy  + Y[i];
   Sumx2 := Sumx2 + (X[i] * X[i]);
   Sumxy := Sumxy + (X[i] * Y[i]);
   End;
Xmean := Sumx/N;
Ymean := Sumy/N;
A1:=(N*Sumxy-Sumx*Sumy)/(N*Sumx2-Sumx*Sumx);
A0 := Ymean - A1*Xmean;
For i := 1 to N do
   Begin
   St := St + sqr(Y[i] - Ymean);
   Sr := Sr + sqr(Y[i]-A1*X[i]+A0);
   End;
R2 := (St-Sr)/St;
Syx := Sqrt(Sr/(N-2));
End;                        {of procedure LinReg}
```

EXAMPLE 11.3   Linear Regression Using the Computer

Problem Statement: A user-friendly computer program to implement linear regression is contained in the Electronic TOOLKIT software package associated with this text. We can use this software to solve a hypothesis-testing problem associated with the falling parachutist discussed in Chap. 1. A theoretical mathematical model for the velocity of the parachutist was given as the following [Eq. (1.10)]:

$$v(t) = \frac{gm}{c}[1 - e^{(-c/m)t}]$$

where v is the velocity in meters per second, g is the gravitational constant of 9.8 m/s^2, m is the mass of the parachutist equal to 68.1 kg, and c is the drag coefficient of 12.5 kg/s. The model predicts the velocity of the parachutist as a function of time, as described in Example 1.1. A plot of the velocity variation was developed in Example 2.1.

An alternative empirical model for the velocity of the parachutist is given by the following

$$v(t) = \frac{gm}{c}\left[\frac{t}{3.75 + t}\right] \qquad\qquad (E11.3.1)$$

Suppose that you would like to test and compare the adequacy of these two mathematical models. This might be accomplished by measuring the actual velocity of the parachutist at known values of time and comparing these results with the predicted velocities according to each model.

Such an experimental-data-collection program was implemented, and the results are listed in column (a) of Table 11.2. Computed velocities for each model are listed in columns (b) and (c).

TABLE 11.2
Measured and calculated velocities for the falling parachutist.

Time, s	Measured v, m/s (a)	Model-calculated v, m/s [Eq. (1.10)] (b)	Model-calculated v, m/s [Eq. (E11.3.1)] (c)
1	10.00	8.953	11.240
2	16.30	16.405	18.570
3	23.00	22.607	23.729
4	27.50	27.769	27.556
5	31.00	32.065	30.509
6	35.60	35.641	32.855
7	39.00	38.617	34.766
8	41.50	41.095	36.351
9	42.90	43.156	37.687
10	45.00	44.872	38.829
11	46.00	46.301	39.816
12	45.50	47.490	40.678
13	46.00	48.479	41.437
14	49.00	49.303	42.110
15	50.00	49.988	42.712

Solution: The adequacy of the models can be tested by plotting the measured velocity versus the model-calculated velocity. Linear regression can be used to calculate a trend line for the plot. This trend line will have a slope of 1 and an intercept of 0 if the model matches the data perfectly. A significant deviation from these values can be used as an indication of the inadequacy of the model.

Figure 11.8a and b are plots of the line and data for the regressions of column (a) versus columns (b) and (c), respectively. These plots indicate that the linear regression between the data and each of the models is highly significant. Both models match the data with a correlation coefficient of greater than 0.99.

However, the model described by Eq. (1.10) conforms to our hypothesis test criteria much better than that described by Eq. (E11.3.1) because the slope and intercept are more nearly equal to 1 and 0. Thus, although each plot is well described by a straight line, Eq. (1.10) appears to be a better model than Eq. (E11.3.1).

Model testing and selection are common and extremely important activities performed in all fields of engineering. The background material provided to you in the present chapter together with your software should allow you to address many practical problems of this type.

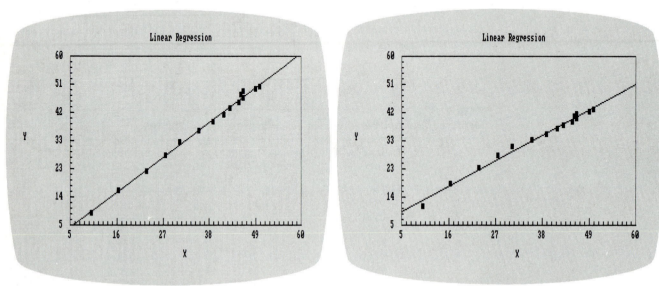

Figure 11.8
(a) Results using linear regression to compare measured values versus model predictions computed with the theoretical Eq. (1.10). (b) Results using linear regression to compare measured values versus model predictions computed with the empirical Eq. (E11.3.1).

11.1.5 Applications of Linear Regression—Linearization of Nonlinear Relationships

Linear regression provides a powerful technique for fitting a "best" line to data. However, it is predicated on the fact that the relationship between the dependent and independent variables is linear. This is not always the case, and the first step in any regression analysis should be to plot and visually inspect the data to ascertain whether a linear model applies. For example, Fig. 11.9 shows some data that is obviously curvilinear. In some cases, techniques such as polynomial regression, which is described in Sec. 11.2, are appropriate. For others, transformations can be used to express the data in a form that is compatible with linear regression.

One example is the *exponential model*

$$y = a_1 e^{b_1 x} \tag{11.12a}$$

where a_1 and b_1 are constants. This model is used in many fields of engineering to characterize quantities that increase (positive b_1) or decrease (negative b_1) at a rate that is directly proportional to their own magnitude. For example, population growth or

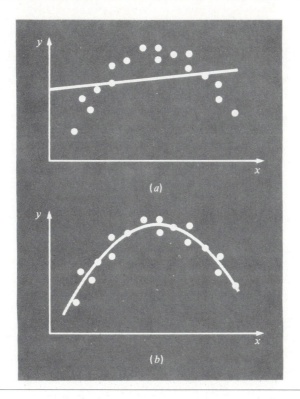

Figure 11.9
(a) Data that is ill-suited for linear least-squares regression. (b) Indication that a parabola is preferable.

radioactive decay can exhibit such behavior. As depicted in Fig. 11.10a, the equation represents a nonlinear relationship (for $b_1 \neq 0$) between y and x.

Another example of a nonlinear model is the simple *power equation*

$$y = a_2 x^{b_2} \tag{11.12b}$$

where a_2 and b_2 are constant coefficients. This model has wide applicability in all fields of engineering. As depicted in Fig. 11.10b, the equation (for $b_2 \neq 0$ or 1) is nonlinear.

A third example of a nonlinear model is the *saturation-growth-rate equation* [recall Eq. (E11.3.1)]

$$y = a_3 \frac{x}{b_3 + x} \tag{11.13}$$

where a_3 and b_3 are constant coefficients. This model, which is particularly well-suited for characterizing population growth rate under limiting conditions, also represents a nonlinear relationship between y and x (Fig. 11.10c) that levels off, or "saturates," as x increases.

Nonlinear regression techniques are available to fit these equations to experimental data directly. (Note that we will discuss nonlinear regression in Sec. 11.5.) However, a simpler alternative is to use mathematical manipulations to transform the equations into a linear form. Then simple linear regression can be employed to fit the equations to data.

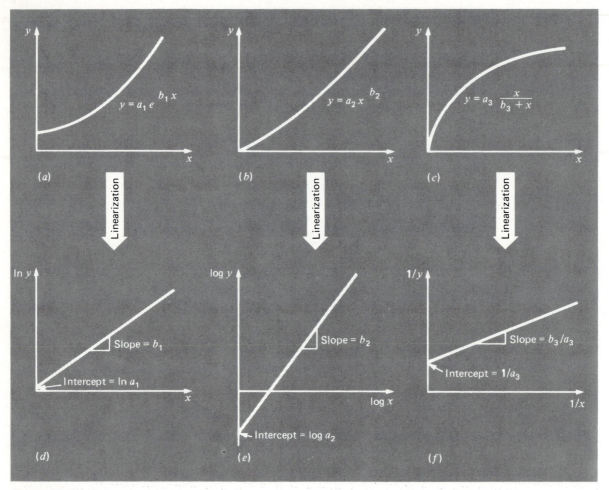

Figure 11.10

(a) The exponential equation, (b) the power equation, and (c) the saturation-growth-rate equation. Parts (d), (e), and (f) are linearized versions of these equations that result from simple transformations.

For example, Eq. (11.11) can be linearized by taking its natural logarithm to yield

$$\ln y = \ln a_1 + b_1 x \ln e$$

But because $\ln e = 1$,

$$\ln y = \ln a_1 + b_1 x \qquad (11.14)$$

Thus a plot of $\ln y$ versus x will yield a straight line with a slope of b_1 and an intercept of $\ln a_1$ (Fig. 11.10d).

Equation (11.12) is linearized by taking its base-10 logarithm to give

$$\log y = b_2 \log x + \log a_2 \qquad (11.15)$$

Thus, a plot of log y versus log x will yield a straight line with a slope of b_2 and an intercept of log a_2 (Fig. 11.10e).

Equation (11.13) is linearized by inverting it to give

$$\frac{1}{y} = \frac{b_3}{a_3}\frac{1}{x} + \frac{1}{a_3} \tag{11.16}$$

Thus, a plot of $1/y$ versus $1/x$ will be linear, with a slope of b_3/a_3 and an intercept of $1/a_3$ (Fig. 11.10f).

In their transformed states, these models are fit using linear regression in order to evaluate the constant coefficients. They could then be transformed back to their original state and used for predictive purposes. Example 11.4 illustrates this procedure for Eq. (11.12b). In addition, Case Studies 14.2 and 14.3 will provide an engineering example of the same sort of computation.

EXAMPLE 11.4   *Linearization of a Power Equation*

Problem Statement: Fit Eq. (11.12b) to the data in Table 11.3 using a logarithmic transformation of the data.

TABLE 11.3
Data to be fit to the power equation.

x	y	log x	log y
1	0.5	0	−0.301
2	1.7	0.301	0.226
3	3.4	0.477	0.534
4	5.7	0.602	0.753
5	8.4	0.699	0.922

Solution: Figure 11.11a is a plot of the original data in its untransformed state. Figure 11.11b shows the plot of the transformed data. A linear regression of the log-transformed data yields the result

$$\log y = 1.75 \log x - 0.300$$

Thus, the intercept, log a_2, equals −0.300, and therefore, by taking the antilogarithm, $a_2 = 10^{-0.3} = 0.5$. The slope is $b_2 = 1.75$. Consequently, the power equation is

$$y = 0.5x^{1.75}$$

This curve, as plotted in Fig. 11.11a, indicates a good fit.

11.1.6   General Comments on Linear Regression

Before proceeding to curvilinear and multiple linear regression, we must emphasize the introductory nature of the foregoing material on linear regression. We have focused on

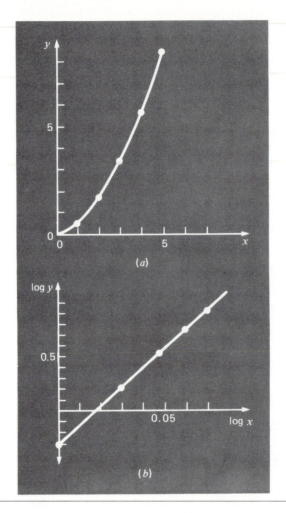

Figure 11.11
(a) Plot of untransformed data with the power equation that fits the data. (b) Plot of transformed data used to determine the coefficients of the power equation.

the simple derivation and practical use of equations to fit data. You should be cognizant of the fact that there are theoretical aspects of regression that are of practical importance but are beyond the scope of this book. For example, some statistical assumptions that are inherent in the linear least-squares procedures are

1. x has a fixed value; it is not random and is measured without error.
2. The y values are independent random variables and all have the same variance.
3. The y values for a given x must be normally distributed.

Such assumptions are relevant to the proper derivation and use of regression. For example, the first assumption means that (1) the x's must be error-free and (2) the regression of y versus x is not the same as x versus y (try Prob. 11.4 at the end of the chapter). You are urged to consult other references such as Draper and Smith (1981) in order to appreciate aspects and nuances of regression that are beyond the scope of this book.

11.2 POLYNOMIAL REGRESSION

In Sec. 11.1, a procedure was developed to derive the equation of a straight line using the least-squares criterion. Some engineering data, although exhibiting a marked pattern such as seen in Fig. 11.9, is poorly represented by a straight line. For these cases, a curve would be better suited to fit the data. As discussed in the previous section, one method to accomplish this objective is to use transformations. Another alternative is to fit polynomials to the data using *polynomial regression*.

The least-squares procedure can be readily extended to fit the data to an *m*th-degree polynomial:

$$y = a_0 + a_1 x + a_2 x^2 + \cdots + a_m x^m + e$$

For this case the sum of the squares of the residuals is [compare with Eq. (11.3)]

$$S_r = \sum_{i=1}^{n} (y_i - a_0 - a_1 x_i - a_2 x_i^2 - \cdots - a_m x_i^m)^2 \tag{11.17}$$

Following the procedure of the previous section, we take the derivative of Eq. (11.17) with respect to each of the unknown coefficients of the polynomial, as in

$$\frac{\partial S_r}{\partial a_0} = -2 \sum (y_i - a_0 - a_1 x_i - a_2 x_i^2 - \cdots - a_m x_i^m)$$

$$\frac{\partial S_r}{\partial a_1} = -2 \sum x_i (y_i - a_0 - a_1 x_i - a_2 x_i^2 - \cdots - a_m x_i^m)$$

$$\frac{\partial S_r}{\partial a_2} = -2 \sum x_i^2 (y_i - a_0 - a_1 x_i - a_2 x_i^2 - \cdots - a_m x_i^m)$$

$$\cdot \qquad \cdot$$
$$\cdot \qquad \cdot$$
$$\cdot \qquad \cdot$$

$$\frac{\partial S_r}{\partial a_m} = -2 \sum x_i^m (y_i - a_0 - a_1 x_i - a_2 x_i^2 - \cdots - a_m x_i^m)$$

These equations can be set equal to zero and rearranged to develop the following set of normal equations:

$$a_0 n + a_1 \sum x_i + a_2 \sum x_i^2 + \cdots + a_m \sum x_i^m = \sum y_i$$

$$a_0 \sum x_i + a_1 \sum x_i^2 + a_2 \sum x_i^3 + \cdots + a_m \sum x_i^{m+1} = \sum x_i y_i$$

$$a_0 \sum x_i^2 + a_1 \sum x_i^3 + a_2 \sum x_i^4 + \cdots + a_m \sum x_i^{m+2} = \sum x_i^2 y_i \tag{11.18}$$

$$\cdot \qquad \cdot \qquad \cdot \qquad \cdot \qquad \cdot$$

. . . . .

. . . . .

$$a_0 \sum x_i^m + a_1 \sum x_i^{m+1} + a_2 \sum x_i^{m+2} + \cdots + a_m \sum x_i^{2m} = \sum x_i^m y_i$$

where all summations are from $i = 1$ through n. Note that the above $m + 1$ equations are linear and have $m + 1$ unknowns: $a_0, a_1, a_2, \ldots, a_m$. The coefficients of the unknowns can be calculated directly from the observed data. Thus, the problem of determining a least-squares polynomial of degree m is equivalent to solving a system of $m + 1$ simultaneous linear equations. Techniques to solve such equations were discussed in Part Three.

Just as for linear regression, the error of polynomial regression can be quantified by a standard error of the estimate,

$$s_{y/x} = \sqrt{\frac{S_r}{n - (m + 1)}} \tag{11.19}$$

where m is the order of the polynomial. This quantity is divided by $n - (m + 1)$ because $m + 1$ data-derived coefficients—$a_0, a_1, \ldots, a_m$—were used to compute S_r; thus, we have lost $m + 1$ degrees of freedom. In addition to the standard error, a coefficient of determination can also be computed for polynomial regression in the same manner as for the linear case:

$$r^2 = \frac{S_t - S_r}{S_t}$$

EXAMPLE 11.5   Polynomial Regression

Problem Statement: Fit a second-order polynomial to the data in the first two columns of Table 11.4.

TABLE 11.4
Computations for an error analysis of the quadratic least-squares fit.

x_i	y_i	$(y_i - \bar{y})^2$	$(y_i - a_0 - a_1 x_i - a_2 x_i^2)^2$
0	2.1	544.44	0.14332
1	7.7	314.47	1.00286
2	13.6	140.03	1.08158
3	27.2	3.12	0.80491
4	40.9	239.22	0.61951
5	61.1	1272.11	0.09439
Σ	152.6	2513.39	3.74657

Solution: From the given data,

$m = 2$	$\sum x_i = 15$	$\sum x_i^4 = 979$
$n = 6$	$\sum y_i = 152.6$	$\sum x_i y_i = 585.6$
$\bar{x} = 2.5$	$\sum x_i^2 = 55$	$\sum x_i^2 y_i = 2488.8$
$\bar{y} = 25.433$	$\sum x_i^3 = 225$	

Therefore, the simultaneous linear equations are

$$6a_0 + 15a_1 + 55a_2 = 152.6$$
$$15a_0 + 55a_1 + 225a_2 = 585.6$$
$$55a_0 + 225a_1 + 979a_2 = 2488.8$$

Solving these equations through a technique such as Gauss elimination gives

$$a_0 = 2.47857$$
$$a_1 = 2.35929$$
$$a_2 = 1.86071$$

Therefore, the least-squares quadratic equation for this case is

$$y = 2.47857 + 2.35929x + 1.86071x^2$$

The standard error of the estimate based on the regression polynomial is [Eq. (11.19)]

$$s_{y/x} = \sqrt{\frac{3.74657}{6 - 3}} = 1.12$$

The coefficient of determination is

$$r^2 = \frac{2513.39 - 3.74657}{2513.39} = 0.99851$$

and the correlation coefficient is

$$r = 0.99925$$

These results indicate that 99.851 percent of the original uncertainty has been explained by the model. This result supports the conclusion that the quadratic equation represents an excellent fit, as is also evident from Fig. 11.12.

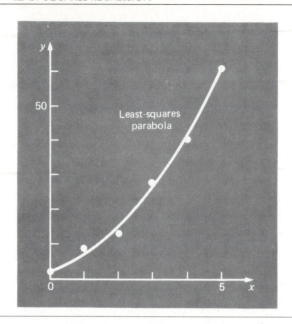

Figure 11.12
Fit of a second-order polynomial.

11.2.1 Algorithm for Polynomial Regression

An algorithm for polynomial regression is delineated in Fig. 11.13. Note that the primary task is the generation of the coefficients of the normal equations [Eq. (11.18)]. (Pseudocode for accomplishing this is presented in Fig. 11.14.) Then, techniques from Part Three can be applied to solve these simultaneous equations for the coefficients.

A potential problem associated with implementing polynomial regression on the computer is that the normal equations are sometimes ill-conditioned. This is particularly true for higher-order versions. For these cases, the computed coefficients may be highly susceptible to round-off error, and consequently the results can be inaccurate. Among other things, this problem is related to the structure of the normal equations and to the fact that for higher-order polynomials the normal equations can have very large and very small coefficients. This is because the coefficients are summations of the data raised to powers.

Figure 11.13
Algorithm for implementation of polynomial and multiple regression.

Step 1: Input order of polynomial to be fit, m.

Step 2: Input number of data points, n.

Step 3: If $n \leq m$, print out an error message that polynomial regression is impossible and terminate the process. If $n > m$, continue.

Step 4: Compute the elements of the normal equation in the form of an augmented matrix.

Step 5: Solve the augmented matrix for the coefficients $a_0, a_1, a_2, \ldots, a_m$, using an elimination method.

Step 6: Print out the coefficients.

```
DOFOR i = 1 to order+1
    DOFOR j = 1 to i
        k = i + j - 2
        sum = 0
        DOFOR l = 1 to n
            sum = sum + x^k_l
        ENDDO
        a_{i,j} = sum
        a_{j,i} = sum
    ENDDO
    sum = 0
    DOFOR l = 1 to n
        sum = sum + y_l·x^{i-1}_l
    ENDDO
    a_{i,order+2} = sum
ENDDO
```

Figure 11.14
Pseudocode to assemble the elements of the normal equations for polynomial regression.

Although the strategies for mitigating round-off error discussed in Part Three, such as pivoting, can help to partially remedy this problem, a simpler alternative is to use a computer with higher precision. This is one area where some personal computers may presently represent a limitation on the effective implementation of this particular numerical method. Fortunately, most practical problems are limited to lower-order polynomials for which round off is usually negligible. In situations where higher-order versions are required, other alternatives are available for certain types of data. However, these techniques (such as orthogonal polynomials) are beyond the scope of this book. The reader should consult texts on regression such as Draper and Smith (1981) for additional information regarding the problem and possible alternatives.

11.3 MULTIPLE LINEAR REGRESSION

A useful extension of linear regression is the case where y is a linear function of two or more variables. For example, y might be a linear function of x_1 and x_2, as in

$$y = a_0 + a_1x_1 + a_2x_2 + e$$

Such an equation is particularly useful when fitting experimental data where the variable being studied is often a function of two other variables. For this two-dimensional case, the regression "line" becomes a "plane" (Fig. 11.15).

As with the previous cases, the "best" values of the coefficients are determined by setting up the sum of the squares of the residuals,

$$S_r = \sum_{i=1}^{n}(y_i - a_0 - a_1x_{1i} - a_2x_{2i})^2$$

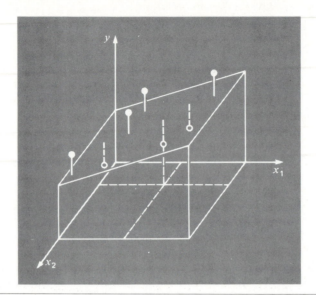

Figure 11.15
Graphical depiction of multiple
linear regression where y is a
linear function of x_1 and x_2.

and differentiating with respect to each of the unknown coefficients,

$$\frac{\partial S_r}{\partial a_0} = -2\sum(y_i - a_0 - a_1 x_{1i} - a_2 x_{2i})$$

$$\frac{\partial S_r}{\partial a_1} = -2\sum x_{1i}(y_i - a_0 - a_1 x_{1i} - a_2 x_{2i}) \qquad (11.20)$$

$$\frac{\partial S_r}{\partial a_2} = -2\sum x_{2i}(y_i - a_0 - a_1 x_{1i} - a_2 x_{2i})$$

The coefficients yielding the minimum sum of the squares of the residuals are obtained
by setting the partial derivatives equal to zero and expressing Eq. (11.20) as a set of
simultaneous linear equations:

$$na_0 + \sum x_{1i} a_1 + \sum x_{2i} a_2 = \sum y_i$$

$$\sum x_{1i} a_0 + \sum x_{1i}^2 a_1 + \sum x_{1i} x_{2i} a_2 = \sum x_{1i} y_i$$

$$\sum x_{2i} a_0 + \sum x_{1i} x_{2i} a_1 + \sum x_{2i}^2 a_2 = \sum x_{2i} y_i$$

or as a matrix:

$$\begin{bmatrix} n & \sum x_{1i} & \sum x_{2i} \\ \sum x_{1i} & \sum x_{1i}^2 & \sum x_{1i} x_{2i} \\ \sum x_{2i} & \sum x_{1i} x_{2i} & \sum x_{2i}^2 \end{bmatrix} \begin{Bmatrix} a_0 \\ a_1 \\ a_2 \end{Bmatrix} = \begin{Bmatrix} \sum y_i \\ \sum x_{1i} y_i \\ \sum x_{2i} y_i \end{Bmatrix} \qquad (11.21)$$

EXAMPLE 11.6   Multiple Linear Regression

Problem Statement: The following data was calculated from the equation $y = 5 + 4x_1 - 3x_2$.

x_1	x_2	y
0	0	5
2	1	10
2.5	2	9
1	3	0
4	6	3
7	2	27

Use multiple linear regression to fit this data.

TABLE 11.5
Computations required to develop the normal equations for Example 11.6.

	y	x_1	x_2	x_1^2	x_2^2	x_1x_2	x_1y	x_2y
	5	0	0	0	0	0	0	0
	10	2	1	4	1	2	20	10
	9	2.5	2	6.25	4	5	22.5	18
	0	1	3	1	9	3	0	0
	3	4	6	16	36	24	12	18
	27	7	2	49	4	14	189	54
Σ	54	16.5	14	76.25	54	48	243.5	100

Solution: The summations required to develop Eq. (11.21) are computed in Table 11.5. They can be substituted into Eq. (11.21) to yield

$$\begin{bmatrix} 6 & 16.5 & 14 \\ 16.5 & 76.25 & 48 \\ 14 & 48 & 54 \end{bmatrix} \begin{Bmatrix} a_0 \\ a_1 \\ a_2 \end{Bmatrix} = \begin{Bmatrix} 54 \\ 243.5 \\ 100 \end{Bmatrix}$$

which can be solved using a method such as Gauss elimination for

$$a_0 = 5 \qquad a_1 = 4 \qquad a_2 = -3$$

which is consistent with the original equation from which the data was derived.

Multiple linear regression can be formulated for the more general case,

$$y = a_0 + a_1x_1 + a_2x_2 + \cdots + a_mx_m + e$$

where the coefficients that minimize the sum of the squares of the residuals are determined by solving

$$
\begin{bmatrix}
n & \sum x_{1i} & \sum x_{2i} & \cdots & \sum x_{mi} \\
\sum x_{1i} & \sum x_{1i}^2 & \sum x_{1i}x_{2i} & \cdots & \sum x_{1i}x_{mi} \\
\sum x_{2i} & \sum x_{2i}x_{1i} & \sum x_{2i}^2 & \cdots & \sum x_{2i}x_{mi} \\
\cdot & \cdot & \cdot & & \cdot \\
\cdot & \cdot & \cdot & & \cdot \\
\cdot & \cdot & \cdot & & \cdot \\
\sum x_{mi} & \sum x_{mi}x_{1i} & \sum x_{mi}x_{2i} & \cdots & \sum x_{mi}^2
\end{bmatrix}
\begin{Bmatrix}
a_0 \\ a_1 \\ a_2 \\ \cdot \\ \cdot \\ \cdot \\ a_m
\end{Bmatrix}
=
\begin{Bmatrix}
\sum y_1 \\ \sum x_{1i}y_i \\ \sum x_{2i}y_i \\ \cdot \\ \cdot \\ \cdot \\ \sum x_{mi}y_i
\end{Bmatrix}
\qquad (11.22)
$$

The standard error of the estimate for multiple linear regression is formulated as

$$
s_{y/x} = \sqrt{\frac{S_r}{n - (m + 1)}}
$$

and the coefficient of determination is computed as in Eq. (11.10).

Although there may be certain cases where a variable is linearly related to two or more other variables, multiple linear regression has additional utility in the derivation of power equations of the general form,

$$
y = a_0 x_1^{a_1} x_2^{a_2} \cdots x_m^{a_m}
$$

Such equations are extremely useful when fitting experimental data. In order to use multiple linear regression, the equation is transformed by taking its logarithm to yield

$$
\log y = \log a_0 + a_1 \log x_1 + a_2 \log x_2 + \cdots + a_m \log x_m
$$

This transformation is similar in spirit to the one used in Sec. 11.1.5 and Example 11.4 to fit a power equation when y was a function of a single variable x. Case Study 14.5 provides an example of such an application for two independent variables.

An algorithm to set up the normal equations is listed in Fig. 11.16.

Figure 11.16
Pseudocode to assemble the elements of the normal equations for multiple regression. Note that aside from storing the independent variables in $x_{1,i}$, $x_{2,i}$, etc., 1's must be stored in $x_{0,i}$ for this algorithm to work.

```
DOFOR i = 1 to order + 1
    DOFOR j = 1 to i
        sum = 0
        DOFOR l = 1 to n
            sum = sum + x_{i-1,l}·x_{j-1,l}
        ENDDO
        a_{i,j} = sum
        a_{j,i} = sum
    ENDDO
    sum = 0
    DOFOR l = 1 to n
        sum = sum + y_l·x_{i-1,l}
    ENDDO
    a_{i,order+2} = sum
ENDDO
```

11.4   GENERAL LINEAR LEAST SQUARES

To this point, we have focused on the mechanics of obtaining least-squares fits of some simple functions to data. Before turning to nonlinear regression, there are several issues that we would like to discuss to enrich your understanding of the preceding material.

11.4.1   General Matrix Formulation for Linear Least Squares

In the preceding pages, we have introduced three types of regression: simple linear, polynomial, and multiple linear. In fact, all three belong to the following general linear least-squares model:

$$y = a_0 z_0 + a_1 z_1 + a_2 z_2 + \cdots + a_m z_m + e \tag{11.23a}$$

where $z_0, z_1, \ldots, z_m$ are $m + 1$ different functions. It can easily be seen how simple and multiple linear regression fall within this model—that is, $z_0 = 1, z_1 = x_1, z_2 = x_2, \ldots, z_m = x_m$. Further, polynomial regression is also included if the z's are simple monomials as in $z_0 = x^0 = 1, z_1 = x, z_2 = x^2, \ldots, z_m = x^m$. Note that the terminology "linear" only refers to the model's dependence on its *parameters*—that is, the a's. As in the case of polynomial regression, the functions themselves can be highly nonlinear.

Equation (11.23a) can be expressed in matrix notation as

$$\{Y\} = [Z]\{A\} + \{E\} \tag{11.23b}$$

where $[Z]$ is a matrix of the observed values for the independent variables,

$$[Z] = \begin{bmatrix} z_{01} & z_{11} & \cdot & \cdot & \cdot & z_{m1} \\ z_{02} & z_{12} & \cdot & \cdot & \cdot & z_{m2} \\ \cdot & & & & & \cdot \\ \cdot & & & & & \cdot \\ \cdot & & & & & \cdot \\ z_{0n} & z_{1n} & \cdot & \cdot & \cdot & z_{mn} \end{bmatrix}$$

where m is the number of variables in the model and n is the number of data points. The column vector $\{Y\}$ contains the observed values of the dependent variable

$$\{Y\}^T = [\,y_1 \;\; y_2 \;\; \cdots \;\; y_n]$$

The column vector $\{A\}$ contains the unknown coefficients

$$\{A\}^T = [a_0 \;\; a_1 \;\; a_2 \;\; \cdots \;\; a_m]$$

and the column vector $\{E\}$ contains the residuals

$$\{E\}^T = [e_1 \;\; e_2 \;\; \cdots \;\; e_n]$$

As was done throughout this chapter, the sum of the squares of the residuals for this model can be defined as

$$S_r = \sum_{i=1}^{n} \left(y_i - \sum_{j=0}^{m} a_j z_{ji} \right)^2$$

This quantity can be minimized by taking its partial derivative with respect to each of the coefficients and setting the resulting equation equal to zero. The outcome of this process is the normal equations which can be expressed concisely in matrix form as

$$[[Z]^T[Z]]\{A\} = [Z]^T\{Y\} \tag{11.24}$$

It will be left as a homework assignment for you to verify that Eq. (11.24) is, in fact, equivalent to the normal equations developed previously for simple linear, polynomial, and multiple linear regression.

Our primary motivation for the foregoing has been to illustrate the unity among the three approaches and to show how they can all be expressed simply in the same matrix notation. It also sets the stage for the next section where we will gain some insights into the preferred strategies for solving Eq. (11.24). The matrix notation will also have relevance when we turn to nonlinear regression in the last section of this chapter.

11.4.2   Solution Techniques

In previous discussions in this chapter, we have glossed over the issue of the specific numerical techniques to solve the normal equations. Now that we have established the unity among the various models, we can explore this question in more detail.

First, it should be clear that Gauss-Seidel cannot be employed because the normal equations are not diagonally dominant. We are thus left with the elimination methods. For the present purposes, we can divide these techniques into three categories: (1) *LU* decomposition methods including Gauss elimination, (2) Cholesky's method, and (3) matrix inversion approaches. There are obviously overlaps involved in this breakdown. For example, Cholesky's method is, in fact, an *LU* decomposition and all the approaches can be formulated so that they can generate the matrix inverse. However, this breakdown has merit in that each category offers benefits regarding the solution of the normal equations.

LU Decomposition.   If you are merely interested in applying a least-squares fit for the case where the appropriate model is known a priori, any of the *LU* decomposition approaches described in Chap. 9 are perfectly acceptable. In fact, the non-*LU*-decomposition formulation of Gauss elimination described in Chap. 7 can also be employed. It is a relatively straightforward programming task to incorporate any of these into an algorithm for linear least squares. In fact, if a modular approach has been followed, it is almost trivial.

Cholesky's Method.   Cholesky's decomposition algorithm has several advantages with regard to the solution of the general linear regression problem. First, it is expressly designed for solving symmetric matrices like the normal equations. Thus, it is fast and requires less storage space to solve such systems. Second, it is ideally suited for cases where the order of the model [that is, the value of m in Eq. (11.23a)] is not known beforehand (see Ralston and Rabinowitz, 1978). A case in point would be polynomial regression. For this case, we might not know a priori whether a linear, quadratic, cubic, or higher-order polynomial is the "best" model to describe our data. Because of the way in which both the normal equations are constructed and the Cholesky algorithm proceeds

(Fig. 9.11), we can develop successively higher-order models in an extremely efficient manner. At each step we could examine the residual sum of the squares error (and a plot!) to examine whether the inclusion of higher-order terms significantly improves the fit.

The analogous situation for multiple linear regression occurs when independent variables are added to the model one at a time. Suppose that the dependent variable of interest is a function of a number of independent variables: say, temperature, moisture content, pressure, etc. We could first perform a linear regression with temperature and compute a residual error. Next, we could include moisture content by performing a two-variable multiple regression and see whether the additional variable results in an improved fit. Cholesky's method makes this process efficient because the decomposition of the linear model would merely be supplemented to incorporate a new variable.

Matrix Inverse Approaches. Recalling Eq. (PT3.6), the matrix inverse can be employed to solve Eq. (11.24) as in

$$\{A\} = \left[[Z]^T [Z] \right]^{-1} [Z]^T \{Y\} \tag{11.25}$$

Each of the elimination methods can be used to determine the inverse (recall Secs. 8.1.2 and 9.5.1) and, thus, can be used to implement Eq. (11.25). However, as we have learned in Part Three, this is an inefficient approach for solving a set of simultaneous equations. Thus, if we were merely interested in solving for the regression coefficients, it is preferrable to employ an *LU* decomposition approach without inversion. However, from a statistical perspective, there are a number of reasons why we might be interested in obtaining the inverse and examining its coefficients. These reasons will be discussed next.

11.4.3 Statistical Aspects of Least-Squares Theory

In Sec. PT4.2.1, we reviewed a number of descriptive statistics that can be used to describe a sample. These included the arithmetic mean, the standard deviation, and the variance.

Aside from yielding a solution for the regression coefficients, the matrix formulation of Eq. (11.25) provides estimates of their statistics. It can be shown (Draper and Smith, 1981) that the diagonal and off-diagonal terms of the matrix $[[Z][Z]^T]^{-1}$ give, respectively, the variances and the covariances* of the a's. If the elements of $[[Z][Z]^T]^{-1}$ are designated as z_{ii}^{-1}, then

$$\operatorname{var}(a_i) = z_{ii}^{-1} s_{y/x}^2$$

and

$$\operatorname{cov}(a_i, a_j) = z_{ij}^{-1} s_{y/x}^2$$

---

*The covariance is a statistic that measures the dependency of one variable on another. Thus, cov (x, y) indicates the dependency of x and y. For example, cov $(x, y) = 0$ would indicate that x and y are totally independent.

These statistics have a number of important applications including hypothesis testing related to the least-squares fit. You can consult Draper and Smith (1981) for additional information on this subject.

11.5   NONLINEAR REGRESSION

There are many cases in engineering where nonlinear models must be fit to data. In the present context, these models are defined as those which have a nonlinear dependence on their parameters. For example,

$$f(x) = a_0(1 - e^{-a_1 x})$$

There is no way that this equation can be manipulated so that it conforms to the general form of Eq. (11.23a).

As with linear least squares, nonlinear regression is based on determining the values of the parameters that minimize the sum of the squares of the residuals. However, for the nonlinear case, the solution must proceed in an iterative fashion. As with other such nonlinear approaches, successful solutions are often highly dependent on good initial guesses for the parameters.

The *Gauss-Newton method* is one algorithm for minimizing the sum of the squares of the residuals between data and nonlinear equations. The key concept underlying the technique is that a Taylor series expansion is used to express the original nonlinear equation in an approximate, linear form. Then, least-squares theory can be used to obtain new estimates of the parameters that move in the direction of minimizing the residual.

To illustrate how this is done, first the relationship between the nonlinear equation and the data can be expressed generally as

$$y_i = f(x_i; a_0, a_1, \ldots, a_m) + e_i$$

where y_i is a measured value of the dependent variable, $f(x_i; a_0, a_1, \ldots, a_m)$ is the equation which is a function of the independent variable x_i and a nonlinear function of the parameters $a_0, a_1, \ldots, a_m$, and e_i is a random error. For convenience, this model can be expressed in abbreviated form by omitting the parameters,

$$y_i = f(x_i) + e_i \tag{11.26}$$

The nonlinear model can be expanded in a Taylor series around the parameter values and curtailed after the first derivatives. For example, for a two-parameter case,

$$f(x_i)_{j+1} = f(x_i)_j + \frac{\partial f(x_i)_j}{\partial a_0} \Delta a_0 + \frac{\partial f(x_i)_j}{\partial a_1} \Delta a_1 \tag{11.27}$$

where j is the initial guess, $j + 1$ is the prediction, $\Delta a_0 = a_{0,j+1} - a_{0,j}$, and $\Delta a_1 = a_{1,j+1} - a_{1,j}$. Thus, we have linearized the original model with respect to the parameters. Equation (11.27) can be substituted into Eq. (11.26) to yield

$$y_i - f(x_i)_j = \frac{\partial f(x_i)_j}{\partial a_0} \Delta a_0 + \frac{\partial f(x_i)_j}{\partial a_1} \Delta a_1 + e_i$$

or in matrix form [compare with Eq. (11.23b)]

$$\{D\} = [Z_j]\{\Delta A\} + \{E\} \tag{11.28}$$

where $[Z_j]$ is the matrix of partial derivatives of the function evaluated at the initial guess, j,

$$[Z_j] = \begin{bmatrix} \dfrac{\partial f_1}{\partial a_0} & \dfrac{\partial f_1}{\partial a_1} \\[2mm] \dfrac{\partial f_2}{\partial a_0} & \dfrac{\partial f_2}{\partial a_1} \\ \cdot & \cdot \\ \cdot & \cdot \\ \cdot & \cdot \\ \dfrac{\partial f_n}{\partial a_0} & \dfrac{\partial f_n}{\partial a_1} \end{bmatrix}$$

where n is the number of data points and $\partial f_i / \partial a_k$ is the partial derivative of the function with respect to the kth parameter evaluated at the ith data point. The vector $\{D\}$ contains the differences between the measurements and the function values

$$\{D\} = \begin{Bmatrix} y_1 - f(x_1) \\ y_2 - f(x_2) \\ \cdot \\ \cdot \\ \cdot \\ y_n - f(x_n) \end{Bmatrix}$$

and the vector $\{\Delta A\}$ contains the changes in the parameter values

$$\{\Delta A\} = \begin{Bmatrix} \Delta a_0 \\ \Delta a_1 \end{Bmatrix}$$

Applying linear least-squares theory to Eq. (11.28) results in the following normal equations [recall Eq. (11.24)]

$$\left[[Z_j]^T [Z_j] \right] \{\Delta A\} = [Z_j]^T \{D\} \tag{11.29}$$

Thus, the approach consists of solving Eq. (11.29) for $\{\Delta A\}$ which can be employed to compute improved values for the parameters as in

$$a_{0,j+1} = a_{0,j} + \Delta a_0$$

and

$$a_{1,j+1} = a_{1,j} + \Delta a_1$$

This procedure is repeated until the solution converges—that is, until

$$|\epsilon_a|_k = \left| \frac{a_{k,j+1} - a_{k,j}}{a_{k,j+1}} \right| 100\% \tag{11.30}$$

falls below an acceptable stopping criterion.

EXAMPLE 11.7   Gauss-Newton Method

Problem Statement: Fit the function $f(x; a_0, a_1) = a_0(1 - e^{-a_1 x})$ to the data:

x	y
0.25	0.28
0.75	0.57
1.25	0.68
1.75	0.74
2.25	0.79

Use initial guesses of $a_0 = 1.0$ and $a_1 = 1.0$ for the parameters. Note that for these guesses the initial sum of the squares of the residuals is 0.0248.

Solution: The partial derivatives of the function with respect to the parameters are

$$\frac{\partial f}{\partial a_0} = 1 - e^{-a_1 x} \tag{E11.7.1}$$

and

$$\frac{\partial f}{\partial a_1} = a_0 x e^{-a_1 x} \tag{E11.7.2}$$

Equations (E11.7.1) and (E11.7.2) can be used to evaluate the matrix

$$[Z_0] = \begin{bmatrix} 0.2212 & 0.1947 \\ 0.5276 & 0.3543 \\ 0.7135 & 0.3581 \\ 0.8262 & 0.3041 \\ 0.8946 & 0.2371 \end{bmatrix}$$

This matrix multiplied by its transpose results in

$$[Z_0]^T[Z_0] = \begin{bmatrix} 2.3193 & 0.9489 \\ 0.9489 & 0.4404 \end{bmatrix}$$

which in turn can be inverted to yield

$$\left[[Z_0]^T[Z_0]\right]^{-1} = \begin{bmatrix} 3.6397 & -7.8421 \\ -7.8421 & 19.1678 \end{bmatrix}$$

The vector $\{D\}$ consists of the differences between the measurements and the model predictions,

$$\{D\} = \begin{Bmatrix} 0.28 - 0.2212 \\ 0.57 - 0.5276 \\ 0.68 - 0.7135 \\ 0.74 - 0.8262 \\ 0.79 - 0.8946 \end{Bmatrix} = \begin{Bmatrix} 0.0588 \\ 0.0424 \\ -0.0335 \\ -0.0862 \\ -0.1046 \end{Bmatrix}$$

It is multiplied by $[Z_0]^T$ to give

$$[Z_0]^T\{D\} = \begin{Bmatrix} -0.1533 \\ -0.0365 \end{Bmatrix}$$

The vector $\{\Delta A\}$ is then calculated by solving Eq. (11.29) for

$$\{\Delta A\} = \begin{Bmatrix} -0.2714 \\ -0.5019 \end{Bmatrix}$$

which can be added to the initial parameter guesses to yield,

$$\begin{Bmatrix} a_0 \\ a_1 \end{Bmatrix} = \begin{Bmatrix} 1.0 \\ 1.0 \end{Bmatrix} + \begin{Bmatrix} -0.2714 \\ 0.5019 \end{Bmatrix} = \begin{Bmatrix} 0.7286 \\ 1.5019 \end{Bmatrix}$$

Thus, the improved estimates of the parameters are $a_0 = 0.7286$ and $a_1 = 1.5019$. The new parameters result in a sum of the squares of the residuals equal to 0.0242. Equation (11.30) can be used to compute ϵ_0 and ϵ_1 equal to 37 and 33 percent, respectively. The computation would then be repeated until these values fell below the prescribed stopping criterion. The final result is $a_0 = 0.79186$ and $a_1 = 1.6751$. These coefficients give a sum of the squares of the residuals of 0.000662.

A potential problem with the Gauss-Newton method as developed to this point is that the partial derivatives of the function may be difficult to evaluate. Consequently, many computer programs use difference equations to approximate the partial derivatives. One method is

$$\frac{\partial f_i}{\partial a_k} \simeq \frac{f(x_i; a_0, \ldots, a_k + h, \ldots, a_m) - f(x_i; a_0, \ldots, a_k, \ldots, a_m)}{h} \tag{11.31}$$

where h is a prespecified small increment.

The Gauss-Newton method has a number of other possible shortcomings:

1. It may converge slowly.
2. It may oscillate widely, i.e., continually change directions.
3. It may not converge at all.

Modifications of the method (Booth and Peterson, 1958; Hartley, 1961) have been developed to remedy the shortcomings. In addition, other methods such as the *steepest-descent* and the *Levenberg-Marquardt* techniques have been developed to fit nonlinear equations. See Draper and Smith (1981) for a discussion of these methods.

PROBLEMS

Hand Calculations

11.1 Given the data

0.95	1.42	1.30	1.55	1.63
1.32	1.15	1.47	1.95	1.66
1.46	1.47	1.92	1.35	1.05
1.85	1.74	1.65	1.78	1.71
2.39	1.82	2.06	2.14	2.27

determine (**a**) the mean, (**b**) the standard deviation, (**c**) the variance, and (**d**) the coefficient of variation.

11.3 Given the data

55	6	18	21	26	28	32
39	22	28	24	27	27	33
2	12	17	34	29	31	38
45	36	41	37	43	38	46

determine (**a**) the mean, (**b**) the standard deviation, (**c**) the variance, and (**d**) the coefficient of variation.
(**e**) Construct a histogram. Use a range from 0 to 55 with increments of 5.
(**f**) Assuming that the distribution is normal and that your estimate of the standard deviation is valid, compute the range (that is, the lower and the upper values) that encompasses 68% of the readings. Determine whether this is a valid estimate for the data in this problem.

11.2 Construct a histogram for the data in Prob. 11.1. Use a range of 0.6 to 2.4 with intervals of 0.2.

11.4 Use least-squares regression to fit a straight line to

x	1	3	5	7	10	12	13	16	18	20
y	4	2	6	5	8	7	10	9	12	11

Along with the slope and intercept, compute the standard error of the estimate and the correlation coefficient. Plot the data and the regression line. Then repeat the problem, but regress x versus y—that is, switch the variables. Interpret your results.

11.5 Use least-squares regression to fit a straight line to

x	4	6	8	10	14	16	20	22	24	28	28	34	36	38
y	30	22	22	28	14	22	16	8	20	8	14	14	0	4

Along with the slope and the intercept, compute the standard error of the estimate and the correlation coefficient. Plot the data and the regression line. If someone made an additional measurement of $x = 30, y = 30$, would you suspect, based on a visual assessment and the standard error, that the measurement was valid or faulty? Justify your conclusion.

11.6 Use least-squares regression to fit a straight line to

x	0	2	4	4	8	12	16	20	24	28	30	34
y	12	12	18	22	20	30	26	30	26	28	22	18

(a) Along with the slope and intercept, compute the standard error of the estimate and the correlation coefficient. Plot the data and the straight line. Assess the fit.

(b) Recompute (a), but use polynomial regression to fit a parabola to the data. Compare the results with those of (a).

11.7 Fit a saturation-growth-rate model to

x	1	2	2.5	4	6	8	8.5
y	0.4	0.7	0.8	1.0	1.2	1.3	1.4

Plot the data and the equation.

11.8 Fit a power equation to the data from Prob. 11.7. Plot the data and the equation.

11.9 Fit a parabola to the data from Prob. 11.7. Plot the data and the equation.

11.10 Fit a power equation to

x	2.5	3.5	5	6	7.5	10	12.5	15	17.5	20
y	5	3.4	2	1.6	1.2	0.8	0.6	0.4	0.3	0.3

Plot y versus x along with the power equation.

11.11 Fit an exponential model to

x	0.05	0.4	0.8	1.2	1.6	2.0	2.4
y	550	750	1000	1400	2000	2700	3750

Plot the data and the equation on both standard and semi-logarithmic graph paper. Discuss your results.

11.12 Fit a power equation to the data in Prob. 11.11. Plot the data and the equation.

11.13 Fit a parabola to the data in Prob. 11.11. Plot the data and the equation.

11.14 Given the data

x	5	10	15	20	25	30	35	40	45	50
y	17	25	30	33	36	38	39	40	41	42

use least-squares regression to fit (a) a straight line, (b) a power equation, (c) a saturation-growth-rate equation, and (d) a parabola. Plot the data along with all the curves. Is any one of the curves superior? If so, justify.

11.15 Fit a parabola to

x	0	2	4	6	9	11	13	15	17	19	23	25	28
y	1.2	0.6	0.4	−0.2	0	−0.6	−0.4	−0.2	−0.4	0.2	0.4	1.2	1.8

Compute the coefficients, the standard error of the estimate, and the correlation coefficient. Plot the results and assess the fit.

11.16 Use multiple linear regression to fit

x_1	0	1	2	0	1	2
x_2	2	2	4	4	6	6
y	19	12	11	24	22	15

Compute the coefficients, the standard error of the estimate, and the correlation coefficient.

11.17 Use multiple linear regression to fit

x_1	1	1	2	2	3	3	4	4
x_2	1	2	1	2	1	2	1	2
y	18	12.8	25.7	20.6	35.0	29.8	45.5	40.3

Compute the coefficients, the standard error of the estimate, and the correlation coefficient.

11.18 Use nonlinear regression to fit a parabola to the data in Prob. 11.11.

11.19 Use nonlinear regression to fit a saturation-growth-rate equation to the data from Prob. 11.14.

Computer-Related-Problems

11.20 Develop a user-friendly computer program for linear regression based on Fig. 11.6 or 11.7. Among other things:
(a) Add statements to document the code.
(b) Make the input and output more descriptive and user-oriented.
(c) (*Optional*) Include a computer plot of the data and the regression line.
(d) (*Optional*) Include a feature that allows exponential, power, and saturation-growth-rate equations to be analyzed.

11.21 Develop a user-friendly computer program for polynomial regression based on Figs. 11.13 and 11.14. Test the program by duplicating the computations from Example 11.5.

11.22 Develop a user-friendly computer program for multiple regression based on Figs. 11.13 and 11.16. Test the program by duplicating the computations from Example 11.6.

11.23 Repeat Probs. 11.4 and 11.5 using the program from Prob. 11.20.

11.24 Use the Electronic TOOLKIT software to solve Probs. 11.4, 11.5, and 11.6a.

11.25 Repeat Probs. 11.9, 11.13, and 11.15 using the program from Prob. 11.21.

11.26 Repeat Probs. 11.16 and 11.17 using the program from Prob. 11.22.

CHAPTER 12
Interpolation

You will frequently have occasion to estimate intermediate values between precise data points. The most common method used for this purpose is polynomial interpolation.

Recall that the general formula for an nth-order polynomial is

$$f(x) = a_0 + a_1 x + a_2 x^2 \cdot \quad \cdot \quad \cdot + a_n x^n \tag{12.1}$$

For $n + 1$ data points, there is one and only one polynomial of order n or less that passes through all the points. For example, there is only one straight line (that is, a first-order polynomial) that connects two points (Fig. 12.1a). Similarly, only one parabola connects a set of three points (Fig. 12.1b). *Polynomial interpolation* consists of determining the unique nth-order polynomial that fits $n + 1$ data points. This polynomial then provides a formula to compute intermediate values.

Although there is one and only one nth-order polynomial that fits $n + 1$ points, there are a variety of mathematical formats in which this polynomial can be expressed. In the present chapter, we will describe two alternatives that are well-suited for computer implementation. These are the Newton and the Lagrange polynomials.

12.1 NEWTON'S DIVIDED-DIFFERENCE INTERPOLATING POLYNOMIALS

As stated above, there are a variety of alternative forms for expressing an interpolating polynomial. *Newton's divided-difference interpolating polynomial* is among the most

Figure 12.1
Examples of interpolating polynomials: (a) first-order (linear) connecting two points, (b) second-order (quadratic or parabolic) connecting three points, and (c) third-order (cubic) connecting four points.

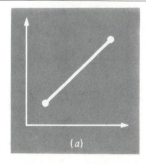
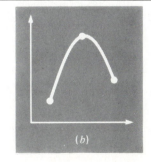
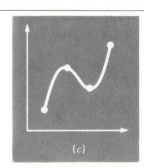

popular and useful forms. Before presenting the general equation, we will introduce the first- and second-order versions because of their simple visual interpretation.

12.1.1   Linear Interpolation

The simplest form of interpolation is to connect two data points with a straight line. This technique, called *linear interpolation,* is depicted graphically in Fig. 12.2. Using similar triangles,

$$\frac{f_1(x) - f(x_0)}{x - x_0} = \frac{f(x_1) - f(x_0)}{x_1 - x_0}$$

which can be rearranged to yield

$$f_1(x) = f(x_0) + \frac{f(x_1) - f(x_0)}{x_1 - x_0}(x - x_0) \tag{12.2}$$

which is a *linear-interpolation formula.* The notation $f_1(x)$ designates that this is a first-order interpolating polynomial. Notice that besides representing the slope of the line connecting the points, the term $[f(x_1) - f(x_0)]/(x_1 - x_0)$ is a finite-divided-difference approximation of the first derivative [recall Eq. (3.28)]. In general, the smaller the

Figure 12.2
Graphical depiction of linear interpolation. The shaded areas indicate the similar triangles that are used to derive the linear-interpolation formula [Eq. (12.2)].

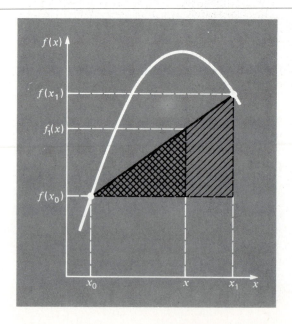

interval between the data points, the better the approximation. This is due to the fact that, as the interval decreases, a continuous function will be better approximated by a straight line. This characteristic is demonstrated in the following example.

EXAMPLE 12.1   *Linear Interpolation*

Problem Statement: Estimate the natural logarithm of 2 using linear interpolation. First, perform the computation by interpolating between $\ln 1 = 0$ and $\ln 6 = 1.7917595$. Then, repeat the procedure, but use a smaller interval from $\ln 1$ to $\ln 4$ (1.3862944). Note that the true value of $\ln 2$ is 0.69314718.

Solution: Using Eq. (12.2), a linear interpolation from $x_0 = 1$ to $x_1 = 6$ gives

$$f_1(2) = 0 + \frac{1.7917595 - 0}{6 - 1}(2 - 1) = 0.35835190$$

which represents a percent error of $\epsilon_t = 48.3$ percent. Using the smaller interval from $x_0 = 1$ to $x_1 = 4$ yields

$$f_1(2) = 0 + \frac{1.3862944 - 0}{4 - 1}(2 - 1) = 0.46209813$$

Thus, using the shorter interval reduces the percent relative error to $\epsilon_t = 33.3$ percent. Both interpolations are shown in Fig. 12.3, along with the true function.

Figure 12.3
Two linear interpolations to estimate ln 2. Note how the smaller interval provides a better estimate.

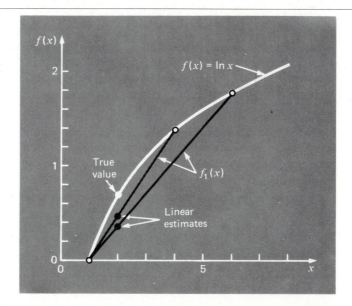

12.1.2 Quadratic Interpolation

The error in Example 12.1 was due to the fact that we approximated a curve with a straight line. Consequently, a strategy for improving the estimate is to introduce some curvature into the line connecting the points. If three data points are available, this can be accomplished with a second-order polynomial (also called a quadratic polynomial or a parabola). A particularly convenient form for this purpose is

$$f_2(x) = b_0 + b_1(x - x_0) + b_2(x - x_0)(x - x_1) \tag{12.3}$$

Note that although Eq. (12.3) might seem to differ from the general polynomial [Eq. (12.1)], the two equations are equivalent. This can be shown by multiplying the terms in Eq. (12.3) to yield

$$f_2(x) = b_0 + b_1 x - b_1 x_0 + b_2 x^2 + b_2 x_0 x_1 - b_2 x x_0 - b_2 x x_1$$

or, collecting terms,

$$f_2(x) = a_0 + a_1 x + a_2 x^2$$

where

$$a_0 = b_0 - b_1 x_0 + b_2 x_0 x_1$$

$$a_1 = b_1 - b_2 x_0 - b_2 x_1$$

$$a_2 = b_2$$

Thus, Eqs. (12.1) and (12.3) are alternative, equivalent formulations of the unique second-order polynomial joining the three points.

A simple procedure can be used to determine the values of the coefficients. For b_0, Eq. (12.3) with $x = x_0$ can be used to compute

$$b_0 = f(x_0) \tag{12.4}$$

Equation (12.4) can be substituted into Eq. (12.3), which can be evaluated at $x = x_1$ for

$$b_1 = \frac{f(x_1) - f(x_0)}{x_1 - x_0} \tag{12.5}$$

Finally, Eqs. (12.4) and (12.5) can be substituted into Eq. (12.3), which can be evaluated at $x = x_2$ and solved (after some algebraic manipulations) for

$$b_2 = \frac{\dfrac{f(x_2) - f(x_1)}{x_2 - x_1} - \dfrac{f(x_1) - f(x_0)}{x_1 - x_0}}{x_2 - x_0} \tag{12.6}$$

Notice that, as was the case with linear interpolation, b_1 still represents the slope of the line connecting points x_0 and x_1. Thus, the first two terms of Eq. (12.3) are equivalent to linear interpolation from x_0 to x_1, as specified previously in Eq. (12.2). The last term, $b_2(x - x_0)(x - x_1)$, introduces the second-order curvature into the formula.

Before illustrating how to use Eq. (12.3), we should examine the form of the coefficient b_2. It is very similar to the finite-divided-difference approximation of the second derivative introduced previously in Eq. (3.35). Thus, Eq. (12.3) is beginning to manifest a structure that is very similar to the Taylor series expansion. This observation will be explored further when we relate Newton's interpolating polynomials to the Taylor series in Sec. 12.1.4. But first, we will show how Eq. (12.3) is used to interpolate between three points.

EXAMPLE 12.2   Quadratic Interpolation

Problem Statement: Fit a second-order polynomial to the three points used in Example 12.1:

$$x_0 = 1 \qquad f(x_0) = 0$$
$$x_1 = 4 \qquad f(x_1) = 1.3862944$$
$$x_2 = 6 \qquad f(x_2) = 1.7917595$$

Use the polynomial to evaluate ln 2.

Solution: Applying Eq. (12.4) yields

$$b_0 = 0$$

Equation (12.5) yields

$$b_1 = \frac{1.3862944 - 0}{4 - 1} = 0.46209813$$

and Eq. (12.6) gives

$$b_2 = \frac{\dfrac{1.7917595 - 1.3862944}{6 - 4} - 0.46209813}{6 - 1} = -0.051873116$$

Substituting these values into Eq. (12.3) yields the quadratic formula

$$f_2(x) = 0 + 0.46209813(x - 1) - 0.051873116(x - 1)(x - 4)$$

which can be evaluated at $x = 2$ for

$$f_2(2) = 0.56584436$$

which represents a percent relative error of $\epsilon_t = 18.4\%$. Thus, the curvature introduced by the quadratic formula (Fig. 12.4) improves the interpolation compared with the result obtained using straight lines in Example 12.1 and Fig. 12.3.

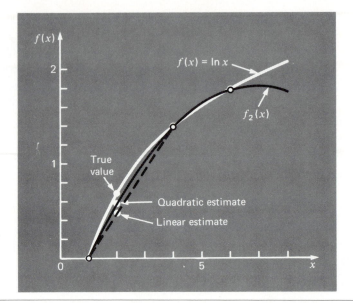

Figure 12.4
The use of quadratic interpolation to estimate ln 2. The linear interpolation from $x = 1$ to 4 is also included for comparison.

12.1.3 General Form of Newton's Interpolating Polynomials

The preceding analysis can be generalized to fit an nth-order polynomial to $n + 1$ data points. The nth-order polynomial is

$$f_n(x) = b_0 + b_1(x - x_0) + \cdots + b_n(x - x_0)(x - x_1) \cdots (x - x_{n-1}) \qquad (12.7)$$

As was done previously with the linear and quadratic interpolations, data points can be used to evaluate the coefficients $b_0, b_1, \ldots, b_n$. For an nth-order polynomial, $n + 1$ data points are required: $x_0, x_1, x_2, \ldots, x_n$. Using these data points, the following equations are used to evaluate the coefficients:

$$b_0 = f(x_0) \qquad (12.8)$$

$$b_1 = f[x_1, x_0] \qquad (12.9)$$

$$b_2 = f[x_2, x_1, x_0] \qquad (12.10)$$

.

.

.

$$b_n = f[x_n, x_{n-1}, \ldots, x_1, x_0] \qquad (12.11)$$

where the bracketed function evaluations are finite divided differences. For example, the *first finite divided difference* is represented generally as

$$f[x_i, x_j] = \frac{f(x_i) - f(x_j)}{x_i - x_j} \qquad (12.12)$$

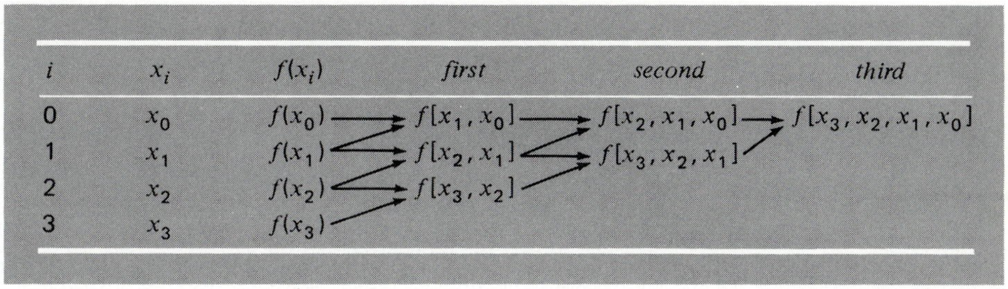

Figure 12.5
Graphical depiction of the recursive nature of finite divided differences.

The *second finite divided difference*, which represents the difference of two first divided differences, is expressed generally as

$$f[x_i, x_j, x_k] = \frac{f[x_i, x_j] - f[x_j, x_k]}{x_i - x_k}$$

(12.13)

Similarly, the *nth finite divided difference* is

$$f[x_n, x_{n-1}, \ldots, x_1, x_0] = \frac{f[x_n, x_{n-1}, \ldots, x_1] - f[x_{n-1}, x_{n-2}, \ldots, x_0]}{x_n - x_0}$$

(12.14)

These differences can be used to evaluate the coefficients in Eqs. (12.8) through (12.11), which can then be substituted into Eq. (12.7) to yield the interpolating polynomial

$$f_n(x) = f(x_0) + (x - x_0)f[x_1, x_0] + (x - x_0)(x - x_1)f[x_2, x_1, x_0]$$

$$+ \cdots + (x - x_0)(x - x_1) \cdots (x - x_{n-1})f[x_n, x_{n-1}, \ldots, x_0]$$

(12.15)

which is called *Newton's divided-difference interpolating polynomial*. It should be noted that it is not necessary that the data points used in Eq. (12.15) be equally spaced or that the abscissa values necessarily be in ascending order, as illustrated in the following example. Also, notice how Eqs. (12.12) through (12.14) are recursive—that is, higher-order differences are composed of lower-order differences (Fig. 12.5). This property will be exploited when we develop an efficient computer program in Sec. 12.1.5 to implement the method.

EXAMPLE 12.3   Newton's Divided-Difference Interpolating Polynomials

Problem Statement: In Example 12.2, data points at $x_0 = 1, x_1 = 4$, and $x_2 = 6$ were used to estimate ln 2 with a parabola. Now, adding a fourth point $[x_3 = 5; f(x_3) = 1.6094379]$, estimate ln 2 with a third-order Newton's interpolating polynomial.

Solution: The third-order polynomial, Eq. (12.7) with $n = 3$, is

$$f_3(x) = b_0 + b_1(x - x_0) + b_2(x - x_0)(x - x_1)$$

$$+ b_3(x - x_0)(x - x_1)(x - x_2)$$

The first divided differences for the problem are [Eq. (12.12)]

$$f[x_1, x_0] = \frac{1.3862944 - 0}{4 - 1} = 0.46209813$$

$$f[x_2, x_1] = \frac{1.7917595 - 1.3862944}{6 - 4} = 0.20273255$$

$$f[x_3, x_2] = \frac{1.6094379 - 1.7917595}{5 - 6} = 0.18232160$$

The second divided differences are [Eq. (12.13)]

$$f[x_2, x_1, x_0] = \frac{0.20273255 - 0.46209813}{6 - 1} = -0.051873116$$

$$f[x_3, x_2, x_1] = \frac{0.18232160 - 0.20273255}{5 - 4} = -0.020410950$$

The third divided difference is [Eq. (12.14) with $n = 3$]

$$f[x_3, x_2, x_1, x_0] = \frac{-0.020410950 - (-0.051873116)}{5 - 1}$$

$$= 0.0078655415$$

The results for $f[x_1, x_0], f[x_2, x_1, x_0]$, and $f[x_3, x_2, x_1, x_0]$ represent the coefficients b_1, b_2, and b_3 of Eq. (12.7). Along with $b_0 = f(x_0) = 0.0$, Eq. (12.7) is

$$f_3(x) = 0 + 0.46209813(x - 1) - 0.051873116(x - 1)(x - 4)$$

$$+ 0.0078655415(x - 1)(x - 4)(x - 6)$$

which can be used to evaluate

$$f_3(2) = 0.62876869$$

which represents a percent relative error of $\epsilon_t = 9.3\%$. The complete cubic polynomial is shown in Fig. 12.6.

12.1.4   Errors of Newton's Interpolating Polynomials

Notice that the structure of Eq. (12.15) is similar to the Taylor series expansion in the sense that terms are added sequentially in order to capture the higher-order behavior of the underlying function. These terms are finite divided differences and, thus, represent approximations of the higher-order derivatives. Consequently, as with the Taylor series, if the true underlying function is an nth-order polynomial, the nth-order interpolating polynomial based on $n + 1$ data points will yield exact results.

Also, as was the case with the Taylor series, a formulation for the truncation error can be obtained. Recall from Eq. (3.17) that the truncation error for the Taylor series could be expressed generally as

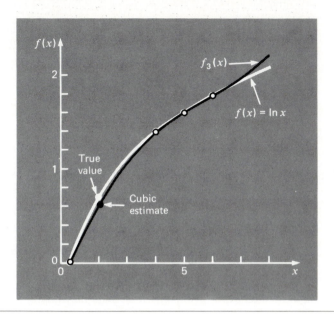

Figure 12.6
The use of cubic interpolation to estimate ln 2.

$$R_n = \frac{f^{(n+1)}(\xi)}{(n+1)!}(x_{i+1} - x_i)^{n+1}$$

where ξ is somewhere in the interval from x_i to x_{i+1}. For an nth-order interpolating polynomial, an analogous relationship for the error is

$$R_n = \frac{f^{(n+1)}(\xi)}{(n+1)!}(x - x_0)(x - x_1) \cdots (x - x_n) \qquad (12.16)$$

where ξ is somewhere in the interval containing the unknown and the data. For this formula to be of use, the function in question must be known and differentiable. This is not usually the case. Fortunately, an alternative formulation is available that does not require prior knowledge of the function. Rather, it uses a finite divided difference to approximate the $(n + 1)$th derivative,

$$R_n = f[x, x_n, x_{n-1}, \ldots, x_0](x - x_0)(x - x_1) \cdots (x - x_n) \qquad (12.17)$$

where $f[x, x_n, x_{n-1}, \ldots, x_0]$ is the $(n + 1)$th finite divided difference. Because Eq. (12.17) contains the unknown $f(x)$, it cannot be solved for the error. However, if an additional data point $f(x_{n+1})$ is available, Eq. (12.17) can be used to estimate the error, as in

$$R_n \simeq f[x_{n+1}, x_n, x_{n-1}, \cdots, x_0](x - x_0)(x - x_1) \cdots (x - x_n) \qquad (12.18)$$

EXAMPLE 12.4   Error Estimation for Newton's Polynomial

Problem Statement: Use Eq. (12.18) to estimate the error for the second-order polynomial interpolation of Example 12.2. Use the additional data point $f(x_3) = f(5) = 1.6094379$ to obtain your results.

Solution: Recall that in Example 12.2, the second-order interpolating polynomial provided an estimate of $f_2(2) = 0.56584436$, which represents an error of $0.69314718 - 0.56584436 = 0.12730282$. If we had not known the true value, as is most usually the case, Eq. (12.18), along with the additional value at x_3, could have been used to estimate the error, as in

$$R_2 = f[x_3, x_2, x_1, x_0](x - x_0)(x - x_1)(x - x_2)$$

or

$$R_2 = 0.0078655415(x - 1)(x - 4)(x - 6)$$

where the value for the third-order finite divided difference is as computed previously in Example 12.3. This relationship can be evaluated at $x = 2$ for

$$R_2 = 0.0078655415(2 - 1)(2 - 4)(2 - 6) = 0.062924332$$

which is of the same order of magnitude as the true error.

From the previous example and from Eq. (12.18), it should be clear that the error estimate for the nth-order polynomial is equivalent to the difference between the $(n + 1)$th-order and the nth-order prediction. That is,

$$R_n = f_{n+1}(x) - f_n(x) \qquad (12.19)$$

In other words, the increment that is added to the nth-order case in order to create the $(n + 1)$th-order case [that is, Eq. (12.18)] is interpreted as an estimate of the nth-order error. This can be clearly seen by rearranging Eq. (12.19) to give

$$f_{n+1}(x) = f_n(x) + R_n$$

The validity of this approach is predicated on the fact that the series is strongly convergent. For such a situation, the $(n + 1)$th-order prediction should be much closer to the true value than the nth-order prediction. Consequently, Eq. (12.19) conforms to our standard definition of error as representing the difference between the truth and an approximation. However, note that whereas all other error estimates for iterative approaches introduced up to this point have been determined as a present prediction minus a previous one, Eq. (12.19) represents a future prediction minus a present one. This means that for a series that is converging rapidly, the error estimate of Eq. (12.19) could be less than the true error. This would represent a highly unattractive quality if the error estimate were being employed as a stopping criterion. However, as will be described in the following section, higher-order interpolating polynomials are highly sensitive to data errors—that is, they are very ill conditioned. When employed for interpolation, they often yield predictions that diverge significantly from the true value. By "looking ahead" to sense errors, Eq. (12.19) is more sensitive to such divergence. As such, it is more valuable for the sort of exploratory data analysis for which Newton's polynomial is best-suited.

12.1.5 Computer Program for Newton's Interpolating Polynomial

Three properties make Newton's interpolating polynomials extremely attractive for computer applications:

1. As in Eq. (12.7), higher-order versions can be developed sequentially by adding a single term to the next lower-order equation. This facilitates the evaluation of several different-order versions in the same program. Such a capability is especially valuable when the order of the polynomial is not known a priori. By adding new terms sequentially, we can determine when a point of diminishing returns is reached—that is, when addition of higher-order terms no longer significantly improves the estimate or in certain situations actually detracts from it. The error equations discussed below in (3) are useful in devising an objective criterion for identifying this point of diminishing terms.

2. The finite divided differences that constitute the coefficients of the polynomial [Eqs. (12.8) through (12.11)] can be computed using a recurrence relationship. That is, as in Eq. (12.14) and Fig. 12.5, lower-order differences are used to compute higher-order differences. By utilizing this previously determined information, the coefficients can be computed efficiently. The programs in Figs. 12.7 and 12.8 contain such a scheme.

3. The error estimate [Eq. (12.18)] can be very simply incorporated into a computer algorithm because of the sequential way in which the prediction is built.

Figure 12.7
An interactive computer program for Newton's interpolating polynomial written in Microsoft BASIC.

```
100 REM NEWTON'S POLYNOMIAL (BASIC VERSION)         225     GOSUB 450 'perform interpolation
105 REM ***********************************         230     GOSUB 500 'output results
110 REM *       DEFINITION OF VARIABLES    *        235     PRINT
115 REM *                                  *        240     PRINT "SOLUTION? (Y OR N)";
120 REM * N     = NUMBER OF DATA POINTS    *        245     INPUT CONTIN$
125 REM * X()   = INDEPENDENT VARIABLES    *        250     CLS
130 REM * F()   = DEPENDENT VARIABLES      *        255 WEND
135 REM * FDD[] = FINITE DIVIDED-          *        260 END
140 REM *                DIFFERENCE TABLE  *        300 REM ******** SUBROUTINE INPUT ********
145 REM * XI    = X-VALUE FOR INTERPOLATION*        305 REM
150 REM ***********************************         310 CLS
155 REM                                             315 INPUT "NUMBER OF POINTS? ", N
160 DIM X(10),F(10),FDD(10,10)                      320 PRINT
165 REM                                             325 FOR I = 0 TO N-1
170 REM ********** MAIN PROGRAM **********           330     PRINT "X(";I;"),F(";I;") = ";
175 REM                                             335     INPUT X(I),F(I)
180 GOSUB 300 'input data                           340 NEXT I
185 GOSUB 350 'compute finite differences           345 RETURN
190 CLS                                             350 REM ***** SUBROUTINE DIFFERENCE ******
195 PRINT "SOLUTION? (Y OR N)";                     355 REM
200 INPUT CONTIN$                                   360 FOR I = 1 TO N
205 WHILE CONTIN$ = "y" OR CONTIN$ = "Y"            365     FDD(I,1) = F(I-1)
210    PRINT                                        370 NEXT I
215    INPUT "INTERPOLATE AT ";XI                   375 FOR J = 2 TO N
220    CLS                                          380     FOR I = 1 TO N-J+1
```

```
385       FDD(I,J)=(FDD(I+1,J-1)-FDD(I,J-1))/(X(I+J-2)-X(I-1))
390    NEXT I
395 NEXT J
400 RETURN
450 REM ******** SUBROUTINE SOLVE ********
455 XTERM=1
460 FA(0)=FDD(1,1)
465 FOR ORDER = 1 TO N-1
470    XTERM=XTERM*(XI-X(ORDER-1))
475    FA2=FA(ORDER-1)+FDD(1,ORDER+1)*XTERM
480    ET(ORDER-1)=FA2-FA(ORDER-1)
485    FA(ORDER)=FA2
490 NEXT ORDER
```

```
495 RETURN
500 REM ******** SUBROUTINE OUTPUT *******
505 REM
510 PRINT "INTERPOLATION AT X = ";XI
515 PRINT
520 PRINT "ORDER        F(X)           ERROR"
525 PRINT
530 FOR I = 0 TO N-2
535    PRINT I,FA(I),ET(I)
540 NEXT I
545 PRINT N-1,FA(N-1)
550 RETURN
```

All the above characteristics can be exploited and incorporated into general computer codes for implementing Newton's polynomial (Figs. 12.7 and 12.8). Note that aside from input-output, the codes consist of two major subroutines or procedures. The first determines the coefficients from Eq. (12.7) by a recurrence relationship. The second determines the predictions and their associated error. The utility of these programs is demonstrated in the following example.

Figure 12.8
Subroutines for Newton's interpolating polynomial written in (a) FORTRAN 77 and (b) Turbo Pascal.

```
      SUBROUTINE NEWTON(X,F,FDD,N,XI,YI)
************************************************
*          DEFINITION OF VARIABLES            *
*                                             *
*    N     = NUMBER OF DATA POINTS            *
*    X()   = INDEPENDENT VARIABLE             *
*    F()   = DEPENDENT VARIABLE               *
*    FDD[] = FINITE DIVIDED-DIFFERENCES       *
*    XI    = X-VALUE FOR INTERPOLATION        *
************************************************
      DIMENSION X(10),F(10),FDD(10,10)
      DIMENSION FA(10),ET(10)
      CALL DIVDIF(X,F,FDD,N)
      XTERM = 1.
      FA(1) = FDD(1,1)
      DO 10 IORD = 1,N-1
         XTERM=XTERM*(XI-X(IORD))
         FA2=FA(IORD)+FDD(1,IORD+1)*XTERM
         ET(IORD)=FA2-FA(IORD)
         FA(IORD+1)=FA2
   10 CONTINUE
      RETURN
      END
************************************************
      SUBROUTINE DIVDIF(X,F,FDD,N)
      DIMENSION X(10),F(10),FDD(10)
      DO 10 I=0,N-1
         FDD(I,1)=F(I)
   10 CONTINUE
```

```
PROCEDURE Newton(VAR Fdd: Matrix;
                 VAR X,F,Fa,Et: Vector;
                 VAR Xi: Real;
                     N: Integer);

{        Driver program type definitions
   Matrix = a 2 dimensional real array
   Vector = a 1 dimensional real array
            (defined with zero subscript)}

{           Definition of variables
   N     = number of data points
   X()   = independent variables
   F()   = dependent variables
   Fdd[] = finite divided-difference table
   Xi    = x-value for interpolation
   Fa    = dependent variable at Xi        }

VAR
   i,j,k : integer;

PROCEDURE DivDiff (VAR Fdd: matrix;
                       X: vector);
VAR
   i,j,k: Integer;
Begin
For i := 1 to N do
```

```
      DO 20 J=2,N
         DO 30 I=1,N-J+1
            FDD(I,J)=
   *              (FDD(I+1,J-1)-FDD(I,J-1))
   *                        /(X(I+J-1)-X(I))
30       CONTINUE
20    CONTINUE
      RETURN
      END
```

```
      Begin
      Fdd[i,1] := F[i-1];
      End;
   For j := 2 to N do
      Begin
      For i := 1 to N-j+1 do
         Begin
         Fdd[i,j] := (Fdd[i+1,j-1]-
              Fdd[i,j-1])/(X[i+j-2]-X[i-1]);
         End;
      End;
   End;                      {of procedure DivDiff}

   PROCEDURE NewtsPoly (XI: real;
                           VAR Fa,Et:vector);
   VAR
      Fa2, Xterm: real;
      order: integer;
   Begin
   Xterm := 1;
   Fa[0] := Fdd[1,1];
   For order := 1 to N-1 do
      Begin
      Xterm := Xterm*(Xi-X[order-1]);
      Fa2:=Fa[order-1]+Fdd[1,order+1]*Xterm;
      Et[order-1] := Fa2-Fa[order-1];
      Fa[order] := Fa2;
      End;
   End;                      {of procedure NewtsPoly}

   Begin                          {procedure Newton}
   DivDiff (Fdd,X);
   NewtsPoly (XI,Fa,Et);
   End;                      {of procedure Newton}
```

EXAMPLE 12.5   Using Error Estimates to Determine the Appropriate Order of Interpolation

Problem Statement: After incorporating the error [Eq. (12.18)], utilize the computer program given in Fig. 12.7 and the following information to evaluate $f(x) = \ln x$ at $x = 2$:

x	f(x) = ln x
1	0
4	1.3862944
6	1.7917595
5	1.6094379
3	1.0986123
1.5	0.40546511
2.5	0.91629073
3.5	1.2527630

Solution: The results of employing the program in Fig. 12.7 to obtain a solution are shown in Fig. 12.9. The error estimates, along with the true error (based on the fact that ln 2 = 0.69314718), are depicted in Fig. 12.10. Note that the estimated and the true error are similar and that their agreement improves as the order increases. From this plot, it can be concluded that the fifth-order version yields a good estimate and that higher-order terms do not significantly enhance the prediction.

This exercise also illustrates the importance of the positioning and ordering of the points. For example, up through the third-order estimate, the rate of improvement is slow because the points that are added (at $x = 4$, 6, and 5) are distant and on one side of the point in question at $x = 2$. The fourth-order estimate shows a somewhat greater improvement because the new point at $x = 3$ is closer to the unknown. However, the most dramatic decrease in the error is associated with the inclusion of the fifth-order term using the data point at $x = 1.5$. Not only is this point close to the unknown but it is also positioned on the opposite side from most of the other points. As a consequence, the error is reduced almost an order of magnitude.

The significance of the position and sequence of the data can also be illustrated by using the same data to obtain an estimate for ln 2, but considering the points in a different sequence. Figure 12.10 shows results for the case of reversing the order of the original data, that is, $x_0 = 3.5, x_1 = 2.5, x_3 = 1.5$, and so forth. Because the initial points for this case are closer to and spaced on either side of ln 2, the error decreases much more rapidly than for the original situation. By the second-order term, the error has been reduced to a percent-relative level of less than $\epsilon_t = 2$ percent. Other combinations could be employed to obtain different rates of convergence.

The foregoing example illustrates the importance of the choice of base points. As should be intuitively obvious, the points should be centered around and as close as

Figure 12.9
The output of the BASIC program to evaluate ln 2.

NUMBER OF POINTS? 8

X(0),F(0) = ? 1,0
X(1),F(1) = ? 4,1.3862944
X(2),F(2) = ? 6,1.7917595
X(3),F(3) = ? 5,1.6094379
X(4),F(4) = ? 3,1.0986123
X(5),F(5) = ? 1.5,0.40546411
X(6),F(6) = ? 2.5,0.91629073
X(7),F(7) = ? 3.5,1.2527630

SOLUTION? (Y OR N)? Y

INTERPOLATION AT X = 2

ORDER	F(X)	ERROR
0	0.000000	0.462098
1	0.462098	0.103746
2	0.565844	0.062924
3	0.628769	0.046953
4	0.675722	0.021792
5	0.697514	-0.003616
6	0.693898	-0.000459
7	0.693439	

SOLUTION? (Y OR N)?

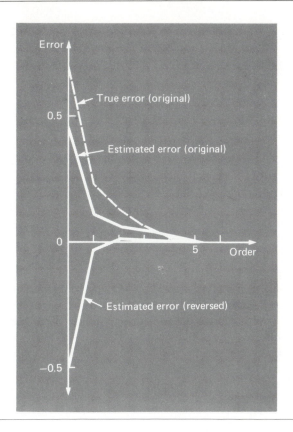

Figure 12.10
Percent relative errors for the prediction of ln 2 as a function of the order of the interpolating polynomial.

possible to the unknown. This observation is also supported by direct examination of the error equation [Eq. (12.17)]. Assuming that the finite divided difference does not vary markedly along the range of the data, the error is proportional to the product: $(x - x_0)(x - x_1) \cdots (x - x_n)$. Obviously, the closer the base points are to x, the smaller the magnitude of this product.

12.2 LAGRANGE INTERPOLATING POLYNOMIALS

The *Lagrange interpolating polynomial* is simply a reformulation of the Newton polynomial that avoids the computation of divided differences. It can be represented concisely as

$$f_n(x) = \sum_{i=0}^{n} L_i(x) f(x_i) \tag{12.20}$$

where

$$L_i(x) = \prod_{\substack{j=0 \\ j \neq i}}^{n} \frac{x - x_j}{x_i - x_j} \tag{12.21}$$

where Π designates the "product of." For example, the linear version ($n = 1$) is

$$f_1(x) = \frac{x - x_1}{x_0 - x_1} f(x_0) + \frac{x - x_0}{x_1 - x_0} f(x_1) \tag{12.22}$$

and the second-order version is

$$f_2(x) = \frac{(x - x_1)(x - x_2)}{(x_0 - x_1)(x_0 - x_2)} f(x_0) + \frac{(x - x_0)(x - x_2)}{(x_1 - x_0)(x_1 - x_2)} f(x_1)$$

$$+ \frac{(x - x_0)(x - x_1)}{(x_2 - x_0)(x_2 - x_1)} f(x_2) \tag{12.23}$$

Equation (12.20) can be derived directly from Newton's polynomial (Box 12.1). However, the rationale underlying the Lagrange formulation can be grasped directly by realizing that each term $L_i(x)$ will be 1 at $x = x_i$ and 0 at all other sample points (Fig. 12.11). Thus, each product $L_i(x)f(x_i)$ takes on the value of $f(x_i)$ at the sample point x_i. Consequently, the summation of all the products designated by Eq. (12.20) is the unique nth-order polynomial that passes exactly through all $n + 1$ data points.

Box 12.1   *Derivation of the Lagrange Form Directly from Newton's Interpolating Polynomial*

The Lagrange interpolating polynomial can be derived directly from Newton's formulation. We will do this for the first-order case [Eq. (12.2)],

$$f_1(x) = f(x_0) + (x - x_0)f[x_1, x_0] \tag{B12.1.1}$$

In order to derive the Lagrange form, we reformulate the divided differences. For example, the first divided difference,

$$f[x_1, x_0] = \frac{f(x_1) - f(x_0)}{x_1 - x_0}$$

can be reformulated as

$$f[x_1, x_0] = \frac{f(x_1)}{x_1 - x_0} + \frac{f(x_0)}{x_0 - x_1} \tag{B12.1.2}$$

which is referred to as the *symmetric form*. Substituting Eq. (B12.1.2) into Eq. (B12.1.1) yields

$$f_1(x) = f(x_0) + \frac{(x - x_0)}{(x_1 - x_0)} f(x_1)$$

$$+ \frac{(x - x_0)}{(x_0 - x_1)} f(x_0)$$

Finally, grouping similar terms and simplifying yields the Lagrange form,

$$f_1(x) = \frac{x - x_1}{x_0 - x_1} f(x_0) + \frac{x - x_0}{x_1 - x_0} f(x_1)$$

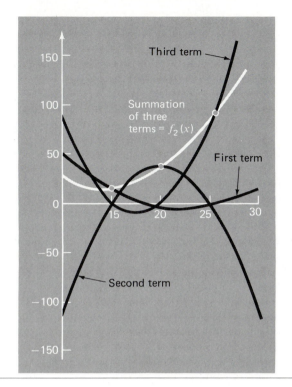

Figure 12.11
A visual depiction of the rationale behind the Lagrange polynomial. This figure shows a second-order case. Each of the three terms in Eq. (12.23) passes through one of the data points and is zero at the other two. The summation of the three terms must, therefore, be the unique second-order polynomial $f_2(x)$ that passes exactly through the three points.

EXAMPLE 12.6   Lagrange Interpolating Polynomials

Problem Statement: Use a Lagrange interpolating polynomial of the first and second order to evaluate ln 2 on the basis of the data given in Example 12.2:

$$x_0 = 1 \qquad f(x_0) = 0$$

$$x_1 = 4 \qquad f(x_1) = 1.3862944$$

$$x_2 = 6 \qquad f(x_2) = 1.7917595$$

Solution: The first-order polynomial [Eq. (12.22)] is

$$f_1(x) = \frac{x - x_1}{x_0 - x_1} f(x_0) + \frac{x - x_0}{x_1 - x_0} f(x_1)$$

and, therefore, the estimate at $x = 2$ is

$$f_1(2) = \frac{2 - 4}{1 - 4} 0 + \frac{2 - 1}{4 - 1} 1.3862944 = 0.4620981$$

In a similar fashion, the second-order polynomial is developed as [Eq. (12.23)]

$$f_2(2) = \frac{(2 - 4)(2 - 6)}{(1 - 4)(1 - 6)} 0 + \frac{(2 - 1)(2 - 6)}{(4 - 1)(4 - 6)} 1.3862944$$

$$+ \frac{(2-1)(2-4)}{(6-1)(6-4)} 1.7917595 = 0.56584437$$

As expected, both these results agree closely with those obtained previously using Newton's interpolating polynomial.

Note that, as with Newton's method, the Lagrange version has an estimated error of [Eq. (12.17)]

$$R_n = f[x, x_n, x_{n-1}, \ldots, x_0] \prod_{i=0}^{n} (x - x_i)$$

Thus, if an additional point is available at $x = x_{n+1}$, an error estimate can be obtained. However, because the finite divided differences are not employed as part of the Lagrange algorithm, this is rarely done.

Equations (12.20) and (12.21) can be very simply programmed for implementation on a computer. Figure 12.12 shows pseudocode that can be employed for this purpose.

Figure 12.12
Pseudocode to implement Lagrange interpolation. This algorithm is set up to compute a single (n)-th order prediction where $n + 1$ is the number of data points.

```
y = 0
DOFOR i = 0 to n
    product = f
             i
    DOFOR j = 0 to n
        IF (i ≠ j)
            product = product·(x - x )/ (x  - x )
                                     j     i     j
        ELSE ENDIF
    ENDDO
    y = y + product
ENDDO
```

In summary, for cases where the order of the polynomial is unknown, the Newton method has advantages because of the insight it provides into the behavior of the different-order formulas. In addition, the error estimate represented by Eq. (12.18) can usually be integrated easily into the Newton computation because the estimate employs a finite difference (Example 12.5). Thus, for single exploratory computations, Newton's method is often preferable.

When only one interpolation is to be performed, the Lagrange and Newton formulations require comparable computational effort. However, the Lagrange version is somewhat easier to program. Because it does not require computation and storage of divided differences, the Lagrange form is often used when the order of the polynomial is known a priori.

EXAMPLE 12.7   Lagrange Interpolation Using the Computer

Problem Statement: A user-friendly computer program to implement Lagrange interpolation is contained in the Electronic TOOLKIT software package associated with this text.

We can use this software to study a trend analysis problem associated with our now-familiar falling parachutist. Assume that we have developed instrumentation to measure the velocity of the parachutist. The measured data obtained for a particular test case is

Time, s	Measured Velocity v, cm/s
1	800
3	2310
5	3090
7	3940
13	4755

Our problem is to estimate the velocity of the parachutist at $t = 10$ s to fill in the large gap in the measurements between $t = 7$ and $t = 13$ s. We are aware that the behavior of interpolating polynomials can be unexpected. Therefore, we will construct polynomials of orders 4, 3, 2, and 1 and compare the results.

Solution: The Electronic TOOLKIT program can be used to construct fourth-, third-, second-, and first-order interpolating polynomials. These results are

Polynomial Order	Coefficient					Estimated Value of v at $t = 10$ s
	4th Order	3d Order	2d Order	1st Order	0 Order	
4	−1.76302	44.87501	−392.87	1813.625	−663.867	5430.195
3		−4.498586	76.09375	1.239258	1742.656	4874.838
2			−36.14584	858.75	−300.1035	4672.812
1				135.8333	2989.167	4347.5

The fourth-order polynomial and the input data can be plotted as shown in Fig. 12.13a. It is evident from this plot that the estimated value of y at x = 10 is higher than the overall trend of the data.

Figure 12.13b through d shows plots of the results of the computations for third-, second-, and first-order interpolating polynomials. It is noted that the lower the order of the interpolating polynomial, the lower the estimated value of the velocity at $t = 10$ s. The plots of the interpolating polynomials indicate that the higher-order polynomials tend to overshoot the trend of the data. This suggests that the first- or second-order polynomials are most appropriate for this particular trend analysis. It should be remembered, however, that because we are dealing with uncertain data, regression would actually be more appropriate.

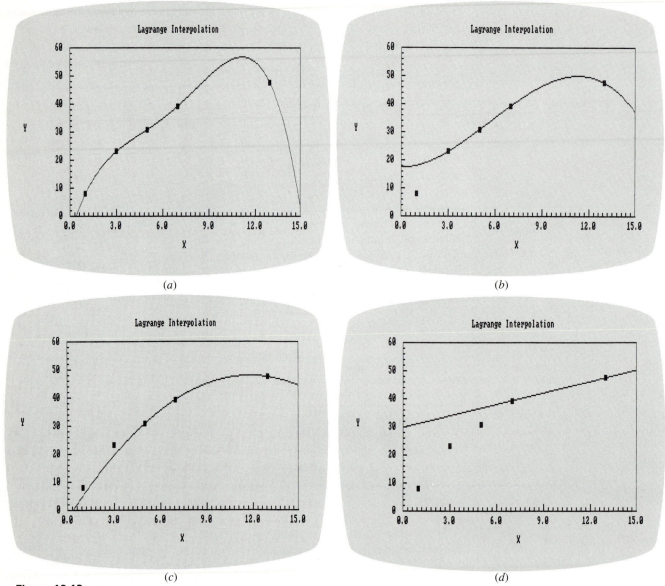

Figure 12.13
Computer-generated plots showing (*a*) fourth-order, (*b*) third-order, (*c*) second-order, and (*d*) first-order interpolations.

12.3   COEFFICIENTS OF AN INTERPOLATING POLYNOMIAL

Although both the Newton and the Lagrange polynomials are well-suited for determining intermediate values between points, they do not provide a convenient polynomial of the conventional form

$$f(x) = a_0 + a_1 x + a_2 x^2 + \cdots + a_n x^n \tag{12.24}$$

A straightforward method for computing the coefficients of this polynomial is based on the fact that $n + 1$ data points are required to determine the $n + 1$ coefficients. Thus, simultaneous linear algebraic equations can be used to calculate the a's. For example, suppose that you desired to compute the coefficients of the parabola

$$f(x) = a_0 + a_1 x + a_2 x^2 \qquad (12.25)$$

Three data points are required: $[x_0, f(x_0)]$, $[x_1, f(x_1)]$, and $[x_2, f(x_2)]$. Each can be substituted into Eq. (12.25) to give

$$f(x_0) = a_0 + a_1 x_0 + a_2 x_0^2$$

$$f(x_1) = a_0 + a_1 x_1 + a_2 x_1^2 \qquad (12.26)$$

$$f(x_2) = a_0 + a_1 x_2 + a_2 x_2^2$$

Thus, for this case, the x's are the knowns and the a's are the unknowns. Because there are the same number of equations as unknowns, Eq. (12.26) could be solved by an elimination method from Part Three.

It should be noted that the foregoing approach is not the most efficient method that is available to determine the coefficients of an interpolating polynomial. Press et al. (1986) provide a discussion and computer codes for more efficient approaches. Whatever technique is employed, a word of caution is in order. Systems such as Eq. (12.26) are notoriously ill-conditioned. Whether they are solved with an elimination method or with a more efficient algorithm, the resulting coefficients can be highly inaccurate, particularly for large n. When used for a subsequent interpolation, they often yield erroneous results.

In summary, if you are interested in determining an intermediate point, employ Newton or Lagrange interpolation. If you must determine an equation of the form of Eq. (12.24), limit yourself to lower-order polynomials and check your results carefully.

12.4 ADDITIONAL COMMENTS

Before proceeding to the next section, we must mention two additional topics: interpolation with equally spaced data and extrapolation.

Because both the Newton and Lagrange polynomials are compatible with arbitrarily spaced data, you might wonder why we address the special case of equally spaced data (Box 12.2). Prior to the advent of digital computers, these techniques had great utility for interpolation from tables with equally spaced arguments. In fact, a computational framework known as a divided-difference table was developed to facilitate the implementation of these techniques. (Figure 12.5 is an example of such tables.)

However, because the formulas are subsets of the computer-compatible Newton and Lagrange schemes and because many tabular functions are available as library subroutines, the need for the equispaced versions has waned. In spite of this, we have included them at this point because of their relevance to later parts of this book. In particular, they can be employed to derive numerical integration formulas which typically employ equispaced data (Chap. 15). Because the numerical integration formulas have relevance to the solution of ordinary differential equations, the material in Box 12.2 also has significance to Chap. 20.

Box 12.2 Interpolation with Equally Spaced Data

If data is equally spaced and in ascending order, then the independent variable assumes values of

$$x_1 = x_0 + h$$

$$x_2 = x_0 + 2h$$

.

.

.

$$x_n = x_0 + nh$$

where h is the interval, or step size, between the data. On this basis, the finite divided differences can be expressed in concise form. For example, the second forward divided difference is

$$f[x_0, x_1, x_2] = \frac{\dfrac{f(x_2) - f(x_1)}{x_2 - x_1} - \dfrac{f(x_1) - f(x_0)}{x_1 - x_0}}{x_2 - x_0}$$

which can be expressed as

$$f[x_0, x_1, x_2] = \frac{f(x_2) - 2f(x_1) + f(x_0)}{2h^2} \qquad \text{(B12.2.1)}$$

because $x_1 - x_0 = x_2 - x_1 = (x_2 - x_0)/2 = h$. Now recall that the second forward difference $\Delta^2 f(x_0)$ is equal to [numerator of Eq. (3.35)]

$$\Delta^2 f(x_0) = f(x_2) - 2f(x_1) + f(x_0)$$

Therefore, Eq. (B12.2.1) can be represented as

$$f[x_0, x_1, x_2] = \frac{\Delta f^2(x_0)}{2!h^2}$$

or, in general,

$$f[x_0, x_1, \dots, x_n] = \frac{\Delta f^n(x_0)}{n!h^n} \qquad \text{(B12.2.2)}$$

Using Eq. (B12.2.2), Newton's interpolating polynomial [Eq. (12.15)] can be expressed for the case of equispaced data as

$$f_n(x) = f(x_0) + \frac{\Delta f(x_0)}{h}(x - x_0)$$

$$+ \frac{\Delta^2 f(x_0)}{2!h^2}(x - x_0)(x - x_0 - h) + \cdots$$

$$+ \frac{\Delta^n f(x_0)}{n!h^n}(x - x_0)(x - x_0 - h) \cdots \qquad \text{(B12.2.3)}$$

$$[x - x_0 - (n - 1)h]$$

$$+ R_n$$

where the remainder is the same as Eq. (12.16). This equation is known as *Newton's formula* or the *Newton-Gregory forward formula*. It can be simplied further by defining a new quantity, α:

$$\alpha = \frac{x - x_0}{h}$$

This definition can be used to develop the following simplified expressions for the terms in Eq. (B12.2.3),

$$x - x_0 = \alpha h$$

$$x - x_0 - h = \alpha h - h = h(\alpha - 1)$$

.

.

.

$$x - x_0 - (n - 1)h = \alpha h - (n - 1)h$$

$$= h(\alpha - n + 1)$$

which can be substituted into Eq. (B12.2.3) to give

$$f_n(x) = f(x_0) + \Delta f(x_0)\alpha + \frac{\Delta^2 f(x_0)}{2!}\alpha(\alpha - 1)$$

$$+ \cdots + \frac{\Delta^n f(x_0)}{n!}\alpha(\alpha - 1) \cdots \qquad \text{(B12.2.4)}$$

$$(\alpha - n + 1) + R_n$$

where

$$R_n = \frac{f^{(n+1)}(\xi)}{(n + 1)!}h^{n+1}\alpha(\alpha - 1)(\alpha - 2) \cdots (\alpha - n)$$

This concise notation will have utility in our derivation and error analyses of the integration formulas in Chap. 15.

In addition to the forward formula, backward and central Newton-Gregory formulas are also available. Carnahan, Luther, and Wilkes (1969) can be consulted for further information regarding interpolation for equally spaced data.

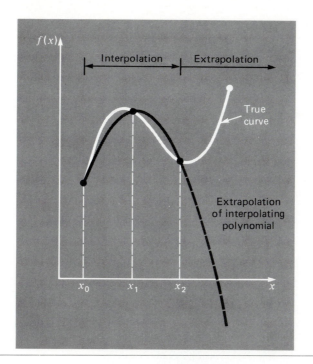

Figure 12.14
Illustration of the possible divergence of an extrapolated prediction. The extrapolation is based on fitting a parabola through the first three known points.

Extrapolation is the process of estimating a value of $f(x)$ that lies outside the range of the known base points, $x_0, x_1, \ldots, x_n$ (Fig. 12.14). In a previous section, we mentioned that the most accurate interpolation is usually obtained when the unknown lies near the center of the base points. Obviously, this is violated when the unknown lies outside the range, and consequently, the error in extrapolation can be very large. As depicted in Fig. 12.14, the open-ended nature of extrapolation represents a step into the unknown because the process extends the curve beyond the known region. As such, the true curve could easily diverge from the prediction. Extreme care should, therefore, be exercised whenever a case arises where one must extrapolate. Case Study 14.1 presents an example of the liabilities involved in projecting beyond the limits of the data.

12.5 SPLINE INTERPOLATION

In the previous sections, nth-order polynomials were used to interpolate between $n + 1$ data points. For example, for eight points, we can derive a perfect seventh-order polynomial. This curve would capture all the meanderings (at least up to and including seventh derivatives) suggested by the points. However, there are cases where these functions can lead to erroneous results. An alternative approach is to apply lower-order polynomials to subsets of data points. Such connecting polynomials are called *spline functions*.

For example, third-order curves employed to connect each pair of data points are called *cubic splines*. These functions have the additional property that the connections between adjacent cubic equations are visually smooth. On the surface, it would seem

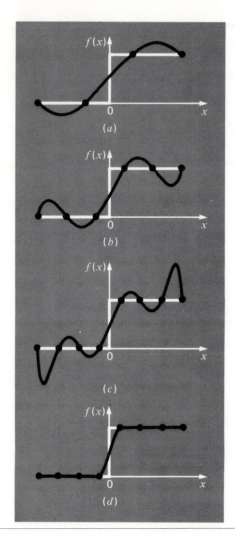

Figure 12.15
A visual representation of a situation where the splines are superior to higher-order interpolating polynomials. The function to be fit undergoes an abrupt increase at $x = 0$. Parts (a) through (c) indicate that the abrupt change induces oscillations in interpolating polynomials. In contrast, because it is limited to third-order curves with smooth transitions, the cubic spline (d) provides a much more acceptable approximation.

that the third-order approximation of the splines would be inferior to the seventh-order expression. You might wonder why a spline would ever be preferable.

Figure 12.15 illustrates a situation where a spline performs better than a higher-order polynomial. This is the case where a function is generally smooth but undergoes an abrupt change somewhere along the region of interest. The step increase depicted in Fig. 12.15 is an extreme example of such a change and serves to illustrate the point.

Figure 12.15a through c illustrates how higher-order polynomials tend to swing through wild oscillations in the vicinity of an abrupt change. In contrast, the spline also connects the points, but because it is limited to third-order changes, the oscillations are kept to a minimum. As such the spline usually provides a superior approximation of the behavior of functions that have local, abrupt changes.

The concept of the spline originated from the drafting technique of using a thin,

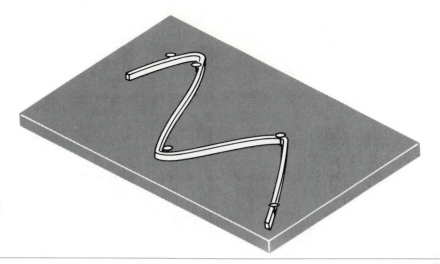

Figure 12.16
The drafting technique of using a spline to draw smooth curves through a series of points. Notice how, at the end points, the spline straightens out. This is called a "natural" spline.

flexible strip (called a *spline*) to draw smooth curves through a set of points. The process is depicted in Fig. 12.16 for a series of five pins (data points). In this technique, the drafter places paper over a wooden board and hammers nails or pins into the paper (and board) at the location of the data points. A smooth cubic curve results from interweaving the strip between the pins. Hence, the name "cubic spline" has been adopted for polynomials of this type.

In the present section, simple linear functions will first be used to introduce some basic concepts and problems associated with spline interpolation. Then we derive an algorithm for fitting quadratic splines to data. Finally, we present material on the cubic spline, which is the most common and useful version in engineering practice.

12.5.1 Linear Splines

The simplest connection between two points is a straight line. The first-order splines for a group of ordered data points can be defined as a set of linear functions,

$$f(x) = f(x_0) + m_0(x - x_0) \qquad x_0 \le x \le x_1$$
$$f(x) = f(x_1) + m_1(x - x_1) \qquad x_1 \le x \le x_2$$

$$.$$

$$.$$

$$.$$

$$f(x) = f(x_{n-1}) + m_{n-1}(x - x_{n-1}) \qquad x_{n-1} \le x \le x_n$$

where m_i is the slope of the straight line connecting the points:

$$m_i = \frac{f(x_{i+1}) - f(x_i)}{x_{i+1} - x_i} \tag{12.27}$$

These equations can be used to evaluate the function at any point between x_0 and x_n by first locating the interval within which the point lies. Then the appropriate equation is used to determine the function value within the interval. The method is obviously identical to linear interpolation.

EXAMPLE 12.8   First-Order Splines

Problem Statement: Fit the data in Table 12.1 with first-order splines. Evaluate the function at $x = 5$.

TABLE 12.1
Data to be fit with spline functions.

x	f(x)
3.0	2.5
4.5	1.0
7.0	2.5
9.0	0.5

Solution: The data can be used to determine the slopes between points. For example, for the interval from $x = 4.5$ to $x = 7$ the slope can be computed using Eq. (12.27):

$$m = \frac{2.5 - 1.0}{7.0 - 4.5} = 0.60$$

The slopes for the other intervals can be computed, and the resulting first-order splines are plotted in Fig. 12.17a. The value at $x = 5$ is 1.3.

Visual inspection of Fig. 12.17a indicates that the primary disadvantage of first-order splines is that they are not smooth. In essence, at the data points where two splines meet (called a *knot*), the slope changes abruptly. In formal terms, the first derivative of the function is discontinuous at these points. This deficiency is overcome by using higher-order polynomial splines that ensure smoothness at the knots by equating derivatives at these points, as discussed in the next section.

12.5.2 Quadratic Splines

In order to ensure that the mth derivatives are continuous at the knots, a spline of at least $m + 1$ order must be used. Third-order polynomials or cubic splines that ensure continuous first and second derivatives are most frequently used in practice. Although third and higher derivatives could be discontinuous when using cubic splines, they usually cannot be detected visually and consequently are ignored.

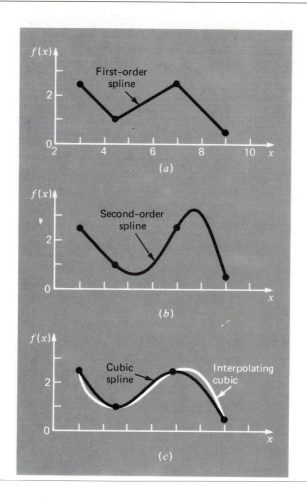

Figure 12.17
Spline fits of a set of four points. (a) Linear spline, (b) quadratic spline, and (c) cubic spline, with a cubic interpolating polynomial also plotted.

Because the derivation of cubic splines is somewhat involved, we have chosen to include them in a subsequent section. We have decided to first illustrate the concept of spline interpolation using second-order polynomials. These "quadratic splines" have continuous first derivatives at the knots. Although quadratic splines do not ensure equal second derivatives at the knots, they serve nicely to demonstrate the general procedure for developing higher-order splines.

The objective in quadratic splines is to derive a second-order polynomial for each interval between data points. The polynomial for each interval can be represented generally as

$$f_i(x) = a_i x^2 + b_i x + c_i \tag{12.28}$$

Figure 12.18 has been included to help clarify the notation. For $n + 1$ data points ($i = 0, 1, 2, \ldots, n$), there are n intervals and, consequently, $3n$ unknown constants (the a's, b's and c's) to evaluate. Therefore, $3n$ equations or conditions are required to evaluate the unknowns. These are:

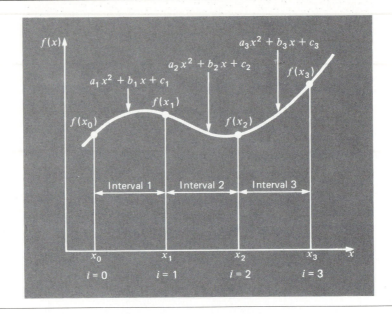

Figure 12.18
Notation used to derive quadratic splines. Notice that there are n intervals and $n + 1$ data points. The examples shown is for $n = 3$.

1. *The function values must be equal at the interior knots.* This condition can be represented as

$$a_{i-1}x_{i-1}^2 + b_{i-1}x_{i-1} + c_{i-1} = f(x_{i-1}) \tag{12.29}$$

$$a_i x_{i-1}^2 + b_i x_{i-1} + c_i = f(x_{i-1}) \tag{12.30}$$

for $i = 2$ to n. Because only interior knots are used, Eqs. (12.29) and (12.30) each provide $n - 1$ for a total of $2n - 2$ conditions.

2. *The first and last functions must pass through the end points.* This adds two additional equations:

$$a_1 x_0^2 + b_1 x_0 + c_1 = f(x_0) \tag{12.31}$$

$$a_n x_n^2 + b_n x_n + c_n = f(x_n) \tag{12.32}$$

for a total of $2n - 2 + 2 = 2n$ conditions.

3. *The first derivatives at the interior knots must be equal.* The first derivative of Eq. (12.28) is

$$f'(x) = 2ax + b$$

Therefore, the condition can be represented generally as

$$2a_{i-1}x_{i-1} + b_{i-1} = 2a_i x_{i-1} + b_i \tag{12.33}$$

for $i = 2$ to n. This provides another $n - 1$ conditions for a total of $2n + n - 1 = 3n - 1$. Because we have $3n$ unknowns, we are one condition short. Unless we have some additional information regarding the functions or their derivatives, we must make an arbitrary choice in order to successfully compute the constants. Although there are a number of different choices that can be made, we select the following:

4. *Assume that the second derivative is zero at the first point.* Because the second derivative of Eq. (12.28) is $2a_i$, this condition can be expressed mathematically as

$$a_1 = 0 \tag{12.34}$$

The visual interpretation of this condition is that the first two points will be connected by a straight line.

EXAMPLE 12.9   Quadratic Splines

Problem Statement: Fit quadratic splines to the same data used in Example 12.8 (Table 12.1). Use the results to estimate the value at $x = 5$.

Solution: For the present problem, we have four data points and $n = 3$ intervals. Therefore, $3(3) = 9$ unknowns must be determined. Equations (12.29) and (12.30) yield $2(3) - 2 = 4$ conditions:

$$20.25a_1 + 4.5b_1 + c_1 = 1.0$$

$$20.25a_2 + 4.5b_2 + c_2 = 1.0$$

$$49a_2 + 7b_2 + c_2 = 2.5$$

$$49a_3 + 7b_3 + c_3 = 2.5$$

Passing the first and last functions through the initial and final values adds 2 more [Eq. (12.31)]:

$$9a_1 + 3b_1 + c_1 = 2.5$$

and [Eq. (12.32)]

$$81a_3 + 9b_3 + c_3 = 0.5$$

Continuity of derivatives creates an additional $3 - 1 = 2$ [Eq. (12.33)]:

$$9a_1 + b_1 = 9a_2 + b_2$$

$$14a_2 + b_2 = 14a_3 + b_3$$

Finally, Eq. (12.34) specifies that $a_1 = 0$. Because this equation specifies a_1 exactly, the problem reduces to solving eight simultaneous equations. These conditions can be expressed in matrix form as

$$
\begin{bmatrix}
4.5 & 1.0 & 0.0 & 0.0 & 0.0 & 0.0 & 0.0 & 0.0 \\
0.0 & 0.0 & 20.25 & 4.5 & 1.0 & 0.0 & 0.0 & 0.0 \\
0.0 & 0.0 & 49.00 & 7.0 & 1.0 & 0.0 & 0.0 & 0.0 \\
0.0 & 0.0 & 0.0 & 0.0 & 0.0 & 49.00 & 7.00 & 1.00 \\
3.0 & 1.0 & 0.0 & 0.0 & 0.0 & 0.0 & 0.0 & 0.0 \\
0.0 & 0.0 & 0.0 & 0.0 & 0.0 & 81.00 & 9.00 & 1.00 \\
1.0 & 0.0 & -9.00 & -1.0 & 0.0 & 0.0 & 0.0 & 0.0 \\
0.0 & 0.0 & 14.00 & 1.0 & 0.0 & -14.00 & -1.00 & 0.0
\end{bmatrix}
\begin{Bmatrix}
b_1 \\
c_1 \\
a_2 \\
b_2 \\
c_2 \\
a_3 \\
b_3 \\
c_3
\end{Bmatrix}
=
\begin{Bmatrix}
1.0 \\
1.0 \\
2.5 \\
2.5 \\
2.5 \\
0.5 \\
0.0 \\
0.0
\end{Bmatrix}
$$

These equations can be solved using techniques from Part Three with the results:

$$a_1 = 0 \qquad b_1 = -1 \qquad c_1 = 5.5$$

$$a_2 = 0.64 \qquad b_2 = -6.76 \qquad c_2 = 18.46$$

$$a_3 = -1.6 \qquad b_3 = 24.6 \qquad c_3 = -91.3$$

which can be substituted into the original quadratic equations to develop the following relationships for each interval:

$$f_1(x) = -x + 5.5 \qquad\qquad 3.0 \le x \le 4.5$$

$$f_2(x) = 0.64x^2 - 6.76x + 18.46 \qquad 4.5 \le x \le 7.0$$

$$f_3(x) = -1.6x^2 + 24.6x - 91.3 \qquad 7.0 \le x \le 9.0$$

The prediction for $x = 5$ is, therefore,

$$f_2(5) = 0.64(5)^2 - 6.76(5) + 18.46 = 0.66$$

The total spline fit is depicted in Fig. 12.17b. Notice that there are two shortcomings that detract from the fit: (1) the straight line connecting the first two points and (2) the spline for the last interval seems to swing too high. The cubic splines in the next section do not exhibit these shortcomings and as a consequence are usually better methods for spline interpolation.

12.5.3 Cubic Splines

The objective in cubic splines is to derive a third-order polynomial for each interval between knots, as in

$$f_i(x) = a_i x^3 + b_i x^2 + c_i x + d_i \tag{12.35}$$

Thus, for $n + 1$ data points ($i = 0, 1, 2, \ldots, n$), there are n intervals and, consequently, $4n$ unknown constants to evaluate. Just as for quadratic splines, $4n$ conditions are required to evaluate the unknowns. These are:

1. *The function values must be equal at the interior knots (2n − 2 conditions).*
2. *The first and last functions must pass through the end points (2 conditions).*
3. *The first derivatives at the interior knots must be equal (n − 1 conditions).*
4. *The second derivatives at the interior knots must be equal (n − 1 conditions).*
5. *The second derivatives at the end knots are zero (2 conditions).*

The visual interpretation of condition 5 is that the function becomes a straight line at the end knots. Specification of such an end condition leads to what is termed a "natural" spline. It is given this name because the drafting spline naturally behaves in this fashion (Fig. 12.16). If the value of the second derivative at the end knots is nonzero (that is,

Box 12.3 Derivation of Cubic Splines

The first step in the derivation (Cheney and Kincaid, 1985) is based on the observation that because each pair of knots is connected by a cubic, the second derivative within each interval is a straight line. Equation (12.35) can be differentiated twice to verify this observation. On this basis, the second derivatives can be represented by a first-order Lagrange interpolating polynomial [Eq. (12.22)]:

$$f_i''(x) = f''(x_{i-1})\frac{x - x_i}{x_{i-1} - x_i} + f''(x_i)\frac{x - x_{i-1}}{x_i - x_{i-1}} \qquad (B12.3.1)$$

where $f_i''(x)$ is the value of the second derivative at any point x within the ith interval. Thus, this equation is a straight line connecting the second derivative at the first knot $f''(x_{i-1})$ with the second derivative at the second knot $f''(x_i)$.

Next, Eq. (B12.3.1) can be integrated twice to yield an expression for $f_i(x)$. However, this expression will contain two unknown constants of integration. These constants can be evaluated by invoking the function-equality conditions—$f(x)$ must equal $f(x_{i-1})$ at x_{i-1} and $f(x)$ must equal $f(x_i)$ at x_i. By performing these evaluations, the following cubic equation results:

$$f_i(x) = \frac{f''(x_{i-1})}{6(x_i - x_{i-1})}(x_i - x)^3$$

$$+ \frac{f''(x_i)}{6(x_i - x_{i-1})}(x - x_{i-1})^3$$

$$+ \left[\frac{f(x_{i-1})}{x_i - x_{i-1}} - \frac{f''(x_{i-1})(x_i - x_{i-1})}{6}\right](x_i - x)$$

$$+ \left[\frac{f(x_i)}{x_i - x_{i-1}} - \frac{f''(x_i)(x_i - x_{i-1})}{6}\right](x - x_{i-1})$$

$$(B12.3.2)$$

Now, admittedly, this relationship is a much more complicated expression for the cubic spline for the ith interval, than, say,

Eq. (12.35). However, notice that it contains only two unknown "coefficients," the second derivatives at the beginning and the end of the interval—$f''(x_{i-1})$ and $f''(x_i)$. Thus, if we can determine the proper second derivative at each knot, Eq. (B12.3.2) is a third-order polynomial that can be used to interpolate within the interval.

The second derivatives can be evaluated by invoking the condition that the first derivatives at the knots must be continuous:

$$f_{i-1}'(x_i) = f_i'(x_i) \qquad (B12.3.3)$$

Equation (B12.3.2) can be differentiated to give an expression for the first derivative. If this is done for both the $(i - 1)$th and the ith intervals and the two results are set equal according to Eq. (B12.3.3), the following relationship results:

$$(x_i - x_{i-1})f''(x_{i-1}) + 2(x_{i+1} - x_{i-1})f''(x_i)$$

$$+ (x_{i+1} - x_i)f''(x_{i+1})$$

$$= \frac{6}{(x_{i+1} - x_i)}[f(x_{i+1}) - f(x_i)] \qquad (B12.3.4)$$

$$+ \frac{6}{(x_i - x_{i-1})}[f(x_{i-1}) - f(x_i)]$$

If Eq. (B12.3.4) is written for all interior knots, $n - 1$ simultaneous equations result with $n + 1$ unknown second derivatives. However, because this is a natural cubic spline, the second derivatives at the end knots are zero and the problem reduces to $n - 1$ equations with $n - 1$ unknowns. In addition, notice that the system of equations will be tridiagonal. Thus, not only have we reduced the number of equations but we have also cast them in a form that is extremely easy to solve (recall Sec. 9.6).

there is some curvature), this information can be used alternatively to supply the two final conditions.

The above five types of conditions provide the total of $4n$ equations required to solve for the $4n$ coefficients. Whereas it is certainly possible to develop cubic splines in this fashion, we will present an alternative technique that only requires the solution of $n - 1$ equations. Although the derivation of this method (Box 12.3) is somewhat less straightforward than that for quadratic splines, the gain in efficiency is well worth the effort.

The derivation from Box 12.3 results in the following cubic equation for each interval:

$$f_i(x) = \frac{f''(x_{i-1})}{6(x_i - x_{i-1})}(x_i - x)^3 + \frac{f''(x_i)}{6(x_i - x_{i-1})}(x - x_{i-1})^3$$

$$+ \left[\frac{f(x_{i-1})}{x_i - x_{i-1}} - \frac{f''(x_{i-1})(x_i - x_{i-1})}{6} \right] (x_i - x) \qquad (12.36)$$

$$+ \left[\frac{f(x_i)}{x_i - x_{i-1}} - \frac{f''(x_i)(x_i - x_{i-1})}{6} \right] (x - x_{i-1})$$

This equation contains only two unknowns—the second derivatives at the end of each interval. These unknowns can be evaluated using the following equation:

$$(x_i - x_{i-1})f''(x_{i-1}) + 2(x_{i+1} - x_{i-1})f''(x_i) + (x_{i+1} - x_i)f''(x_{i+1})$$

$$= \frac{6}{(x_{i+1} - x_i)}[f(x_{i+1}) - f(x_i)] + \frac{6}{(x_i - x_{i-1})}[f(x_{i-1}) - f(x_i)] \qquad (12.37)$$

If this equation is written for all the interior knots, $n - 1$ simultaneous equations result with $n - 1$ unknowns. (Remember, the second derivatives at the end knots are zero.) The application of these equations is illustrated in the following example.

EXAMPLE 12.10   Cubic Splines

Problem Statement: Fit cubic splines to the same data used in Examples 12.8 and 12.9 (Table 12.1). Utilize the results to estimate the value at $x = 5$.

Solution: The first step is to employ Eq. (12.37) to generate the set of simultaneous equations that will be utilized to determine the second derivatives at the knots. For example, for the first interior knot, the following data is used:

$x_0 = 3 \qquad f(x_0) = 2.5$

$x_1 = 4.5 \qquad f(x_1) = 1$

$x_2 = 7 \qquad f(x_2) = 2.5$

These values can be substituted into Eq. (12.37) to yield

$(4.5 - 3)f''(3) + 2(7 - 3)f''(4.5) + (7 - 4.5)f''(7)$

$$= \frac{6}{7 - 4.5}(2.5 - 1) + \frac{6}{4.5 - 3}(2.5 - 1)$$

Because of the natural spline condition, $f''(3) = 0$, and the equation reduces to

$8f''(4.5) + 2.5f''(7) = 9.6$

In a similar fashion, Eq. (12.37) can be applied to the second interior point to give

$2.5f''(4.5) + 9f''(7) = -9.6$

These two equations can be solved simultaneously for

$$f''(4.5) = 1.67909$$

$$f''(7) \quad = -1.53308$$

These values can then be substituted into Eq. (12.36), along with values for the x's and the $f(x)$'s, to yield

$$f_1(x) = \frac{1.67909}{6(4.5 - 3)}(x - 3)^3 + \frac{2.5}{4.5 - 3}(4.5 - x)$$

$$+ \left[\frac{1}{4.5 - 3} - \frac{1.67909(4.5 - 3)}{6} \right](x - 3)$$

or

$$f_1(x) = 0.186566(x - 3)^3 + 1.666667(4.5 - x) + 0.246894(x - 3)$$

This equation is the cubic spline for the first interval. Similar substitutions can be made to develop the equations for the second and third intervals:

$$f_2(x) = 0.111939(7 - x)^3 - 0.102205(x - 4.5)^3$$

$$- 0.299621(7 - x) + 1.638783(x - 4.5)$$

and

$$f_3(x) = -0.127757(9 - x)^3 + 1.761027(9 - x) + 0.25(x - 7)$$

The three equations can then be employed to compute values within each interval. For example, the value at $x = 5$, which falls within the second interval, is calculated as

$$f_2(5) = 0.111939(7 - 5)^3 - 0.102205(5 - 4.5)^3$$

$$- 0.299621(7 - 5) + 1.638783(5 - 4.5) = 1.102886$$

Other values are computed and the results are plotted in Fig. 12.17c.

The results of Examples 12.8 through 12.10 are summarized in Fig. 12.17. Notice the progressive improvement of the fit as we move from linear to quadratic to cubic splines. We have also superimposed a cubic interpolating polynomial on Fig. 12.17c. Although the cubic spline consists of a series of third-order curves, the resulting fit differs from that obtained using the third-order polynomial. This is due to the fact that the natural spline requires zero second derivatives at the end knots, whereas the cubic polynomial has no such constraint.

12.5.4 Computer Algorithm for Cubic Splines

The method for calculating cubic splines outlined in the previous section is ideal for computer implementation. Recall that by some clever manipulations, the method reduced to solving $n - 1$ simultaneous equations. An added benefit of the derivation is that, as specified by Eq. (12.37), the system of equations is *tridiagonal*. As described in Sec. 9.6, algorithms are available to solve such systems in an extremely efficient manner. Figure 12.19 outlines a computational framework that incorporates these features.

Step 1: Input data

Step 2: Apply Eq. (12.37) in order to generate a tridiagonal system of equations.

Step 3: Solve the system of equations for the unknown second derivatives using an algorithm of the sort described in Sec. 9.6.

Figure 12.19
Algorithm for cubic spline interpolation.

Step 4: Use Eq. (12.36) to interpolate at a given value of x.

Step 5: If you require another interpolation, return to step 4. If not, end the computation.

PROBLEMS

Hand Calculations

12.1 Estimate the logarithm of 4 to the base 10 (log 4) using linear interpolation.
 (a) Interpolate between log 3 = 0.4771213 and log 5 = 0.6989700.
 (b) Interpolate between log 3 and log 4.5 = 0.6532125. For each of the interpolations, compute the percent relative error based on the true value of log 4 = 0.6020600.

12.2 Fit a second-order Newton's interpolating polynomial to estimate log 4 using the data from Prob. 12.1. Compute the true percent relative error.

12.3 Fit a third-order Newton's interpolating polynomial to estimate log 4 using the data from Prob. 12.1 along with the additional point, log 3.5 = 0.5440680. Compute the true percent relative error.

12.4 Given the data

x	0	0.5	1.0	1.5	2.0	2.5
$f(x)$	1	2.119	2.910	3.945	5.720	8.695

 (a) Calculate $f(1.6)$ using Newton's interpolating polynomials of order 1 through 3. Choose the sequence of the points for your estimates to attain good accuracy.
 (b) Utilize Eq. (12.18) to estimate the error for each prediction.

12.5 Given the data

x	1	2	3	5	6
$f(x)$	4.75	4	5.25	19.75	36

Calculate $f(3.5)$ using Newton's interpolating polynomials of order 1 through 4. Choose your base points to attain good accuracy. What do your results indicate regarding the order of the polynomial used to generate the data in the table?

12.6 Repeat Probs. 12.1 through 12.3 using the Lagrange polynomial.

12.7 Repeat Prob. 12.4a using the Lagrange polynomial.

12.8 Repeat Prob. 12.5 using the Lagrange polynomial of order 1 through 3.

12.9 Develop quadratic splines for the data in Prob. 12.5 and predict $f(3.5)$.

12.10 Develop cubic splines for the data in Prob. 12.5 and predict $f(3.5)$.

12.11 Determine the coefficients of the parabola that passes through the first three points in Prob. 12.4.

12.12 Determine the coefficients of the cubic equation that passes through the last four points in Prob. 12.5.

Computer-Related Problems

12.13 Reprogram either Fig. 12.7 or 12.8 so that it is user-friendly. Among other things:

(**a**) Place documentation statements throughout the program to identify what each section is intended to accomplish.

(**b**) Label the input and output.

12.14 Test the program you developed in Prob. 12.13 by duplicating the computation from Example 12.5.

12.15 Use the program you developed in Prob. 12.13 to solve Probs. 12.1 through 12.3.

12.16 Use the program you developed in Prob. 12.13 to solve Probs. 12.4 and 12.5. In Prob. 12.4, utilize all the data to develop first- through fifth-order polynomials. For both problems, plot the estimated error versus order.

12.17 Repeat Probs. 12.14 and 12.15, except use the Electronic TOOLKIT software associated with the text.

12.18 Use the Electronic TOOLKIT software to duplicate Examples 12.6 and 12.7.

12.19 Develop a user-friendly program for Lagrange interpolation. Test it by duplicating Example 12.7.

12.20 Develop a user-friendly program for cubic spline interpolation based on Fig. 12.19 and Sec. 12.5.4. Test the program by duplicating Example 12.10.

12.21 Use the software developed in Prob. 12.20 to fit cubic splines through the data in Probs. 12.4 and 12.5. For both cases, predict $f(2.25)$.

CHAPTER 13
Fourier Approximation

To this point, our presentation of curve fitting has emphasized standard polynomials—that is, linear combinations of the monomials $1, x, x^2, \ldots, x^m$ (Fig. 13.1a). We now turn to another class of functions that has immense importance in engineering. These are the trigonometric functions $1, \cos x, \cos 2x, \ldots, \cos nx, \sin x, \sin 2x, \ldots, \sin mx$ (Fig. 13.1b).

Engineers often deal with systems that oscillate or vibrate. As might be expected, trigonometric functions play a fundamental role in modeling such problem contexts. *Fourier approximation* represents a systematic framework for using trigonometric series for this purpose.

One of the hallmarks of a Fourier analysis is that it deals with both the time and the frequency domains. Unfortunately, many engineers are not comfortable with the latter. For this reason, we have devoted a large fraction of the subsequent material to a general overview of Fourier approximation. An important aspect of this overview will be to familiarize you with the frequency domain. This orientation is then followed by an introduction to numerical methods for computing discrete Fourier transforms.

13.1 CURVE FITTING WITH SINUSOIDAL FUNCTIONS

A periodic function $f(t)$ is one for which

$$f(t) = f(t + T) \tag{13.1}$$

where T is a constant called the *period* which is the smallest value for which Eq. (13.1) holds. Common examples include waveforms such as square and sawtooth waves (Fig. 13.2). The most fundamental are sinusoidal functions.

In the present discussion, we will use the term *sinusoid* to represent any waveform that can be described as a sine or cosine. There is no clear-cut convention for choosing either function, and in any case the results will be identical. For the present chapter we will use the cosine which is expressed generally as

$$f(t) = A_0 + C_1 \cos (\omega_0 t + \theta) \tag{13.2}$$

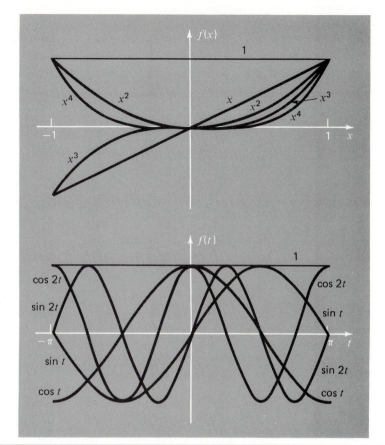

Figure 13.1
The first five (*a*) monomials and (*b*) trigonometric functions. Note that for the intervals shown, both types of function range in value between −1 and 1. However, notice that the peak values for the monomials all occur at the extremes whereas for the trigonometric functions the peaks are more uniformly distributed across the interval.

Thus, four parameters serve to characterize the sinusoid (Fig. 13.3). The *mean value* A_0 sets the average height above the abscissa. The *amplitude* C_1 specifies the height of the oscillation. The *angular frequency* ω_0 characterizes how often the cycles occur. Finally, the *phase angle* or *phase shift* θ parameterizes the extent to which the sinusoid is shifted horizontally. It can be measured as the distance in radians from $t = 0$ to the point at which the cosine function begins a new cycle. As depicted in Fig. 13.4*a*, a negative value is referred to as a *lagging phase angle* because the curve $\cos(\omega_0 t - \theta)$ begins a new cycle θ radians after $\cos(\omega_0 t)$. Thus, $\cos(\omega_0 t - \theta)$ is said to lag $\cos(\omega_0 t)$. Conversely, as in Fig. 13.4*b*, a positive value is referred to as a *leading phase angle*.

Note that the angular frequency (in radians/time) is related to frequency f (in cycles/time) by

$$\omega_0 = 2\pi f \tag{13.3}$$

and frequency in turn is related to period T (in units of time) by

$$f = \frac{1}{T} \tag{13.4}$$

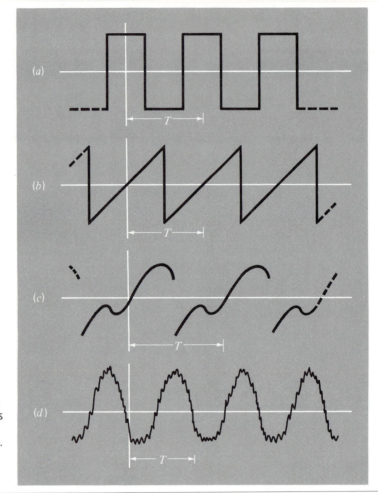

Figure 13.2
Aside from trigonometric functions such as sines and cosines, periodic functions include waveforms such as (a) the square wave and (b) the sawtooth wave. Beyond these idealized forms, periodic signals in nature can be (c) nonideal and (d) contaminated by noise. The trigonometric functions can be used to represent and to analyze all of these cases.

Although Eq. (13.2) is an adequate mathematical characterization of a sinusoid, it is awkward to work with from the standpoint of curve fitting because the phase shift is included in the argument of the cosine function. This deficiency can be overcome by invoking the trigonometric identity

$$C_1 \cos (\omega_0 t + \theta) = C_1[\cos (\omega_0 t) \cos (\theta) - \sin (\omega_0 t) \sin (\theta)] \tag{13.5}$$

Substituting Eq. (13.5) into Eq. (13.2) and collecting terms gives (Fig. 13.3b)

$$f(t) = A_0 + A_1 \cos (\omega_0 t) + B_1 \sin (\omega_0 t) \tag{13.6}$$

where

$$A_1 = C_1 \cos (\theta) \qquad B_1 = -C_1 \sin (\theta) \tag{13.7}$$

Dividing the two parts of Eq. (13.7) gives

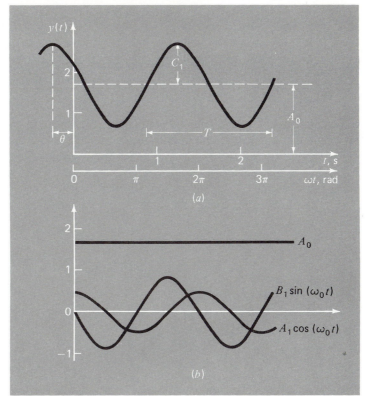

Figure 13.3

(a) A plot of the sinusoidal function $y(t) = A_0 + C_1 \cos(\omega_0 t + \theta)$. For this case, $A_0 = 1.7$, $C_1 = 1$, $\omega_0 = 2\pi/T = 2\pi/(1.5)$ s, $\theta = \pi/3$ radians $= 1.0472\ (=0.25$ s). Other parameters used to describe the curve are the frequency $f = \omega_0/(2\pi)$, which for this case is 1 cycle/(1.5 s) and the period $T = 1.5$ s. (b) An alternative expression of the same curve is $y(t) = A_0 + A_1 \cos(\omega_0 t) + B_1 \sin(\omega_0 t)$. The three components of this function are depicted in (b) where $A_1 = 0.5$ and $B_1 = -0.866$. The summation of the three curves in (b) yields the single curve in (a).

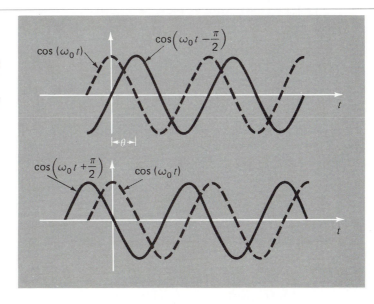

Figure 13.4

Graphical depictions of (a) a lagging phase angle and (b) a leading phase angle. Note that the lagging curve in (a) can be alternatively described as $\cos(\omega_0 t + 3\pi/2)$. In other words, if a curve lags by an angle of α, it can also be representing as leading by $2\pi - \alpha$.

$$\theta = \arctan\left(-\frac{B_1}{A_1}\right) \tag{13.8}$$

where, if $A_1 < 0$, add π to θ. Squaring and summing Eq. (13.7) leads to

$$C_1 = \sqrt{A_1^2 + B_1^2} \tag{13.9}$$

Thus, Eq. (13.6) represents an alternative formulation of Eq. (13.2) that still requires four parameters but which is cast in the format of a general linear model [recall Eq. (11.23)]. As we will discuss in the next section, it can be simply applied as the basis for a least-squares fit.

Before proceeding to the next section, however, we should stress that we could have employed a sine rather than a cosine as our fundamental model of Eq. (13.2). For example,

$$f(t) = A_0 + C_1 \sin(\omega_0 t + \delta)$$

could have been used. Simple relationships can be applied to convert between the two forms

$$\sin(\omega_0 t + \delta) = \cos\left(\omega_0 t + \delta - \frac{\pi}{2}\right)$$

and

$$\cos(\omega_0 t + \theta) = \sin\left(\omega_0 t + \theta + \frac{\pi}{2}\right) \tag{13.10}$$

In other words, $\theta = \delta - \pi/2$. The only important consideration is that one or the other format should be used consistently. Thus, we will use the cosine version throughout our discussion.

13.1.1 Least-Squares Fit of a Sinusoid

Equation (13.6) can be thought of as a linear least-squares model

$$y = A_0 + A_1 \cos(\omega_0 t) + B_1 \sin(\omega_0 t) + e \tag{13.11}$$

which is just another example of the general model [recall Eq. (11.23a)]

$$y = a_0 z_0 + a_1 z_1 + a_2 z_2 + \cdots + a_m z_m + e$$

where $z_0 = 1, z_1 = \cos(\omega_0 t), z_2 = \sin(\omega_0 t)$, and all other z's $= 0$. Thus, our goal is to determine coefficient values that minimize

$$S_r = \sum_{i=1}^{N} \{y_i - [A_0 + A_1 \cos(\omega_0 t_i) + B_1 \sin(\omega_0 t_i)]\}^2$$

The normal equations to accomplish this minimization can be expressed in matrix form as [recall Eq. (11.24)]

$$\begin{bmatrix} N & \Sigma \cos(\omega_0 t) & \Sigma \sin(\omega_0 t) \\ \Sigma \cos(\omega_0 t) & \Sigma \cos^2(\omega_0 t) & \Sigma \cos(\omega_0 t)\sin(\omega_0 t) \\ \Sigma \sin(\omega_0 t) & \Sigma \sin(\omega_0 t)\cos(\omega_0 t) & \Sigma \sin^2(\omega_0 t) \end{bmatrix} \begin{Bmatrix} A_0 \\ A_1 \\ B_1 \end{Bmatrix}$$

$$= \left\{ \begin{array}{c} \Sigma y \\ \Sigma y \cos (\omega_0 t) \\ \Sigma y \sin (\omega_0 t) \end{array} \right\} \qquad (13.12)$$

These equations can be employed to solve for the unknown coefficients. However, rather than do this, we can examine the special case where there are N observations equispaced at intervals of Δt and with a total record length of $T = (N - 1)\Delta t$. For this situation, the following average values can be determined (see Prob. 13.3):

$$\frac{\Sigma \sin (\omega_0 t)}{N} = 0 \qquad \frac{\Sigma \cos (\omega_0 t)}{N} = 0$$

$$\frac{\Sigma \sin^2 (\omega_0 t)}{N} = \frac{1}{2} \qquad \frac{\Sigma \cos^2 (\omega_0 t)}{N} = \frac{1}{2} \qquad (13.13)$$

$$\frac{\Sigma \cos (\omega_0 t) \sin (\omega_0 t)}{N} = 0$$

Thus, for equispaced points the normal equations become

$$\begin{bmatrix} N & 0 & 0 \\ 0 & N/2 & 0 \\ 0 & 0 & N/2 \end{bmatrix} \left\{ \begin{array}{c} A_0 \\ A_1 \\ B_1 \end{array} \right\} = \left\{ \begin{array}{c} \Sigma y \\ \Sigma y \cos (\omega_0 t) \\ \Sigma y \sin (\omega_0 t) \end{array} \right\}$$

The inverse of a diagonal matrix is merely another diagonal matrix whose elements are the reciprocals of the original. Thus, the coefficients can be determined as

$$\left\{ \begin{array}{c} A_0 \\ A_1 \\ B_1 \end{array} \right\} = \begin{bmatrix} 1/N & 0 & 0 \\ 0 & 2/N & 0 \\ 0 & 0 & 2/N \end{bmatrix} \left\{ \begin{array}{c} \Sigma y \\ \Sigma y \cos (\omega_0 t) \\ \Sigma y \sin (\omega_0 t) \end{array} \right\}$$

or

$$A_0 = \frac{\Sigma y}{N} \qquad (13.14)$$

$$A_1 = \frac{2}{N} \Sigma y \cos (\omega_0 t) \qquad (13.15)$$

$$B_1 = \frac{2}{N} \Sigma y \sin (\omega_0 t) \qquad (13.16)$$

EXAMPLE 13.1   *Least-Squares Fit of a Sinusoid*

Problem Statement: The curve in Fig. 13.3 is described by $y = 1.7 + \cos (4.189t + 1.0472)$. Generate 10 discrete values for this curve at intervals of $\Delta t = 0.15$ for the range from t $= 0$ to 1.35. Use this information to evaluate the coefficients of Eq. (13.11) by a least-squares fit.

Solution: The data required to evaluate the coefficients are

t	y	$y \cos (\omega_0 t)$	$y \sin (\omega_0 t)$
0	2.200	2.200	0.000
0.15	1.595	1.291	0.938
0.30	1.031	0.319	0.980
0.45	0.722	−0.223	0.687
0.60	0.786	−0.636	0.462
0.75	1.200	−1.200	0.000
0.90	1.805	−1.460	−1.061
1.05	2.369	−0.732	−2.253
1.20	2.678	0.829	−2.547
1.35	2.614	2.114	−1.536
$\Sigma =$	17.000	2.502	−4.330

These results can be used to determine [Eqs. (13.14) to (13.16)]

$$A_0 = \frac{17.000}{10} = 1.7 \qquad A_1 = \frac{2}{10} 2.502 = 0.500 \qquad B_1 = \frac{2}{10}(-4.330) = -0.866$$

Thus, the least-squares fit is

$$y = 1.7 + 0.500 \cos (\omega_0 t) - 0.866 \sin (\omega_0 t)$$

The model can also be expressed in the format of Eq. (13.2) by calculating [Eq. (13.8)]

$$\theta = \arctan \left(-\frac{-0.866}{0.500} \right) = 1.0472$$

and [Eq. (13.9)]

$$C_1 = \sqrt{(0.5)^2 + (-0.866)^2} = 1.00$$

to give

$$y = 1.7 + \cos (\omega_0 t + 1.0472)$$

or alternatively as a sine by using Eq. (13.10)

$$y = 1.7 + \sin (\omega_0 t + 2.618)$$

The foregoing analysis can be extended to the general model

$$y(t) = A_0 + A_1 \cos (\omega_0 t) + B_1 \sin (\omega_0 t) + A_2 \cos (2\omega_0 t)$$
$$+ B_2 \sin (2\omega_0 t) + \cdots + A_m \cos (m\omega_0 t) + B_m \sin (m\omega_0 t)$$

where, for equally spaced data, the coefficients can be evaluated by

$$A_0 = \frac{\sum y}{N}$$

$$\left.\begin{array}{l} A_j = \dfrac{2}{N} \sum y \cos (j\omega_0 t) \\[2em] B_j = \dfrac{2}{N} \sum y \sin (j\omega_0 t) \end{array}\right\} \quad j = 1, 2, \ldots, m$$

Although these relationships can be used to fit data in the regression sense (i.e., $N > 2m + 1$), an alternative application is to employ them for interpolation or collocation. That is, to use them for the case where the number of unknowns, $2m + 1$, is equal to the number of data points, N. This is the approach used in the continuous Fourier series as described next.

13.2 THE CONTINUOUS FOURIER SERIES

In the course of studying heat-flow problems, Fourier showed that an arbitrary periodic function can be represented by an infinite series of sinusoids of harmonically related frequencies. For a function with period T, a continuous Fourier series can be written as[*]

$$f(t) = a_0 + a_1 \cos (\omega_0 t) + b_1 \sin (\omega_0 t) + a_2 \cos (2\omega_0 t)$$
$$+ b_2 \sin (2\omega_0 t) + \cdots$$

or more concisely

$$f(t) = a_0 + \sum_{k=1}^{\infty} [a_k \cos (k\omega_0 t) + b_k \sin (k\omega_0 t)] \tag{13.17}$$

where $\omega_0 = 2\pi/T$ is called the *fundamental frequency* and its constant multiples $2\omega_0$, $3\omega_0$, etc., are called *harmonics*. Thus, Eq. (13.17) expresses $f(t)$ as a linear combination of the basis functions: 1, $\cos (\omega_0 t)$, $\sin (\omega_0 t)$, $\cos (2\omega_0 t)$, $\sin (2\omega_0 t)$, ...

As described in Box 13.1, the coefficients of Eq. (13.17) can be computed via

$$a_k = \frac{2}{T} \int_0^T f(t) \cos (k\omega_0 t)\, dt \tag{13.18}$$

and

$$b_k = \frac{2}{T} \int_0^T f(t) \sin (k\omega_0 t)\, dt \tag{13.19a}$$

for $k = 1, 2, \ldots$ and

[*]The existence of the Fourier series is predicated on the *Dirichlet conditions*. These specify that the periodic function have a finite number of finite maxima and minima and that there be a finite number of jump discontinuities. In general, all physically derived periodic functions satisfy these conditions.

Box 13.1 *Determination of the coefficients of the continuous Fourier series.*

As was done for the discrete data of Sec. 13.1.1, the following relationships can be established.

$$\int_0^T \sin(k\omega_0 t)\, dt = \int_0^T \cos(k\omega_0 t)\, dt = 0 \qquad (B13.1.1)$$

$$\int_0^T \cos(k\omega_0 t) \sin(g\omega_0 t)\, dt = 0 \qquad (B13.1.2)$$

$$\int_0^T \sin(k\omega_0 t) \sin(g\omega_0 t)\, dt = 0 \qquad (B13.1.3)$$

$$\int_0^T \cos(k\omega_0 t) \cos(g\omega_0 t)\, dt = 0 \qquad (B13.1.4)$$

$$\int_0^T \sin^2(k\omega_0 t)\, dt = \int_0^T \cos^2(k\omega_0 t)\, dt = \frac{T}{2} \qquad (B13.1.5)$$

To evaluate its coefficients, each side of Eq. (13.17) can be integrated to give

$$\int_0^T f(t)\, dt = \int_0^T a_0\, dt + \int_0^T \sum_{k=1}^\infty [a_k \cos(k\omega_0 t) + b_k \sin(k\omega_0 t)]\, dt$$

Because every term in the summation is of the form of Eq. (B13.1.1), the equation becomes

$$\int_0^T f(t)\, dt = a_0 T$$

which can be solved for

$$a_0 = \frac{\int_0^t f(t)\, dt}{T}$$

Thus, a_0 is simply the average value of the function over the period.

To evaluate one of the cosine coefficients, for example, a_m, Eq. (13.17) can be multiplied by $\cos(m\omega_0 t)$ and integrated to give

$$\int_0^T f(t) \cos(m\omega_0 t)\, dt = \int_0^T a_0 \cos(m\omega_0 t)\, dt$$

$$+ \int_0^T \sum_{k=1}^\infty a_k \cos(k\omega_0 t) \cos(m\omega_0 t)\, dt \quad (B13.1.6)$$

$$+ \int_0^T \sum_{k=1}^\infty b_k \sin(k\omega_0 t) \cos(m\omega_0 t)\, dt$$

From Eqs. (B13.1.1), (B13.1.2), and (B13.1.4), we see that every term on the right-hand side is zero with the exception of the case where $k = m$. This latter case can be evaluated by Eq. (B13.1.5) and, therefore, Eq. (B13.1.6) can be solved for a_m, or more generally [Eq. (13.18)]

$$a_k = \frac{2}{T} \int_0^T f(t) \cos(k\omega_0 t)\, dt$$

for $k = 1, 2, \ldots$

In a similar fashion Eq. (13.17) can be multiplied by $\sin(m\omega_0 t)$, integrated, and manipulated to yield Eq. (13.19a).

$$a_0 = \frac{1}{T} \int_0^T f(t)\, dt \qquad\qquad (13.19b)$$

EXAMPLE 13.2 Continuous Fourier Series Approximation

Problem Statement: Use the continuous Fourier series to approximate the square or rectangular wave function (Fig. 13.5)

$$f(t) = \begin{cases} -1 & -T/2 < t < -T/4 \\ 1 & -T/4 < t < T/4 \\ -1 & T/4 < t < T/2 \end{cases}$$

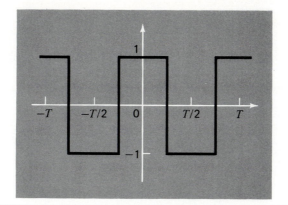

Figure 13.5
A square or rectangular
waveform with a height of
2 and a period $T = 2\pi/\omega_0$.

Solution: Because the average height of the wave is zero, a value of $a_0 = 0$ can be obtained directly. The remaining coefficients can be evaluated as [Eq. (13.18)]

$$a_k = \frac{2}{T}\int_{-T/2}^{T/2} f(t)\cos(k\omega_0 t)\,dt$$

$$= \frac{2}{T}\left[-\int_{-T/2}^{-T/4}\cos(k\omega_0 t)\,dt + \int_{-T/4}^{T/4}\cos(k\omega_0 t)\,dt - \int_{T/4}^{T/2}\cos(k\omega_0 t)\,dt\right]$$

The integrals can be evaluated to give

$$a_k = \begin{cases} \dfrac{4}{k\pi} & \text{for } k = 1, 5, 9, \ldots \\[2mm] -\dfrac{4}{k\pi} & \text{for } k = 3, 7, 11, \ldots \\[2mm] 0 & \text{for } k = \text{even integers} \end{cases}$$

Similarly, it can be determined that all the b's $= 0$. Therefore, the Fourier series approximation is

$$f(t) = \frac{4}{\pi}\cos(\omega_0 t) - \frac{4}{3\pi}\cos(3\omega_0 t) + \frac{4}{5\pi}\cos(5\omega_0 t) - \frac{4}{7\pi}\cos(7\omega_0 t) + \cdots$$

The results up to the first three terms are shown in Fig. 13.6.

It should be mentioned that the square wave in Fig. 13.5 is called an *even function* because $f(t) = f(-t)$. Another example of an even function is cos (t). It can be shown (Van Valkenburg, 1974) that the b's in the Fourier series always equal zero for even functions. Note also that *odd functions* are those for which $f(t) = -f(-t)$. The function sin (t) is an odd function. For this case, the a's will equal zero.

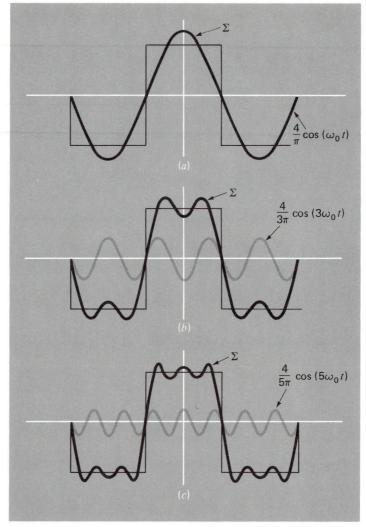

Figure 13.6
The Fourier series approximation of the square wave from Fig. 13.5. The series of plots shows the summation up to and including the (a) first, (b) second, and (c) third terms. The individual terms that were added at each stage are also shown.

Aside from the trigonometric format of Eq. (13.17), the Fourier series can be expressed in terms of exponential functions as (see Box 13.2 and App. B)

$$f(t) = \sum_{k=-\infty}^{\infty} \tilde{c}_k e^{ik\omega_0 t} \tag{13.20}$$

Box 13.2   Complex Form of the Fourier Series

The trigonometric form of the continuous Fourier series is

$$f(t) = a_0 + \sum_{k=1}^{\infty} [a_k \cos (k\omega_0 t) + b_k \sin (k\omega_0 t)] \quad \text{(B13.2.1)}$$

From Euler's identity, the sine and cosine can be expressed in exponential form as

$$\sin x = \frac{e^{ix} - e^{-ix}}{2i} \quad \text{(B13.2.2)}$$

$$\cos x = \frac{e^{ix} + e^{-ix}}{2} \quad \text{(B13.2.3)}$$

which can be substituted into Eq. (B13.2.1) to give

$$f(t) = a_0 + \sum_{k=1}^{\infty} \left(e^{ik\omega_0 t}\frac{a_k - ib_k}{2} + e^{-ik\omega_0 t}\frac{a_k + ib_k}{2} \right) \quad \text{(B13.2.4)}$$

because $1/i = -i$. We can define a number of constants

$$\tilde{c}_0 = a_0$$

$$\tilde{c}_k = \frac{a_k - ib_k}{2} \quad \text{(B13.2.5)}$$

$$\tilde{c}_{-k} = \frac{a_{-k} - ib_{-k}}{2} = \frac{a_k + ib_k}{2}$$

where because of the odd and even properties of the sine and cosine, $a_k = a_{-k}$ and $b_k = -b_{-k}$ Equation (B13.2.4) can therefore, be reexpressed as

$$f(t) = \tilde{c}_0 + \sum_{k=1}^{\infty} \tilde{c}_k e^{ik\omega_0 t} + \sum_{k=1}^{\infty} \tilde{c}_{-k} e^{-ik\omega_0 t}$$

or

$$f(t) = \sum_{k=0}^{\infty} \tilde{c}_k e^{ik\omega_0 t} + \sum_{k=1}^{\infty} \tilde{c}_{-k} e^{-ik\omega_0 t}$$

To simplify further, instead of summing the second series from 1 to ∞, perform the sum from -1 to $-\infty$

$$f(t) = \sum_{k=0}^{\infty} \tilde{c}_k e^{ik\omega_0 t} + \sum_{k=-1}^{-\infty} \tilde{c}_k e^{ik\omega_0 t}$$

or

$$f(t) = \sum_{k=-\infty}^{\infty} \tilde{c}_k e^{ik\omega_0 t} \quad \text{(B13.2.6)}$$

where the summation includes a term for $k = 0$.

To evaluate the $\tilde{c}_k$'s, Eqs. (13.18) and (13.19a) can be substituted into Eq. (B13.2.5) to yield

$$\tilde{c}_k = \frac{1}{T}\int_{-T/2}^{T/2} f(t) \cos (k\omega_0 t)\, dt - i\frac{1}{T}\int_{-T/2}^{T/2} f(t) \sin (k\omega_0 t)\, dt$$

Employing Eqs. (B13.2.2) and (B13.2.3) and simplifying gives

$$\tilde{c}_k = \frac{1}{T}\int_{-T/2}^{T/2} f(t)\, e^{-ik\omega_0 t}\, dt \quad \text{(B13.2.7)}$$

Therefore, Eqs. (B13.2.6) and (B13.2.7) are the complex versions of Eqs. (13.17) through (13.19b). Note that App. B includes a summary of the interrelationships among all the formats of the Fourier series introduced in this chapter.

where

$$\tilde{c}_k = \frac{1}{T}\int_{-T/2}^{T/2} f(t) e^{-ik\omega_0 t}\, dt \quad \text{(13.21)}$$

This alternative formulation will have utility throughout the remainder of the chapter.

13.3   FREQUENCY AND TIME DOMAINS

To this point, our discussion of Fourier approximation has been limited to the *time domain*. We have done this because most of us are fairly comfortable conceptualizing

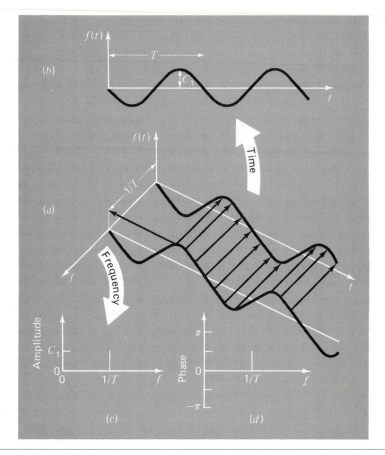

Figure 13.7

(a) A depiction of how a sinusoid can be portrayed in the time and the frequency domains. The time projection is reproduced in (b), whereas the amplitude-frequency projection is reproduced in (c). The phase-frequency projection is shown in (d).

a function's behavior in this dimension. Although it is not as familiar, the *frequency domain* provides an alternative perspective for characterizing the behavior of oscillating functions.

Thus, just as amplitude can be plotted versus time so also can it be plotted versus frequency. Both types of expression are depicted in Fig. 13.7a where we have drawn a three-dimensional graph of a sinusoidal function

$$f(t) = C_1 \cos \left(t + \frac{\pi}{2} \right)$$

In this plot, the magnitude or amplitude of the curve, $f(t)$, is the dependent variable and time t and frequency $f = \omega_0/2\pi$ are the independent variables. Thus, the amplitude and the time axes form a *time plane* and the amplitude and the frequency axes form a *frequency plane*. The sinusoid can, therefore, be conceived of as existing a distance $1/T$ out along the frequency axis and running parallel to the time axes. Consequently, when we speak about the behavior of the sinusoid in the time domain, we mean the projection of the curve onto the time plane (Fig. 13.7b). Similarly, the behavior in the frequency domain is merely its projection onto the frequency plane.

As in Fig. 13.7*c*, this projection is a measure of the sinusoid's maximum positive amplitude C_1. The full peak-to-peak swing is unnecessary because of the symmetry. Together with the location $1/T$ along the frequency axis, Fig. 13.7*c* now defines the amplitude and frequency of the sinusoid. This is enough information to reproduce the shape and size of the curve in the time domain. However, one more parameter, namely the phase angle, is required to position the curve relative to $t = 0$. Consequently, a phase diagram, as shown in Fig. 13.7*d*, must also be included. The phase angle is determined as the distance (in radians) from zero to the point at which the positive peak occurs. If the peak occurs after zero, it is said to be delayed (recall our discussion of lags and leads in Sec. 13.1) and, by convention, the phase angle is given a negative sign. Conversely, a peak before zero is said to be advanced and the phase angle is positive. Thus, for Fig. 13.7, the peak leads zero and the phase angle is plotted as $+\pi/2$. Figure 13.8 depicts some other possibilities.

We can now see that Figs. 13.7*c* and *d* provide an alternative way to present or summarize the pertinent features of the sinusoid in Fig. 13.7*a*. They are referred to as *line spectra*. Admittedly, for a single sinusoid they are not very interesting. However, when applied to a more complicated situation, say a Fourier series, their true power and value is revealed. For example, Fig. 13.9 shows the amplitude and phase line spectra for the square-wave function from Example 13.2.

Figure 13.8
Various phases of a sinusoid showing the associated phase line spectra.

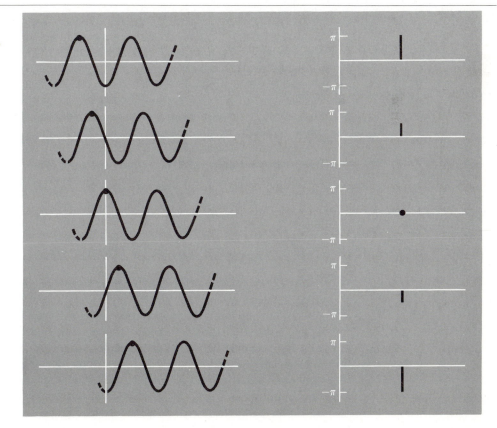

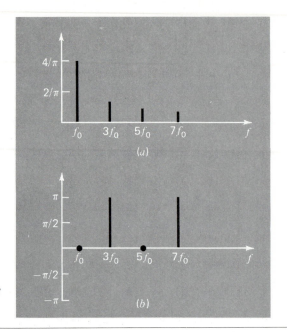

Figure 13.9
The (a) amplitude and (b) phase line spectra for the square wave from Fig. 13.5.

Such spectra provide information that would not be apparent from the time domain. This can be seen by contrasting Figs. 13.6 and 13.9. Figure 13.6 presents two alternative time-domain perspectives. The first, the original square wave, tells us nothing about the sinusoids that comprise it. The alternative is to display these sinusoids—that is, $(4/\pi)$ cos $(\omega_0 t)$, $-(4/3\pi)$ cos $(3\omega_0 t)$, $(4/5\pi)$ cos $(5\omega_0 t)$, etc. This alternative does not provide an adequate visualization of the structure of these harmonics. In contrast, Figs. 13.9a and b provide a graphic display of this structure. As such, the line spectra represent "fingerprints" that can help us to characterize and understand a complicated waveform. They are particularly valuable for nonidealized cases where they sometimes allow us to discern structure in otherwise obscure signals. In the next section, we will describe the Fourier transform that will allow us to extend such analyses to nonperiodic waveforms.

13.4   FOURIER INTEGRAL AND TRANSFORM

Although the Fourier series is a useful tool for investigating the spectrum of a periodic function, there are many waveforms that do not repeat themselves regularly. For example, a lightning bolt occurs only once (or at least it will be a long time until it occurs again), but it will cause interference with receivers operating on a broad range of frequencies—for example, TVs, radios, shortwave receivers, etc. Such evidence suggests that a nonrecurring signal such as that produced by lightning exhibits a continuous frequency spectrum. Because such phenomena are of great interest to engineers, an alternative to the Fourier series would be valuable for analyzing these aperiodic waveforms.

The *Fourier integral* is the primary tool that is available for this purpose. It can be derived from the exponential form of the Fourier series

$$f(t) = \sum_{k=-\infty}^{\infty} \tilde{c}_k e^{ik\omega_0 t} \tag{13.22}$$

where

$$\tilde{c}_k = \frac{1}{T} \int_{-T/2}^{T/2} f(t) e^{-ik\omega_0 t} \, dt \tag{13.23}$$

where $\omega_0 = 2\pi/T$ and $k = 0, 1, 2, \ldots$.

The transition from a periodic to a nonperiodic function can be effected by allowing the period to approach infinity. In other words, as T becomes infinite, the function never repeats itself and thus becomes aperiodic. If this is allowed to occur, it can be demonstrated (for example, Van Valkenburg, 1974, Hayt and Kemmerly, 1986) that the Fourier series reduces to

$$f(t) = \frac{1}{2\pi} \int_{-\infty}^{\infty} F(i\omega_0) e^{i\omega_0 t} \, d\omega_0 \tag{13.24}$$

and the coefficients become a continuous function of the frequency variable ω as in

$$F(i\omega_0) = \int_{-\infty}^{\infty} f(t) e^{-i\omega_0 t} \, dt \tag{13.25}$$

The function $F(i\omega_0)$, as defined by Eq. (13.25), is called the *Fourier integral* of $f(t)$. In addition, Eqs. (13.24) and (13.25) are collectively referred to as the *Fourier transform pair*. Thus, along with being called the Fourier integral, $F(i\omega_0)$ is also called the *Fourier transform* of $f(t)$. In the same spirit, $f(t)$, as defined by Eq. (13.24), is referred to as the *inverse Fourier transform* of $F(i\omega_0)$. Thus, the pair allows us to transform back and forth between the time and the frequency domains for an aperiodic signal.

The distinction between the Fourier series and transform should now be quite clear. The major difference is that each applies to a different class of function—the series to periodic and the transform to nonperiodic waveforms. Beyond this major distinction, the two approaches differ in how they move between the time and the frequency domains. The Fourier series converts a continuous, periodic time-domain function to frequency-domain magnitudes at discrete frequencies. In contrast, the Fourier transform converts a continuous time-domain function to a continuous frequency-domain function. Thus, the discrete frequency spectrum generated by the Fourier series is analogous to a continuous frequency spectrum generated by the Fourier transform.

The shift from a discrete to a continuous spectrum can be illustrated graphically. In Fig. 13.10a, we can see a pulse train of rectangular waves with pulse widths equal to one-half the period along with its associated discrete spectrum. This is the same

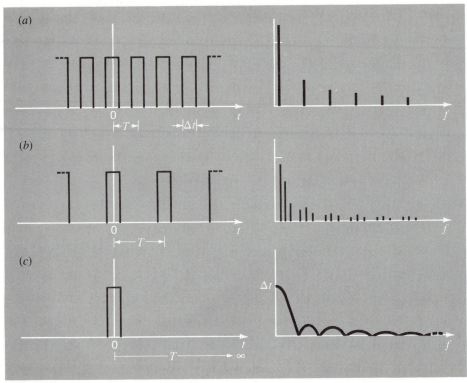

Figure 13.10
Illustration of how the discrete frequency spectrum of a Fourier series for a pulse train (*a*) approaches a continuous frequency spectrum of a Fourier integral (*c*) as the period is allowed to approach infinity.

function as was investigated previously in Example 13.2 with the exception that it is shifted vertically.

In Fig. 13.10*b* a doubling of the pulse train's period has two effects on the spectrum. First, two additional frequency lines are added on either side of the original components. Second, the amplitudes of the components are reduced.

As the period is allowed to approach infinity, these effects continue as more and more spectral lines are packed together until the spacing between lines goes to zero. At the limit, the series converges on the continuous Fourier integral which is depicted in Fig. 13.10*c*.

Now that we have introduced a way to analyze an aperiodic signal, we will take the final step in our development. In the next section, we will acknowledge the fact that a signal is rarely characterized as a continuous function of the sort needed to implement Eq. (13.25). Rather, the data is invariably in a discrete form. Thus, we will now show how to compute a Fourier transform for such discrete measurements.

13.5 DISCRETE FOURIER TRANSFORM (DFT)

In engineering, functions are often represented by finite sets of discrete values. Additionally, data is often collected in or converted to such a discrete format. As depicted in Fig. 13.11, an interval from 0 to T can be divided into N equispaced subintervals with widths of $\Delta t = T/N$. The subscript n is employed to designate the discrete times at which samples are taken. Thus, f_n designates a value of the continuous function $f(t)$ taken at t_n.

Note that the data points are specified at $n = 0, 1, 2, \ldots, N-1$. A value is not included at $n = N$. (See Ramirez 1985 for the rationale for excluding f_N.)

For the system in Fig. 13.11, a discrete Fourier transform can be written as

$$F_k = \sum_{n=0}^{N-1} f_n e^{-ik\omega_0 n} \qquad \text{for } k = 0 \text{ to } N-1 \tag{13.26}$$

and the inverse Fourier transform as

$$f_n = \frac{1}{N} \sum_{k=0}^{N-1} F_k e^{ik\omega_0 n} \qquad \text{for } n = 0 \text{ to } N-1 \tag{13.27}$$

where $\omega_0 = 2\pi/N$.

Equations (13.26) and (13.27) represent the discrete analogs of Eqs. (13.25) and (13.24), respectively. As such, they can be employed to compute both a direct and an inverse Fourier transform for discrete data. Although such calculations can be performed by hand, they are extremely arduous. As expressed by Eq. (13.26), the DFT requires

Figure 13.11

The sampling points of the discrete Fourier series.

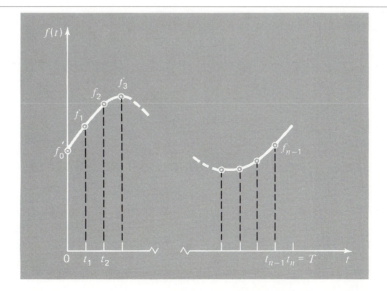

N^2 complex operations. Thus, we will now develop a computer algorithm to implement the DFT.

Computer Algorithm for the DFT. Note that the factor $1/N$ in Eq. (13.27) is merely a scale factor that can be included in either Eq. (13.26) or (13.27) but not both. For our computer algorithm, we will shift it to Eq. (13.26) so that the first coefficient F_0 (which is the analog of the continuous coefficient a_0) is equal to the arithmetic mean of the samples. Also, in order to develop an algorithm that can be implemented in languages that do not accommodate complex variables, we can use Euler's identity

$$e^{\pm ia} = \cos a \pm i \sin a$$

to reexpress Eqs. (13.26) and (13.27) as

$$F_k = \frac{1}{N} \sum_{n=0}^{N-1} [f_n \cos (k\omega_0 n) - i f_n \sin (k\omega_0 n)] \tag{13.28}$$

and

$$f_n = \sum_{k=0}^{N-1} [F_k \cos (k\omega_0 n) + i F_k \sin (k\omega_0 n)] \tag{13.29}$$

Pseudocode to implement Eq. (13.28) is listed in Fig. 13.12. This algorithm can be developed into a computer program to compute the DFT. The output from such a program is listed in Fig. 13.13 for the analysis of a cosine function. Note that additional information on interpreting Fig 13.13 and DFTs in general is included in Case Study 14.4.

Figure 13.12
Pseudocode for computing the DFT.

```
DOFOR k = 0 to N - 1
    DOFOR n = 0 to N - 1
        angle = kω₀n
        real_k = real_k + f_n cos(angle)/N
        imaginary_k = imaginary_k - f_n sin(angle)/N
    ENDDO
ENDDO
```

13.6   THE FAST FOURIER TRANSFORM (FFT)

Although the algorithm described in the previous section adequately calculates the DFT, it is computationally burdensome because N^2 operations are required. Consequently, for data samples of even moderate size, the direct determination of the DFT can be extremely time-consuming.

The *fast Fourier transform,* or *FFT,* is an algorithm that has been developed to compute the DFT in an extremely economical fashion. Its speed stems from the fact that

INDEX	f(t)	REAL	IMAGINARY
0	1.000	0.0000	0.0000
1	0.707	0.0000	0.0000
2	-0.000	0.0000	-0.0000
3	-0.707	0.0000	-0.0000
4	-1.000	0.5000	0.0000
5	-0.707	0.0000	-0.0000
6	0.000	0.0000	0.0000
7	0.707	0.0000	0.0000
8	1.000	-0.0000	0.0000
9	0.707	0.0000	0.0000
10	-0.000	-0.0000	-0.0000
11	-0.707	0.0000	0.0000
12	-1.000	-0.0000	0.0000
13	-0.707	0.0000	-0.0000
14	0.000	0.0000	0.0000
15	0.707	0.0000	-0.0000
16	1.000	-0.0000	0.0000
17	0.707	0.0000	0.0000
18	-0.000	-0.0000	0.0000
19	-0.707	-0.0000	-0.0000
20	-1.000	-0.0000	0.0000
21	-0.707	0.0000	0.0000
22	0.000	0.0000	0.0000
23	0.707	-0.0000	-0.0000
24	1.000	-0.0000	0.0000
25	0.707	-0.0000	0.0000
26	-0.000	-0.0000	0.0000
27	-0.707	-0.0000	-0.0000
28	-1.000	0.5000	-0.0000
29	-0.707	0.0000	-0.0000
30	0.000	0.0000	0.0000
31	0.707	0.0000	0.0000

Figure 13.13

Output of a program based on the algorithm from Fig. 13.12 for the DFT of data generated by a cosine function, $f(t) = \cos[2\pi(12.5)t]$ at 32 points with $\Delta t = 0.01$ s.

it utilizes the results of previous computations to reduce the number of operations. In particular, it exploits the periodicity and symmetry of trigonometric functions to compute the transform with approximately $N \log_2 N$ operations (Fig. 13.14). Thus, for $N = 50$ samples, the FFT is about 10 times faster than the standard DFT. For $N = 1000$, it is about 100 times faster.

The first FFT algorithm was developed by Gauss in the early nineteenth century (Heideman et al., 1984). Other major contributions were made by Runge, Danielson, and Lanczos and others in the early twentieth century. However, because discrete transforms often took days to weeks to calculate by hand, they did not attract broad interest prior to the development of the modern digital computer.

In 1965, J. W. Cooley and J. W. Tukey published a key paper in which they outlined an algorithm for calculating the FFT. This scheme, which is similar to those of Gauss and other earlier investigators, is called the *Cooley-Tukey algorithm*. Today, there are a host of other approaches that are offshoots of this method.

The basic idea behind each of these algorithms is that a DFT of length N is

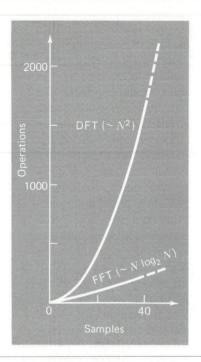

Figure 13.14
Plot of number of operations vs. sample size for the standard DFT and the FFT.

decomposed or "decimated" into successively smaller DFTs. There are a variety of different ways to implement this principle. For example, the Cooley-Tukey algorithm is a member of what are called *decimation-in-time* techniques. In the present section, we will describe an alternative approach called the *Sande-Tukey algorithm*. This method is a member of another class of algorithms called *decimation-in-frequency* techniques. The distinction between the two classes will be discussed after we have elaborated on the method.

13.6.1 The Sande-Tukey Algorithm.

In the present case, N will be assumed to be an integral power of 2,

$$N = 2^M \tag{13.30}$$

where M is an integer. This constraint is introduced to simplify the resulting algorithm. Now, recall that the DFT can be generally represented as

$$F_k = \sum_{n=0}^{N-1} f_n e^{-i(2\pi/N)nk} \qquad \text{for } k = 0, 1, 2, \ldots, N-1 \tag{13.31}$$

where $2\pi/N = \omega_0$. Equation (13.31) can also be expressed as

$$F_k = \sum_{n=0}^{N-1} f_n W^{nk}$$

where W is a complex-valued weighting function defined as

$$W = e^{-i(2\pi/N)} \tag{13.32}$$

Suppose now that we divide the sample in half and express Eq. (13.31) in terms of the first and last $N/2$ points:

$$F_k = \sum_{n=0}^{(N/2)-1} f_n e^{-i(2\pi/N)kn} + \sum_{n=N/2}^{N-1} f_n e^{-i(2\pi/N)kn}$$

where $k = 0, 1, 2, \ldots, N-1$. A new variable, $m = n - N/2$, can be created so that the range of the second summation is consistent with the first,

$$F_k = \sum_{n=0}^{(N/2)-1} f_n e^{-i(2\pi/N)kn} + \sum_{m=0}^{(N/2)-1} f_{m+N/2} e^{-i(2\pi/N)k(m+N/2)}$$

or

$$F_k = \sum_{n=0}^{(N/2)-1} (f_n + e^{-i\pi k} f_{n+N/2}) e^{-i2\pi kn/N} \tag{13.33}$$

Next, recognize that the factor $e^{-i\pi k} = (-1)^k$. Thus, for even points it is equal to 1 and for odd points it is equal to -1. Therefore, the next step in the method is to separate Eq. (13.33) according to even values and odd values of k. For the even values

$$F_{2k} = \sum_{n=0}^{(N/2)-1} (f_n + f_{n+N/2}) e^{-i2\pi(2k)n/N}$$

$$= \sum_{n=0}^{(N/2)-1} (f_n + f_{n+N/2}) e^{-i2\pi kn/(N/2)}$$

and for the odd values

$$F_{2k+1} = \sum_{n=0}^{(N/2)-1} (f_n - f_{n+N/2}) e^{-i2\pi(2k+1)n/N}$$

$$= \sum_{n=0}^{(N/2)-1} (f_n - f_{n+N/2}) e^{-i2\pi n/N} e^{-i2\pi kn/(N/2)}$$

for $k = 0, 1, 2, \ldots, (N/2) - 1$.

These equations can also be expressed in terms of Eq. (13.32). For the even values

$$F_{2k} = \sum_{n=0}^{(N/2)-1} (f_n + f_{n+N/2}) W^{2kn}$$

and for the odd values

$$F_{2k+1} = \sum_{n=0}^{(N/2)-1} (f_n - f_{n+N/2}) W^n W^{2kn}$$

Now, a key insight can be made. These even and odd expressions can be interpreted as being equal to the transforms of the $N/2$-length sequences

$$g_n = f_n + f_{n+N/2} \tag{13.34}$$

and

$$h_n = (f_n - f_{n+N/2}) W^n \tag{13.35}$$

for $n = 0, 1, 2, \ldots, (N/2) - 1$. Thus, it directly follows that

$$F_{2k} = G_k$$

and

$$F_{2k+1} = H_k$$

for $k = 0, 1, 2, \ldots, (N/2) - 1$.

In other words, one N-point computation has been replaced by two $(N/2)$-point computations. Because each of the latter requires approximately $(N/2)^2$ complex multiplications and additions, the approach produces a factor-of-2 savings—that is, N^2 versus $2(N/2)^2 = N^2/2$.

The scheme is depicted in Fig. 13.15 for $N = 8$. The DFT is computed by first forming the sequence g_n and h_n and then computing the $N/2$ DFTs to obtain the even- and odd-numbered transforms. The weights W^n are sometimes called *twiddle factors*.

Figure 13.15
Flow graph of the first stage in a decimation-in-frequency decomposition of an N-point DFT into two $N/2$-point DFTs for $N = 8$.

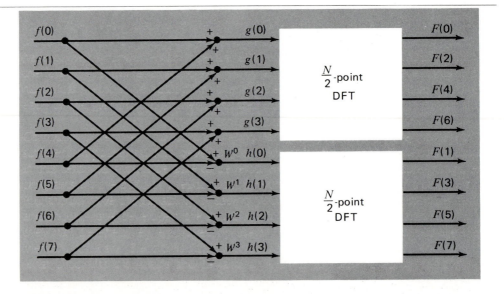

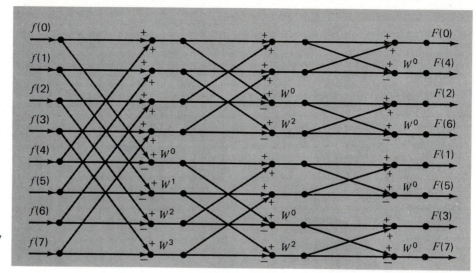

Figure 13.16
Flow graph of the complete decimation-in-frequency decomposition of an 8-point DFT.

Now it is clear that this "divide-and-conquer" approach can be repeated at the second stage. Thus, we can compute the $(N/4)$-point DFTs of the four $N/4$ sequences composed of the first and last $N/4$ points of Eqs. (13.34) and (13.35).

The strategy is continued to its inevitable conclusion when $N/2$ two-point DFTs are computed (Fig. 13.16). The total number of calculations for the entire computation is on the order of $N \log_2 N$. The contrast between this level of effort and that of the standard DFT (Fig. 13.14) illustrates why the FFT is so important.

Computer Algorithm. It is a relatively straightforward proposition to express Fig. 13.16 as an algorithm. As was the case for the DFT algorithm of Fig. 13.12, we will use Euler's identity

$$e^{\pm ia} = \cos a \pm i \sin a$$

to allow the algorithm to be implemented in languages that do not explicitly accommodate complex variables.

Close inspection of Fig. 13.16 indicates that its fundamental computational molecule is the so-called *butterfly network* depicted in Fig. 13.17a. Pseudocode to implement one of these molecules is shown in Fig. 13.17b.

Figure 13.17
(a) A butterfly network that represents the fundamental computation of Fig. 13.16. (b) Pseudocode to implement (a).

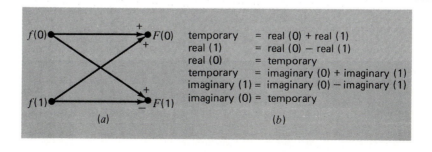

Pseudocode for the FFT is listed in Fig. 13.18. The first part essentially consists of three nested loops to implement the computation embodied in Fig. 13.16. Note that the real-valued data is originally stored in the array x. Also note that the outer loop steps through the M stages [recall Eq. (13.30)] of the flowgraph.

(a)
```
N2 = N
DOFOR k = 1 to M
    N1 = N2
    N2 = N2/2
    angle = 0
    argument = 2π/N1
    DOFOR j = 0 to N2-1
        c = cos(angle)
        s = -sin(angle)
        DOFOR i = j to N-1 step N1
            l = i + N2
            x_t = x_i - x_l
            x_i = x_i + x_l
            y_t = y_i - y_l
            y_i = y_i + y_l
            x_l = x_t c - y_t s
            y_l = y_t c + x_t s
        ENDDO
        angle = (j + 1)·argument
    ENDDO
ENDDO
```

(b)
```
j = 0
DOFOR i = 0 to N-2
    IF (i < j)
        x_t = x_j
        x_j = x_i
        x_i = x_t
        y_t = y_j
        y_j = y_i
        y_i = y_t
    ELSE ENDIF
    k = N/2
    DOWHILE (k < j+1)
        j = j - k
        k = k/2
    ENDDO
    j = j + k
ENDDO
DOFOR i = 0 to N-1
    x_i = x_i /N
    y_i = y_i /N
ENDDO
```

Figure 13.18
Pseudocode to implement a decimation-in-frequency FFT. Note that the pseudocode is composed of two parts: (a) the FFT itself and (b) a bit-reversal routine to unscramble the order of the resulting Fourier coefficients.

After this first part is executed, the DFT will have been computed but in a scrambled order (see the right-hand side of Fig. 13.16). These Fourier coefficients can be unscrambled by a procedure called *bit reversal*. If the subscripts 0 through 7 are expressed in binary, the correct ordering can be obtained by reversing these bits (Fig. 13.19). The second part of the algorithm implements this procedure.

13.6.2 The Cooley-Tukey Algorithm

Figure 13.20 shows a flow network to implement the Cooley-Tukey algorithm. For this case, the sample is initially divided into odd- and even-numbered points, and the final results are in correct order.

This approach is called a *decimation in time*. It is the reverse of the Sande-Tukey algorithm described in the previous section. Although the two classes of method differ in organization, they both exhibit the $N \log_2 N$ operations that are the strength of the FFT approach.

Scrambled order (decimal)		Scrambled order (binary)		Bit-reversed order (binary)		Final result (decimal)
$F(0)$		$F(000)$		$F(000)$		$F(0)$
$F(4)$		$F(100)$		$F(001)$		$F(1)$
$F(2)$		$F(010)$		$F(010)$		$F(2)$
$F(6)$	$\Rightarrow$	$F(110)$	$\Rightarrow$	$F(011)$	$\Rightarrow$	$F(3)$
$F(1)$		$F(001)$		$F(100)$		$F(4)$
$F(5)$		$F(101)$		$F(101)$		$F(5)$
$F(3)$		$F(011)$		$F(110)$		$F(6)$
$F(7)$		$F(111)$		$F(111)$		$F(7)$

FIGURE 13.19
Depiction of the bit-reversal process.

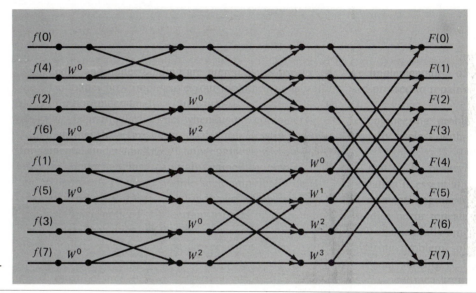

Figure 13.20
Flowgraph of a decimation-in-time FFT of an 8-point DFT.

Additional Information. The foregoing has been a brief introduction to Fourier approximation and the FFT. Additional information on the former can be found in Van Valkenburg (1974), Chirlian (1969), and Hayt and Kemmerly (1986). References on the FFT include Davis and Rabinowitz (1975), Cooley, Lewis, and Welch (1977), and Brigham (1974). Nice introductions to both can be found in Ramirez (1985), Oppenheim and Schafer (1975), and Gabel and Roberts (1987). Finally, we have devoted Case Study 14.4 to an application of the methods introduced in this chapter.

PROBLEMS

Hand Calculations

13.1 Use least-squares regression to fit Eq. (13.11) to

t	0	2	4	5	7	8.5	12	15	19	20	22	24
y	11	13	20.4	21	17	16	14	11	11.5	9.2	10.7	10.2

13.2 The solar radiation for Grand Junction, Colorado, in 1971 has been tabulated as

Time, month	J	F	M	A	M	J	J	A	S	O	N	D
Radiation, langley	227	315	448	517	609	630	—	—	531	361	239	185

Assuming each month is 30 days long, fit a sinusoid to this data. Use the resulting equation to predict the radiation in mid-August.

13.3 The average values of a function can be determined by

$$\overline{f(x)} = \frac{\int_0^x f(x)\, dx}{x}$$

Use this relationship to verify the results of Eq. (13.13).

13.4 Use a continuous Fourier series to approximate the sawtooth wave in Fig. P13.4. Plot the first three terms.

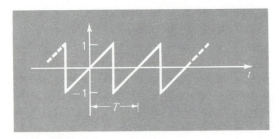

Figure P13.4
A sawtooth wave.

13.5 Use a continuous Fourier series to approximate the half-wave rectifier in Fig. P13.5. Plot the first three terms.

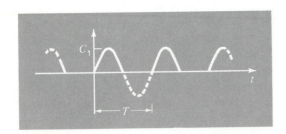

Figure P13.5
\ half-wave rectifier.

13.6 Construct amplitude and phase line spectra for Prob. 13.4.

13.7 Construct amplitude and phase line spectra for Prob. 13.5.

13.8 A triangular wave can be characterized by

$$y(t) = \frac{8C_1}{\pi^2} \left[\cos(\omega_0 t) + \frac{1}{9} \cos(3\omega_0 t) + \frac{1}{25} \cos(5\omega_0 t) + \cdots \right]$$

where C_1 is the amplitude of the wave. Plot the first three terms.

13.9 Construct amplitude and phase line spectra for Prob. 13.8.

Computer-Related Problems

13.10 Develop a user-friendly computer program for the DFT based on the algorithm from Fig. 13.12. Test it by duplicating Fig. 13.13.

13.11 Use the program from Prob. 13.10 to compute a DFT for the triangular wave from Prob. 13.8. Sample the wave from $t = 0$ to $4T$. Use 32, 64, and 128 sample points. Time each run and plot execution versus N to verify Fig. 13.14.

13.12 Develop a user-friendly program for the FFT based on the algorithm from Fig. 13.18. Test it by duplicating Fig. 13.13.

13.13 Repeat Prob. 13.11 using the software you developed in Prob. 13.12.

CHAPTER 14
Case Studies: Curve Fitting

The purpose of this chapter is to use the numerical methods for curve fitting to solve some engineering problems. As with the other case study chapters, the first example is taken from the general area of engineering economics and management. This is followed by case studies from the four major fields of engineering: chemical, civil, electrical, and mechanical. Finally, a case study is included on curve fitting using spreadsheets.

Case Study 14.1 deals with a trend analysis of computer sales data. The example illustrates two important points regarding curve fitting: (1) interpolating polynomials are ill-suited for fitting imprecise data and (2) extrapolation is a particularly tenuous proposition when the cause-effect relationships underlying a trend are unknown.

Case Study 14.2, which is taken from chemical engineering, demonstrates how a nonlinear model can be linearized and fit to data using linear regression. *Case Study 14.3* uses a similar approach but also employs polynomial interpolation in order to determine the stress-strain relationship for a civil engineering structures problem.

Case Study 14.4 illustrates how a fast Fourier transform can be employed to analyze a signal by determining its harmonics. *Case Study 14.5* demonstrates how multiple linear regression is used to analyze experimental data for a fluids problem taken from mechanical engineering. Finally, *Case Study 14.6* illustrates how a spreadsheet can be employed to interpolate with Newton's polynomial.

CASE STUDY 14.1   ENGINEERING PRODUCT SALES MODEL (GENERAL ENGINEERING)

Background: Engineers concerned with the design and manufacture of products such as automobiles, TV sets, and computers may become involved with many other aspects of the business. These involvements include sales, marketing, and distribution of the product.

You are an engineer working for the Ultimate Computer Company (see Case Study 6.1). Planning and resource allocation considerations require that you be able to predict how long your computers will stay on the market as a function of time. In this case study, you are provided with measured data that describes the number of computers on the market for various times up to 60 days (Fig. 14.1). You are asked to examine this data and, using extrapolation techniques, estimate how many computers will be available at day 90. The data is listed in Table 14.1.

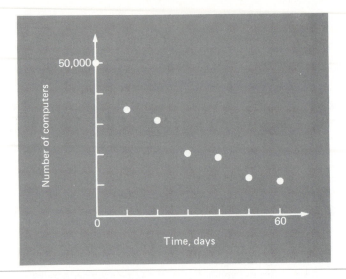

Figure 14.1
Number of computers on the
market versus time.

TABLE 14.1 **Number of computers on the market as a function of time.**

Time, days	Number of Computers on the Market
0	50,000
10	35,000
20	31,000
30	20,000
40	19,000
50	12,050
60	11,000

Solution: This trend analysis and extrapolation problem will be addressed using first-through sixth-order interpolating polynomials as well as with first- through sixth-order polynomial regression. The resulting curves will be used to make predictions at days 55, 65, and 90 in order to contrast interpolation and extrapolation.

Inspection of Fig. 14.1 indicates that the data is not smooth. Although the number of computers decreases with time, the rate of decrease varies from interval to interval in a seemingly random fashion. Thus, even before the analysis begins, we might suspect that extrapolation of this data could pose difficulties.

The results in Table 14.2 bear this suspicion out. Notice that there is a great deal of discrepancy between the predictions using the various techniques. In order to quantify this discrepancy, we have computed a mean, a standard deviation, and a coefficient of variation for the predictions. The coefficient of variation, which is the mean divided by the standard deviation (multiplied by 100 percent), provides a relative measure of the variability for each set of predictions [Eq. (PT4.5)]. Notice how the coefficient of variation is lowest for the interpolated value at day 55. Also, notice how the highest discrepancy results for day 90, which represents the farthest extrapolation.

TABLE 14.2 Results of fitting various interpolating and least-squares polynomials to the data in Table 14.1. An interpolation at t = 55 and extrapolations at t = 65 and 90 are shown. Notice that because the normal equations are ill-conditioned, the sixth-order least-squares polynomial differs from the more precise interpolating polynomial (recall Sec. 11.2.1).

	Interpolation	Extrapolation	
	$t = 55$	$t = 65$	$t = 90$
Interpolating polynomials			
First-order	11,525	10,475	7,850
Second-order	10,788	12,688	43,230
Third-order	10,047	16,391	161,750
Fourth-order	8,961	23,992	578,750
Fifth-order	7,300	38,942	1,854,500
Sixth-order	4,660	67,975	5,458,100
Mean	8,880	28,411	1,350,700
Standard deviation	2,542	21,951	2,128,226
Coefficient of variation	29%	77%	157%
Least-squares polynomials			
First-order	9,820	3,573	−12,045
Second-order	11,829	10,939	16,529
Third-order	12,040	8,872	−3,046
Fourth-order	11,101	12,733	83,104
Fifth-order	11,768	9,366	−78,906
Sixth-order	4,261	71,266	5,768,460
Mean	10,203	18,910	910,623
Standard deviation	2,834	24,233	2,228,408
Coefficient of variation	28%	128%	245%

In addition, the results for the interpolating polynomials deteriorate as the order is increased, to the point that the sixth-order case yields the ridiculous prediction that 5,458,100 computers would be available on day 90. The reason for this poor performance is illustrated by Fig. 14.2, which shows the sixth-order polynomial. Because the trend suggested by the data is not smooth, the higher-order polynomial loops up and down in order to intersect each point. These "oscillations" can lead to both spurious interpolations and extrapolations of the type manifested in Fig. 14.2.

Because it is not constrained to pass through every point, regression is sometimes useful in remedying this situation. Figure 14.3, which shows the quadratic and cubic regression results, suggests that this is true for lower-order regression. Within the range of the data ($t = 0$ to 60 days), the results of the two regressions yield fairly consistent results. However, when they are extrapolated beyond this range, the predictions diverge. At $t = 90$, the second-order regression yields the absurd result that the number of computers has increased, whereas the third-order version yields the equally ridiculous projection that there will be a negative number of computers.

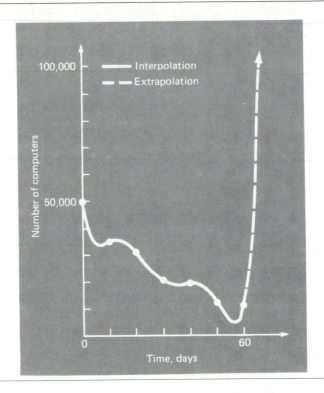

Figure 14.2
Plot of the sixth-order interpolating polynomial used to interpolate and extrapolate computer sales data.

The primary reason that both interpolation and regression are ill-suited for the present case is that neither is based on a model of physical reality. In both instances, the behavior of the predictions is purely an artifact of the behavior of the numbers. For example, neither model takes account of the simple fact that beyond $t = 60$ the number of computers must lie somewhere between 0 and 11,000. Thus, if you were interested in a quick estimate of the number of computers on the market at a future time, an "eyeball"

Figure 14.3
Plot of the second- and third-order regression curves to interpolate and extrapolate computer sales data.

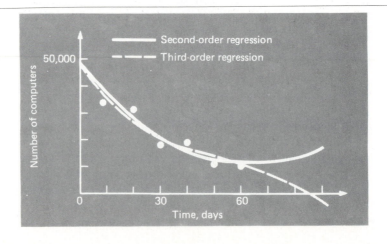

fit and extrapolation of the data would actually provide more realistic results. This is due to the fact that you are cognizant of the physical constraints of the problem and would, therefore, incorporate these constraints into your simple graphical solution. In Case Study 22.1, we will use a differential equation to develop a model that has a theoretical basis and, consequently, yields more satisfactory projections for the present problem context. Another alternative might use an exponential model (see Prob. 14.2).

On the positive side, it should be noted that this example illustrates how regression has some utility for interpolation between noisy data points. However, the primary conclusion of the present case study is that extrapolation should always be performed with care and caution.

CASE STUDY 14.2   LINEAR REGRESSION AND POPULATION MODELS
(CHEMICAL ENGINEERING)

Background: Population growth models are important in many fields of engineering. Fundamental to many of the models is the assumption that the rate of change of the population (dp/dt) is proportional to the actual population (p) at any time (t), or, in equation form,

$$\frac{dp}{dt} = kp \tag{14.1}$$

where k is a proportionality factor called the specific growth rate and has units of time^{-1}. If k is a constant, then the solution of Eq. (14.1) can be obtained from the theory of differential equations:

$$p(t) = p_0 e^{kt} \tag{14.2}$$

where p_0 is the population when $t = 0$. It is observed that $p(t)$ in Eq. (14.2) approaches infinity as t becomes large. This behavior is clearly impossible for real systems. Therefore, the model must be modified to make it more realistic.

Solution: First, it must be recognized that the specific growth rate k cannot be constant as the population becomes large. This is the case because, as p approaches infinity, the organism being modeled will become limited by factors such as food shortages and toxic waste production. One way to express this mathematically is to use a saturation-growth-rate model such that

$$k = k_{max} \frac{f}{K + f} \tag{14.3}$$

where k_{max} is the maximum attainable growth rate for large values of food (f) and K is the half-saturation constant. The plot of Eq. (14.3) in Fig. 14.4 shows that when $f = K, k = k_{max}/2$. Therefore, K is that amount of available food which supports a population growth rate equal to one-half the maximum rate.

The constants K and k_{max} are empirical values based on experimental measurements of k for various values of f. As an example, suppose the population p represents a yeast employed in the commercial production of beer and f is the concentration of the carbon source to be fermented. Measurements of k versus f for the yeast are shown in Table

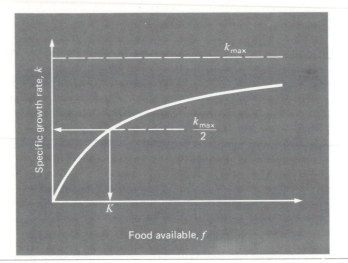

Figure 14.4
Plot of specific growth rate versus available food for the saturation-growth-rate model used to characterize microbial kinetics. The value K is called a half-saturation constant because it conforms to the concentration where the specific growth rate is half its maximum value.

14.3. It is required to calculate k_{max} and K from this empirical data. This is accomplished by inverting Eq. (14.3) in a manner similar to Eq. (11.16) to yield

TABLE 14.3   Data used to evaluate the constants for a saturation-growth-rate model to characterize microbial kinetics.

f, mg/L	k, day^{-1}	$1/f$, L/mg	$1/k$, day
7	0.29	0.14286	3.448
9	0.37	0.11111	2.703
15	0.48	0.06666	2.083
25	0.65	0.04000	1.538
40	0.80	0.02500	1.250
75	0.97	0.01333	1.031
100	0.99	0.01000	1.010
150	1.07	0.00666	0.935

$$\frac{1}{k} = \frac{K+f}{k_{max}f} = \frac{K}{k_{max}}\frac{1}{f} + \frac{1}{k_{max}} \tag{14.4}$$

By this manipulation, we have transformed Eq. (14.3) into a linear form; that is, $1/k$ is a linear function of $1/f$, with slope K/k_{max} and intercept $1/k_{max}$. These values are plotted in Fig. 14.5.

Because of this transformation, the linear least-squares procedures described in Chap. 11 can be used to determine $k_{max} = 1.23$ days^{-1} and $K = 22.18$ mg/L. These results combined with Eq. (14.3) are compared to the untransformed data in Fig. 14.6, and when substituted into the model in Eq. (14.1) give

$$\frac{dp}{dt} = 1.23\frac{f}{22.18 + f}p \tag{14.5}$$

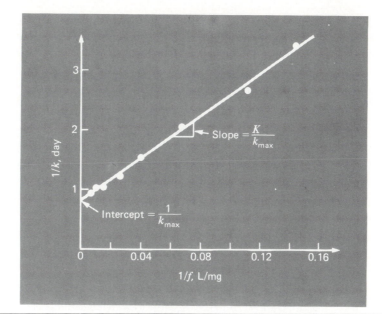

Figure 14.5
Linearized version of the saturation-growth-rate model. The line is a least-squares fit that is used to evaluate the model coefficients $k_{max} = 1.23$ days $^{-1}$ and $K = 22.18$ mg/L for a yeast that is used to produce beer.

Note that the fit yields a sum of the squares of the residuals (as computed for the untransformed data) of 0.001305.

Equation (14.5) can be solved using the theory of differential equations or using numerical methods discussed in Chap. 19 when $f(t)$ is known. If f approaches zero as p becomes large, then dp/dt approaches zero and the population stabilizes.

The linearization of Eq. (14.3) is one way to evaluate the constants k_{max} and K. An alternative approach, which fits the relationship in its original form, is the *nonlinear regression* described in Sec. 11.5. Using the initial guesses of $k_{max} = 1.23$ and $K = 22.18$ obtained from the linearized approach, the Gauss-Newton method can be employed to develop new estimates of $k_{max} = 1.23$ and $K = 22.23$ with an $S_r = 0.001296$. Thus, although as expected the nonlinear regression yields a slightly better fit, the results are

Figure 14.6
Fit of the saturation-growth-rate model to a yeast employed in the commercial production of beer.

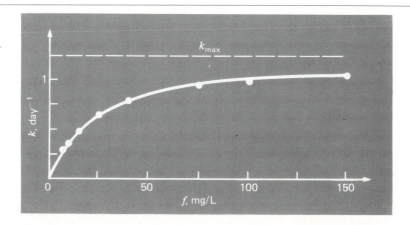

almost identical. In other applications, this may not be true (or the function may not be compatible with linearization) and nonlinear regression could represent the only feasible option for obtaining a least-squares fit.

CASE STUDY 14.3 CURVE FITTING TO DESIGN A SAILBOAT MAST (CIVIL ENGINEERING)

Background: The mast of a racing sailboat (see Case Study 18.3 for more details) has a cross-sectional area of 0.876 in^2 and is constructed of an experimental aluminum alloy. Tests were performed to define the relationship between *stress* (force per area) applied to the material and *strain* (deflection per unit length). These test results are shown in Fig. 14.7 and summarized in Table 14.4. It is necessary to estimate the change in length of the mast due to stress caused by wind force. The stress caused by wind can be computed using

$$\text{Stress} = \frac{\text{force in mast}}{\text{cross-sectional area of mast}}$$

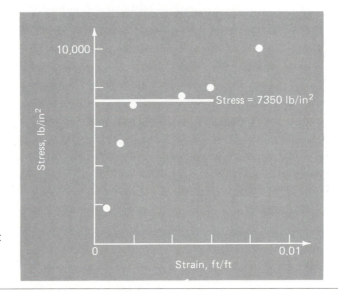

Figure 14.7
Stress-strain curve for an experimental aluminum alloy. For Case Study 14.3, we must obtain a strain estimate from this data conforming to a stress of 7350 lb/in^2.

TABLE 14.4   **Stress-strain data ordered so that the points used for interpolation are always closest to the stress of 7350 lb/in^2.**

Point Number	Stress, lb/in^2	Strain, ft/ft
1	7,200	0.0020
2	7,500	0.0045
3	8,000	0.0060
4	5,200	0.0013
5	10,000	0.0085
6	1,800	0.0005

For this case, a wind force of 6440.6 lb is given (note that in Case Study 18.3 we will use numerical methods to determine this value directly from wind data), and stress is calculated as

$$\text{Stress} = \frac{6440.6}{0.876} = 7350 \text{ lb/in}^2$$

This stress can then be used to determine a strain from Fig. 14.7, which, in turn, can be substituted into Hooke's law to compute the change in length of the mast:

$$\Delta L = (\text{strain})(\text{length}) \tag{14.6}$$

where length refers to the height of the mast. Thus, the problem reduces to determining the value of strain from the data in Fig. 14.7. Because no data point is available at the given stress value of 7350, the problem calls for some sort of curve fit. In the present case study, we will use two approaches: polynomial interpolation and least-squares regression.

Solution: The first approach will be to use interpolating polynomials of order 0 through 5 to estimate the strain at a stress of 7350 lb/in^2. To do this, the data points are ordered so that the interpolation will always use information that is closest to the unknown point (Table 14.4). Newton's interpolating polynomial can then be applied, with the results summarized in Table 14.5.

TABLE 14.5 **Results of using Newton's interpolating polynomial to predict a strain conforming to a stress of 7350 lb/in^2 on the basis of the information from Table 14.4.**

Order of Polynomial (n)	nth-Order Coefficient	Strain (at Stress = 7350)
0	2×10^{-3}	2×10^{-3}
1	8.33×10^{-6}	3.27×10^{-3}
2	-6.67×10^{-9}	3.42×10^{-3}
3	-3.62×10^{-12}	3.36×10^{-3}
4	1.198×10^{-15}	3.40×10^{-3}
5	2.292×10^{-19}	3.38×10^{-3}

All but the zero-order polynomial yield results in fairly close agreement. On the basis of the analysis, we would conclude that a strain of approximately 3.4×10^{-3} ft/ft is a reasonable estimate.

However, a cautionary note is in order. It is actually fortuitous that the estimates of strain are in such close agreement. This can be seen by examining Fig. 14.8, where the fifth-order polynomial is shown along with the data. Notice that because three points are in close proximity to the unknown value of 7350, the interpolation would not be expected to vary significantly at this point. However, if estimates were required at other stresses, oscillations of the polynomial could lead to wild results.

The foregoing results illustrate that higher-order interpolation is ill-suited for uncertain or "noisy" data of the sort in this problem. Regression provides an alternative that is usually more appropriate for such situations.

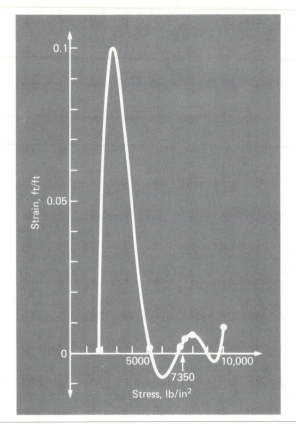

Figure 14.8

Plot of the fifth-order interpolating polynomial that perfectly fits the data from Table 14.4. Note that although the curve passes nicely through the three points in the vicinity of the stress of 7350, the curve oscillates wildly in other parts of the data range.

For example, a linear regression can be used to fit a straight line through the data. The line of best fit is

$$\text{Strain} = -0.002527 + 9.562 \times 10^{-7} \text{ (stress)} \tag{14.7}$$

The line along with the data is shown in Fig. 14.9. Substituting stress $= 7350$ lb/in^2 into Eq. (14.7) yields a prediction of 4.5×10^{-3} in/in.

One problem with a linear regression is that it yields the unrealistic result that strain is negative at a stress of zero. An alternative form of regression that avoids this unrealistic result is to fit a straight line to the logarithm (base 10) of strain versus the logarithm of stress (recall Sec. 11.1.5). The result for this case is

$$\log \text{ (strain)} = -8.565 + 1.586 \log \text{ (stress)}$$

This equation can be transformed back to a form that predicts strain by taking its antilogarithm to yield

$$\text{strain} = 2.723 \times 10^{-9} \text{ (stress)}^{1.586} \tag{14.8}$$

This curve is also superimposed on Fig. 14.9 where it can be seen that this version has the more physically realistic result that strain is equal to zero for a stress of zero.

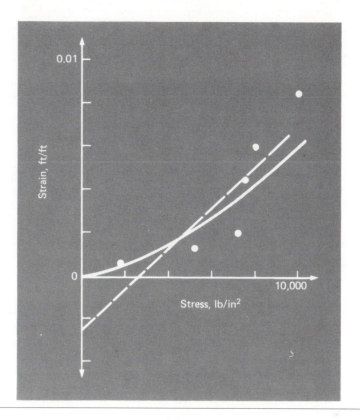

Figure 14.9
Plots of a linear regression line and a log-linear regression line for the strain-stress data for the sailboat mast.

The curve is also somewhat more realistic because it captures some of the curvature suggested by the data. Substituting stress = 7350 into Eq. (14.8) yields a prediction of strain = 3.7×10^{-3} in/in.

Thus, polynomial interpolation and the two types of regression all yield different predictions for strain. Because of its physical realism and its more satisfactory behavior across the entire range of data, we will opt for Eq. (14.8) as providing the best prediction. Using a value of length = 30 ft and Eq. (14.6) yields the following result for the change in length of the mast:

$$\Delta L = (3.7 \times 10^{-3} \text{ ft/ft})(30 \text{ ft}) = 0.11 \text{ ft}$$

CASE STUDY 14.4　　FOURIER ANALYSIS (ELECTRICAL ENGINEERING)

Problem Statement: In Example 4.2, computer graphics were employed to determine the roots of the function (see Fig. 4.4a).

$$f(t) = \cos(3t) + \sin(10t) \tag{14.9}$$

For the present case study, suppose that you were provided with discrete values generated with this function. As displayed in Fig. 14.10, the discrete values are sampled at equal intervals of $\Delta t = 2\pi/32$ over the range from $t = 0$ to 2π. Analyze and interpret this

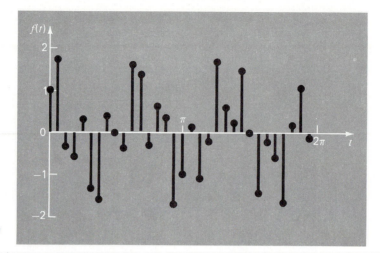

Figure 14.10
Discrete values sampled from the function $f(t) = \cos(3t) + \sin(10t)$ at equal intervals of $t = 2\pi/32$ over the range from $t = 0$ to 2π.

data using the fast Fourier transform. Note that the transforms of the cosine and the sine functions can be analytically determined (Gabel and Roberts, 1987) as in Fig. 14.11. A unit cosine results in two positive spectral components located at $\pm 1/T$ on the real part of the frequency domain. A unit sine results in a positive and a negative spectral component at $-1/T$ and $1/T$, respectively, on the imaginary part of the frequency domain.

Figure 14.11
The (a) sine and cosine functions and (b) their corresponding Fourier transforms.

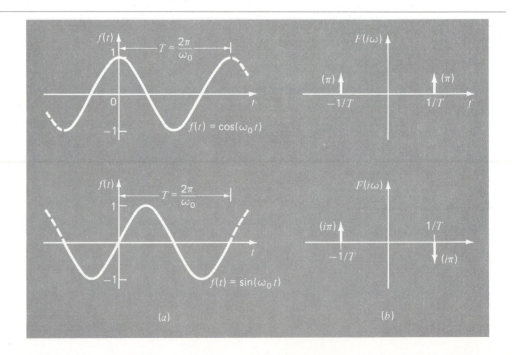

Solution: A computer program based on Fig. 13.18 is used to generate the FFT for the data from Fig. 14.10. An annotated version of the output is shown in Fig. 14.12.

First, notice that the program does not generate the coefficients in a "normal" order. The coefficients for $k = 16$ through 31 are actually for the negative frequency domain and are in descending order. Thus, these results could have been printed before the coefficients from $k = 0$ through 15 in order that the frequencies would be displayed in a more conventional manner—that is, in ascending order from negative to positive values. Most FFT programs incorporate the additional programming steps needed to obtain such output.

Figure 14.12
Output of a computer program based on the algorithm from Fig. 13.18. These results were obtained when the program was used to analyze the data from Fig. 14.10.

index	$f(t)$	real	imaginary	frequency	
0	1.000	−0.000	0.000	0	constant or "DC" term
1	1.755	−0.000	0.000	$1/2\pi$	
2	−0.324	0.000	−0.000	$2/2\pi$	
3	−0.578	0.500	0.000	$3/2\pi$	
4	0.293	−0.000	−0.000	$4/2\pi$	
5	−1.363	−0.000	−0.000	$5/2\pi$	
6	−1.631	−0.000	−0.000	$6/2\pi$	
7	0.368	−0.000	−0.000	$7/2\pi$	
8	−0.000	−0.000	−0.000	$8/2\pi$	positive frequencies
9	−0.368	−0.000	−0.000	$9/2\pi$	
10	1.631	0.000	−0.500	$10/2\pi$	
11	1.363	−0.000	0.000	$11/2\pi$	
12	−0.293	−0.000	0.000	$12/2\pi$	
13	0.578	−0.000	0.000	$13/2\pi$	
14	0.324	−0.000	0.000	$14/2\pi$	
15	−1.755	−0.000	0.000	$15/2\pi$	
16	−1.000	−0.000	0.000	$-16/2\pi$	Nyquist frequency
17	0.092	−0.000	−0.000	$-15/2\pi$	
18	−1.090	−0.000	−0.000	$-14/2\pi$	
19	−0.188	−0.000	−0.000	$-13/2\pi$	
20	1.707	−0.000	−0.000	$-12/2\pi$	
21	0.598	−0.000	−0.000	$-11/2\pi$	
22	0.217	0.000	0.500	$-10/2\pi$	
23	1.479	−0.000	0.000	$-9/2\pi$	
24	−0.000	−0.000	0.000	$-8/2\pi$	negative frequencies
25	−1.479	−0.000	0.000	$-7/2\pi$	
26	−0.217	−0.000	0.000	$-6/2\pi$	
27	−0.598	−0.000	0.000	$-5/2\pi$	
28	−1.707	−0.000	0.000	$-4/2\pi$	
29	0.188	0.500	−0.000	$-3/2\pi$	
30	1.090	0.000	0.000	$-2/2\pi$	
31	−0.092	−0.000	−0.000	$-1/2\pi$	

Next, the results can be displayed graphically as in Fig. 14.13. Note that the real part of the frequency domain exhibits two positive spectral components at $f = \pm 3/2\pi$ (Fig. 14.13a). In addition, the imaginary part (Fig. 14.13b) shows a positive and a negative component at $f = -10/2\pi$ and $f = 10/2\pi$, respectively. Thus, we can deduce that the data is composed of the harmonics: cos $(3t)$ and sin $(10t)$. Consequently, the Fourier transform has provided us with insight into the underlying structure of the original signal. Such insight would be highly unlikely without the benefit of this analytical tool.

In addition to the above, a few more aspects of the analysis bear mentioning. The *Nyquist frequency* $(f = -16/2\pi$ in Fig. 14.12) is the highest-frequency sinusoid that can be detected at a given sampling rate. For equally-spaced samples at intervals of Δt, the Nyquist frequency can be calculated as $1/(2\Delta t)$. Another way of stating this is that there must be at least two samples per cycle for any frequency component that you wish to detect. The practical importance of this fact is that if less than two samples per cycle are taken, an effect called *aliasing* can occur. In essence, this means that different frequency components can yield the same sample. Thus, a distortion is introduced because the analysis will no longer reliably distinguish between different frequency components. This and a variety of other issues must be considered to effectively utilize the power of Fourier approximation. We urge you to consult the many fine references that are available on the subject (see the end of Chap. 13) so that you can successfully employ this tool in your work.

Figure 14.13
Graphical display of the Fourier transform obtained in Fig. 14.12. The plots represent the (a) real and the (b) imaginary components.

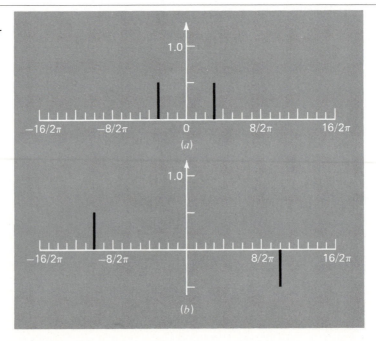

MULTIPLE LINEAR REGRESSION FOR ANALYSIS OF EXPERIMENTAL DATA (MECHANICAL ENGINEERING)

Background: Engineering design variables are often dependent on several independent variables. Often this functional dependence is best characterized by multivariate power equations. As discussed in Sec. 11.3, a multiple linear regression of log-transformed data provides a means to evaluate such relationships.

For example, a mechanical engineering study indicates that fluid flow through a pipe is related to pipe diameter and slope (Table 14.6). Use multiple linear regression to analyze this data. Then use the resulting model to predict the flow for a pipe with a diameter of 2.5 ft and a slope of 0.025 ft/ft.

TABLE 14.6 Experimental data for diameter, slope, and flow of concrete circular pipes.

Experiment	Diameter, ft	Slope, ft/ft	Flow, ft^3/s
1	1	0.001	1.4
2	2	0.001	8.3
3	3	0.001	24.2
4	1	0.01	4.7
5	2	0.01	28.9
6	3	0.01	84.0
7	1	0.05	11.1
8	2	0.05	69.0
9	3	0.05	200.0

Solution: The power equation to be evaluated is

$$Q = a_0 D^{a_1} S^{a_2} \tag{14.10}$$

where Q is flow (in cubic feet per second), S is slope (in feet per foot), D is pipe diameter (in feet), and a_0, a_1, and a_2 are coefficients. Taking the logarithm of this equation yields

$$\log Q = \log a_0 + a_1 \log D + a_2 \log S$$

In this form, the equation is suited for multiple linear regression because $\log Q$ is a linear function of $\log S$ and $\log D$. Using the logarithm (base 10) of the data in Table 14.6, we can generate the following normal equations expressed in matrix form [recall Eq. (11.21)]:

$$\begin{bmatrix} 9 & 2.334 & -18.903 \\ 2.334 & 0.954 & -4.903 \\ -18.903 & -4.903 & 44.079 \end{bmatrix} \begin{Bmatrix} \log a_0 \\ a_1 \\ a_2 \end{Bmatrix} = \begin{Bmatrix} 11.691 \\ 3.945 \\ -22.207 \end{Bmatrix}$$

This system can be solved using Gauss elimination for

$$\log a_0 = 1.7475$$

$$a_1 = 2.62$$

$$a_2 = 0.54$$

If $\log a_0 = 1.7475$, $a_0 = 55.9$ and Eq. (14.10) is

$$Q = 55.9 D^{2.62} S^{0.54} \tag{14.11}$$

Equation (14.11) can be used to predict flow for the case of $D = 2.5$ ft and $S = 0.025$ ft/ft, as in

$$Q = 55.9(2.5)^{2.62}(0.025)^{0.54} = 84.1 \text{ ft}^3/\text{s}$$

It should be noted that Eq. (14.11) can be used for other purposes besides computing flow. For example, the slope is related to head loss h_L and pipe length L by $S = h_L/L$. If this relationship is substituted into Eq. (14.11) and the resulting formula solved for h_L, the following equation can be developed:

$$h_L = \frac{L}{1721} Q^{1.85} D^{4.85}$$

This relationship is called the *Hazen-Williams equation*.

CASE STUDY 14.6   SPREADSHEET SOLUTION OF NEWTON'S POLYNOMIAL

Background: This case study uses the spreadsheet to perform Newton's divided difference interpolation. A spreadsheet to implement the method is contained on the supplementary disk associated with the text. Insert the disk into the computer, type NUMMET following the A prompt, advance to the main menu, and activate the third selection. This causes Fig. 14.14 to appear on the screen.

Solution: The spreadsheet allows you to approximate the function f(X) = LOG(X) (base e) with various polynomials up to order 4. You can input any five base points to develop the divided difference table and a value of X where the interpolation is to be performed. The active cells of the spreadsheet are C3 and A7 through A11. Study the contents of all the cells to be sure you understand the computations.

Observe the formation of the first through fourth divided differences based on five data points and corresponding values of f(X) = LOG(X). The bottom part of the spreadsheet shows zero through fourth-order polynomial approximations of f(X) = LOG(X) along with the percent error. Note how the error decreases as the order of the polynomial increases. Now change the value of X to be interpolated (C3) to 1.5 and strike [F3]. Note that all the polynomials are exact as expected. Change C3 to 4 and strike [F3]. Observe that the zero-order polynomial has error but higher-order polynomials are exact. Repeat changing C3 to 6, 5, and then 3. Note that X = 3 can only be fit exactly when a fourth-order polynomial is used.

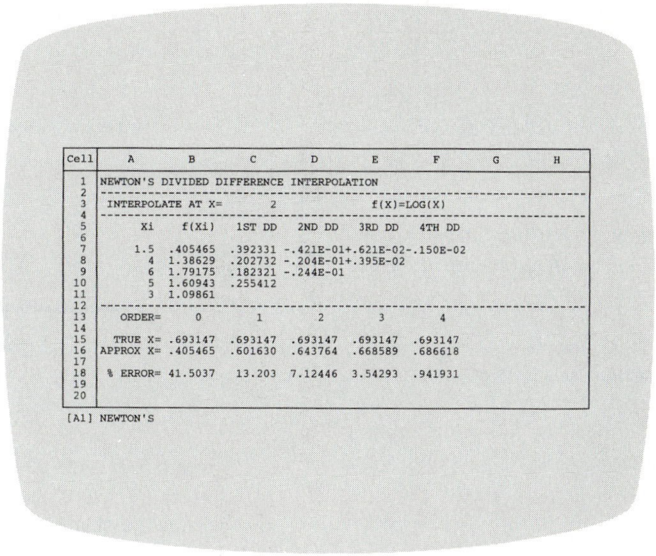

```
Cell    A       B       C       D       E       F       G       H
  1  NEWTON'S DIVIDED DIFFERENCE INTERPOLATION
  2  ---------------------------------------------------------------------
  3  INTERPOLATE AT X=        2                    f(X)=LOG(X)
  4  ---------------------------------------------------------------------
  5      Xi    f(Xi)    1ST DD   2ND DD   3RD DD   4TH DD
  6
  7     1.5   .405465  .392331 -.421E-01+.621E-02-.150E-02
  8      4   1.38629   .202732 -.204E-01+.395E-02
  9      6   1.79175   .182321 -.244E-01
 10      5   1.60943   .255412
 11      3   1.09861
 12  ---------------------------------------------------------------------
 13    ORDER=   0       1        2        3        4
 14
 15   TRUE X= .693147  .693147  .693147  .693147  .693147
 16  APPROX X= .405465  .601630  .643764  .668589  .686618
 17
 18  % ERROR= 41.5037  13.203   7.12446  3.54293  .941931
 19
 20

[A1] NEWTON'S
```

Figure 14.14

Let's use the spreadsheet to see how the accuracy of the fourth-order interpolating polynomial changes as X varies between 3 and 4. The results are presented in Table 14.7. Note how error increases in the middle of the interval and decreases near the points where the polynomial matches the given data exactly.

TABLE 14.7 Interpolation results for $f(x) = \log x$ using a fourth-order polynomial.

x	ϵ_t, %
3.0	0.0000
3.1	0.0268
3.2	0.0439
3.3	0.0528
3.4	0.0549
3.5	0.0517
3.6	0.0447
3.7	0.0348
3.8	0.0235
3.9	0.0116
4.0	0.0000

This example shows how the spreadsheet can be used as an efficient tool to perform interpolation. It has the advantage of displaying *all* the computations involved in performing the interpolation. This encourages numerical experimentation and facilitates understanding the methods and interpreting the results.

PROBLEMS

General Engineering

14.1 Reproduce the computations performed in Case Study 14.1 using your own software.

14.2 Perform the same computation as in Case Study 14.1, but use an exponential model of the type described in Sec. 11.1.5.

14.3 If a sum of money is deposited at a certain interest rate, economic tables can be used to determine the accumulated sum at a later time. For example, the following information is contained in an economics table for the future worth of a deposit after 20 years.

Interest Rate, %	F/P (n = 20 years)
15	16.366
20	38.337
25	86.736
30	190.05

where F/P is the ratio of the future worth to the present value. Thus, if $P = \$10,000$ was deposited, after 20 years at 20 percent interest it would be worth:

$$F = (F/P)P = 38.337(10,000) = \$383,370$$

Use linear, quadratic, and cubic interpolation to determine the future value of $25,000 deposited at 23.6 percent interest. Interpret your results from the perspective of the lending institution.

14.4 Use the information given in Prob. 14.3, but suppose that you have invested $40,000 and are told that 20 years later the lender will pay you back $2,800,000. Use linear, quadratic, and cubic interpolation to determine what interest rate you are being given.

14.5 Suppose you win a sweepstakes and are given the choice of accepting $2 million now or $700,000 per year for the next 5 years. The relationship between a present value P and an annual series of payments A is given by the following information from an economics table:

Interest Rate, %	A/P (n = 5 years)
15	0.29832
20	0.33438
25	0.37185
30	0.41058

where A/P is the ratio of the annual payment to the present worth. Thus, if the interest rate is 15 percent, the five annual payments A that are equivalent to a single present payment ($P = \$2$ million) is calculated as

$$A = (A/P)P = 0.29832(2,000,000) = \$596,640$$

Use interpolation to determine the interest rate at which taking the $2 million becomes the better decision.

14.6 You are performing a study to determine the relationship between the upward drag force and velocity for the falling parachutist. A number of experiments yields the following information for velocity (v in centimeters per second) and upward drag force (F_u in 10^6 dynes):

v	1000	2000	3000	4000	5000
F_u	5	15.3	29.3	46.4	66.3

Plot F_u versus v and then use regression to determine the relationship between upward drag and velocity.

14.7 The population (p) of a small community on the outskirts of a city grows rapidly over a 20-year period:

t	0	5	10	15	20
p	100	212	448	949	2009

As an engineer working for a utility company, you must forecast the population 5 years into the future in order to anticipate the demand for power. Employ an exponential model and linear regression to make this prediction.

14.8 An experiment is performed to determine the percent elongation of a material as a function of temperature. The resulting data is

Temperature, °F	400	500	600	700	800	900	1000	1100
Elongation, %	11	13	13	15	17	19	20	23

Predict the percent elongation for a temperature of 780°F.

14.9 The distance required to stop an automobile is a function of its speed. The following experimental data was collected to quantify this relationship:

Speed, mi/h	15	20	25	30	40	50	60	70
Stopping distance, ft	15	21	45	46	65	90	111	98

Estimate the stopping distance for a car traveling at 50 mi/h.

14.10 You are provided with the following stress-strain data for an aluminum alloy:

Stress	1	2	3	4	5	6	7	8	9	10	11	12
Strain	2	4	6	6	6	7	8	7.5	7	7.5	8	7.5

Use linear regression to determine the strain corresponding to a stress of 7.4.

14.11 An experiment is performed to define the relationship between applied stress and the time to fracture for a stainless steel. Eight different values of stress are applied, and the resulting data is

Applied stress, x, kg/mm^2	5	10	15	20	25	30	35	40
Fracture time, y, h	40	30	25	40	18	20	22	15

(a) Plot the data.
(b) Fit a straight line to the data with linear regression. Superimpose this line on your plot.
(c) Use the best-fit equation to predict the fracture time for an applied stress of 33 kg/mm^2.

14.12 The acceleration due to gravity at an altitude y above the surface of the earth is given by

y, m	0	20,000	40,000	60,000	80,000
g, m/s^2	9.8100	9.7487	9.6879	9.6278	9.5682

Compute g at $y = 55,000$ m to four decimal places of accuracy.

14.13 Bessel functions often arise in advanced engineering analysis. These functions are usually not amenable to straightforward evaluation and, therefore, are often compiled in standard mathematical tables. For example,

x	$J_0(x)$
1.8	0.3400
2.0	0.2239
2.2	0.1104
2.4	0.0025
2.6	0.0968

Estimate $J_0(2.1)$. Note that the true value is 0.1666.

14.14 A great deal of the quantitative data used by engineers is in tabulated form. In the past, linear interpolation was employed to estimate intermediate points in such tables. Today, this situation is changing because of the availability of computers. Write a structured program to perform a table lookup and interpolation on data. Aside from entering the data and the independent variable where the interpolation is to be performed, the program must search through the table to locate the interval where the independent variable lies. One common method for accomplishing this objective is the *binary search*. Suppose that you are given an array A(I) of length N and must determine whether the value V is included in this array. After sorting the array in ascending order, you employ the following algorithm to conduct a binary search:

Compare V with the middle term A(M). (*a*) If V = A(M), the search is successful; (*b*) if V < A(M), V is in the first half; and (*c*) if V > A(M), V is in the second half. In the case of (*b*) or (*c*), the search is repeated in the half in which V is known to lie. Then V is again compared with A(M), where M is now redefined as the middle value of the smaller list. The process is repeated until V is located or the array is exhausted.

After locating the proper interval, use a third-order Lagrange polynomial to interpolate. Be careful if the interval falls at either end of the table. Test your program by generating a table of $\log_{10} x$ from $x = 1$ to 10 using intervals of 1. Use your program to determine estimates at points spaced 0.2 apart across the entire range. Calculate the true percent relative error for each estimate and plot the errors versus x. Discuss your results.

14.15 Repeat Prob. 14.14 but use cubic splines to perform the interpolation.

Chemical Engineering

14.16 Reproduce the computations performed in Case Study 14.2 using your own software.

14.17 Perform the same computation as in Case Study 14.2, but use polynomial regression to fit a parabola to the data. Discuss your results.

14.18 Perform the same computation as in Case Study 14.2, but use linear regression and transformations to fit the data with a power equation. Ignore the first point when fitting the equation.

14.19 You perform experiments and determine the following values of heat capacity c at various temperatures T for a metal:

T	−50	−20	10	70	100	120
c	0.125	0.128	0.134	0.144	0.150	0.155

Use regression to determine a model to predict c as a function of T.

14.20 The saturation concentration of dissolved oxygen in water as a function of temperature and chloride concentration is listed in Table P14.20. Use interpolation to estimate the dissolved oxygen level for $T = 22.4°C$ with chloride = 10,000 mg/L.

TABLE P14.20   **Dependency of dissolved oxygen concentration on temperature and chloride concentration.**

Temperature, °C	Dissolved Oxygen (mg/L) for Stated Concentrations of Chloride		
	Chloride = 0 mg/L	Chloride = 10,000 mg/L	Chloride = 20,000 mg/L
5	12.8	11.6	10.5
10	11.3	10.3	9.2
15	10.0	9.1	8.2
20	9.0	8.2	7.4
25	8.2	7.4	6.7
30	7.4	6.8	6.1

14.21 For the data in Table P14.20, use polynomial interpolation to derive a predictive equation for dissolved oxygen concentration as a function of temperature for the case where chloride concentration is equal to 20,000 mg/L.

14.22 Use polynomial regression to perform the same task as in Prob. 14.21.

14.23 Use multiple linear regression and logarithmic transformations to derive an equation to predict dissolved oxygen concentration as a function of temperature and chloride concentration. Evaluate your results.

14.24 It is known that the tensile strength of a plastic increases as a function of the time it is heat-treated. The following data is collected:

Time	10	15	20	30	40	50	55	60	75
Tensile strength	4	20	18	50	33	48	80	105	78

Fit a straight line to this data and use the equation to determine the tensile strength at a time of 70 min.

14.25 The following data was gathered to determine the relationship between pressure and temperature of a fixed volume of 1 kg of nitrogen. The volume is 10 m^3.

T, °C	−20	0	20	40	50	70	100	120
p, N/m^3	7500	8104	8700	9300	9620	10,200	10,500	11,700

Employ the ideal gas law $pV = nRT$ to determine R on the basis of this data. Note that for the law T must be expressed in kelvins.

14.26 The specific volume of a superheated steam is listed in steam tables for various temperatures. For example, at a pressure of 2950 lb/in^2, absolute:

T, °F	700	720	740	760	780
v	0.1058	0.1280	0.1462	0.1603	0.1703

Determine v at $T = 728$°F.

14.27 A reactor is thermally stratified as in the following table:

Depth, m	0	0.5	1.0	1.5	2.0	2.5	3.0
Temperature, °C	70	70	66	52	18	11	10

As depicted in Fig. P14.27, the tank can be idealized as two zones separated by a strong temperature gradient or *thermocline*. The depth of this gradient can be defined as the inflection point of the temperature-depth curve — that is, the point at which $d^2T/dz^2 = 0$. At this depth, the heat flux from the surface to the bottom layer can be computed with Fourier's law,

$$J = -k\frac{dT}{dz}$$

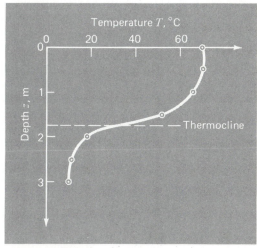

Figure P14.27

Use a cubic spline fit of this data to determine the thermocline depth. If $k = 0.01$ cm^2/s, compute the flux across this interface. Note that this is a dilute aqueous solution, so °C ≃ cal/cm^3.

Civil Engineering

14.28 Reproduce the computations performed in Case Study 14.3 using your own software.

14.29 Perform the same computations as in Case Study 14.3, but use second-order polynomial regression to relate strain to stress.

14.30 Perform the same computations as in Case Study 14.3, but use an exponential formulation to relate strain to stress.

14.31 Perform the same computations as in Case Study 14.3, but use polynomial interpolation to evaluate the ΔL if the stress is 7700 lb/in^2.

14.32 The shear stress, in kips per square foot (ksf), of nine specimens taken at various depths in a clay stratum are

Depth, m	1.9	3.1	4.2	5.1	5.8	6.9	8.1	9.3	10.0
Stress, ksf	0.3	0.6	0.5	0.8	0.7	1.1	1.5	1.3	1.6

Estimate the shear stress at a depth of 7.5 m.

14.33 A transportation engineering study was conducted to determine the proper design of bike lanes. Data was gathered on bike-lane widths and average distance between bikes and passing cars. The data from 11 streets is

Lane width x, ft	5	10	7	7.5	7	6	10	9	5	5.5	8
Distance y, ft	3	8	5	8	6	6	10	10	4	5	7

(a) Plot the data.

(b) Fit a straight line to the data with linear regression. Add this line to the plot.

(c) If the minimum safe average distance between bikes and passing cars is considered to be 6 ft, determine the corresponding minimum lane width.

14.34 In water-resources engineering the sizing of reservoirs depends on accurate estimates of water flow in the river that is being impounded. For some rivers, long-term historical records of such flow data are difficult to obtain. In contrast, meteorological data on precipitation is often available for many years past. Therefore it is often useful to determine a relationship between flow and precipitation. This relationship can then be used to estimate flows for years when only precipitation measurements were made. The following data is available for a river that is to be dammed:

Annual precipitation x, in	35	40	41	55	52	37	46	48	39	45
Annual water flow y, ft^3/s	4050	6075	5400	9500	7290	5700	6210	8440	4590	8000

(a) Plot the data.

(b) Fit a straight line to the data with linear regression. Superimpose this line on your plot.

(c) Use the best-fit line to predict the annual water flow if the precipitation is 50 in.

14.35 The concentration of total phosphorus (p in mg/m^3) and chlorophyll a (c in mg/m^3) for each of the Great Lakes is

	p	c
Lake Superior	4.5	0.8
Lake Michigan	8.0	2.0
Lake Huron	5.5	1.2
Lake Erie: west basin	39.0	11.0
central basin	19.5	4.4
east basin	17.5	3.3
Lake Ontario	21.0	5.5

Chlorophyll a is a parameter that indicates how much plant life is suspended in the water. As such, it indicates how unclear and unsightly the water appears. Use the above data to determine a relationship to predict c as a function of p. Use this equation to predict the level of chlorophyll that can be expected if waste treatment is used to lower the phosphorus concentration of western Lake Erie to 10 mg/m^3.

14.36 The vertical stress σ_z under the corner of a rectangular area subjected to a uniform load of intensity q is given by the solution of Boussinesq's equation:

$$\sigma_z = \frac{q}{4\pi}\left[\frac{2mn\sqrt{m^2+n^2+1}}{m^2+n^2+1+m^2n^2}\frac{m^2+n^2+2}{m^2+n^2+1}\right.$$
$$\left. + \sin^{-1}\left(\frac{2mn\sqrt{m^2+n^2+1}}{m^2+n^2+1+m^2n^2}\right)\right]$$

Because this equation is inconvenient to solve manually, it has been reformulated as

$$\sigma_z = qf_z(m,n)$$

where $f_z(m,n)$ is called the *influence value* and m and n are dimensionless ratios, with $m = a/z$ and $n = b/z$ and a and b as defined in Fig. P14.36. The influence value is then listed in a table, a portion of which is given here:

m	$n = 1.2$	$n = 1.4$	$n = 1.6$
0.1	0.02926	0.03007	0.03058
0.2	0.05733	0.05894	0.05994
0.3	0.08323	0.08561	0.08709
0.4	0.10631	0.10941	0.11135
0.5	0.12626	0.13003	0.13241
0.6	0.14309	0.14749	0.15027
0.7	0.15703	0.16199	0.16515
0.8	0.16843	0.17389	0.17739

If $a = 5.6$ and $b = 14$, compute σ_z at a depth 10 m below the corner of a rectangular footing that is subject to a total load of 100 t (metric tons). Express your answer in tonnes per square meter. Note that q is equal to the load per area.

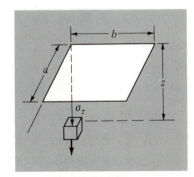

Figure P14.36

14.37 An experiment is used to determine the ultimate moment capacity of a concrete beam as a function of cross-sectional area. The experiment, which is performed with great precision, yields the following data:

Capacity, in·Klb	932.3	1785.2	2558.6	3252.7	3867.4	4402.6	4858.4	5234.8
Area, in^2	1	2	3	4	5	6	7	8

Determine the ultimate moment capacity (to one decimal place) for an area of 5.4 in.

Electrical Engineering

14.38 Reproduce the computations performed in Case Study 14.4 using your own software.

14.39 Perform the same computations as in Case Study 14.4, but analyze data generated with $f(t) = 5 \cos (7t) - 2 \sin (4t) + 6$.

14.40 You measure the voltage drop v across a resistor for a number of different values of current i. The results are

i	0.25	0.75	1.25	1.5	2.0
v	−0.23	−0.33	0.70	1.88	6.00

Use polynomial interpolation to estimate the voltage drop for $i = 0.9$. Interpret your results.

14.41 Duplicate the computation for Prob. 14.40, but use polynomial regression to derive a cubic equation to fit the data. Plot and evaluate your results.

14.42 The current in a wire is measured with great precision as a function of time:

t	0	0.1250	0.2500	0.3750	0.5000
i	0	6.2402	7.7880	4.8599	0.0000

Determine i at $t = 0.32$.

14.43 The following data was taken from an experiment that measured the current in a wire for various imposed voltages:

Voltage, V	0	2	3	4	5	7	10
Current, A	0	5.2	7.8	10.7	13	19.3	26.5

On the basis of a linear regression of this data, determine current for a voltage of 5 V. Plot the line and the data and evaluate the fit.

14.44 It is known that the voltage drop across an inductor follows Faraday's law:

$$V_L = L \frac{di}{dt}$$

where V_L is the voltage drop (in volts), L is inductance (in henrys; 1 H = 1 V·s/A), and i is current (in amperes). Employ the following data to estimate L:

di/dt, A/s	1	2	4	6	8	10
V_L, V	5	11	19	31	39	50

What is the meaning, if any, of the intercept of the regression equation derived from this data?

14.45 Ohm's law states that the voltage drop V across an ideal resistor is linearly proportional to the current i flowing through the resistor as in $V = iR$, where R is the resistance. However, real resistors may not always obey Ohm's law. Suppose that you performed some very precise experiments to measure the voltage drop and corresponding current for a resistor. The results, as listed in Table P14.45, suggest a curvilinear relationship rather than the straight line represented by Ohm's law. In order to quantify this relationship, a curve must be fit to the data. Because of measurement error, regression would typically be the preferred method of curve fitting for analyzing such experimental data. However, the smoothness of the relationship, as well as the precision of the experimental methods, suggests that interpolation might be appropriate. Use Newton's interpolating polynomial to fit the data.

TABLE P14.45

Experimental data for voltage drop across a resistor subjected to various levels of current.

i	V
−1.00	−200.0
−0.50	−40.0
−0.25	−12.5
0.25	12.5
0.50	40.0
1.00	200.0

14.46 Repeat Prob. 14.45 but determine the coefficients of the cubic equation (Sec. 12.3) that fit the data in Table P14.45.

Mechanical Engineering

14.47 Reproduce the computations performed in Case Study 14.5 using your own software.

14.48 Based on Table 14.6, use linear and quadratic interpolation to compute Q for $D = 1.23$ ft and $S = 0.01$ ft/ft. Compare your results with the same value computed with the formula derived in Case Study 14.5.

14.49 Reproduce Case Study 14.5, but develop an equation to predict diameter as a function of slope and flow. Compare your results with the formula from Case Study 14.5 and discuss your results.

14.50 Kinematic viscosity of water, v, is related to temperature in the following manner:

T, °F	40	50	60	70	80
v, 10^{-5} ft²/s	1.66	1.41	1.22	1.06	0.93

Plot this data. Use interpolation to predict v at $T = 62$°F.

14.51 Repeat Prob. 14.50, but use regression.

14.52 Hooke's law, which holds when a spring is not stretched too far, signifies that the extension of the spring and the applied force are linearly related. The proportionality is parameterized by the spring constant k. A value for this parameter can be established experimentally by placing known weights onto the spring and measuring the resulting compression. Such data is contained in Table P14.52 and plotted in Fig. P14.52. Notice that above a weight of 40×10^4 N, the linear relationship between the force and displacement breaks down. This sort of behavior is typical of what is termed a "hardening spring." Employ linear regression to determine a value of k for the linear portion of this system.

14.53 Repeat Prob. 14.52 but fit a power curve to all the data in Table P14.52. Comment on your results.

TABLE P14.52

Experimental values for elongation x and force F for the spring on an automobile suspension system.

Displacement, m	Force, 10^4 N
0.10	10
0.17	20
0.24	30
0.34	40
0.39	50
0.42	60
0.43	70
0.44	80

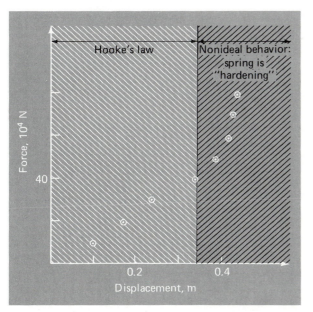

Figure P14.52

Plot of force (in 10^4 newtons) versus displacement (in meters) for the spring from the automobile suspension system.

Spreadsheets

14.54 Use an active spreadsheet that may be available to you (like Lotus 123 or the one on the Electronic TOOLKIT) to perform Newton's divided difference interpolation for up to fourth-order approximations. Try it for the following functions.

(**a**) $f(x) = e^x$
(**b**) $f(x) = \sin x$
(**c**) $f(x) = x^6$
(**d**) $f(x) = \log_{10} x$

Try various base points to approximate at $x = 1.5$. Discuss your results.

14.55 Repeat Prob. 14.54 but use Lagrange interpolation.

14.56 Activate the spreadsheet associated with the text. Change the base points to $X = 10, 20, 21, 30$, and 40, and interpolate at $X = 25$ and record the error associated with the fourth-order polynomial. Now change the base points to $X = 22, 24, 26, 28$, and 30, and compare the error. Discuss and analyze the results on the basis of your knowledge of the theoretical relationship between error and step size.

14.57 Repeat Prob. 14.56 interpolating at X = 20 using base points of X = 2, 3, 4, 5, and 30. Explain the results using plots of the interpolating polynomials. Verify the results by Lagrange interpolation, using the Electronic TOOLKIT or your own software. (Hint: See Fig. 14.14.)

14.58 Repeat Prob. 14.57 interpolating at X = 15 using base points of X = 2, 2.5, 3, 3.5, and 20.

Miscellaneous

14.59 Read all the case studies in Chap. 14. On the basis of your reading and experience, make up your own case study for any one of the fields of engineering. This may involve modifying or reexpressing one of our case studies. However, it can also be totally original. As with our examples, it must be drawn from an engineering problem context and must demonstrate the use of numerical methods for curve fitting. Write up your results using our case studies as models.

EPILOGUE: PART FOUR

PT4.4 TRADE-OFFS

Table PT4.4 provides a summary of the trade-offs involved in curve fitting. The techniques are divided into two broad categories depending on the uncertainty of the data. For imprecise measurements, regression is used to develop a "best" curve that fits the overall trend of the data without necessarily passing through any of the individual points. For precise measurements, interpolation is used to develop a curve that passes directly through each of the points.

TABLE PT4.4 Comparison of the characteristics of alternative methods for curve fitting.

Method	Error Associated with Data	Match of Individual Data Points	Number of Points Matched Exactly	Programming Effort	Comments
Regression					
Linear regression	Large	Approximate	0	Easy	
Polynomial regression	Large	Approximate	0	Moderate	Round-off error becomes pronounced for higher-order versions
Multiple linear regression	Large	Approximate	0	Moderate	
Nonlinear regression	Large	Approximate	0	Difficult	
Interpolation					
Newton's divided-difference polynomials	Small	Exact	$n + 1$	Easy	Usually preferred for exploratory analyses
Lagrange polynomials	Small	Exact	$n + 1$	Easy	Usually preferred when order is known
Cubic splines	Small	Exact	Piecewise fit of data points	Moderate	First and second derivatives equal at knots

All the regression methods are designed to fit functions that minimize the sum of the squares of the residuals between the data and the function. Such methods are termed least-squares regression. Linear least-squares regression is used for cases where a dependent and an independent variable are related to each other in a linear fashion. For situations where a dependent and an independent variable exhibit a curvilinear relationship, several options are available. In some cases, transformations can be used to linearize the relationship. In these instances, linear regression can be applied to the transformed variables in order to determine the best straight line. Alternatively, polynomial regression can be employed to fit a curve directly to the data.

Multiple linear regression is utilized when a dependent variable is a linear function of two or more independent variables. Logarithmic transformations can also be applied to this type of regression for some cases where the multiple dependency is curvilinear.

Polynomial and multiple linear regression (note that simple linear regression is a member of both) belong to a more general class of linear least-squares models. They are classified in this way because they are linear with respect to their coefficients. These models are typically implemented using linear algebraic systems that are sometimes ill-conditioned. However, in many engineering applications (that is, for lower-order fits), this does not come into play. For cases where it is a problem, alternative approaches are available. For example, a technique called orthogonal polynomials is available to perform polynomial regression (See Sec. PT4.6).

Equations that are not linear with respect to their coefficients are called nonlinear. Special regression techniques are available to fit such equations. These are approximate methods that start with initial parameter estimates and then iteratively home in on values that minimize the sum of the squares.

Polynomial interpolation is designed to fit a unique nth-order polynomial that passes exactly through $n + 1$ precise data points. This polynomial is presented in two alternative formats. Newton's divided-difference interpolating polynomial is ideally suited for those cases where the proper order of the polynomial is unknown. Newton's polynomial is appropriate for such situations because it is easily programmed in a format to compare results with different orders. In addition, an error estimate can be simply incorporated into the technique. Thus, you can compare and choose from results using several different-order polynomials.

The Lagrange interpolating polynomial is an alternative formulation that is appropriate when the order is known a priori. For these situations, the Lagrange version is somewhat simpler to program and does not require the computation and storage of finite divided differences.

Another approach to curve fitting is spline interpolation. This technique fits a low-order polynomial to each interval between data points. The fit is made smooth by setting the derivatives of adjacent polynomials to the same value at their connecting points. The cubic spline is the most common version. Splines are of great utility when fitting data that is generally smooth but exhibits local areas of roughness. Such data tends to induce wild oscillations in higher-order interpolating polynomials. Cubic splines are less prone to these oscillations because they are limited to third-order variations.

The final method covered in this part of the book is Fourier approximation. This area deals with using trigonometric functions to approximate waveforms. In contrast to the other techniques, the major emphasis of this approach is not to fit a curve to data points. Rather, the curve fit is employed to analyze the frequency characteristics of a signal. In particular, a fast Fourier transform is available to very efficiently transform a function from the time to the frequency domain in order to elucidate its underlying harmonic structure.

PT4.5 IMPORTANT RELATIONSHIPS AND FORMULAS

Table PT4.5 summarizes important information that was presented in Part Four. This table can be consulted to quickly access important relationships and formulas.

PT4.6 ADVANCED METHODS AND ADDITIONAL REFERENCES

Although polynomial regression with normal equations is adequate for many engineering applications, there are problem contexts where its sensitivity to round-off error can represent a serious limitation. An alternative approach based on *orthogonal polynomials* can mitigate these effects. It should be noted that this approach does not yield a best-fit equation but rather yields individual predictions for given values of the independent variable. Information on orthogonal polynomials can be found in Shampine and Allen (1973) and Guest (1961).

Whereas the orthogonal polynomial technique is helpful for developing a polynomial regression, it does not represent a solution to the instability problem for the general linear regression model [Eq. (11.23)]. An alternative approach based on *single-value decomposition* and called the SVD method is available for this purpose. Forsythe et al. (1977), Lawson and Hanson (1974), and Press et al. (1986) contain information on this approach.

In addition to the *Gauss-Newton* algorithm, there are a number of methods which can be used to directly develop a least-squares fit for a nonlinear equation. These nonlinear regression techniques include *Marquardt's* and the *steepest descent methods*. Information regarding these techniques and regression in general can be found in Draper and Smith (1981).

All the methods in Part Four have been couched in terms of fitting a curve to data points. In addition, you may also desire to fit a curve to another curve. The primary motivation for such *functional approximation* is to represent a complicated function by a simpler version that is easier to manipulate. One way to do this is to

TABLE PT4.5 Summary of important information presented in Part Four.

Method	Formulation	Graphical Interpretation	Errors
Linear regression	$y = a_0 + a_1x$ where $a_1 = \dfrac{n\sum x_iy_i - \sum x_i\sum y_i}{n\sum x_i^2 - (\sum x_i)^2}$ $a_0 = \bar{y} - a_1\bar{x}$		$s_{y/x} = \sqrt{\dfrac{S_r}{n-2}}$ $r^2 = \dfrac{S_t - S_r}{S_t}$
Polynomial regression	$y = a_0 + a_1x + \cdots + a_mx^m$ (Evaluation of a's equivalent to solution of $m+1$ linear algebraic equations)		$s_{y/x} = \sqrt{\dfrac{S_r}{n-(m+1)}}$ $r^2 = \dfrac{S_t - S_r}{S_t}$
Multiple linear regression	$y = a_0 + a_1x_1 + \cdots + a_mx_m$ (Evaluation of a's equivalent to solution of $m+1$ linear algebraic equations)		$s_{y/x} = \sqrt{\dfrac{S_r}{n-(m+1)}}$ $r^2 = \dfrac{S_t - S_r}{S_t}$
Newton's divided-difference interpolating polynomial*	$f_2(x) = b_0 + b_1(x-x_0) + b_2(x-x_0)(x-x_1)$ where $b_0 = f(x_0)$ $ b_1 = f[x_1, x_0]$ $ b_2 = f[x_2, x_1, x_0]$		$R_2 = (x-x_0)(x-x_1)(x-x_2)\dfrac{f^{(3)}(\xi)}{6}$ or $R_2 = (x-x_0)(x-x_1)(x-x_2)f[x_3,x_2,x_1,x_0]$
Lagrange interpolating polynomial*	$f_2(x) = f(x_0)\left(\dfrac{x-x_1}{x_0-x_1}\right)\left(\dfrac{x-x_2}{x_0-x_2}\right)$ $\quad + f(x_1)\left(\dfrac{x-x_0}{x_1-x_0}\right)\left(\dfrac{x-x_2}{x_1-x_2}\right)$ $\quad + f(x_2)\left(\dfrac{x-x_0}{x_2-x_0}\right)\left(\dfrac{x-x_1}{x_2-x_1}\right)$		$R_2 = (x-x_0)(x-x_1)(x-x_2)\dfrac{f^{(3)}(\xi)}{6}$ or $R_2 = (x-x_0)(x-x_1)(x-x_2)f[x_3,x_2,x_1,x_0]$
Cubic splines	A cubic: $\quad a_ix^3 + b_ix^2 + c_ix + d_i$ is fit to each interval between knots. First and second derivatives are equal at each knot		

*Note, for simplicity, second-order versions are shown.

use the complicated function to generate a table of values. Then the techniques discussed in this part of the book can be used to fit polynomials to these discrete values.

Beyond this approach, there are a variety of alternative, and usually preferable, methods for functional approximation. For example, if the function is continuous and differentiable, it can be fit with a truncated Taylor series. However, this strategy is flawed because the truncation error increases as we move away from the base point of the expansion. Thus, we will have a very good prediction at one point in the interval and poorer predictions elsewhere.

An alternative approach is based on the *minimax principle* (recall Fig. 11.2c). This principle specifies that the coefficients of the approximating polynomial be chosen so that the maximum discrepancy is as small as possible. Thus, although the approximation may not be as good as that given by the Taylor series at the base point, it is generally better across the entire range of the fit. *Chebyshev economization* is an example of an approach for functional approximation based on such a strategy (Ralston and Rabinowitz, 1978; Gerald and Wheatley, 1984; and Carnahan, Luther, and Wilkes, 1969).

An important area in curve fitting is the combining of splines with least-squares regression. Thus, a cubic spline is generated that does not intercept every point but rather minimizes the sum of the squares of the residuals between the data points and the spline curves. The approach involves using the so-called *B splines* as basis functions. These are bell-shaped functions, hence the name B splines. Such curves are consistent with a spline approach in that their value and their first and second derivatives would have continuity at their extremes. Thus, continuity of $f(x)$ and its lower derivatives at the knots is ensured. Wold (1974) and Prenter (1974) present discussions of this approach.

In summary, the foregoing is intended to provide you with avenues for deeper exploration of the subject. Additionally, all the above references provide descriptions of the basic techniques covered in Part Four. We urge you to consult these alternative sources to broaden your understanding of numerical methods for curve fitting.[*]

---

[*]Books are referenced only by author here; a complete bibliography will be found at the back of this text.

PART FIVE

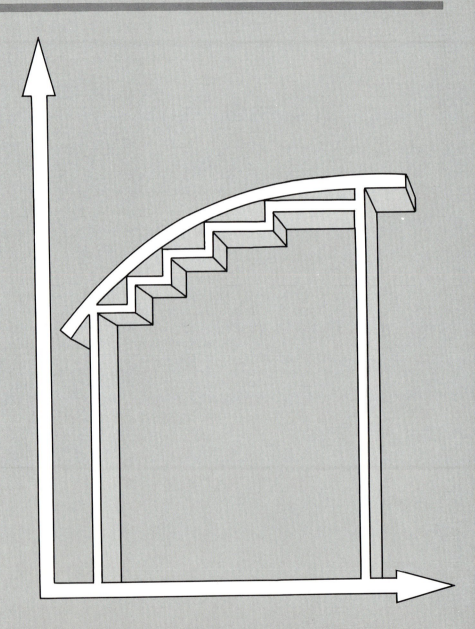

NUMERICAL DIFFERENTIATION AND INTEGRATION

PT5.1   MOTIVATION

Calculus is the mathematics of change. Because engineers must continuously deal with systems and processes that change, calculus is an essential tool of our profession. Standing at the heart of calculus are the related mathematical concepts of differentiation and integration.

According to the dictionary definition, to *differentiate* means "to mark off by differences; distinguish; . . . to perceive the difference in or between." Mathematically, the *derivative,* which serves as the fundamental vehicle for differentiation, represents the rate of change of a dependent variable with respect to an independent variable. As depicted in Fig. PT5.1, the mathematical definition of the derivative begins with a difference approximation:

$$\frac{\Delta y}{\Delta x} = \frac{f(x_i + \Delta x) - f(x_i)}{\Delta x} \tag{PT5.1}$$

where *y* and *f(x)* are alternative representatives for the dependent variable and *x* is the independent variable. If Δx is allowed to approach zero, as occurs in moving from Fig. PT5.1*a* to *c*, the difference becomes a derivative

$$\frac{dy}{dx} = \lim_{\Delta x \to 0} \frac{f(x_i + \Delta x) - f(x_i)}{\Delta x}$$

where *dy/dx* [which can also be designated as *y'* or *f'(x_i)*] is the first derivative of *y* with respect to *x* evaluated at x_i. As seen in the visual depiction of Fig. PT5.1*c*, the derivative is the slope of the tangent to the curve at x_i.

The inverse process to differentiation in calculus is integration. According to the dictionary definition, to *integrate* means "to bring together, as parts, into a whole; to unite; to indicate the total amount . . . ." Mathematically, integration is represented by

$$I = \int_a^b f(x)\, dx \tag{PT5.2}$$

which stands for the integral of the function *f(x)* with respect to the independent variable *x*, evaluated between the limits $x = a$ to $x = b$. The function *f(x)* in Eq. PT5.2 is referred to as the *integrand.*

As suggested by the dictionary definition, the "meaning" of Eq. (PT5.2) is the *total*

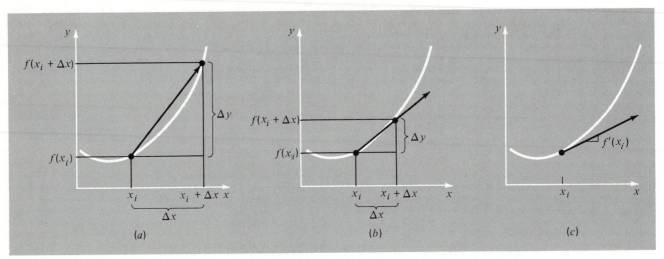

Figure PT5.1
The graphical definition of a derivative: as Δx approaches zero in going from (a) to (c), the difference approximation becomes a derivative.

value, or *summation,* of $f(x)\ dx$ over the range from $x = a$ to b. In fact, the symbol $\int$ is actually a stylized capital S that is intended to signify the close connection between integration and summation.

Figure PT5.2 represents a graphical manifestation of the concept. For functions lying above the x axis, the integral expressed by Eq. (PT5.2) corresponds to the area under the curve of $f(x)$ between $x = a$ and b.[*]

As outlined above, the "marking off or discrimination" of differentiation and the "bringing together" of integration are closely linked processes that are, in fact, inversely related (Fig. PT5.3). For example, if we are given a function $y(t)$ that specifies an object's position as a function of time, differentiation provides a means to determine its velocity as in (Fig. PT5.3a).

$$v(t) = \frac{d}{dt}\,y(t)$$

Conversely, if we are provided with velocity as a function of time, integration can be used to determine its position (Fig. PT5.3b).

$$y(t) = \int_0^t v(t)\,dt$$

[*]It should be noted that the process represented by Eq. (PT5.2) and Fig. PT5.2 is called *definite integration*. There is another type called indefinite integration in which the limits a and b are unspecified. As will be discussed in Part Six, *indefinite integration* deals with determining a function whose derivative is given.

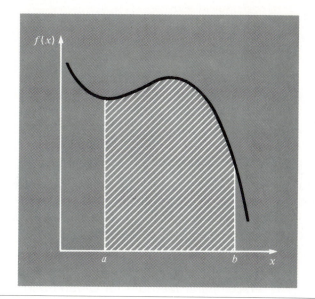

Figure PT5.2
Graphical representation of the integral of $f(x)$ between the limits $x = a$ to b. The integral is equivalent to the area under the curve.

Figure PT5.3
The contrast between (a) differentiation and (b) integration.

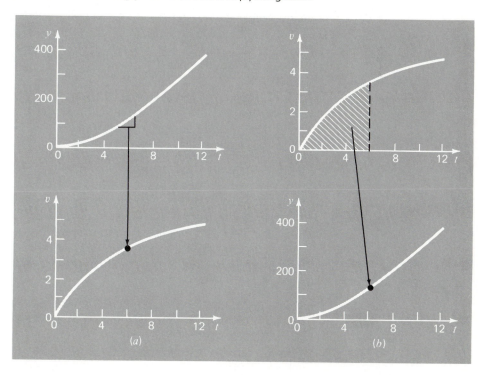

Thus, we can make the general claim that the evaluation of the integral

$$I = \int_a^b f(x)\,dx$$

is equivalent to solving the differential equation

$$\frac{dy}{dx} = f(x)$$

for $y(b)$ given the initial condition $y(a) = 0$.

Because of this close relationship, we have opted to devote this part of the book to both processes. Among other things, this will provide the opportunity to highlight their similarities and differences from a numerical perspective. In addition, our discussion will have relevance to the next part of the book where we will cover differential equations.

PT5.1.1 Precomputer Methods for Differentiation and Integration

The function to be differentiated or integrated will typically be in one of the following three forms:

1. A simple continuous function such as a polynomial, an exponential, or a trigonometric function.

2. A complicated continuous function that is difficult or impossible to differentiate or integrate directly.

3. A tabulated function where values of x and $f(x)$ are given at a number of discrete points, as is often the case with experimental or field data.

In the first case, the derivative or integral of a simple function may be evaluated analytically using calculus. For the second case, analytical solutions are often impractical, and sometimes impossible, to obtain. In these instances, as well as in the third case of discrete data, approximate methods must be employed.

A precomputer method for determining derivatives from data is called *equal-area graphical differentiation.* In this method, the *(x, y)* data is tabulated and, for each interval, a simple divided difference $\Delta y/\Delta x$ is employed to estimate the slope. Then these values are plotted as a stepped curve versus x (Fig. PT5.4). Next, a smooth curve is drawn that attempts to approximate the area under the stepped curve. That is, it is drawn so that visually, the positive and negative areas are balanced. The rates at given values of x can then be read from the curve.

In the same spirit, visually oriented approaches were employed to integrate tabulated data and complicated functions in the precomputer era. A simple intuitive approach is to plot the function on a grid (Fig. PT5.5) and count the number of boxes that approximate the area. This number multiplied by the area of each box provides a rough estimate of the total area under the curve. This estimate can be refined, at the expense of additional effort, by using a finer grid.

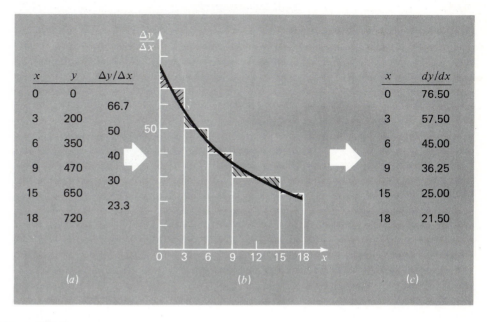

x	y	$\Delta y/\Delta x$
0	0	
		66.7
3	200	
		50
6	350	
		40
9	470	
		30
15	650	
		23.3
18	720	

x	dy/dx
0	76.50
3	57.50
6	45.00
9	36.25
15	25.00
18	21.50

(a) $\qquad\qquad\qquad (b) \qquad\qquad\qquad (c)$

Figure PT5.4
Equal-area differentiation. (a) Centered finite divided differences are used to estimate the derivative for each interval between the data points. (b) The derivative estimates are plotted as a bar graph. A smooth curve is superimposed on this plot to approximate the area under the bar graph. This is accomplished by drawing the curve so that equal positive and negative areas are balanced. (c) Values of dy/dx can then be read off the smooth curve.

Figure PT5.5
The use of a grid to approximate an integral.

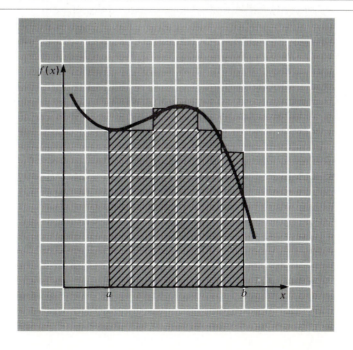

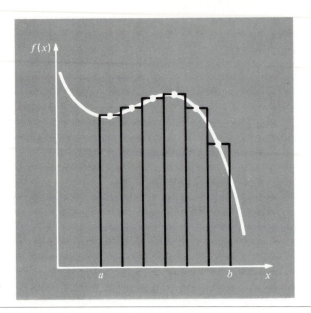

Figure PT5.6
The use of rectangles, or strips, to approximate the integral.

Another commonsense approach is to divide the area into vertical segments, or strips, with a height equal to the function value at the midpoint of each strip (Fig. PT5.6). The area of the rectangles can then be calculated and summed to estimate the total area. In this approach, it is assumed that the value at the midpoint provides a valid approximation of the average height of the function for each strip. As with the grid method, refined estimates are possible by using more (and thinner) strips to approximate the integral.

Although such simple approaches have utility for quick estimates, alternative numerical techniques are available for the same purpose. Not surprisingly, the simplest of these methods are similar in spirit to the precomputer techniques.

For differentiation, the most fundamental numerical techniques use finite divided differences to estimate derivatives. For data with error, an alternative approach is to fit a smooth curve to the data with a technique such as least-squares regression and then differentiate this curve to obtain derivative estimates.

In a similar spirit, numerical integration or *quadrature* methods are available to obtain integrals. These methods, which are actually easier to implement than the grid approach, are similar in spirit to the strip method. That is, function heights are multiplied by strip widths and summed in order to estimate the integral. However, through clever choices of weighting factors, the resulting estimate can be made more accurate than that from the simple "strip method."

As in the simple strip method, numerical integration and differentiation techniques utilize data at discrete points. Because tabulated information is already in such a form, it is naturally compatible with many of the numerical approaches. Although continuous functions are not originally in discrete form, it is usually a simple

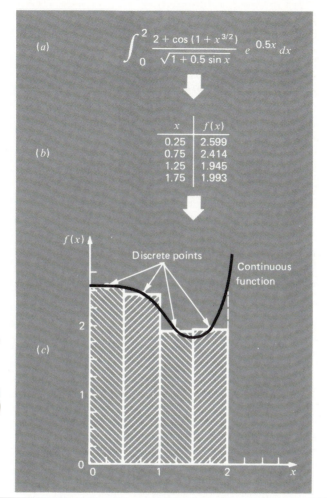

(a)

$$\int_0^2 \frac{2 + \cos\,(1 + x^{3/2})}{\sqrt{1 + 0.5 \sin x}}\ e^{\,0.5x}\,dx$$

(b)

x	$f(x)$
0.25	2.599
0.75	2.414
1.25	1.945
1.75	1.993

(c)

Figure PT5.7
Application of a numerical integration method: (*a*) A complicated, continuous function. (*b*) Table of discrete values of $f(x)$ generated from the function. (*c*) Use of a numerical method (the strip method here) to estimate the integral on the basis of the discrete points. For a tabulated function, the data is already in tabular form (*b*); therefore, step (*a*) is unnecessary.

proposition to use the given equation to generate a table of values. As depicted in Fig. PT5.7, this table can then be evaluated with a numerical method.

PT5.1.2   Numerical Differentiation and Integration in Engineering

The differentiation and integration of a function has so many engineering applications that you were required to take differential and integral calculus in your first year at college. Many specific examples of such applications could be given in all fields of engineering.

Differentiation is commonplace in engineering because so much of our work involves characterizing the changes of variables in both time and space. In fact many of the laws and other generalizations that figure so prominently in our work are based on the predictable ways in which change manifests itself in the physical world. A prime example is Newton's second law which is not couched in terms of the position of an object but rather its change of position with respect to time.

Aside from such temporal examples, numerous laws governing the spatial behavior of variables are expressed in terms of derivatives. Among the most common of these are those laws involving potentials or gradients. For example, Fourier's law of heat conduction quantifies the observation that heat flows from regions of high to low temperature. For the one-dimensional case, this can be expressed mathematically as

$$\text{Heat flux} = -k' \frac{dT}{dx}$$

Thus, the derivative provides a measure of the intensity of the temperature change or *gradient* that drives the transfer of heat. Similar laws provide workable models in many other areas of engineering including the modeling of fluid dynamics, mass transfer, chemical reaction kinetics, and electromagnetic flux. The ability to accurately estimate derivatives is an important facet of our capability to work effectively in these areas.

Just as accurate estimates of derivatives are important in engineering, the calculation of integrals is equally valuable. A number of examples relate directly to the idea of the integral as the area under a curve. Figure PT5.8 depicts a few cases where integration is used for this purpose.

Other common applications relate to the analogy between integration and summation. For example, a common application is to determine the mean of con-

Figure PT5.8
Examples of how integration is used to evaluate the areas of surfaces in engineering. (*a*) A surveyor might need to know the area of a field bounded by a meandering stream and two roads. (*b*) A water-resource engineer might need to know the cross-sectional area of a river. (*c*) A structural engineer might need to determine the net force due to a nonuniform wind blowing against the side of a skyscraper.

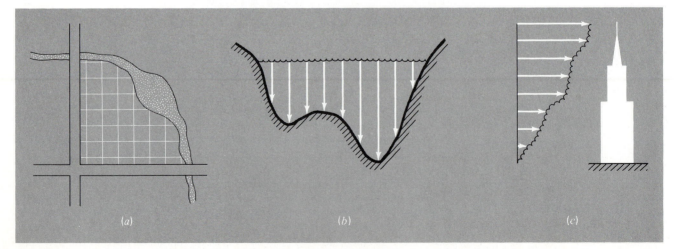

tinuous functions. In Part Four, you were introduced to the concept of the mean of *n discrete* data points [recall Eq. (PT4.1)]:

$$\text{Mean} = \frac{\displaystyle\sum_{i=1}^{n} y_i}{n} \qquad\qquad\text{(PT5.3)}$$

where y_i are individual measurements. The determination of the mean of discrete points is depicted in Fig. PT5.9a.

In contrast, suppose that y is a *continuous function* of an independent variable x, as depicted in Fig. PT5.9b. For this case, there are an infinite number of values between a and b. Just as Eq. (PT5.3) can be applied to determine the mean of the discrete readings, you might also be interested in computing the mean or average of the continuous function $y = f(x)$ for the interval from a to b. Integration is used for this purpose, as specified by the formula

$$\text{Mean} = \frac{\displaystyle\int_{a}^{b} f(x)\,dx}{b - a} \qquad\qquad\text{(PT5.4)}$$

Figure PT5.9
An illustration of the mean for (*a*) discrete and (*b*) continuous data.

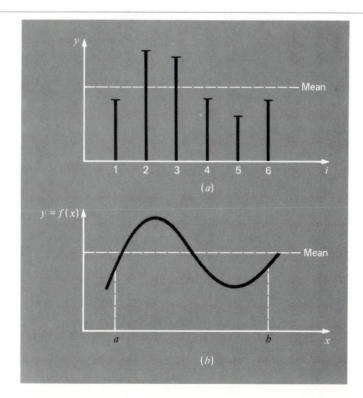

This formula has hundreds of engineering applications. For example, it is used to calculate the center of gravity of irregular objects in mechanical and civil engineering and to determine the root-mean-square current in electrical engineering.

Integrals are also employed by engineers to evaluate the total amount or quantity of a given physical variable. The integral may be evaluated over a line, an area, or a volume. For example, the total mass of chemical contained in a reactor is given as the product of the concentration of chemical and the reactor volume, or

$$\text{Mass} = \text{concentration} \times \text{volume}$$

where concentration has units of mass per volume. However, suppose that concentration varies from location to location within the reactor. In this case, it is necessary to sum the products of local concentrations c_i and corresponding elemental volumes ΔV_i:

$$\text{Mass} = \sum_{i=1}^{n} c_i \, \Delta V_i$$

where n is the number of discrete volumes. For the continuous case, where $c(x, y, z)$ is a known function and x, y, and z are independent variables designating position in cartesian coordinates, integration can be used for the same purpose:

$$\text{Mass} = \int \int \int c(x, y, z) \, dx \, dy \, dz$$

or

$$\text{Mass} = \int \int \int_{V} c(V) \, dV$$

which is referred to as a *volume integral.* Notice the strong analogy between summation and integration.

Similar examples could be given in other fields of engineering. For example, the total rate of energy transfer across a plane where the flux (in calories per square centimeter per second) is a function of position is given by

$$\text{Heat transfer} = \int \int_{A} \text{flux} \, dA$$

which is referred to as an *areal integral* where $A = $ *area.*

Similarly, for the one-dimensional case, the total weight of a variable-density rod is given by

$$w = A \int_{0}^{L} \rho(x) \, dx$$

where w is the total weight (in pounds), L is the length of the rod (in feet), $\rho(x)$ is the known density (in pounds per cubic foot) as a function of length x (in feet), and A is the cross-sectional area of the rod (in square feet).

Finally, integrals are used to evaluate rate equations. Suppose the velocity of a particle is a known continuous function of time $v(t)$. The total distance d traveled by this particle over a time t is given by (Fig. PT5.3b)

$$d = \int_0^t v(t)\, dt \tag{PT5.5}$$

These are just a few of the applications of differentiation and integration that you might face regularly in the pursuit of your profession. When the functions to be analyzed are simple, you will normally choose to evaluate them analytically. For example, in the falling parachutist problem, we determined the solution for velocity as a function of time [Eq. (1.10)]. This relationship could be substituted into Eq. (PT5.5), which could then be integrated easily to determine how far the parachutist fell over a time period t. For this case, the integral is simple to evaluate. However, it is difficult or impossible when the function is complicated, as is typically the case in more realistic examples. In addition, the underlying function is often unknown and defined only by measurement at discrete points. For both these cases, you must have the ability to obtain approximate values for derivatives and integrals using numerical techniques. Several such techniques will be discussed in this part of the book.

PT5.2 MATHEMATICAL BACKGROUND

In high school or during your first year of college, you were introduced to *differential* and *integral calculus*. There you learned techniques to obtain analytical or exact derivatives and integrals.

When we differentiate a function analytically, we generate a second function that can be used to compute the derivative for different values of the independent variable. General rules are available for this purpose. For example, in the case of the monomial

$$y = x^n$$

the following simple rule applies ($n \neq 0$)

$$\frac{dy}{dx} = nx^{n-1}$$

which is the expression of the more general rule for

$$y = u^n$$

where u is a function of x. For this equation, the derivative is computed via

$$\frac{dy}{dx} = nu^{n-1}\frac{du}{dx}$$

Two other useful formulas apply to the products and quotients of functions. For example, if the product of two functions of x (u and v) is represented as $y = uv$, then the derivative can be computed as

$$\frac{dy}{dx} = u\frac{dv}{dx} + v\frac{du}{dx}$$

For the division, $y = u/v$, the derivative can be computed as

$$\frac{dy}{dx} = \frac{v\dfrac{du}{dx} - u\dfrac{dv}{dx}}{v^2}$$

Other useful formulas are summarized in Table PT5.1.

TABLE PT5.1 Some commonly used derivatives.

$\dfrac{d}{dx}\sin x = \cos x$	$\dfrac{d}{dx}\cot x = -\csc^2 x$
$\dfrac{d}{dx}\cos x = -\sin x$	$\dfrac{d}{dx}\sec x = \sec x \tan x$
$\dfrac{d}{dx}\tan x = \sec^2 x$	$\dfrac{d}{dx}\csc x = -\csc x \cot x$
$\dfrac{d}{dx}\ln x = \dfrac{1}{x}$	$\dfrac{d}{dx}\log_a x = \dfrac{1}{x \ln a}$
$\dfrac{d}{dx}e^x = e^x$	$\dfrac{d}{dx}a^x = a^x \ln a$

Similar formulas are available for definite integration, which deals with determining an integral between specified limits, as in

$$I = \int_a^b f(x)\,dx \tag{PT5.6}$$

According to the *fundamental theorem* of integral calculus, Eq. (PT5.6) is evaluated as

$$\int_a^b f(x)\,dx = F(x)\bigg|_a^b$$

where $F(x)$ is the integral of $f(x)$—that is, any function such that $F'(x) = f(x)$. The nomenclature on the right-hand side stands for

$$F(x)\Big|_a^b = F(b) - F(a) \tag{PT5.7}$$

An example of a definite integral is

$$I = \int_0^{0.8} (0.2 + 25x - 200x^2 + 675x^3 - 900x^4 + 400x^5)\,dx \tag{PT5.8}$$

For this case, the function is a simple polynomial that can be integrated analytically by evaluating each term according to the rule

$$\int_a^b x^n\,dx = \frac{x^{n+1}}{n+1}\Big|_a^b \tag{PT5.9}$$

where n cannot equal -1. Applying this rule to each term in Eq. (PT5.8) yields

$$I = 0.2x + 12.5x^2 - \frac{200}{3}x^3 + 168.75x^4 - 180x^5 + \frac{400}{6}x^6 \Big|_0^{0.8}$$

which can be evaluated according to Eq. (PT5.7) as $I = 1.64053334$. This value is equal to the area under the original polynomial [Eq. (PT5.8)] between $x = 0$ and 0.8.

The foregoing integration depends on knowledge of the rule expressed by Eq. (PT5.9). Other functions follow different rules. These "rules" are all merely instances of antidifferentiation, that is, finding $F(x)$ so that $F'(x) = f(x)$. Consequently, analytical integration depends on prior knowledge of the answer. Such knowledge is acquired by training and experience. Many of the rules are summarized in handbooks and in tables of integrals. We list some commonly encountered integrals in Table PT5.2. However, many functions of practical importance are too complicated to be contained in such tables. One reason why the techniques in the present part of the book are so valuable is that they provide a means to evaluate relationships such as Eq. (PT5.8) without knowledge of the rules.

PT5.3 ORIENTATION

Before proceeding to the numerical methods for integration, some further orientation might be helpful. The following is intended as an overview of the material discussed in Part Five. In addition, we have formulated some objectives to help focus your efforts when studying the material.

TABLE PT5.2   Some simple integrals that are used in Part Five. The *a* and *b* in this table are constants and should not be confused with the limits of integration discussed in the text.

$$\int u \, dv = uv - \int v \, du$$

$$\int u^n \, du = \frac{u^{n+1}}{n+1} + C \qquad n \neq -1$$

$$\int a^{bx} \, dx = \frac{a^{bx}}{b \ln a} + C \qquad a > 0, \, a \neq 1$$

$$\int \frac{dx}{x} = \ln |x| + C$$

$$\int \sin (ax + b) = -\frac{1}{a} \cos (ax + b) + C$$

$$\int \cos (ax + b) \, dx = \frac{1}{a} \sin (ax + b) + C$$

$$\int \ln |x| \, dx = x \ln |x| - x + C$$

$$\int e^{ax} \, dx = \frac{e^{ax}}{a} + C$$

$$\int xe^{ax} \, dx = \frac{e^{ax}}{a^2} (ax - 1) + C$$

$$\int \frac{dx}{a + bx^2} = \frac{1}{\sqrt{ab}} \tan^{-1} \frac{\sqrt{ab}}{a} x + C$$

PT5.3.1 Scope and Preview

Figure PT5.10 provides an overview of Part Five. *Chapter 15* is devoted to the most common approaches for numerical integration—the *Newton-Cotes formulas*. These relationships are based on replacing a complicated function or tabulated data with a simple polynomial that is easy to integrate. Three of the most widely used Newton-Cotes formulas are discussed in detail: the *trapezoidal rule, Simpson's 1/3 rule,* and *Simpson's 3/8 rule*. All of these formulas are designed for cases where the data to be integrated is evenly spaced. In addition, we also include a discussion of numerical integration of unequally spaced data. This is a very important topic because many real-world applications deal with data that is in this form.

All the above material relates to closed integration, where the function values at the ends of the limits of integration are known. At the end of Chap. 15, we present *open integration formulas,* where the integration limits extend beyond the range of the known data. Although they are not commonly used for definite integration,

Figure PT5.10
Schematic of the organization of material in Part Five: Numerical Integration and Differentiation.

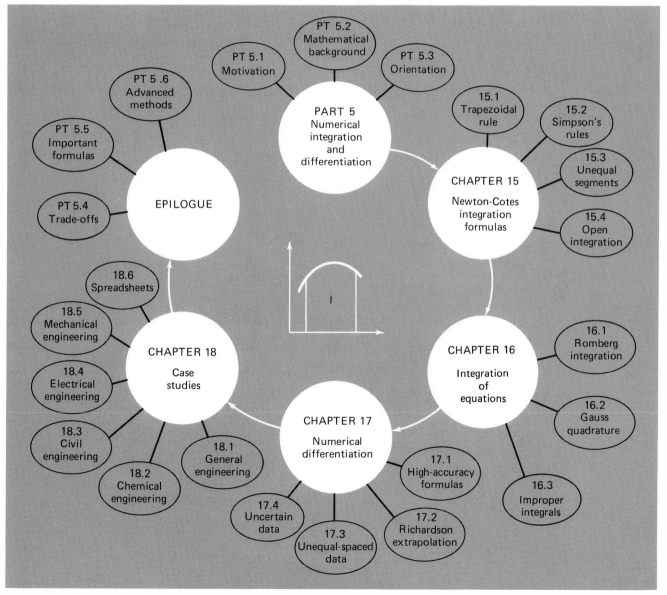

open integration formulas are presented here because they are utilized extensively in the solution of ordinary differential equations in Part Six.

The formulations covered in Chap. 15 can be employed to analyze both tabulated data and equations. *Chapter 16* deals with two techniques that are expressly designed to integrate equations: *Romberg integration* and *Gauss quadrature*. Computer algorithms are provided for both of these methods. In addition, methods for evaluating improper integrals are discussed.

In *Chap. 17*, we present additional information on *numerical differentiation* to supplement the introductory material from Chap. 3. Topics include high-accuracy finite-difference formulas, Richardson's extrapolation, and the differentiation of unequally spaced data. The effect of errors on both numerical differentiation and integration is discussed at the chapter's end.

Chapter 18 demonstrates how the methods can be applied for problem solving. As with other parts of the book, case studies are drawn from all fields of engineering. A case study on spreadsheet applications is also included.

A review section, or *epilogue,* is included at the end of Part Five. This review includes a discussion of trade-offs that are relevant to implementation in engineering practice. In addition, important formulas are summarized. Finally, we present a short review of advanced methods and alternative references that will facilitate your further studies of numerical integration.

PT5.3.2   Goals and Objectives

Study Objectives.   After completing Part Five you should be able to solve many numerical integration and differentiation problems and appreciate their application for engineering problem solving. You should strive to master several techniques and assess their reliability. You should understand the trade-offs involved in selecting the "best" method (or methods) for any particular problem. In addition to these general objectives, the specific concepts listed in Table PT5.3 should be assimilated and mastered.

Computer Objectives.   You have been provided with software, simple computer programs, and algorithms to implement the techniques discussed in Part Five. All have utility as learning tools.

The Electronic TOOLKIT personal computer software is user-friendly. It employs the trapezoidal rule to evaluate the integral of either continuous functions or tabular data. The graphics associated with this software will enable you to easily visualize your problem and the associated mathematical operations as the area between the curve and the *x* axis. The software is very easy to apply to solve many practical problems and can be used to check the results of any computer programs you may develop yourself.

Alternatively, BASIC, FORTRAN, and Pascal programs for the trapezoidal rule and

TABLE PT5.3 Specific study objectives for Part Five.

1. Understand the derivation of the Newton-Cotes formulas; know how to derive the trapezoidal rule and how to set up the derivation of both of Simpson's rules; recognize that the trapezoidal and Simpson's 1/3 and 3/8 rules represent the areas under first-, second-, and third-order polynomials, respectively

2. Know the formulas and error equations for
   (a) The trapezoidal rule
   (b) The multiple-application trapezoidal rule
   (c) Simpson's 1/3 rule
   (d) Simpson's 3/8 rule
   (e) The multiple-application Simpson's rule
   Be able to choose the "best" among these formulas for any particular problem context

3. Recognize that Simpson's 1/3 rule is fourth-order accurate even though it is based on only three points; realize that all the even-segment–odd-point Newton-Cotes formulas have similar enhanced accuracy

4. Know how to evaluate the integral and derivative of unequally spaced data

5. Recognize the difference between open and closed integration formulas

6. Understand the theoretical basis of Richardson extrapolation and how it is applied in the Romberg integration algorithm and for numerical differentiation

7. Understand the fundamental difference between Newton-Cotes and Gauss quadrature formulas

8. Recognize why both Romberg integration and Gauss quadrature have utility when integrating equations (as opposed to tabular or discrete data)

9. Know how open integration formulas are employed to evaluate improper integrals

10. Understand the application of high-accuracy numerical-differentiation formulas

11. Recognize the differing effects of data error on the processes of numerical integration and differentiation

numerical differentiation are supplied directly in the text. In addition, general algorithms are provided for most of the methods in Part Five. This information will allow you to expand your software library to include techniques beyond the trapezoidal rule. For example, you may find it useful from a professional viewpoint to develop software that can handle numerical integration and differentiation of unequally spaced data. You may also want to develop your own software for Simpson's rules, Romberg integration, and Gauss quadrature, which are usually more efficient and accurate than the trapezoidal rule.

Finally, a diskette for spreadsheet applications is included in the back of the book. This software is designed to accompany and illustrate Case Study 18.6.

CHAPTER 15
Newton-Cotes Integration Formulas

The *Newton-Cotes formulas* are the most common numerical integration schemes. They are based on the strategy of replacing a complicated function or tabulated data with some approximating function that is easy to integrate:

$$I = \int_a^b f(x)\, dx \simeq \int_a^b f_n(x)\, dx \tag{15.1}$$

where $f_n(x)$ is a polynomial of the form

$$f_n(x) = a_0 + a_1 + \cdots + a_{n-1}x^{n-1} + a_n x^n$$

where n is the order of the polynomial. For example, in Fig. 15.1a, a first-order

Figure 15.1
The approximation of an integral by the area under (a) a single straight line and (b) a single parabola.

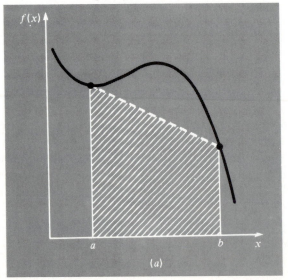
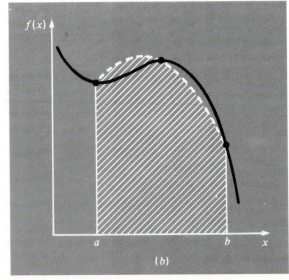

(a)

(b)

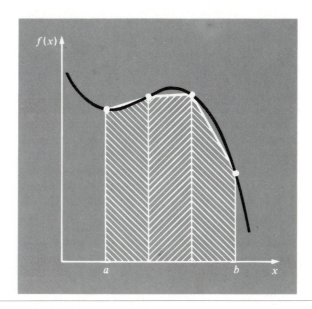

Figure 15.2
The approximation of an
integral by the area under
three straight-line segments.

polynomial (a straight line) is used as an approximation. In Fig. 15.1*b*, a parabola is employed for the same purpose.

The integral can also be approximated using a series of polynomials applied piecewise to the function or data over segments of constant length. For example, in Fig. 15.2, three straight-line segments are used to approximate the integral. Higher-order polynomials can be utilized for the same purpose. With this background, we now recognize that the "strip method" in Fig. PT5.6 employed a series of zero-order polynomials (that is, constants) to approximate the integral.

Closed and open forms of the Newton-Cotes formulas are available. The *closed forms* are those where the data points at the beginning and end of the limits of integration are known (Fig. 15.3*a*). The *open forms* have integration limits that extend beyond the range of the data (Fig. 15.3*b*). In this sense, they are akin to extrapolation as

Figure 15.3
The difference between (*a*)
closed and (*b*) open integration
formulas.

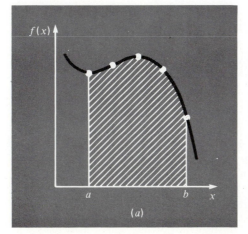

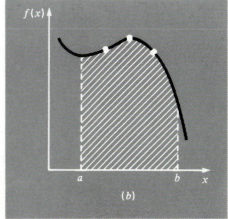

discussed in Sec. 12.4. Open Newton-Cotes formulas are not generally used for definite integration. However, they are utilized for evaluating improper integrals and for the solution of ordinary differential equations. The present chapter emphasizes the closed forms. However, material on open Newton-Cotes formulas is briefly introduced at the end of this chapter.

15.1 THE TRAPEZOIDAL RULE

The *trapezoidal rule* is the first of the Newton-Cotes closed integration formulas. It corresponds to the case where the polynomial in Eq. (15.1) is first-order.

$$I = \int_a^b f(x)\,dx \simeq \int_a^b f_1(x)\,dx$$

Recall from Chap. 12 that a straight line can be represented as [Eq. (12.2)]

$$f_1(x) = f(a) + \frac{f(b) - f(a)}{b - a}(x - a) \tag{15.2}$$

The area under this straight line is an estimate of the integral of $f(x)$ between the limits a and b:

$$I \simeq \int_a^b \left[f(a) + \frac{f(b) - f(a)}{b - a}(x - a) \right] dx$$

The result of the integration (see Box 15.1 for details) is

$$I \simeq (b - a)\frac{f(a) + f(b)}{2} \tag{15.3}$$

which is called the *trapezoidal rule*.

Geometrically, the trapezoidal rule is equivalent to approximating the area of the trapezoid under the straight line connecting $f(a)$ and $f(b)$ in Fig. 15.4. Recall from

Figure 15.4
Graphical depiction of the trapezoidal rule.

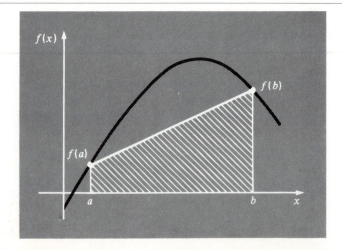

Box 15.1 Derivation of Trapezoidal Rule

Before integration, Eq. (15.2) can be expressed as

$$f_1(x) = \frac{f(b) - f(a)}{b - a} x + f(a) - \frac{af(b) - af(a)}{b - a}$$

Grouping the last two terms gives

$$f_1(x) = \frac{f(b) - f(a)}{b - a} x$$

$$+ \frac{bf(a) - af(a) - af(b) + af(a)}{b - a}$$

or

$$f_1(x) = \frac{f(b) - f(a)}{b - a} x + \frac{bf(a) - af(b)}{b - a}$$

which can be integrated between $x = a$ and $x = b$ to yield

$$I \simeq \frac{f(b) - f(a)}{b - a} \frac{x^2}{2} + \frac{bf(a) - af(b)}{b - a} x \bigg|_a^b$$

This result can be evaluated to give

$$I \simeq \frac{f(b) - f(a)}{b - a} \frac{(b^2 - a^2)}{2}$$

$$+ \frac{bf(a) - af(b)}{b - a} (b - a)$$

Now, since $b^2 - a^2 = (b - a)(b + a)$,

$$I \simeq [f(b) - f(a)] \frac{(b + a)}{2} + bf(a) - af(b)$$

Multiplying and collecting terms yields

$$I \simeq (b - a) \frac{f(a) + f(b)}{2}$$

which is the formula for the trapezoidal rule.

geometry that the formula for computing the area of a trapezoid is the height times the average of the bases (Fig. 15.5a). In our case, the concept is the same but the trapezoid is on its side (Fig. 15.5b). Therefore, the integral estimate can be represented as

$$I \simeq \text{width} \times \text{average height} \tag{15.4}$$

or

$$I \simeq (b - a) \times \text{average height} \tag{15.5}$$

Figure 15.5
(a) The formula for computing the area of a trapezoid: height times the average of the bases. (b) For the trapezoidal rule, the concept is the same but the trapezoid is on its side.

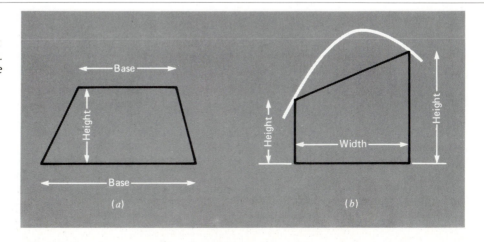
(a)   (b)

where for the trapezoidal rule, the average height is the average of the function values at the end points, or $[f(a) + f(b)]/2$.

All the Newton-Cotes closed formulas can be expressed in the general format of Eq. (15.5). In fact, they only differ with respect to the formulation of the average height.

Box 15.2 Derivation and Error Estimate of the Trapezoidal Rule Based on Integrating the Forward Newton-Gregory Interpolating Polynomial

---

An alternative derivation of the trapezoidal rule is possible by integrating the forward Newton-Gregory interpolating polynomial. Recall that for the first-order version with error term, the integral would be (Box 12.2)

$$I = \int_a^b \left[f(a) + \Delta f(a)\alpha \right.$$
$$\left. + \frac{f''(\xi)}{2}\alpha(\alpha - 1)h^2 \right] dx \qquad \text{(B15.2.1)}$$

In order to simplify the analysis, realize that because $\alpha = (x - a)/h$,

$$dx = h\,d\alpha$$

Inasmuch as $h = b - a$ (for the one-segment trapezoidal rule), the limits of integration, a and b, correspond to 0 and 1, respectively. Therefore, Eq. (B15.2.1) can be expressed as

$$I = h \int_0^1 \left[f(a) + \Delta f(a)\,\alpha \right.$$
$$\left. + \frac{f''(\xi)}{2}\alpha(\alpha - 1)\,h^2 \right] d\alpha$$

If it is assumed that, for small h, the term $f''(\xi)$ is approximately constant, this equation can be integrated:

$$I = h\left[\alpha f(a) + \frac{\alpha^2}{2}\Delta f(a) + \left(\frac{\alpha^3}{6} - \frac{\alpha^2}{4} \right) f''(\xi)h^2 \right]_0^1$$

and evaluated as

$$I = h\left[f(a) + \frac{\Delta f(a)}{2} \right] - \frac{1}{12}f''(\xi)h^3$$

Because $\Delta f(a) = f(b) - f(a)$, the result can be written as

$$I = h\underbrace{\frac{f(a) + f(b)}{2}}_{\text{Trapezoidal rule}} - \underbrace{\frac{1}{12}f''(\xi)h^3}_{\text{Truncation error}}$$

Thus, the first term is the trapezoidal rule and the second is an approximation for the error.

15.1.1 Error of the Trapezoidal Rule

When we employ the integral under a straight-line segment to approximate the integral under a curve, we obviously can incur an error that may be substantial (Fig. 15.6). An estimate for the local truncation error of a single application of the trapezoidal rule is (Box. 15.2)

$$E_t = -\frac{1}{12}f''(\xi)(b - a)^3 \qquad (15.6)$$

where ξ lies somewhere in the interval from a to b. Equation (15.6) indicates that if the function being integrated is linear, the trapezoidal rule will be exact. Otherwise, for functions with second- and higher-order derivatives (that is, with curvature), some error can occur.

EXAMPLE 15.1   Single Application of the Trapezoidal Rule

Problem Statement: Use Eq. (15.3) to numerically integrate

$$f(x) = 0.2 + 25x - 200x^2 + 675x^3 - 900x^4 + 400x^5$$

from $a = 0$ to $b = 0.8$. Recall from PT5.2 that the exact value of the integral can be determined analytically to be 1.64053334.

Solution: The function values

$$f(0) = 0.2$$

$$f(0.8) = 0.232$$

can be substituted into Eq. (15.3) to yield

$$I \simeq 0.8 \frac{0.2 + 0.232}{2} = 0.1728$$

which represents an error of

$$E_t = 1.64053334 - 0.1728 = 1.46773334$$

which corresponds to a percent relative error of $\epsilon_t = 89.5\%$. The reason for this large error is evident from the graphical depiction in Fig. 15.6. Notice that the area under the straight line neglects a significant portion of the integral lying above the line.

Figure 15.6
Graphical depiction of the use of a single application of the trapezoidal rule to approximate the integral of $f(x) = 0.2 + 25x - 200x^2 + 675x^3 - 900x^4 + 400x^5$ from $x = 0$ to 0.8.

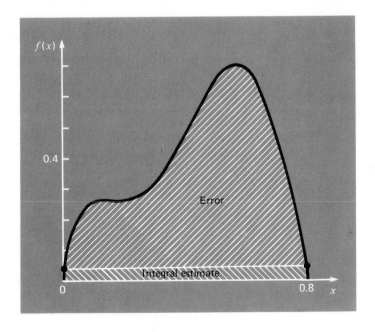

In actual situations, we would have no foreknowledge of the true value. Therefore, an approximate error estimate is required. To obtain this estimate, the function's second derivative over the interval can be computed by differentiating the original function twice to give

$$f''(x) = -400 + 4050x - 10{,}800x^2 + 8000x^3$$

The average value of the second derivative can be computed using Eq. (PT5.4):

$$\bar{f}'' = \frac{\displaystyle\int_0^{0.8} (-400 + 4050x - 10{,}800x^2 + 8000x^3)\,dx}{0.8 - 0} = -60$$

which can be substituted into Eq. (15.6) to yield

$$E_a = -\frac{1}{12}(-60)(0.8)^3 = 2.56$$

which is of the same order of magnitude and sign as the true error. A discrepancy does exist, however, because of the fact that for an interval of this size, the average second derivative is not necessarily an accurate approximation of $f''(\xi)$. Thus, we denote that the error is approximate by using the notation E_a, rather than exact by using E_t.

15.1.2 The Multiple-Application Trapezoidal Rule

One way to improve the accuracy of the trapezoidal rule is to divide the integration interval from a to b into a number of segments and apply the method to each segment (Fig. 15.7). The areas of individual segments can then be added to yield the integral for the entire interval. The resulting equations are called *multiple-application*, or *composite*, *integration formulas*.

Figure 15.8 shows the general format and nomenclature we will use to characterize multiple-application integrals. There are $n + 1$ equally spaced base points $(x_0, x_1, x_2, \ldots, x_n)$. Consequently, there are n segments of equal width:

$$h = \frac{b - a}{n} \tag{15.7}$$

If a and b are designated as x_0 and x_n, respectively, the total integral can be represented as

$$I = \int_{x_0}^{x_1} f(x)\,dx + \int_{x_1}^{x_2} f(x)\,dx + \cdots + \int_{x_{n-1}}^{x_n} f(x)\,dx$$

Substituting the trapezoidal rule for each integral yields

$$I \simeq h\,\frac{f(x_0) + f(x_1)}{2} + h\,\frac{f(x_1) + f(x_2)}{2} + \cdots + h\,\frac{f(x_{n-1}) + f(x_n)}{2} \tag{15.8}$$

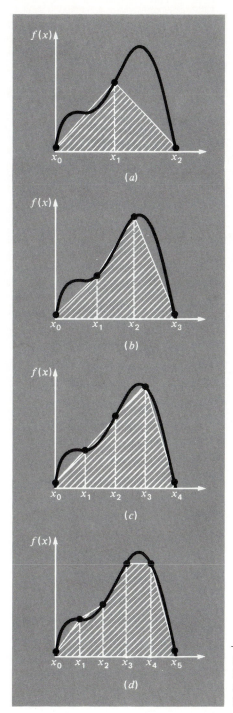

Figure 15.7
Illustration of the multiple-application trapezoidal rule. (*a*) Two segments, (*b*) three segments, (*c*) four segments, and (*d*) five segments.

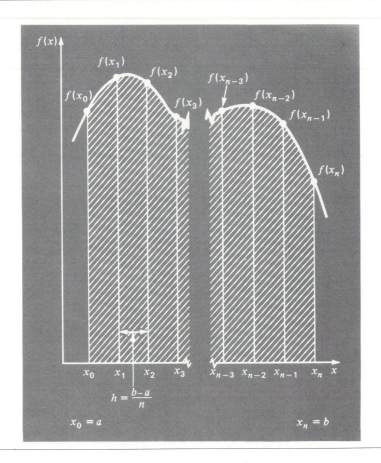

Figure 15.8
The general format and no-menclature for multiple-application integrals.

or, grouping terms,

$$I \simeq \frac{h}{2} \left[f(x_0) + 2 \sum_{i=1}^{n-1} f(x_i) + f(x_n) \right] \tag{15.9}$$

or, using Eq. (15.7) to express Eq. (15.9) in the general form of Eq. (15.5),

$$I \simeq \underbrace{(b - a)}_{\text{Width}} \underbrace{\frac{f(x_0) + 2 \sum_{i=1}^{n-1} f(x_i) + f(x_n)}{2n}}_{\text{Average height}} \tag{15.10}$$

Because the summation of the coefficients of $f(x)$ in the numerator divided by $2n$ is equal to 1, the average height represents a *weighted average* of the function values. According to Eq. (15.10), the interior points are given twice the weight of the two end points $f(x_0)$ and $f(x_n)$.

An error for the multiple-application trapezoidal rule can be obtained by summing the individual errors for each segment to give

$$E_t = -\frac{(b-a)^3}{12n^3} \sum_{i=1}^{n} f''(\xi_i) \tag{15.11}$$

where $f''(\xi_i)$ is the second derivative at a point ξ_i located in segment i. This result can be simplified by estimating the mean or average value of the second derivative for the entire interval as [Eq. (PT5.3)]

$$\bar{f}'' \simeq \frac{\sum_{i=1}^{n} f''(\xi_i)}{n} \tag{15.12}$$

Therefore, $\Sigma f''(\xi_i) \simeq n\bar{f}''$ and Eq. (15.11) can be rewritten as

$$E_a = -\frac{(b-a)^3}{12n^2}\bar{f}'' \tag{15.13}$$

Thus, if the number of segments is doubled, the truncation error will be quartered. Note that Eq. (15.13) is an approximate error because of the approximate nature of Eq. (15.12).

EXAMPLE 15.2   Multiple-Application Trapezoidal Rule

Problem Statement: Use the two-segment trapezoidal rule to estimate the integral of

$$f(x) = 0.2 + 25x - 200x^2 + 675x^3 - 900x^4 + 400x^5$$

from $a = 0$ to $b = 0.8$. Employ Eq. (15.13) to estimate the error. Recall from PT5.2 that the correct value for the integral is 1.64053334.

Solution: $n = 2 \, (h = 0.4)$:

$$f(0) = 0.2$$

$$f(0.4) = 2.456$$

$$f(0.8) = 0.232$$

$$l \simeq 0.8\frac{0.2 + 2(2.456) + 0.232}{4} = 1.0688$$

$$E_t = 1.64053334 - 1.0688 = 0.57173 \qquad \epsilon_t = 34.9\%$$

$$E_a = -\frac{0.8^3}{12(2)^2}(-60) = 0.64$$

where -60 is the average second derivative determined previously in Example 15.1.

The results of the previous example, along with three- through ten-segment applications of the trapezoidal rule, are summarized in Table 15.1. Notice how the error decreases as the number of segments increases. However, also notice that the rate of decrease is gradual. This is because the error is inversely related to the square of n [Eq. (15.13)]. Therefore, doubling the number of segments quarters the error. In subsequent sections we develop higher-order formulas that are more accurate and that converge more quickly on the true integral as the segments are increased. However, before investigating these formulas, we will first discuss a computer program to implement the trapezoidal rule.

TABLE 15.1   Results for multiple-application trapezoidal rule to estimate the integral of $f(x) = 0.2 + 25x - 200x^2 + 675x^3 - 900x^4 + 400x^5$ from $x = 0$ to 0.8. The exact value is 1.64053334.

n	h	I	ϵ_t, %
2	0.4	1.0688	34.9
3	0.2667	1.3695	16.5
4	0.2	1.4848	9.5
5	0.16	1.5399	6.1
6	0.1333	1.5703	4.3
7	0.1143	1.5887	3.2
8	0.1	1.6008	2.4
9	0.0889	1.6091	1.9
10	0.08	1.6150	1.6

15.1.3   Computer Program for the Multiple-Application Trapezoidal Rule

Simple computer programs to implement the trapezoidal rule are listed in Figs. 15.9 and 15.10. These programs have a number of shortcomings. First, they are limited to data that is in tabulated form. A general program should have the capability to evaluate known functions or equations as well. Additionally, the programs are not user-friendly; they are designed strictly to come up with the answer. In Prob. 15.21, you will have the task of making these skeletal computer codes easier to use and understand. Also, you will have the optional task of modifying the programs so that they are capable of evaluating the integral of known functions.

An example of a user-friendly program for the trapezoidal rule is included in the supplementary Electronic TOOLKIT software associated with this text. This software can evaluate the integrals of either tabulated data or user-defined functions. The following example demonstrates its utility for evaluating integrals. It also provides a good reference for assessing and testing your own software.

```
100 REM     TRAPEZOIDAL RULE (BASIC VERSION)
105 REM
110 REM *********************************
115 REM *        DEFINITION OF VARIABLES      *
120 REM *                                     *
125 REM * N   = NUMBER OF SEGMENTS            *
130 REM * X() = INDEPENDENT VARIABLE          *
135 REM * F() = DEPENDENT VARIABLE            *
140 REM *********************************
145 REM
150 DIM X(100),F(100)
155 REM
160 REM ********* MAIN PROGRAM **********
165 REM
170 GOSUB 200 'input data
175 GOSUB 300 'perform integration
180 GOSUB 350 'output results
185 END
200 REM ******** SUBROUTINE INPUT *********
205 CLS
210 INPUT "INTEGRATION INTERVAL (A,B)? ",A,B
215 PRINT
220 INPUT "NUMBER SEGMENTS? ", N
225 H = (B-A)/N
230 PRINT
235 X(0) = A
240 PRINT "F(AT X = ";X(0);") = ";
245 INPUT F(0)
250 FOR I = 1 TO N
255     X(I) = X(I-1)+H
260     PRINT "F(AT X = ";X(I);") = ";
265     INPUT F(I)
270 NEXT I
275 RETURN
300 REM *****  SUBROUTINE INTEGRATION  *****
305 REM
310 SUM = 0
315 FOR I = 1 TO N-1
320     SUM = SUM + F(I)
325 NEXT I
330 INTEGRAL = (B-A)*(F(0)+2*SUM+F(N))/(2*N)
335 RETURN
350 REM ******* SUBROUTINE OUTPUT *******
355 REM
360 CLS
365 PRINT:PRINT
370 PRINT "INTEGRAL = ";INTEGRAL
375 RETURN
```

Figure 15.9
An interactive computer program for the multiple-application trapezoidal rule written in Microsoft BASIC.

Figure 15.10
Subroutines for the multiple-application trapezoidal rule written in FORTRAN 77 and Turbo Pascal.

```
      SUBROUTINE TRAP(F,N,A,B,INTEGR)
*********************************************
*          DEFINITION OF VARIABLES         *
*                                          *
*    N      = NUMBER OF SEGMENTS           *
*    F()    = DEPENDENT VARIABLE           *
*    A,B    = INTEGRATION LIMITS           *
*    INTEGR = INTEGRAL                     *
*********************************************
      DIMENSION F(100)
      REAL INTEGR
      SUM = 0.0
      DO 10 I=2,N
         SUM = SUM + F(I)
   10 CONTINUE
      INTEGR = (B-A)*(F(1)+2*SUM*F(N+1))/(2*N)
      RETURN
      END
```

```
PROCEDURE TrapRule (F: Vector;
                    A,B : real;
                    N :integer;
                    VAR Integral: real);

{      Driver program type definitions
   Vector = a 1 dimensional real array
              (defined with zero subscript)}

{          Definition of variables
   F() = dependent variable
   N   = number of segments
   A,B = interval bounds                 }

VAR
   Sum : real;
   i : integer;
Begin
Sum := 0.0;
For i := 1 to N-1 do
   Begin
   Sum := Sum + F[i];
   End;
Integral := (B - A)*(F[0]+2*Sum+F[N])/(2*N);
End;                  {of procedure TrapRule}
```

EXAMPLE 15.3   Evaluating Integrals with the Computer

Problem Statement: A user-friendly computer program to implement the multiple-application trapezoidal rule is contained in the Electronic TOOLKIT software associated with the text. We can use this software to solve a problem related to our friend the falling parachutist. As you recall from Example 1.1, the velocity of the parachutist is given as the following function of time:

$$v(t) = \frac{gm}{c}[1 - e^{-(c/m)t}] \tag{E15.3.1}$$

where v is the velocity in meters per second, g is the gravitational constant of 9.8 m/s², m is the mass of the parachutist equal to 68.1 kg, and c is the drag coefficient of 12.5 kg/s. The model predicts the velocity of the parachutist as a function of time as described in Example 1.1. A plot of the velocity variation was developed in Example 2.1.

Suppose we would like to know how far the parachutist has fallen after a certain time t. This distance is given by [Eq. (PT5.5)]

$$d = \int_0^t v(t)\, dt$$

where d is the distance in meters. Substituting Eq. (E15.3.1) and setting $t = 10$ s,

$$d = \frac{gm}{c}\int_0^{10}[1 - e^{-(c/m)t}]\, dt$$

Figure 15.11
Computer screens showing (a) the entry of integration parameters and integration results and (b) a graph of the integral as the area between the function and the x axis.

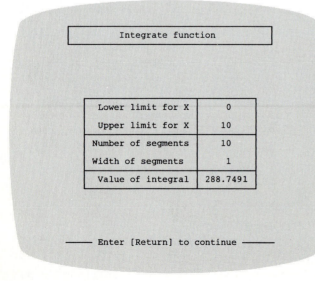

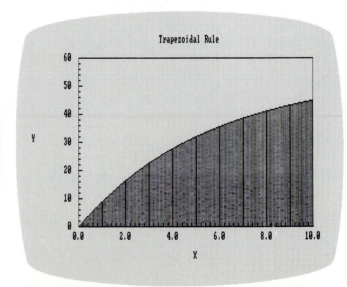

Performing the integration and substituting known parameter values results in

$$d = 289.435147 \text{ m}$$

This exact result can be used to study the efficiency of the multiple-application trapezoidal rule. Figure 15.11*a* shows the computer screen that requests the upper and lower limits of integration and the step size. After the computations are complete, the integral is printed as 288.7491. The integral is equivalent to the area under the $v(t)$ versus t curve, as shown in Fig. 15.11*b*. Visual inspection confirms that the integral is the width of the interval (10 s) times the average height (about 29 m/s).

Other numbers of segments can be easily tested by repeating the computations. The results indicate how the estimated distance the parachutist falls approaches the exact result as the segment size decreases:

Segments	Segment size	Estimated d, m	ϵ_t, %
10	1.0	288.7491	0.237
20	0.5	289.2636	0.0593
50	0.2	289.4076	9.51×10^{-3}
100	0.1	289.4282	2.40×10^{-3}
200	0.05	289.4336	5.37×10^{-4}
500	0.02	289.4348	1.15×10^{-4}
1,000	0.01	289.4360	-2.96×10^{-4}
2,000	0.005	289.4369	-5.92×10^{-4}
5,000	0.002	289.4337	5.16×10^{-4}
10,000	0.001	289.4317	1.20×10^{-3}

Thus, the multiple-application trapezoidal rule attains excellent accuracy. However, notice how the error changes sign and begins to increase in absolute value beyond the 500-segment case. This is due to the intrusion of round-off error because of the great number of computations for this many segments. Thus, the level of precision is limited, and we would never reach the exact result of 289.435147 obtained analytically. This limitation will be discussed in further detail in Chap. 16.

15.2 SIMPSON'S RULES

Aside from applying the trapezoidal rule with finer segmentation, another way to obtain a more accurate estimate of an integral is to use higher-order polynomials to connect the points. For example, if there is an extra point midway between $f(a)$ and $f(b)$, the three points can be connected with a parabola (Fig. 15.12*a*). If there are two points equally spaced between $f(a)$ and $f(b)$, the four points can be connected with a third-order polynomial (Fig. 15.12*b*). The formulas that result from taking the integrals under these polynomials are called *Simpson's rules*.

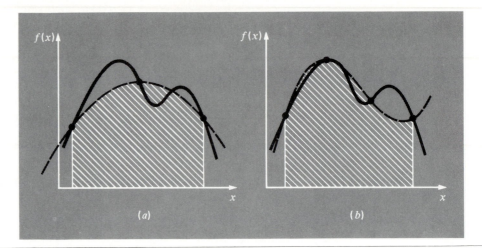

Figure 15.12
(*a*) Graphical depiction of
Simpson's 1/3 rule: it consists
of taking the area under
a parabola connecting three
points. (*b*) Graphical depiction
of Simpson's 3/8 rule: it consists of taking the area under
a cubic equation connecting
four points.

15.2.1 Simpson's 1/3 Rule

Simpson's 1/3 rule results when a second-order interpolating polynomial is substituted
into Eq. (15.1):

$$I = \int_a^b f(x)\, dx \simeq \int_a^b f_2(x)\, dx$$

If a and b are designated as x_0 and x_2 and $f_2(x)$ is represented by a second-order Lagrange
polynomial [Eq. (12.23)], the integral becomes

$$I \simeq \int_{x_0}^{x_2} \left[\frac{(x - x_1)(x - x_2)}{(x_0 - x_1)(x_0 - x_2)} f(x_0) + \frac{(x - x_0)(x - x_2)}{(x_1 - x_0)(x_1 - x_2)} f(x_1) \right.$$

$$\left. + \frac{(x - x_0)(x - x_1)}{(x_2 - x_0)(x_2 - x_1)} f(x_2) \right] dx$$

After integration and algebraic manipulation, the following formula results:

$$I \simeq \frac{h}{3}[f(x_0) + 4f(x_1) + f(x_2)] \qquad\qquad (15.14)$$

where, for this case, $h = (b - a)/2$. This equation is known as *Simpson's 1/3 rule*. It
is the second Newton-Cotes closed integration formula. The label "1/3" stems from the
fact that h is divided by 3 in Eq. (15.14). An alternative derivation is shown in Box 15.3
where the Newton-Gregory polynomial is integrated to obtain the same formula (recall
Box 12.2).

Simpson's 1/3 rule can also be expressed using the format of Eq. (15.5):

$$I \simeq \underbrace{(b - a)}_{\text{Width}} \underbrace{\frac{f(x_0) + 4f(x_1) + f(x_2)}{6}}_{\text{Average height}} \qquad\qquad (15.15)$$

Box 15.3   *Derivation and Error Estimate of Simpson's Rule Based on Integrating the Forward Newton-Gregory Interpolating Polynomial*

As was done in Box 15.2 for the trapezoidal rule, Simpson's 1/3 rule can be derived by integrating the forward Newton-Gregory interpolating polynomial:

$$I = \int_{x_0}^{x_2} \left[f(x_0) + \Delta f(x_0)\,\alpha + \frac{\Delta^2 f(x_0)}{2}\,\alpha\,(\alpha-1) \right.$$
$$+ \frac{\Delta^3 f(x_0)}{6}\,\alpha\,(\alpha-1)(\alpha-2)$$
$$\left. + \frac{f^{(4)}(\xi)}{24}\,\alpha(\alpha-1)(\alpha-2)(\alpha-3)h^4 \right] dx$$

Notice that we have written the polynomial up to the fourth-order term rather than to the third-order term as would be expected. The reason for this will be apparent shortly. Also notice that the limits of integration are from x_0 to x_2. Therefore, when the simplifying substitutions are made (recall Box 15.2), the integral is from $\alpha = 0$ to 2:

$$I = h \int_{0}^{2} \left[f(x_0) + \Delta f(x_0)\,\alpha + \frac{\Delta^2 f(x_0)}{2}\,\alpha\,(\alpha-1) \right.$$
$$+ \frac{\Delta^3 f(x_0)}{6}\,\alpha\,(\alpha-1)(\alpha-2)$$
$$\left. + \frac{f^{(4)}(\xi)}{24}\,\alpha\,(\alpha-1)(\alpha-2)(\alpha-3)\,h^4 \right] d\alpha$$

which can be integrated to yield

$$I = h \left[\alpha f(x_0) + \frac{\alpha^2}{2}\,\Delta f(x_0) \right.$$
$$+ \left(\frac{\alpha^3}{6} - \frac{\alpha^2}{4} \right) \Delta^2 f(x_0)$$
$$+ \left(\frac{\alpha^4}{24} - \frac{\alpha^3}{6} + \frac{\alpha^2}{6} \right) \Delta^3 f(x_0)$$
$$\left. + \left(\frac{\alpha^5}{120} - \frac{\alpha^4}{16} + \frac{11\alpha^3}{72} - \frac{\alpha^2}{8} \right) f^{(4)}(\xi)\,h^4 \right]_{0}^{2}$$

and evaluated for the limits to give

$$I = h \left[2f(x_0) + 2\Delta f(x_0) + \frac{\Delta^2 f(x_0)}{3} \right.$$
$$\left. + (0)\Delta^3 f(x_0) - \frac{1}{90}\,f^{(4)}(\xi)\,h^4 \right] \qquad \text{(B15.3.1)}$$

Notice the significant result that the coefficient of the third divided difference is zero. Because $\Delta f(x_0) = f(x_1) - f(x_0)$ and $\Delta^2 f(x_0) = f(x_2) - 2f(x_1) + f(x_0)$, Eq. (B15.3.1) can be rewritten as

$$I = \underbrace{\frac{h}{3}\,[f(x_0) + 4f(x_1) + f(x_2)]}_{\text{Simpson's 1/3 rule}} - \underbrace{\frac{1}{90}\,f^{(4)}(\xi)\,h^5}_{\text{Truncation error}}$$

Thus, the first term is Simpson's 1/3 rule and the second is the truncation error. Because the third divided difference dropped out, we obtain the significant result that the formula is third-order accurate.

---

where $a = x_0, b = x_2$, and x_1 is the point midway between a and b which is given by $(b + a)/2$. Notice that according to Eq. (15.15), the middle point is weighted by two-thirds and the two end points by one-sixth.

It can be shown that a single-segment application of Simpson's 1/3 rule has a truncation error of (Box 15.3)

$$E_t = -\frac{1}{90}h^5 f^{(4)}(\xi)$$

or, because $h = (b - a)/2$,

$$E_t = -\frac{(b - a)^5}{2880}\,f^{(4)}(\xi) \qquad\qquad (15.16)$$

where ξ lies somewhere in the interval from a to b. Thus, Simpson's 1/3 rule is more accurate than the trapezoidal rule. However, comparison with Eq. (15.6) indicates that it is *much* more accurate than expected. Rather than being proportional to the third derivative, the error is proportional to the fourth derivative. This is because, as shown in Box 15.3, the coefficient of the third-order term goes to zero during the integration of the interpolating polynomial. Consequently, Simpson's 1/3 rule is third-order accurate even though it is based on only three points. In other words, it yields exact results for cubic polynomials even though it is derived from a parabola!

EXAMPLE 15.4   Single Application of Simpson's 1/3 Rule

Problem Statement: Use Eq. (15.15) to integrate

$$f(x) = 0.2 + 25x - 200x^2 + 675x^3 - 900x^4 + 400x^5$$

from $a = 0$ to $b = 0.8$. Recall that the exact integral is 1.64053334.

Solution:

$$f(0) = 0.2 \qquad f(0.4) = 2.456 \qquad f(0.8) = 0.232$$

Therefore, Eq. (15.15) can be used to compute

$$I \simeq 0.8 \frac{0.2 + 4(2.456) + 0.232}{6} = 1.36746667$$

which represents an exact error of

$$E_t = 1.64053334 - 1.36746667 = 0.27306666 \qquad \epsilon_t = 16.6\%$$

which is approximately 5 times more accurate than for a single application of the trapezoidal rule (Example 15.1).

The estimated error is [Eq. (15.16)]

$$E_a = -\frac{(0.8)^5}{2880}(-2400) = 0.27306667$$

where -2400 is the average fourth derivative for the interval as obtained using Eq. (PT5.4). As was the case in Example 15.1, the error is approximate (E_a) because the average fourth derivative is not an exact estimate of $f^{(4)}(\xi)$. However, because this case deals with a fifth-order polynomial, the discrepancy is actually due to roundoff.

15.2.2 The Multiple-Application Simpson's 1/3 Rule

Just as with the trapezoidal rule, Simpson's rule can be improved by dividing the integration interval into a number of segments of equal width (Fig. 15.13):

$$h = \frac{b - a}{n} \tag{15.17}$$

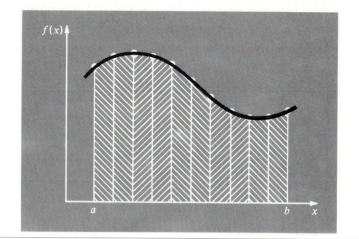

Figure 15.13
Graphical representation of
the multiple application of
Simpson's 1/3 rule. Note that
the method can be employed
only if the number of segments is even.

The total integral can be represented as

$$I = \int_{x_0}^{x_2} f(x)\, dx + \int_{x_2}^{x_4} f(x)\, dx + \cdots + \int_{x_{n-2}}^{x_n} f(x)\, dx$$

Substituting Simpson's rule for the individual integral yields

$$I \simeq 2h\frac{f(x_0) + 4f(x_1) + f(x_2)}{6}$$

$$+\ 2h\frac{f(x_2) + 4f(x_3) + f(x_4)}{6}$$

$$+\ \cdots\ +\ 2h\frac{f(x_{n-2}) + 4f(x_{n-1}) + f(x_n)}{6}$$

or, combining terms and using Eq. (15.17),

$$I \simeq (b - a)\ \underbrace{\frac{f(x_0) + 4\sum\limits_{i=1,3,5}^{n-1} f(x_i) + 2\sum\limits_{j=2,4,6}^{n-2} f(x_j) + f(x_n)}{3n}}_{\text{Average height}} \qquad (15.18)$$

$$\underbrace{}_{\text{Width}}$$

Notice that, as illustrated in Fig. 15.13, an even number of segments must be utilized
to implement the method. In addition, the coefficients "4" and "2" in Eq. (15.18) might
seem peculiar at first glance. However, they follow naturally from Simpson's 1/3 rule.
The odd points represent the middle term for each application and hence carry the weight
of 4 from Eq. (15.15). The even points are common to adjacent applications and hence
are counted twice.

An error estimate for the multiple-application Simpson's rule is obtained in the same fashion as for the trapezoidal rule by summing the individual errors for the segments and averaging the derivative to yield

$$E_a = -\frac{(b-a)^5}{180n^4}\bar{f}^{(4)}$$
(15.19)

where $\bar{f}^{(4)}$ is the average fourth derivative for the interval.

EXAMPLE 15.5   **Multiple-Application Version of Simpson's 1/3 Rule**

Problem Statement: Use Eq. (15.18) with $n = 4$ to estimate the integral of

$$f(x) = 0.2 + 25x - 200x^2 + 675x^3 - 900x^4 + 400x^5$$

from $a = 0$ to $b = 0.8$. Recall that the exact integral is 1.64053334.

Solution: $n = 4\,(h = 0.2)$:

$$f(0) = 0.2 \qquad f(0.2) = 1.288$$

$$f(0.4) = 2.456 \qquad f(0.6) = 3.464$$

$$f(0.8) = 0.232$$

from Eq. (15.18)

$$I = 0.8\frac{0.2 \quad + 4(1.288 + 3.464) + 2(2.456) + 0.232}{12}$$

$$= 1.62346667$$

$$E_t = 1.64053334 - 1.62346667 = 0.01706667 \qquad \epsilon_t = 1.04\%$$

The estimated error [Eq. (15.19)] is

$$E_a = -\frac{(0.8)^5}{180(4)^4}(-2400) = 0.01706667$$

The previous example illustrates that the multiple-application version of Simpson's 1/3 rule yields very accurate results. For this reason, it is considered superior to the trapezoidal rule for most applications. However, as mentioned previously, it is limited to cases where the values are equispaced. Further, it is limited to situations where there is an even number of segments and an odd number of points. Consequently, as discussed in the next section, an odd-segment–even-point formula known as Simpson's 3/8 rule is used in conjunction with the 1/3 rule to permit evaluation of both even and odd numbers of segments.

15.2.3   Simpson's 3/8 Rule

In a similar manner to the derivation of the trapezoidal and Simpson's 1/3 rule, a third-order Lagrange polynomial can be fit to four points and integrated:

$$I = \int_a^b f(x)\, dx \simeq \int_a^b f_3(x)\, dx$$

to yield

$$I \simeq \frac{3h}{8}[f(x_0) + 3f(x_1) + 3f(x_2) + f(x_3)]$$

where $h = (b - a)/3$. This equation is called *Simpson's 3/8 rule* because h is multiplied by 3/8. It is the third Newton-Cotes closed integration formula. The 3/8 rule can also be expressed in the form of Eq. (15.5):

$$I \simeq \underbrace{(b - a)}_{\text{Width}} \underbrace{\frac{f(x_0) + 3f(x_1) + 3f(x_2) + f(x_3)}{8}}_{\text{Average height}} \tag{15.20}$$

Thus, the two interior points are given weights of three-eighths, whereas the end points are weighted with one-eighth. Simpson's 3/8 rule has an error of

$$E_t = -\frac{3}{80}h^5 f^{(4)}(\xi)$$

or, because $h = (b - a)/3$,

$$E_t = -\frac{(b - a)^5}{6480} f^{(4)}(\xi) \tag{15.21}$$

Because the denominator of Eq. (15.21) is larger than for Eq. (15.16), the 3/8 rule is somewhat more accurate than the 1/3 rule.

Simpson's 1/3 rule is usually the method of preference because it attains third-order accuracy with three points rather than the four points required for the 3/8 version. However, the 3/8 rule has utility when the number of segments is odd. For instance, in Example 15.5 we used Simpson's rule to integrate the function for four segments. Suppose that you desired an estimate for five segments. One option would be to use a multiple-application version of the trapezoidal rule as was done in Examples 15.2 and 15.3. This may not be advisable, however, because of the large truncation error associated with this method. An alternative would be to apply Simpson's 1/3 rule to the first two segments and Simpson's 3/8 rule to the last three (Fig. 15.14). In this way, we could obtain an estimate with third-order accuracy across the entire interval.

EXAMPLE 15.6   Simpson's 3/8 Rule

Problem Statement:
(a) Use Simpson's 3/8 rule to integrate

$$f(x) = 0.2 + 25x - 200x^2 + 675x^3 - 900x^4 + 400x^5$$

from $a = 0$ to $b = 0.8$.

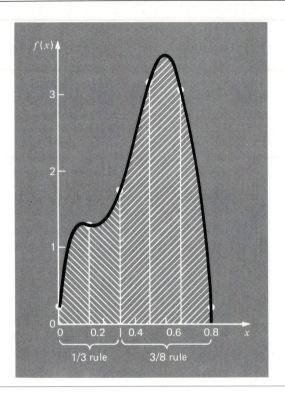

Figure 15.14
Illustration of how Simpson's
1/3 and 3/8 rules can be
applied in tandem to handle
multiple applications with odd
numbers of intervals.

(b) Use it in conjunction with Simpson's 1/3 rule to integrate the same function for five segments.

Solution:

(a) A single application of Simpson's 3/8 rule requires four equally spaced points:

$$f(0) = 0.2 \qquad\qquad f(0.2667) = 1.43272428$$

$$f(0.5333) = 3.48717696 \qquad\qquad f(0.8) = 0.232$$

Using Eq. (15.20),

$$I \simeq 0.8\frac{0.2 + 3(1.43272428 + 3.48717696) + 0.232}{8}$$

$$= 1.51917037$$

$$E_t = 1.64053334 - 1.51917037 = 0.12136297 \qquad \epsilon_t = 7.4\%$$

$$E_a = -\frac{(0.8)^5}{6480}(-2400) = 0.12136296$$

(b) The data needed for a five-segment application ($h = 0.16$) is

$$f(0) = 0.2 \qquad\qquad f(0.16) = 1.29691904$$

$$f(0.32) = 1.74339328 \qquad f(0.48) = 3.18601472$$

$$f(0.64) = 3.18192896 \qquad f(0.80) = 0.232$$

The integral for the first two segments is obtained using Simpson's 1/3 rule:

$$I \simeq 0.32\frac{0.2 + 4(1.29691904) + 1.74339328}{6} = 0.38032370$$

For the last three segments, the 3/8 rule can be used to obtain

$$I \simeq 0.48\frac{1.74339328 + 3(3.18601472 + 3.18192896) + 0.232}{8}$$

$$= 1.26475346$$

The total integral is computed by summing the two results:

$$I = 0.38032370 + 1.26475346 = 1.64507716$$

$$E_t = 1.64053334 - 1.64507716 = -0.00454383 \qquad \epsilon_t = -0.28\%$$

15.2.4 Computer Algorithm for Simpson's Rule

Pseudocode for Simpson's rule is outlined in Fig. 15.15. Notice that the program is set up so that either an even or odd number of segments may be used. For the former case, Simpson's 1/3 rule is applied to each pair of segments, and the results are summed to compute the total integral. For the latter, Simpson's 3/8 rule is applied to the last three segments and the 1/3 rule is applied to all the previous segments.

Figure 15.15
Pseudocode for multiple-application version of Simpson's rule. Note that n must be ≥ 1.

```
INPUT a,b (integration limits)
INPUT n (number of segments)
h = (b-a)/n ; m = n
INPUT f(xᵢ), for i = 1 to n
IF (n is odd) and (n > 1)
```
$$\text{int} = \text{int} + 3h\,\frac{f(x_{n-3}) + 3[f(x_{n-2}) + f(x_{n-1})] + f(x_n)}{8}$$
```
   m = n - 3
ENDIF
IF (m > 1)
```
$$\text{int} = \text{int} + h\,\frac{f(x_0) + 4\sum_{i=1,3,5}^{m-1} f(x_i) + 2\sum_{j=2,4,6}^{m-2} f(x_j) + f(x_m)}{3}$$
```
ENDIF
OUTPUT integral
```

15.2.5 Higher-Order Newton-Cotes Closed Formulas

As noted previously, the trapezoidal rule and both of Simpson's rules are members of a family of integrating equations known as the *Newton-Cotes closed integration formulas*. Some of the formulas are summarized in Table 15.2, along with their truncation-error estimates.

TABLE 15.2
Newton-Cotes closed integration formulas. The formulas are presented in the format of Eq. (15.5) so that the weighting of the data points to estimate the average height is apparent. The step size is given by $h = (b - a)/n$.

Segments (n)	Points	Name	Formula	Truncation Error
1	2	Trapezoidal rule	$(b - a)\dfrac{f(x_0) + f(x_1)}{2}$	$-(1/12)h^3 f''(\xi)$
2	3	Simpson's 1/3 rule	$(b - a)\dfrac{f(x_0) + 4f(x_1) + f(x_2)}{6}$	$-(1/90)h^5 f^{(4)}(\xi)$
3	4	Simpson's 3/8 rule	$(b - a)\dfrac{f(x_0) + 3f(x_1) + 3f(x_2) + f(x_3)}{8}$	$-(3/80)h^5 f^{(4)}(\xi)$
4	5	Boole's rule	$(b - a)\dfrac{7f(x_0) + 32f(x_1) + 12f(x_2) + 32f(x_3) + 7f(x_4)}{90}$	$-(8/945)h^7 f^{(6)}(\xi)$
5	6		$(b - a)\dfrac{19f(x_0) + 75f(x_1) + 50f(x_2) + 50f(x_3) + 75f(x_4) + 19f(x_5)}{288}$	$-(275/12,096)h^7 f^{(6)}(\xi)$

Notice that, as was the case with Simpson's 1/3 and 3/8 rules, the five- and six-point formulas have the same order error. This general characteristic holds for the higher-point formulas and leads to the result that the even-segment–odd-point formulas (for example, 1/3 rule and Boole's rule) are usually the methods of preference.

However, it must also be stressed that in engineering practice, the higher-order (that is, greater than four-point) formulas are rarely used. Simpson's rules are sufficient for most applications. Accuracy can be improved by using the multiple-application version rather than opting for the higher-point formulas. Furthermore, when the function is known and high accuracy is required, methods such as Romberg integration or Gauss quadrature, described in Chap. 16, offer viable and attractive alternatives.

15.3 INTEGRATION WITH UNEQUAL SEGMENTS

To this point, all formulas for numerical integration have been based on equally spaced data points. In practice, there are many situations where this assumption does not hold and we must deal with unequal-sized segments. For example, experimentally-derived data is often of this type. For these cases, one method is to apply the trapezoidal rule to each segment and sum the results:

$$I = h_1\frac{f(x_1) + f(x_0)}{2} + h_2\frac{f(x_2) + f(x_1)}{2} + \cdots + h_n\frac{f(x_n) + f(x_{n-1})}{2} \tag{15.22}$$

where h_i is the width of segment i. Note that this was the same approach used for the multiple-application trapezoidal rule. The only difference between Eqs. (15.8) and (15.22) is that the h's in the former are constant. Consequently, Eq. (15.8) could be simplified by grouping terms to yield Eq. (15.9). Although this simplification cannot be applied to Eq. (15.22), a computer program can be easily developed to accommodate unequal-sized segments. Before describing such a program, we will illustrate in the following example how Eq. (15.22) is applied to evaluate an integral.

EXAMPLE 15.7   Trapezoidal Rule with Unequal Segments

Problem Statement: The information in Table 15.3 was generated using the same polynomial employed in Example 15.1. Use Eq. (15.22) to determine the integral for this data. Recall that the correct answer is 1.64053334.

TABLE 15.3
Data for $f(x) = 0.2 + 25x - 200x^2 + 675x^3 - 900x^4 + 400x^5$, with unequally spaced values of x.

x	f(x)	x	f(x)
0.0	0.20000000	0.44	2.84298496
0.12	1.30972928	0.54	3.50729696
0.22	1.30524128	0.64	3.18192896
0.32	1.74339328	0.70	2.36300000
0.36	2.07490304	0.80	0.23200000
0.40	2.45600000		

Solution: Applying Eq. (15.22) to the data in Table 15.3 yields

$$I = 0.12\frac{1.30972928 + 0.2}{2} + 0.10\frac{1.30524128 + 1.30972928}{2}$$

$$+ \cdots + 0.10\frac{0.232 + 2.363}{2}$$

$$= 0.09058376 + 0.13074853 + \cdots + 0.12975$$

$$= 1.56480098$$

which represents an absolute percent relative error of $\epsilon_t = 4.6\%$.

The data from Example 15.7 is depicted in Fig. 15.16. Notice that some adjacent segments are of equal width and, consequently, could have been evaluated using Simp-

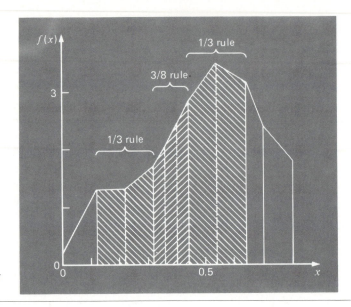

Figure 15.16
Use of the trapezoidal rule to determine the integral of unevenly spaced data. Notice how the shaded segments could be evaluated with Simpson's rule to attain higher accuracy.

son's rules. This usually leads to more accurate results, as illustrated by the following example.

EXAMPLE 15.8   *Inclusion of Simpson's Rules in the Evaluation of Uneven Data*

Problem Statement: Recompute the integral for the data in Table 15.3, but use Simpson's rules for those segments where they are appropriate.

Solution: The first segment is evaluated with the trapezoidal rule:

$$I = 0.12\frac{1.30972928 + 0.2}{2} = 0.09058376$$

Because the next two segments from $x = 0.12$ to 0.32 are of equal length, their integral can be computed with Simpson's 1/3 rule:

$$I = 0.2\frac{1.74339328 + 4(1.30524128) + 1.30972928}{6}$$

$$= 0.27580292$$

The next three segments are also equal and, as such, may be evaluated with the 3/8 rule to give $I = 0.27268631$. Similarly, the 1/3 rule can be applied to the two segments from $x = 0.44$ to 0.64 to yield $I = 0.66847006$. Finally, the last two segments, which are of unequal length, can be evaluated with the trapezoidal rule to give values of 0.16634787 and 0.12975000, respectively. The area of these individual segments can be summed to yield a total integral of 1.60364092. This represents an error of $\epsilon_t = 2.2\%$, which is superior to the result using the trapezoidal rule in Example 15.7.

Computer Program for Unequally Spaced Data. It is a fairly simple proposition to program Eq. (15.22). However, as demonstrated in Example 15.8, the approach is enhanced if it implements Simpson's rules wherever possible. For this reason, we have developed a computer algorithm that incorporates this capability.

As depicted in Fig. 15.17, the algorithm checks the length of adjacent segments. If two consecutive segments are of equal length, Simpson's 1/3 rule is applied. If three are equal, the 3/8 rule is used. When adjacent segments are of unequal length, the trapezoidal rule is implemented.

Figure 15.17
Pseudocode for integrating unequally spaced data.

```
INPUT n (number of segments)
INPUT x_i , f(x_i), for i = 0 to n
h = x_1 - x_0
k = 1
DOFOR j = 1 to n
    h_future = x_{j+1} - x_j
    IF (h = h_future)
        IF (k = 3)
            I = I + 2h [f(x_{j-1}) + 4f(x_{j-2}) + f(x_{j-3})] / 6
            k = k-1
        ELSE
            k = k+1
        ENDIF
    ELSE
        IF (k = 1)
            I = I + h [f(x_j) + f(x_{j-1})] / 2
        ELSE
            IF (k = 2)
                I = I + 2h [f(x_j) + 4f(x_{j-1}) + f(x_{j-2})] / 6
            ELSE
                I = I + 3h [f(x_j) + 3[f(x_{j-1}) + f(x_{j-2})] + f(x_{j-3})] / 8
            ENDIF
            k = 1
        ENDIF
    ENDIF
    h = h_future
ENDDO
```

We suggest that you develop your own computer program from this pseudocode. Not only does it allow evaluation of unequal segment data, but if equally spaced information is used, it reduces to Simpson's rules. As such, it represents a basic, all-purpose algorithm for the determination of the integral of tabulated data.

15.4   OPEN INTEGRATION FORMULAS

Recall from Fig. 15.3b that open integration formulas have limits that extend beyond the range of the data. Table 15.4 summarizes the *Newton-Cotes open integration formulas*. The formulas are expressed in the form of Eq. (15.5) so that the weighting factors are evident. As with the closed versions, successive pairs of the formulas have the same order error. The even-segment–odd-point formulas are usually the methods of preference because they require fewer points to attain the same accuracy as the odd-segment–even-point formulas.

The open formulas are not often used for integration. However, as discussed in Chap. 16, they have utility for analyzing improper integrals. In addition, they will have relevance to our discussion of multistep methods for solving ordinary differential equations in Chap. 20.

TABLE 15.4
Newton-Cotes open integration formulas. The formulas are presented in the format of Eq. (15.5) so that the weighting of the data points to estimate the average height is apparent. The step size is given by $h = (b - a)/n$.

Segments (n)	Points	Name	Formula	Truncation Error
2	1	Midpoint method	$(b - a) f(x_1)$	$(1/3)h^3 f''(\xi)$
3	2		$(b - a)\dfrac{f(x_1) + f(x_2)}{2}$	$(3/4)h^3 f''(\xi)$
4	3		$(b - a)\dfrac{2f(x_1) - f(x_2) + 2f(x_3)}{3}$	$(14/45)h^5 f^{(4)}(\xi)$
5	4		$(b - a)\dfrac{11f(x_1) + f(x_2) + f(x_3) + 11f(x_4)}{24}$	$(95/144)h^5 f^{(4)}(\xi)$
6	5		$(b - a)\dfrac{11f(x_1) - 14f(x_2) + 26f(x_3) - 14f(x_4) + 11f(x_5)}{20}$	$(41/140)h^7 f^{(6)}(\xi)$

PROBLEMS

Hand Calculations

15.1 Use analytical means to evaluate

(a) $\int_0^{10} (10 + 2x - 6x^2 + 5x^4)\, dx$

(b) $\int_{-3}^{5} (1 - x - 4x^3 + 3x^5)\, dx$

(c) $\int_0^{\pi} (8 + 5 \sin x)\, dx$

15.2 Use a single application of the trapezoidal rule to evaluate the integrals from Prob. 15.1.

15.3 Evaluate the integrals from Prob. 15.1 with a multiple-application trapezoidal rule, with $n = 2$, 4, and 6.

15.4 Evaluate the integrals from Prob. 15.1 with a single application of Simpson's 1/3 rule.

15.5 Evaluate the integrals from Prob. 15.1 with a multiple-application Simpson's 1/3 rule, with $n = 4$ and 6.

15.6 Evaluate the integrals from Prob. 15.1 with a single application of Simpson's 3/8 rule.

15.7 Evaluate the integrals from Prob. 15.1, but use a multiple-application Simpson's rule, with $n = 5$.

15.8 Integrate the following function both analytically and using the trapezoidal rule, with $n = 1$, 2, 3, and 4:

$$\int_0^{3\pi/20} [\sin (5x + 1)]\, dx$$

Compute percent relative errors to evaluate the accuracy of the trapezoidal approximations.

15.9 Integrate the following function both analytically and using Simpson's rules, with $n = 4$ and 5:

$$\int_{-4}^{6} [(4x + 8)^3]\, dx$$

Discuss the results.

15.10 Integrate the following function both analytically and numerically. Use both the trapezoidal and Simpson's 1/3 rules to numerically integrate the function. For both cases, use the multiple-application version, with $n = 4$.

$$\int_0^4 xe^{2x}\, dx$$

Compute percent relative errors for the numerical results.

15.11 Integrate the following function both analytically and numerically. Use a single application of the trapezoidal rule, Simpson's 1/3 and 3/8 rules, and Boole's rule (see Table 15.2).

$$\int_0^1 15.3^{2.5x}\, dx$$

Compute percent relative errors for the numerical results.

15.12 Evaluate the integral

$$\int_0^{\pi} (4 + 2 \sin x)\, dx$$

(a) Analytically.
(b) By single application of the trapezoidal rule.
(c) By multiple application of the trapezoidal rule ($n = 5$).
(d) By single application of Simpson's 1/3 rule.
(e) By single application of Simpson's 3/8 rule.
(f) By multiple application of Simpson's rules ($n = 5$).
For (b) through (f), compute the percent relative error (ϵ_t) based on (a).

15.13 Evaluate the integral of the following tabular data with the trapezoidal rule:

x	0	0.1	0.2	0.3	0.4	0.5
$f(x)$	1	7	4	3	5	9

15.14 Perform the same evaluation as in Prob. 15.13, but use Simpson's rules.

15.15 Evaluate the integral of the following tabular data using the trapezoidal rule:

x	-3	-1	1	3	5	7	9	11
$f(x)$	1	-4	-5	2	4	8	6	-3

15.16 Perform the same evaluation as in Prob. 15.15, but use Simpson's rules.

15.17 Determine the mean value of the function

$$f(x) = -46 + 45.4x - 13.8x^2 + 1.71x^3 - 0.0729x^4$$

between $x = 2$ and 10 by
(a) Graphing the function and visually estimating the mean value.
(b) Using Eq. (PT5.4) and the analytical evaluation of the integral.
(c) Using Eq. (PT5.4) and a four-segment version of the trapezoidal rule to estimate the integral.
(d) Using Eq. (PT5.4) and a four-segment version of Simpson's 1/3 rule.

15.18 The function

$$f(x) = 10 - 38.6x + 74.07x^2 - 40.1x^3$$

can be used to generate the following table of unequally spaced data:

x	0	0.1	0.3	0.5	0.7	0.95	1.2
$f(x)$	10	6.84	4	4.20	5.51	5.77	1

Evaluate the integral from $a = 0$ to $b = 1.2$ using
(a) Analytical means.
(b) The trapezoidal rule.
(c) A combination of the trapezoidal and Simpson's rules; employ Simpson's rules wherever possible to obtain the highest accuracy.
For (b) and (c), compute the percent relative error (ϵ_t).

15.19 Evaluate the following double integral:

$$\int_{-2}^{2}\int_{0}^{4} (x^3 - 3y^2 + xy^3)\, dx\, dy$$

(a) Analytically.
(b) Using a multiple-application trapezoidal rule ($n = 2$).
(c) Using single applications of Simpson's 1/3 rule.
For (b) and (c) compute the percent relative error (ϵ_t).

15.20 Evaluate the triple integral

$$\int_{-4}^{4}\int_{0}^{6}\int_{-1}^{3} (x^4 - 2yz)\, dx\, dy\, dz$$

(a) Analytically.

(b) Using single applications of Simpson's 1/3 rule.
For (b), compute the percent relative error (ϵ_t).

Computer-Related Problems

15.21 Develop a user-friendly computer program for the multiple-application trapezoidal rule based on Fig. 15.9 or 15.10. Among other things,
(a) Add documentation statements to the code.
(b) Make the input and output more descriptive and user-oriented.
(c) (*Optional*) Modify the program so that it is capable of evaluating given functions in addition to tabular data.

Test your program by duplicating the computation from Example 15.2.

15.22 Develop a user-friendly computer program for the multiple-application version of Simpson's rule based on Fig. 15.15. Test it by duplicating the computations from Examples 15.5 and 15.6.

15.23 Develop a user-friendly computer program for integrating unequally spaced data based on Fig. 15.17. Test it by duplicating the computation from Example 15.8.

15.24 Use the TRAPEZOIDAL RULE program on the Electronic TOOLKIT disk (or your own program from Prob. 15.21) to repeat (a) Prob. 15.2, (b) Prob. 15.3, (c) Prob. 15.8, (d) Prob. 15.10, and (e) Prob. 15.13. Use the graphical option to help you visualize the concept that $I = \int_a^b f(x)\, dx$ is the area between the $f(x)$ curve and the axis. Try several different step sizes for each problem.

15.25 Make up five of your own functions. Use the Electronic TOOLKIT software (or your own program) to calculate the integral of each function over some limits based on your inputs. Try step sizes of $h = (b - a)/n$ for $n = 1, \ldots, n = 10$. Plot I as a function of n.

15.26 Use the Electronic TOOLKIT software (or your own program) to calculate the integral of tabular data. Make up your own data for both x and $f(x)$. Use negative values and zero values for both x and $f(x)$. Look at your function using the plot option and convince yourself that the Electronic TOOLKIT software is working properly.

CHAPTER 16
Integration of Equations

In the introduction to Part Five, we noted that functions to be integrated numerically will typically be of two forms: a table of values or an equation. The form of the data has an important influence on the approaches that can be used to evaluate the integral. For tabulated information, you are limited by the number of points that are given. In contrast, if the function is available in equation form, you can generate as many values of $f(x)$ as are required to attain acceptable accuracy (recall Fig. PT5.7).

The present chapter is devoted to two techniques that are expressly designed to analyze cases where the function is given. Both capitalize on the ability to generate function values in order to develop efficient schemes for numerical integration. The first is based on *Richardson's extrapolation,* which is a method for combining two numerical integral estimates in order to obtain a third, more accurate value. The computational algorithm for implementing Richardson's extrapolation in a highly efficient manner is called *Romberg integration.* This technique is recursive and can be used to generate an integral estimate within a prespecified error tolerance.

The second method is called *Gauss quadrature.* Recall that in the last chapter, $f(x)$ values for the Newton-Cotes formulas were determined at specified values of x. For example, if we used the trapezoidal rule to determine an integral, we were constrained to take the weighted average of $f(x)$ at the ends of the interval. Gauss-quadrature formulas employ x values that are positioned between a and b in such a manner that a much more accurate integral estimate results.

In addition to these two standard techniques, we devote a final section to the evaluation of improper integrals. In this discussion, we focus on integrals with infinite limits and show how a change of variable and open integration formulas prove useful for such cases.

16.1 ROMBERG INTEGRATION

In Chap. 15, we presented multiple-application versions of the trapezoidal rule and Simpson's rules. For an analytical (as opposed to a tabular) function, the error equations [Eqs. (15.13) and (15.19)] indicate that increasing the number of segments n will result in more accurate integral estimates. This observation is borne out by Fig. 16.1, which is a plot of true error versus n for the integral of $f(x) = 0.2 + 25x - 200x^2 + 675x^3 -$

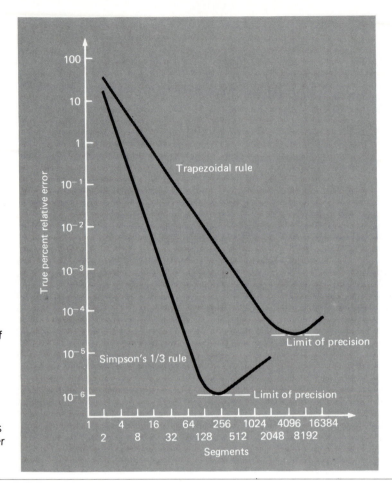

Figure 16.1
Absolute value of the true percent relative error versus number of segments for the determination of the integral of $f(x) = 0.2 + 25x - 200x^2 + 675x^3 - 900x^4 + 400x^5$, evaluated from $a = 0$ to $b = 0.8$ using the multiple-application trapezoidal rule and the multiple-application Simpson's 1/3 rule. Note that both results indicate that for a large number of segments, round-off errors limit precision.

$900x^4 + 400x^5$. Notice how the error drops as n increases. However, also notice that at large values of n, the error starts to increase as round-off errors begin to dominate. Also observe that a very large number of segments (and, hence, computational effort) is required to attain high levels of accuracy. As a consequence of these shortcomings, the multiple-application trapezoidal rule and Simpson's rules are sometimes inadequate for problem contexts where high efficiency and low errors are needed.

Romberg integration is one technique that is designed to obviate these shortcomings. It is quite similar to the techniques discussed in Chap. 15, in the sense that it is based on successive application of the trapezoidal rule. However, through mathematical manipulations, superior results are attained for less effort.

16.1.1   Richardson's Extrapolation

Recall that in Sec. 8.2.3, we used iterative refinement to improve the solution of a set of simultaneous linear equations. Error-correction techniques are also available to improve the results of numerical integration on the basis of the integral estimates themselves. Generally called *Richardson's extrapolation,* these methods use two estimates of an integral to compute a third, more accurate approximation.

The estimate and error associated with a multiple-application trapezoidal rule can be represented generally as

$$I = I(h) + E(h)$$

where I is the exact value of the integral, $I(h)$ is the approximation from an n-segment application of the trapezoidal rule with step size $h = (b - a)/n$, and $E(h)$ is the truncation error. If we make two separate estimates using step sizes of h_1 and h_2 and have exact values for the error,

$$I(h_1) + E(h_1) = I(h_2) + E(h_2) \tag{16.1}$$

Now recall that the error of the multiple-application trapezoidal rule can be represented approximately by Eq. (15.13) [with $n = (b - a)/h$]:

$$E \simeq -\frac{b - a}{12} h^2 \bar{f}'' \tag{16.2}$$

If it is assumed that $\bar{f}''$ is constant regardless of step size, Eq. (16.2) can be used to determine that the ratio of the two errors will be

$$\frac{E(h_1)}{E(h_2)} \simeq \frac{h_1^2}{h_2^2} \tag{16.3}$$

This calculation has the important effect of removing the term $\bar{f}''$ from the computation. In so doing, we have made it possible to utilize the information embodied by Eq. (16.2) without prior knowledge of the function's second derivative. To do this, we rearrange Eq. (16.3) to give

$$E(h_1) \simeq E(h_2)\left(\frac{h_1}{h_2}\right)^2$$

which can be substituted into Eq. (16.1):

$$I(h_1) + E(h_2)\left(\frac{h_1}{h_2}\right)^2 \simeq I(h_2) + E(h_2)$$

which can be solved for

$$E(h_2) \simeq \frac{I(h_1) - I(h_2)}{1 - (h_1/h_2)^2}$$

Thus, we have developed an estimate of the truncation error in terms of the integral estimates and their step sizes. This estimate can then be substituted into

$$I = I(h_2) + E(h_2)$$

to yield an improved estimate of the integral:

$$I \simeq I(h_2) + \left[\frac{1}{(h_1/h_2)^2 - 1}\right][I(h_2) - I(h_1)] \tag{16.4}$$

It can be shown (Ralston and Rabinowitz, 1978) that the error of this estimate is $O(h^4)$. Thus, we have combined two trapezoidal rule estimates of $O(h^2)$ to yield a new estimate of $O(h^4)$. For the special case where the interval is halved ($h_2 = h_1/2$), this equation becomes

$$I \simeq I(h_2) + \frac{1}{2^2 - 1}\, [I(h_2) - I(h_1)]$$

or, collecting terms,

$$I \simeq \frac{4}{3}\, I(h_2) - \frac{1}{3}\, I(h_1) \tag{16.5}$$

EXAMPLE 16.1   Error Corrections of the Trapezoidal Rule

Problem Statement: In the previous chapter (Example 15.1 and Table 15.1), single and multiple applications of the trapezoidal rule yielded the following results:

Segments	h	Integral	ϵ_t, %
1	0.8	0.1728	89.5
2	0.4	1.0688	34.9
4	0.2	1.4848	9.5

Use this information along with Eq. (16.5) to compute improved estimates of the integral.

Solution: The estimates for one and two segments can be combined to yield

$$I \simeq \frac{4}{3}\, (1.0688) - \frac{1}{3}\, (0.1728) = 1.36746667$$

The error of the improved integral is

$$E_t = 1.64053334 - 1.36746667 = 0.27306667 \qquad \epsilon_t = 16.6\%$$

which is superior to the estimates upon which it was based.

In the same manner, the estimates for two and four segments can be combined to give

$$I \simeq \frac{4}{3}\, (1.4848) - \frac{1}{3}\, (1.0688) = 1.62346667$$

which represents an error of

$$E_t = 1.64053334 - 1.62346667 = 0.01706667 \qquad \epsilon_t = 1.0\%$$

Equation (16.4) provides a way to combine two applications of the trapezoidal rule with error $O(h^2)$ in order to compute a third estimate with error $O(h^4)$. This approach is a subset of a more general method for combining integrals to obtain improved estimates. For instance, in Example 16.1, we computed two improved integrals of $O(h^4)$ on the basis of three trapezoidal rule estimates. These two improved estimates can, in turn, be combined to yield an even better value with $O(h^6)$. For the special case where the original trapezoidal estimates are based on successive halving of the step size, the equation used for $O(h^6)$ accuracy is

$$I = \frac{16}{15} I_m - \frac{1}{15} I_l \tag{16.6}$$

where I_m and I_l are the more and less accurate estimates, respectively. Similarly, two $O(h^6)$ results can be combined to compute an integral that is $O(h^8)$ using

$$I = \frac{64}{63} I_m - \frac{1}{63} I_l \tag{16.7}$$

EXAMPLE 16.2   Higher-Order Error Correction of Integral Estimates

Problem Statement: In Example 16.1, we used Richardson's extrapolation to compute two integral estimates of $O(h^4)$. Utilize Eq. (16.6) to combine these estimates to compute an integral with $O(h^6)$.

Solution: The two integral estimates of $O(h^4)$ obtained in Example 16.1 were 1.36746667 and 1.62346667. These values can be substituted into Eq. (16.6) to yield

$$I = \frac{16}{15}(1.62346667) - \frac{1}{15}(1.36746667) = 1.64053334$$

which is the correct answer to the nine significant figures that are carried in this example.

16.1.2   The Romberg Integration Algorithm

Notice that the coefficients in each of the extrapolation equations [Eqs. (16.5), (16.6), and (16.7)] add up to 1. Thus, they represent weighting factors that, as accuracy increases, place relatively greater weight on the superior integral estimate. These formulations can be expressed in a general form that is well-suited for computer implementation:

$$I_{j,k} \simeq \frac{4^{k-1} I_{j+1,k-1} - I_{j,k-1}}{4^{k-1} - 1} \tag{16.8}$$

where $I_{j+1,k-1}$ and $I_{j,k-1}$ are the more and less accurate integrals, respectively, and $I_{j,k}$ is the improved integral. The index k signifies the level of the integration where $k = 1$ corresponds to the original trapezoidal rule estimates, $k = 2$ corresponds to $O(h^4)$, $k = 3$ to $O(h^6)$, and so forth. The index j is used to distinguish between the more ($j + 1$) and the less (j) accurate estimates. For example, for $k = 2$ and $j = 1$, Eq. (16.8) becomes

$$I_{1,2} \simeq \frac{4I_{2,1} - I_{1,1}}{3}$$

which is equivalent to Eq. (16.5).

The general form represented by Eq. (16.8) is attributed to Romberg, and its systematic application to evaluate integrals is known as *Romberg integration.* Figure 16.2 is a graphical depiction of the sequence of integral estimates generated using this approach. Each matrix corresponds to a single iteration. The first column contains the trapezoidal rule evaluations that are designated $I_{j,1}$, where $j = 1$ is for a single-segment application (step size is $b - a$); $j = 2$ is for a two-segment application [step size is $(b - a)/2$]; $j = 3$ is for a four-segment application [step size is $(b - a)/4$]; and so forth. The other columns of the matrix are generated by systematically applying Eq. (16.8) to obtain successively better estimates of the integral.

For example, the first iteration (Fig. 16.2a) involves computing the one- and two-segment trapezoidal rule estimates ($I_{1,1}$ and $I_{2,1}$). Equation (16.8) is then used to compute the element $I_{1,2} = 1.36746667$, which has an error of $O(h^4)$.

Now, we must check to determine whether this result is adequate for our needs. As in other approximate methods in this book, a termination, or stopping, criterion is required to assess the accuracy of the results. One method that can be employed for the present purposes is [Eq. (3.5)]

$$|\epsilon_a| = \left| \frac{I_{1,k} - I_{1,k-1}}{I_{1,k}} \right| 100\% \tag{16.9}$$

where ϵ_a is an estimate of the percent relative error. Thus, as was done previously in other iterative processes, we compare the new estimate with a previous value. When the change between the old and new values as represented by ϵ_a is below a prespecified error criterion ϵ_s, the computation is terminated. For Fig. 16.2a, this evaluation indicates an 87.4 percent change over the course of the first iteration.

The object of the second iteration (Fig. 16.2b) is to obtain the $O(h^6)$ estimate—$I_{1,3}$. To do this, an additional trapezoidal rule estimate, $I_{3,1} = 1.4848$, is determined. Then it is combined with $I_{2,1}$ using Eq. (16.8) to generate $I_{2,2} = 1.62346667$. The result is, in turn, combined with $I_{1,2}$ to yield $I_{1,3} = 1.64053334$. Equation (16.9) can be applied to determine that this result represents a change of 16.6 percent when compared with the previous result $I_{1,2}$.

The third iteration (Fig. 16.2c) continues the process in the same fashion. In this

Figure 16.2
Graphical depiction of the sequence of integral estimates generated using Romberg integration.

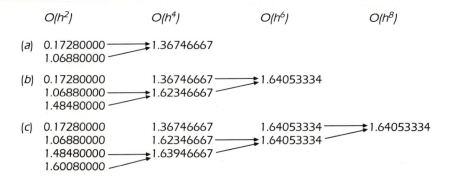

case, a trapezoidal estimate is added to the first column and then Eq. (16.8) is applied to compute successively more accurate integrals along the lower diagonal. After only three iterations, the result, $I_{1,4} = 1.64053334$, is known to be accurate to at least nine significant figures.

Romberg integration is more efficient than the trapezoidal rule and Simpson's rules discussed in Chap. 15. For example, for determination of the integral as shown in Fig. 16.1, Simpson's 1/3 rule would require a 256-segment application to yield an estimate of 1.64053332. Finer approximations would not be possible because of round-off error. In contrast, Romberg integration yields an exact result (to nine significant figures) based on combining one-, two-, four-, and eight-segment trapezoidal rules.

Figure 16.3 presents pseudocode for Romberg integration. By using loops, this algorithm implements the method in an efficient manner. Remember that Romberg integration is designed for cases where the function to be integrated is known. This is because knowledge of the function permits the evaluations required for the initial implementations of the trapezoidal rule. Tabulated data is rarely in the form needed to make the necessary successive halvings.

Figure 16.3
Pseudocode for Romberg integration.

```
INPUT a,b (integration limits)
INPUT ε_s (stopping criterion)
INPUT maxit (maximum iterations)
n = 1
CALL traprule(n,a,b,integral)
I_1,1 = integral
ε_a = 1.1ε_s
i = 0
DOWHILE (ε_a > ε_s) and (i < maxit)
    i = i + 1
    n = 2^i
    CALL traprule(n,a,b,integral)
    I_i+1,1 = integral
    DOFOR k = 2 to i + 1
        j = 2 + i - k
```

$$I_{j,k} = \frac{4^{k-1}\,I_{j+1,k-1} - I_{j,k-1}}{4^{k-1} - 1}$$

```
    ENDDO
```

$$\epsilon_a = \left| \frac{I_{1,i+1} - I_{1,1}}{I_{1,i+1}} \right| \cdot 100$$

```
ENDDO
```

16.2 GAUSS QUADRATURE

In Chap. 15, we studied the group of numerical integration or quadrature formulas known as the Newton-Cotes equations. A characteristic of these formulas (with the exception of the special case of Sec. 15.3) was that the integral estimate was based on evenly spaced

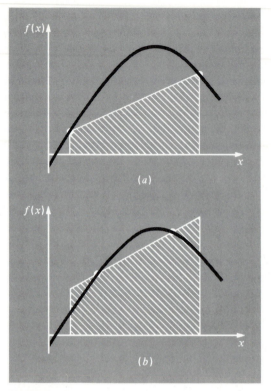

Figure 16.4
(*a*) Graphical depiction of the trapezoidal rule as the area under the straight line joining fixed end points. (*b*) An improved integral estimate obtained by taking the area under the straight line passing through two intermediate points. By positioning these points wisely, the positive and negative errors are balanced, and an improved integral estimate results.

function values. Consequently, the location of the base points used in these equations was predetermined or fixed.

For example, as depicted in Fig. 16.4*a*, the trapezoidal rule is based on taking the area under the straight line connecting the function values at the ends of the integration interval. The formula that is used to compute this area is

$$I \simeq (b - a)\frac{f(a) + f(b)}{2} \tag{16.10}$$

where a and b are the limits of integration and $b - a$ is the width of the integration interval. Because the trapezoidal rule must pass through the end points, there are cases such as Fig. 16.4*a* where the formula results in a large error.

Now, suppose that the constraint of fixed base points was removed and we were free to evaluate the area under a straight line joining *any* two points on the curve. By positioning these points wisely, we could define a straight line that would balance the positive and negative errors. Hence, as in Fig. 16.4*b*, we would arrive at an improved estimate of the integral.

Gauss quadrature is the name for one class of techniques to implement such a strategy. The particular Gauss quadrature formulas described in the present section are called *Gauss-Legendre formulas*. Before describing the approach, we will show how numerical integration formulas such as the trapezoidal rule can be derived using the method of undetermined coefficients. This method will then be employed to develop the Gauss-Legendre formulas.

16.2.1 Method of Undetermined Coefficients

In Chap. 15, we derived the trapezoidal rule by integrating a linear interpolating polynomial and by geometrical reasoning. The *method of undetermined coefficients* offers a third approach that also has utility in deriving other integration techniques such as Gauss quadrature.

In order to illustrate the approach, Eq. (16.10) is expressed as

$$I \simeq c_0 f(a) + c_1 f(b) \tag{16.11}$$

where the c's are constants. Now realize that the trapezoidal rule should yield exact results when the function being integrated is a constant or a straight line. Two simple equations that represent these cases are $y = 1$ and $y = x$. Both are illustrated in Fig. 16.5. Thus, the following equalities should hold

$$c_0 + c_1 = \int_{-(b-a)/2}^{(b-a)/2} 1 \, dx$$

$$-c_0 \frac{b-a}{2} + c_1 \frac{b-a}{2} = \int_{-(b-a)/2}^{(b-a)/2} x \, dx$$

or, evaluating the integrals,

$$c_0 + c_1 = b - a$$

$$-c_0 \frac{b-a}{2} + c_1 \frac{b-a}{2} = 0$$

These are two equations with two unknowns that can be solved for

$$c_0 = c_1 = \frac{b-a}{2}$$

which, when substituted back into Eq. (16.11), gives

$$I \simeq \frac{b-a}{2} f(a) + \frac{b-a}{2} f(b)$$

which is equivalent to the trapezoidal rule.

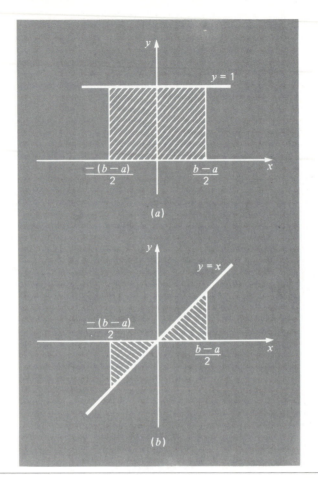

Figure 16.5
Two integrals that should
be evaluated exactly by the
trapezoidal rule: (a) a constant
and (b) a straight line.

16.2.2   Derivation of the Two-Point Gauss-Legendre Formula

Just as was the case for the above derivation of the trapezoidal rule, the object of Gauss
quadrature is to determine the coefficients of an equation of the form

$$I \simeq c_0 f(x_0) + c_1 f(x_1) \tag{16.12}$$

where the c's are the unknown coefficients. However, in contrast to the trapezoidal rule
that used fixed end points a and b, the function arguments x_0 and x_1 are not fixed at the
end points, but are unknowns (Fig. 16.6). Thus, we now have a total of four unknowns
that must be evaluated, and consequently, we require four conditions to determine them
exactly.

Just as for the trapezoidal rule, we can obtain two of these conditions by assuming
that Eq. (16.12) fits the integral of a constant and a linear function exactly. Then, to
arrive at the other two conditions, we merely extend this reasoning by assuming that it
also fits the integral of a parabolic ($y = x^2$) and a cubic ($y = x^3$) function. By doing this,

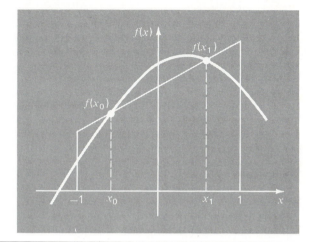

Figure 16.6
Graphical depiction of the unknown variables x_0 and x_1 for integration by Gauss quadrature.

we determine all four unknowns and in the bargain derive a linear two-point integration formula that is exact for cubics. The four equations to be solved are

$$c_0 f(x_0) + c_1 f(x_1) = \int_{-1}^{1} 1 \, dx = 2 \tag{16.13}$$

$$c_0 f(x_0) + c_1 f(x_1) = \int_{-1}^{1} x \, dx = 0 \tag{16.14}$$

$$c_0 f(x_0) + c_1 f(x_1) = \int_{-1}^{1} x^2 \, dx = \frac{2}{3} \tag{16.15}$$

$$c_0 f(x_0) + c_1 f(x_1) = \int_{-1}^{1} x^3 \, dx = 0 \tag{16.16}$$

Equations (16.13) through (16.16) can be solved simultaneously for

$$c_0 = c_1 = 1$$

$$x_0 = \frac{-1}{\sqrt{3}} = -0.577350629\ldots$$

$$x_1 = \frac{1}{\sqrt{3}} = 0.577350269\ldots$$

which can be substituted into Eq. (16.12) to yield the two-point Gauss-Legendre formula

$$I \simeq f\left(\frac{-1}{\sqrt{3}}\right) + f\left(\frac{1}{\sqrt{3}}\right) \tag{16.17}$$

Thus, we arrive at the interesting result that the simple addition of the function values at $x = 1/\sqrt{3}$ and $-1/\sqrt{3}$ yields an integral estimate that is third-order accurate.

Notice that the integration limits in Eqs. (16.13) through (16.16) are from -1 to 1. This was done to simplify the mathematics and to make the formulation as general as possible. A simple change of variable can be used to translate other limits of integration into this form. This is accomplished by assuming that a new variable x_d is related to the original variable x in a linear fashion, as in

$$x = a_0 + a_1 x_d \tag{16.18}$$

If the lower limit, $x = a$, corresponds to $x_d = -1$, these values can be substituted into Eq. (16.18) to yield

$$a = a_0 + a_1(-1) \tag{16.19}$$

Similarly, the upper limit, $x = b$, corresponds to $x_d = 1$, to give

$$b = a_0 + a_1(1) \tag{16.20}$$

Equations (16.19) and (16.20) can be solved simultaneously for

$$a_0 = \frac{b + a}{2} \tag{16.21}$$

and

$$a_1 = \frac{b - a}{2} \tag{16.22}$$

which can be substituted into Eq. (16.18) to yield

$$x = \frac{(b + a) + (b - a)x_d}{2} \tag{16.23}$$

This equation can be differentiated to give

$$dx = \frac{b - a}{2}\, dx_d \tag{16.24}$$

Equations (16.23) and (16.24) can be substituted for x and dx, respectively, in the equation to be integrated. These substitutions effectively transform the integration interval without changing the value of the integral. The following example illustrates how this is done in practice.

EXAMPLE 16.3   Two-Point Gauss-Legendre Formula

Problem Statement: Use Eq. (16.17) to evaluate the integral of

$$f(x) = 0.2 + 25x - 200x^2 + 675x^3 - 900x^4 + 400x^5$$

between the limits $x = 0$ to 0.8. Recall that this was the same problem that we solved in Chap. 15 using a variety of Newton-Cotes formulations. The exact value of the integral is 1.64053334.

Solution: Before integrating the function, we must perform a change of variable so that the limits are from -1 to $+1$. To do this, we substitute $a = 0$ and $b = 0.8$ into Eq. (16.23) to yield

$$x = 0.4 + 0.4x_d$$

The derivative of this relationship is [Eq. (16.24)]

$$dx = 0.4dx_d$$

Both of these can be substituted into the original equation to yield

$$\int_0^{0.8} (0.2 + 25x - 200x^2 + 675x^3 - 900x^4 + 400x^5)\, dx$$

$$= \int_{-1}^{1} \{[0.2 + 25(0.4 + 0.4x_d) - 200(0.4 + 0.4x_d)^2$$

$$+ 675(0.4 + 0.4x_d)^3 - 900(0.4 + 0.4x_d)^4$$

$$+ 400(0.4 + 0.4x_d)^5]0.4\}\ dx_d$$

Therefore, the right-hand side is in the form that is suitable for evaluation using Gauss quadrature. The transformed function can be evaluated at $-1/\sqrt{3}$ to be equal to 0.51674055 and at $1/\sqrt{3}$ to be equal to 1.30583723. Therefore, the integral according to Eq. (16.17) is

$$I \simeq 0.51674055 + 1.30583723 = 1.82257778$$

which represents a percent relative error of -11.1 percent. This result is comparable in magnitude to a four-segment application of the trapezoidal rule (Table 15.1) or a single application of Simpson's 1/3 and 3/8 rules (Examples 15.4 and 15.6). This latter result is to be expected because Simpson's rules are also third-order accurate. However, because of the clever choice of base points, Gauss quadrature attains this accuracy on the basis of only two function evaluations.

16.2.3 Higher-Point Formulas

Beyond the two-point formula described in the previous section, higher-point versions can be developed in the general form

$$I \simeq c_0 f(x_0) + c_1 f(x_1) + \cdots + c_{n-1} f(x_{n-1}) \tag{16.25}$$

where n is the number of points. Values for c's and x's for up to and including the six-point formula are summarized in Table 16.1.

EXAMPLE 16.4   *Three-Point Gauss-Legendre Formula*

Problem Statement: Use the three-point formula from Table 16.1 to estimate the integral for the same function as in Example 16.3.

Solution: According to Table 16.1, the three-point formula is

$$I = 0.555555556 f(-0.774596669) + 0.888888889 f(0)$$

$$+ 0.555555556 f(0.774596669)$$

which is equal to

$$I = 0.281301290 + 0.873244444 + 0.485987599 = 1.64053334$$

which is exact.

Because Gauss quadrature requires function evaluations at nonuniformly spaced points within the integration interval, it is not appropriate for cases where the function is unknown. Thus, it is not suited for the many engineering problems that deal with tabulated data. However, where the function is known, its efficiency can be a decided advantage. This is particularly true when numerous integral evaluations must be performed.

TABLE 16.1   Weighting factors c and function arguments x used in Gauss-Legendre formulas.

Points	Weighting Factors	Function Arguments	Truncation Error
2	$c_0 = 1.000000000$ $c_1 = 1.000000000$	$x_0 = -0.577350269$ $x_1 = 0.577350269$	$\simeq f^{(4)}(\xi)$
3	$c_0 = 0.555555556$ $c_1 = 0.888888889$ $c_2 = 0.555555556$	$x_0 = -0.774596669$ $x_1 = 0.0$ $x_2 = 0.774596669$	$\simeq f^{(6)}(\xi)$
4	$c_0 = 0.347854845$ $c_1 = 0.652145155$ $c_2 = 0.652145155$ $c_3 = 0.347854845$	$x_0 = -0.861136312$ $x_1 = -0.339981044$ $x_2 = 0.339981044$ $x_3 = 0.861136312$	$\simeq f^{(8)}(\xi)$
5	$c_0 = 0.236926885$ $c_1 = 0.478628670$ $c_2 = 0.568888889$ $c_3 = 0.478628670$ $c_4 = 0.236926885$	$x_0 = -0.906179846$ $x_1 = -0.538469310$ $x_2 = 0.0$ $x_3 = 0.538469310$ $x_4 = 0.906179846$	$\simeq f^{(10)}(\xi)$
6	$c_0 = 0.171324492$ $c_1 = 0.360761573$ $c_2 = 0.467913935$ $c_3 = 0.467913935$ $c_4 = 0.360761573$ $c_5 = 0.171324492$	$x_0 = -0.932469514$ $x_1 = -0.661209386$ $x_2 = -0.238619186$ $x_3 = 0.238619186$ $x_4 = 0.661209386$ $x_5 = 0.932469514$	$\simeq f^{(12)}(\xi)$

EXAMPLE 16.5   Applying Gauss Quadrature to the Falling Parachutist Problem

Problem Statement: In Example 15.3, we used the multiple-application trapezoidal rule to evaluate

$$d = \frac{gm}{c} \int_0^{10} [1 - e^{-(c/m)t}] \, dt$$

where $g = 9.8$, $c = 12.5$, and $m = 68.1$. The exact value of the integral was determined by calculus to be 289.435147. Recall that the best estimate obtained using a 500-segment trapezoidal rule was 289.4348 with an $|\epsilon_t| \simeq 1.15 \times 10^{-4}$ percent. Repeat this computation using Gauss quadrature.

Solution: After modifying the function, the following results are obtained:

Two-point estimate = 290.014478

Three-point estimate = 289.439297

Four-point estimate = 289.435162

Five-point estimate = 289.435147

Six-point estimate = 289.435147

Thus, the five- and six-point estimates yield results that are exact to nine significant figures.

16.2.4   Error Analysis for Gauss Quadrature

The error for the Gauss-Legendre formulas is specified generally by (Carnahan et al., 1969)

$$E_t = \frac{2^{2n+3}[(n + 1)!]^4}{(2n + 3)[(2n + 2)!]^3} f^{(2n+2)}(\xi) \tag{16.26}$$

where n is the number of points minus one and $f^{(2n+2)}(\xi)$ is the $(2n + 2)$th derivative of the function after the change of variable and ξ is located somewhere on the interval from -1 to 1. Comparison of Eq. (16.26) with Table 15.2 indicates the superiority of Gauss quadrature to Newton-Cotes formulas, provided the higher-order derivatives do not increase substantially with increasing n. Problem 16.8 at the end of this chapter illustrates a case where the Gauss-Legendre formulas perform poorly. In these situations, the multiple-application Simpson's rule or Romberg integration would be preferable. However, for many functions confronted in engineering practice, Gauss quadrature provides an efficient means for evaluating integrals.

16.3 IMPROPER INTEGRALS

To this point, we have dealt exclusively with integrals having finite limits and bounded integrands. Although these types are commonplace in engineering, there will be times when improper integrals must be evaluated. In the present section we will focus on one type of improper integral—that is, one with a lower limit of $-\infty$ or an upper limit of $+\infty$.

Such integrals usually can be evaluated by making a change of variable that transforms the infinite range to one that is finite. The following identity serves this purpose and works for any function that decreases towards zero at least as fast as $1/x^2$ as x approaches infinity:

$$\int_a^b f(x)\,dx = \int_{1/b}^{1/a} \frac{1}{t^2} f\left(\frac{1}{t}\right) dt \tag{16.27}$$

for $ab > 0$. Therefore, it can only be used when a is positive and b is ∞ or when a is $-\infty$ and b is negative. For cases where the limits are from $-\infty$ to a positive value or from a negative value to ∞, the integral can be implemented in two steps. For example,

$$\int_{-\infty}^b f(x)\,dx = \int_{-\infty}^{-A} f(x)\,dx + \int_{-A}^b f(x)\,dx \tag{16.28}$$

where $-A$ is chosen as a sufficiently large negative value so that the function has begun to approach zero asymptotically at least as fast as $1/x^2$. After the integral has been divided into two parts, the first can be evaluated with Eq. (16.27) and the second with a Newton-Cotes closed formula such as Simpson's 1/3 rule.

One problem with using Eq. (16.27) to evaluate an integral is that the transformed function will be singular at one of the limits. The open integration formulas can be used to circumvent this dilemma as they allow evaluation of the integral without employing data at the end points of the integration interval. To allow the maximum flexibility, a multiple-application version of one of the open formulas from Table 15.4 is required.

Multiple-application versions of the open formulas can be concocted by using closed formulas for the interior segments and open formulas for the ends. For example, the multiple-segment trapezoidal rule and the midpoint rule can be combined to give

$$\int_{x_0}^{x_n} f(x)\,dx = h\left[\frac{3}{2}f(x_1) + \sum_{i=2}^{n-2} f(x_i) + \frac{3}{2}f(x_{n-1})\right]$$

In addition, semiopen formulas can be developed for cases where one or the other end of the interval is closed. For example, a formula that is open at the lower limit and closed at the upper limit is given as

$$\int_{x_0}^{x_n} f(x)\,dx = h\left[\frac{3}{2}f(x_1) + \sum_{i=2}^{n-1} f(x_i) + \frac{1}{2}f(x_n)\right]$$

Although these relationships can be used, a preferred formula is (Press et al., 1986)

$$\int_{x_0}^{x_n} f(x)\, dx = h[\, f(x_{1/2}) + f(x_{3/2}) + \cdots f(x_{n-3/2}) + f(x_{n-1/2})] \tag{16.29}$$

which is called the *extended midpoint rule*. Notice that this formula is based on limits of integration that are $h/2$ after and before the first and last data points (Fig. 16.7).

Figure 16.7
Placement of data points relative to integration limits for the extended midpoint rule.

EXAMPLE 16.6   Evaluation of an Improper Integral

Problem Statement: The *cumulative normal distribution* is an important formula in statistics (see Fig. 16.8):

$$N(x) = \int_{-\infty}^{x} \frac{1}{\sqrt{2\pi}}\, e^{-x^2/2}\, dx \tag{E16.6.1}$$

where $x = (y - \bar{y})/s_y$ is called the *normalized standard deviate*. It represents a change of variable to scale the normal distribution so that it is centered on zero and the distance along the abscissa is measured in multiples of the standard deviation (Fig. 16.8b).

Equation (E16.6.1) represents the probability that an event will be less than x. For example, if $x = 1$, Eq. (E16.6.1) can be used to determine that the probability that an event will occur that is less than one standard deviation above the mean is $N(1) = 0.8413$. In other words, if 100 events occur, approximately 84 will be less than the mean plus one standard deviation. Because Eq. (E16.6.1) cannot be evaluated in a simple functional form, it is solved numerically and listed in statistical tables. Use Eq. (16.28) in conjunction with Simpson's 1/3 rule and the extended midpoint rule to determine $N(1)$ numerically.

Solution: Equation (E16.6.1) can be reexpressed in terms of Eq. (16.28) as

$$N(x) = \frac{1}{\sqrt{2\pi}} \left(\int_{-\infty}^{-2} e^{-x^2/2}\, dx + \int_{-2}^{1} e^{-x^2/2}\, dx \right)$$

The first integral can be evaluated by applying Eq. (16.27) to give

$$\int_{-\infty}^{-2} e^{-x^2/2}\, dx = \int_{-1/2}^{0} \frac{1}{t^2}\, e^{-1/(2t^2)}\, dt$$

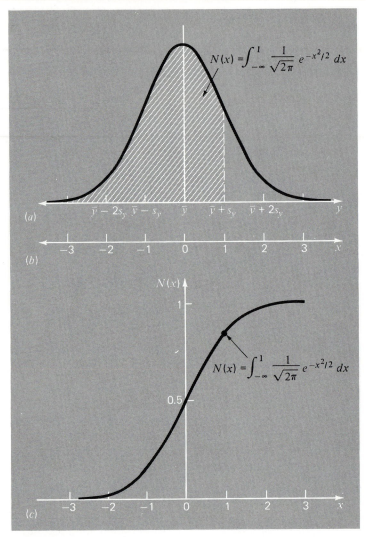

Figure 16.8
(a) The normal distribution, (b) the transformed abscissa in terms of the standardized normal deviate, and (c) the cumulative normal distribution. The shaded area in (a) and the point in (c) represent the probability that a random event will be less than the mean plus one standard deviation.

Then the extended midpoint rule with $h = 1/8$ can be employed to estimate

$$\int_{-1/2}^{0} \frac{1}{t^2} e^{-1/(2t^2)} dt \simeq \frac{1}{8} [f(x_{-7/16}) + f(x_{-5/16}) + f(x_{-3/16}) + f(x_{-1/16})]$$

$$= \frac{1}{8} (0.3833 + 0.0612 + 0 + 0) = 0.0556$$

Simpson's 1/3 rule with $h = 0.5$ can be used to estimate the second integral as

$$\int_{-2}^{1} e^{-x^2/2}\, dx =$$

$$[1 - (-2)]\ \frac{0.1353 + 4(0.3247 + 0.8825 + 0.8825) + 2(0.6065 + 1) + 0.6065}{3(6)}$$

$$= 2.0523$$

Therefore, the final result can be computed as

$$N(1) \simeq \frac{1}{\sqrt{2\pi}}\,(0.0556 + 2.0523) = 0.8409$$

which represents an error of $\epsilon_t = 0.046$ percent.

The foregoing computation can be improved in a number of ways. First, higher-order interpolation formulas could be used. For example, a Romberg integration could be employed. Second, more points could be used. Press et al. (1986) explore both options in depth.

Aside from infinite limits, there are other ways in which an integral can be improper. Common examples include cases where the integral is singular at either the limits or at a point within the integral. Press et al. (1986) provide a nice discussion of ways to handle these situations.

PROBLEMS

Hand Calculations

16.1 Use Romberg integration to evaluate

$$\int_{0}^{3\pi/20} [\sin(5x + 1)]\, dx$$

to an accuracy of $\epsilon_s = 0.5\%$. Your results should be presented in the form of Fig. 16.2. Compute the analytical solution and use it to determine the true error ϵ_t of the result obtained with Romberg integration. Check that ϵ_t is less than the stopping criterion ϵ_s.

16.2 Perform the same computations as in Prob. 16.1, but for the integral

$$\int_{0}^{4} xe^{2x}\, dx$$

16.3 Use Romberg integration to evaluate

$$\int_{0}^{3} \frac{e^x \sin x}{1 + x^2}\, dx$$

to an accuracy of 0.1%. Your results should be presented in the form of Fig. 16.2.

16.4 Obtain an estimate of the integral from Prob. 16.1, but using two-, three-, and four-point Gauss-Legendre formulas. Compute ϵ_t for each case on the basis of the analytical solution.

16.5 Obtain an estimate of the integral from Prob. 16.2, but using two-, three-, and four-point Gauss-Legendre formulas. Compute ϵ_t for each case on the basis of the analytical solution.

16.6 Obtain an estimate of the integral from Prob. 16.3, but using two- through five-point Gauss-Legendre formulas.

16.7 Perform the computation in Examples 15.3 and 16.5 for the falling parachutist, but use Romberg integration ($\epsilon_s = 0.01\%$).

16.8 Employ analytical methods (recall Table PT5.2) and two- through six-point Gauss-Legendre formulas to solve

$$\int_{-3}^{3} \frac{2}{1 + 2x^2} \, dx$$

Interpret your results in light of Eq. (16.26).

16.9 Use numerical integration to evaluate the following

(a) $\displaystyle\int_{0}^{\infty} e^{-y} \sin^2 y \, dy$

(b) $\displaystyle\int_{0}^{\infty} \frac{1}{(1 + y^2)(1 + y^2/2)} \, dy$

(c) $\displaystyle\int_{-1}^{\infty} ye^{-y} \, dy$

(d) $\displaystyle\int_{-\infty}^{\infty} \frac{1}{\sqrt{2\pi}} \, e^{-x^2/2} \, dx$

Note that (d) is the normal distribution.

Computer-Related Problems

16.10 Develop a user-friendly computer program for Romberg integration based on Fig. 16.3. Test it by duplicating the computation depicted in Fig. 16.2.

16.11 Develop a user-friendly computer program for Gauss quadrature. Test it by duplicating the results of Examples 16.3 and 16.4.

16.12 Use the program developed in Prob. 16.10 to solve Probs. 16.1, 16.2, and 16.3.

16.13 Use the program developed in Prob. 16.11 to solve Probs. 16.4, 16.5, and 16.6.

16.14 Develop a program to implement the extended midpoint rule iteratively. Start the iterations with an initial estimate based on a single point and the midpoint rule from Table 15.4. Then successively apply Eq. (16.29) with the interval divided by 3 at each stage; that is, $h = (b - a)/3$, $(b - a)/9$, $(b - a)/27$, etc. Note that this means that one-third of the function estimates will have been determined in the previous iteration. Develop your algorithm so that it capitalizes on this property. Perform iterations until an approximate error estimate ϵ_a falls below a prespecified stopping criteria ϵ_s. Test the program by evaluating Example 16.6.

CHAPTER 17
Numerical Differentiation

We have already introduced the notion of numerical differentiation in Chap. 3. Recall that we employed Taylor series expansions to derive finite-divided-difference approximations of derivatives. In Chap. 3, we developed forward, backward, and centered difference approximations of first and higher derivatives. Recall that, at best, these estimates had errors that were $O(h^2)$—that is, their errors were proportional to the square of the step size. This level of accuracy is due to the number of terms of the Taylor series that were retained during the derivation of these formulas. We will now illustrate how to develop more accurate formulas by retaining more terms.

17.1 HIGH-ACCURACY DIFFERENTIATION FORMULAS

As noted above, high-accuracy divided-difference formulas can be generated by including additional terms from the Taylor series expansion. For example, the forward Taylor series expansion can be written as [Eq. (3.32)].

$$f(x_{i+1}) = f(x_i) + f'(x_i)h + \frac{f''(x_i)}{2} h^2 + \cdots \tag{17.1}$$

which can be solved for

$$f'(x_i) = \frac{f(x_{i+1}) - f(x_i)}{h} - \frac{f''(x_i)}{2} h + O(h^2) \tag{17.2}$$

In Chap. 3, we truncated this result by excluding the second- and higher-derivative terms and were thus left with a final result of

$$f'(x_i) = \frac{f(x_{i+1}) - f(x_i)}{h} + O(h) \tag{17.3}$$

In contrast to this approach, we now retain the second-derivative term by substituting the following approximation of the second derivative [recall Eq. (3.35)]:

$$f''(x_i) = \frac{f(x_{i+2}) - 2f(x_{i+1}) + f(x_i)}{h^2} + O(h)$$

525

into Eq. (17.2) to yield

$$f'(x_i) = \frac{f(x_{i+1}) - f(x_i)}{h} - \frac{f(x_{i+2}) - 2f(x_{i+1}) + f(x_i)}{2h^2} h + O(h^2)$$

or by collecting terms

$$f'(x_i) = \frac{-f(x_{i+2}) + 4f(x_{i+1}) - 3f(x_i)}{2h} + O(h^2)$$

Notice that inclusion of the second-derivative term has improved the accuracy to $O(h^2)$. Similar improved versions can be developed for the backward and centered formulas as well as for the approximations of the higher derivatives. The formulas are summarized in Figs. 17.1 through 17.3 along with all the results from Chap. 3. The following example illustrates the utility of these formulas for estimating derivatives.

Figure 17.1
Forward finite-divided-difference formulas: two versions are presented for each derivative. The latter version incorporates more terms of the Taylor series expansion and is, consequently, more accurate.

First Derivative   Error

$$f'(x_i) = \frac{f(x_{i+1}) - f(x_i)}{h} \qquad\qquad O(h)$$

$$f'(x_i) = \frac{-f(x_{i+2}) + 4f(x_{i+1}) - 3f(x_i)}{2h} \qquad\qquad O(h^2)$$

Second Derivative

$$f''(x_i) = \frac{f(x_{i+2}) - 2f(x_{i+1}) + f(x_i)}{h^2} \qquad\qquad O(h)$$

$$f''(x_i) = \frac{-f(x_{i+3}) + 4f(x_{i+2}) - 5f(x_{i+1}) + 2f(x_i)}{h^2} \qquad\qquad O(h^2)$$

Third Derivative

$$f'''(x_i) = \frac{f(x_{i+3}) - 3f(x_{i+2}) + 3f(x_{i+1}) - f(x_i)}{h^3} \qquad\qquad O(h)$$

$$f'''(x_i) = \frac{-3f(x_{i+4}) + 14f(x_{i+3}) - 24f(x_{i+2}) + 18f(x_{i+1}) - 5f(x_i)}{2h^3} \qquad\qquad O(h^2)$$

Fourth Derivative

$$f''''(x_i) = \frac{f(x_{i+4}) - 4f(x_{i+3}) + 6f(x_{i+2}) - 4f(x_{i+1}) + f(x_i)}{h^4} \qquad\qquad O(h)$$

$$f''''(x_i) = \frac{-2f(x_{i+5}) + 11f(x_{i+4}) - 24f(x_{i+3}) + 26f(x_{i+2}) - 14f(x_{i+1}) + 3f(x_i)}{h^4} \qquad\qquad O(h^2)$$

First Derivative   Error

$$f'(x_i) = \frac{f(x_i) - f(x_{i-1})}{h}$$ $O(h)$

$$f'(x_i) = \frac{3f(x_i) - 4f(x_{i-1}) + f(x_{i-2})}{2h}$$ $O(h^2)$

Second Derivative

$$f''(x_i) = \frac{f(x_i) - 2f(x_{i-1}) + f(x_{i-2})}{h^2}$$ $O(h)$

$$f''(x_i) = \frac{2f(x_i) - 5f(x_{i-1}) + 4f(x_{i-2}) - f(x_{i-3})}{h^2}$$ $O(h^2)$

Third Derivative

$$f'''(x_i) = \frac{f(x_i) - 3f(x_{i-1}) + 3f(x_{i-2}) - f(x_{i-3})}{h^3}$$ $O(h)$

$$f'''(x_i) = \frac{5f(x_i) - 18f(x_{i-1}) + 24f(x_{i-2}) - 14f(x_{i-3}) + 3f(x_{i-4})}{2h^3}$$ $O(h^2)$

Fourth Derivative

$$f''''(x_i) = \frac{f(x_i) - 4f(x_{i-1}) + 6f(x_{i-2}) - 4f(x_{i-3}) + f(x_{i-4})}{h^4}$$ $O(h)$

$$f''''(x_i) = \frac{3f(x_i) - 14f(x_{i-1}) + 26f(x_{i-2}) - 24f(x_{i-3}) + 11f(x_{i-4}) - 2f(x_{i-5})}{h^4}$$ $O(h^2)$

Figure 17.2
Backward finite-divided-difference formulas: two versions are presented for each derivative. The latter version incorporates more terms of the Taylor series expansion and is, consequently, more accurate.

EXAMPLE 17.1   High-Accuracy Differentiation Formulas

Problem Statement: Recall that in Example 3.13 we estimated the derivative of

$$f(x) = -0.1x^4 - 0.15x^3 - 0.5x^2 - 0.25x + 1.2$$

at $x = 0.5$ using finite divided differences and a step size of $h = 0.25$,

	Forward $O(h)$	Backward $O(h)$	Centered $O(h^2)$
Estimate	−1.155	−0.714	−0.934
ϵ_t, %	−26.5	21.7	−2.4

where the errors were computed on the basis of the true value of -0.9125. Repeat this computation, but employ the high-accuracy formulas from Figs. 17.1 through 17.3.

Solution: The data needed for this example is

$$x_{i-2} = 0 \qquad f(x_{i-2}) = 1.2$$

$$x_{i-1} = 0.25 \qquad f(x_{i-1}) = 1.103515625$$

$$x_i = 0.5 \qquad f(x_i) = 0.925$$

$$x_{i+1} = 0.75 \qquad f(x_{i+1}) = 0.636328125$$

$$x_{i+2} = 1 \qquad f(x_{i+2}) = 0.2$$

The forward difference of accuracy $O(h^2)$ is computed as (Fig. 17.1)

$$f'(0.5) = \frac{-0.2 + 4(0.636328125) - 3(0.925)}{2(0.25)} = -0.859375 \qquad \epsilon_t = 5.82\%$$

The backward difference of accuracy $O(h^2)$ is computed as (Fig. 17.2)

$$f'(0.5) = \frac{3(0.925) - 4(1.103515625) + 1.2}{2(0.25)} = -0.878125 \qquad \epsilon_t = 3.77\%$$

The centered difference of accuracy $O(h^4)$ is computed as (Fig. 17.3)

$$f'(0.5) = \frac{-0.2 + 8(0.636328125) - 8(1.103515625) + 1.2}{12(0.25)}$$

$$= -0.9125 \qquad \epsilon_t = 0$$

As expected, the errors for the forward and backward differences are considerably more accurate than the results from Example 3.13. However, surprisingly, the centered difference yields a perfect result. This is due to the fact that the formulas based on the Taylor series are equivalent to passing polynomials through the data points.

17.1.1 Computer Programs

Figures 17.4 and 17.5 provide programs to compute finite-difference approximations of derivatives for equispaced discrete data. The codes employ centered differences of $O(h^2)$ for the interior points and forward and backward differences of $O(h)$ for the first and last points, respectively. Figure 17.6 shows the results of applying the programs to estimate derivatives from tabulated data generated via the function $y = 2x^2$. The derivative can be evaluated analytically as $y' = 4x$. Therefore, the estimates for the interior points

First Derivative   Error

$$f'(x_i) = \frac{f(x_{i+1}) - f(x_{i-1})}{2h}$$ $O(h^2)$

$$f'(x_i) = \frac{-f(x_{i+2}) + 8f(x_{i+1}) - 8f(x_{i-1}) + f(x_{i-2})}{12h}$$ $O(h^4)$

Second Derivative

$$f''(x_i) = \frac{f(x_{i+1}) - 2f(x_i) + f(x_{i-1})}{h^2}$$ $O(h^2)$

$$f''(x_i) = \frac{-f(x_{i+2}) + 16f(x_{i+1}) - 30f(x_i) + 16f(x_{i-1}) - f(x_{i-2})}{12h^2}$$ $O(h^4)$

Third Derivative

$$f'''(x_i) = \frac{f(x_{i+2}) - 2f(x_{i+1}) + 2f(x_{i-1}) - f(x_{i-2})}{2h^3}$$ $O(h^2)$

$$f'''(x_i) = \frac{-f(x_{i+3}) + 8f(x_{i+2}) - 13f(x_{i+1}) + 13f(x_{i-1}) - 8f(x_{i-2}) + f(x_{i-3})}{8h^3}$$ $O(h^4)$

Fourth Derivative

$$f''''(x_i) = \frac{f(x_{i+2}) - 4f(x_{i+1}) + 6f(x_i) - 4f(x_{i-1}) + f(x_{i-2})}{h^4}$$ $O(h^2)$

$$f''''(x_i) = \frac{-f(x_{i+3}) + 12f(x_{i+2}) - 39f(x_{i+1}) + 56f(x_i) - 39f(x_{i-1}) + 12f(x_{i-2}) - f(x_{i-3})}{6h^4}$$ $O(h^4)$

Figure 17.3
Centered finite-divided-difference formulas: two versions are presented for each derivative. The latter version incorporates more terms of the Taylor series expansion and is, consequently, more accurate.

Figure 17.4
An interactive computer program for numerical differentiation written in Microsoft BASIC.

```
100 REM  NUMERICAL DIFFERENTIATION (BASIC)        200 REM ********** MAIN PROGRAM **********
105 REM *******************************           205 REM
110 REM *        DEFINITION OF VARIABLES    *     210 GOSUB 250 'input data
115 REM *                                   *     215 GOSUB 350 'compute derivatives
120 REM * X  = INDEPENDENT VARIABLE         *     220 GOSUB 400 'output results
125 REM * F  = DEPENDENT VARIABLE           *     225 END
130 REM * X1 = INITIAL VALUE OF X           *     250 REM ******** SUBROUTINE INPUT ********
135 REM * N  = NUMBER OF POINTS             *     255 REM
140 REM * H  = STEP SIZE                    *     260 CLS
145 REM * D  = DERIVATIVES                  *     265 INPUT "NUMBER OF POINTS? ", N
165 REM *******************************           270 PRINT:INPUT "STEP SIZE? ", H
170 REM                                           275 PRINT:INPUT "FIRST VALUE OF X? ", X1
```

(*Continues* **Figure 17.4**)

```
280 X=X1
285 PRINT
290 FOR I = 1 TO N
295    PRINT "F(AT X = ";X;") = ";
300    INPUT F(I)
305    X = X + H
310 NEXT I
315 PRINT
320 RETURN
350 REM **** SUBROUTINE DIFFERENTIATE ****
355 D(1)=(F(2)-F(1))/H
360 FOR I = 2 TO N-1
365    D(I)=(F(I+1)-F(I-1))/(2*H)
```

```
370 NEXT I
375 D(N)=(F(N)-F(N-1))/H
380 RETURN
400 REM ******* SUBROUTINE OUTPUT ********
405 REM
410 PRINT "X-VALUE     Y-VALUE     DERIVATIVE"
415 X=X1
420 FOR I = 1 TO N
425    PRINT X,F(I),D(I)
430    X = X + H
435 NEXT I
440 RETURN
```

(determined with the centered differences) are perfect. This is because the function is a simple second-order monomial. In contrast, the first and last points have errors of $|E_t| = 2$ because they were estimated with the lower-accuracy forward and backward differences. Note that other functions such as $y = \cos x$ or $y = 5e^{-2x}$ would indicate errors for all points but with less discrepancy for the centered-difference approximations.

Figure 17.5
Subroutines for numerical differentiation written in (*a*) FORTRAN 77 and (*b*) Turbo Pascal.

```
      SUBROUTINE NUMDIF(F,D,N,H)
**********************************************
*          DEFINITION OF VARIABLES
*
*    N       = NUMBER OF POINTS
*    D()     = DERIVATIVES
*    F()     = DEPENDENT VARIABLE
*    H       = STEP-SIZE
**********************************************
      DIMENSION D(100),F(100)
      D(1) = (F(2) - F(1))/H
      DO 10 I=1,N-1
         D(I) = (F(I+1) - F(I-1))/(2*H)
   10 CONTINUE
      D(N) = (F(N) - F(N-1))/H
      RETURN
      END
```

```
PROCEDURE NumDif (VAR F,D: Vector;
                          N :integer;
                          H : Real);

{      Driver program type definitions
    Vector = a 1 dimensional real array
                 (defined with zero subscript)}

{          Definition of variables
    N   = number of points
    H   = step-size
    F() = dependent variable
    D() = finite divided difference        }

VAR
    i : integer;
Begin
D[1] := (F[2] - F[1])/H;
For i := 2 to N-1 do
  Begin
    D[i] := (F[i+1] - F[i-1])/(2*H);
  End;
D[N] := (F[N] - F[N-1])/H;
End                    {of procedure NumDif}
```

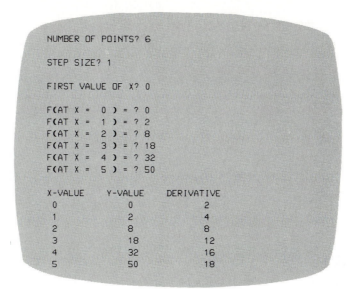

```
NUMBER OF POINTS? 6

STEP SIZE? 1

FIRST VALUE OF X? 0

F(AT X =   0  ) = ? 0
F(AT X =   1  ) = ? 2
F(AT X =   2  ) = ? 8
F(AT X =   3  ) = ? 18
F(AT X =   4  ) = ? 32
F(AT X =   5  ) = ? 50

X-VALUE      Y-VALUE      DERIVATIVE
   0            0             2
   1            2             4
   2            8             8
   3            18            12
   4            32            16
   5            50            18
```

Figure 17.6
Application of the program from Fig. 17.4 to estimate the derivative of $y = 2x^2$ at equispaced points with $\Delta x = 1$ on the interval from $x = 0$ to 5.

17.2 RICHARDSON EXTRAPOLATION

To this point, we have seen that there are two ways to improve derivative estimates when employing finite divided differences: (1) decrease the step size or (2) use a higher-order formula that employs more points. A third approach, based on Richardson extrapolation, uses two derivative estimates to compute a third, more accurate approximation.

Recall from Sec. 16.1.1 that Richardson extrapolation provided a means to obtain an improved integral estimate I by the formula [Eq. (16.4)]

$$I \simeq I(h_2) + \frac{1}{(h_1/h_2)^2 - 1} \left[I(h_2) - I(h_1) \right] \tag{17.5a}$$

where $I(h_1)$ and $I(h_2)$ are integral estimates using two step sizes h_1 and h_2. Because of its convenience when expressed as a computer algorithm, this formula is usually written for the case where $h_2 = h_1/2$ as in

$$I \simeq \frac{4}{3} I(h_2) - \frac{1}{3} I(h_1) \tag{17.5a}$$

In a similar fashion, Eq. (17.5a) can be written for derivatives as

$$D \simeq \frac{4}{3} D(h_2) - \frac{1}{3} D(h_1) \tag{17.5b}$$

For centered difference approximations with $O(h^2)$, the application of this formula will yield a new derivative estimate of $O(h^4)$.

EXAMPLE 17.2   Richardson Extrapolation

Problem Statement: Using the same function as in Example 17.1, estimate the first-derivative at $x = 0.5$ employing step sizes of $h_1 = 0.5$ and $h_2 = 0.25$. Then use Eq. (17.5b) to compute an improved estimate with Richardson extrapolation. Recall that the true value is -0.9125.

Solution: The first-derivative estimates can be computed with centered differences as

$$D(0.5) = \frac{0.2 - 1.2}{1.0} = -1.0 \qquad \epsilon_t = -9.6\%$$

and

$$D(0.25) = \frac{0.63632813 - 1.10351563}{0.5} = -0.934375 \qquad \epsilon_t = -2.4\%$$

The improved estimate can be determined by applying Eq. (17.5b) to give

$$D = \frac{4}{3}(-0.934375) - \frac{1}{3}(-1) = -0.9125$$

which for the present case is a perfect result.

The previous example yielded a perfect result because the function being analyzed was a fourth-order polynomial. The perfect outcome was due to the fact that Richardson extrapolation is actually equivalent to fitting a higher-order polynomial through the data and then evaluating the derivatives by centered divided differences. Thus, the present case matched the derivative of the fourth-order polynomial precisely. For most other functions, of course, this would not occur and our derivative estimate would be improved but not perfect. Consequently, as was the case for the application of Richardson extrapolation, the approach can be applied iteratively using a Romberg algorithm until the result falls below an acceptable error criterion.

17.3   DERIVATIVES OF UNEQUALLY SPACED DATA

The approaches discussed to this point are primarily designed to determine the derivative of a given function. For the finite-divided-difference approximations of Sec. 17.1, the data had to be evenly spaced. For the Richardson extrapolation technique of Sec. 17.2, the data had to be evenly spaced and generated for successively halved intervals. Such control of data spacing is usually available only in cases where we can use a function to generate a table of values.

In contrast, empirically derived information—that is, data from experiments or field studies—is often collected at unequal intervals. Such information cannot be analyzed with the techniques discussed to this point.

One way to handle nonequispaced data is to fit a second-order Lagrange interpolating polynomial [recall Eq. (12.23)] to each set of three adjacent points. Remember that this polynomial does not require that the points be equispaced. The second-order polynomial can be differentiated analytically to give

$$
\begin{aligned}
f'(x) \simeq\ & f(x_{i-1}) \frac{2x - x_i - x_{i+1}}{(x_{i-1} - x_i)(x_{i-1} - x_{i+1})} \\
& + f(x_i) \frac{2x - x_{i-1} - x_{i+1}}{(x_i - x_{i-1})(x_i - x_{i+1})} \\
& + f(x_{i+1}) \frac{2x - x_{i-1} - x_i}{(x_{i+1} - x_{i-1})(x_{i+1} - x_i)}
\end{aligned}
\tag{17.6}
$$

where x is the value at which you want to estimate the derivative. Although this equation is certainly more complicated than the first-derivative approximations from Figs. 17.1 through 17.3, it has some important advantages. First, it can be used to estimate the derivative anywhere within the range prescribed by the three points. Second, the points themselves do not have to be equally spaced. Third, the derivative estimate is of the same accuracy as the centered difference [Eq. (3.33)]. In fact, for equispaced points, Eq. (17.6) evaluated at $x = x_i$ reduces to Eq. (3.33).

17.4 DERIVATIVE AND INTEGRAL ESTIMATES FOR DATA WITH ERRORS

Aside from unequal spacing, the other problem related to differentiating empirical data is that it usually includes measurement error. A shortcoming of numerical differentiation is that it tends to amplify errors in the data. Figure 17.7a shows smooth, error-free data that when numerically differentiated yields a smooth result (Fig. 17.7b). In contrast, Fig. 17.7c uses the same data, but with some points raised and some lowered slightly. This minor modification is barely apparent from Fig. 17.7c. However, the resulting effect in Fig. 17.7d is significant because the process of differentiation amplifies errors.

As might be expected, the primary approach for determining derivatives for imprecise data is to use least-squares regression to fit a smooth, differentiable function to the data. In the absence of any other information, a lower-order polynomial regression might be a good first choice. Obviously, if the true functional relationship between the dependent and independent variable is known, this relationship should form the basis for the least-squares fit.

17.4.1 Differentiation versus Integration of Uncertain Data

Just as curve-fitting techniques like regression can be used to differentiate uncertain data, a similar process can be employed for integration. However, because of the difference in stability between differentiation and integration, this is rarely done.

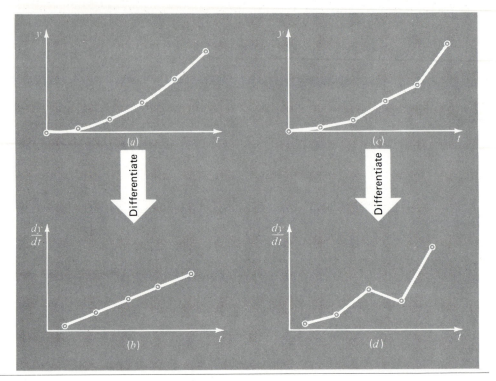

Figure 17.7
Illustration of how even small data errors are amplified by numerical differentiation: (*a*) data with no error, (*b*) the resulting numerical differentiation, (*c*) data modified slightly, (*d*) the resulting differentiation manifesting increased variability. In contrast, the reverse operation of integration [moving from (*d*) to (*c*) by taking the area under (*d*)] tends to attenuate or smooth data errors.

As depicted in Fig. 17.7, differentiation tends to be unstable—that is, it amplifies errors. In contrast, the fact that integration is a summing process tends to make it very forgiving with regard to uncertain data. In essence, as points are summed to form an integral, random positive and negative errors tend to cancel out. In contrast, because differentiation is subtractive, random positive and negative errors tend to add.

PROBLEMS

Hand Calculations

17.1 Compute forward and backward difference approximations of $O(h)$ and $O(h^2)$, and central difference approximations of $O(h^2)$ and $O(h^4)$ for the first derivative of $y = \sin x$ at $x = \pi/3$ using a value of $h = \pi/12$. Estimate the true percent relative error ϵ_t for each approximation.

17.2 Repeat Prob. 17.1 but for $y = \log x$ evaluated at $x = 4$ with $h = 0.2$.

17.3 Use centered difference approximations to estimate the first and second derivatives of $y = e^x$ at $x = 2$ for $h = 0.25$. Employ both $O(h^2)$ and $O(h^4)$ formulas for your estimates.

17.4 Use Richardson extrapolation to estimate the first derivative of $y = \sin x$ at $x = \pi/3$ using step sizes of $h_1 = \pi/3$ and $h_2 = \pi/6$. Employ centered differences of $O(h^2)$ for the initial estimates.

17.5 Repeat Prob. 17.4 but for the first derivative of $\ln x$ at $x = 5$ using $h_1 = 2$ and $h_2 = 1$.

17.6 Employ Eq. (17.6) to determine the first derivative of $y = 3x^4 - 7x^3 - 12x - 3$ at $x = 1$ based on values at $x_0 = 0$, $x_1 = 2$, and $x_2 = 2.2$. Compare this result with the true value and with an estimate obtained using a centered difference approximation based on $h = 1$.

17.7 Prove that for equispaced data points, Eq. (17.6) reduces to Eq. (3.33) at $x = x_i$.

17.8 Compute the central difference approximations of $O(h^4)$ for each of the following functions at the specified location and for the specified step size:

    (a) $y = x^3 + 2x - 10$ at $x = 0$, $h = 0.5$
    (b) $y = x^3 + 2x - 10$ at $x = 0$, $h = 0.1$
    (c) $y = \tan x$ at $x = 4$, $h = 0.2$
    (d) $y = \sin (0.5 \sqrt{x})$ at $x = 1$, $h = 0.25$
    (e) $y = e^x$ at $x = 0$, $h = 0.5$

17.9 The following data was collected for the distance traveled versus time for a rocket:

t	0	1	2	3	4	5
y	0	2.1	7.8	18.2	31.9	50.3

Use numerical differentiation to estimate the rocket's velocity and acceleration at each time.

Computer-Related Problems

17.10 Develop a user-friendly computer program for numerical differentiation based on Fig. 17.4 or 17.5 and the material in this chapter. Among other things modify the algorithm so that the program computes second derivatives.

17.11 Use the program developed in Prob. 17.10 to solve Prob. 17.9.

17.12 Develop a user-friendly program to apply a Romberg algorithm to estimate the derivative of a given function.

17.13 Develop a user-friendly program to obtain first-derivative estimates for unequally spaced data. Test it with the following data:

x	1.2	3	3.2	5	7
$f(x)$	1.807	0.7468	0.6522	0.1684	0.03192

where $f(x) = 5e^{-x}x$. Compare your results with the true derivatives.

CHAPTER 18
Case Studies: Numerical Integration and Differentiation

The purpose of this chapter is to apply the methods of numerical integration and differentiation discussed in Part Five to practical engineering problems. Two situations are most frequently encountered. In the first case, the function under study can be expressed in analytic form but is too complicated to be readily evaluated using the methods of calculus. Numerical methods are applied to situations of this type by using the analytic expression to generate a table of arguments and function values. In the second case, the function to be evaluated is inherently tabular in nature. This type of function usually represents a series of measurements, observations, or some other empirical information. Data for either case is directly compatible with several schemes discussed in this part of the book.

Case Study 18.1, which analyzes cash flow for a computer company, is an example of the analysis of tabular data. The trapezoidal rule, Simpson's 1/3 rule, and numerical differentiation are used to determine cash flow. *Case Study 18.2*, which deals with heat calculations from chemical engineering, involves equations. In this case study, an analytic function is integrated numerically to determine the heat required to raise the temperature of a material.

Case Studies 18.3 and 18.4 also involve functions that are available in equation form. *Case Study 18.3*, which is taken from civil engineering, uses numerical integration to determine the total wind force acting on the mast of a racing sailboat. *Case Study 18.4* determines the root-mean-square current for an electric circuit. This example is used to demonstrate the utility of Romberg integration and Gauss quadrature.

Case Study 18.5 returns to the analysis of tabular information to determine the work required to move a block. Although this example has a direct connection with mechanical engineering, it is germane to all other areas of engineering. Among other things, we use this case study to illustrate the integration of unequally spaced data.

Finally, *Case Study 18.6* illustrates how spreadsheets can be employed to integrate data. Examples include the trapezoidal and Simpson's rules, Romberg integration, and Gauss quadrature.

CASE STUDY 18.1   CASH-FLOW ANALYSIS (GENERAL ENGINEERING)

Background: An important part of any engineering or business project is cash-flow analysis. Available cash may affect many aspects of the problem, for instance, resource allocation. Your engineering position at Ultimate Computer Company requires that you calculate the total cash generated from computer sales for the first 60 days following the introduction of the new computer on the market. (See Table 18.1 for computer sales data.)

TABLE 18.1　**Computer sales data and cash-flow analysis. Column (*c*) is calculated using numerical differentiation of the information in column (*b*). The first and last values in column (*c*) are determined using forward and backward differences of order h^2; the middle values are determined by centered differences of order h^2.**

Numbers of Computers Available on the Market (*a*)	Number of Computers Sold (*b*)	Computer Sales Rate per day (*c*)	Cost per Computer, $ [Based on Column (*a*) and Eq. (18.1)] (*d*)	Cash Generated per day, $ [(*c*) × (*d*)] (*e*)	Time, days (*f*)
50,000	0	2050.0	1542	3,161,100	0
35,000	15,000	950.0	1639	1,557,050	10
31,000	19,000	1500.0	1677	2,515,500	20
20,000	30,000	600.0	1833	1,099,800	30
19,000	31,000	397.5	1853	736,568	40
12,050	37,950	400.0	2040	816,000	50
11,000	39,000	−190.0	2083	−395,770	60

Your problem is complicated by the fact that the cost of the computer is very sensitive to the supply or availability. Your sales and marketing research teams have given you the information that the base sales price at infinite supply is $1250 per computer. As the supply diminishes, the price increases to a maximum of $3000 per computer. Furthermore, the continuous variation of cost with supply N is defined by the empirically derived equation

$$\text{Cost per computer (\$)} = 3000 - 1750\frac{N}{10,000 + N} \qquad (18.1)$$

which is plotted in Fig. 18.1.

Solution: The total cash generated is given by

$$\text{Total cash} = \int_0^{60} (\text{cash generated per day}) \, dt$$

or

$$\text{Total cash} = \int_0^{60} (\text{sales rate} \times \text{unit cost}) \, dt$$

In this case, the rate of sales on days 0 through 60 is given by column (*c*) in Table 18.1. The rate is determined using finite divided differences to estimate the first derivative

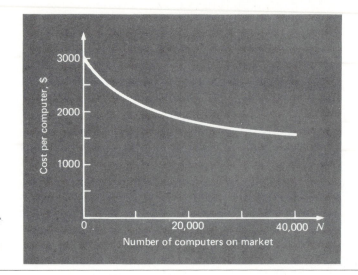

Figure 18.1
Cost of a computer versus the number of computers on the market. The curve is based on Eq. (18.1).

of column (*b*). Notice how, because of the noise in the data, the derivative estimates in column (*c*) vary greatly. In fact, even though the number of computers sold always increases, the noise in the data results in a negative sales rate at day 60. This outcome is due to the fact that numerical estimates of derivatives are highly sensitive to data noise.

The cost per computer on each day is computed on the basis of Eq. (18.1) and the number of computers available as listed in column (*a*) in Table 18.1. The cost per computer for each day 0 through 60 is given in column (*d*). Column (*e*) gives the cash generated per day. This data can be used in conjunction with the numerical integration procedures discussed in Chap. 15.

Table 18.2 gives the results of applying the trapezoidal rule and Simpson's 1/3 rule to this problem. Notice how the results vary widely depending on how many segments are employed for the analysis. In particular, the estimate from the three-segment version of the trapezoidal rule is much higher than other estimates because of the selective inclusion of the high cash-flow estimate on day 20.

TABLE 18.2 Results of applying the trapezoidal rule and Simpson's 1/3 rule to calculate the cash generated from computer sales.

Technique	Segments	Cash Generated, $
Trapezoidal rule	1	82,959,900
	2	74,473,950
	3	96,294,660
	6	81,075,830
Simpson's 1/3 rule	2	71,645,300
	6	77,202,887

On the basis of this analysis, we can conclude that cash flow is approximately $77 million. However, the results indicate that care should be exercised when applying numerical integration formulas and that estimates from tabular data can usually be improved only by obtaining additional information. This conclusion is reiterated in Case Study 18.5 where we demonstrate how the number of data points can have a significant effect on the final outcome of an integral estimate.

CASE STUDY 18.2   THE USE OF INTEGRALS TO DETERMINE THE TOTAL QUANTITY OF HEAT OF MATERIALS (CHEMICAL ENGINEERING)

Background: Heat calculations are employed routinely in chemical engineering as well as in many other fields of engineering. The present case study provides a simple but useful example of such computations.

One problem that is often encountered is the determination of the quantity of heat required to raise the temperature of a material. The characteristic that is needed to carry out this computation is the heat capacity c. This parameter represents the quantity of heat required to raise a unit mass by a unit temperature. If c is constant over the range of temperatures being examined, the required heat ΔH (in calories) can be calculated by

$$\Delta H = mc \, \Delta T \tag{18.2}$$

where c has units of calories per gram per degree Celsius, m is mass (in grams), and ΔT is change in temperature (in degrees Celsius). For example, the amount of heat required to raise 20 g of water from 5 to 10°C is equal to

$$\Delta H = 20(1)(10 - 5) = 100 \text{ cal}$$

where the heat capacity of water is approximately 1 cal/(g·°C). Such a computation is adequate when the ΔT is small. However, for large ranges of temperature, the heat capacity is not constant and, in fact, varies as a function of temperature. For example, the heat capacity of a material could increase with temperature according to a relationship such as

$$c(T) = 0.132 + 1.56 \times 10^{-4}T + 2.64 \times 10^{-7}T^2 \tag{18.3}$$

In this instance you are asked to compute the heat required to raise 1000 g of this material from -100 to 200°C.

Solution: Equation (PT5.4) provides a way to calculate the average value of $c(T)$:

$$\bar{c}(T) = \int_{T_1}^{T_2} \frac{c(T)}{T_2 - T_1} \, dT \tag{18.4}$$

which can be substituted into Eq. (18.2) to yield

$$\Delta H = m \int_{T_1}^{T_2} c(T) \, dT \tag{18.5}$$

where $\Delta T = T_2 - T_1$. Now because, for the present case, $c(T)$ is a simple quadratic, ΔH can be determined analytically. Equation (18.3) is substituted into Eq. (18.5) and the result integrated to yield an exact value of $\Delta H = 42,732$ cal. It is useful and instructive to compare this result with the numerical methods developed in Chap. 15. To accomplish this, it is necessary to generate a table of values of c for various values of T:

T, °C	c, cal/(g·°C)
−100	0.11904
−50	0.12486
0	0.13200
50	0.14046
100	0.15024
150	0.16134
200	0.17376

These points can be used in conjunction with a six-segment Simpson's 1/3 rule to compute an integral estimate of 42.732. This result can be substituted into Eq. (18.5) to yield a value of $\Delta H = 42,732$ cal, which agrees exactly with the analytical solution. This exact agreement would occur no matter how many segments were used. This is to be expected because c is a quadratic function and Simpson's rule is exact for polynomials of the third order or less (see Sec. 15.2.1).

    The results using the trapezoidal rule are listed in Table 18.3. It is seen that the trapezoidal rule is also capable of estimating the total heat very accurately. However, a small step (<10°C) is required for five-place accuracy. This example is a good illustration of why Simpson's rule is very popular. It is easy to perform with either a hand calculator or, better yet, a personal computer. In addition, it is usually sufficiently accurate for relatively large step sizes and is exact for polynomials of the third order or less.

TABLE 18.3   **Results using the trapezoidal rule with various step sizes.**

Step Size, °C	ΔH	ϵ_t, %
300	96,048	125
150	43,029	0.7
100	42,864	0.3
50	42,765	0.07
25	42,740	0.018
10	42,733.3	<0.01
5	42,732.3	<0.01
1	42,732.01	<0.01
0.05	42,732.0003	<0.01

CASE STUDY 18.3   EFFECTIVE FORCE ON THE MAST OF A RACING SAILBOAT
(CIVIL ENGINEERING)

Background: A cross section of a racing sailboat is shown in Fig. 18.2a. Wind forces (f) exerted per foot of mast from the sails vary as a function of distance above the deck of the boat (z) as in Fig. 18.2b. Calculate the tensile force T in the left mast support

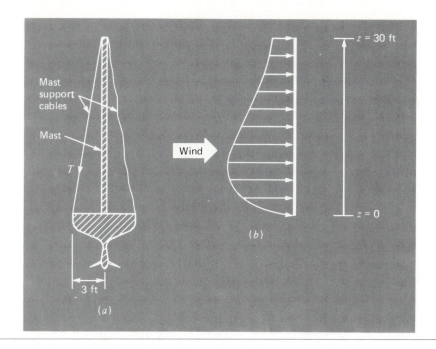

Figure 18.2
(*a*) Cross section of a racing sailboat. (*b*) Wind forces *f* exerted per foot of mast as a function of distance *z* above the deck of the boat.

cable, assuming that the right support cable is completely slack and the mast joins the deck in a manner that transmits horizontal or vertical forces but no moments. Assume that the mast remains vertical.

Solution: In order to proceed with this problem, it is required that the distributed force *f* be converted to an equivalent total force *F* and that its effective location above the deck *d* be calculated (Fig. 18.3). This computation is complicated by the fact that the force exerted per foot of mast varies with the distance above the deck. The total force exerted on the mast can be expressed as the integral of a continuous function:

$$F = \int_0^{30} 200\left(\frac{z}{5+z}\right) e^{-2z/30} dz \tag{18.6}$$

This nonlinear integral is difficult to evaluate analytically. Therefore, it is convenient to employ numerical approaches such as Simpson's rule and the trapezoidal rule for this problem. This is accomplished by calculating $f(z)$ for various values of z and then using Eqs. (15.10) or (15.18). For example, Table 18.4 has values of $f(z)$ for a step size of 3 ft that provide data for Simpson's 1/3 rule or the trapezoidal rule. Results for several step sizes are given in Table 18.5. It is observed that both methods give a value of $F = 1480.6$ lb as the step size becomes small. In this case, step sizes of 0.05 ft for the trapezoidal rule and 0.5 ft for Simpson's rule provide good results.

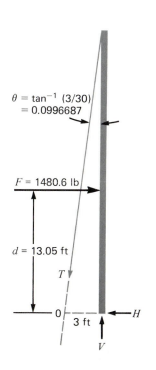

$\theta = \tan^{-1}(3/30)$
$\quad = 0.0996687$

$F = 1480.6$ lb

$d = 13.05$ ft

T

0

3 ft

H

V

Figure 18.3
Free-body diagram of the forces exerted on the mast of a sailboat.

TABLE 18.4 Values of $f(z)$ for a step size of 3 ft that provide data for the trapezoidal rule and Simpson's 1/3 rule.

z, ft	$f(z)$, lb/ft
0	0
3	61.40
6	73.13
9	70.56
12	63.43
15	55.18
18	47.14
21	39.83
24	33.42
27	27.89
30	23.20

TABLE 18.5 Values of F computed on the basis of various versions of the trapezoidal rule and Simpson's 1/3 rule.

Technique	Step Size, ft	Segments	F, lb
Trapezoidal	15	2	1001.7
rule	10	3	1222.3
	6	5	1372.3
	3	10	1450.8
	1	30	1477.1
	0.5	60	1479.7
	0.25	120	1480.3
	0.1	300	1480.5
	0.05	600	1480.6
Simpson's 1/3	15	2	1219.6
rule	5	6	1462.9
	3	10	1476.9
	1	30	1480.5
	0.5	60	1480.6

The effective line of action of F (Fig. 18.3) can be calculated by evaluation of the integral

$$d = \frac{\int_0^{30} z f(z)\, dz}{\int_0^{30} f(z)\, dz} \tag{18.7}$$

or

$$d = \frac{\int_0^{30} 200z[z/(5+z)]e^{-2z/30}\, dz}{1480.6} \tag{18.8}$$

This integral can be evaluated using methods similar to the above. For example, Simpson's 1/3 rule with a step size of 0.5 gives

$$d = \frac{19,326.9}{1480.6} = 13.05 \text{ ft}$$

With F and d known from numerical methods, a free-body diagram is used to develop force and moment balance equations. This free-body diagram is shown in Fig. 18.3. Summing forces in the horizontal and vertical direction and taking moments about point 0 gives

$$\sum F_H = 0 = F - T \sin \theta - H \tag{18.9}$$

$$\sum F_V = 0 = V - T \cos \theta \tag{18.10}$$

$$\sum M_0 = 0 = 3V - Fd \tag{18.11}$$

where T is the tension in the cable. H and V are the unknown reactions on the mast transmitted by the deck. The direction, as well as the magnitude, of H and V is unknown. Equation (18.11) can be solved directly for V because F and d are known.

$$V = \frac{Fd}{3} = \frac{(1480.6)(13.05)}{3} = 6440.6 \text{ lb}$$

Therefore, from Eq. (18.10),

$$T = \frac{V}{\cos \theta} = \frac{6440.6}{0.995} = 6473 \text{ lb}$$

and from Eq. (18.9),

$$H = F - T \sin \theta = 1480.6 - (6473)(0.0995) = 836.54 \text{ lb}$$

These forces now enable you to proceed with other aspects of the structural design of the boat such as the cables and the deck support system for the mast. This problem illustrates nicely two uses of numerical integration that may be encountered during the engineering design of structures. It is seen that both the trapezoidal rule and Simpson's 1/3 rule are easy to apply and are practical problem-solving tools. Simpson's 1/3 rule is more accurate than the trapezoidal rule for the same step size and thus may often be preferred.

CASE STUDY 18.4   DETERMINATION OF THE ROOT-MEAN-SQUARE CURRENT
                   BY NUMERICAL INTEGRATION (ELECTRICAL ENGINEERING)

Background: The average value of an oscillating electric current over one period may be zero. For example, suppose that the current is described by a simple sinusoid: $i(t) = \sin(2\pi t/T)$ where T is the period. The average value of this function can be determined by the following equation:

$$i = \frac{\int_0^T \sin\left(\frac{2\pi t}{T}\right) dt}{T - 0} = \frac{-\cos 2\pi + \cos 0}{T} = 0$$

Despite the fact that the net result is zero, such current is capable of performing work and generating heat. Therefore, electical engineers often characterize such current by

$$I_{RMS} = \sqrt{\frac{1}{T} \int_0^T i^2(t)\, dt} \tag{18.12}$$

where $i(t)$ is the instantaneous current. Calculate the RMS current of the waveform shown in Fig. 18.4 using the trapezoidal rule, Simpson's 1/3 rule, Romberg integration, and Gauss quadrature for $T = 1$ s.

Solution: Integral estimates for various applications of the trapezoidal rule and Simpson's 1/3 rule are listed in Table 18.6. Notice that Simpson's rule is more accurate than the trapezoidal rule.

Figure 18.4
A periodically varying electric current.

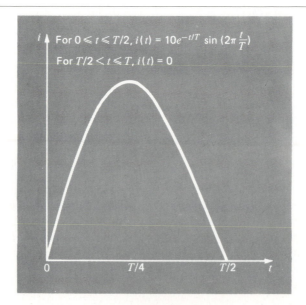

TABLE 18.6 Values for the integral calculated using various numerical schemes.
The percent relative error ϵ_t is based on a true value of 15.4126081.

Technique	Segments	Integral	ϵ_t, %
Trapezoidal	1	0.0	100
rule	2	15.1632665	1.62
	4	15.4014291	0.0725
	8	15.4119584	4.21×10^{-3}
	16	15.4125682	2.59×10^{-4}
	32	15.4126056	1.62×10^{-5}
	64	15.4126079	1.30×10^{-6}
	128	15.4126081	0
Simpson's 1/3	2	20.2176887	−31.2
rule	4	15.4808166	−0.443
	8	15.4154681	−0.0186
	16	15.4127714	-1.06×10^{-3}
	32	15.4126081	0

The exact value for the integral is 15.4126081. This result is obtained using a 128-segment trapezoidal rule or a 32-segment Simpson's rule. The same estimate is also determined using Romberg integration (Fig. 18.5).

In addition, Gauss quadrature can be used to make the same estimate. The determination of the root-mean-square current involves the evaluation of the integral ($T = 1$)

$$I = \int_0^{1/2} (10e^{-t} \sin 2\pi t)^2 \, dt \qquad (18.13)$$

First, a change in variable is performed by applying Eqs. (16.23) and (16.24) to yield

$$t = \frac{1}{4} + \frac{1}{4} t_d$$

Figure 18.5
Result of using Romberg integration to estimate the RMS current.

$O(h^2)$	$O(h^4)$	$O(h^6)$	$O(h^8)$	$O(h^{10})$	$O(h^{12})$	$O(h^{14})$
0	20.2176887	15.1650251	15.4150177	15.4126058	15.4126081	15.4126081
15.1632665	15.4808166	15.4111116	15.4126152	15.4126081	15.4126081	
15.4014291	15.4154682	15.4122517	15.4126081	15.4126081		
15.4119584	15.4127715	15.4126078	15.4126081			
15.4125682	15.4126180	15.4126081				
15.4126056	15.4126087					
15.4126079						

and

$$dt = \frac{1}{4} dt_d$$

These relationships can be substituted into Eq. (18.13) to yield

$$I = \int_{-1}^{1} \left[10e^{-[1/4+(1/4)t_d]} \sin 2\pi \left(\frac{1}{4} + \frac{1}{4}t_d \right) \right]^2 \frac{1}{4} dt_d$$

For the two-point Gauss-Legendre formula, this function is evaluated at $t_d = -1/\sqrt{3}$ and $1/\sqrt{3}$, with the results being 7.6840962 and 4.3137280, respectively. These values can be substituted into Eq. (16.17) to yield an integral estimate of 11.9978242, which represents an error of $\epsilon_t = 22.1\%$.

The three-point formula is (Table 16.1)

$$I = 0.555555556\,(1.237449345) + 0.888888889\,(15.16326649)$$

$$+ \ 0.555555556\,(2.684914679)$$

$$= 15.65755021 \qquad \epsilon_t = 1.6\%$$

The results of using the higher-point formulas are summarized in Table 18.7.

The integral estimate of 15.4126081 can be substituted into Eq. (18.12) to compute an I_{RMS} of 3.9258895 A. This result could then be employed to guide other aspects of the design and operation of the circuit.

TABLE 18.7   **Results of using various-point Gauss quadrature formulas to approximate the integral.**

Points	Estimate	ϵ_t, %
2	11.9978243	22.1
3	15.6575502	−1.59
4	15.4058023	4.42×10^{-2}
5	15.4126391	-2.01×10^{-4}
6	15.4126109	-1.82×10^{-5}

CASE STUDY 18.5   USE OF NUMERICAL INTEGRATION TO COMPUTE WORK
(MECHANICAL ENGINEERING)

Background: Many engineering problems involve the calculation of work. The general formula is

Work = force × distance

When you were introduced to this concept in high school physics, simple applications were presented using forces that remained constant throughout the displacement. For example, if a force of 10 lb was used to pull a block a distance of 15 ft, the work would be calculated as 150 ft·lb.

Although such a simple computation is useful for introducing the concept, realistic problem settings are usually more complex. For example, suppose that the force varies during the course of the calculation. In such cases, the work equation is reexpressed as

$$W = \int_{x_0}^{x_n} F(x) \, dx \tag{18.14}$$

where W is work in foot pounds, x_0 and x_n are the initial and final positions, respectively, and $F(x)$ is a force which varies as a function of position. If $F(x)$ is easy to integrate, Eq. (18.14) can be evaluated analytically. However, in a realistic problem setting, the force might not be expressed in such a manner. In fact, when analyzing measured data, the force might only be available in tabular form. For such cases, numerical integration is the only viable option for the evaluation.

Further complexity is introduced if the angle between the force and the direction of movement also varies as a function of position (Fig. 18.6). The work equation can be modified further to account for this effect, as in

$$W = \int_{x_0}^{x_n} F(x) \cos \, [\theta(x)] \, dx \tag{18.15}$$

Again if $F(x)$ and $\theta(x)$ are simple functions, Eq. (18.15) might be solved analytically. However, as in Fig. 18.6, it is more likely that the functional relationship is complicated. For this situation, numerical methods provide the only alternative for determining the integral.

Suppose that you have to perform the computation for the situation depicted in Fig. 18.6. Although the figure shows the continuous values for $F(x)$ and $\theta(x)$, assume that because of experimental constraints, you are provided only with discrete measurements at $x = 5$-ft intervals (Table 18.8). Use single- and multiple-application versions of the trapezoidal rule and Simpson's 1/3 and 3/8 rules to compute work for this data.

TABLE 18.8   **Data for force $F(x)$ and angle $\theta(x)$ as a function of position x.**

x, ft	$F(x)$, lb	θ, rad	$F(x) \cos \theta$
0	0.0	0.50	0.0000
5	9.0	1.40	1.5297
10	13.0	0.75	9.5120
15	14.0	0.90	8.7025
20	10.5	1.30	2.8087
25	12.0	1.48	1.0881
30	5.0	1.50	0.3537

Solution: The results of the analysis are summarized in Table 18.9. A percent relative error ϵ_t was computed in reference to a true value of the integral of 129.52 that was estimated on the basis of values taken from Fig. 18.6 at 1-ft intervals.

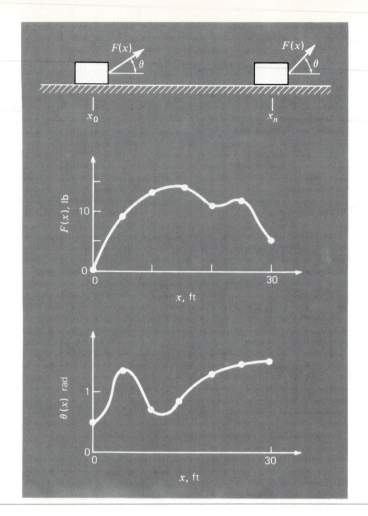

Figure 18.6
The case of a variable force acting on a block. For this case, the angle, as well as the magnitude, of the force varies.

TABLE 18.9 Estimates of work calculated using the trapezoidal rule and Simpson's rules. The percent relative error ϵ_t was computed in reference to a true value of the integral (129.52 ft·lb) that was estimated on the basis of values at 1-ft intervals.

Technique	Segments	Work	ϵ_t, %
Trapezoidal	1	5.31	95.9
	2	133.19	−2.84
	3	124.98	3.51
	6	119.09	8.05
Simpson's 1/3 rule	2	175.82	−35.75
	6	117.13	9.57
Simpson's 3/8 rule	3	139.93	−8.04

The results are interesting because the most accurate result occurs for the simple two-segment trapezoidal rule. More refined estimates using more segments, as well as Simpson's rules, yield less accurate results.

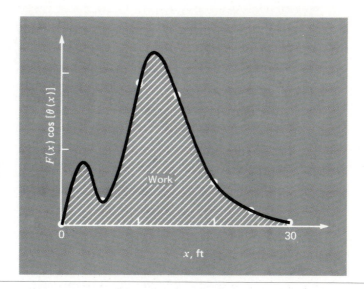

Figure 18.7
A continuous plot of $F(x)$ $\cos [\theta(x)]$ versus position with the seven discrete points used to develop the numerical integration estimates in Table 18.9. Notice how the use of seven points to characterize this continuously varying function misses two peaks at $x = 2.5$ and 12.5 ft.

The reason for this apparently counterintuitive result is that the coarse spacing of the points is not adequate to capture the variations of the forces and angles. This is particularly evident in Fig. 18.7, where we have plotted the continuous curve for the product of $F(x)$ and $\cos [\theta(x)]$. Notice how the use of seven points to characterize the continuously varying function misses the two peaks at $x = 2.5$ and 12.5 ft. The omission of these two points effectively limits the accuracy of the numerical integration estimates in Table 18.9. The fact that the two-segment trapezoidal rule yields the most accurate result is due to the chance positioning of the points for this particular problem (Fig. 18.8).

Figure 18.8
Graphical depiction of why the two-segment trapezoidal rule yields a good estimate of the integral for this particular case. By chance, the use of two trapezoids happens to lead to an even balance between positive and negative errors.

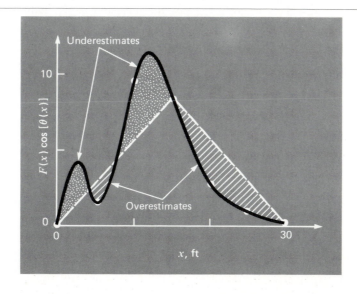

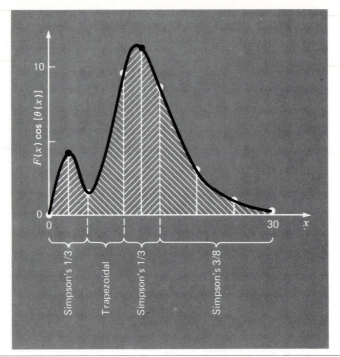

Figure 18.9
The unequal segmentation scheme that results from the inclusion of two additional points at $x - 2.5$ and 12.5 in the data in Table 18.8. The numerical integration formulas applied to each set of segments are shown.

The conclusion to be drawn from Fig. 18.7 is that an adequate number of measurements must be made in order to accurately compute integrals. For the present case, if data were available at $F(2.5) \cos [\theta(2.5)] = 4.3500$ and $F(12.5) \cos [\theta(12.5)] = 11.3600$, we could determine an integral estimate using the algorithm for unequally spaced data described previously in Sec. 15.3. Figure 18.9 illustrates the unequal segmentation for this case. Including the two additional points yields an improved integral estimate of 126.9 ($\epsilon_t = 2.02\%$). Thus, the inclusion of the additional data would incorporate the peaks that were missed previously and, as a consequence, lead to better results.

CASE STUDY 18.6   SPREADSHEET APPLICATION OF INTEGRATION OF KNOWN FUNCTIONS

This case study uses the spreadsheet to integrate known functions. A spreadsheet to implement Romberg integration is contained on the supplementary disk associated with the text. Insert the disk, type NUMMET following the prompt, advance to the main menu, and activate the fourth selection. See Fig. 18.10 for what will appear on the screen.

The spreadsheet allows you to approximate the integral of the function $f(x) = x^N$, where N is an input parameter, using Romberg integration. Three levels of extrapolation [$k = 4$ in Eq. (16.8)] are automatically computed. The active cells of the spreadsheet are B3, B4, and B5. Modification of these cells allows you to change the order of the monomial to be integrated (N) and the range of the integration. Study the contents of all the cells carefully to be sure you understand the computations.

Observe how the trapezoidal rule applications are contained in the k = 1 column.

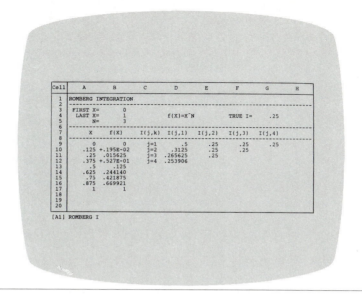

Figure 18.10
Application of Romberg integration using a spreadsheet. This spreadsheet is designed to evaluate the integral of a simple monomial.

Note that the error decreases as j increases but that it does this slowly. Observe that for k = 2 integration is exact. This is consistent with the theoretical error formulation. Now change the value of B5 to 4 and strike [F3]. Note that the k = 2 column is not exact but that the error is zero for the k = 3 and k = 4 columns. Continue with N = 6 and N = 8 and note the results. Observe that the computational efficiency is better as one moves across the table horizontally increasing k compared to increasing the number of applications of the trapezoidal rule as one moves down the first column.

The spreadsheet has the advantage that all the computation results are readily available. This increases your insight into the behavior of the algorithm. One disadvantage is that the limited size of the spreadsheet restricts the number of columns that can be generated. This, of course, would not be the case for professional spreadsheets such as Lotus 123.

Another application of the spreadsheet is to compare the efficiency of the trapezoidal rule, Simpson's 1/3 rule, and Gauss quadrature methods to integrate functions. A spreadsheet to implement these methods is contained on the supplementary disk associated with the text. Advance to the main menu and activate the fifth selection. See Fig. 18.11 for what will appear on the screen.

The spreadsheet allows you to approximate the integral of the function $f(x) = x^N$ where N is an input parameter. It uses three integration techniques that require exactly the same computational effort. The range of the integration is fixed between -1 and 1. The only active cell in the spreadsheet is B3 which is the value of N. Study the contents of all the cells carefully to be sure you understand the methods and confirm that the computational effort is the same for all three methods.

Note that all three methods require five values of x and five function evaluations. The trapezoidal rule and Simpson's rule utilize the same values of x with only the relative weighting of each different. Gauss quadrature uses different values of x from the other two methods. However, note that the error for N = 4 for Gauss quadrature is much smaller than for the other methods. This is consistent with the theoretical error

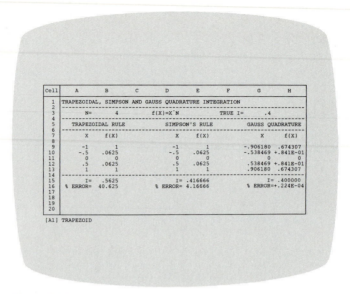

Cell	A	B	C	D	E	F	G	H
1	TRAPEZOIDAL, SIMPSON AND GAUSS QUADRATURE INTEGRATION							
2	--							
3	N=	4		f(X)=X^N		TRUE I=	.4	
4	--							
5	TRAPEZOIDAL RULE			SIMPSON'S RULE			GAUSS QUADRATURE	
6	--							
7	X	f(X)		X	f(X)		X	f(X)
8								
9	-1	1		-1	1		-.906180	.674307
10	-.5	.0625		-.5	.0625		-.538469	+.841E-01
11	0	0		0	0		0	0
12	.5	.0625		.5	.0625		.538469	+.841E-01
13	1	1		1	1		.906180	.674307
14								
15	I=	.5625		I=	.416666		I=	.400000
16	% ERROR=	40.625		% ERROR=	4.16666		% ERROR=	+.224E-04
17								
18								
19								
20								

[A1] TRAPEZOID

Figure 18.11
Application of the trapezoidal rule, Simpson's 1/3 rule, and Gauss quadrature to integrate a
simple monomial.

formulation. Next move the cursor to B3, change the value to 2, strike [F3], and observe
the results. Note that the error for both Simpson's and the Gauss quadrature methods
is essentially zero except for roundoff error. Change N = 3, strike [F3] and note that
an error message (*error*) is displayed because TRUE I = 0. Try N = 8 and note the
large difference among the errors.

This spreadsheet example helps make it clear how much more accurate Gauss
quadrature is compared to the trapezoidal rule and Simpson's rule for exactly the same
computational effort.

PROBLEMS

General Engineering:

18.1 Reproduce the computations in Case Study 18.1 using your
own software.

18.2 Perform the same computations as in Case Study 18.1, but
instead of using Eq. (18.1) use the following alternative
formulation:

$$\text{Cost per computer (\$)} = 1250 + 1750e^{-5.5 \times 10^{-5} N}$$

18.3 You are doing a study of an assembly line at an automobile
plant. Over the course of a 24-h period, you visit two
points on the line, and at different times during the day
you spot-check the number of autos that pass through in a
minute. The data is

Point A		Point B	
Time	**Cars/min**	**Time**	**Cars/min**
Midnight	3	Midnight	3
2 A.M.	3	1 A.M.	3
3 A.M.	5	4 A.M.	5
6 A.M.	4	5 A.M.	2
9 A.M.	5	7 A.M.	1
11 A.M.	6	10 A.M.	4
2 P.M.	2	1 P.M.	3
5 P.M.	1	3 P.M.	4
6 P.M.	1	9 P.M.	6
7 P.M.	3	10 P.M.	1
8 P.M.	4	11 P.M.	3
Midnight	6	Midnight	6

Use numerical integration and Eq. (PT5.4) to determine the total number of cars per day that pass through each point.

18.4 The data listed in Table P18.4 gives hourly measurements of heat flux q at the surface of a solar collector. Estimate the total heat absorbed by a 150,000-cm^2 collector panel during a 14-h period. The panel has an absorption efficiency e_{ab} of 45%. The total heat absorbed is given by

$$H = e_{ab} \int_0^t q \, A \, dt$$

where A is area and q is heat flux.

TABLE P18.4

Measurements of solar heat flux.

Time, h	Heat flux, q, cal/(cm$^2 \cdot$h)
0	0.1
1	1.62
2	5.32
3	6.29
4	7.8
5	8.81
6	8.00
7	8.57
8	8.03
9	7.04
10	6.27
11	5.56
12	3.54
13	1.0
14	0.2

18.5 The *heat current H* is the quantity of heat flowing through a material per unit time. It can be computed with

$$H = -kA \frac{dT}{dx}$$

where H has units of joules (or kg·m^2/s^2) per second; k, a coefficient of thermal conductivity that parameterizes the heat-conducting properties of the material, has units of joules per second per meter per degree Celsius; A is the cross-sectional area perpendicular to the path of heat flow; T is temperature (in degrees Celsius); and x is distance (in centimeters) along the path of heat flow. If the temperature outside a house is $-20°C$ and inside is $21°C$, what is the rate of heat loss through a 6-in-thick, 1000-ft^2 wall made of

Material	k
Wood	0.08
Red brick	0.60
Concrete	0.80
Insulating brick	0.15

18.6 The following data was collected when a large oil tanker was loading:

t, min	0	15	30	45	60	90	120
V, 10^6 barrels	0.5	0.65	0.73	0.88	1.03	1.14	1.30

Calculate the flow rate Q (that is, dV/dt) for each time.

18.7 A jet fighter's position on an aircraft carrier's runway was timed during landing:

t, s	0	0.51	1.03	1.74	2.36	3.24	3.82
x, m	154	186	209	250	262	272	274

where x is the distance from the end of the carrier. Estimate (a) velocity (dx/dt) and (b) acceleration (dv/dt) using numerical differentiation.

18.8 Employ the multiple-application trapezoidal rule to evaluate the vertical distance traveled by a rocket if the vertical velocity is given by

$$\begin{array}{ll} v = 10t^2 & 0 \le t \le 10 \\ v = 1000 - 5t & 10 < t \le 20 \\ v = 45t + 2(t - 20)^2 & 20 < t \le 30 \end{array}$$

18.9 You normally jog down a country road, and you want to determine how far you run. Although the odometer on your car is broken, you can use the speedometer to record the car's velocity at 1-min intervals as you follow your jogging route:

Time, min	0	1	2	3	4	5	6	6.33
Velocity, mi/h	0	35	40	48	50	55	32	36

Use this data to determine how many miles you jog. Be careful of units.

18.10 The upward velocity of a rocket can be computed by the following formula:

$$v = u \ln \left(\frac{m_0}{m_0 - qt} \right) - gt$$

where v is the upward velocity, u is the velocity at which fuel is expelled relative to the rocket, m_0 is the initial mass of the rocket at time $t = 0$, q is the fuel consumption rate, and g is the downward acceleration of gravity (assumed constant $= 9.8$ m/s^2). If $u = 2200$ m/s, $m_0 = 160,000$ kg, and $q = 2680$ kg/s, use a numerical method to determine how high the rocket will fly in 30 s.

18.11 If the velocity of a roller coaster in the horizontal direction is given by

$$v(t) = (60 - t)^2 + (60 - t) \sin (t^{1/2})$$

determine the horizontal distance traveled in 60 s using the multiple-application trapezoidal rule.

Chemical Engineering

18.12 Reproduce the computations in Case Study 18.2 using your own software.

18.13 Perform the same computation as in Case Study 18.2, but compute the amount of heat required to raise the temperature of 2000 g of the material from -200 to $100°C$. Use Simpson's rule for your computation, with values of T at $50°C$ increments.

18.14 Repeat Prob. 18.13, but use Romberg integration to $\epsilon_s = 0.01\%$.

18.15 Repeat Prob. 18.13, but use a two- and a three-point Gauss-Legendre formula. Interpret your results.

18.16 Integration provides a means to compute how much mass enters or leaves a reactor over a specified time period, as in

$$M = \int_{t_1}^{t_2} Qc \, dt$$

where t_1 and t_2 are the initial and final times. This formula makes intuitive sense if you recall the analogy between integration and summation. Thus the integral represents the summation of the product of flow times concentration to give the total mass entering or leaving from t_1 to t_2. If the flow rate is constant, Q can be moved outside the integral:

$$M = Q \int_{t_1}^{t_2} c \, dt \qquad \text{(P18.16)}$$

Use numerical integration to evaluate this equation for the data in Table P18.16. Note that $Q = 5$ m^3/min.

TABLE P18.16
Values of concentration measured in the exit pipe of a reactor.

t, min	c, mg/m^3
0	10.00
5	21.62
10	35.00
15	47.05
20	54.73
25	56.31
30	52.16
35	44.59
40	37.07
45	32.91
50	34.06

18.17 The outflow chemical concentration from a completely mixed reactor is measured as:

t, min	0	5	10	15	20	30	40	50	60
c, mg/m^3	10	20	30	40	60	80	70	50	60

For an outflow of $Q = 10$ m^3/min, estimate the mass of chemical that exits the reactor from $t = 0$ to 60 min.

18.18 *Fick's first diffusion law* states that

$$\text{Mass flux} = -D\,\frac{dc}{dx} \qquad (P18.18)$$

where mass flux is the quantity of mass that passes across a unit area per unit time (in grams per square centimeter per second), D is a diffusion coefficient (in square centimeters per second), c is concentration, and x is distance (in centimeters). An environmental engineer measures the following concentration of a pollutant in the sediments underlying a lake ($x = 0$ at the sediment-water interface and increases downward):

x, cm	0	1	2.5
c, 10^{-6}g/cm^3	0.1	0.4	0.75

Use the best numerical differentiation technique available to estimate the derivative at $x = 0$. Employ this estimate in conjunction with Eq. (P18.18) to compute the mass flux of pollutant out of the sediments and into the overlying waters ($D = 10^{-6}$ cm/s^2). For a lake with 10^6 m^2 of sediments, how much pollutant would be transported into the lake over a year's time?

Civil Engineering

18.19 Reproduce the computations in Case Study 18.3 using your own software.

18.20 Perform the same computation as in Case Study 18.3, but use Romberg integration to evaluate the integral. Employ a stopping criterion of $\epsilon_s = 0.25\%$.

18.21 Perform the same computation as in Case Study 18.3, but use Gauss quadrature to evaluate the integral.

18.22 Perform the same computation as in Case Study 18.3, but change the integral to

$$F = \int_0^{30} \frac{250z}{4 + z}\, e^{-2z/30}\, dz$$

18.23 Stream cross-sectional areas (A) are required for a number of tasks in water resources engineering, including flood forecasting and reservoir design. Unless electronic sounding devices are available to obtain continuous profiles of the channel bottom, the engineer must rely on discrete depth measurements to compute A. An example of a typical stream cross section is shown in Fig. P18.23. The data points represent locations where a boat was anchored and depth readings taken. Use two trapezoidal rule applications ($h = 4$ and 2 m), and Simpson's 1/3 rule to estimate the cross-sectional area from this data.

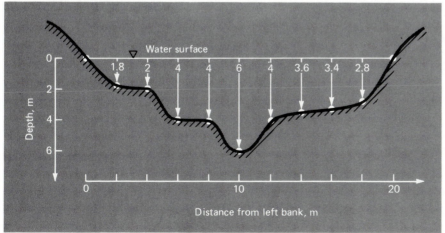

Figure P18.23
A stream cross section.

18.24 During a survey, you are required to compute the area of the field shown in Fig. P18.24. Use Simpson's rules to determine the area.

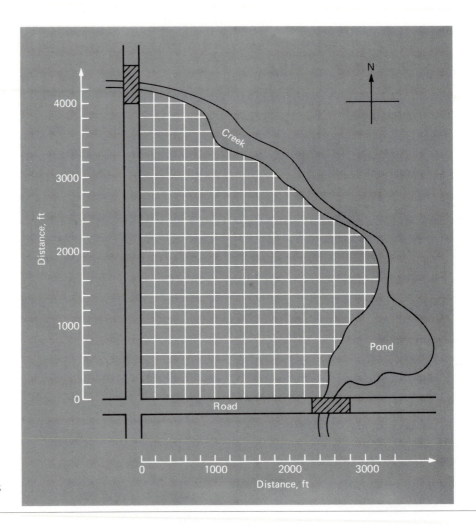

Figure P18.24
A field bounded by two roads and a creek.

18.25 A transportation engineering study requires the calculation of the total number of cars that pass through an intersection over a 24-h period. An individual visits the intersection at various times during the course of a day and counts the number of cars that pass through the intersection in a minute. Utilize this data, which is summarized in Table P18.25 to estimate the total number of cars that pass through the intersection per day. (Be careful of units.)

TABLE P18.25

Traffic flow rate for an intersection measured at various times over a 24-h period.

Time	Rate, cars/min	Time	Rate, cars/min
12:00 midnight	2	12:30 P.M.	15
2:00 A.M.	2	2:00 P.M.	7
4:00 A.M.	0	4:00 P.M.	9
5:00 A.M.	2	5:00 P.M.	20
6:00 A.M.	5	6:00 P.M.	19
7:00 A.M.	8	7:00 P.M.	10
8:00 A.M.	20	8:00 P.M.	9
9:00 A.M.	12	9:00 P.M.	11
10:30 A.M.	5	10:00 P.M.	8
11:30 A.M.	10	11:00 P.M.	5
		12:00 P.M.	3

18.26 A wind force distributed against the side of a skyscraper is measured as

Height l, ft	Force $F(l)$, lb/ft
0	0
100	50
200	155
300	200
400	220
500	400
600	450
700	475
750	490

Compute the net force and the line of action due to this distributed wind.

18.27 Water exerts pressure on the upstream face of a dam as shown in Fig. P18.27. The pressure can be characterized by

$$p(z) = \rho g(D - z) \qquad \text{(P18.27)}$$

where $p(z)$ is pressure in pascals (or newtons per square meter) exerted at an elevation z meters above the reservoir bottom; ρ is the density of water, which for the present problem is assumed to be a constant value of 10^3 kg/m³; g is the acceleration due to gravity (9.8 m/s²); and D is the elevation (in meters) of the water surface above the reservoir bottom. According to Eq. (P18.27), pressure increases linearly with depth, as depicted in Fig. P18.27a. Omitting atmospheric pressure (because it works against both sides of the dam face and essentially cancels out), the total force f_t can be determined by multiplying pressure times the area of the dam face (as shown in Fig. P18.27b). Because both pressure and area vary with elevation, the total force is obtained by evaluating

$$f_t = \int_0^D \rho g w(z)(D - z)\, dz$$

where $w(z)$ is the width of the dam face in meters at elevation z (Fig. P18.27b). The line of action can also be obtained by evaluating

$$d = \frac{\int_0^D \rho g z w(z)(D - z)\, dz}{\int_0^D \rho g w(z)(D - z)\, dz}$$

Use Simpson's rule to compute f_t and d. Check the results with your computer program for the trapezoidal rule.

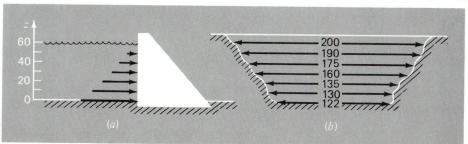

Figure P18.27
Water exerting pressure on the upstream face of a dam: (a) side view showing force increasing linearly with depth; (b) front view showing width of dam in meters.

18.28 In order to estimate the size of a new dam you have to determine the total volume of water (in gallons) that flows down a river in a year's time. You have available to you the following long-term average data for the river:

Date	mid-Jan.	mid-Feb.	mid-Mar.	mid-Apr.	mid-June	mid-Aug.	mid-Oct.	mid-Nov.	mid-Dec.
Flow rate, ft³/s	1100	1300	2200	4200	3600	700	750	800	1000

Determine the volume. Be careful of units. Note that there are 7.481 gal/ft³ and take care to make a proper estimate of flow at the end points.

Electrical Engineering

18.29 Reproduce the computations in Case Study 18.4 using your own software.

18.30 Perform the same computation as in Case Study 18.4, but for the current as specified by

$$i(t) = 5e^{-2t} \sin 2\pi t \qquad \text{for } 0 \le t \le T/2$$

$$i(t) = 0 \qquad \text{for } T/2 \le t \le T$$

where $T = 1$ s. Use a 16-segment Simpson's 1/3 rule to estimate the integral.

18.31 Repeat Prob. 18.30, but use Gauss quadrature.

18.32 Repeat Prob. 18.30, but use Romberg integration to $\epsilon_s = 0.1\%$.

18.33 *Faraday's law* characterizes the voltage drop across an inductor as

$$V_L = L \frac{di}{dt}$$

where V_L is the voltage drop (in volts), L is the inductance (in henrys; 1 H = 1 V·s/A), i is the current (in amperes), and t is time (in seconds). Determine the voltage drop as a function of time from the following data:

t	0	0.1	0.2	0.4	0.6	0.8
i	0	0.1	0.2	0.6	1.3	2.7

The inductance is equal to 5 H.

Mechanical Engineering

18.34 Reproduce the computations in Case Study 18.5 using your own software.

18.35 Perform the same computation as in Case Study 18.5, but use the following equation to compute:

$$F(x) = 1.17x - 0.035x^2$$

Employ the values of θ from Table 18.8.

18.36 Perform the same computation as in Case Study 18.5, but use the following equation to compute:

$$\theta(x) = 0.5 + 0.1375x - 0.01x^2 + (2.5 \times 10^{-4})x^3$$

Employ the equation from Prob. 18.35 for $F(x)$. Use four-, eight-, and sixteen-segment trapezoidal rules to compute the integral.

18.37 Repeat Prob. 18.36, but use Simpson's 1/3 rule.

18.38 Repeat Prob. 18.36, but use Romberg integration to $\epsilon_s = 0.1\%$.

18.39 Repeat Prob. 18.36, but use Gauss quadrature.

18.40 The work done on an object is equal to the force times the distance moved in the direction of the force. The velocity of an object in the direction of a force is given by

$$v = 5t \qquad\qquad 0 \le t \le 7$$

$$v = 35 + (7 - t)^2 \qquad 7 < t \le 14$$

Employ the multiple-application trapezoidal rule to determine the work if a constant force of 200 lb is applied for all t.

18.41 The rate of cooling of a body (Fig. P18.41) can be expressed as

$$\frac{dT}{dt} = -k(T - T_a)$$

where T is the temperature of the body (in degrees Celsius), T_a is the temperature of the surrounding medium (in degrees Celsius), and k is a proportionality constant (per minute). Thus this equation (which is called *New-*

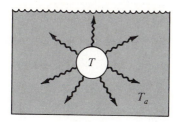

Figure P18.41

ton's law of cooling) specifies that the rate of cooling is proportional to the difference in the temperatures of the body and of the surrounding medium. If a metal ball heated to 90°C is dropped into water that is held constant at $T_a = 20$°C, the temperature of the ball changes as in

Time, min	0	5	10	15	20	25
Temperature, °C	90	62.5	45.8	35.6	29.5	25.8

Utilize numerical differentiation to determine dT/dt at each value of time. Plot dT/dt versus $T - T_a$ and employ linear regression to evaluate k.

18.42 A rod subject to an axial load (Fig. P18.42a) will be deformed as shown in the stress-strain curve in Fig. P18.42b. The area under the curve from zero stress out to the point of rupture is called the *modulus of toughness* of the material. It provides a measure of the energy per unit volume required to cause the material to rupture. As such, it is representative of the material's ability to withstand an impact load. Use numerical integration to compute the modulus of toughness for the stress-strain curve seen in Fig. P18.42b.

18.43 If the velocity distribution of a fluid flowing through a pipe is known (Fig. P18.43), the flow rate Q (that is, the volume of water passing through the pipe per unit time) can be computed by $Q = \int v \, dA$, where v is the velocity and A is the pipe's cross-sectional area. (To grasp the meaning of this relationship physically, recall the close connection between summation and integration). For a circular pipe, $A = \pi r^2$ and $dA = 2\pi r \, dr$. Therefore

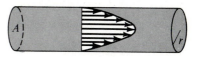

Figure P18.43

$$Q = \int_0^r v(2\pi r) \, dr$$

where r is the radial distance measured outward from the center of the pipe. If the velocity distribution is given by

$$v = 3.0\left(1 - \frac{r}{r_0}\right)^{1/7}$$

where r_0 is the total radius (in this case, 3 in), compute Q using the multiple-application trapezoidal rule.

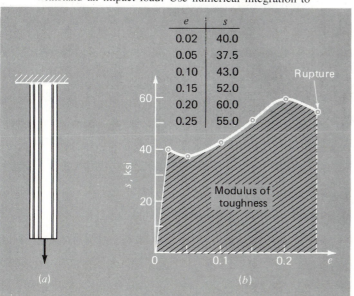

e	s
0.02	40.0
0.05	37.5
0.10	43.0
0.15	52.0
0.20	60.0
0.25	55.0

Figure P18.42
(*a*) A rod under axial loading and (*b*) the resulting stress-strain curve where stress is in kips per square inch (10^3 lb/in^2) and strain is dimensionless.

18.44 Using the following data, calculate the work done by stretching a spring that has a spring constant of $k \simeq 3 \times 10^2$ N/m to $x = 0.4$ m.

F, 10^3 N	0	0.01	0.028	0.046	0.063	0.082	0.11	0.13	0.18
x, m	0	0.05	0.1	0.15	0.2	0.25	0.3	0.35	0.4

Spreadsheet

18.45 Use an active spreadsheet that may be available to you (like the one on the Electronic TOOLKIT or Lotus 123) to perform four levels of Romberg integration for the following functions:

(a) $f(x) = \sin (x)$ $x = 0$ to 10
(b) $f(x) = \exp (x)$ $x = 0$ to 2
(c) $f(x) = \log (x)$ $x = 3$ to 5

Plot the error versus k and discuss the results and compare with the theoretical behavior of the error.

18.46 Activate the spreadsheet associated with the text (Fig. 18.10). Change the value of cell B5 (N) to 1. Record the value of I(1,2) and compute the percent error. Repeat the procedure for N = 2, . . . , 10. Plot the percent error versus N.

18.47 Repeat Prob. 18.46 for I(1,3).

18.48 Integrate the function

$$f(x) = x^5$$

between x = 5 and x = 10 using the Romberg integration program on the supplementary software disk. Observe the value of I(1,2) and the percent error. Use interval halving and the trapezoidal rule to obtain an equivalent error. Compare the computational efficiency of the two methods.

18.49 Use an active spreadsheet that may be available to you (like the one on the Electronic TOOLKIT or Lotus 123) to perform seven-point Gauss quadrature integration of the following functions:

(a) $f(x) = \exp (x)$ $x = -1$ to $x = 1$
(b) $f(x) = \sin (x)$ $x = -1$ to $x = 1$
(c) $f(x) = 1/(5 + x)$ $x = -1$ to $x = 1$

18.50 Repeat Prob. 18.49 except use three applications of Simpson's 1/3 rule.

18.51 Repeat Prob. 18.49 except use two applications of Simpson's 3/8 rule.

18.52 In Case Study 18.6 the error associated with Gauss quadrature for N = 8 is a small number. Is this value consistent with the theoretical error formulation? Discuss.

18.53 For the spreadsheet form Fig. 18.11, make a plot of percent error versus N for the trapezoidal rule, Simpson's rule, and Gauss quadrature for N = 0 to N = 16. Discuss.

Miscellaneous

18.54 Read all the case studies in Chap. 18. On the basis of your reading and experience make up your own case study for any one of the fields of engineering. This may involve modifying or reexpressing one of our case studies. However, it can also be totally original. As with our examples, it must be drawn from an engineering problem context and must demonstrate the use of numerical methods for integration and/or differentiation. Write up your results using our case studies as models.

EPILOGUE: PART FIVE

PT5.4 TRADE-OFFS

Table PT5.4 provides a summary of the trade-offs involved in numerical integration or quadrature. Most of these methods are based on the simple physical interpretation of an integral as the area under a curve. These techniques are designed to evaluate the integral of two different cases: (1) a continuous mathematical function and (2) discrete data in tabular form.

The Newton-Cotes formulas are the primary methods discussed in Chap. 15. They are applicable to both continuous and discrete functions. Both closed and open versions of these formulas are available. The open forms, which have integration limits that extend beyond the range of the data, are rarely used for the evaluation of definite integrals. However, they have utility for the solution of ordinary differential equations and for evaluating improper integrals.

The closed Newton-Cotes formulas are based on replacing a mathematical function or tabulated data by an interpolating polynomial that is easy to integrate. The simplest version is the trapezoidal rule, which is based on taking the area below a straight line joining adjacent values of the function. One way to improve the accuracy of the trapezoidal rule is to divide the integration interval from a to b into a number of segments and apply the method to each segment.

Aside from applying the trapezoidal rule with finer segmentation, another way to obtain a more accurate estimate of the integral is to use higher-order polynomials

TABLE PT5.4 Comparison of the characteristics of alternative methods for numerical integration. The comparisons are based on general experience and do not account for the behavior of special functions.

Method	Data Points Required for One Application	Data Points Required for n Applications	Truncation Error	Application	Programming Effort	Comments
Trapezoidal rule	2	$n + 1$	$\simeq h^3 f''(\xi)$	Wide	Easy	
Simpson's 1/3 rule	3	$2n + 1$	$\simeq h^5 f^{(4)}(\xi)$	Wide	Easy	
Simpson's rule (1/3 and 3/8)	3 or 4	≥ 3	$\simeq h^5 f^{(4)}(\xi)$	Wide	Easy	
Higher-order Newton-Cotes	≥ 5	N/A	$\geq h^7 f^{(6)}(\xi)$	Rare	Easy	
Romberg integration	3			Requires $f(x)$ be known	Moderate	Inappropriate for tabular data
Gauss quadrature	≥ 2	N/A		Requires $f(x)$ be known	Easy	Inappropriate for tabular data

to connect the points. If a quadratic equation is employed, the result is Simpson's 1/3 rule. If a cubic equation is used, the result is Simpson's 3/8 rule. Because they are much more accurate than the trapezoidal rule, these formulas are usually preferred. Multiple-application versions are available. For situations with an even number of segments, the multiple application of the 1/3 rule is recommended. For an odd number of segments, the 3/8 rule can be applied to the last three segments and the 1/3 rule to the remaining segments.

Higher-order Newton-Cotes formulas are also available. However, they are rarely used in practice. Where high accuracy is required, Romberg integration and Gauss quadrature formulas are available. It should be noted that both Romberg integration and Gauss quadrature are of practical value only in cases where the function is available in equation form. These techniques are ill-suited for tabulated data.

PT5.5 IMPORTANT RELATIONSHIPS AND FORMULAS

Table PT5.5 summarizes important information that was presented in Part Five. This table can be consulted to quickly access important relationships and formulas.

PT5.6 ADVANCED METHODS AND ADDITIONAL REFERENCES

Although we have reviewed a number of numerical integration techniques, there are other methods that have utility in engineering practice. For example, *adaptive Simpson's integration* is based on dividing the integration interval into a series of subintervals of width h. Then Simpson's 1/3 rule is used to evaluate the integral of each subinterval by halving the step size in an iterative fashion, that is, with a step size of h, $h/2$, $h/4$, $h/8$, and so forth. The iterations are continued for each subinterval until an approximate error estimate ϵ_a [Eq. (3.5)] falls below a prespecified stopping criterion ϵ_s. The total integral is then computed as the summation of the integral estimates for the subintervals. This technique is especially valuable for complicated functions that have regions exhibiting both lower- and higher-order variations. Discussions of adaptive integration may be found in Gerald and Wheatley (1984) and Rice (1983). In addition, adaptive schemes for solving ordinary differential equations can be used to evaluate complicated integrals as was mentioned in PT5.1 and as will be discussed in Chap. 20.

Another method for obtaining integrals is to fit *cubic splines* to the data. The resulting cubic equations can be integrated easily (Forsythe et al., 1977). A similar approach is also sometimes used for differentiation. Finally, aside from the Gauss-Legendre formulas discussed in Sec. 16.2, there are a variety of other *quadrature formulas*. Carnahan, Luther, and Wilkes (1969) and Ralston and Rabinowitz (1978) summarize many of these approaches.

In summary, the foregoing is intended to provide you with avenues for deeper exploration of the subject. Additionally, all the above references describe basic techniques covered in Part Five. We urge you to consult these alternative sources to broaden your understanding of numerical methods for integration.

TABLE PT5.5 Summary of important information presented in Part Five.

Method	Formulation	Graphic Interpretations	Error
Trapezoidal rule	$I \cong (b - a)\dfrac{f(a) + f(b)}{2}$		$-\dfrac{(b - a)^3}{12}f''(\xi)$
Multiple-application trapezoidal rule	$I \cong (b - a)\dfrac{f(x_0) + 2\sum\limits_{i=1}^{n-1} f(x_i) + f(x_n)}{2n}$		$-\dfrac{(b - a)^3}{12n^2}\bar{f}''$
Simpson's 1/3 rule	$I \cong (b - a)\dfrac{f(x_0) + 4f(x_1) + f(x_2)}{6}$		$-\dfrac{(b - a)^5}{2880}f^{(4)}(\xi)$
Multiple-application Simpson's 1/3 rule	$I \cong (b - a)\dfrac{f(x_0) + 4\sum\limits_{i=1,3}^{n-1} f(x_i) + 2\sum\limits_{j=2,4}^{n-2} f(x_j) + f(x_n)}{3n}$		$-\dfrac{(b - a)^5}{180n^4}\overline{f^{(4)}}$
Simpson's 3/8 rule	$I \cong (b - a)\dfrac{f(x_0) + 3f(x_1) + 3f(x_2) + f(x_3)}{8}$		$-\dfrac{(b - a)^5}{6480}f^{(4)}(\xi)$
Romberg integration	$I_{j,k} = \dfrac{4^{k-1} I_{j+1,\,k-1} - I_{j,k-1}}{4^{k-1} - 1}$		$O(h^{2k})$
Gauss quadrature	$I \cong c_0 f(x_0) + c_1 f(x_1) + \cdots + c_{n-1} f(x_{n-1})$		$\cong f^{(2n+2)}(\xi)$

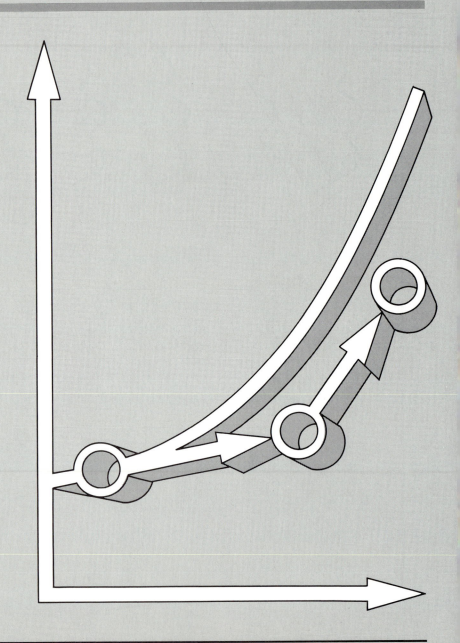

PART SIX

ORDINARY DIFFERENTIAL EQUATIONS

PT6.1 MOTIVATION

In the first chapter of this book, we derived the following equation based on Newton's second law to compute the velocity v of a falling parachutist as a function of time t [recall Eq. (1.9)]:

$$\frac{dv}{dt} = g - \frac{c}{m}v \tag{PT6.1}$$

where g is the gravitational constant, m is the mass, and c is a drag coefficient. Such equations, which are composed of an unknown function and its derivatives, are called *differential equations*. Equation (PT6.1) is sometimes referred to as a *rate equation* because it expresses the rate of change of a variable as a function of variables and parameters. Such equations play a fundamental role in engineering because many physical phenomena are best formulated mathematically in terms of their rate of change.

In Eq. (PT6.1), the quantity being differentiated, v, is called the *dependent variable*. The quantity with respect to which v is differentiated, t, is called the *independent variable*. When the function involves one independent variable, the equation is called an *ordinary differential equation* (or *ODE*). This is in contrast to a *partial differential equation* (or *PDE*) that involves two or more independent variables.

Differential equations are also classified as to their order. For example, Eq. (PT6.1) is called a *first-order equation* because the highest derivative is a first derivative. A *second-order equation* would include a second derivative. For example, the equation describing the position x of a mass-spring system with damping is the second-order equation (recall Case Study 6.5),

$$m\frac{d^2x}{dt^2} + c\frac{dx}{dt} + kx = 0 \tag{PT6.2}$$

where c is a damping coefficient and k is a spring constant. Similarly, an nth-order equation would include an nth derivative.

Higher-order equations can be reduced to a system of first-order equations. For Eq. (PT6.2), this is done by defining a new variable y where

$$y = \frac{dx}{dt} \tag{PT6.3}$$

which itself can be differentiated to yield

$$\frac{dy}{dt} = \frac{d^2x}{dt^2} \tag{PT6.4}$$

Equations (PT6.3) and (PT6.4) can then be substituted into Eq. (PT6.2) to give

$$m\frac{dy}{dt} + cy + kx = 0 \tag{PT6.5}$$

or

$$\frac{dy}{dt} = -\frac{cy + kx}{m} \tag{PT6.6}$$

Thus, Eqs. (PT6.3) and (PT6.6) are a pair of first-order equations that are equivalent to the original second-order equation. Because other nth-order differential equations can be similarly reduced, the present part of our book focuses on the solution of first-order equations. Some of the case studies in Chap. 22 deal with the solution of second-order ODEs by reduction to a pair of first-order equations.

PT6.1.1 Precomputer Methods for Solving ODEs

Before the computer age, ODEs were usually solved with analytical integration techniques. For example, Eq. (PT6.1) could be multiplied by dt and integrated to yield

$$v = \int \left(g - \frac{c}{m}v\right)\, dt \tag{PT6.7}$$

The right-hand side of this equation is called an *indefinite integral* because the limits of integration are unspecified. This is in contrast to the definite integrals discussed previously in Part Five [compare Eq. (PT6.7) with Eq. (PT5.6)].

An analytical solution for Eq. (PT6.7) is obtained if the indefinite integral can be evaluated exactly in equation form. For example, recall that for the falling parachutist problem, Eq. (PT6.7) was solved analytically by Eq. (1.10): (assuming $v = 0$ at $t = 0$)

$$v = \frac{gm}{c}[1 - e^{-(c/m)t}] \tag{PT6.8}$$

The mechanics of deriving such analytical solutions will be discussed in Sec. PT6.2. For the time being, the important fact is that, as was the case for definite integration, the analytical evaluation of indefinite integrals usually hinges on foreknowledge of the answer. Unfortunately, exact solutions for many ODEs of practical importance are not available. As is true for most situations discussed in other parts of this book, numerical methods offer the only viable alternative for these cases. Because these numerical methods usually require computers, engineers in the precomputer era were somewhat limited in the scope of their investigations.

One very important method that engineers and applied mathematicians developed to overcome this dilemma was *linearization*. A linear ordinary differential equation is one that fits the general form

$$a_n(x)y^{(n)} + \cdots + a_1(x)y' + a_0(x)y = f(x) \tag{PT6.9}$$

where $y^{(n)}$ is the nth derivative of y with respect to x and the a's and f's are specified functions of x. This equation is called *linear* because there are no products or nonlinear functions of the dependent variable y and its derivatives. The practical importance of linear ODEs is that they can be solved analytically. In contrast, most nonlinear equations cannot be solved exactly. Thus, in the precomputer era, one tactic to solve nonlinear equations was to linearize them.

A simple example is the application of ODEs to predict the motion of a swinging pendulum (Fig. PT6.1). In a manner similar to the derivation of the falling parachutist problem, Newton's second law can be used to develop the following differential equation (see Case Study 22.5 for the complete derivation):

$$\frac{d^2\theta}{dt^2} + \frac{g}{l}\sin\theta = 0 \tag{PT6.10}$$

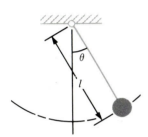

Figure PT6.1
The swinging pendulum.

where θ is the angle of displacement of the pendulum, g is the gravitational constant, and l is the pendulum length. This equation is nonlinear because of the term $\sin\theta$. One way to obtain an analytical solution is to realize that for small displacements of the pendulum from equilibrium (that is, for small values of θ),

$$\sin\theta \simeq \theta \tag{PT6.11}$$

Thus, if it is assumed that we are only interested in cases where θ is small, Eq. (PT6.11) can be substituted into Eq. (PT6.10) to give

$$\frac{d^2\theta}{dt^2} + \frac{g}{l}\theta = 0 \tag{PT6.12}$$

We have, therefore, transformed Eq. (PT6.10) into a linear form that is easy to solve analytically.

Although linearization remains a very valuable tool for engineering problem solving, there are cases where it cannot be invoked. For example, suppose that we were interested in studying the behavior of the pendulum for large displacements from equilibrium. In such instances, numerical methods offer a viable option for obtaining solutions. Today, the widespread availability of computers places this option within reach of all practicing engineers.

PT6.1.2 ODEs and Engineering Practice

The fundamental laws of physics, mechanics, electricity, and thermodynamics are usually based on empirical observations that explain variations in physical properties and states of systems. Rather than describing the *state* of physical systems directly, the laws are usually couched in terms of spatial and temporal *changes*.

Several examples are listed in Table PT6.1. These laws define mechanisms of change. When combined with continuity laws for energy, mass, or momentum, differential equations result. Subsequent integration of these differential equations results in mathematical functions that describe the spatial and temporal state of a system in terms of energy, mass, or velocity variations.

TABLE PT6.1 **Examples of fundamental laws that are written in terms of the rate of change of variables (t = time and x = position).**

Law	Mathematical Expression	Variables and Parameters
Newton's second law of motion	$\dfrac{dv}{dt} = \dfrac{F}{m}$	Velocity (v), force (F), and mass (m)
Fourier's heat law	Heat flux $= -k\dfrac{dT}{dx}$	Thermal conductivity (k) and temperature (T)
Fick's law of diffusion	Mass flux $= -D\dfrac{dc}{dx}$	Diffusion coefficient (D) and concentration (c)
Faraday's law (describes voltage drop across an inductor)	Voltage drop $= L\dfrac{di}{dt}$	Inductance (L) and current (i)

The falling parachutist problem introduced in Chap. 1 is an example of the derivation of an ordinary differential equation from a fundamental law. Recall that Newton's second law was used to develop an ODE describing the rate of change of velocity of a falling parachutist. By integrating this relationship, we obtained an equation to predict fall velocity as a function of time (Fig. PT6.2). This equation could be utilized in a number of different ways, including design purposes.

In fact, such mathematical relationships are the basis of the solution for a great number of engineering problems. However, as described in the previous section, many of the differential equations of practical significance cannot be solved using the analytical methods of calculus. Thus, the methods discussed in the following chapters are extremely important in all fields of engineering.

PT6.2 MATHEMATICAL BACKGROUND

A *solution* of an ordinary differential equation is a specific function of the independent variable and parameters that satisfies the original differential equation.

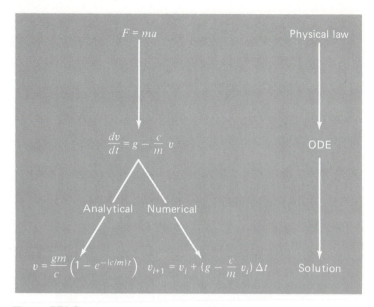

Figure PT6.2
The sequence of events in the application of ODEs for engineering problem solving. The example shown is the velocity of a falling parachutist.

In order to illustrate this concept, let us start with a given function

$$y = -0.5x^4 + 4x^3 - 10x^2 + 8.5x + 1 \tag{PT6.13}$$

which is a fourth-order polynomial (Fig. PT6.3a). Now, if we differentiate Eq. (PT6.13), we obtain an ODE:

$$\frac{dy}{dx} = -2x^3 + 12x^2 - 20x + 8.5 \tag{PT6.14}$$

This equation also describes the behavior of the polynomial but in a different manner than does Eq. (PT6.13). Rather than explicitly representing the values of y for each value of x, Eq. (PT6.14) gives the rate of change of y with respect to x (that is, the slope) at every value of x. Figure PT6.3 shows both the function and the derivative plotted versus x. Notice how the zero values of the derivatives correspond to the point at which the original function is flat—that is, has a zero slope. Also, the maximum absolute values of the derivatives are at the ends of the interval where the slopes of the function are greatest.

Although, as just demonstrated, we can determine a differential equation given the original function, the object here is to determine the original function given the differential equation. The original function then represents the solution. For the present case, we can determine this solution analytically by integrating Eq. (PT6.14):

$$y = \int [-2x^3 + 12x^2 - 20x + 8.5] \, dx$$

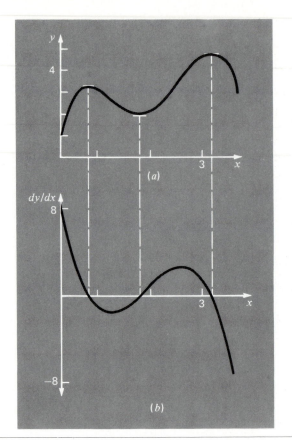

Figure PT6.3
Plots of (a) y versus x and (b) dy/dx versus x for the function $y = -0.5x^4 + 4x^3 - 10x^2 + 8.5x + 1$.

Applying the integration rule (recall Table PT5.2)

$$\int u^n\, du = \frac{u^{n+1}}{n+1} + C \qquad n \neq -1$$

to each term of the equation gives the solution:

$$y = -0.5x^4 + 4x^3 - 10x^2 + 8.5x + C \tag{PT6.15}$$

which is identical to the original function with one notable exception. In the course of differentiating and then integrating, we lost the constant value of 1 in the original equation and gained the value C. This C is called a *constant of integration*. The fact that such an arbitrary constant appears indicates that the solution is not unique. In fact, it is but one of an infinite number of possible functions (corresponding to an infinite number of possible values of C) that satisfy the differential equation. For example, Fig. PT6.4 shows six possible functions that satisfy Eq. (PT6.15).

Therefore, in order to specify the solution completely, a differential equation is usually accompanied by *auxiliary conditions*. For first-order ODEs, a type of

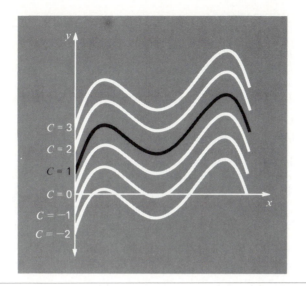

Figure PT6.4
Six possible solutions for the integral of $-2x^3 + 12x^2 - 20x + 8.5$. Each conforms to a different value of the constant of integration C.

auxiliary condition called an *initial value* is required to determine the constant and obtain a unique solution. For example, Eq. (PT6.14) could be accompanied by the initial condition that at $x = 0$, $y = 1$. These values could be substituted into Eq. (PT6.15):

$$1 = -0.5(0)^4 + 4(0)^3 - 10(0)^2 + 8.5(0) + C \tag{PT6.16}$$

to determine $C = 1$. Therefore, the unique solution that satisfies both the differential equation and the specified initial condition is obtained by substituting $C = 1$ into Eq. (PT6.15) to yield

$$y = -0.5x^4 + 4x^3 - 10x^2 + 8.5x + 1 \tag{PT6.17}$$

Thus, we have "pinned down" Eq. (PT6.15) by forcing it to pass through the initial condition, and in so doing, we have developed a unique solution to the ODE and have come full circle to the original function [Eq. (PT6.13)].

Initial conditions usually have very tangible interpretations for differential equations derived from physical problem settings. For example, in the falling parachutist problem the initial condition was reflective of the physical fact that at time zero the vertical velocity was zero. If the parachutist had already been in vertical motion at time zero, the solution would have been modified to account for this initial velocity.

When dealing with an *n*th-order differential equation, *n* conditions are required to obtain a unique solution. If all conditions are specified at the same value of the independent variable (for example, at *x* or *t* = 0), then the problem is called an *initial-value problem*. This is in contrast to *boundary-value problems* where specification of conditions occurs at different values of the independent variable. Chapters 19 and 20 will focus on initial-value problems. Boundary-value problems are covered in Chap. 21 along with eigenvalues.

PT6.3   ORIENTATION

Before proceeding to numerical methods for solving ordinary differential equations, some orientation might be helpful. The following material is intended to provide you with an overview of the material discussed in Part Six. In addition, we have formulated objectives to focus your studies of the subject area.

PT6.3.1   Scope and Preview

Figure PT6.5 provides an overview of Part Six. Two broad categories of numerical methods for initial-value problems will be discussed in the present part of this book. One-step methods, which are covered in Chap. 19, permit the calculation of y_{i+1}, given the differential equation and y_i. Multistep methods, which are covered in Chap. 20, require additional values of y other than at i.

With all but a minor exception, the *one-step methods* in *Chap. 19* belong to what are called Runge-Kutta techniques. Although the chapter might have been organized around this theoretical notion, we have opted for a more graphical, intuitive approach to introduce the methods. Thus, we begin the chapter with *Euler's method* which has a very straightforward graphical interpretation. Then, we use visually oriented arguments to develop two improved versions of Euler's method—the *Heun* and the *improved polygon* techniques. After this introduction, we formally develop the concept of *Runge-Kutta* (or RK) approaches and demonstrate how the foregoing techniques are actually first- and second-order RK methods. This is followed by a discussion of the higher-order RK formulations that are frequently used for engineering problem solving. The chapter ends with the application of one-step methods to *systems of ODEs*.

As presented in Chap. 19, the open methods employ a constant step size. In contrast, the techniques discussed in *Chap. 20* use what is called *adaptive step size control*. These algorithms automatically adjust the step size in response to the truncation error of the computation. This is done in two ways. First, the one-step methods presented in Chap. 19 are modified to include an adaptive step size control strategy. Second, *multistep methods* can be derived. These algorithms retain information of previous steps in order to more effectively capture the trajectory of the solution. They also yield the truncation error estimates that are needed to implement adaptive step size control. In this section, we initially take a visual, intuitive approach by using a simple method—the *non-self-starting Heun*—to introduce all the essential features of the multistep approaches. We also review higher-order methods.

In *Chap. 21* we turn to *boundary-value* and *eigenvalue problems*. For the former, we introduce both shooting and finite-difference methods. For the latter we discuss several approaches including the *polynomial* and the *power methods*.

Chapter 22 is devoted to case studies from all the fields of engineering as well as a case study using a spreadsheet to solve an ODE. Finally, a short review section is included at the end of Part Six. This *epilogue* summarizes and compares

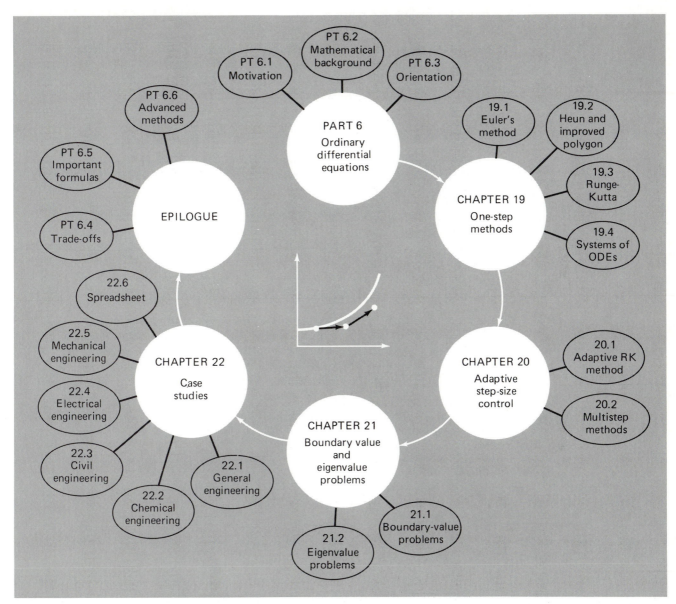

Figure PT6.5
Schematic representation of the organization of Part Six: Ordinary differential equations.

the important formulas and concepts related to ODEs. This comparison includes a discussion of trade-offs that are relevant to their implementation in engineering practice. The epilogue also summarizes important formulas and includes references for advanced topics.

PT6.3.2 Goals and Objectives

Study Objectives. After completing Part Six, you should have greatly enhanced your capability to confront and solve ordinary differential equations. General study goals should include mastering the techniques, having the capability to assess the reliability of the answers, and being able to choose the "best" method (or methods) for any particular problem. In addition to these general objectives, the specific study objectives in Table PT6.2 should be mastered.

Computer Objectives. You have been provided with software, simple computer programs, and algorithms, to implement the techniques discussed in Part Six. All have utility as learning tools.

The Electronic TOOLKIT personal computer software, which utilizes Euler's method, is user-friendly. The solution can be displayed in either graphical or tabular form. The graphical output will enable you to easily visualize your problem and solution. You can study the efficiency of the method by examining several different step sizes. The software is very easy to apply and can be used to check the results of any computer programs you may develop yourself.

Alternatively, BASIC, FORTRAN, and Pascal programs for Euler's method are supplied directly in the text. In addition, algorithms or flowcharts are provided for most of the other methods discussed in Part Six. This information will allow you to expand your software library to include techniques beyond Euler's method. For example, you may find it useful from a professional viewpoint to have software that employs the fourth-order Runge-Kutta method, which is usually more efficient and accurate than Euler's method. You may also want to develop software to accommodate systems of ordinary differential equations and to solve ODEs with an adaptive step-size approach.

Finally, spreadsheet software is included on a diskette that comes with this book. It is used to implement the computation performed in Case Study 22.6.

TABLE PT6.2   Specific study objectives for Part Six.

1. Understand the visual representations of Euler's, Heun's, and the improved polygon methods

2. Know the relationship of Euler's method to the Taylor series expansion and the insight it provides regarding the error of the method

3. Understand the difference between local and global truncation errors and how they relate to the choice of a numerical method for a particular problem

4. Know the order and the step-size dependency of the global truncation errors for all the methods described in Part Six; understand how these errors bear on the accuracy of the techniques

5. Understand the basis of predictor-corrector methods. In particular, realize that the efficiency of the corrector is highly dependent on the accuracy of the predictor

6. Know the general form of the Runge-Kutta methods. Understand the derivation of the second-order RK method and how it relates to the Taylor series expansion; realize that there are an infinite number of possible versions for second- and higher-order RK methods

7. Know how to apply any of the RK methods to systems of equations; be able to reduce an nth-order ODE to a system of n first-order ODEs

8. Understand the difference between initial-value and boundary-value problems

9. Know the difference between multistep and one-step methods; realize that all multistep methods are predictor-correctors but that not all predictor-correctors are multistep methods

10. Understand the connection between integration formulas and predictor-corrector methods

11. Recognize the fundamental difference between Newton-Cotes and Adams integration formulas

12. Recognize the type of problem context where step size adjustment is important

13. Understand how adaptive step size control is integrated into a fourth-order RK method

14. Know the rationale behind the polynomial and the power methods for determining eigenvalues; in particular, recognize their strengths and limitations

15. Understand how Hoteller's deflation allows the power method to be used to compute intermediate eigenvalues

CHAPTER 19
One-Step Methods

The present chapter is devoted to solving ordinary differential equations of the form

$$\frac{dy}{dx} = f(x, y)$$

In Chap. 1 we used a numerical method to solve such an equation for the velocity of the falling parachutist. Recall that the equation used to solve this problem was of the general form

New value = old value + slope × step size

or, in mathematical terms,

$$y_{i+1} = y_i + \phi h \tag{19.1}$$

According to this equation, the slope estimate of ϕ is used to extrapolate from an old value y_i to a new value y_{i+1} over a distance h (Fig. 19.1). This formula can be applied step by step to compute out into the future and, hence, trace out the trajectory of the solution.

Figure 19.1
Graphical depiction of a one-step method.

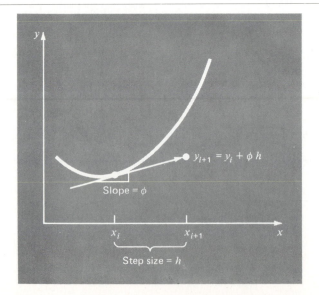

All one-step methods can be expressed in this general form, with the only difference being the manner in which the slope is estimated. As in the falling parachutist problem, the simplest approach is to use the differential equation to estimate the slope in the form of the first derivative at x_i. In other words, the slope at the beginning of the interval is taken as an approximation of the average slope over the whole interval. This approach, which is called *Euler's method,* is discussed in the first part of this chapter. This is followed by other one-step methods that employ alternative slope estimates that result in more accurate predictions.

19.1 EULER'S METHOD

The first derivative provides a direct estimate of the slope at x_i (Fig. 19.2):

$$\phi = f(x_i, y_i)$$

where $f(x_i, y_i)$ is the differential equation evaluated at x_i and y_i. This estimate can be substituted into Eq. (19.1):

$$y_{i+1} = y_i + f(x_i, y_i)h \qquad (19.2)$$

This formula is referred to as *Euler's* (or the *Euler-Cauchy* or the *point-slope*) *method.* A new value of y is predicted using the slope (equal to the first derivative at the original value of x) to extrapolate linearly over the step size h (Fig. 19.2).

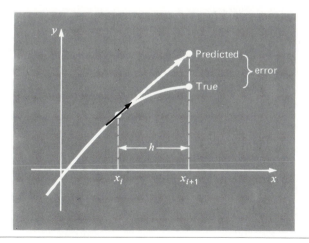

Figure 19.2
Euler's method.

EXAMPLE 19.1   Euler's Method

Problem Statement: Use Euler's method to numerically integrate Eq. (PT6.14):

$$f(x, y) = -2x^3 + 12x^2 - 20x + 8.5$$

from $x = 0$ to $x = 4$ with a step size of 0.5. The initial condition at $x = 0$ is $y = 1$. Recall that the exact solution is given by Eq. (PT6.17):

$$y = -0.5x^4 + 4x^3 - 10x^2 + 8.5x + 1$$

Solution: Equation (19.2) can be used to implement Euler's method:

$$y(0.5) = y(0) + f(0, 1)0.5$$

where $y(0) = 1$ and the slope estimate at $x = 0$ is

$$f(0, 1) = -2(0)^3 + 12(0)^2 - 20(0) + 8.5 = 8.5$$

Therefore,

$$y(0.5) = 1.0 + 8.5(0.5) = 5.25$$

The true solution at $x = 0.5$ is

$$y(0.5) = -0.5(0.5)^4 + 4(0.5)^3 - 10(0.5)^2 + 8.5(0.5) + 1$$

$$= 3.21875$$

Thus, the error is

$$E_t = \text{true} - \text{approximate} = 3.21875 - 5.25 = -2.03125$$

or, expressed as percent relative error, $\epsilon_t = -63.1\%$. For the second step

$$y(1.0) = y(0.5) + f(0.5, 5.25)0.5$$

$$= 5.25 + [-2(0.5)^3 + 12(0.5)^2 - 20(0.5) + 8.5]0.5$$

$$= 5.875$$

The true solution at $x = 1.0$ is 3.0, and therefore the percent relative error is -95.8%. The computation is repeated, and the results are compiled in Table 19.1 and Fig. 19.3. Note that, although the computation captures the general trend of the true solution, the error is considerable. As discussed in the next section, this error can be reduced by using a smaller step size.

19.1.1 Error Analysis for Euler's Method

The numerical solution of ODEs involves two types of error (recall Chap. 3):

1. *Truncation* or *discretization errors* caused by the nature of the techniques employed to approximate values of y.
2. *Round-off errors* caused by the limited numbers of significant digits that can be retained by a computer

The truncation errors are composed of two parts. The first is a *local truncation error* that results from an application of the method in question over a single step. The second is a *propagated truncation error* that results from the approximations produced during the previous steps. The sum of the two is the total or *global truncation error*.

TABLE 19.1 Comparison of true and approximate values of the integral of $y' = -2x^3 + 12x^2 - 20x + 8.5$, with the initial condition that $y = 1$ at $x = 0$. The approximate values were computed using Euler's method with a step size of 0.5. The local error refers to the error incurred over a single step. The global error is the total discrepancy due to the past steps as well as the present.

| x | y_{true} | y_{Euler} | ϵ_t, Percent Relative Error | |
			Global	Local
0.0	1.00000	1.00000		
0.5	3.21875	5.25000	−63.1	−63.1
1.0	3.00000	5.87500	−95.8	−28.0
1.5	2.21875	5.12500	−131.0	−1.41
2.0	2.00000	4.50000	−125.0	20.5
2.5	2.71875	4.75000	−74.7	17.3
3.0	4.00000	5.87500	−46.9	4.0
3.5	4.71875	7.12500	−51.0	−11.3
4.0	3.00000	7.00000	−133.3	−53.0

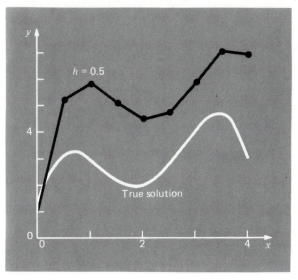

Figure 19.3
Comparison of the true solution with a numerical solution using Euler's method for the integral of $y' = -2x^3 + 12x^2 - 20x + 8.5$ from $x = 0$ to $x = 4$ with a step size of 0.5. The initial condition at $x = 0$ is $y = 1$.

Insight into the magnitude and properties of the truncation error can be gained by deriving Euler's method directly from the Taylor series expansion. In order to do this, realize that the differential equation being integrated will be of the general form

$$y' = f(x, y) \tag{19.3}$$

where $y' = dy/dx$ and x and y are the independent and the dependent variables, respectively. If the solution—that is, the function describing the behavior of y—has continuous derivatives, it can be represented by a Taylor series expansion about a starting value (x_i, y_i), as in [recall Eq. (3.18)]

$$y_{i+1} = y_i + y_i'h + \frac{y_i''}{2}h^2 + \cdots + \frac{y_i^{(n)}}{n!}h^n + R_n \tag{19.4}$$

where $h = x_{i+1} - x_i$ and R_n is the remainder term defined as

$$R_n = \frac{y^{(n+1)}(\xi)}{(n+1)!}h^{n+1} \tag{19.5}$$

where ξ lies somewhere in the interval from x_i to x_{i+1}. An alternative form can be developed by substituting Eq. (19.3) into Eqs. (19.4) and (19.5) to yield

$$y_{i+1} = y_i + f(x_i, y_i)h + \frac{f'(x_i, y_i)}{2}h^2$$

$$+ \cdots + \frac{f^{(n-1)}(x_i, y_i)}{n!}h^n + O(h^{n+1}) \tag{19.6}$$

where $O(h^{n+1})$ specifies that the local truncation error is proportional to the step size raised to the $(n + 1)$th power.

By comparing Eqs. (19.2) and (19.6), it can be seen that Euler's method corresponds to the Taylor series up to and including the term $f(x_i, y_i)h$. Additionally, the comparison indicates that the truncation error is due to the fact that we approximate the true solution using a finite number of terms from the Taylor series. We thus truncate, or leave out, a part of the true solution. For example, the truncation error in Euler's method is attributable to the remaining terms in the Taylor series expansion that were not included in Eq. (19.2). Subtracting Eq. (19.2) from Eq. (19.6) yields

$$E_t = \frac{f'(x_i, y_i)}{2}h^2 + \cdots + O(h^{n+1}) \tag{19.7}$$

where E_t is the true local truncation error. For sufficiently small h, the errors in the terms in Eq. (19.7) usually decrease as the order increases (recall Example 3.11 and the accompanying discussion), and the result is often represented as

$$E_a = \frac{f'(x_i, y_i)}{2}h^2 \tag{19.8}$$

or

$$E_a = O(h^2) \tag{19.9}$$

where E_a is the approximate local truncation error.

EXAMPLE 19.2   Taylor Series Estimate for the Error of Euler's Method

Problem Statement: Use Eq. (19.7) to estimate the error of the first step of Example 19.1. Also use it to determine the error due to each higher-order term of the Taylor series expansion.

Solution: Because we are dealing with a polynomial, we can use the Taylor series to obtain exact estimates of the errors in Euler's method. Equation (19.7) can be written as

$$E_t = \frac{f'(x_i, y_i)}{2} h^2 + \frac{f''(x_i, y_i)}{3!} h^3 + \frac{f'''(x_i, y_i)}{4!} h^4 \qquad \text{(E19.2.1)}$$

where $f'(x_i, y_i)$ is the first derivative of the differential equation (that is, the second derivative of the solution). For the present case, this is

$$f'(x_i, y_i) = -6x^2 + 24x - 20 \qquad \text{(E19.2.2)}$$

and $f''(x_i, y_i)$ is the second derivative of the ODE

$$f''(x_i, y_i) = -12x + 24 \qquad \text{(E19.2.3)}$$

and $f'''(x_i, y_i)$ is the third derivative of the ODE

$$f'''(x_i, y_i) = -12 \qquad \text{(E19.2.4)}$$

We can omit additional terms (that is, fourth derivatives and higher) from Eq. (E19.2.1) because for this particular case they equal zero. It should be noted that for other functions (for example, transcendental functions such as sinusoids or exponentials) this would not necessarily be true, and higher-order terms would have nonzero values. However, for the present case, Eqs. (E19.2.1) through (E19.2.4) completely define the truncation error for a single application of Euler's method.

For example, the error due to truncation of the second-order term can be calculated as

$$E_{t,2} = \frac{-6(0.0)^2 + 24(0.0) - 20}{2} (0.5)^2 = -2.5 \qquad \text{(E19.2.5)}$$

For the third-order term:

$$E_{t,3} = \frac{-12(0.0) + 24}{6} (0.5)^3 = 0.5$$

and the fourth-order term:

$$E_{t,4} = \frac{-12}{24} (0.5)^4 = -0.03125$$

These three results can be added to yield the total truncation error:

$$E_t = E_{t,2} + E_{t,3} + E_{t,4} = -2.5 + 0.5 - 0.03125 = -2.03125$$

which is exactly the error that was incurred in the initial step of Example 19.1. Note how $E_{t,2} > E_{t,3} > E_{t,4}$, which supports the approximation represented by Eq. (19.8).

As illustrated in Example 19.2, the Taylor series provides a means of quantifying the error in Euler's method. However, there are a number of limitations associated with its use for this purpose:

1. The Taylor series only provides an estimate of the local truncation error—that is, the error created during a single step of the method. It does not provide a measure of the propagated and, hence, the global truncation error. In Table 19.1, we have included the local and global truncation errors for Example 19.1. The local error was computed for each time step with Eq. (19.2) but using the true value of y_i (the second column of the table) to compute each y_{i+1} rather than the approximate value (the third column), as is done in the Euler method. As expected, the average absolute local truncation error (25 percent) is less than the average global error (90 percent). The only reason that we can make these exact error calculations is that we know the true value a priori. Such would not be the case in an actual problem. Consequently, as discussed below, you must usually apply techniques such as Euler's method using a number of different step sizes to obtain an indirect estimate of the errors involved.

2. As mentioned above, in actual problems we usually deal with functions that are more complicated than simple polynomials. Consequently, the derivatives that are needed to evaluate the Taylor series expansion would not always be easy to obtain.

Although these limitations preclude exact error analysis for most practical problems, the Taylor series still provides valuable insight into the behavior of Euler's method. According to Eq. (19.8), we see that the local error is proportional to the square of the step size and the first derivative of the differential equation. It can also be demonstrated that the global truncation error is $O(h)$; that is, it is proportional to the step size (Carnahan et al., 1969). These observations lead to some useful conclusions:

1. The error can be reduced by decreasing the step size.
2. The method will provide error-free predictions if the underlying function (that is, the solution of the differential equation) is linear, because for a straight line the second derivative would be zero.

This latter conclusion makes intuitive sense because Euler's method uses straight-line segments to approximate the solution. Hence, Euler's method is referred to as a *first-order method*.

EXAMPLE 19.3   *Effect of Reduced Step Size on Euler's Method*

Problem Statement: Repeat the computation of Example 19.1, but use a step size of 0.25.

Solution: The computation is repeated, and the results are compiled in Fig. 19.4a. Halving the step size reduces the absolute value of the average global error to 40 percent and the absolute value of the local error to 6.4 percent. This is compared to global and local errors for Example 19.1 of 90 percent and 24.8 percent. Thus, as expected, the local error is quartered and the global error is halved.

Also, notice how the local error changes sign for intermediate values along the range. This is due primarily to the fact that the first derivative of the differential equation is a parabola that changes sign [recall Eq. (E19.2.2) and see Fig. 19.4b]. Because the local error is proportional to this function, the net effect of the oscillation in sign is to keep the global error from continuously growing as the calculation proceeds. Thus, from $x = 0$ to $x = 1.25$, the local errors are all negative, and consequently, the global error increases over this interval. In the intermediate section of the range, positive local errors begin to reduce the global error. Near the end, the process is reversed and the global error again inflates. If the local error continuously changes sign over the computation interval,

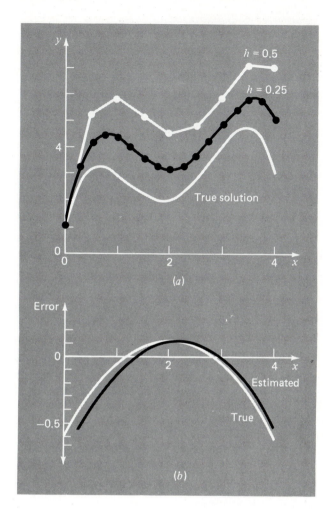

Figure 19.4
(a) Comparison of two numerical solutions with Euler's method using step sizes of 0.5 and 0.25. (b) Comparison of true and estimated local truncation error for the case where the step size is 0.5. Note that the "estimated" error is based on Eq. (E19.2.5).

the net effect is usually to reduce the global error. However, where the local errors are of the same sign, the numerical solution may diverge farther and farther from the true solution as the computation proceeds. Such results are said to be *unstable*.

The effect of further step-size reductions on the global truncation error of Euler's method is illustrated in Fig. 19.5. This plot shows the absolute percent relative error at $x = 5$ as a function of step size for the problem we have been examining in Examples 19.1 through 19.3. Notice that even when h is reduced to 0.001, the error still exceeds 0.1 percent. Because this step size corresponds to 5000 steps to proceed from $x = 0$ to $x = 5$, the plot suggests that a first-order technique such as Euler's method demands grea computational effort to obtain acceptable error levels. The following sections presen higher-order techniques that attain much better accuracy for the same computationa

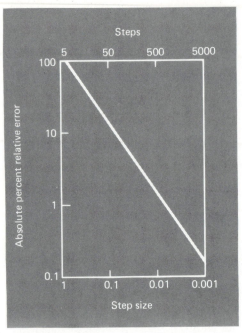

Figure 19.5
Effect of step size on the global truncation error of Euler's method for the integral of $y' = -2x^3 + 12x^2 - 20x + 8.5$. The plot shows the absolute percent relative global error at $x = 5$ as a function of step size.

effort. However, it should be noted that despite its inefficiency, the simplicity of Euler's method makes it an extremely attractive option for many engineering problems. Because it is very easy to program, the technique is particularly useful for quick initial computations prior to full-scale analysis. In the next section, a computer program for Euler's method is developed.

19.1.2 Computer Programs for Euler's Method

Algorithms for one-step techniques such as Euler's method are extremely simple to program on a personal computer. As specified previously at the beginning of this chapter, all one-step methods have the general form:

$$\text{New value} = \text{old value} + \text{slope} \times \text{step size} \tag{19.10}$$

The only way in which the methods differ is in the calculation of the slope.

Although the programs in Figs. 19.6 and 19.7 are specifically designed to implement Euler's method, they are cast in the general form of Eq. (19.10). All that is required to apply these programs to the other one-step methods is to modify the computation of the slope.

```
100 REM      EULER'S METHOD (BASIC VERSION)        280 PRINT:INPUT "STEP SIZE? ", H
105 REM ******************************             285 RETURN
110 REM *      DEFINITION OF VARIABLES    *        300 REM ****** SUBROUTINE INTEGRATE ******
115 REM *                                 *        305 REM
120 REM * X   = INDEPENDENT VARIABLE      *        310 NP = (X1-X0)/TP
125 REM * Y   = DEPENDENT VARIABLE        *        315 NC=TP/H
130 REM * X0  = INITIAL VALUE OF X        *        320 X=X0
135 REM * Y0  = INITIAL VALUE OF Y        *        325 Y=Y0
140 REM * H   = STEP SIZE                 *        330 XP(1) = X
145 REM * TP  = PRINT INTERVAL            *        335 YP(1) = Y
150 REM * NP  = NUMBER OF PRINT STEPS     *        340 FOR I = 1 TO NP
155 REM * NC  = NUMBER OF CALCULATION STEPS*       345    FOR J = 1 TO NC
160 REM ******************************             350       Y = Y + FNF(Y)*H
165 REM                                            355       X = X + H
170 DEF FNF(Y)=-2*X^3 + 12*X^2 - 20*X + 8.5        360    NEXT J
175 REM                                            365    XP(I+1) = X
180 REM ********* MAIN PROGRAM **********           370    YP(I+1) = Y
185 REM                                            375 NEXT I
190 GOSUB 250 'input data                          380 RETURN
195 GOSUB 300 'integrate                           400 REM ******* SUBROUTINE OUTPUT ********
200 GOSUB 400 'output results                      405 REM
205 END                                            410 PRINT:PRINT
250 REM ******** SUBROUTINE INPUT ********         415 PRINT "   X                 Y"
255 REM                                            420 PRINT
260 CLS                                            425 FOR I = 1 TO NP+1
265 PRINT:INPUT "INITIAL,FINAL X? ", X0,X1         430    PRINT XP(I),YP(I)
270 PRINT:INPUT "INITIAL Y? ", Y0                  435 NEXT I
275 PRINT:INPUT "PRINT INTERVAL? ", TP             440 RETURN
```

Figure 19.6
An interactive computer program for Euler's method written in Microsoft BASIC.

The programs in Figs. 19.6 and 19.7 are not user-friendly; they are designed strictly to come up with the answer. In Prob. 19.12 you will have the task of making this skeletal computer code easier to use and understand. An example of a user-friendly program for Euler's method is included in the supplementary Electronic TOOLKIT software associated with this text. The following example demonstrates the use of this software for solving ODEs. It also provides a reference for assessing and testing your own software.

EXAMPLE 19.4   Solving ODEs with the Computer

Problem Statement: A user-friendly computer program to implement Euler's method is contained in the Electronic TOOLKIT software associated with the text. We can use this software to solve another problem associated with the falling parachutist. You recall from Part One that our mathematical model for the velocity was based on Newton's second law in the form

$$\frac{dv}{dt} = g - \frac{c}{m} v$$

```
SUBROUTINE EULER(XO,X1,YO,TP,H,XP,YP)
*************************************************
*          DEFINITION OF VARIABLES          *
*                                              *
*     X    = INDEPENDENT VARIABLE             *
*     Y    = DEPENDENT VARIABLE               *
*     XO   = INITIAL VALUE OF X               *
*     X1   = FINAL VALUE OF X                 *
*     YO   = INITIAL VALUE OF Y               *
*     XP() = OUTPUT VALUES OF X               *
*     YP() = OUTPUT VALUES OF Y               *
*     H    = STEP-SIZE                         *
*     TP   = PRINT INTERVAL                    *
*************************************************
      DIMENSION XP(100),YP(100)
      NP = (X1 - XO)/TP
      NC = TP/H
      X = XO
      Y = YO
      DO 10 I=1,NP
         DO 20 J=1,NC
            Y = Y + DYDX(X,Y)*H
            X = X + H
20       CONTINUE
         XP(I) = X
         YP(I) = Y
10    CONTINUE
      RETURN
      END
```

```
PROCEDURE Euler (VAR Xp,Yp : Vector;
                 XO,X1,YO,Tp,H: Real);

{         Driver program type definitions
      Vector = a 1 dimensional real array
               (defined with zero subscript)}

{           Definition of variables
    X      = independent variable
    Y      = dependent variable
    XO     = initial value of X
    X1     = final value of X
    YO     = initial value of Y
    Xp()   = output values of X
    Yp()   = output values of Y
    H      = step size
    Tp     = print interval
    slope  = derivitive evaluated at x and y }

VAR
   x,y,slope: real;
   np,nc,i,j : integer;
Begin
np := Trunc((X1-XO)/Tp);
nc := Trunc(Tp/H);
X := XO;
Y := YO;
Xp[1] := X;
Yp[1] := Y;
For i := 1 to np do
   Begin
   For j := 1 to nc do
      Begin
      {slope is computed with a user
      supplied function of x and y}
      slope := dydx(x,y);
      Y := Y + slope * H;
      X := X + H;
      End;
   Xp[i+1] := X;
   Yp[i+1] := Y;
   End;
End;                    { of procedure Euler}
```

Figure 19.7
Subroutines for Euler's method written in FORTRAN 77 and
Turbo Pascal.

This differential equation was solved both analytically (Example 1.1) and numerically using Euler's method (Example 1.2). The objective of the present example is to repeat these numerical computations employing a more complicated model for the velocity based on a more complete mathematical description of the drag force caused by wind resistance. This model is given by

$$\frac{dv}{dt} = g - \frac{c}{m}\left[v + a\left(\frac{v}{v_{\max}}\right)^b\right]$$

(E19.4.1)

where a, b, and v_{max} are empirical constants which for this case are equal to 8.30, 2.20, and 46.00, respectively. Note that this model is more capable of accurately fitting empirical measurements of drag forces versus velocity than is the simple linear model of Example 1.1. However, this increased flexibility is gained at the expense of evaluating three coefficients rather than one. Furthermore, the resulting mathematical model is more difficult to solve analytically. In this case, Euler's method provides a convenient alternative to obtain an approximate numerical solution.

Solution: We will use the Electronic TOOLKIT to tackle Eq. (E19.4.1). Figure 19.8*a* shows the solution of the model with an integration step size of 0.1 s. The plot in Fig. 19.8*b* also shows an overlay of the solution of the linear model for comparison purposes. Note that the software can plot only one solution at a time.

The results of the two simulations indicate how increasing the complexity of the formulation of the drag force affects the velocity of the parachutist. In this case, the terminal velocity is lowered because of resistance caused by the higher-order terms in Eq. (E19.4.1).

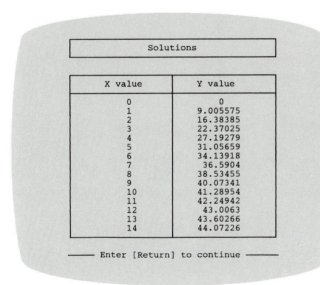
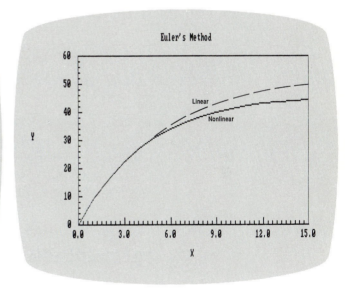

Figure 19.8
(*a*) Tabular results of the computation and (*b*) graphical results for the solution of the nonlinear ODE [Eq. (E19.4.1)]. Notice that (*b*) also shows the solution for the linear model for comparative purposes. In fact, the software is not designed to superimpose plots in this manner.

Alternative models could be tested in a similar fashion. The combination of the Electronic TOOLKIT software and your personal computer makes this an easy and efficient task. This convenience should allow you to devote more of your time to considering creative alternatives and holistic aspects of the problem rather than to tedious manual computations.

19.1.3 Higher-Order Taylor Series Methods

One way to reduce the error of Euler's method would be to include higher-order terms of the Taylor series expansion in the solution. For example, including the second-order term from Eq. (19.6) yields

$$y_{i+1} = y_i + f(x_i, y_i)h + \frac{f'(x_i, y_i)}{2} h^2 \tag{19.11}$$

with a local truncation error of

$$E_a = \frac{f''(x_i, y_i)}{6} h^3$$

Although the incorporation of higher-order terms is simple enough to implement for polynomials, their inclusion is not so trivial when the ODE is more complicated. In particular, ODEs that are a function of both the dependent and independent variable require chain-rule differentiation. For example, the first derivative of $f(x, y)$ is

$$f'(x, y) = \frac{\partial f(x, y)}{\partial x} + \frac{\partial f(x, y)}{\partial y} \frac{dy}{dx}$$

The second derivative is

$$f''(x, y) = \frac{\partial[\partial f/\partial x + (\partial f/\partial y)(dy/dx)]}{\partial x}$$

$$+ \frac{\partial[\partial f/\partial x + (\partial f/\partial y)(dy/dx)]}{\partial y} \frac{dy}{dx}$$

Higher-order derivatives become increasingly more complicated.

Consequently, as described in the following sections, alternative one-step methods have been developed. These schemes are comparable in performance to the higher-order Taylor-series approaches but require only the calculation of first derivatives.

19.2   MODIFICATIONS AND IMPROVEMENTS OF EULER'S METHOD

A fundamental source of error in Euler's method is that the derivative at the beginning of the interval is assumed to apply across the entire interval. Two simple modifications are available to help circumvent this shortcoming. As will be demonstrated in Sec. 19.3, both modifications actually belong to a larger class of solution techniques called Runge-Kutta methods. However, because they have a very straightforward graphical interpretation, we will present them prior to their formal derivation as Runge-Kutta methods.

19.2.1 Heun's Method

One method to improve the estimate of the slope involves the determination of two derivatives for the interval—one at the initial point and another at the end point. The two derivatives are then averaged to obtain an improved estimate of the slope for the entire interval. This approach, called *Heun's method,* is depicted graphically in Fig. 19.9.

Recall that in Euler's method, the slope at the beginning of an interval

$$y_i' = f(x_i, y_i) \tag{19.12}$$

is used to extrapolate linearly to y_{i+1}:

$$y_{i+1}^0 = y_i + f(x_i, y_i) h \tag{19.13}$$

For the standard Euler method we would stop at this point. However, in Heun's method the y_{i+1}^0 calculated in Eq. (19.13) is not the final answer but an intermediate prediction. This is why we have distinguished it with a superscripted 0. Equation (19.13) is called a *predictor equation.* It provides an estimate of y_{i+1} that allows the calculation of an estimated slope at the end of the interval:

$$y_{i+1}' = f(x_{i+1}, y_{i+1}^0) \tag{19.14}$$

Thus, the two slopes [Eqs. (19.12) and (19.14)] can be combined to obtain an average slope for the interval:

$$\bar{y}' = \frac{y_i' + y_{i+1}'}{2} = \frac{f(x_i, y_i) + f(x_{i+1}, y_{i+1}^0)}{2}$$

Figure 19.9
Graphical depiction of Heun's method. (*a*) Predictor and (*b*) corrector.

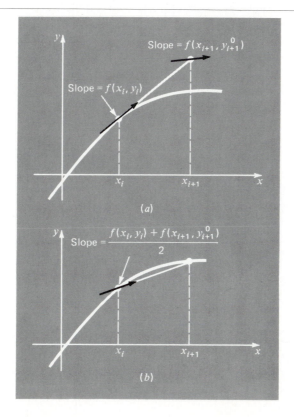

This average slope is then used to extrapolate linearly from y_i to y_{i+1} using Euler's method:

$$y_{i+1} = y_i + \frac{f(x_i, y_i) + f(x_{i+1}, y_{i+1}^0)}{2} h$$

which is called a *corrector equation*.

The Heun method is a *predictor-corrector approach*. All the multistep methods to be discussed subsequently in Chap. 20 are of this type. The Heun method is the only one-step, predictor-corrector method described in this book. As derived above, it can be expressed concisely as

Predictor (Fig. 19.9a): $y_{i+1}^0 = y_i + f(x_i, y_i) h$ (19.15)

Corrector (Fig. 19.9b): $y_{i+1} = y_i + \dfrac{f(x_i, y_i) + f(x_{i+1}, y_{i+1}^0)}{2} h$ (19.16)

Note that because Eq. (19.16) has y_{i+1} on both sides of the equal sign, it can be applied to "correct" in an iterative fashion. That is, an old estimate can be used repeatedly to provide an improved estimate of y_{i+1}. The process is depicted in Fig. 19.10. It should be understood that this iterative process will not necessarily converge on the true answer but will converge on an estimate with a finite truncation error, as demonstrated in the following example.

As with similar iterative methods discussed in previous sections of the book, a termination criterion for convergence of the corrector is provided by [recall Eq. (3.5)]

$$|\epsilon_a| = \left| \frac{y_{i+1}^j - y_{i+1}^{j-1}}{y_{i+1}^j} \right| 100\%$$ (19.17)

Figure 19.10

Graphical representation of iterating the corrector of Heun's method to obtain an improved estimate.

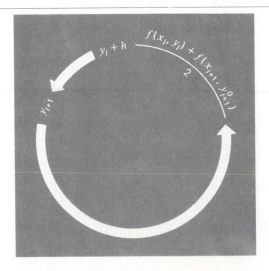

where y_{i+1}^{j-1} and y_{i+1}^{j} are the result from the prior and the present iteration of the corrector, respectively.

EXAMPLE 19.5   Heun's Method

Problem Statement: Use Heun's method to integrate $y' = 4e^{0.8x} - 0.5y$ from $x = 0$ to $x = 4$ with a step size of 1. The initial condition at $x = 0$ is $y = 2$.

Solution: Before solving the problem numerically, calculus can be used to determine the following analytical solution:

$$y = \frac{4}{1.3}(e^{0.8x} - e^{-0.5x}) + 2e^{-0.5x} \qquad (E19.5.1)$$

This formula can be used to generate the true-solution values in Table 19.2.

The numerical solution is obtained by using the predictor [Eq. (19.15)] to obtain an estimate of y at 1.0:

$$y_1^0 = 2 + [4e^0 - 0.5(2)] 1 = 5$$

Note that this is the result that would be obtained by the standard Euler method. The true value in Table 19.2 shows that it corresponds to a percent relative error of 19.3 percent.

The slope at (x_0, y_0) is

$$y_0' = 4e^0 - 0.5(2) = 3$$

This result is quite different from the actual average slope for the interval from 0 to 1.0, which is equal to 4.1946, as calculated from the original differential equation using Eq.

TABLE 19.2
Comparison of true and approximate values of the integral of $y' = 4e^{0.8x} - 0.5y$ with the initial condition that $y = 2$ at $x = 0$. The approximate values were computed using the Heun method with a step size of 1. Two cases, corresponding to different numbers of corrector iterations, are shown, along with the absolute percent relative error.

| | | Iterations of Heun's Method | | | |
| | | 1 | | 15 | |
x	y_{true}	y_{heun}	$\lvert \epsilon_t \rvert$, %	y_{heun}	$\lvert \epsilon_t \rvert$, %
0	2.00000000	2.00000000	0.00	2.00000000	0.00
1	6.19463138	6.70108186	8.18	6.36086549	2.68
2	14.8439219	16.3197819	9.94	15.3022367	3.09
3	33.6771718	37.1992489	10.46	34.7432761	3.17
4	75.3389626	83.3377674	10.62	77.7350962	3.18

(PT5.4). Therefore, in order to improve the estimate of the slope, we use the value y_1^0 to predict the slope at the end of the interval:

$$y_1' = f(x_1, y_1^0) = 4e^{0.8(1)} - 0.5(5) = 6.40216371$$

which can be combined with the initial slope to yield

$$y' = \frac{3 + 6.40216371}{2} = 4.70108186$$

which is closer to the true average slope of 4.1946. This result can then be substituted into the corrector [Eq. (19.16)] to give the prediction at $x = 1$:

$$y_1 = 2 + (4.70108186)(1) = 6.70108186$$

which represents a percent relative error of -8.18 percent. Thus, the Heun method reduces the absolute value of the error by a factor of 2.4 as compared with Euler's method.

Now this estimate can be used to refine or correct the prediction of y_1 by substituting the new result back into the right-hand side of Eq. (19.16):

$$y_1 = 2 + \frac{[3 + 4e^{0.8(1)} - 0.5(6.70108186)]}{2} 1 = 6.27581139$$

which represents an absolute percent relative error of 1.31 percent. This result, in turn, can be substituted back into Eq. (19.16) to further correct y_1:

$$y_1 = 2 + \frac{[3 + 4e^{0.8(1)} - 0.5(6.27581139)]}{2} 1 = 6.38212901$$

which represents an $|\epsilon_t|$ of 3.03%. Notice how the errors sometimes grow as the iterations proceed. For example, for three iterations the error grows to 3.03 percent. Such increases can occur, especially for large step sizes, and they prevent us from drawing the general conclusion that an additional iteration will always improve the result. However, for a sufficiently small step size, the iterations should eventually converge on a single value. For our case, 6.36086549, which represents a relative error of 2.68 percent, is attained after 15 iterations. Table 19.2 shows results for the remainder of the computation using the method with 1 and 15 iterations per step.

In the previous example, the derivative is a function of both the dependent variable y and the independent variable x. For cases such as polynomials, where the ODE is solely a function of the independent variable, the predictor step [Eq. (19.15)] is not required and the corrector is only applied once for each iteration. For such cases, the technique is expressed concisely as

$$y_{i+1} = y_i + \frac{f(x_i) + f(x_{i+1})}{2} h \tag{19.18}$$

Notice the similarity between the right-hand side of Eq. (19.18) and the trapezoidal rule [Eq. (15.3)]. The connection between the two methods can be formally demonstrated by starting with the ordinary differential equation

$$\frac{dy}{dx} = f(x)$$

This equation can be solved for y by integration:

$$\int_{y_i}^{y_{i+1}} dy = \int_{x_i}^{x_{i+1}} f(x)\, dx \tag{19.19}$$

which yields

$$y_{i+1} - y_i = \int_{x_i}^{x_{i+1}} f(x)\, dx \tag{19.20}$$

or

$$y_{i+1} = y_i + \int_{x_i}^{x_{i+1}} f(x)\, dx \tag{19.21}$$

Now, recall from Sec. 15.1 that the trapezoidal rule [Eq. (15.3)] is defined as

$$\int_a^b f(x)\, dx \simeq (b - a)\frac{f(b) + f(a)}{2}$$

or, for the present case,

$$\int_{x_i}^{x_{i+1}} f(x)\, dx \simeq \frac{f(x_i) + f(x_{i+1})}{2} h \tag{19.22}$$

where $h = x_{i+1} - x_i$. Substituting Eq. (19.22) into Eq. (19.21) yields

$$y_{i+1} = y_i + \frac{f(x_i) + f(x_{i+1})}{2} h \tag{19.23}$$

which is equivalent to Eq. (19.18).

Because Eq. (19.23) is a direct expression of the trapezoidal rule, the local truncation error is given by [recall Eq. (15.6)]

$$E_t = -\frac{f''(\xi)}{12} h^3 \tag{19.24}$$

where ξ is between x_i and x_{i+1}. Thus, the method is second order because the second derivative of the ODE is zero when the true solution is a quadratic. In addition, the local and global errors are $O(h^3)$ and $O(h^2)$, respectively. Therefore, decreasing the step size decreases the error at a faster rate than for Euler's method. Figure 19.11, which shows the result of using Heun's method to solve the polynomial from Example 19.1, demonstrates this behavior.

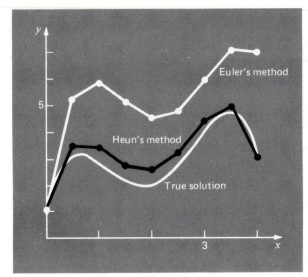

Figure 19.11
Comparison of the true solution
with a numerical solution using
Euler's and Heun's method for
the integral of $y' = -2x^3 + 12x^2 - 20x + 8.5$.

19.2.2   The Improved Polygon Method (Modified Euler)

Figure 19.12 illustrates another simple modification of Euler's method. Called the *improved polygon* (or the *modified Euler*), this technique uses Euler's method to predict a value of y at the midpoint of the interval (Fig. 19.12a):

$$y_{i+1/2} = y_i + f(x_i, y_i)\frac{h}{2} \tag{19.25}$$

Then this predicted value is used to estimate a slope at the midpoint:

$$y'_{i+1/2} = f(x_{i+1/2}, y_{i+1/2}) \tag{19.26}$$

which is assumed to represent a valid approximation of the average slope for the entire interval. This slope is then used to extrapolate linearly from x_i to x_{i+1} using Euler's method (Fig. 19.12b):

$$y_{i+1} = y_i + f(x_{i+1/2}, y_{i+1/2})h \tag{19.27}$$

Notice that because y_{i+1} is not on both sides, the corrector [Eq. (19.27)] cannot be applied iteratively to improve the solution.

   The improved polygon method is superior to Euler's method because it utilizes a slope estimate at the midpoint of the prediction interval. Recall from our discussion of numerical differentiation in Sec. 3.5.4 that centered finite divided differences are better approximations of derivatives than either forward or backward versions. In the same sense, a centered approximation such as Eq. (19.26) has a local truncation error

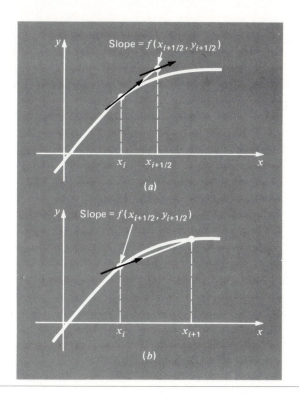

Figure 19.12
Graphical depiction of the improved polygon method. (a) Equation (19.25) and (b) Eq. (19.27).

of $O(h^2)$ in comparison with the forward approximation of Euler's method which has an error of $O(h)$. Consequently, the local and global errors of the improved polygon method are $O(h^3)$ and $O(h^2)$, respectively.

19.2.3 Computer Algorithms for Heun and Modified Euler Methods

Both the Heun method with a single corrector and the improved polygon method can be easily programmed using the general structure depicted in Figs. 19.6 and 19.7. It is a relatively straightforward task to modify the program in order to compute the slope in conformance with these methods.

However, when the iterative version of the Heun method is to be implemented, the modifications are a bit more involved. We have developed pseudocode for this purpose in Fig. 19.13. This algorithm can be combined with Fig. 19.6 or 19.7 to develop software for the iterative Heun method.

19.2.4 Summary

By tinkering with Euler's method, we have derived two new second-order techniques. Even though these versions require more computational effort to determine the slope, the accompanying reduction in error will allow us to conclude in a subsequent section (Sec. 19.3.4) that the improved accuracy is usually worth the effort. Although there are certain cases where easily programmable techniques such as Euler's method can be applied to

```
s₁ = dydx(x,y)
x₂ = x + h
y₂ = y + s₁·h
iter = 0
εₐ = 1.1 εₛ
DOWHILE (εₐ > εₛ) and (iter < maxit)
    iter = iter + 1
    s₂ = dydx(x₂,y₂)
    slope = (s₁ + s₂)/2
    y₂,new = y + slope·h

    εₐ =  | y₂,new - y₂ | ·100
         | ————————— |
         |   y₂,new   |

    y₂ = y₂,new
ENDDO
IF (εₐ > εₛ) THEN
    OUTPUT "nonconvergence"
ELSE ENDIF
```

Figure 19.13
Pseudocode to implement the iterative Heun method.

advantage, the Heun and improved polygon methods are generally superior and should be implemented if consistent with the problem objectives.

As noted at the beginning of this section, the Heun (without iterations), the improved polygon method, and, in fact, the Euler technique itself are versions of a broader class of one-step approaches called Runge-Kutta methods. We now turn to a formal derivation of these techniques.

19.3 RUNGE-KUTTA METHODS

Runge-Kutta (RK) methods achieve the accuracy of a Taylor series approach without requiring the calculation of higher derivatives. Many variations exist but all can be cast in the generalized form of Eq. (19.1):

$$y_{i+1} = y_i + \phi(x_i, y_i, h)\, h \tag{19.28}$$

where $\phi(x_i, y_i, h)$ is called an *increment function* which can be interpreted as a representative slope over the interval. The increment function can be written in general form as

$$\phi = a_1 k_1 + a_2 k_2 + \cdots + a_n k_n \tag{19.29}$$

where the a's are constants and the k's are

$$k_1 = f(x_i, y_i) \tag{19.29a}$$

$$k_2 = f(x_i + p_1 h, y_i + q_{11} k_1 h) \tag{19.29b}$$

$$k_3 = f(x_i + p_2 h, y_i + q_{21} k_1 h + q_{22} k_2 h) \tag{19.29c}$$

.

.

$$k_n = f(x_i + p_{n-1}h,\ y_i + q_{n-1,1}\ k_1h + q_{n-1,2}\ k_2h$$

$$+\ \cdots\ +\ q_{n-1,\,n-1}\ k_{n-1}h) \tag{19.29d}$$

Notice that the k's are recurrence relationships. That is, k_1 appears in the equation for k_2, which appears in the equation for k_3, and so forth. This recurrence makes RK methods efficient for computer calculations.

Various types of Runge-Kutta methods can be devised by employing different numbers of terms in the increment function as specified by n. Note that the first-order RK method with $n = 1$ is, in fact, Euler's method. Once n is chosen, values for the a's, p's, and q's are evaluated by setting Eq. (19.28) equal to terms in a Taylor series expansion (Box 19.1). Thus, at least for the lower-order versions, the number of terms n usually represents the order of the approach. For example, in the next section, second-order RK methods use an increment function with two terms ($n = 2$). These second-order methods will be exact if the solution to the differential equation is quadratic. In addition, because terms with h^3 and higher are dropped during the derivation, the local truncation error is $O(h^3)$ and the global error is $O(h^2)$. In subsequent sections, the third- and fourth-order RK methods ($n = 3$ and 4) are developed. For these cases, the global truncation errors are $O(h^3)$ and $O(h^4)$, respectively.

19.3.1 Second-Order Runge-Kutta Methods

The second-order version of Eq. (19.28) is

$$y_{i+1} = y_i + (a_1k_1 + a_2k_2)\,h \tag{19.30}$$

where

$$k_1 = f(x_i,\ y_i) \tag{19.30a}$$

$$k_2 = f(x_i + p_1h,\ y_i + q_{11}k_1h) \tag{19.30b}$$

As described in Box 19.1, values for a_1, a_2, p_1, and q_{11} are evaluated by setting Eq. (19.30) equal to a Taylor series expansion to the second-order term. By doing this, we derive three equations to evaluate the four unknown constants. The three equations are

$$a_1 + a_2 = 1 \tag{19.31}$$

$$a_2\,p_1 = \frac{1}{2} \tag{19.32}$$

$$a_2q_{11} = \frac{1}{2} \tag{19.33}$$

Because we have three equations with four unknowns, we must assume a value of one of the unknowns in order to determine the other three. Suppose that we specify a value for a_2. Then Eqs. (19.31) through (19.33) can be solved simultaneously for

$$a_1 = 1 - a_2 \tag{19.34}$$

Box 19.1   *Derivation of the Second-Order Runge-Kutta Methods*

The second-order version of Eq. (19.28) is

$$y_{i+1} = y_1 + (a_1 k_1 + a_2 k_2)h \tag{B19.1.1}$$

where

$$k_1 = f(x_i, y_i) \tag{B19.1.2}$$

and

$$k_2 = f(x_i + p_1 h, y_i + q_{11} k_1 h) \tag{B19.1.3}$$

In order to use Eq. (B19.1.1) we have to determine values for the constants a_1, a_2, p_1, and q_{11}. To do this, we recall that the second-order Taylor series for y_{i+1} in terms of y_i and $f(x_i, y_i)$ is written as [Eq. (19.11)]

$$y_{i+1} = y_i + f(x_i, y_i)h + f'(x_i, y_i)\frac{h^2}{2} \tag{B19.1.4}$$

where $f'(x_i, y_i)$ must be determined by chain-rule differentiation (Sec. 19.1.3):

$$f'(x_i, y_i) = \frac{\partial f}{\partial x} + \frac{\partial f}{\partial y}\frac{dy}{dx} \tag{B19.1.5}$$

Substituting Eq. (B19.1.5) into (B19.1.4) gives

$$y_{i+1} = y_i + f(x_i, y_i)h$$
$$+ \left(\frac{\partial f}{\partial x} + \frac{\partial f}{\partial y}\frac{dy}{dx}\right)\frac{h^2}{2} \tag{B19.1.6}$$

The basic strategy underlying Runge-Kutta methods is to use algebraic manipulations to solve for values of a_1, a_2, p_1, and q_{11} that make Eqs. (B19.1.1) and (B19.1.6) equivalent.

To do this, we first use a Taylor series to expand Eq. (19.1.3). The Taylor series for a two-variable function is defined as [recall Eq. (3.37)]

$$g(x + r, y + s) = g(x, y) + r\frac{\partial g}{\partial x} + s\frac{\partial g}{\partial y} + \cdots$$

Applying this method to expand Eq. (B19.1.3) gives

$$f(x_i + p_1 h, y_i + q_{11} k_1 h)$$

$$= f(x_i, y_i) + p_1 h \frac{\partial f}{\partial x} + q_{11} k_1 h \frac{\partial f}{\partial y} + O(h^2)$$

This result can be substituted along with Eq. (B19.1.2) into Eq. (B19.1.1) to yield

$$y_{i+1} = y_i + a_1 h f(x_i, y_i) + a_2 h f(x_i, y_i)$$

$$+ a_2 p_1 h^2 \frac{\partial f}{\partial x} + a_2 q_{11} h^2 f(x_i, y_i)\frac{\partial f}{\partial y}$$

$$+ O(h^3)$$

or, by collecting terms,

$$y_{i+1} = y_i + [a_1 f(x_i, y_i) + a_2 f(x_i, y_i)]h$$

$$+ \left[a_2 p_1 \frac{\partial f}{\partial x} + a_2 q_{11} f(x_i, y_i)\frac{\partial f}{\partial y}\right]h^2$$

$$+ O(h^3) \tag{B19.1.7}$$

Now, comparing like terms in Eqs. (B19.1.6) and (B19.1.7), we determine that in order for the two equations to be equivalent, the following must hold:

$$a_1 + a_2 = 1$$

$$a_2 p_1 = \frac{1}{2}$$

$$a_2 q_{11} = \frac{1}{2}$$

These three simultaneous equations contain the four unknown constants. Because there is one more unknown than the number of equations, there is no unique set of constants that satisfy the equations. However, by assuming a value for one of the constants, we can determine the other three. Consequently, there is a family of second-order methods rather than a single version.

$$p_1 = q_{11} = \frac{1}{2a_2} \tag{19.35}$$

Because we can choose an infinite number of values for a_2, there are an infinite number of second-order RK methods. Every version would yield exactly the same results if the solution to the ODE were quadratic, linear, or a constant. However, they yield different results when (as is typically the case) the solution is more complicated. We present three of the most commonly used and preferred versions:

Heun Method with a Single Corrector $(a_2 = 1/2)$. If a_2 is assumed to be 1/2, Eqs. (19.34) and (19.35) can be solved for $a_1 = 1/2$ and $p_1 = q_{11} = 1$. These parameters, when substituted into Eq. (19.30), yield

$$y_{i+1} = y_i + \left(\frac{1}{2}k_1 + \frac{1}{2}k_2 \right) h \qquad (19.36)$$

where

$$k_1 = f(x_i, y_i) \qquad (19.36a)$$

$$k_2 = f(x_i + h, y_i + hk_1) \qquad (19.36b)$$

Note that k_1 is the slope at the beginning of the interval and k_2 is the slope at the end of the interval. Consequently, this second-order Runge-Kutta method is actually Heun's technique with a single iteration of the corrector.

The Improved Polygon Method $(a_2 = 1)$. If a_2 is assumed to be 1, then $a_1 = 0$, $p_1 = q_{11} = 1/2$, and Eq. (19.30) becomes

$$y_{i+1} = y_i + k_2 h \qquad (19.37)$$

where

$$k_1 = f(x_i, y_i) \qquad (19.37a)$$

$$k_2 = f\left(x_i + \frac{1}{2}h, y_i + \frac{1}{2}hk_1 \right) \qquad (19.37b)$$

This is the improved polygon method.

Ralston's Method $(a_2 = 2/3)$. Ralston (1962) and Ralston and Rabinowitz (1978) determined that choosing $a_2 = 2/3$ provides a minimum bound on the truncation error for the second-order RK algorithms. For this version, $a_1 = 1/3$ and $p_1 = q_{11} = 3/4$:

$$y_{i+1} = y_i + \left(\frac{1}{3}k_1 + \frac{2}{3}k_2 \right) h \qquad (19.38)$$

where

$$k_1 = f(x_i, y_i) \qquad (19.38a)$$

$$k_2 = f\left(x_i + \frac{3}{4}h, y_i + \frac{3}{4}hk_1 \right) \qquad (19.38b)$$

EXAMPLE 19.6   Comparison of Various Second-Order RK Schemes

Problem Statement: Use the improved polygon [Eq. (19.37)] and Ralston's method [Eq. (19.38)] to numerically integrate Eq. (PT6.14):

$$f(x, y) = -2x^3 + 12x^2 - 20x + 8.5$$

from $x = 0$ to $x = 4$ using a step size of 0.5. The initial condition at $x = 0$ is $y = 1$. Compare the results with the values obtained using another second-order RK algorithm: the Heun method with one corrector iteration (Fig. 19.11 and Table 19.3).

Solution: The first step in the improved polygon method is to use Eq. (19.37a) to compute

$$k_1 = -2(0)^3 + 12(0)^2 - 20(0) + 8.5 = 8.5$$

However, because the ODE is a function of x only, this result has no bearing on the second step—the use of Eq. (19.37b) to compute

$$k_2 = -2(0.25)^3 + 12(0.25)^2 - 20(0.25) + 8.5 = 4.21875$$

Notice that this estimate of the slope is much closer to the average value for the interval (4.4375) than the slope at the beginning of the interval (8.5) that would have been used for Euler's approach. The slope at the midpoint can then be substituted into Eq. (19.37) to predict

$$y(0.5) = 1 + 4.21875(0.5) = 3.109375 \qquad \epsilon_t = 3.4\%$$

The computation is repeated, and the results are summarized in Fig. 19.14 and Table 19.3.

Figure 19.14
Comparison of the true solution with numerical solutions using three second-order RK methods and Euler's method.

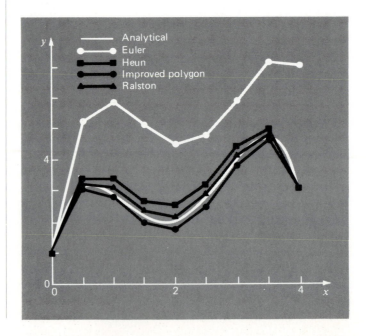

TABLE 19.3
Comparison of true and approximate values of the integral of $y' = -2x^3 + 12x^2 - 20x + 8.5$, with the initial condition that $y = 1$ at $x = 0$. The approximate values were computed using three versions of second-order RK methods with a step size of 0.5.

x	y_{true}	Single-Corrector Heun		Improved Polygon		Second-Order Ralston RK	
		y	$\lvert\epsilon_t\rvert$, %	y	$\lvert\epsilon_t\rvert$, %	y	$\lvert\epsilon_t\rvert$, %
0.0	1.00000	1.00000	0	1.00000	0	1.00000	0
0.5	3.21875	3.43750	6.8	3.109375	3.4	3.27734375	1.8
1.0	3.00000	3.37500	12.5	2.81250	6.3	3.1015625	3.4
1.5	2.21875	2.68750	21.1	1.984375	10.6	2.34765625	5.8
2.0	2.00000	2.50000	25.0	1.75	12.5	2.140625	7.0
2.5	2.71875	3.18750	17.2	2.484375	8.6	2.85546875	5.0
3.0	4.00000	4.37500	9.4	3.81250	4.7	4.1171875	2.9
3.5	4.71875	4.93750	4.6	4.609375	2.3	4.80078125	1.7
4.0	3.00000	3.00000	0	3	0	3.03125	1.0

For Ralston's method, k_1 for the first interval also equals 8.5 and [Eq. (19.38b)]

$$k_2 = -2(0.375)^3 + 12(0.375)^2 - 20(0.375) + 8.5 = 2.58203125$$

The average slope is computed by

$$\phi = \frac{1}{3}(8.5) + \frac{2}{3}(2.58203125) = 4.5546875$$

which can be used to predict

$$y(0.5) = 1 + 4.5546875(0.5) = 3.27734375 \qquad \epsilon_t = -1.82\%$$

The computation is repeated, and the results are summarized in Fig. 19.14 and Table 19.3. Notice how all the second-order RK methods are superior to Euler's method.

19.3.2 Third-Order Runge-Kutta Methods

For $n = 3$, a derivation similar to the one for the second-order method can be performed. The result of this derivation is six equations with eight unknowns. Therefore, values for two of the unknowns must be specified a priori in order to determine the remaining parameters. One common version that results is

$$y_{i+1} = y_i + \left[\frac{1}{6}(k_1 + 4k_2 + k_3)\right]h \qquad (19.39)$$

where

$$k_1 = f(x_i, y_i) \tag{19.39a}$$

$$k_2 = f\left(x_i + \frac{1}{2}h, y_i + \frac{1}{2}hk_1\right) \tag{19.39b}$$

$$k_3 = f(x_i + h, y_i - hk_1 + 2hk_2) \tag{19.39c}$$

Note that if the derivative is a function of x only, this third-order method reduces to Simpson's 1/3 rule. Ralston (1962) and Ralston and Rabinowitz (1978) have developed an alternative version that provides a minimum bound on the truncation error. In any case, the third-order RK methods have local and global errors of $O(h^4)$ and $O(h^3)$, respectively, and yield exact results when the solution is a cubic. As shown in the following example, when dealing with polynomials, Eq. (19.39) will also be exact when the differential equation is cubic and the solution is quartic. This is because Simpson's 1/3 rule provides exact integral estimates for cubics (recall Box 15.3).

EXAMPLE 19.7 *Third-Order RK Method*

Problem Statement: Use Eq. (19.39) to integrate
(*a*) An ODE that is solely a function of x [Eq. (PT6.14)]:

$$\frac{dy}{dx} = -2x^3 + 12x^2 - 20x + 8.5$$

with $y(0) = 1$ and a step size of 0.5.
(*b*) An ODE that is a function of both x and y:

$$\frac{dy}{dx} = 4e^{0.8x} - 0.5y$$

with $y(0) = 2$ from $x = 0$ to 1 with a step size of 1.

Solution: (*a*) Equation (19.39a) through (19.39c) can be used to compute

$$k_1 = -2(0)^3 + 12(0)^2 - 20(0) + 8.5 = 8.5$$
$$k_2 = -2(0.25)^3 + 12(0.25)^2 - 20(0.25) + 8.5 = 4.21875$$
$$k_3 = -2(0.5)^3 + 12(0.5)^2 - 20(0.5) + 8.5 = 1.25$$

which can be substituted into Eq. (19.39) to yield

$$y(0.5) = 1 + \left\{\frac{1}{6}[8.5 + 4(4.21875) + 1.25]\right\}0.5 = 3.21875$$

which is exact. Thus, because the true solution is a fourth-order polynomial [Eq. (PT6.17)], Simpson's 1/3 rule provides an exact result.
(*b*) Equation (19.39a) through (19.39c) can be used to compute

$$k_1 = 4e^{0.8(0)} - 0.5(2) = 3$$

$$k_2 = 4e^{0.8(0.5)} - 0.5[2 + 0.5(1)3] = 4.21729879$$

$$k_3 = 4e^{0.8(1.0)} - 0.5[2 - 1(3) + 2(1)4.21729879] = 5.184864924$$

which can be substituted into Eq. (19.39) to yield

$$y(1.0) = 2 + \left\{ \frac{1}{6}[3 + 4(4.21729879) + 5.184864924] \right\} 1$$

$$= 6.175676681$$

which represents an $\epsilon_t = 0.31$ percent (true value $= 6.19463138$), which is far superior to the results obtained previously with the second-order RK method (that is, Heun without iterations) in Example 19.5.

19.3.3 Fourth-Order Runge-Kutta Methods

The most popular RK methods are fourth order. As with the second-order approaches, there are an infinite number of versions. The following is sometimes called the *classical fourth-order RK method:*

$$y_{i+1} = y_i + \left[\frac{1}{6}(k_1 + 2k_2 + 2k_3 + k_4) \right] h \tag{19.40}$$

where

$$k_1 = f(x_i, y_i) \tag{19.40a}$$

$$k_2 = f\left(x_i + \frac{1}{2}h, \ y_i + \frac{1}{2}hk_1 \right) \tag{19.40b}$$

$$k_3 = f\left(x_i + \frac{1}{2}h, \ y_i + \frac{1}{2}hk_2 \right) \tag{19.40c}$$

$$k_4 = f(x_i + h, \ y_i + hk_3) \tag{19.40d}$$

Notice that for ODEs that are a function of x alone, the classical fourth-order RK method is also equivalent to Simpson's 1/3 rule.

EXAMPLE 19.8   Classical Fourth-Order RK Method

Problem Statement: Use the classical fourth-order RK method [Eq. (19.40)] to integrate

$$f(x, y) = -2x^3 + 12x^2 - 20x + 8.5$$

using a step size of 0.5 and an initial condition of $y = 1$ at $x = 0$.

Solution: Equation (19.40a) through (19.40d) can be used to compute

$$k_1 = -2(0)^3 + 12(0)^2 - 20(0) + 8.5 = 8.5$$

$$k_2 = -2(0.25)^3 + 12(0.25)^2 - 20(0.25) + 8.5 = 4.21875$$

$$k_3 = 4.21875$$

$$k_4 = -2(0.5)^3 + 12(0.5)^2 - 20(0.5) + 8.5 = 1.25$$

which can be substituted into Eq. (19.40) to yield

$$y(0.5) = 1 + \left\{ \frac{1}{6}[8.5 + 2(4.21875) + 2(4.21875) + 1.25] \right\} 0.5$$

$$= 3.21875$$

which is exact. Thus, because the true solution is a quartic [Eq. (PT6.17)], the fourth-order method gives an exact result.

19.3.4 Higher-Order Runge-Kutta Methods

Where more accurate results are required, *Butcher's* (1964) *fifth-order RK method* is recommended:

$$y_{i+1} = y_i + h\left[\frac{1}{90}(7k_1 + 32k_3 + 12k_4 + 32k_5 + 7k_6) \right] \tag{19.41}$$

where

$$k_1 = f(x_i, y_i) \tag{19.41a}$$

$$k_2 = f\left(x_i + \frac{1}{4}h, \ y_i + \frac{1}{4}hk_1 \right) \tag{19.41b}$$

$$k_3 = f\left(x_i + \frac{1}{4}h, \ y_i + \frac{1}{8}hk_1 + \frac{1}{8}hk_2 \right) \tag{19.41c}$$

$$k_4 = f\left(x_i + \frac{1}{2}h, \ y_i - \frac{1}{2}hk_2 + hk_3 \right) \tag{19.41d}$$

$$k_5 = f\left(x_i + \frac{3}{4}h, \ y_i + \frac{3}{16}hk_1 + \frac{9}{16}hk_4 \right) \tag{19.41e}$$

$$k_6 = f\left(x_i + h, \ y_i - \frac{3}{7}hk_1 + \frac{2}{7}hk_2 + \frac{12}{7}hk_3 - \frac{12}{7}hk_4 + \frac{8}{7}hk_5 \right) \tag{19.41f}$$

Note the similarity between Butcher's method and Boole's Rule in Table 15.2. Higher-order RK formulas such as Butcher's method are available, but, in general, beyond fourth-order methods the gain in accuracy is offset by the added computational effort and complexity.

EXAMPLE 19.9   Comparison of Runge-Kutta Methods

Problem Statement: Use first- through fifth-order RK methods to solve

$$\frac{dy}{dx} = 4e^{0.8x} - 0.5y$$

with $y(0) = 2$ from $x = 0$ to $x = 4$ with various step sizes. Compare the accuracy of the various methods for the result at $x = 4$ based on the exact answer of $y(4) = 75.33896261$.

Solution: The computation is performed using Euler's, the uncorrected Heun, the third-order RK [Eq. (19.39)], the classical fourth-order RK, and Butcher's fifth-order RK methods. The results are presented in Fig. 19.15, where we have plotted the absolute value of the percent relative error versus the computational effort. This latter quantity is equivalent to the number of function evaluations required to attain the result, as in

$$\text{Effort} = n_f \frac{b - a}{h} \tag{E19.9.1}$$

where n_f is the number of function evaluations involved in the particular RK computation. For orders ≤ 4, n_f is equal to the order of the method. However, note that Butcher's fifth-order technique requires six function evaluations [Eq. (19.41a) through (19.41f)]. The quantity $(b - a)/h$ is the total integration interval divided by the step size—that is, it is the number of applications of the RK technique required to obtain the result. Thus, because the function evaluations are usually the primary time-consuming steps, Eq. (E19.9.1) provides a rough measure of the run time required to attain the answer.

Figure 19.15
Comparison of percent relative error versus computational effort for first-through fifth-order RK methods.

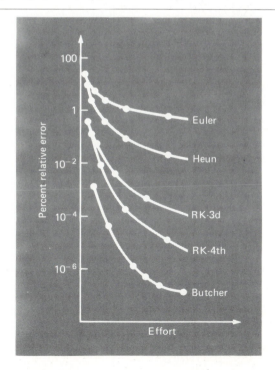

Inspection of Fig. 19.15 leads to a number of conclusions: first, that the higher-order methods attain better accuracy for the same computational effort and, second, that the gain in accuracy for the additional effort tends to diminish after a point. (Notice that the curves drop rapidly at first and then tend to level off.)

Example 19.9 and Fig. 19.15 might lead one to conclude that higher-order RK techniques are always the methods of preference. However, other factors such as programming costs and the accuracy requirements of the problem also must be considered when choosing a solution technique. Such trade-offs will be explored in detail in the case studies in Chap. 22 and in the epilogue for Part Six.

19.3.5 Computer Algorithms for Runge-Kutta Methods

As with all the methods covered in this chapter, the RK techniques fit nicely into the general algorithm embodied in Figs. 19.6 and 19.7. Figure 19.16 presents pseudocode to determine the slope of Ralston's second-order RK method [Eq. (19.38)]. Subroutines to compute slopes for all the other versions can be easily programmed in a similar fashion.

Figure 19.16
Pseudocode to determine the slope of Ralston's second-order RK method.

```
k₁ = dydx(x,y)
x₂ = x + (3/4)·h
y₂ = y + k₁(3/4)h
k₂ = dydx(x₂,y₂)
slope = (1/3)k₁ + (2/3)k₂
```

19.4 SYSTEMS OF EQUATIONS

Many practical problems in engineering and science require the solution of a system of simultaneous differential equations rather than a single equation. Such systems may be represented generally as

$$\frac{dy_1}{dx} = f_1(x, y_1, y_2, \ldots, y_n)$$

$$\frac{dy_2}{dx} = f_2(x, y_1, y_2, \ldots, y_n)$$

$$\vdots$$

$$(19.42)$$

$$\frac{dy_n}{dx} = f_n(x, y_1, y_2, \ldots, y_n)$$

The solution of such a system requires that n initial conditions be known at the starting value of x.

All the methods discussed in this chapter for single equations can be extended to the system shown above. Engineering applications can involve several hundred simultaneous equations. In each case, the procedure for solving a system of equations simply involves applying the one-step technique for every equation at each step before proceeding to the next step. This is best illustrated by the following example.

EXAMPLE 19.10   Solving Systems of ODEs Using Euler's method

Problem Statement: Solve the following set of differential equations using Euler's method, assuming that at $x = 0$, $y_1 = 4$ and $y_2 = 6$. Integrate to $x = 2$ with a step size of 0.5.

$$\frac{dy_1}{dx} = -0.5y_1$$

$$\frac{dy_2}{dx} = 4 - 0.3y_2 - 0.1y_1$$

Solution: Euler's method is implemented as in Eq. (19.2):

$$y_1(0.5) = 4 + [-0.5(4)]0.5 = 3$$

$$y_2(0.5) = 6 + [4 - 0.3(6) - 0.1(4)]0.5 = 6.9$$

Note that $y_1(0) = 4$ is used in the second equation rather than the $y_1(0.5) = 3$ computed with the first equation. Proceeding in a like manner gives

x	y_1	y_2
0	4	6
0.5	3	6.9
1.0	2.25	7.715
1.5	1.6875	8.44525
2.0	1.265625	9.0940875

19.4.1 Computer Algorithm for Solving Systems of ODEs

The computer code for solving a single ODE with Euler's method (Figs. 19.6 and 19.7) can be easily extended to systems of equations. The modifications include:

1. Inputting the number of equations, n
2. Inputting the initial values for each of the n dependent variables
3. Modifying the subroutine so that it computes slopes for each of the dependent variables
4. Including additional functions to compute derivative values for each of the ODEs
5. Including additional equations to compute a new value for each dependent variable

Note that any of the one-step methods in the present chapter could be used for such an algorithm. The only difference would be the formulation of the subroutine to compute the slopes. The classic fourth-order RK method is a good choice for this purpose as it provides excellent accuracy yet is relatively easy to program. An important feature of a computer program to solve systems of ODEs with an RK method is the sequencing of the calculation of the k's as demonstrated in the following example.

EXAMPLE 19.11   Solving Systems of ODEs Using the Fourth-Order RK Method

Problem Statement: Use the fourth-order RK method to solve the ODEs from Example 19.10.

Solution: First, we must solve for all the k_1's:

$$k_{1,1} = f(0, 4, 6) = -0.5(4) = -2$$

$$k_{1,2} = f(0, 4, 6) = 4 - 0.3(6) - 0.1(4) = 1.8$$

where $k_{i,j}$ is the ith value of k for the jth dependent variable. Next, we must calculate the values of y_1 and y_2 that are needed to determine the k_2's:

$$y_1 + \frac{1}{2}hk_{1,1} = 4 + \frac{1}{2}(0.5)(-2) = 3.5$$

$$y_2 + \frac{1}{2}hk_{1,2} = 6 + \frac{1}{2}(0.5)(1.8) = 6.45$$

which can be used to compute

$$k_{2,1} = f(0.25, 3.5, 6.45) = -1.75$$

$$k_{2,2} = f(0.25, 3.5, 6.45) = 1.715$$

The process can be continued to calculate the remaining k's:

$$k_{3,1} = f(0.25, 3.5625, 6.42875) = -1.78125$$

$$k_{3,2} = f(0.25, 3.5625, 6.42875) = 1.715125$$

$$k_{4,1} = f(0.5, 3.109375, 6.8575625) = -1.5546875$$

$$k_{4,2} = f(0.5, 3.109375, 6.8575625) = 1.63179375$$

The values of k can then be used to compute [Eq. (19.40)]:

$$y_1(0.5) = 4 + \frac{1}{6}[-2 + 2(-1.75 - 1.78125) - 1.5546875]0.5$$

$$= 3.11523438$$

$$y_2(0.5) = 6 + \frac{1}{6}[1.8 + 2(1.715 + 1.715125) + 1.63179375]0.5$$

$$= 6.85767032$$

Proceeding in a like manner for the remaining steps yields

x	y_1	y_2
0	4	6
0.5	3.1152344	6.8576703
1.0	2.4261713	7.6321057
1.5	1.8895231	8.3268860
2.0	1.4715768	8.9468651

PROBLEMS

Hand Calculations

19.1 Solve the following initial-value problem analytically over the interval from $x = 0$ to $x = 2$:

$$\frac{dy}{dx} = yx^2 - y$$

where $y(0) = 1$. Plot the solution.

19.2 Use Euler's method with $h = 0.5$ and 0.25 to solve Prob. 19.1. Plot the results on the same graph to visually compare the accuracy for the two step sizes.

19.3 Use Heun's method with $h = 0.5$ and 0.25 to solve Prob. 19.1. Iterate the corrector to $\epsilon_s = 1\%$. Plot the results on the same graph to visually compare the accuracy for the two step sizes with the analytical solution. Interpret your results.

19.4 Use the improved polygon method with $h = 0.5$ and 0.25 to solve Prob. 19.1.

19.5 Use Ralston's second-order RK method with $h = 0.5$ to solve Prob. 19.1.

19.6 Use the classical fourth-order RK method with $h = 0.5$ to solve Prob. 19.1.

19.7 Repeat Probs. 19.1 through 19.6 but for the following initial-value problem over the interval from $x = 0$ to $x = 1$:

$$\frac{dy}{dx} = x\sqrt{y} \qquad y(0) = 1$$

19.8 Use a numerical method to solve

$$\frac{d^2y}{dx^2} + 7y = 0$$

where $y(0) = 2$ and $y'(0) = 0$. Solve from $x = 0$ to 4.

19.9 Solve the following problem numerically:

$$\frac{d^2y}{dx^2} + \frac{dy}{dx} + 5y = 0$$

where $y(0) = 3$ and $y'(0) = 0$. Solve from $x = 0$ to 5.

19.10 Solve the following problem numerically:

$$\frac{dy}{dt} = y \cos (t)$$

where $y(0) = 1$.

19.11 Use a numerical method to solve

$$\frac{dy}{dt} = y + t$$

where $y(0) = 1$.

Computer-Related Problems

19.12 Reprogram Fig. 19.6 or 19.7 so that it is user-friendly. Among other things,
  (a) Place documentation statements throughout the program to identify what each section is intended to accomplish.
  (b) Label the input and output.
  (c) If you choose to implement Fig. 19.7, develop a main program to input and output information and drive the subroutine.

19.13 Test the program you developed in Prob. 19.12 by duplicating the computations from Examples 19.1, 19.3, and 19.4.

19.14 Use the program you developed in Prob. 19.12 to repeat Probs. 19.1 and 19.2.

19.15 Repeat Probs. 19.13 and 19.14, except use the Electronic TOOLKIT software available with the text.

19.16 Develop a user-friendly program for the Heun method with an iterative corrector. Test the program by duplicating the results in Table 19.2.

19.17 Develop a user-friendly computer program for Ralston's second-order RK method. Test the program by duplicating Example 19.6.

19.18 Develop a user-friendly computer program for the classical fourth-order RK method. Test the program by duplicating Example 19.8 and Prob. 19.6.

19.19 Develop a user-friendly computer program for systems of equations using Euler's method. Base your program on the discussion in Sec. 19.4.1. Use this program to duplicate the computation in Example 19.10.

19.20 Repeat Prob. 19.19, but use the fourth-order RK method. For this problem, duplicate the computation from Example 19.11.

CHAPTER 20
Adaptive Step Size Control

To this point, we have presented methods for solving ODEs that employ a constant step size. For a significant number of problems, this can represent a serious limitation. For example, suppose that we are integrating an ODE with a solution of the type depicted in Fig. 20.1. For most of the range, the solution changes gradually. Such behavior suggests that a fairly large step size could be employed to obtain adequate results. However, for a localized region from $x = 1.75$ to $x = 2.25$, the solution undergoes an abrupt change. The practical consequence of dealing with such functions is that a very small step size would be required to accurately capture the impulsive behavior. If a constant step size algorithm were employed, the smaller step size required for the region of abrupt change would have to be applied to the entire computation. As a consequence, a much smaller step size than necessary—and, therefore, many more calculations—would be wasted on the regions of gradual change.

Figure 20.1
An example of a solution of an ODE that exhibits an abrupt change. Automatic step size adjustment has great advantages for such cases.

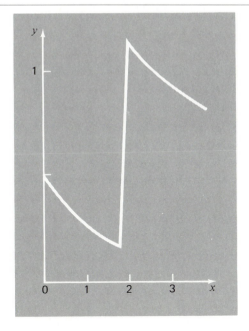

Algorithms that automatically adjust the step size can avoid such overkill and hence be of great advantage. Because they "adapt" to the solution's trajectory, they are said to have *adaptive step size control*. Implementation of such approaches requires that an estimate of the local truncation error be obtained at each step. This error estimate can then serve as a basis for either lengthening or decreasing the step size.

In the present chapter, we will present two such approaches. The first involves adding an adaptive step size control mechanism to the one-step methods from the previous chapter. The second deals with multistep methods.

Before proceeding, we should mention that aside from solving ODEs, the methods described in the present chapter can also be used to evaluate definite integrals. As mentioned previously in the introduction to Part Five, the evaluation of the integral

$$I = \int_a^b f(x)\, dx$$

is equivalent to solving the differential equation

$$\frac{dy}{dx} = f(x)$$

for $y(b)$ given the initial condition $y(a) = 0$. Thus, the following techniques can be employed to efficiently evaluate definite integrals involving functions that are generally smooth but exhibit regions of abrupt change.

20.1   ADAPTIVE RUNGE-KUTTA METHODS

There are two primary approaches to incorporate adaptive step size control into one-step methods. In the first, the local truncation error is estimated as the difference between two predictions using different-order RK methods. We will discuss such an approach in Sec. 20.1.4. In the second approach, the error is estimated as the difference between two predictions using the same-order RK method but with different step sizes. This approach, which is called step doubling, is described next.

20.1.1   Step Doubling

Step doubling involves taking each step twice, once as a full step and independently as two half steps. The difference in the two results represents an estimate of the local truncation error. If y_1 designates the single-step prediction and y_2 designates the prediction using the two half steps, the error Δ can be represented as

$$\Delta = y_2 - y_1 \tag{20.1}$$

In addition to providing a criterion for step size control, Eq. (20.1) can also be used to correct the y_2 prediction. For the fourth-order RK version, the correction is

$$y_2 \leftarrow y_2 + \frac{\Delta}{15} \qquad (20.2)$$

This estimate is fifth-order accurate.

EXAMPLE 20.1   Adaptive Fourth-Order RK Method

Problem Statement: Use the adaptive fourth-order RK method to integrate $y' = 4e^{0.8x} - 0.5y$ from $x = 0$ to 2 using $h = 2$ and an initial condition of $y(0) = 2$. This is the same differential equation that was solved previously in Example 19.5. Recall that the true solutions were $y(1) = 6.19463138$ and $y(2) = 14.8439219$.

Solution: The single prediction with a step of h is computed as

$$y(2) = 2 + \frac{1}{6} [3 + 2(6.40216 + 4.70108) + 14.11105]2 = 15.10584$$

The two half-step predictions are

$$y(1) = 2 + \frac{1}{6} [3 + 2(4.21730 + 3.91297) + 5.94568]1 = 6.20104$$

and

$$y(2) = 6.20104 + \frac{1}{6} [5.80164 + 2(8.72954 + 7.99756) + 12.71283]1 = 14.86249$$

Therefore, the approximate error is

$$E_a = \frac{14.86249 - 15.10584}{15} = -0.01622$$

which compares favorably with the true error of

$$E_t = 14.8439219 - 14.86249 = -0.01857$$

The error estimate can also be used to correct the prediction

$$y(2) = 14.86249 - 0.01622 = 14.84627$$

which has $E_t = -0.00235$.

20.1.2 Step Size Control

Now that we have an error estimate, it can be used to adjust the step size. In general, the strategy is to increase the step size if the error is too small and decrease it if the error is too large. Press et al. (1986) have suggested the following criterion to accomplish this:

$$h_{new} = h_{present} \left| \frac{\Delta_{new}}{\Delta_{present}} \right|^{\alpha} \tag{20.3}$$

where $h_{present}$ and h_{new} are the present and the new step sizes, $\Delta_{present}$ is the present accuracy as computed by Eq. (20.1), Δ_{new} is the desired accuracy, and α is a constant power that is equal to 0.2 when the step size is increased (i.e., when $\Delta_{present} \leq \Delta_{new}$) and 0.25 when the step size is decreased ($\Delta_{present} > \Delta_{new}$).

The key parameter in Eq. (20.3) is obviously Δ_{new} because it is your vehicle for specifying the desired accuracy. One way to do this would be to relate Δ_{new} to a relative error level. Although this works well when only positive values occur, it can cause problems for solutions that pass through zero. For example, you might be simulating an oscillating function that repeatedly passes through zero but is bounded by maximum absolute values. For such a case, you might want these maximum values to figure in the desired accuracy.

A more general way to handle such cases is to determine Δ_{new} as

$$\Delta_{new} = \epsilon \, y_{scale}$$

where ϵ is an overall tolerance level. Your choice of y_{scale} will then determine how the error is scaled. For example, if $y_{scale} = y$, the accuracy will be couched in terms of fractional relative errors. If you are dealing with a case where you desire constant errors relative to a prescribed maximum bound, set y_{scale} equal to that bound. A trick suggested by Press et al. (1986) to obtain the constant relative errors except very near zero crossings is

$$y_{scale} = |y| + \left| h \frac{dy}{dx} \right|$$

This is the version we will use in our algorithm.

20.1.3 Computer Algorithm

Figure 20.2 outlines pseudocode to implement the algorithm delineated in Sec. 20.1.1 and 20.1.2. As illustrated in the following example, the pseudocode can be expressed as a program for solving an ODE.

EXAMPLE 20.2   Computer Application of an Adaptive Fourth-Order RK Scheme

Problem Statement: The following nonhomogeneous ordinary differential equation is well-suited for an adaptive integration scheme

$$\frac{dy}{dx} + 0.6y = 10e^{-(x-2)^2/[2(0.075)^2]} \tag{E20.2.1}$$

Notice for the initial condition, $y(0) = 0.5$, the general solution is

$$y = 0.5e^{-0.6x}$$

```
INPUT maxstep (max number of steps)
INPUT x_i , x_f (initial and final x)
INPUT y_i (initial y)
INPUT h_i (initial step size)
INPUT ε (error tolerance)
OUTPUT x_i , y_i
x = x_i ; y = y_i ; h = h_i
istep = 0
DOWHILE (istep ≤ maxstep) and (x < x_f)
    istep = istep + 1
    y_scale = |y| + |h dy/dx|
    IF (x + h > x_f ) THEN h = x_f - x
    CALL ADAPT (x,y,h,y_scale,ε,h_next)
    OUTPUT x,y
    h = h_next
ENDDO
END
```

```
SUBROUTINE ADAPT (x,y,h,y_scale,ε,h_next)
safety = 0.9 ; ε_con = 6 × 10^-4
ε_max = 1.1
h_1 = _____h_____
        (ε_max^-0.25) safety
DOWHILE (ε_max > 1)
    h_1 = safety·ε_max^-0.25 ·h_1
    h_2 = 0.5 h_1
    CALL RK4 (x, y, h_1, y_1)
    CALL RK4 (x, y, h_2, y_2)
    CALL RK4 (x + h_2, y, h_2, y_2)
    ε_max = |(y_2 - y_1)/ y_scale |/ε
ENDDO
y = y_2 ; x = x + h_1
IF (ε_max > ε_con)
    h_next = safety·ε_max^-0.2 ·h_1
ELSE
    h_next = 4h_1
ENDIF
y = y + (y_2 - y_1)/15
RETURN
```

Figure 20.2
Pseudocode for an adaptive fourth-order RK method to solve a single ODE.

which is a smooth curve that gradually approaches zero as x increases. In contrast, the particular solution undergoes an abrupt transition in the vicinity of $x = 2$ due to the nature of the forcing function (Fig. 20.3a). Use a standard fourth-order RK scheme to solve Eq. (E20.2.1) from $x = 0$ to 4. Then employ the adaptive scheme described in this section to perform the same computation.

Solution: First, the classical fourth-order scheme is used to compute the solid curve in Fig. 20.3b. For this computation, a step size of 0.1 is used so that $4/(0.1) = 40$ applications of the technique are made. Then, the calculation is repeated with a step size of 0.05 for a total of 80 applications. The major discrepancy between the two results occurs in the region from 1.8 to 2.0. The magnitude of the discrepancy is about 0.1 to 0.2 percent.

Next, the algorithm in Fig. 20.2 is developed into a computer program and used to solve the same problem. An initial step size of 0.5 and an $ε = 0.001$ were chosen. The results were superimposed on Fig. 20.3b. Notice how large steps are taken in the regions of gradual change. Then, in the vicinity of $x = 2$ the steps are decreased to accommodate the abrupt nature of the forcing function. A minimum step size of 0.0685 is employed at this point. As with the constant step size version, the error incurred at this

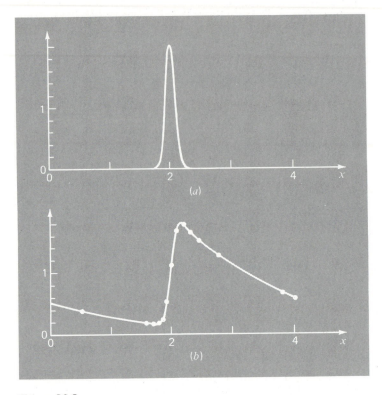

Figure 20.3
(a) A bell-shaped forcing function that induces an abrupt change in the solution of an ODE [Eq. (E20.2.1)]. (b) The solution. The points indicate the predictions of an adaptive step size routine.

point is on the order of 0.1 to 0.2 percent. Note that even though only 14 final steps are taken, additional steps are incurred during step size adjustment. Three RK evaluations are required for each step, and a total of 54 applications are made. Thus, the effort involved is roughly equivalent to the constant step size applications.

The utility of an adaptive integration scheme obviously depends on the nature of the functions being modeled. It is particularly advantageous for those solutions with long smooth stretches and short regions of abrupt change. In addition, it has utility in those situations where the correct step size is not known a priori. For these cases, an adaptive routine will "feel" its way through the solution while keeping the results within the desired tolerance. Thus, it will tiptoe through the regions of abrupt change and step out briskly when the variations become more gradual.

Aside from the scheme outlined in this section, there are a variety of other one-step algorithms involving adaptive step size control. For example, Press et al. (1986) present computer code for an adaptive fourth-order RK solver for systems of ODEs. Additional information can be found in Gear (1971).

20.1.4 Runge-Kutta Fehlberg

Aside from step doubling, an alternative approach for obtaining an error estimate involves computing two RK predictions of different order. The results can then be subtracted to obtain an estimate of the local truncation error. One shortcoming of this approach is that it greatly increases the computational overhead. For example, a fourth- and fifth-order prediction amount to a total of 10 function evaluations per step. The *Runge-Kutta Fehlberg* method cleverly circumvents this problem by using a fifth-order RK method that employs the function evaluations from the accompanying fourth-order RK method. Thus, the approach yields the error estimate on the basis of only six function evaluations!

$$y_{i+1} = y_i + \left(\frac{25}{216} k_1 + \frac{1408}{2565} k_3 + \frac{2197}{4104} k_4 - \frac{1}{5} k_5 \right) h \tag{20.4}$$

along with a fifth-order formula:

$$y_{i+1} = y_i + \left(\frac{16}{135} k_1 + \frac{6656}{12,825} k_3 + \frac{28,561}{56,430} k_4 - \frac{9}{50} k_5 + \frac{2}{55} k_6 \right) h \tag{20.5}$$

where

$$k_1 = 2f(x_i, y_i)$$

$$k_2 = f\left(x_i + \frac{1}{4} h, \; y_i + \frac{1}{4} hk_1 \right)$$

$$k_3 = f\left(x_i + \frac{3}{8} h, \; y_i + \frac{3}{32} hk_1 + \frac{9}{32} hk_2 \right)$$

$$k_4 = f\left(x_i + \frac{12}{13} h, \; y_i + \frac{1932}{2197} hk_1 - \frac{7200}{2197} hk_2 + \frac{7296}{2197} hk_3 \right)$$

$$k_5 = f\left(x_i + h, \; y_i + \frac{439}{216} hk_1 - 8hk_2 + \frac{3680}{513} hk_3 - \frac{845}{4104} hk_4 \right)$$

$$k_6 = f\left(x_i + \frac{1}{2} h, \; y_i - \frac{8}{27} hk_1 + 2hk_2 - \frac{3544}{2565} hk_3 \right.$$
$$\left. + \frac{1859}{4104} hk_4 - \frac{11}{40} hk_5 \right)$$

The error estimate is obtained by subtracting Eq. (20.4) from Eq. (20.5) to yield

$$E_a = \left(\frac{1}{360} k_1 - \frac{128}{4275} k_3 - \frac{2197}{75,240} k_4 + \frac{1}{50} k_5 + \frac{2}{55} k_6 \right) h \tag{20.6}$$

Thus, the ODE can be solved with Eq. (20.4) and the error estimated from Eq. (20.6). Note, that after each step, Eq. (20.6) can be added to Eq. (20.4) to make the result fifth-order [Eq. (20.5)]. Further information on this approach and software for its implementation can be found in Maron (1982), Gerald and Wheatley (1984), and Rice (1983).

20.2   MULTISTEP METHODS

The *one-step methods* described in the previous sections utilize information at a single point x_i to predict a value of the dependent variable y_{i+1} at a future point x_{i+1} (Fig. 20.4a). Alternative approaches, called *multistep methods* (Fig. 20.4b), are based on the insight that, once the computation has begun, valuable information from previous points is at our command. The curvature of the lines connecting these previous values provides information regarding the trajectory of the solution. The multistep methods explored in the present chapter exploit this information in order to solve ODEs and evaluate their error. Before describing the higher-order versions, we will present a simple second-order method that serves to demonstrate the general characteristics of multistep approaches.

20.2.1   The Non-Self-Starting Heun Method

Recall that the Heun approach uses *Euler's method* as a *predictor* [Eq. (19.15)]:

$$y_{i+1}^0 = y_i + f(x_i, y_i)h \tag{20.7}$$

and the *trapezoidal rule* as a *corrector* [Eq. (19.16)]:

$$y_{i+1} = y_i + \frac{f(x_i, y_i) + f(x_{i+1}, y_{i+1}^0)}{2} h \tag{20.8}$$

Thus, the predictor and the corrector have local truncation errors of $O(h^2)$ and $O(h^3)$, respectively. This suggests that the predictor is the weak link in the method because it

Figure 20.4
Graphical depiction of the fundamental difference between (a) one-step and (b) multistep methods for solving ODEs.

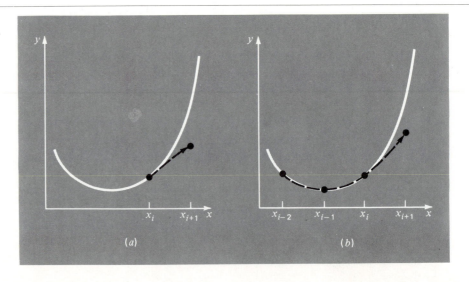

has the greatest error. This weakness is significant because the efficiency of the iterative corrector step depends on the accuracy of the initial prediction. Consequently, one way to improve Heun's method is to develop a predictor that has a local error of $O(h^3)$. This can be accomplished by using Euler's method and the slope at y_i, but making the prediction from a previous point y_{i-1}, as in

$$y_{i+1}^0 = y_{i-1} + f(x_i, y_i)2h \tag{20.9}$$

Notice that Eq. (20.9) attains $O(h^3)$ at the expense of employing a larger step size, $2h$. In addition, note that Eq. (20.9) is not self-starting because it involves a previous value of the dependent variable y_{i-1}. Such a value would not be available in a typical initial-value problem. Because of this fact, Eqs. (20.8) and (20.9) are called the *non-self-starting Heun method*.

Notice that, as depicted in Fig. 20.5, the derivative estimate in Eq. (20.9) is now located at the midpoint rather than at the beginning of the interval over which the prediction is made. As demonstrated subsequently, this centering improves the error of the predictor to $O(h^3)$. However, before proceeding to a formal derivation of the

Figure 20.5
A graphical depiction of the non-self-starting Heun method. (*a*) The midpoint method that is used as a predictor. (*b*) The trapezoidal rule that is employed as a corrector.

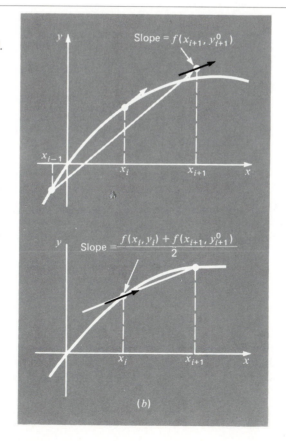

non-self-starting Heun, we will summarize the method and express it using a slightly modified nomenclature:

$$\text{Predictor:} \quad y_{i+1}^0 = y_{i-1}^m + f(x_i, y_i^m) 2h \tag{20.10}$$

$$\text{Corrector:} \quad y_{i+1}^j = y_i^m + \frac{f(x_i, y_i^m) + f(x_{i+1}, y_{i+1}^{j-1})}{2} h \tag{20.11}$$

$$\text{(for } j = 1, 2, \ldots, m)$$

where the superscripts have been added to denote that the corrector is applied iteratively from $j = 1$ to m in order to obtain refined solutions. Note that y_i^m and y_{i-1}^m are the final results of the corrector iterations at the previous time steps. The iterations are terminated at any time step on the basis of the stopping criterion

$$|\epsilon_a| = \left| \frac{y_{i+1}^j - y_{i+1}^{j-1}}{y_{i+1}^j} \right| 100\% \tag{20.12}$$

When ϵ_a is less than a prespecified error tolerance ϵ_s, the iterations are terminated. At this point, $j = m$. The use of Eqs. (20.10) through (20.12) to solve an ODE is demonstrated in the following example.

EXAMPLE 20.3   Non-Self-Starting Heun Method

Problem Statement: Use the non-self-starting Heun method to perform the same computations as were performed previously in Example 19.5 using Heun's method. That is, integrate $y' = 4e^{0.8x} - 0.5y$ from $x = 0$ to $x = 4$ using a step size of 1.0. As with Example 19.5, the initial condition at $x = 0$ is $y = 2$. However, because we are now dealing with a multistep method, we require the additional information that y is equal to -0.392995325 at $x = -1$.

Solution: The predictor [Eq. (20.10)] is used to extrapolate linearly from $x = -1$ to $x = 1$:

$$y_1^0 = -0.392995325 + [4e^{0.8(0)} - 0.5(2)]2 = 5.607004675$$

The corrector [Eq. (20.11)] is then used to compute the value:

$$y_1^1 = 2 + \frac{[4e^{0.8(0)} - 0.5(2) + 4e^{0.8(1)} - 0.5(5.607004675)]}{2} 1$$

$$= 6.549330688$$

which represents a percent relative error of -5.73 percent (true value $= 6.194631377$). This error is somewhat smaller than the value of -8.18 percent incurred in the self-starting Heun.

Now Eq. (20.11) can be applied iteratively to improve the solution:

$$y_1^2 = 2 + \frac{3 + 4e^{0.8(1)} - 0.5(6.549330688)}{2} 1 = 6.313749185$$

which represents an ϵ_t of -1.92%. An approximate estimate of the error can also be determined using Eq. (20.12):

$$|\epsilon_a| = \left| \frac{6.313749185 - 6.549330688}{6.313749185} \right| 100\% = 3.7\%$$

Equation (20.11) can be applied iteratively until ϵ_a falls below a prespecified value of ϵ_s. As was the case with the Heun method (recall Example 19.5), the iterations converge on a value of 6.36086549 ($\epsilon_t = -2.68\%$). However, because the initial predictor value is more accurate, the multistep method converges at a somewhat faster rate.

For the second step, the predictor is

$$y_2^0 = 2 + [4e^{0.8(1)} - 0.5(6.36086549)]\,2$$

$$= 13.4434619 \qquad \epsilon_t = 9.43\%$$

which is superior to the prediction of 12.08259646 ($\epsilon_t = 18\%$) that was computed with the original Heun method. The first corrector yields 15.95539553 ($\epsilon_t = -6.8\%$), and subsequent iterations converge on the same result as was obtained with the self-starting Heun method: 15.3022367 ($\epsilon_t = -3.1\%$). As with the previous step, the rate of convergence of the corrector is somewhat improved because of the better initial prediction.

Derivation and Error Analysis of Predictor-Corrector Formulas. We have just employed graphical concepts to derive the non-self-starting Heun. We will now show how the same equations can be derived mathematically. This derivation is particularly interesting because it ties together ideas from curve fitting, numerical integration, and ODEs. The exercise is also useful because it provides a simple procedure for developing higher-order multistep methods and estimating their errors.

The derivation is based on solving the general ODE

$$\frac{dy}{dx} = f(x, y)$$

This equation can be solved by multiplying both sides by dx and integrating between limits at i and $i + 1$:

$$\int_{y_i}^{y_{i+1}} dy = \int_{x_i}^{x_{i+1}} f(x, y)\, dx$$

The left side can be integrated and evaluated using the fundamental theorem [recall Eq. (19.21)]:

$$y_{i+1} = y_i + \int_{x_i}^{x_{i+1}} f(x, y)\, dx \tag{20.13}$$

Equation (20.13) represents a solution to the ODE if the integral can be evaluated. That is, it provides a means to compute a new value of the dependent variable y_{i+1} on the basis of a prior value y_i and the differential equation.

Numerical integration formulas such as those developed in Chap. 15 provide one way to make this evaluation. For example, the trapezoidal rule [Eq. (15.3)] can be used to evaluate the integral, as in

$$\int_{x_i}^{x_{i+1}} f(x, y) \, dx \simeq \frac{f(x_i, y_i) + f(x_{i+1}, y_{i+1})}{2} h \tag{20.14}$$

where $h = x_{i+1} - x_i$ is the step size. Substituting Eq. (20.14) into Eq. (20.13) yields

$$y_{i+1} = y_i + \frac{f(x_i, y_i) + f(x_{i+1}, y_{i+1})}{2} h$$

which is the corrector equation for the Heun method. Because this equation is based on the trapezoidal rule, the truncation error can be taken directly from Table 15.2.

$$E_c = -\frac{1}{12} h^3 y'''(\xi_c) = -\frac{1}{12} h^3 f''(\xi_c) \tag{20.15}$$

where the subscript c designates that this is the error of the corrector.

A similar approach can be used to derive the predictor. For this case, the integration limits are from $i - 1$ to $i + 1$:

$$\int_{y_{i-1}}^{y_{i+1}} dy = \int_{x_{i-1}}^{x_{i+1}} f(x, y) \, dx$$

which can be integrated and rearranged to yield

$$y_{i+1} = y_{i-1} + \int_{x_{i-1}}^{x_{i+1}} f(x, y) \, dx \tag{20.16}$$

Now, rather than using a closed formula from Table 15.2, the first Newton-Cotes open integration formula (see Table 15.4) can be used to evaluate the integral, as in

$$\int_{x_{i-1}}^{x_{i+1}} f(x, y) \, dx = 2h f(x_i, y_i) \tag{20.17}$$

which is called the *midpoint method*. Substituting Eq. (20.17) into Eq. (20.16) yields

$$y_{i+1} = y_{i-1} + f(x_i, y_i) 2h$$

which is the predictor for the non-self-starting Heun. As with the corrector, the local truncation error can be taken directly from Table 15.4:

$$E_p = \frac{1}{3} h^3 y'''(\xi_p) = \frac{1}{3} h^3 f''(\xi_p) \tag{20.18}$$

where the subscript p designates that this is the error of the predictor.

Thus, the predictor and the corrector for the non-self-starting Heun method have truncation errors of the same order. Aside from upgrading the accuracy of the predictor, this fact has additional benefits related to error analysis, as elaborated in the next section.

Error Estimates.   If the predictor and the corrector of a multistep method are of the same order, the local truncation error may be estimated during the course of a computation. This is a tremendous advantage because it establishes a criterion for adjustment of the step size.

The local truncation error for the predictor is estimated by Eq. (20.18). This error estimate can be combined with the estimate of y_{i+1} from the predictor step to yield [recall our basic definition of Eq. (3.1)]

$$\text{True value} = y_{i+1}^0 + \frac{1}{3} h^3 y'''(\xi_p) \tag{20.19}$$

Using a similar approach, the error estimate for the corrector [Eq. (20.15)] can be combined with the true value and the corrector result y_{i+1} to give

$$\text{True value} = y_{i+1}^m - \frac{1}{12} h^3 y'''(\xi_c) \tag{20.20}$$

Equation (20.19) can be subtracted from Eq. (20.20) to yield

$$0 = y_{i+1}^m - y_{i+1}^0 - \frac{5}{12} h^3 y'''(\xi) \tag{20.21}$$

where ξ is between x_{i-1} and x_{i+1}. Now dividing Eq. (20.21) by 5 and rearranging the result gives

$$\frac{y_{i+1}^0 - y_{i+1}^m}{5} = -\frac{1}{12} h^3 y'''(\xi) \tag{20.22}$$

Notice that the right-hand sides of Eqs. (20.15) and (20.22) are identical, with the exception of the argument of the third derivative. If the third derivative does not vary appreciably over the interval in question, we can assume that the right-hand sides are equal, and, therefore, the left-hand sides should also be equivalent, as in

$$E_c \simeq -\frac{y_{i+1}^m - y_{i+1}^0}{5} \tag{20.23}$$

Thus, we have arrived at a relationship that can be used to estimate the per-step truncation error on the basis of two quantities—the predictor (y_{i+1}^0) and the corrector (y_{i+1}^m), which are routine by-products of the computation.

EXAMPLE 20.4   Estimate of Per-Step Truncation Error for the Non-Self-Starting Heun Method

Problem Statement: Use Eq. (20.23) to estimate the per-step truncation error of Example 20.3. Note that the true values at $x = 1$ and 2 are 6.19463138 and 14.8439219, respectively.

Solution: At $x_{i+1} = 1$, the predictor gives 5.607004675 and the corrector yields 6.36086549. These values can be substituted into Eq. (20.23) to give

$$E_c = -\frac{6.36086549 - 5.607004675}{5} = -0.150772163$$

which compares well with the exact error,

$$E_t = 6.19463138 - 6.36086549 = -0.166234110$$

At $x_{i+1} = 2$, the predictor gives 13.4434619 and the corrector yields 15.3022367, which can be used to compute

$$E_c = -\frac{15.3022367 - 13.4434619}{5} = -0.371754960$$

which also compares favorably with the exact error, $E_t = 14.8439219 - 15.3022367 = -0.4583148$.

The ease with which the error can be estimated using Eq. (20.23) provides a rational basis for step-size adjustment during the course of a computation. For example, if Eq. (20.23) indicates that the error is greater than an acceptable level, the step size could be decreased.

Modifiers. Before developing computer algorithms, we must note two other ways in which the non-self-starting Heun method can be made more accurate and efficient. First, you should realize that besides providing a criterion for step size adjustment, Eq. (20.23) represents a numerical estimate of the discrepancy between the final corrected value at each step y_{i+1} and the true value. Thus, it can be added directly to y_{i+1} to refine the estimate further:

$$y_{i+1}^m \leftarrow y_{i+1}^m - \frac{y_{i+1}^m - y_{i+1}^0}{5} \tag{20.24}$$

Equation (20.24) is called a *corrector modifier*. (The symbol $\leftarrow$ is read "is replaced by.") The left-hand side is the modified value of y_{i+1}^m.

A second improvement, one that relates more to program efficiency, is a *predictor modifier,* which is designed to adjust the predictor result so that it is closer to the final convergent value of the corrector. This is advantageous because, as noted previously at the beginning of this section, the number of iterations of the corrector is highly dependent on the accuracy of the initial prediction. Consequently, if the prediction is modified properly, we might reduce the number of iterations required to converge on the ultimate value of the corrector.

Such a modifier can be derived simply by assuming that the third derivative is relatively constant from step to step. Therefore, using the result of the previous step at i, Eq. (20.22) can be solved for

$$h^3 y'''(\xi) = -\frac{12}{5}(y_i^0 - y_i^m) \tag{20.25}$$

which, assuming that $y'''(\xi) \simeq y'''(\xi_p)$, can be substituted into Eq. (20.18) to give

$$E_p = \frac{4}{5}(y_i^m - y_i^0) \tag{20.26}$$

which can then be used to modify the predictor result:

$$y_{i+1}^0 \leftarrow y_{i+1}^0 + \frac{4}{5}(y_i^m - y_i^0) \tag{20.27}$$

EXAMPLE 20.5   Effect of Modifiers on Predictor-Corrector Results

Problem Statement: Recompute Example 20.3 using the modifiers as specified in Fig. 20.6.

Solution: As in Example 20.3, the initial predictor result is 5.607004675. Because the predictor modifier [Eq. (20.27)] requires values from a previous iteration, it cannot be employed to improve this initial result. However, Eq. (20.24) can be used to modify the corrected value of 6.36086549 ($\epsilon_t = -2.684\%$), as in

$$y_1^m = 6.36086549 - \frac{6.36086549 - 5.607004675}{5} = 6.210093327$$

which represents an $\epsilon_t = -0.25\%$. Thus, the error is reduced over an order of magnitude.
    For the next iteration, the predictor [Eq. (20.10)] is used to compute

$$y_2^0 = 2 + [4e^{0.8(1)} - 0.5(6.210093327)]2$$

$$= 13.59423410 \qquad \epsilon_t = 8.42\%$$

which is about half the error of the predictor for the second iteration of Example 20.3, which was $\epsilon_t = 18.6\%$. This improvement is due to the fact that we are using a superior estimate of y (6.210093327, as opposed to 6.36086549) in the predictor. In other words, the propagated and global errors are reduced by the inclusion of the corrector modifier.
    Now because we have information from the prior iteration, Eq. (20.27) can be employed to modify the predictor, as in

$$y_2^0 = 13.59423410 + \frac{4}{5}(6.36086549 - 5.607004675)$$

$$= 14.19732275 \qquad \epsilon_t = -4.36\%$$

which, again, halves the error.
    This modification has no effect on the final outcome of the subsequent corrector step. Regardless of whether the unmodified or modified predictors are used, the corrector will ultimately converge on the same answer. However, because the rate or efficiency of convergence depends on the accuracy of the initial prediction, the modification can reduce the number of iterations required for convergence.
    Implementing the corrector yields a result of 15.21177723 ($\epsilon_t = -2.48\%$) which represents an improvement over Example 20.3 because of the reduction of global error. Finally, this result can be modified using Eq. (20.24):

$$y_2^m = 15.21177723 - \frac{15.21177723 - 13.59423410}{5}$$

$$= 14.88826860 \qquad \epsilon_t = -0.30\%$$

Again, the error has been reduced an order of magnitude.

Predictor:

$$y_{i+1}^0 = y_{i-1}^m + f(x_i, y_i^m)2h$$

(Save result as $y_{i+1,u}^0 = y_{i+1}^0$ where the subscript u designates that the variable is unmodified.)

Predictor modifier:

$$y_{i+1}^0 \leftarrow y_{i+1,u}^0 + \frac{4}{5}(y_{i,u}^m - y_{i,u}^0)$$

Corrector:

$$y_{i+1}^j = y_i^m + \frac{f(x, y_i^m) + f(x_{i+1}, y_{i+1}^{j-1})}{2}h$$

(for $j = 1$ to maximum iterations m)

Error check:

$$|\epsilon_a| = \left|\frac{y_{i+1}^j - y_{i+1}^{j-1}}{y_{i+1}^j}\right| 100\%$$

(If $|\epsilon_a| >$ error criterion, set $j = j + 1$ and repeat corrector; if $\epsilon_a \leq$ error criterion, save result as $y_{i+1,u}^m = y_{i+1}^m$.)

Corrector error estimate:

$$E_c = -\frac{1}{5}(y_{i+1,u}^m - y_{i+1,u}^0)$$

(If computation is to continue, set $i = i + 1$ and return to predictor.)

Figure 20.6
The sequence of formulas used to implement the non-self-starting Heun method. Note that the corrector error estimates can be used to modify the corrector. However, because this can affect the corrector's stability, the modifier is not included in this algorithm. The corrector error estimate is included because of its utility for step-size adjustment.

As in the previous example, the addition of the modifiers increases both the efficiency and accuracy of multistep methods. In particular, the corrector modifier effectively increases the order of the technique. Thus, the non-self-starting Heun with modifiers is third order rather than second order as is the case for the unmodified version. However, it should be noted that there are situations where the corrector modifier will affect the stability of the corrector iteration process. As a consequence, the modifier is not included in the algorithm for the non-self-starting Heun delineated in Fig. 20.6. Nevertheless, the corrector modifier can still have utility for step size control as discussed next.

20.2.2 Step Size Control and Computer Programs

Constant Step Size.   It is relatively simple to develop a constant step size version of the non-self-starting Heun method. About the only complication is that a one-step method is required to generate the extra point required to start the computation.

Additionally, because a constant step size is employed, a value for h must be chosen prior to the computation. In general, experience indicates that an optimal step size should be small enough to ensure convergence within two iterations of the corrector (Hull and Creemer, 1963). In addition, it must be small enough to yield a sufficiently small truncation error. At the same time, the step size should be as large as possible to minimize run-time cost and round-off error. As with other methods for ODEs, the only practical way to assess the magnitude of the global error is to compare the results for the same problem but with a halved step size.

Variable Step Size. Two criteria are typically used to decide whether a change in step size is warranted. First, if Eq. (20.23) is greater than some prespecified error criterion, the step size is decreased. Second, the step size is chosen so that the convergence criterion of the corrector is satisfied in two iterations. This criterion is intended to account for the trade-off between the rate of convergence and the total number of steps in the calculation. For smaller values of h, convergence will be more rapid but more steps are required. For larger h, convergence is slower but fewer steps result. Experience (Hull and Creemer, 1963) suggests that the total steps will be minimized if h is chosen so that the corrector converges within two iterations. Therefore, if over two iterations are required, the step size is decreased, and if less than two iterations are required, the step size is increased.

Although the above strategy specifies when step size modifications are in order, it does not indicate *how* they should be changed. This is a critical question because multistep methods by definition require several points in order to compute a new point. Once the step size is changed, a new set of points must be determined. One approach is to restart the computation and use the one-step method to generate a new set of starting points.

A more efficient strategy that makes use of presently existing information is to increase and decrease by doubling and halving the step size. As depicted in Fig. 20.7*b*, if a sufficient number of previous values have been generated, increasing the step size by doubling is a relatively straightforward task (Fig. 20.7*c*). All that is necessary is to keep track of subscripts so that old values of x and y become the appropriate new values. Halving the step size is somewhat more complicated because some of the new values will be unavailable (Fig. 20.7*a*). However, interpolating polynomials of the type developed in Chap. 12 can be used to determine these intermediate values.

In any event, the decision to incorporate step size control represents a trade-off between initial investment in program complexity versus the long-term return because of increased efficiency. Obviously, the magnitude and importance of the problem itself will have a strong bearing on this trade-off.

20.2.3 Integration Formulas

The non-self-starting Heun method is characteristic of most multistep methods. It employs an open integration formula (the midpoint method) to make an initial estimate. This predictor step requires a previous data point. Then, a closed integration formula (the trapezoidal rule) is applied iteratively to improve the solution.

It should be obvious that a strategy for improving multistep methods would be to use higher-order integration formulas as predictors and correctors. For example, the higher-order Newton-Cotes formulas developed in Chap. 15 could be used for this purpose.

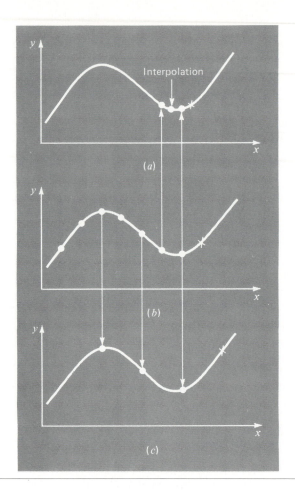

Figure 20.7
A plot indicating how a halving-doubling strategy allows the use of (*b*) previously calculated values for a third-order multistep method. (*a*) Halving; (*c*) doubling.

Before describing these methods, we will review the most common integration formulas upon which they are based. As mentioned above, the first of these are the Newton-Cotes formulas. However, there is a second class called the Adams formulas which we will also review and which are often preferred. As depicted in Fig. 20.8, the fundamental difference between the Newton-Cotes and the Adams formulas relates to the manner in which the integral is applied to obtain the solution. As depicted in Fig. 20.8*a*, the Newton-Cotes formulas estimate the integral over an interval spanning several points. This integral is then used to project from the beginning of the interval to the end. In contrast, the Adams formulas (Fig. 20.8*b*) use a set of points from an interval to estimate the integral solely for the last segment in the interval. This integral is then used to project across this last segment.

Newton-Cotes Formulas.   Some of the most common formulas for solving ordinary differential equations are based on fitting an *n*th-degree interpolating polynomial to $n + 1$ known values of *y* and then using this equation to compute the integral. As discussed

$$y_{i+1} = y_i + \frac{h}{2}(f_i + f_{i+1})$$

which is equivalent to the trapezoidal rule. For $n = 2$,

$$y_{i+1} = y_{i-1} + \frac{h}{3}(f_{i-1} + 4f_i + f_{i+1}) \tag{20.32}$$

which is equivalent to Simpson's 1/3 rule. Equation (20.32) is depicted in Fig. 20.9*b*.

Adams Formulas. The other type of integration formulas that can be used to solve ODEs are the Adams formulas. Many popular computer algorithms for multistep solution of ODEs are based on these methods.

Open Formulas (Adams-Bashforth). The Adams formulas can be derived in a variety of ways. One technique is to write a forward Taylor series expansion around x_i:

$$y_{i+1} = y_i + f_i h + \frac{f_i'}{2} h^2 + \frac{f_i''}{3!} h^3 + \cdots$$

which can also be written as

$$y_{i+1} = y_i + h\left(f_i + \frac{h}{2} f_i' + \frac{h^2}{6} f_i'' + \cdots\right) \tag{20.33}$$

Recall from Sec. 3.5.4 that a backward difference can be used to approximate the derivative:

$$f_i' = \frac{f_i - f_{i-1}}{h} + \frac{f_i''}{2} h + O(h^2)$$

which can be substituted into Eq. (20.33) to yield

$$y_{i+1} = y_i + h\left\{f_i + \frac{h}{2}\left[\frac{f_i - f_{i-1}}{h} + \frac{f_i''}{2} h + O(h^2)\right] + \frac{h^2}{6} f_i'' + \cdots\right\}$$

or, collecting terms,

$$y_{i+1} = y_i + h\left(\frac{3}{2} f_i - \frac{1}{2} f_{i-1}\right) + \frac{5}{12} h^3 f_i'' + O(h^4) \tag{20.34}$$

This formula is called the *second-order open Adams formula*. Open Adams formulas are also referred to as *Adams-Bashforth* formulas. Consequently, Eq. (20.34) is sometimes called the second Adams-Bashforth formula.

Higher-order Adams-Bashforth formulas can be developed by substituting higher-difference approximations into Eq. (20.33). The nth-order open Adams formula can be represented generally as

$$y_{i+1} = y_i + h \sum_{k=0}^{n-1} \beta_k f_{i-k} + O(h^{n+1}) \tag{20.35}$$

The coefficients β_k are compiled in Table 20.1. The fourth-order version is depicted in Fig. 20.10a. Notice that the first-order version is Euler's method.

TABLE 20.1 Coefficients and truncation error for Adams-Bashforth predictors.

Order	β_0	β_1	β_2	β_3	β_4	β_5	Local Truncation Error
1	1						$\dfrac{1}{2} h^2 f'(\xi)$
2	$\dfrac{3}{2}$	$-\dfrac{1}{2}$					$\dfrac{5}{12} h^3 f''(\xi)$
3	$\dfrac{23}{12}$	$-\dfrac{16}{12}$	$\dfrac{5}{12}$				$\dfrac{9}{24} h^4 f^{(3)}(\xi)$
4	$\dfrac{55}{24}$	$-\dfrac{59}{24}$	$\dfrac{37}{24}$	$-\dfrac{9}{24}$			$\dfrac{251}{720} h^5 f^{(4)}(\xi)$
5	$\dfrac{1901}{720}$	$-\dfrac{2774}{720}$	$\dfrac{2616}{720}$	$-\dfrac{1274}{720}$	$\dfrac{251}{720}$		$\dfrac{475}{1440} h^6 f^{(5)}(\xi)$
6	$\dfrac{4277}{720}$	$-\dfrac{7923}{720}$	$\dfrac{9982}{720}$	$-\dfrac{7298}{720}$	$\dfrac{2877}{720}$	$-\dfrac{475}{720}$	$\dfrac{19,087}{60,480} h^7 f^{(6)}(\xi)$

Closed Formulas (Adams-Moulton). A backward Taylor series around x_{i+1} can be written as

$$y_i = y_{i+1} - f_{i+1}h + \frac{f'_{i+1}}{2} h^2 - \frac{f''_{i+1}}{3!} h^3 + \cdots$$

Solving for y_{i+1} yields

$$y_{i+1} = y_i + h \left(f_{i+1} - \frac{h}{2} f'_{i+1} + \frac{h^2}{6} f''_{i+1} - \cdots \right) \tag{20.36}$$

A difference can be used to approximate the derivative:

$$f'_{i+1} = \frac{f_{i+1} - f_i}{h} + \frac{f''_{i+1}}{2} h + O(h^2)$$

which can be substituted into Eq. (20.36) to yield

$$y_{i+1} = y_i + h \left\{ f_{i+1} - \frac{h}{2} \left[\frac{f_{i+1} - f_i}{h} + \frac{f''_{i+1}}{2} h + O(h^2) \right] + \frac{h^2}{6} f''_{i+1} - \cdots \right\}$$

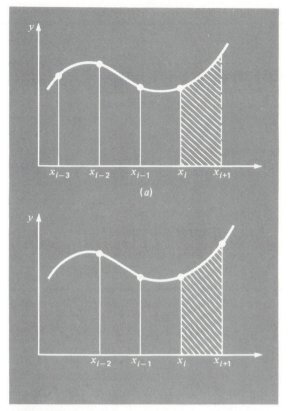

Figure 20.10
Graphical depiction of open and closed Adams integration formulas. (*a*) The fourth Adams-Bashforth open formula and (*b*) the fourth Adams-Moulton closed formula.

or, collecting terms,

$$y_{i+1} = y_i + h\left(\frac{1}{2}f_{i+1} + \frac{1}{2}f_i\right) - \frac{1}{12}h^3 f''_{i+1} - O(h^4)$$

This formula is called the *second-order closed Adams formula* or the *second Adams-Moulton formula*. Also, notice that it is the trapezoidal rule.

The *n*th-order closed Adams formula can be written generally as

$$y_{i+1} = y_i + h \sum_{k=0}^{n-1} \beta_k f_{i+1-k} + O(h^{n+1})$$

The coefficients β_k are listed in Table 20.2. The fourth-order method is depicted in Fig. 20.10*b*.

TABLE 20.2   Coefficients and truncation error for Adams-Moulton correctors.

Order	β_0	β_1	β_2	β_3	β_4	β_5	Local Truncation Error
2	$\dfrac{1}{2}$	$\dfrac{1}{2}$					$-\dfrac{1}{12} h^3 f''(\xi)$
3	$\dfrac{5}{12}$	$\dfrac{8}{12}$	$-\dfrac{1}{12}$				$-\dfrac{1}{24} h^4 f^{(3)}(\xi)$
4	$\dfrac{9}{24}$	$\dfrac{19}{24}$	$-\dfrac{5}{24}$	$\dfrac{1}{24}$			$-\dfrac{19}{720} h^5 f^{(4)}(\xi)$
5	$\dfrac{251}{720}$	$\dfrac{646}{720}$	$-\dfrac{264}{720}$	$\dfrac{106}{720}$	$-\dfrac{19}{720}$		$-\dfrac{27}{1440} h^6 f^{(5)}(\xi)$
6	$\dfrac{475}{1440}$	$\dfrac{1427}{1440}$	$-\dfrac{798}{1440}$	$\dfrac{482}{1440}$	$-\dfrac{173}{1440}$	$\dfrac{27}{1440}$	$-\dfrac{863}{60,480} h^7 f^{(6)}(\xi)$

20.2.4   Higher-Order Multistep Methods

Now that we have formally developed the Newton-Cotes and Adams integration formulas, we can use them to derive higher-order multistep methods. As was the case with the non-self-starting Heun method, the integration formulas are applied in tandem as predictor-corrector methods. In addition, if the open and closed formulas have local truncation errors of the same order, modifiers of the type listed in Fig. 20.6 can be incorporated to improve accuracy and allow step size control. Box 20.1 provides general equations for these modifiers. In the following section, we present two of the most common higher-order multistep approaches: Milne's method and the fourth-order Adams method.

Milne's Method.   *Milne's method* is the most common multistep method based on Newton-Cotes integration formulas. It uses the three-point Newton-Cotes open formula as a predictor:

$$y_{i+1}^0 = y_{i-3}^m + \frac{4h}{3} (2f_i^m - f_{i-1}^m + 2f_{i-2}^m) \qquad (20.37)$$

and the three-point Newton-Cotes closed formula (Simpson's 1/3 rule) as a corrector:

$$y_{i+1}^j = y_{i-1}^m + \frac{h}{3} (f_{i-1}^m + 4f_i^m + f_{i+1}^{j-1}) \qquad (20.38)$$

The predictor and corrector modifiers for Milne's method can be developed from the formulas in Box 20.1 and the error coefficients in Tables 15.2 and 15.4:

$$E_p \simeq \frac{28}{29} (y_i^m - y_i^0) \qquad (20.39)$$

Box 20.1 *Derivation of General Relationships for Modifiers*

The relationship between the true value, the approximation, and the error of a predictor can be represented generally as

$$\text{True value} = y_{i+1}^0 + \frac{\eta_p}{\delta_p} h^{n+1} y^{(n+1)}(\xi_p) \qquad (B20.1.1)$$

where η_p and δ_p are the numerator and denominator of the constant of the truncation error for either an open Newton-Cotes (Table 15.4) or an Adams-Bashforth (Table 20.1) predictor and n is the order.

A similar relationship can be developed for the corrector

$$\text{True value} = y_{i+1}^m - \frac{\eta_c}{\delta_c} h^{n+1} y^{(n+1)}(\xi_c) \qquad (B20.1.2)$$

where η_c and δ_c are the numerator and denominator of the constant of the truncation error for either a closed Newton-Cotes (Table 15.2) or an Adams-Moulton (Table 20.2) corrector. As was done in the derivation of Eq. (20.21), Eq. (B20.1.1) can be subtracted from Eq. (B20.1.2) to yield

$$0 = y_{i+1}^m - y_{i+1}^0$$

$$- \frac{\eta_c + \eta_p \delta_c / \delta_p}{\delta_c} h^{n+1} y^{(n+1)}(\xi) \qquad (B20.1.3)$$

Now dividing the equation by $\eta_c + \eta_p \delta_c / \delta_p$, multiplying the last

term by δ_p / δ_p, and rearranging provides an estimate of the local truncation error of the corrector:

$$E_c \simeq \frac{y_{i+1}^0 - y_{i+1}^m}{\eta_c + \eta_p \delta_c / \delta_p}$$

$$= -\frac{\eta_c \delta_p}{\eta_c \delta_p + \eta_p \delta_c} (y_{i+1}^m - y_{i+1}^0) \qquad (B20.1.4)$$

For the predictor modifier, Eq. (B20.1.3) can be solved at the previous step for

$$h^n y^{(n+1)}(\xi) = -\frac{\delta_c \delta_p}{\eta_c \delta_p + \eta_p \delta_c} (y_i^0 - y_i^m)$$

which can be substituted into the error term of Eq. (B20.1.1) to yield

$$E_p = \frac{\eta_p \delta_c}{\eta_c \delta_p + \eta_p \delta_c} (y_i^m - y_i^0) \qquad (B20.1.5)$$

Equations (B20.1.4) and (B20.1.5) are general versions of modifiers that can be used to improve multistep algorithms. For example, Milne's method has $\eta_p = 14$, $\delta_p = 45$, $\eta_c = 1$, and $\delta_c = 90$. Substituting these values into Eqs. (B20.1.4) and (B20.1.5) yields Eqs. (20.39) and (20.40). Similar modifiers can be developed for other pairs of open and closed formulas that have local truncation errors of the same order.

and

$$E_c \simeq -\frac{1}{29} (y_{i+1}^m - y_{i+1}^0) \qquad (20.40)$$

EXAMPLE 20.6   **Milne's Method**

Problem Statement: Use Milne's method to integrate $y' = 4e^{0.8x} - 0.5y$ from $x = 0$ to $x = 4$ using a step size of 1. The initial condition at $x = 0$ is $y = 2$. Because we are dealing with a multistep method, previous points are required. In an actual application, a one-step method such as a fourth-order RK would be used to compute the required points. For the present example, we will use the analytical solution [recall Eq. (E19.5.1) from Example 19.5] to compute exact values at $x_{i-3} = -3$, $x_{i-2} = -2$, and $x_{i-1} = -1$ of $y_{i-3} = -4.547302219$, $y_{i-2} = -2.306160375$, and $y_{i-1} = -0.392995325$, respectively.

Solution: The predictor [Eq. (20.37)] is used to calculate a value at $x = 1$:

$$y_1^0 = -4.547302219 + \frac{4(1)}{3}[2(3) - 1.993813519 + 2(1.960666259)]$$

$$= 6.02272313 \qquad \epsilon_t = 2.8\%$$

The corrector [Eq. (20.38)] is then employed to compute

$$y_1^1 = -0.392995325 + \frac{1}{3}[1.993813519 + 4(3) + 5.890802157]$$

$$= 6.235209902 \qquad \epsilon_t = -0.66\%$$

This result can be substituted back into Eq. (20.38) to iteratively correct the estimate. This process converges on a final corrected value of 6.20485465 ($\epsilon_t = -0.17\%$).

This value is more accurate than the comparable estimate of 6.36086549 ($\epsilon_t = -2.68\%$) obtained previously with the non-self-starting Heun method (Examples 20.3 through 20.5). The results for the remaining steps are $y(2) = 14.8603072(\epsilon_t = -0.11\%)$, $y(3) = 33.7242601$ ($\epsilon_t = -0.14\%$), and $y(4) = 75.4329487$ ($\epsilon_t = -0.12\%$).

As in the previous example, Milne's method usually yields results of high accuracy. However, there are certain cases where it performs poorly (see Ralston and Rabinowitz, 1978). Before elaborating on these cases, we will describe another higher-order multistep approach—the fourth-order Adams method.

Fourth-Order Adams Method.　A popular multistep method based on the Adams integration formulas uses the fourth-order Adams-Bashforth formula (Table 20.1) as the predictor:

$$y_{i+1}^0 = y_i^m + h\left(\frac{55}{24}f_i^m - \frac{59}{24}f_{i-1}^m + \frac{37}{24}f_{i-2}^m - \frac{9}{24}f_{i-3}^m\right) \qquad (20.41)$$

and the fourth-order Adams-Moulton formula (Table 20.2) as the corrector:

$$y_{i+1}^j = y_i^m + h\left(\frac{9}{24}f_{i+1}^{j-1} + \frac{19}{24}f_i^m - \frac{5}{24}f_{i-1}^m + \frac{1}{24}f_{i-2}^m\right) \qquad (20.42)$$

The predictor and the corrector modifiers for the fourth-order Adams method can be developed from the formulas in Box 20.1 and the error coefficients in Tables 20.1 and 20.2 as

$$E_p \simeq \frac{251}{270}(y_i^m - y_i^0) \qquad (20.43)$$

and

$$E_c \simeq -\frac{19}{270}(y_{i+1}^m - y_{i+1}^0) \qquad (20.44)$$

EXAMPLE 20.7　*Fourth-Order Adams Method*

Problem Statement: Use the fourth-order Adams method to solve the same problem as in Example 20.6.

Solution: The predictor [Eq. (20.41)] is used to compute a value at $x = 1$.

$$y_i^0 = 2 + 1\left(\frac{55}{24}3 - \frac{59}{24}1.993813519 + \frac{37}{24}1.960666259\right.$$

$$\left. - \frac{9}{24}2.649382908\right)$$

$$= 6.002716992 \qquad \epsilon_t = 3.1\%$$

which is comparable to but somewhat less accurate than the result using the Milne method. The corrector [Eq. (20.42)] is then employed to calculate

$$y_1^1 = 2 + 1\left(\frac{9}{24}5.900805218 + \frac{19}{24}3 - \frac{5}{24}1.993813519\right.$$

$$\left. + \frac{1}{24}1.960666259\right)$$

$$= 6.254118568 \qquad \epsilon_t = -0.96\%$$

which again is comparable to but less accurate than the result using Milne's method. This result can be substituted back into Eq. (20.42) to iteratively correct the estimate. The process converges on a final corrected value of 6.214423582 ($\epsilon_t = 0.32\%$) which is an accurate result but again somewhat inferior to that obtained with the Milne method.

Stability of Multistep Methods. The superior accuracy of the Milne method exhibited in Examples 20.6 and 20.7 would be anticipated on the basis of the error terms for the predictors [Eqs. (20.39) and (20.43)] and the correctors [Eqs. (20.40) and (20.44)]. The coefficients for the Milne method, $\frac{14}{45}$ and $\frac{1}{90}$, are smaller than for the fourth-order Adams, $\frac{251}{720}$ and $\frac{19}{720}$. Additionally, the Milne method employs fewer function evaluations to attain these higher accuracies. At face value, these results might lead to the conclusion that the Milne method is superior and, therefore, preferable to the fourth-order Adams. Although this conclusion holds for many cases, there are instances where the Milne method performs unacceptably. Such behavior is exhibited in the following example.

EXAMPLE 20.8 **Stability of Milne's and Fourth-Order Adams Methods**

Problem Statement: Employ Milne's and the fourth-order Adams methods to solve

$$\frac{dy}{dx} = -y$$

with the initial condition that $y = 1$ at $x = 0$. Solve this equation from $x = 0$ to $x = 10$ using a step size of $h = 0.5$. Note that the analytical solution is $y = e^{-x}$.

Solution: The results, as summarized in Fig. 20.11, indicate problems with Milne's method. Shortly after the onset of the computation, the errors begin to grow and oscillate in sign. By $t = 10$, the relative error has inflated to 2831 percent and the predicted value itself has started to oscillate in sign.

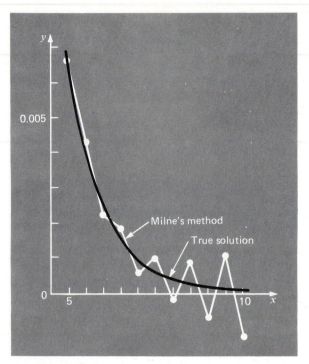

Figure 20.11
Graphical depiction of the instability of Milne's method.

In contrast, the results for the Adams method would be much more acceptable. Although the error also grows, it would do so at a slow rate. Additionally, the discrepancies would not exhibit the wild swings in sign exhibited by the Milne method.

The unacceptable behavior manifested in the previous example by the Milne method is referred to as *instability*. Although it does not always occur, its possibility leads to the conclusion that Milne's approach should be avoided. Thus, the fourth-order Adams method is normally preferred.

The instability of Milne's method is due to the corrector. Consequently, attempts have been made to rectify the shortcoming by developing stable correctors. One commonly used alternative that employs this approach is *Hamming's method*, which uses the Milne predictor and a stable corrector:

$$y_{i+1}^{j} = \frac{9y_i^m - y_{i-2}^m + 3h(f_{i+1}^{j-1} + 2f_i^m - f_{i-1}^m)}{8}$$

which has a local truncation error:

$$E_c = \frac{1}{40} h^5 f^{(4)}(\xi_c)$$

Hamming's method also includes modifiers of the form

$$E_p \simeq \frac{9}{121} (y_i^m - y_i^0)$$

and

$$E_c \simeq -\frac{112}{121} (y_{i+1}^m - y_{i+1}^0)$$

The reader can obtain additional information on this and other multistep methods elsewhere (Hamming, 1973; Lapidus and Seinfield, 1971).

PROBLEMS

Hand Calculations

20.1 Compute the first step of Example 20.2 using the adaptive fourth-order RK method. Verify whether step size adjustment is in order.

20.2 If $\epsilon = 0.0005$, determine whether step size adjustment is required for Example 20.1.

20.3 Develop your own second-order adaptive RK method. Employ the Heun method with single corrector as your basic solver. Test it by repeating Prob. 20.1.

20.4 Solve the following initial-value problem over the interval from $x = 2$ to $x = 3$:

$$\frac{dy}{dx} = -0.5y$$

Use the non-self-starting Heun method with a step size of 0.5 and initial conditions of $y(1.5) = 4.72367$ and $y(2.0) = 3.67879$. Iterate the corrector to $\epsilon_s = 1\%$. [*Note:* The exact results obtained analytically are $y(2.5) = 2.86505$ and $y(3.0) = 2.23130$.] Compute the true percent relative errors ϵ_t for your results.

20.5 Repeat Prob. 20.4, but use Milne's method. [*Note:* $y(0.5) = 7.78801$ and $y(1.0) = 6.06531$.] Iterate the corrector to $\epsilon_s = 0.01\%$.

20.6 Repeat Prob. 20.5, but use the fourth-order Adams method ($\epsilon_s = 0.01\%$).

20.7 Solve the following initial-value problem from $x = 4$ to $x = 5$:

$$\frac{dy}{dx} = -\frac{y}{x}$$

Use a step size of 0.5 and initial values of $y(2.5) = 1.2$, $y(3) = 1$, $y(3.5) = 0.857142857$, and $y(4) = 0.75$. Obtain your solutions using the following techniques: (*a*) the non-self-starting Heun Method ($\epsilon_s = 1\%$) (*b*) Milne's method ($\epsilon_s = 0.01\%$), and (*c*) the fourth-order Adams method ($\epsilon_s = 0.01\%$). [*Note:* The exact answers obtained analytically are $y(4.5) = 0.66666667$ and $y(5) = 0.6$.] Compute the true percent relative errors ϵ_t for your results.

20.8 Solve the following initial-value problem from $x = 0$ to $x = 0.5$:

$$\frac{dy}{dx} = yx^2 - y$$

Use the non-self-starting Heun method with a step size of 0.25. If $y(-0.25) = 1.277355170$, employ a fourth-order RK method with a step size of 0.25 to predict the starting value at $y(0)$.

20.9 Solve the following initial-value problem from $x = 1.5$ to $x = 2.5$:

$$\frac{dy}{dx} = \frac{-y}{1 + x}$$

Use the fourth-order Adams method. Employ a step size of 0.5 and the fourth-order RK method to predict the start-up values if $y(0) = 2$.

20.10 Repeat Prob. 20.9, but use Milne's method.

20.11 Determine the predictor, corrector, and modifiers for a second-order Adams method. Use it to solve Prob. 20.4.

20.12 Determine the predictor, corrector, and modifiers for a third-order Adams method. Use it to solve Prob. 20.7.

Computer-Related Problems

20.13 Develop a user-friendly program for the adaptive fourth-order RK method outlined in Fig. 20.2. Test it by duplicating the results shown in Fig. 20.3.

20.14 Use the program developed in Prob. 20.13 to solve the equation in Prob. 20.4 from $x = 0$ to 5 with $y(0) = 1$, initial $h = 1$, and $\epsilon = 0.0001$.

20.15 Use the program developed in Prob. 20.13 to solve the equation in Prob. 20.7 from $x = 1$ to 5 with $y(0) = 0.1$, initial $h = 0.2$, and $\epsilon = 0.001$.

20.16 Develop a user-friendly program for the non-self-starting Heun method with modifiers. Employ a fourth-order RK method to compute starter values. Test the program by duplicating Example 20.5.

20.17 Use the program developed in Prob. 20.16 to solve Prob. 20.8.

20.18 Develop a user-friendly program for the fourth-order Adams method with modifiers. Use a fourth-order RK method to compute starter values. Test the program by duplicating Example 20.7.

20.19 Employ the program developed in Prob. 20.18 to solve Prob. 20.9.

CHAPTER 21

Boundary-Value and Eigenvalue Problems

Recall from our discussion at the beginning of Part Six that an ordinary differential equation is accompanied by auxiliary conditions. These conditions are used to evaluate the constants of integration that result during the solution of the equation. For an nth-order equation, n conditions are required. If all the conditions are specified at the same value of the independent variable, then we are dealing with an *initial-value problem* (Fig. 21.1a). To this point, the material in Part Six has been devoted to this type of problem.

In contrast, there is another application for which the conditions are not known at a single point but rather at different values of the independent variable. Because these values are often specified at the extreme points or boundaries of a system, they are customarily referred to as *boundary-value problems* (Fig. 21.1b). A variety of significant engineering applications fall within this class. In the present chapter, we discuss two general approaches for obtaining their solution: the *shooting method* and the *finite-difference approach*. Additionally, we present techniques to approach a special type of boundary-value problem: the determination of *eigenvalues*.

21.1 GENERAL METHODS FOR BOUNDARY-VALUE PROBLEMS

The conservation of heat can be used to develop a heat balance for a long, thin rod (Fig. 21.2). If the rod is not insulated along its length and the system is at a steady state, the equation that results is

$$\frac{d^2T}{dx^2} + h'\,(T_a - T) = 0 \tag{21.1}$$

where h' is a radiative heat loss coefficient (cm^{-2}) that parameterizes the rate of heat dissipation to the surrounding air and T_a is the temperature of the surrounding air (°C).

To obtain a solution for Eq. (21.1) there must be appropriate boundary conditions. A simple case is where the temperatures at the ends of the bar are held at fixed values. These can be expressed mathematically as

$$T(0) = T_1$$

$$T(L) = T_2$$

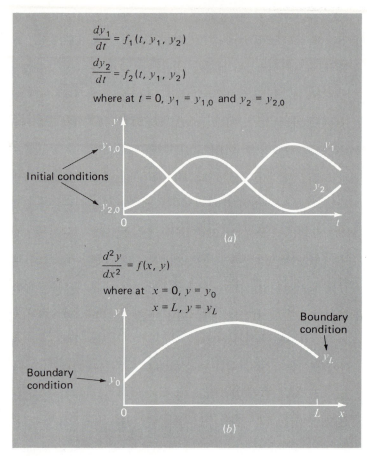

$$\frac{dy_1}{dt} = f_1(t, y_1, y_2)$$

$$\frac{dy_2}{dt} = f_2(t, y_1, y_2)$$

where at $t = 0$, $y_1 = y_{1,0}$ and $y_2 = y_{2,0}$

Initial conditions

$y_{1,0}$

$y_{2,0}$

y_1

y_2

(a)

$$\frac{d^2 y}{dx^2} = f(x, y)$$

where at $x = 0$, $y = y_0$

$x = L$, $y = y_L$

Boundary condition

Boundary condition

y_0

y_L

(b)

Figure 21.1
Initial-value versus boundary-value problems. (a) An initial-value problem where all of the conditions are specified at the same value of the independent variable. (b) A boundary-value problem where the conditions are specified at different values of the independent variable.

Figure 21.2
A noninsulated uniform rod positioned between two bodies of constant but different temperature. For this case $T_1 > T_2$ and $T_2 > T_a$.

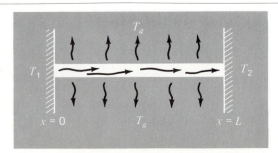

T_a

T_1

T_2

$x = 0$

T_a

$x = L$

With these conditions, Eq. (21.1) can be solved analytically using calculus. For a 10-m rod with $T_a = 20$, $T_1 = 40$, $T_2 = 200$, and $h' = 0.01$, the solution is

$$T = 73.4523e^{0.1x} - 53.4523e^{-0.1x} + 20 \qquad (21.2)$$

In the following sections, the same problem will be solved using numerical approaches.

21.1.1 The Shooting Method

The *shooting method* is based on converting the boundary-value problem into an equivalent initial-value problem. A trial-and-error approach is then implemented to solve the initial-value version. The approach can be illustrated by an example.

EXAMPLE 21.1   The Shooting Method

Problem Statement: Use the shooting method to solve Eq. (21.1) for a 10-m rod, with $h' = 0.01$ m^{-2}, $T_a = 20$ and the boundary conditions

$$T(0) = 40 \qquad T(10) = 200$$

Solution: Using the same approach as was employed to transform Eq. (PT6.2) into Eqs. (PT6.3) and (PT6.6), the second-order equation can be expressed as two first-order ODEs:

$$\frac{dT}{dx} = z \qquad (E21.1.1)$$

$$\frac{dz}{dx} = h'(T - T_a) \qquad (E21.1.2)$$

In order to solve these equations, we require an initial value for z. For the shooting method, we guess a value—say $z(0) = 10$. The solution is then obtained by integrating Eq. (E21.1.1) and (E21.1.2) simultaneously. For example, using a fourth-order RK method with a step size of 2, we obtain a value at the end of the interval of $T(10) = 168.3797$ (Fig. 21.3a), which differs from the boundary condition of $T(10) = 200$. Therefore, we make another guess, $z(0) = 20$, and perform the computation again. This time, the result of $T(10) = 285.8980$ is obtained (Fig. 21.3b).

Now, because the original ODE is linear, the values

$$z(0) = 10 \qquad T(10) = 168.3797$$

and

$$z(0) = 20 \qquad T(10) = 285.8980$$

are linearly related. As such, they can be used to compute the value of $z(0)$ that yields $T(10) = 200$. A linear interpolation formula [recall Eq. (12.2)] can be employed for this purpose:

$$z(0) = 10 + \frac{20 - 10}{285.8980 - 168.3797}(200 - 168.3797) = 12.6907$$

This value can then be used to determine the correct solution as depicted in Fig. 21.3c.

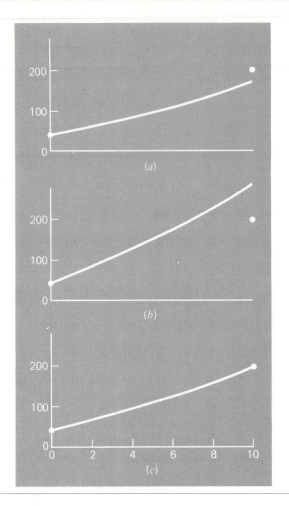

Figure 21.3
The shooting method: (a) the first "shot," (b) the second "shot," and (c) the final exact "hit."

For nonlinear boundary-value problems, linear interpolation or extrapolation through two solution points will not necessarily result in an accurate estimate of the required boundary condition to attain an exact solution. An alternative is to perform three applications of the shooting method and use a quadratic interpolating polynomial to estimate the proper boundary condition. However, it is unlikely that such an approach would yield the exact answer, and additional iterations would be necessary to obtain the solution.

The shooting method also becomes arduous for higher-order equations where the necessity to assume two or more conditions makes the approach inefficient. For these reasons, alternative methods are available as described next.

21.1.2 Finite-Difference Methods

The most common alternatives to the shooting method are finite difference approaches. In these techniques, finite divided differences are substituted for the derivatives in the original equation. Thus, the differential equation is transformed into a set of simultaneous algebraic equations that can be solved using the methods from Part Three.

For the case of Fig. 21.2, the finite-divided-difference approximation for the second derivative is (recall Fig. 17.3)

$$\frac{d^2T}{dx^2} = \frac{T_{i+1} - 2T_i + T_{i-1}}{\Delta x^2}$$

This approximation can be substituted into Eq. (21.1) to give

$$\frac{T_{i+1} - 2T_i + T_{i+1}}{\Delta x^2} - h'(T_i - T_a) = 0$$

Collecting terms gives

$$-T_{i-1} + (2 + h'\Delta x^2)T_i - T_{i+1} = h'\Delta x^2 T_a \tag{21.2a}$$

This equation applies for each of the interior nodes of the rod. For the first and last interior nodes, T_{i-1} and T_{i+1}, respectively, are specified by the boundary conditions. Therefore, the resulting set of linear algebraic equations will be tridiagonal. As such, it can be solved with the efficient algorithms that are available for such systems (Sec. 9.6).

EXAMPLE 21.2 Finite-Difference Approximation of Boundary-Value Problems

Problem Statement: Use the finite-difference approach to solve the same problem as in Example 21.1.

Solution: Employing the parameters in Example 21.1, Eq. (21.2a) can be written for the rod from Fig. 21.2. Using four interior nodes with a segment length of $\Delta x = 2$ m results in the following equations:

$$\begin{bmatrix} 2.04 & -1 & & \\ -1 & 2.04 & -1 & \\ & -1 & 2.04 & -1 \\ & & -1 & 2.04 \end{bmatrix} \begin{Bmatrix} T_1 \\ T_2 \\ T_3 \\ T_4 \end{Bmatrix} = \begin{Bmatrix} 40.8 \\ 0.8 \\ 0.8 \\ 200.8 \end{Bmatrix}$$

which can be solved for

$$\{T\}^T = [65.9698 \quad 93.7785 \quad 124.5382 \quad 159.4795]$$

Table 21.1 provides a comparison between the analytical solution [Eq. (21.2)] and the numerical solutions obtained in Examples 21.1 and 21.2. Note that there are some discrepancies among the approaches. For both numerical methods, these errors can be mitigated by decreasing their respective step sizes. Although both techniques perform well for the present case, the finite-difference approach is preferred because of the ease with which it can be applied to more complex cases such as nonlinear and higher-order systems.

TABLE 21.1 Comparison of the exact analytical solution with the shooting and finite-difference methods.

x	True	Shooting Method	Finite Difference
0	40	40	40
2	65.9518	65.9520	65.9698
4	93.7478	93.7481	93.7785
6	124.5036	124.5039	124.5382
8	159.4534	159.4538	159.4795
10	200	200	200

Aside from the shooting and finite-difference methods, there are other techniques available for solving boundary-value problems. Some of these will be described in Part Seven. These include steady-state (Chap. 23) and transient (Chap. 24) solution of two-dimensional boundary-value problems using finite differences and steady-state solutions of the one-dimensional problem with the finite-element approach (Chap. 25).

21.2 EIGENVALUE PROBLEMS

Eigenvalue, or *characteristic-value, problems* are a special class of boundary-value problems that are common in engineering problem contexts involving vibrations, elasticity, and other oscillating systems. Before describing numerical methods for solving these problems, we will present some general background information. This includes discussion of both the mathematics and the engineering significance of eigenvalues.

21.2.1 Mathematical Background

Part Three dealt with methods for solving sets of linear algebraic equations of the general form

$$[A]\{X\} = \{C\}$$

Such systems are called *nonhomogeneous* because of the presence of the vector $\{C\}$ on the right-hand side of the equality. If the equations comprising such a system are linearly independent (that is, they have a nonzero determinant), they will have a unique solution. In other words, there is one set of x values that will make the equations balance.

In contrast, a homogeneous linear algebraic system has the general form

$$[A]\{X\} = 0$$

Although nontrivial solutions (that is, solutions other than all x's $= 0$) of such systems are possible, they are generally not unique. Rather, the simultaneous equations establish relationships among the x's that can be satisfied by various combinations of values.

Eigenvalue problems associated with engineering are typically of the general form

$$(a_{11} - \lambda)x_1 + \quad a_{12}x_2 + \cdots + \quad a_{1n}x_n = 0$$
$$a_{21}x_1 + (a_{22} - \lambda)x_2 + \cdots + \quad a_{2n}x_n = 0$$

$$\vdots \qquad\qquad \vdots \qquad\qquad\qquad \vdots \qquad \vdots$$

$$a_{n1}x_1 + \quad a_{n2}x_2 + \cdots + (a_{nn} - \lambda)x_n = 0$$

where λ is an unknown parameter called the *eigenvalue*, or *characteristic, value*. A solution $\{X\}$ for such a system is referred to as an *eigenvector*. The above set of equations may also be expressed concisely as

$$\big[[A] - \lambda[I]\big]\{X\} = 0 \tag{21.3}$$

The solution of Eq. (21.3) hinges on determining λ. One way to accomplish this is based on the fact that the determinant of the matrix $[[A] - \lambda[I]]$ must equal zero for nontrivial solutions to be possible. Expanding the determinant yields a polynomial in λ. The roots of this polynomial are the solutions for the eigenvalues. An example of this approach will be provided in the next section.

21.2.2 Physical Background

The mass-spring system in Fig. 21.4a is a simple context to illustrate how eigenvalues occur in physical problem settings. It also will help to illustrate some of the mathematical concepts introduced in the previous section.

To simplify the analysis, assume that each mass has no external or damping forces acting on it. In addition, assume that each spring has the same natural length l and the same spring constant k. Finally, assume that the displacement of each spring is measured

Figure 21.4
Positioning the masses away from equilibrium creates forces in the springs that upon release lead to oscillations of the masses. The positions of the masses can be referenced to local coordinates with origins at their respective equilibrium positions.

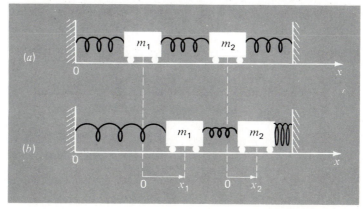

relative to its own local coordinate system with an origin at the spring's equilibrium position (Fig. 21.4b). Under these assumptions, Newton's second law can be employed to develop a force balance for each mass (recall Case Study 10.5),

$$m_1 \frac{d^2x_1}{dt^2} = -kx_1 + k(x_2 - x_1)$$

and

$$m_2 \frac{d^2x_2}{dt^2} = -k(x_2 - x_1) - kx_2$$

where x_i is the displacement of mass i away from its equilibrium position (Fig. 21.4b). These equations can be expressed as

$$m_1 \frac{d^2x_1}{dt^2} - k(-2x_1 + x_2) = 0 \qquad (21.4a)$$

$$m_2 \frac{d^2x_2}{dt^2} - k(x_1 - 2x_2) = 0 \qquad (21.4b)$$

From vibration theory, it is known that solutions to Eq. (21.4) can take the form

$$x_i = A_i \sin (\omega t) \qquad (21.5)$$

where A_i is the amplitude of the vibration of mass i and ω is the frequency of the vibration which is equal to

$$\omega = \frac{2\pi}{T_p} \qquad (21.6)$$

where T_p is the period. From Eq. (21.5) it follows that

$$x_i'' = -A_i \omega^2 \sin (\omega t) \qquad (21.7)$$

Equations (21.5) and (21.7) can be substituted into Eq. (21.4), which, after collection of terms, can be expressed as

$$\left(\frac{2k}{m_1} - \omega^2\right) A_1 - \frac{k}{m_1} A_2 = 0 \qquad (21.8a)$$

$$-\frac{k}{m_2} A_1 + \left(\frac{2k}{m_2} - \omega^2\right) A_2 = 0 \qquad (21.8b)$$

Comparison of Eq. (21.8) with Eq. (21.3) indicates that at this point, the solution has been reduced to an eigenvalue problem.

EXAMPLE 21.3   Eigenvalues and Eigenvectors for a Mass-Spring System

Problem Statement: Evaluate the eigenvalues and the eigenvectors of Eq. (21.8) for the case where $m_1 = m_2 = 40$ kg and $k = 200$ N/m.

Solution: Substituting the parameter values into Eq. (21.8) yields

$$(10 - \omega^2)A_1 - 5A_2 = 0$$
$$-5A_1 + (10 - \omega^2)A_2 = 0$$

The determinant of this system is [recall Eq. (7.3)]

$$(\omega^2)^2 - 20\omega^2 + 75 = 0$$

which can be solved by the quadratic formula for $\omega^2 = 15$ and 5 s^{-2}. Therefore, the frequencies for the vibrations of the masses are $\omega = 3.873$ s^{-1} and 2.236 s^{-1}, respectively. These values can be used to determine the periods for the vibrations with Eq. (21.6). For the first mode, $T_p = 1.62$ s and for the second, $T_p = 2.81$ s.

As stated in Sec. 21.2.1, a unique set of values cannot be obtained for the unknowns. However, their ratios can be specified by substituting the eigenvalues back into the equations. For example, for the first mode ($\omega^2 = 15$ s^{-2}), it can be determined that $A_1 = -A_2$. For the second mode ($\omega^2 = 5$ s^{-2}), $A_1 = A_2$.

This example provides valuable information regarding the behavior of the system in Fig. 21.4. Aside from its period, we know that if the system is vibrating in the first mode, the amplitude of the second mass will be equal but of opposite sign to the amplitude of the first. As in Fig. 21.5a, the masses vibrate apart and then together indefinitely.

Figure 21.5
The principal modes of vibration of two equal masses connected by three identical springs between fixed walls.

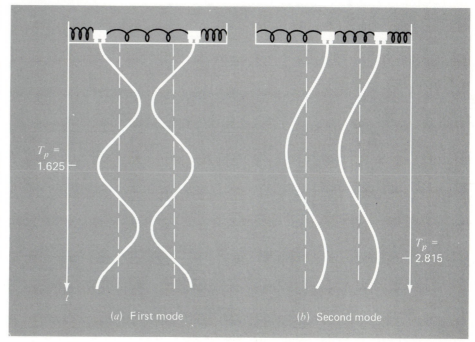

$T_p = 1.625$

$T_p = 2.815$

(a) First mode   (b) Second mode

In the second mode, the two masses have equal amplitudes at all times. Thus, as in Fig. 21.5b, they vibrate back and forth in unison. It should be noted that the configuration of the amplitudes provides guidance on how to set their initial values to attain pure motion in either of the two modes. Any other configuration will lead to additional harmonics (recall Chap. 13).

21.2.3 A Boundary-Value Problem

Now that you have been introduced to eigenvalues, we turn to the type of problem that is the subject of the present chapter: boundary-value problems for ordinary differential equations. Figure 21.6 shows a physical system that can serve as a context for examining this type of problem.

The curvature of a slender column subject to an axial load P can be modeled by

$$\frac{d^2y}{dx^2} = \frac{M}{EI} \tag{21.9}$$

where d^2y/dx^2 specifies the curvature, M is the bending moment, E is the modulus of elasticity, and I is the moment of inertia of the cross section about its neutral axis. Considering the free body in Fig. 21.6b, it is clear that the bending moment at x is $M = -Py$. Substituting this value into Eq. (21.9) gives

$$\frac{d^2y}{dx^2} + p^2y = 0 \tag{21.10}$$

where

$$p^2 = \frac{P}{EI} \tag{21.11}$$

Figure 21.6
(a) A slender rod. (b) A free-body diagram of a rod.

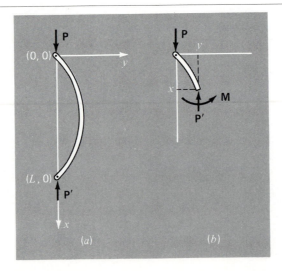

For the system in Fig. 21.6, subject to the boundary conditions

$$y(0) = 0 \qquad (21.12a)$$

$$y(L) = 0 \qquad (21.12b)$$

the general solution for Eq. (21.10) is

$$y = A \sin (px) + B \cos (px) \qquad (21.13)$$

where A and B are arbitrary constants that are to be evaluated via the boundary conditions. According to the first condition [Eq. (21.12a)],

$$0 = A \sin (0) + B \cos (0)$$

Therefore, we conclude that $B = 0$.

According to the second condition [Eq. (21.12b)],

$$0 = A \sin (pL) + B \cos (pL)$$

but since $B = 0$,

$$A \sin (pL) = 0$$

Because $A = 0$ represents a trivial solution, we conclude that

$$\sin (pL) = 0$$

In order for this equality to hold

$$pL = n\pi \qquad \text{for } n = 1, 2, 3, \ldots \qquad (21.14)$$

Thus, there are an infinite number of values that meet the boundary condition. Equation (21.14) can be solved for

$$p = \frac{n\pi}{L} \qquad \text{for } n = 1, 2, 3, \ldots \qquad (21.15)$$

which are the eigenvalues for the column.

Figure 21.7, which shows the solution for the first four eigenvalues, can provide insight into the physical significance of the results. Each eigenvalue corresponds to a way in which the column buckles. Combining Eqs. (21.11) and (21.15) gives

$$P = \frac{n^2 \pi^2 EI}{L^2} \qquad \text{for } n = 1, 2, 3, \ldots \qquad (21.16)$$

These can be thought of as *buckling loads* because they represent the levels at which the column moves into each succeeding buckling configuration. In a practical sense, it is usually the first value that is of interest because failure will usually occur when the column first buckles. Thus, a critical load can be defined as

$$P_{\text{crit}} = \frac{\pi^2 EI}{L^2}$$

which is formally known as *Euler's formula*.

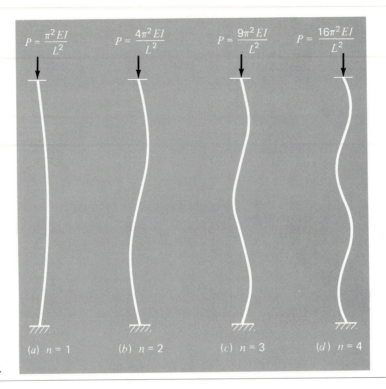

Figure 21.7
The first four eigenvalues for the slender rod from Fig. 21.6.

EXAMPLE 21.4   Eigenvalue Analysis of an Axially Loaded Column

Problem Statement: An axially loaded wooden column has the following characteristics: $E = 10 \times 10^9$ Pa, $I = 1.25 \times 10^{-5}$ m^4, and $L = 3$ m. Determine the first eight eigenvalues and the corresponding buckling loads.

Solution: Equations (21.15) and (21.16) can be used to compute

n	p, m^{-2}	P, kN
1	1.0472	137.078
2	2.0944	548.311
3	3.1416	1233.701
4	4.1888	2193.245
5	5.2360	3426.946
6	6.2832	4934.802
7	7.3304	6716.814
8	8.3776	8772.982

The critical buckling load is, therefore, 137.078 kN.

Although analytical solutions of the sort obtained above are useful, they are often difficult or impossible to obtain. This is usually true when dealing with nonlinear systems or those with nonhomogeneous properties. In such cases, numerical methods of the sort described next are the only practical alternative.

21.2.4 The Polynomial Method

Equation (21.10) can be solved numerically by substituting a central finite-divided-difference approximation (Table 17.3) for the second derivative to give

$$\frac{y_{i+1} - 2y_i + y_{i-1}}{h^2} + p^2 y_i = 0$$

which can be expressed as

$$y_{i-1} - (2 - h^2 p^2) y_i + y_{i+1} = 0 \tag{21.17}$$

Writing this equation for a series of nodes along the axis of the column yields a homogeneous system of equations. For example, if the column is divided into five segments (that is, four interior nodes), the result is

$$\begin{bmatrix} (2 - h^2 p^2) & -1 & & \\ -1 & (2 - h^2 p^2) & -1 & \\ & -1 & (2 - h^2 p^2) & -1 \\ & & -1 & (2 - h^2 p^2) \end{bmatrix} \begin{Bmatrix} y_1 \\ y_2 \\ y_3 \\ y_4 \end{Bmatrix} = 0 \tag{21.18}$$

Expansion of the determinant of the system yields an eighth-order polynomial, the roots of which are the eigenvalues. This approach, called the *polynomial method,* is performed in the following example.

EXAMPLE 21.5 *The Polynomial Method*

Problem Statement: Employ the polynomial method to determine the eigenvalues for the axially loaded column from Example 21.4 using (*a*) one, (*b*) two, (*c*) three, and (*d*) four interior nodes.

Solution: (*a*) Writing Eq. (21.17) for one interior node yields ($h = 3/2$)

$$-(2 - 2.25p^2)y_1 = 0$$

Thus, for this simple case, the eigenvalue is analyzed by setting the determinant equal to zero

$$2 - 2.25p^2 = 0$$

and solving for

$$p = \pm 0.9428$$

which is about 10 percent less than the exact value of 1.0472 obtained in Example 21.4.
   (*b*) For two interior nodes ($h = 3/3$), Eq. (21.17) is written as

$$\begin{bmatrix} (2 - p^2) & -1 \\ -1 & (2 - p^2) \end{bmatrix} \begin{Bmatrix} y_1 \\ y_2 \end{Bmatrix} = \{0\}$$

Expansion of the determinant gives

$$(2 - p^2)^2 - 1 = 0$$

which can be solved for

$$p = \pm 1$$

and

$$p = \pm 1.73205$$

Thus, the first eigenvalue is now about 4.5 percent low and a second eigenvalue is obtained that is about 17 percent low.

(c) For three interior points ($h = 3/4$), Eq. (21.17) yields

$$\begin{bmatrix} 2 - 0.5625p^2 & -1 & 0 \\ -1 & 2 - 0.5625p^2 & -1 \\ 0 & -1 & 2 - 0.5625p^2 \end{bmatrix} \begin{Bmatrix} y_1 \\ y_2 \\ y_3 \end{Bmatrix} = 0 \qquad \text{(E21.5.1)}$$

The determinant can be set equal to zero and expanded to give

$$(2 - 0.5625p^2)^3 - 2(2 - 0.5625p^2) = 0$$

For this equation to hold,

$$2 - 0.5625p^2 = 0$$

and

$$2 - 0.5625p^2 = \sqrt{2}$$

which can be solved for the first three eigenvalues

$$p = \pm 1.0205 \qquad (|\epsilon_t| = 2.5\%)$$

$$p = \pm 1.8856 \qquad (|\epsilon_t| = 10\%)$$

$$p = \pm 2.4637 \qquad (|\epsilon_t| = 22\%)$$

(d) For four interior points ($h = 3/5$), the result is Eq. (21.18) with $2 - 0.36p^2$ on the diagonal. Setting the determinant equal to zero and expanding it gives

$$(2 - 0.36p^2)^4 - 3(2 - 0.36p^2)^2 + 1 = 0$$

which can be solved for the first four eigenvalues

$$p = \pm 1.0301 \qquad (|\epsilon_t| = 1.6\%)$$

$$p = \pm 1.9593 \qquad (|\epsilon_t| = 6.5\%)$$

$$p = \pm 2.6967 \qquad (|\epsilon_t| = 14\%)$$

$$p = \pm 3.1702 \qquad (|\epsilon_t| = 24\%)$$

Table 21.2, which summarizes the results of this example, illustrates some fundamental aspects of the polynomial method. As the segmentation is made more refined, additional eigenvalues are determined and the previously determined values become progressively more accurate. Thus, the approach is best suited for cases where the lower eigenvalues are required.

TABLE 21.2 The results of applying the polynomial method to an axially loaded column. The numbers in parentheses represent the absolute value of the true percent relative error.

Eigenvalue	True	Polynomial Method			
		$h = 3/2$	$h = 3/3$	$h = 3/4$	$h = 3/5$
1	1.0472	0.9428 (10%)	1.0000 (4.5%)	1.0205 (2.5%)	1.0301 (1.6%)
2	2.0944		1.7321 (21%)	1.8856 (10%)	1.9593 (6.5%)
3	3.1416			2.4637 (22%)	2.6967 (14%)
4	4.1888				3.1702 (24%)

As in Example 21.5, the polynomial method consists of two major steps: (1) expansion of the determinant to yield a polynomial and (2) the calculation of the roots of the polynomial. We now turn to a computer method for generating the polynomial.

The Fadeev-Leverrier Method. The *Fadeev-Leverrier method* is an efficient approach for generating the coefficients p_i of the polynomial

$$(-1)^n (\lambda^n - p_{n-1}\lambda^{n-1} - \cdots - p_1\lambda - p_0) = 0 \tag{21.19}$$

which result from the expansion of the determinant of the system

$$\left| [A] - \lambda[I] \right| \{X\} = 0$$

Note that the term $(-1)^n$ and the negative signs are included so that the terms of the polynomial have the same signs as they would have if they were generated by expanding the determinant with minors. The method has the additional advantage that the matrix inverse $[A]^{-1}$ can be generated efficiently in the process.

The method consists of generating a sequence of matrices $[B]$ that can be employed to determine the p_i values. For example,

$$[B]_{n-1} = [A] \tag{21.20}$$

and

$$p_{n-1} = \text{tr } [B]_{n-1} \tag{21.21}$$

where tr $[B]_{n-1}$ is the *trace* of the matrix $[B]_{n-1}$ which is (recall PT3.2.2) the sum of the diagonal coefficients.

The method can be continued by generating the matrix

$$[B]_{n-2} = [A]\big[[B]_{n-1} - p_{n-1}[I]\big] \tag{21.22}$$

which can be used to calculate

$$p_{n-2} = \frac{1}{2} \text{ tr } [B]_{n-2} \tag{21.23}$$

and

$$[B]_{n-3} = [A]\big[[B]_{n-2} - p_{n-2}[I]\big] \tag{21.24}$$

which can be used to calculate

$$p_{n-3} = \frac{1}{3} \text{ tr } [B]_{n-3} \tag{21.25}$$

and so on until p_0 is determined from

$$[B]_0 = [A]\big[[B]_1 - p_1[I]\big] \tag{21.26}$$

and

$$p_0 = \frac{1}{n} \text{ tr } [B]_0 \tag{21.27}$$

The matrix inverse can then be simply computed as

$$[A]^{-1} = \frac{1}{p_0}\big[[B]_1 - p_1[I]\big] \tag{21.28}$$

EXAMPLE 21.6   The Fadeev-Leverrier Method

Problem Statement: Employ the Fadeev-Leverrier method to determine the coefficients for the matrix in part (c) of Example 21.5.

Solution: The matrix can be expressed in the general form of Eq. (21.2) by dividing each row of Eq. (E21.5.1) by h^2. For the present case $[h^2 = (3/4)^2 = 0.5625]$, this gives

$$[A] - \lambda[I] = \begin{bmatrix} 3.556 - \lambda & -1.778 & 0 \\ -1.778 & 3.556 - \lambda & -1.778 \\ 0 & -1.778 & 3.556 - \lambda \end{bmatrix}$$

where $\lambda = p^2$. Before proceeding, the determinant can be expanded by minors to give

$$-\lambda^3 + 10.667\,\lambda^2 - 31.607\,\lambda + 22.487$$

The same evaluation can be implemented with the Fadeev-Leverrier method. According to Eq. (21.20)

$$[B]_2 = \begin{bmatrix} 3.556 & -1.778 & 0 \\ -1.778 & 3.556 & -1.778 \\ 0 & -1.778 & 3.556 \end{bmatrix}$$

Therefore, according to Eq. (21.21)

$$p_2 = 3.556 + 3.556 + 3.556 = 10.667 \tag{E21.6.1}$$

This value can be substituted into Eq. (21.22) to give

$$[B]_1 = \begin{bmatrix} 3.556 & -1.778 & 0 \\ -1.778 & 3.556 & -1.778 \\ 0 & -1.778 & 3.556 \end{bmatrix} \begin{bmatrix} -7.111 & -1.778 & 0 \\ -1.778 & -7.111 & -1.778 \\ 0 & -1.778 & -7.111 \end{bmatrix}$$

which can be evaluated to yield

$$[B]_1 = \begin{bmatrix} -22.123 & 6.321 & 3.160 \\ 6.321 & -18.963 & 6.321 \\ 3.160 & 6.321 & -22.123 \end{bmatrix}$$

Therefore, Eq. (21.23) can be employed to compute

$$p_1 = \frac{1}{2}(-22.123 - 18.963 - 22.123) = -31.605 \tag{E21.6.2}$$

This result in turn can be substituted into Eq. (21.26) to give

$$[B]_0 = \begin{bmatrix} 3.556 & -1.778 & 0 \\ -1.778 & 3.556 & -1.778 \\ 0 & -1.778 & 3.556 \end{bmatrix} \begin{bmatrix} 9.481 & 6.321 & 3.160 \\ 6.321 & 12.642 & 6.321 \\ 3.160 & 6.321 & 9.481 \end{bmatrix}$$

which can be evaluated to yield

$$[B]_0 = \begin{bmatrix} 22.475 & 0 & 0 \\ 0 & 22.475 & 0 \\ 0 & 0 & 22.475 \end{bmatrix}$$

Therefore, Eq. (21.27) can be employed to compute

$$p_0 = \frac{1}{3}(22.475 + 22.475 + 22.475) = 22.475 \tag{E21.6.3}$$

Substituting Eqs. (E21.6.1) through (E21.6.2) into Eq. (21.19) gives the resulting polynomial

$$-\lambda^3 + 10.667\lambda^2 - 31.605\lambda + 22.475 = 0$$

which, aside from slight differences due to round-off error, is identical to the result obtained by expanding the determinant using minors.

The matrix inverse can also be computed with Eq. (21.28)

$$[A]^{-1} = \frac{1}{22.475}$$

$$\times \begin{bmatrix} -22.123 - (-31.605) & 6.321 & 3.160 \\ 6.321 & -18.963 - (-31.605) & 6.321 \\ 3.160 & 6.321 & -22.123 - (-31.605) \end{bmatrix}$$

$$[A]^{-1} = \begin{bmatrix} 0.422 & 0.281 & 0.141 \\ 0.281 & 0.562 & 0.281 \\ 0.141 & 0.281 & 0.422 \end{bmatrix}$$

which when multiplied by $[A]$ yields $[I]$ (see Prob. 21.14 at the end of the chapter).

The Fadeev-Leverrier method can be programmed concisely. Figure 21.8 lists pseudocode for this purpose. Notice that a subroutine is required to compute the trace.

```
Pn = -1
[B] = [A]
CALL trace([B],trace n)
Pn-1 = trace
DOFOR ii = n-2 to 0 step -1
    DOFOR i = 1 to n
        bi,i = bi,i - Pii+1
    ENDDO
    [B] = [A]·[B]
    CALL trace([B],trace n)
    Pii = trace/(n-ii)
ENDDO
```

Figure 21.8
Code for the Fadeev-Leverrier method. Notice that a subroutine is required to compute the trace.

Determination of the Eigenvalues as the Roots of the Characteristic Polynomial. After applying the Fadeev-Leverrier technique, the roots of the resulting polynomial must be evaluated. Although the techniques from Part Two could be employed, they have two major disadvantages in this regard. First, they all require good initial guesses for each root. Second, they are not designed to locate complex roots. Although the Newton-Raphson method can be modified to overcome the second shortcoming, it requires good complex guesses to attain convergence. Consequently, other approaches, that exploit the mathematical properties of polynomials, are usually the method of choice. We will now briefly describe one of these approaches—Bairstow's method.

Before launching into a mathematical description of the technique, some orientation might be useful. Recall from Chap. 12 that polynomials are typically formulated as

$$f_n(x) = a_0 + a_1 x + a_2 x^2 + \cdots + a_n x^n \tag{21.29}$$

Although this is a familiar format, it is not necessarily the best expression to understand the polynomial's mathematical behavior. For example, a fifth-order polynomial might be expressed alternatively as

$$f_5(x) = (x + 1)(x - 4)(x - 5)(x + 7)(x - 2) \qquad (21.29a)$$

If this version were multiplied out and like terms collected, a polynomial of the form of Eq. (21.29) would be obtained. However, the format of Eq. (21.29a) has the advantage that it clearly indicates the function's roots. Thus, it is apparent that $x = -1, 4, 5, -7$, and 2 are all roots because each causes Eq. (21.29a) to become zero.

Now suppose that we divide this fifth-order polynomial by any of its factors, for example, $x + 7$. For this case the result would be a fourth-order polynomial with a remainder of zero. In contrast, if we divided by a factor that is not a root (for example, $x + 6$), the quotient would again be a fourth-order polynomial. However, for this case, a remainder would result.

On the basis of the above, we can elaborate an algorithm for determining a root of a polynomial: (1) Guess a value for the root, $x = t$, (2) divide the polynomial by the factor $x - t$, and (3) determine whether there is a remainder. If not, the guess was perfect and the root is equal to t. If there is a remainder, the guess can be systematically adjusted and the procedure repeated until the remainder disappears and a root is located. After this is accomplished, the entire procedure can be repeated for the quotient to locate another root.

Bairstow's method is generally based on this approach. Consequently, it hinges on the mathematical process of dividing a polynomial by a factor. Recall from algebra that synthetic division involves dividing a polynomial by a factor, $x - t$. For example, the general polynomial [Eq. (21.29)]

$$f_n(x) = a_0 + a_1x + a_2x^2 + \cdots + a_nx^n$$

can be divided by the factor $x - t$ to yield a second polynomial that is one order lower,

$$f_{n-1}(x) = b_1 + b_2x + b_3x^2 + \cdots + b_nx^{n-1}$$

with a remainder $R = b_0$ where the coefficients can be calculated by the recurrence relationship

$$b_n = a_n$$

$$b_i = b_{i+1}t + a_i \qquad \text{for } i = n - 1 \text{ to } 0$$

Note that if t were a root of the original polynomial, the remainder, b_0, would equal zero.

In order to permit the evaluation of complex roots, Bairstow's method divides the polynomial by a quadratic factor, $x^2 - rx - s$. If this is done to Eq. (21.29), the result is a new polynomial,

$$f_{n-2}(x) = b_nx^{n-2} + b_{n-1}x^{n-3} + \cdots + b_2$$

with a remainder

$$R = b_1(x - r) + b_0 \qquad (21.30)$$

As with normal synthetic division, a simple recurrence relationship can be used to perform the division by the quadratic factor:

$$b_n = a_n \qquad (21.31a)$$

$$b_{n-1} = a_{n-1} + rb_n \tag{21.31b}$$

$$b_i = a_i + rb_{i+1} + sb_{i+2} \qquad \text{for } i = n - 2 \text{ to } 0 \tag{21.31c}$$

The quadratic factor is introduced to allow the determination of complex roots. This relates to the fact that, if the coefficients of the original polynomial are real, the complex roots occur in conjugate pairs. If $x^2 - rx - s$ is an exact divisor of the polynomial, complex roots can be determined by the quadratic formula. Thus, the method reduces to determining the values of r and s that make the quadratic factor an exact divisor. In other words, the values that make the remainder term equal to zero.

Inspection of Eq. (21.30) leads us to conclude that in order for the remainder to be zero, b_1 and b_0 must be zero. Because it is unlikely that our initial guesses at the values of r and s will lead to this result, we must determine a systematic way to modify our guesses so that b_0 and b_1 approach zero. To do this, Bairstow's method uses a strategy similar to the Newton-Raphson approach. Because both b_0 and b_1 are functions of both r and s, they can be expanded using a Taylor series as in [recall Eq. (3.37)]

$$b_1(r + \Delta r, s + \Delta s) \simeq b_1 + \frac{\partial b_1}{\partial r} \Delta r + \frac{\partial b_1}{\partial s} \Delta s$$
$$\tag{21.32}$$
$$b_0(r + \Delta r, s + \Delta s) \simeq b_0 + \frac{\partial b_0}{\partial r} \Delta r + \frac{\partial b_0}{\partial s} \Delta s$$

where the values on the right-hand side are all evaluated at r and s. Notice that second- and higher-order terms have been neglected. This represents an implicit assumption that Δr and Δs are small enough that the higher-order terms are negligible. Another way of expressing this assumption is to say that the initial guesses are adequately close to the values of r and s at the roots.

The changes, Δr and Δs, needed to improve our guesses can be estimated by setting Eq. (21.32) equal to zero to give

$$\frac{\partial b_1}{\partial r} \Delta r + \frac{\partial b_1}{\partial s} \Delta s = -b_1 \tag{21.33}$$

$$\frac{\partial b_0}{\partial r} \Delta r + \frac{\partial b_0}{\partial s} \Delta s = -b_0 \tag{21.34}$$

If the partial derivatives of the b's can be determined, these are a system of two equations that can be solved simultaneously for the two unknowns—Δr and Δs. Bairstow showed that the partial derivatives can be obtained by a synthetic division of the b's in a fashion similar to the way in which the b's were derived themselves,

$$c_n = b_n \tag{21.35a}$$

$$c_{n-1} = b_{n-1} + rc_n \tag{21.35b}$$

$$c_i = b_i + rc_{i+1} + sc_{i+2} \qquad \text{for } i = n - 2 \text{ to } 1 \tag{21.35c}$$

where $\partial b_0 / \partial r = c_1$, $\partial b_0 / \partial s = \partial b_1 / \partial r = c_2$, and $\partial b_1 / \partial s = c_3$. Thus, the partial derivatives are obtained by synthetic division of the b's. Then the partial derivatives can be substituted into Eqs. (21.33) and (21.34) along with the b's to give

$$c_2 \Delta r + c_3 \Delta s = -b_1$$

$$c_1 \Delta r + c_2 \Delta s = -b_0$$

These equations can be solved for Δr and Δs, which can in turn be employed to improve the initial guesses of r and s. At each step, an approximate error in r and s can be estimated as in

$$|\epsilon_{a,r}| = \left| \frac{\Delta r}{r} \right| 100\% \tag{21.36}$$

and

$$|\epsilon_{a,s}| = \left| \frac{\Delta s}{s} \right| 100\% \tag{21.37}$$

When both of these error estimates fall below a prespecified stopping criterion, ϵ_s, the values of the roots can be determined by

$$x = \frac{r \pm \sqrt{r^2 + 4s}}{2} \tag{21.38}$$

At this point, three possibilities exist:

1. *The quotient is a third-order polynomial or greater.* For this case, Bairstow's method would be applied to the quotient to evaluate new values for r and s. The previous values of r and s can serve as the starting guesses for this application.
2. *The quotient is a quadratic.* For this case, the remaining two roots could be evaluated directly with Eq. (21.38).
3. *The quotient is a first-order polynomial.* For this case, the remaining single root can be evaluated simply as

$$x = -\frac{s}{r} \tag{21.39}$$

EXAMPLE 21.7   Bairstow's Method

Problem Statement: Employ Bairstow's method to determine the roots of the polynomial

$$f(x) = x^5 - 3.5 x^4 + 2.75 x^3 + 2.125 x^2 - 3.875 x + 1.25$$

Use initial guesses of $r = s = -1$ and iterate to a level of $\epsilon_s = 1\%$.

Solution: Equations (21.31) and (21.35) can be applied to compute

$$b_5 = 1 \quad b_4 = -4.5 \quad b_3 = 6.25 \quad b_2 = 0.375 \quad b_1 = -10.5 \quad b_0 = 11.375$$

$$c_5 = 1 \quad c_4 = -5.5 \quad c_3 = 10.75 \quad c_2 = -4.875 \quad c_1 = -16.375$$

Thus, the simultaneous equations to solve for Δr and Δs are

$$-4.875 \Delta r + 10.75 \Delta s = 10.5$$

$$-16.375 \Delta r - 4.875 \Delta s = -11.375$$

which can be solved for $\Delta r = 0.3558$ and $\Delta s = 1.1381$. Therefore, our original guesses can be corrected to

$$r = -1 + 0.3558 = -0.6442$$

$$s = -1 + 1.1381 = 0.1381$$

and the approximate errors can be evaluated by Eq. (21.36)

$$|\epsilon_{a,r}| = \left| \frac{0.3558}{-0.6442} \right| 100\% = 55.23\%$$

and Eq. (21.37)

$$|\epsilon_{a,s}| = \left| \frac{1.1381}{0.1381} \right| 100\% = 824.1\%$$

Next, the computation is repeated using the revised values for r and s. Applying Eqs. (21.31) and (21.35) yields

$$b_5 = 1 \quad b_4 = -4.1442 \quad b_3 = 5.5578 \quad b_2 = -2.0276 \quad b_1 = -1.8013 \quad b_0 = 2.1304$$

$$c_5 = 1 \quad c_4 = -4.7884 \quad c_3 = 8.7806 \quad c_2 = -8.3454 \quad c_1 = 4.7874$$

Therefore, we must solve

$$-8.3454\,\Delta r + 8.7806\,\Delta s = 1.8013$$

$$4.7874\,\Delta r - 8.3454\,\Delta s = -2.1304$$

for $\Delta r = 0.1331$ and $\Delta s = 0.3316$, which can be used to correct the root estimates as in

$$r = -0.6442 + 0.1331 = -0.5111$$

$$s = 0.1381 + 0.3316 = 0.4697$$

where

$$|\epsilon_{a,r}| = \left| \frac{0.1331}{-0.5111} \right| 100\% = 26.0\%$$

and

$$|\epsilon_{a,s}| = \left| \frac{0.3316}{0.4697} \right| 100\% = 70.6\%$$

The computation can be continued with the result that after four iterations the method converges on values of $r = -0.5\,(|\epsilon_{a,r}| = 0.063\%)$ and $s = 0.5\,(|\epsilon_{a,s}| = 0.040\%)$. Equation (21.38) can then be employed to evaluate the roots as

$$x = \frac{-0.5 \pm \sqrt{(-0.5)^2 + 4(0.5)}}{2} = 0.5, -1.0$$

At this point, the quotient is the cubic equation

$$f(x) = x^3 - 4x^2 + 5.25x - 2.5$$

Bairstow's method can be applied to this polynomial using the results of the previous step, $r = -0.5$ and $s = 0.5$, as starting guesses to yield

$$b_3 = 1 \qquad b_2 = -4.5 \qquad b_1 = 8 \qquad b_0 = -8.75$$

$$c_3 = 1 \qquad c_2 = -5 \qquad c_1 = 11$$

Thus, the simultaneous equations to solve for Δr and Δs are

$$-5\,\Delta r + 1\,\Delta s = -8$$

$$11\,\Delta r - 5\,\Delta s = 8.75$$

which can be solved for $\Delta r = 2.232$ and $\Delta s = 3.161$. Therefore, the initial values can be corrected to give

$$r = -0.5 + 2.232 = 1.732$$

$$s = 0.5 + 3.161 = 3.661$$

which represent approximate errors of 128.9% and 86.3%, respectively. Continuing the procedure for four additional iterations yields estimates of $r = 2$ and $s = -1.249$, which can be used to compute

$$x = \frac{2 \pm \sqrt{2^2 + 4(-1.249)}}{2} = 1 \pm 0.499i$$

At this point, the quotient is a first-order polynomial that can be directly evaluated by Eq. (21.39) to determine the fifth root: 2.

Note that the heart of Bairstow's method is the evaluation of the b's and c's via Eqs. (21.31) and (21.35). One of the method's primary strengths is the concise way in which these recurrence relationships can be programmed.

Figure 21.9 lists pseudocode to implement Bairstow's method. The heart of the algorithm consists of the loop to evaluate the b's and c's. Also notice that the code to solve the simultaneous equations checks to prevent division by zero. If this is the case, the values of r and s are perturbed slightly and the procedure is begun again. In addition, the algorithm places a user-defined upper limit on the number of iterations (MAXIT) and should be designed to avoid division by zero while calculating the error estimates.

EXAMPLE 21.8   Roots of the Characteristic Polynomial

Problem Statement: Determine the roots of the characteristic polynomial that was generated in Example 21.6, and compare the resulting eigenvalues with those calculated in part (c) of Example 21.5.

Solution: The polynomial

$$-\lambda^3 + 10.667\,\lambda^2 - 31.605\,\lambda + 22.475 = 0$$

```
INPUT n (order of polynomial)
INPUT a_i , i = 0 to n (coefficients)
INPUT r,s (initial guesses for roots)
INPUT ε_s (stopping criterion)
INPUT maxit (maximum iterations)
DOWHILE (n ≥ 3) and (iter < maxit)
    iter = 0
    ε_{a,1} = 1.1 ε_s; ε_{a,2} = 1.1ε_s
    DOWHILE (ε_{a,1} > ε_s or ε_{a,2} > ε_s) and (iter < maxit)
        iter = iter + 1
        b_n = a_n
        b_{n-1} = a_{n-1} + rb_n
        c_n = b_n
        c_{n-1} = b_{n-1} + rc_n
        DOFOR i = n-2 to 0 step -1
            b_i = a_i + rb_{i+1} + sb_{i+2}
            c_i = b_i + rc_{i+1} + sc_{i+2}
        ENDDO
        determinant = c_2^2 - c_3 c_1
        IF (determinant ≠ 0)
            Δr = (-b_1 c_2 + b_0 c_3 )/determinant
            Δs = (-b_0 c_2 + b_1 c_1 )/determinant
            r = r + Δr
            s = s + Δs
            ε_{a,1} = (Δr/r)·100
            ε_{a,2} = (Δs/s)·100
        ELSE
            r = r+1
            s = s+1
            iter = 0
        ENDIF
    ENDDO
    CALL roots (determine roots of x^2 - rx - s)
    n = n - 2
    DOFOR ii = 0 to n
        a_{ii} = b_{ii+2}
    ENDDO
ENDDO
IF (iter < maxit)
    IF (n = 2)
        r = a_1 /a_2 , s = a_0 /a_2
        CALL roots
    ELSE
        single root = -a_0 /a_1
    ENDIF
ELSE
    OUTPUT "maximum iterations exceeded"
ENDIF
```

Figure 21.9
Algorithm for implementing
Bairstow's method.

can be evaluated with Bairstow's method to yield

$$\lambda = 1.041,\ 3.555,\ 6.071$$

Because $\lambda = p^2$, the square root of these values can be taken to give

$$p = 1.020,\ 1.885,\ 2.464$$

which are the same results as in Example 21.5*c* (with small discrepancies due to round-off error).

Computer Program for the Polynomial Method. A computer program for the polynomial method can be simply produced by combining the code for the Fadeev-Leverrier method (Fig. 21.8) with that for Bairstow's method (Fig. 21.9). This will be left as a homework exercise.

21.2.5 The Power Method

The *power method* is an iterative approach that can be employed to determine the largest eigenvalue. With slight modification, it can also be employed to determine the smallest and the intermediate values. It has the additional benefit that the corresponding eigenvector is obtained as a by-product of the method.

Determination of the Largest Eigenvalue. In order to implement the power method, the system being analyzed must be expressed in the form

$$[A]\{X\} = \lambda\{X\} \tag{21.40}$$

As illustrated by the following example, the approach will eventually yield the highest eigenvalue and its associated eigenvector.

EXAMPLE 21.9   *Power Method for Highest Eigenvalue*

Problem Statement: Employ the power method to determine the highest eigenvalue for part (*c*) of Example 21.5.

Solution: The system is first written in the form of Eq. (21.40)

$$
\begin{aligned}
3.556x_1 - 1.778x_2 &= \lambda x_1 \\
-1.778x_1 + 3.556x_2 - 1.778x_3 &= \lambda x_2 \\
-1.778x_2 + 3.556x_3 &= \lambda x_3
\end{aligned}
$$

Then, assuming the *x*'s on the left-hand side of the equation are equal to 1,

$$
\begin{aligned}
3.556(1) - 1.778(1) &= 1.778 \\
-1.778(1) + 3.556(1) - 1.778(1) &= 0 \\
-1.778(1) + 3.556(1) &= 1.778
\end{aligned}
$$

Next, the right-hand side is normalized by 1.778 to make the largest element equal to 1.

$$\begin{Bmatrix} 1.778 \\ 0 \\ 1.778 \end{Bmatrix} = 1.778 \begin{Bmatrix} 1 \\ 0 \\ 1 \end{Bmatrix}$$

Thus, the first estimate of the eigenvalue is 1.778. This iteration can be expressed concisely in matrix form as

$$\begin{bmatrix} 3.556 & -1.778 & 0 \\ -1.778 & 3.556 & -1.778 \\ 0 & -1.778 & 3.556 \end{bmatrix} \begin{Bmatrix} 1 \\ 1 \\ 1 \end{Bmatrix} = \begin{Bmatrix} 1.778 \\ 0 \\ 1.778 \end{Bmatrix} = 1.778 \begin{bmatrix} 1 \\ 0 \\ 1 \end{bmatrix}$$

The next iteration consists of multiplying $[A]$ by $[1 \quad 0 \quad 1]^T$ to give

$$\begin{bmatrix} 3.556 & -1.778 & 0 \\ -1.778 & 3.556 & -1.778 \\ 0 & -1.778 & 3.556 \end{bmatrix} \begin{Bmatrix} 1 \\ 0 \\ 1 \end{Bmatrix} = \begin{Bmatrix} 3.556 \\ -3.556 \\ 3.556 \end{Bmatrix} = 3.556 \begin{Bmatrix} 1 \\ -1 \\ 1 \end{Bmatrix}$$

Therefore, the eigenvalue estimate for the second iteration is 3.556 which can be employed to determine the error estimate

$$|\epsilon_a| = \left| \frac{3.556 - 1.778}{3.556} \right| \times 100\% = 50\%$$

The process can then be repeated.
Third iteration:

$$\begin{bmatrix} 3.556 & -1.778 & 0 \\ -1.778 & 3.556 & -1.778 \\ 0 & -1.778 & 3.556 \end{bmatrix} \begin{Bmatrix} 1 \\ -1 \\ 1 \end{Bmatrix} = \begin{Bmatrix} 5.334 \\ -7.112 \\ 5.334 \end{Bmatrix} = -7.112 \begin{Bmatrix} -0.75 \\ 1 \\ -0.75 \end{Bmatrix}$$

where $|\epsilon_a| = 150\%$ (which is high because of the sign change).
Fourth iteration:

$$\begin{bmatrix} 3.556 & -1.778 & 0 \\ -1.778 & 3.556 & -1.778 \\ 0 & -1.778 & 3.556 \end{bmatrix} \begin{Bmatrix} -0.75 \\ 1 \\ -0.75 \end{Bmatrix} = \begin{Bmatrix} -4.445 \\ 6.223 \\ -4.445 \end{Bmatrix} = 6.223 \begin{Bmatrix} -0.714 \\ 1 \\ -0.714 \end{Bmatrix}$$

where $|\epsilon_a| = 214\%$ (again inflated because of sign change).
Fifth iteration:

$$\begin{bmatrix} 3.556 & -1.778 & 0 \\ -1.778 & 3.556 & -1.778 \\ 0 & -1.778 & 3.556 \end{bmatrix} \begin{Bmatrix} -0.714 \\ 1 \\ -0.714 \end{Bmatrix} = \begin{Bmatrix} -4.317 \\ 6.095 \\ -4.317 \end{Bmatrix} = 6.095 \begin{Bmatrix} -0.708 \\ 1 \\ -0.708 \end{Bmatrix}$$

where $|\epsilon_a| = 2.1\%$.

Sixth iteration:

$$\begin{bmatrix} 3.556 & -1.778 & 0 \\ -1.778 & 3.556 & -1.778 \\ 0 & -1.778 & 3.556 \end{bmatrix} \begin{Bmatrix} -0.708 \\ 1 \\ -0.708 \end{Bmatrix} = \begin{Bmatrix} -4.296 \\ 6.074 \\ -4.296 \end{Bmatrix} = 6.074 \begin{Bmatrix} -0.707 \\ 1 \\ -0.707 \end{Bmatrix}$$

where $|\epsilon_a| = 0.35\%$.

Thus, the normalizing factor is converging on the value of 6.070 ($= 2.4637^2$) obtained in part (c) of Example 21.5.

Note that there are some instances where the power method will converge to the second-largest eigenvalue instead of to the largest. James, Smith, and Wolford (1985) provide an illustration of such a case. Other special cases are discussed in Fadeev and Fadeeva (1963).

Determination of the Smallest Eigenvalue. There are often cases in engineering where we are interested in determining the smallest eigenvalue. Such was the case for the rod in Fig. 21.6 where the smallest eigenvalue could be used to identify a critical buckling load. This can be done by applying the power method to the matrix inverse of [A]. For this case, the power method will converge on the largest value of $1/\lambda$—in other words, the smallest value of λ.

EXAMPLE 21.10   **Power Method for Lowest Eigenvalue**

Problem Statement: Employ the power method to determine the lowest eigenvalue for part (c) of Example 21.5.

Solution: Recall from Example 21.6 that the matrix inverse is

$$[A]^{-1} = \begin{bmatrix} 0.422 & 0.281 & 0.141 \\ 0.281 & 0.562 & 0.281 \\ 0.141 & 0.281 & 0.422 \end{bmatrix}$$

Using the same format as in Example 21.9, the power method can be applied to this matrix.

First iteration:

$$\begin{bmatrix} 0.422 & 0.281 & 0.141 \\ 0.281 & 0.562 & 0.281 \\ 0.141 & 0.281 & 0.422 \end{bmatrix} \begin{Bmatrix} 1 \\ 1 \\ 1 \end{Bmatrix} = \begin{Bmatrix} 0.844 \\ 1.124 \\ 0.844 \end{Bmatrix} = 1.124 \begin{Bmatrix} 0.751 \\ 1.000 \\ 0.751 \end{Bmatrix}$$

Second iteration:

$$\begin{bmatrix} 0.422 & 0.281 & 0.141 \\ 0.281 & 0.562 & 0.281 \\ 0.141 & 0.281 & 0.422 \end{bmatrix} \begin{Bmatrix} 0.751 \\ 1 \\ 0.751 \end{Bmatrix} = \begin{Bmatrix} 0.704 \\ 0.984 \\ 0.704 \end{Bmatrix} = 0.984 \begin{Bmatrix} 0.751 \\ 1.000 \\ 0.715 \end{Bmatrix}$$

where $|\epsilon_a| = 14.6\%$.

Third iteration:

$$\begin{bmatrix} 0.422 & 0.281 & 0.141 \\ 0.281 & 0.562 & 0.281 \\ 0.141 & 0.281 & 0.422 \end{bmatrix} \begin{Bmatrix} 0.715 \\ 1 \\ 0.715 \end{Bmatrix} = \begin{Bmatrix} 0.684 \\ 0.964 \\ 0.684 \end{Bmatrix} = 0.964 \begin{Bmatrix} 0.709 \\ 1.000 \\ 0.709 \end{Bmatrix}$$

where $|\epsilon_a| = 4\%$.

Thus, after only three iterations, the result is converging on the value of 0.955, which is the reciprocal of the smallest eigenvalue, 1.0472, obtained in Example 21.4.

Determination of Intermediate Eigenvalues. After finding the largest eigenvalue, it is possible to determine the next highest by replacing the original matrix by one that includes only the remaining eigenvalues. The process of removing the largest known eigenvalue is called *deflation*. The technique outlined here, *Hotelling's method,* is designed for symmetric matrices. This is because it exploits the orthogonality of the eigenvectors of such matrices which can be expressed as

$$\{X\}_i^T \{X\}_j = \begin{cases} 0 & \text{for } i \neq j \\ 1 & \text{for } i = j \end{cases} \tag{21.41}$$

where the components of the eigenvector $\{X\}$ have been normalized so that $\{X\}^T\{X\} = 1$; that is, so that the sum of the squares of the components equals 1. This can be accomplished by dividing each of the elements by the normalizing factor

$$\sqrt{\sum_{k=1}^{n} x_k^2}$$

Now a new matrix $[A]_2$ can be computed as

$$[A]_2 = [A]_1 - \lambda_1 \{X\}_1 \{X\}_1^T \tag{21.42}$$

where $[A]_1$ is the original matrix and λ_1 is the largest eigenvalue. If the power method is applied to this matrix, the iteration process will converge to the second largest eigenvalue, λ_2. To show this, first postmultiply Eq. (21.42) by $\{X\}_1$

$$[A]_2\{X\}_1 = [A]_1\{X\}_1 - \lambda_1\{X\}_1\{X\}_1^T\{X\}_1$$

Invoking the orthogonality principle converts this equation to

$$[A]_2\{X\}_1 = [A]_1\{X\}_1 - \lambda_1\{X\}_1 = 0$$

where the right-hand side is equal to zero according to Eq. (21.40). Thus, $[A]_2\{X\}_1 = 0$. Consequently, $\lambda = 0$ and $\{X\} = \{X\}_1$ is a solution to $[A]_2\{X\} = \lambda\{X\}$. In other words, the $[A]_2$ has eigenvalues of 0, λ_2, λ_3, . . . , λ_n. The largest eigenvalue, λ_1, has been replaced by a 0 and, therefore, the power method will converge on the next biggest λ_2.

The above process can be repeated by generating a new matrix $[A]_3$, etc. Although in theory this process could be continued to determine the remaining eigenvalues, it is limited by the fact that errors in the eigenvectors are passed along at each step. Thus, it is only of value in determining several of the highest eigenvalues. Although this is somewhat of a shortcoming, such information is precisely what is required in many engineering problems.

21.2.6 Other Methods

A wide variety of additional methods are available for solving eigenvalue problems. Most are based on a two-step process. The first step involves transforming the original matrix to a simpler form (e.g., tridiagonal) which retains all the original eigenvalues. Then, iterative methods are used to determine these eigenvalues.

Many of these approaches are designed for special types of matrices. In particular, a variety of techniques are devoted to symmetric systems. For example, *Jacobi's* method transforms a symmetric matrix to a diagonal matrix by eliminating off-diagonal terms in a systematic fashion. Unfortunately, the method requires an infinite number of operations because the removal of each nonzero element often creates a new nonzero value at a previous zero element. Although an infinite time is required to create all nonzero off-diagonal elements, the matrix will eventually tend toward a diagonal form. Thus, the approach is iterative in that it is repeated until the off-diagonal terms are "sufficiently" small.

Given's method also involves transforming a symmetric matrix into a simpler form. However, in contrast to the Jacobi method, the simpler form is tridiagonal. In addition, it differs in that the zeros that are created in off-diagonal positions are retained. Consequently, it is finite and, thus, more efficient than Jacobi.

Householder's method also transforms a symmetric matrix into a tridiagonal form. It is a finite method and is more efficient than Given's approach in that it reduces whole rows and columns of off-diagonal elements to zero.

Once a tridiagonal system is obtained from Given's or Householder's method, the remaining step involves finding the eigenvalues. A direct way to do this is to expand the determinant. The result is a sequence of polynomials which can be evaluated iteratively for the eigenvalues.

Aside from symmetric matrices, there are also techniques that are available when all eigenvalues of a general matrix are required. These include the *LR method* of Rutishauser and the *QR method* of Francis. Although the QR method is less efficient, it is usually the preferred approach because it is more stable. As such, it is considered to be the best general-purpose solution method.

Finally, it should be mentioned that the aforementioned techniques are often used in tandem to capitalize on their respective strengths. For example, it should be noted that Given's and Householder's methods can also be applied to nonsymmetric systems. The result will not be tridiagonal but rather a special type called the *Hessenberg form*. One approach is to exploit the speed of Householder's approach by employing it to transform the matrix to this form and then use the stable QR algorithm to find the eigenvalues. Additional information on these and other issues related to eigenvalues can be found in Ralston and Rabinowitz (1978), Wilkinson (1965), Fadeev and Fadeeva (1963), and Householder (1953). Computer codes can be found in a number of sources including Press et al. (1987). Rice (1983) discusses available software packages.

PROBLEMS

Hand Calculations

21.1 A steady-state heat balance for a rod can be represented as

$$\frac{d^2T}{dx^2} + h'(T_a - T) = 0$$

Obtain an analytical solution for a 20-m rod with $h' = 0.05$, $T_a = 30$, $T(0) = 100$, and $T(20) = 50$.

21.2 Use the shooting method to solve Prob. 21.1.

21.3 Use the finite-difference approach with $\Delta x = 2$ and $\Delta x = 1$ to solve Prob. 21.1.

21.4 Use the shooting method to solve

$$8\frac{d^2y}{dx^2} + 16\frac{dy}{dx} - 4y = 20$$

with the boundary conditions $y(0) = 5$ and $y(20) = 2$.

21.5 Solve Prob. 21.4 with the finite-difference approach using $\Delta x = 4$.

21.6 Repeat Example 21.3 but for three masses. Produce a plot like Fig. 21.5 to identify the principle modes of vibration.

21.7 Repeat Example 21.5 but for five interior points ($h = 3/6$).

21.8 Use minors to expand the determinant of

$$\begin{bmatrix} 2-\lambda & 8 & 10 \\ 8 & 3-\lambda & 4 \\ 10 & 4 & 7-\lambda \end{bmatrix}$$

Employ the Fadeev-Leverrier method to perform the same computation. Also compute the matrix inverse and verify that it is correct.

21.9 Employ synthetic division to divide

$$f(x) = x^4 - 5x^3 + 5x^2 + 5x - 6$$

by the factor $x - 2$. Interpret your results.

21.10 Employ synthetic division to divide

$$f(x) = x^5 - 6x^4 + x^3 - 7x^2 - 7x + 12$$

by the factor $x - 2$.

21.11 Use Bairstow's method to determine the roots of

(a) $f(x) = -2 + 6.2x - 4x^2 + 0.7x^3$

(b) $f(x) = 9.34 - 21.97x + 16.3x^2 - 3.704x^3$

(c) $f(x) = x^4 - 8.5x^3 - 35.5x^2 + 465x - 1000$

(d) $f(x) = x^3 - 6x^2 + 11x - 5.9$

(e) $f(x) = x^3 - 4.8x^2 + 7.56x - 3.92$

21.12 Use the power method to determine the highest eigenvalue and corresponding eigenvector for Prob. 21.8.

21.13 Use the power method to determine the lowest eigenvalue and corresponding eigenvector for Prob. 21.8.

21.14 Verify that the matrix inverse in Example 21.6 when multiplied by $[A]$ yields the identity matrix I.

Computer-Related Problems

21.15 Develop a user-friendly computer program to implement the shooting method for a linear, second-order ODE. Test the program by duplicating Example 21.1.

21.16 Use the program developed in Prob. 21.15 to solve Probs. 21.2 and 21.4.

21.17 Develop a user-friendly computer program to implement the finite-difference approach for solving a linear, second-order ODE. Test it by duplicating Example 21.2.

21.18 Use the program developed in Prob. 21.17 to solve Probs. 21.3 and 21.5.

21.19 Develop a subroutine to implement the Fadeev-Leverrier approach. Test it by duplicating Example 21.6.

21.20 Use the subroutine developed in Prob. 21.19 to solve Prob. 21.8.

21.21 Develop a subroutine to implement Bairstow's method. Test it by duplicating Example 21.7.

21.22 Use the subroutine developed in Prob. 21.21 to determine the roots of the equations in Prob. 21.11.

21.23 Develop a user-friendly program to implement the polynomial method. Test it by duplicating Example 21.5c.

21.24 Develop a user-friendly program to solve for the largest eigenvalue with the power method. Test it by duplicating Example 21.8.

21.25 Develop a user-friendly program to solve for the smallest eigenvalue with the power method. Test it by duplicating Example 21.9.

CHAPTER 22

Case Studies: Ordinary Differential Equations

The purpose of this chapter is to solve some ordinary differential equations using the numerical methods presented in Part Six. The equations originate from practical engineering applications. Many of these applications result in nonlinear differential equations that cannot be solved using analytic techniques. Therefore, numerical methods are usually required. Thus, the techniques for the numerical solution of ordinary differential equations are fundamental capabilities that characterize good engineering practice. The problems in this chapter illustrate some of the trade-offs associated with various methods developed in Part Six.

In *Case Study 22.1,* a differential equation is used to predict trends of computer sales. Among other things, this example illustrates how a parameter of a mathematical model is calibrated to data. The fourth-order RK method is used for this application.

Case Study 22.2 originates from a chemical-engineering problem context. It demonstrates how the transient behavior of chemical reactors can be simulated.

Case Studies 22.3, 22.4, and 22.5, which are taken from civil, electrical, and chemical engineering, respectively, deal with the solution of systems of equations. In *Case Study 22.3,* Euler's method is used because the problem does not require highly accurate results. *Case Study 22.4,* on the other hand, demands high accuracy, and as a consequence, a fourth-order RK scheme is used. In addition, both case studies deal with determining eigenvalues for engineering systems.

Case Study 22.5 employs a variety of different approaches to investigate the behavior of a swinging pendulum. This problem also utilizes two simultaneous equations. An important aspect of this example is that it illustrates how numerical methods allow nonlinear effects to be incorporated easily into an engineering analysis. Finally, *Case Study 22.6* illustrates how spreadsheets can be applied to solve ODEs numerically. Among other things, the efficiency of first- and second-order RK methods is compared.

CASE STUDY 22.1 MATHEMATICAL MODEL FOR COMPUTER SALES PROJECTIONS (GENERAL ENGINEERING)

Background: Operations and profitability at a computer company are very dependent on management's knowledge of the number of computers available on the market at any time. The extrapolation techniques discussed in Case Study 14.1 have proved unreliable and inaccurate. You have, therefore, been asked to derive a mathematical model that is capable of simulating and predicting the number of unsold computers available on the market as a function of time t. An ordinary differential equation can be developed for this purpose.

The marketing department of the company has determined from long experience and empirical observation that the expected sales rate of the computers can be described by

$$\text{Sales rate (computers sold per day)} \propto \frac{\text{number of computers on the market}}{\text{cost of individual computer}} \qquad (22.1)$$

That is, the more computers that are exposed to the public, the faster they sell; and the higher their cost, the slower they sell. Furthermore, the cost of an individual computer is related to the number of individual computers on the market, as in [recall Eq. (18.1)]

$$\text{Cost per computer(\$)} = 3000 - 1750 \frac{N}{10,000 + N} \qquad (22.2)$$

where N is the number of computers.

The time rate of change of the number of computers remaining on the market is equal to the negative of the sales rate:

$$\frac{dN}{dt} = -\text{sales rate} \qquad (22.3)$$

where the sales rate is derived by combining Eqs. (22.1) and (22.2):

$$\text{Sales rate} = k \frac{N}{3000 - 1750N/(10,000 + N)} \qquad (22.4)$$

where k is a proportionality constant having units of dollars per time. Substituting Eq. (22.4) into Eq. (22.3) yields

$$\frac{dN}{dt} = -k \frac{N}{3000 - 1750N/(10,000 + N)} \qquad (22.5)$$

Planning considerations require that an estimate be obtained on how long 50,000 new computers will remain on the market as a function of time. You have at your disposal the data from Table 14.1. Use this information to estimate the parameter k. Then use a fourth-order Runge-Kutta method to solve Eq. (22.5) from $t = 0$ to $t = 90$.

Solution: The first step in this analysis will be to determine a value for k. To do this, we can solve Eq. (22.5) for

$$k = -\frac{dN}{dt} \frac{3 \times 10^7 + 1250N}{N(10,000 + N)}$$

Thus, we could evaluate k if we had an estimate of dN/dt. This can be done on the basis of the data in Table 22.1, using finite divided differences to estimate dN/dt, as in (recall Sec. 3.5.4),

$$\frac{dN}{dt}\Big|_i \simeq \frac{N_{i+1} - N_{i-1}}{2\Delta t}$$

The results are contained in Table 22.1 and can be used to determine a mean value of $k = \$49.3$ per day.

TABLE 22.1 **Estimates of k obtained from computer sales data. The mean of k values is 49.3.**

t, days	N	dN/dt	k
0	50,000		
10	35,000	−950	44.5
20	31,000	−750	40.6
30	20,000	−600	55.0
40	19,000	−397.5	38.8
50	12,050	−400	67.8
60	11,000		

Now this value can be substituted into Eq. (22.5) to yield

$$\frac{dN}{dt} = -49.3 \frac{N}{3000 - 1750[N/(10,000 + N)]}$$

which can be integrated using a fourth-order RK method with an initial condition of $N = 50,000$ and a time step of 1 day. Note that we also performed the simulation using a time step of 0.5 day and obtained almost identical results, indicating that the accuracy using a step size of 1.0 is acceptable.

The results are depicted in Fig. 22.1 along with the data. Just as in regression, we can compute the sum of the squares of the residuals to quantify the goodness of fit. The result is 2.85×10^7. Although the fit appears to be satisfactory, we perform the computation again using values of k that are ± 20 percent of the original value of $\$49.3$ per day. Using these k's of 59.2 and 39.4 results in residual sum of the squares of 1.05×10^8 and 5.35×10^7, respectively. These simulations are also depicted in Fig. 22.1.

Next we plot the sum of the squares of the residuals versus the k values (Fig. 22.2) and fit a parabola through the points using an interpolating polynomial. We then determine the k corresponding to the minimum sum of the squares by differentiating the second-order equation, setting it equal to zero, and solving for k. The resulting value of $k = \$46.8$ per day can then be substituted into Eq. (22.5) to give

$$\frac{dN}{dt} = -46.8 \frac{N}{3000 - 1750[N/(10,000 + N)]}$$

This model yields a sum of the squares of the residuals of 2.24×10^7. It can then be used for predictive purposes. The predictions are shown along with the original data in Fig.

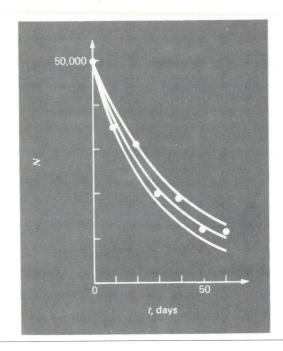

Figure 22.1
Plot of the number of computers N on the market versus time t in days. Three simulations using an ordinary differential equation model [Eq. (22.5)] are depicted for the case where $N = 50,000$ at $t = 0$. The three simulations correspond to different values of a model parameter k.

22.3. The results for $t = 55$, 65, and 90 are 11,720, 9383, and 5596, respectively. This information, which is superior to that obtained by curve fitting in Chap. 14, can then be utilized by management to guide decisions regarding the marketing of these computers.

Figure 22.2
Plot of the sum of the squares of the residuals (S_r) versus values of the model parameter k. The curve is a parabola that was fit to the three points. The point of zero slope of this curve represents an estimate of the k value ($\$46.8$/day) that corresponds to a minimum value of S_r.

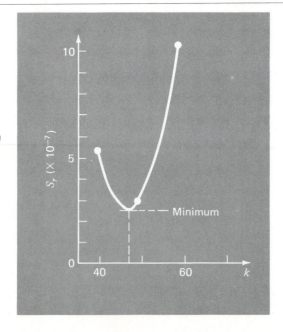

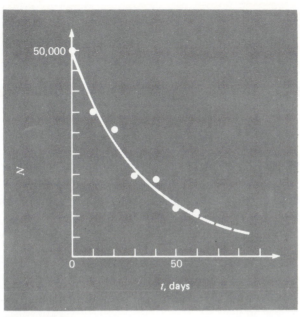

Figure 22.3
Model predictions using Eq. (22.5) with a k of $46.8/day.

CASE STUDY 22.2 | USING ODES TO ANALYZE THE TRANSIENT RESPONSE OF A REACTOR (CHEMICAL ENGINEERING)

Background: In Case Study 10.2, we analyzed the steady state of a series of reactors. In addition to steady-state computations, we might also be interested in the transient response of a completely mixed reactor. To do this, we have to develop a mathematical expression for the accumulation term in Eq. (10.6). Because accumulation represents the change in mass in the reactor per change in time, accumulation can be simply formulated as

$$\text{Accumulation} = \frac{\Delta M}{\Delta t} \tag{22.6}$$

where M is the mass of chemical in the reactor. By convention, concentration is defined as mass per unit volume, or

$$c = \frac{M}{V}$$

This equation can be solved for mass and the result ($M = cV$) can be substituted into Eq. (22.6) to reexpress accumulation in terms of concentration:

$$\text{Accumulation} = \frac{\Delta cV}{\Delta t}$$

If we assume that the volume of liquid in the reactor is constant, V can be moved outside the difference:

$$\text{Accumulation} = V \frac{\Delta c}{\Delta t}$$

Finally, as Δt approaches zero, the difference can be reexpressed as a derivative:

$$\text{Accumulation} = V \frac{dc}{dt} \tag{22.7}$$

Thus a mathematical formulation for accumulation is volume times the derivative of c with respect to t.

Solution: Equations (22.7) and (10.6) can be used to represent the mass balance for a single reactor such as the one shown in Fig. 22.4:

$$V \frac{dc}{dt} = Q c_{\text{in}} - Q c \tag{22.8}$$

This rate equation can be used to determine transient or time-variable solutions for the reactor. For example, if $c = c_0$ at $t = 0$, calculus can be employed to analytically solve Eq. (22.8) for

$$c = c_{\text{in}}(1 - e^{-(Q/V)t}) + c_0 e^{-(Q/V)t}$$

If $c_{\text{in}} = 50$ mg/m^3, $Q = 5$ m^3/min, $V = 100$ m^3, and $c_0 = 10$ mg/m^3, the equation is

$$c = 50(1 - e^{-0.05t}) + 10e^{-0.05t}$$

Figure 22.5 shows this exact, analytical solution.

Euler's method provides an alternative approach for solving Eq. (22.8). Figure 22.5 includes two solutions with different step sizes. As the step size is decreased, the numerical solution converges on the analytical solution. Thus, for this case, the numerical method can be used to check the analytical result.

Figure 22.4
A single, completely mixed reactor with an inflow and an outflow.

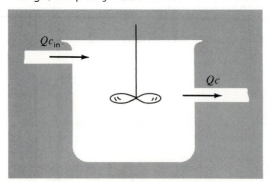

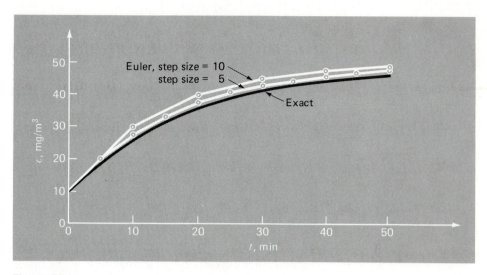

Figure 22.5
Plot of analytical and numerical solutions of Eq. 22.8. The numerical solutions are obtained with Euler's method using different step sizes.

Besides checking the results of an analytical solution, numerical solutions have added value in those situations where analytical solutions are impossible or so difficult that they are impractical. Suppose, for example, that the concentration of the inflow to the reactor is not constant but rather varies sinusoidally with time, as in (see Fig. 22.6a).

$$c_{in} = \bar{c}_{in} + c_a \sin \frac{2\pi t}{T} \tag{22.9}$$

where $\bar{c}_{in}$ is the base-inflow concentration, c_a is the amplitude of the oscillation, and T is its period (that is, the time required for a complete cycle). For this case, the mass balance equation is

$$V \frac{dc}{dt} = Q \left(\bar{c}_{in} + c_a \sin \frac{2\pi t}{T} \right) - Qc \tag{22.10}$$

It is possible to solve this equation by calculus, but the solution is time-consuming and complicated:

$$c = c_o e^{-(Q/V)t} + \bar{c}_{in}(1 - e^{-(Q/V)t})$$

$$+ \frac{Qc_a}{V\sqrt{(Q/V)^2 + (2\pi/T)^2}} \left[\sin \left(\frac{2\pi t}{T} - \tan^{-1} \frac{2\pi V}{TQ} \right) \right.$$

$$\left. + e^{-(Q/V)t} \sin \left(-\tan^{-1} \frac{2\pi V}{TQ} \right) \right] \tag{22.11}$$

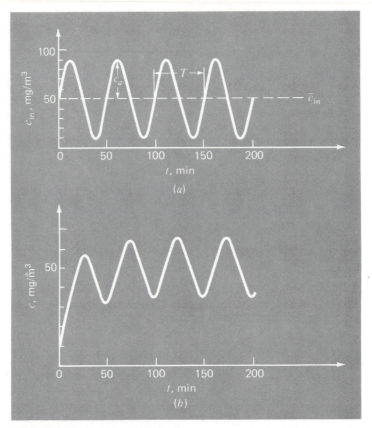

Figure 22.6
(*a*) Sinusoidal inflow concentration and (*b*) the resulting concentration in the reactor, as computed with Euler's method.

For such cases, Euler's approach offers a much easier means of obtaining the solution. Figure 22.6*b* shows the result corresponding to the situation in Fig. 22.6*a* for $c_0 = 10$ mg/m^3. The solution starts at the initial condition and then oscillates upward.

Aside from a single reactor, numerical methods have utility when simulating the dynamics of systems of reactors. For example, ODEs can be written for the five coupled reactors in Fig. 10.4. For the first reactor

$$V_1 \frac{dc_1}{dt} = Q_{01}c_{01} + Q_{31}c_3 - Q_{12}c_1 - Q_{15}c_1$$

or

$$\frac{dc_1}{dt} = \frac{50 + c_3 - 6c_1}{V_1}$$

Similarly, balances can be developed for the other reactors as

$$\frac{dc_2}{dt} = \frac{3c_1 - 3c_2}{V_2}$$

$$\frac{dc_3}{dt} = \frac{160 + c_2 - 9c_3}{V_3}$$

$$\frac{dc_4}{dt} = \frac{c_2 + 8c_3 - 11c_4 + 2c_5}{V_4}$$

$$\frac{dc_5}{dt} = \frac{3c_1 + c_2 - 4c_5}{V_5}$$

Suppose that at $t = 0$, all the concentrations in the reactors are zero. Compute how long it will take each of them to reach 90 percent of their steady-state values (from Case Study 10.2), if their volumes are

$$V_1 = 50 \qquad V_3 = 40 \qquad V_5 = 100$$

$$V_2 = 20 \qquad V_4 = 80$$

The results are depicted in Fig. 22.7. Notice that all the reactors eventually reach the steady-state concentrations previously calculated in Case Study 10.2. This serves as an independent check on both computations. In addition to their steady-state behavior, each of the reactors shows a different transient response to the introduction of chemical. These responses can be parameterized by a 90 percent response time t_{90}, which measures the time required for each reactor to reach 90 percent of its ultimate steady-state level. The times range from about 10 min for reactor 3 to about 70 min for reactor 5. The response times of reactors 4 and 5 are of particular concern because the two outflow streams for the system exit these tanks. Thus, a chemical engineer designing the system might change the flows or volumes of the reactors to speed up the response of these tanks while still maintaining the desired outputs. Numerical methods of the sort described in this part of the book can prove useful in these design calculations.

CASE STUDY 22.3   DEFLECTION OF A SAILBOAT MAST (CIVIL ENGINEERING)

Background: A sailboat similar to that of Case Studies 14.3 and 18.3 is shown in Fig. 22.8, with a uniform force f distributed along the mast. In this case, the cables supporting the mast have been removed, but the mast is mounted solidly to the deck for support.

The wind force causes the mast to deflect as depicted in Fig. 22.9. The deflection is similar to that of a cantilever beam. The following differential equation, based on the laws of mechanics, can be used to characterize this deflection:

$$\frac{d^2y}{dz^2} = \frac{f}{2EI} (L - z)^2 \tag{22.12}$$

where E is the modulus of elasticity, L is the height of the mast, and I is the moment of

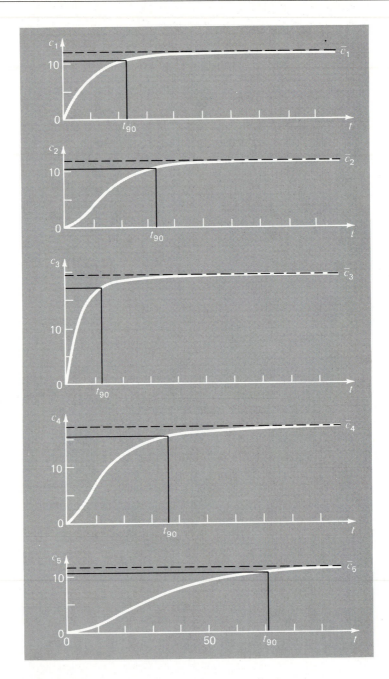

Figure 22.7
Plots of transient or dynamic response of the network of reactors from Fig. 10.4. Note that all the reactors eventually approach their steady-state concentrations c_i, previously computed in Case Study 10.2. In addition, the time to steady state is parameterized by the 90 percent response time t_{90}.

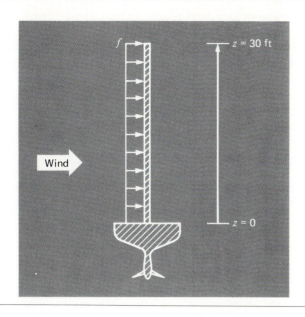

Figure 22.8
Sailboat mast subject to a
uniform force *f*.

inertia. At $z = 0$, $y = 0$ and $dy/dz = 0$. Calculate the deflection of the top of the mast where $z = L$ using both analytical and numerical methods. Assume that the hull does not rotate.

Solution: Equation (22.12) can be solved analytically for the deflection at $z = L$:

$$y(z = L) = \frac{fL^4}{8EI} \tag{22.13}$$

Figure 22.9
Deflection of a mast subjected
to a uniform force.

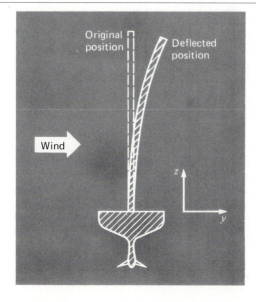

This problem involves a simple differential equation that has a solution with smooth characteristics. Furthermore, the integration interval is relatively short, and the mast deflection is small. Also, the values of f and E are based on experimental data that is variable and difficult to measure accurately. Therefore, it seems satisfactory to use a low-order, simple method to solve the differential equation. Only one starting value will be needed, and we can probably use a small step size without accumulation of excessive round-off error.

Equation (22.12) can be written as a system of two first-order equations by transforming variables. Let

$$\frac{dy}{dz} = u \qquad (22.14)$$

Therefore, Eq. (22.12) can be expressed as

$$\frac{du}{dz} = \frac{f}{2EI}(L - z)^2 \qquad (22.15)$$

This pair of differential equations can be solved simultaneously using Euler's method.

First, however, we can obtain the analytical solution for comparison. Given a uniform load of $f = 50$ lb/ft, $L = 30$ ft, $E = 1.5 \times 10^8$ lb/ft^2, and $I = 0.06$ ft^4, Eq. (22.13) can be solved for

$$y(30) = \frac{50(30)^4}{8(1.5 \times 10^8)(0.06)} = 0.5625 \text{ ft}$$

Next we can solve Eqs. (22.14) and (22.15) using Euler's method. The results for a number of step sizes are

$y(30)$	Euler's Step Size
0.5744	1.0
0.5637	0.1
0.5631	0.05

Therefore, the answers obtained appear satisfactory. The deflection of the mast is depicted in Fig. 22.10.

This computation can be employed for design purposes. It is especially valuable for cases where the wind stress is not constant but varies in a complicated manner as a function of the height above the deck. Problem 22.17 provides an example of such a situation.

Numerical methods can be employed to investigate the dynamics of the mast, in addition to the steady-state deflection. The following differential equation (Meirovitch, 1986) provides a model of its flexural vibrations

$$-\frac{\partial^2}{\partial z^2}\left[EI(z)\frac{\partial^2 y(z,\, t)}{\partial z^2}\right] + f(z,\, t) = m(z)\frac{\partial^2 y(z,\, t)}{\partial t^2} \qquad (22.16)$$

Figure 22.10
Plot of deflection of a sailboat mast as computed with Euler's method.

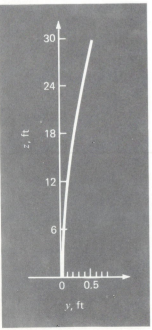

where $f(z, t)$ is force as a function of time t and height z above the deck and $m(z)$ is mass per unit length. Given the appropriate boundary conditions, this fourth-order partial differential equation can be solved for displacement as a function of both time and space.

Now, aside from the transient solution, we can develop an eigenvalue solution to identify the free vibrations of the mast. By "free vibrations" we mean for the case where $f(z, t) = 0$. In this situation, the solution of Eq. (22.16) becomes separable in space and time and, thus, can be represented as

$$y(z, t) = Y(z)F(t)$$

where $Y(z)$ and $F(t)$ are the spatial and the temporal dependence of the solution. If the frequency of $F(t)$ is denoted by ω, the eigenvalue formulation can be represented as

$$\frac{d^2}{dz^2}\left[EI(z)\,\frac{d^2Y(z)}{dz^2}\right] = \omega^2 m(x)Y(x) \tag{22.17}$$

which is subject to the boundary conditions at the deck $(z = 0)$,

$$Y(0) = 0 \tag{22.18}$$

$$\frac{dY(0)}{dy} = 0 \tag{22.19}$$

and at the top of the mast $(z = L)$

$$\frac{d^2Y(L)}{dz^2} = 0 \tag{22.20}$$

$$\frac{d^3Y(L)}{dz^3} = 0 \tag{22.21}$$

For a mast of constant thickness $(m, E, \text{ and } I = \text{constants})$, ω can be determined analytically by

$$\omega = p^2\sqrt{\frac{EI}{m}} \tag{22.22}$$

where p is a root of

$$f(p) = \cos(pL)\cosh(pL) + 1$$

For $L = 30$ ft, the first three roots are 0.062879, 0.15647, and 0.26183 ft^{-1}. If $m = 15$ lb$_m$/ft, then Eq. (22.22) can be used to compute the first three eigenvalues as 17.4, 108, and 301 s^{-1} (note that a conversion of 1 lb$_f$ = 32.2 lb$_m$ ft/s^2 must be used to obtain these results). The corresponding modes are depicted in Fig. 22.11.

Numerical methods can be used to obtain the same results. As in Sec. 21.2.3, a finite divided difference (Fig. 17.3) can be substituted for the fourth derivative in Eq. (22.17) to give

$$EI\,\frac{y_{i-2} - 4y_{i-1} + 6y_i - 4y_{i+1} + y_{i+2}}{h^4} = \omega^2 m y_i$$

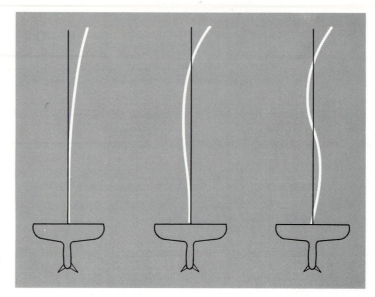

Figure 22.11
The first three vibration modes of the sailboat mast. The displacements are exaggerated to make the shapes more apparent.

or

$$y_{i-2} - 4y_{i-1} + (6 - \beta)y_i - 4y_{i+1} + y_{i+2} = 0 \tag{22.23}$$

where

$$\beta = \frac{\omega^2 m h^4}{EI} \tag{22.24}$$

The first approximation will be for the case shown in Fig. 22.12a. Because one of the boundary conditions [Eq. (22.18)] specifies that $y_0 = 0$, Eq. (22.23) has to be applied only to nodes 1 and 2. For node 1, the result is

$$y_{-1} - 4y_0 + (6 - \beta)y_1 - 4y_2 + y_3 = 0 \tag{22.25}$$

Figure 22.12
Segmentation schemes used in the numerical computation of the eigenvalues of the sailboat mast. The filled circles represent nodes for which the eigenvalue formulation is written in finite-difference format. The open circles represent nodes that are removed using boundary conditions.

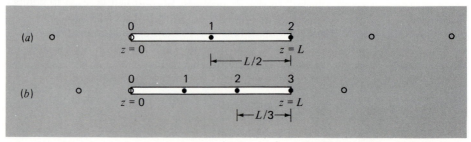

and for node 2,

$$y_0 - 4y_1 + (6 - \beta)y_2 - 4y_3 + y_4 = 0 \tag{22.26}$$

At this point, these equations cannot be solved because they contain three unknowns that lie outside the system: y_{-1}, y_3, and y_4. However, the boundary conditions can be employed to purge these unknowns from the equations.

First, we can remove y_0 because according to the boundary conditions, it is equal to zero. Then, the fact that the first derivative is zero at $z = 0$ can be employed to remove y_{-1}. This can be accomplished by realizing that the first derivative at $z = 0$ can be approximated by a centered difference

$$\frac{dy(0)}{dz} \simeq \frac{y_1 - y_{-1}}{2h}$$

which can be solved for

$$y_{-1} = y_1 - 2h\frac{dy(0)}{dz}$$

However, because the derivative is zero, $y_{-1} = y_1$. Substituting this result along with $y_0 = 0$ into Eq. (22.25) gives

$$(7 - \beta)y_1 - 4y_2 + y_3 = 0 \tag{22.27}$$

A similar approach can be used to remove the y_3. In this case, the second derivative at node 2 can be approximated by a centered difference

$$\frac{d^2y(L)}{dz^2} = \frac{y_3 - 2y_2 + y_1}{h^2}$$

which can be solved for

$$y_3 = 2y_2 - y_1 + h^2\frac{d^2y(L)}{dz^2}$$

but because the second derivative at the top of the mast is zero

$$y_3 = 2y_2 - y_1 \tag{22.28}$$

This result can be substituted into Eq. (22.27) to yield

$$(6 - \beta)y_1 - 2y_2 = 0$$

Equation (22.28) can also be substituted for the y_3 in Eq. (22.26):

$$(-2 - \beta)y_2 + y_4 = 0 \tag{22.29}$$

Finally, the y_4 can be removed from this equation by approximating the third derivative at node 2 by a centered difference (Fig. 17.3)

$$\frac{d^3y(L)}{dz^3} \simeq \frac{-y_0 + 2y_1 - 2y_3 + y_4}{2h^3}$$

which can be solved for

$$y_4 = y_0 - 2y_1 + 2y_3 + 2h^3 \frac{d^3y(L)}{dz^3}$$

Substituting Eq. (22.28) for y_3 and setting the third derivative and y_0 equal to zero gives

$$y_4 = -4y_1 + 4y_2$$

which can be substituted into Eq. (22.29) to give

$$4y_1 + (2 - \beta)y_2 = 0$$

Therefore, we now have an eigenvalue problem with two unknowns,

$$(6 - \beta)y_1 - 2y_2 = 0$$

$$4y_1 + (2 - \beta)y_2 = 0$$

Expansion of the determinant of this system gives

$$\beta^2 - 8\beta + 4 = 0$$

which can be solved for $\beta = 0.53590$ and 10.928 which, in turn, can be used to evaluate [using Eq. (22.24)] $\omega = 14.3$ and 64.6 s^{-1}. These represent errors of 18 and 40 percent when compared to the true values of 17.4 and 108 s^{-1}.

To improve these estimates, the rod can be divided into three equispaced segments as in Fig. 22.12b. The eigenvalue equation can be written in finite-difference form for nodes 1 through 3. After boundary conditions are used to remove unknowns at external nodes, the resulting simultaneous equations are

$$\begin{bmatrix} (7 - \beta) & -4 & 1 \\ -4 & (5 - \beta) & -2 \\ 2 & -4 & (2 - \beta) \end{bmatrix} \begin{Bmatrix} y_1 \\ y_2 \\ y_3 \end{Bmatrix} = \{0\}$$

The determinant can be expanded using the Fadeev-Leverrier method to give

$$-\beta^3 + 14\beta^2 - 33\beta + 4 = 0$$

which can be evaluated with Bairstow's method for $\beta = 0.1281, 2.827, 11.05$. Substituting these results into Eq. (22.24) yields the first three eigenvalue estimates of 15.73 ($\epsilon_t = 9.6\%$), 73.90 ($\epsilon_t = 32\%$), and 146.1 ($\epsilon_t = 51\%$).

CASE STUDY 22.4   SIMULATING TRANSIENT CURRENT FOR AN ELECTRIC CIRCUIT
                    (ELECTRICAL ENGINEERING)

Background: Electric circuits where the current is time-variable rather than constant are common. A transient current is established in the right-hand loop of the circuit shown in Fig. 22.13 when the switch is suddenly closed.

Equations that describe the transient behavior of the circuit in Fig. 22.13 are based on Kirchhoff's law, which states that the algebraic sum of the voltage drops around a closed loop is zero (recall Case Study 6.4). Thus

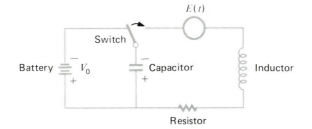

Figure 22.13
An electric circuit where the
current varies with time.

$$L\frac{di}{dt} + Ri + \frac{q}{c} - E(t) = 0 \tag{22.30}$$

where $L(di/dt)$ is the voltage drop across the inductor, L is inductance (in henrys), R is resistance (in ohms), q is the charge on the capacitor (in coulombs), C is capacitance (in farads), $E(t)$ is the time-variable voltage source (in volts), and

$$i = \frac{dq}{dt} \tag{22.31}$$

Equations (22.30) and (22.31) are a pair of first-order linear differential equations that can be solved analytically. For example, if $E(t) = E_0 \sin wt$ and $R = 0$,

$$q(t) = \frac{-E_0}{L(p^2 - w^2)} \frac{w}{p} \sin pt + \frac{E_0}{L(p^2 - w^2)} \sin wt \tag{22.32}$$

where $p = 1/\sqrt{LC}$. The values of q and dq/dt are zero for $t = 0$. Use a numerical approach to solve Eqs. (22.30) and (22.31) and compare the results with Eq. (22.32).

Solution: This problem involves a rather long integration range and demands the use of a highly accurate scheme to solve the differential equation if good results are expected. Let's assume that $L = 1$ H, $E_0 = 1$ V, $C = 0.25$ C, and $w^2 = 3.5$ s^2. This gives $p = 2$, and Eq. (22.32) becomes

$$q(t) = -1.8708 \sin 2t + 2 \sin (1.8708t)$$

for the analytical solution. This function is plotted in Fig. 22.14. The rapidly changing nature of the function places a severe requirement on any numerical procedure to find $q(t)$. Furthermore, because the function exhibits a slowly varying periodic nature as well as a rapidly varying component, long integration ranges are necessary to portray the solution. Thus, we expect that a high-order method is preferred for this problem.

However, we can try both Euler and fourth-order Runge-Kutta methods and compare the results. Using a step size of 0.1 s gives a value for q at $t = 10$ s of -6.638 with Euler's method and a value of -1.9897 with the fourth-order Runge-Kutta method. These results compare to an exact solution of -1.996 C.

Figure 22.15 shows the results of Euler integration every 1.0 s compared to the exact solution. Note that only every tenth output point is plotted. It is seen that the global error increases as t increases. This divergent behavior intensifies as t approaches infinity.

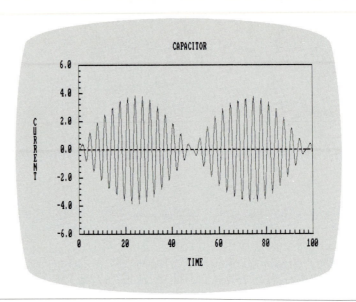

Figure 22.14
Computer screen showing the plot of the function represented by Eq. (22.32).

In addition to directly simulating a network's transient response, numerical methods can also be used to determine its eigenvalues. For example, Fig. 22.16 shows an *LC* network for which Kirchhoff's voltage law can be employed to develop the following system of ODEs:

$$-L_1 \frac{di_1}{dt} - \frac{1}{C_1} \int_{-\infty}^{t} (i_1 - i_2) \, dt = 0$$

Figure 22.15
Results of Euler integration versus exact solution. Note that only every tenth output point is plotted.

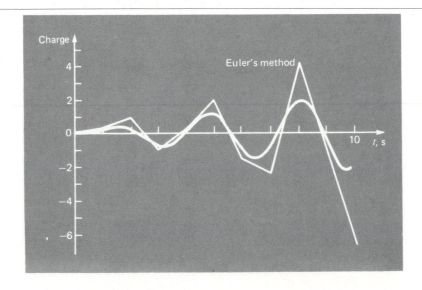

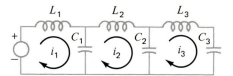

Figure 22.16
An *LC* network.

$$-L_2 \frac{di_2}{dt} - \frac{1}{C_2} \int_{-\infty}^{t} (i_2 - i_3)\, dt + \frac{1}{C_1} \int_{-\infty}^{t} (i_1 - i_2)\, dt = 0$$

$$-L_3 \frac{di_3}{dt} - \frac{1}{C_3} \int_{-\infty}^{t} i_3\, dt + \frac{1}{C_2} \int_{-\infty}^{t} (i_2 - i_3)\, dt = 0$$

Notice that we have represented the voltage drop across the capacitor as

$$V_C = \frac{1}{C} \int_{-\infty}^{t} i\, dt$$

This is an alternative and equivalent expression to the relationship used in Eq. (22.30) and introduced in Case Study 6.4.

The system of ODEs can be differentiated and rearranged to yield

$$L_1 \frac{d^2 i_1}{dt^2} + \frac{1}{C_1} (i_1 - i_2) = 0$$

$$L_2 \frac{d^2 i_2}{dt^2} + \frac{1}{C_2} (i_2 - i_3) - \frac{1}{C_1} (i_1 - i_2) = 0$$

$$L_3 \frac{d^2 i_3}{dt^2} + \frac{1}{C_3} i_3 - \frac{1}{C_2} (i_2 - i_3) = 0$$

Comparison of this system with the one in Eq. (21.4) indicates an analogy between a spring-mass system and an *LC* circuit. As was done with Eq. (21.4), the solution can be assumed to be of the form

$$i_j = A_j \sin (\omega t)$$

This solution along with its second derivative can be substituted into the simultaneous ODEs. After simplification, the result is

$$\left(\frac{1}{C_1} - L_1 \omega^2 \right) A_1 \qquad\qquad - \frac{1}{C_2} A_2 \qquad\qquad\qquad\qquad = 0$$

$$-\frac{1}{C_1} A_1 \qquad + \left(\frac{1}{C_1} + \frac{1}{C_2} - L_2 \omega^2 \right) A_2 \qquad\qquad - \frac{1}{C_2} A_3 \qquad\qquad = 0$$

$$-\frac{1}{C_2} A_2 \qquad\qquad + \left(\frac{1}{C_2} + \frac{1}{C_3} - L_3 \omega^2 \right) A_3 \quad = 0$$

Thus, we have formulated an eigenvalue problem. Further simplification results for the special case where the C's and L's are constant. For this situation, the system can be expressed in matrix form as

$$
\begin{bmatrix}
1 - \lambda & -1 & \\
-1 & 2 - \lambda & -1 \\
 & -1 & 2 - \lambda
\end{bmatrix}
\begin{Bmatrix}
i_1 \\
i_2 \\
i_3
\end{Bmatrix} = \{0\}
\tag{22.33}
$$

where

$$
\lambda = LC\omega^2
\tag{22.34}
$$

Numerical methods can be employed to determine values for $\lambda = 0.1981$, 1.555, and 3.247. These values in turn can be substituted into Eq. (22.34) to solve for the natural circular frequencies of the system

$$
\omega = \begin{Bmatrix}
0.4451 / \sqrt{LC} \\
1.2470 / \sqrt{LC} \\
1.8019 / \sqrt{LC}
\end{Bmatrix}
\tag{22.35}
$$

Aside from providing the natural frequencies, the eigenvalues can be substituted into Eq. (22.33) to gain further insight into the circuit's physical behavior. For example, substituting $\lambda = 0.1981$ yields

$$
\begin{bmatrix}
0.8019 & -1 & \\
-1 & 1.8019 & -1 \\
 & -1 & 1.8019
\end{bmatrix}
\begin{Bmatrix}
i_1 \\
i_2 \\
i_3
\end{Bmatrix} = \{0\}
$$

Although this system does not have a unique solution, it will be satisfied if the currents are in fixed ratios as in

$$
0.8019i_1 = i_2 = 1.8019i_3
$$

Thus, as depicted in Fig. 22.17a, they oscillate in the same direction. In a similar fashion, the second eigenvalue of $\lambda = 1.555$ can be substituted and the result evaluated to yield

$$
-1.8018i_1 = i_2 = 2.247i_3
$$

As depicted in Fig. 22.17b, the first loop oscillates in the opposite direction from the second and third. Finally, the third mode can be determined as

$$
-0.445i_1 = i_2 = -0.8718i_3
$$

Consequently, as in Fig. 22.17c, the first and third loops oscillate in the opposite direction from the second.

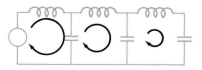 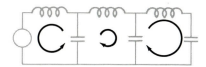

(a) $\omega = \dfrac{0.4451}{\sqrt{LC}}$

(b) $\omega = \dfrac{1.2470}{\sqrt{LC}}$

(c) $\omega = \dfrac{1.8019}{\sqrt{LC}}$

Figure 22.17
A visual representation of the natural modes of oscillation of the *LC* network for Fig. 22.16. Note that the diameters of the circular arrows are proportional to the magnitudes of the currents for each loop.

CASE STUDY 22.5   THE SWINGING PENDULUM (MECHANICAL ENGINEERING)

Background: Mechanical engineers (as well as all other engineers) are frequently faced with problems concerning the periodic motion of free bodies. The engineering approach to such problems ultimately requires that the position and velocity of the body be known as a function of time. These functions of time invariably are the solution of ordinary differential equations. The differential equations are usually based on Newton's laws of motion.

As an example, consider the simple pendulum shown previously in Fig. PT6.1. The particle of weight W is suspended on a weightless rod of length l. The only forces acting on the particle are its weight and the tension R in the rod. The position of the particle at any time is completely specified in terms of the angle θ and l.

The free-body diagram in Fig. 22.18 shows the forces on the particle and the acceleration. It is convenient to apply Newton's laws of motion in the x direction tangent to the path of the particle:

$$\sum F = -W \sin \theta = \frac{W}{g} a$$

Figure 22.18
A free-body diagram of the swinging pendulum showing the forces on the particle and the acceleration.

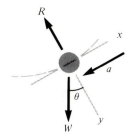

where g is the gravitational constant (32.2 ft/s^2) and a is the acceleration in the x direction. The angular acceleration of the particle (α) becomes

$$\alpha = \frac{a}{l}$$

Therefore, in polar coordinates ($\alpha = d^2\theta/dt^2$),

$$-W \sin \theta = \frac{Wl}{g} \alpha = \frac{Wl}{g} \frac{d^2\theta}{dt^2}$$

or

$$\frac{d^2\theta}{dt^2} + \frac{g}{l} \sin \theta = 0 \tag{22.36}$$

This apparently simple equation is a second-order nonlinear differential equation. In general, such equations are difficult or impossible to solve analytically. You have two choices regarding further progress. First, the differential equation might be reduced to a form that can be solved analytically (recall Sec. PT6.1.1), or second, a numerical approximation technique can be used to solve the differential equation directly. We will examine both of these alternatives in this example.

Solution: Proceeding with the first approach, we note that the series expansion for $\sin \theta$ is given by

$$\sin \theta = \theta - \frac{\theta^3}{3!} + \frac{\theta^5}{5!} - \frac{\theta^7}{7!} + \cdots \tag{22.37}$$

For small angular displacements, $\sin \theta$ is approximately equal to θ when expressed in radians. Therefore, for small displacements, Eq. (22.36) becomes

$$\frac{d^2\theta}{dt^2} + \frac{g}{l} \theta = 0 \tag{22.38}$$

which is a second-order linear differential equation. This approximation is very important because Eq. (22.38) is easy to solve analytically. The solution, based on the theory of differential equations, is given by

$$\theta(t) = \theta_0 \cos \sqrt{\frac{g}{l}} t \tag{22.39}$$

where θ_0 is the displacement at $t = 0$ and where it is assumed that the velocity ($v = d\theta/dt$) of the particle is zero at $t = 0$. The time required for the particle to complete one cycle of oscillation is called the period and is given by

$$T = 2\pi \sqrt{\frac{l}{g}}$$

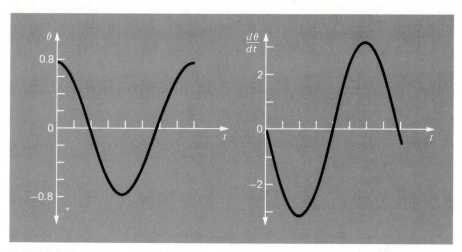

Figure 22.19

Plot of displacement θ and velocity $d\theta/dt$ as a function of time t, as calculated from Eq. (22.39).
θ_0 is $\pi/4$ and length is 2 ft.

Figure 22.19 shows a plot of the displacement (θ) and velocity ($d\theta/dt$) as a function of time, as calculated from Eq. (22.39) with $\theta_0 = \pi/4$ and $l = 2$ ft. The period, as calculated from Eq. (22.39a), is 1.5659 s.

The above calculations essentially are a complete solution of the motion of the particle. However, you must also consider the accuracy of the results because of the assumptions inherent in Eq. (22.38). In order to evaluate the accuracy, it is necessary to obtain a numerical solution for Eq. (22.36), which is a more complete physical representation of the motion. Any of the methods discussed in Chaps. 19 and 20 could be used for this purpose—for example, the Euler and fourth-order Runge-Kutta methods. Equation (22.36) must be transformed into two first-order equations to be compatible with the above methods. This is accomplished as follows. The velocity v is defined by

$$\frac{d\theta}{dt} = v \tag{22.40}$$

and, therefore, Eq. (22.36) can be expressed as

$$\frac{dv}{dt} = -\frac{g}{l} \sin \theta \tag{22.41}$$

Equations (22.40) and (22.41) are a coupled system of two ordinary differential equations. The numerical solutions by the Euler method and the fourth-order Runge-Kutta method give the results shown in Table 22.2. Table 22.2 compares the analytic solution for the linear equation of motion [Eq. (22.39)] in column (a) with the numerical solutions in columns (b), (c), and (d).

TABLE 22.2 Comparison of a linear analytical solution of the swinging pendulum problem with three nonlinear numerical solutions.

Time, s	Linear Analytical Solution (a)	Nonlinear Numerical Solutions		
		Euler (h = 0.05) (b)	4th-Order RK (h = 0.05) (c)	4th-Order RK (h = 0.01) (d)
0.0	0.785398	0.785398	0.785398	0.785398
0.2	0.545784	0.615453	0.566582	0.566579
0.4	−0.026852	0.050228	0.021895	0.021882
0.6	−0.583104	−0.639652	−0.535802	−0.535820
0.8	−0.783562	−1.050679	−0.784236	−0.784242
1.0	−0.505912	−0.940622	−0.595598	−0.595583
1.2	0.080431	−0.299819	−0.065611	−0.065575
1.4	0.617698	0.621700	0.503352	0.503392
1.6	0.778062	1.316795	0.780762	0.780777

The Euler and fourth-order RK methods yield different results and both disagree with the analytic solution, although the fourth-order Runge-Kutta method for the nonlinear case is closer to the analytic solution than is the Euler method. To properly evaluate the difference between the linear and nonlinear models it is important to determine the accuracy of the numerical results. This is accomplished in three ways. First, the Euler numerical solution is easily recognized as inadequate because it overshoots the initial condition at $t = 0.8$ s. This clearly violates conservation of energy. Second, columns (c) and (d) in Table 22.2 show the solution of the fourth-order Runge-Kutta method for step sizes of 0.05 and 0.01. Because these vary in the fourth decimal place, it is reasonable to assume that the solution with a step size of 0.01 is also accurate with this degree of certainty. Third, for the 0.01-s step-size case, θ obtains a local maximum value of 0.785385 at $t = 1.63$ s (not shown in Table 22.2). This indicates that the particle returns to its original position with four-place accuracy with a period of 1.63 s. These considerations allow you to safely assume that the difference between columns (a) and (d) in Table 22.2 truly represents the difference between the linear and nonlinear model.

Another way to characterize the difference between the linear and the nonlinear model is on the basis of period. Table 22.3 shows the period of oscillation as calculated by the linear model and nonlinear model for three different initial displacements. It is seen that the calculated periods agree closely when θ is small because θ is a good approximation for sin θ in Eq. (22.37). This approximation deteriorates when θ becomes large.

These analyses are typical of cases you will routinely encounter as an engineer. The utility of the numerical techniques becomes particularly significant in nonlinear problems, and in many cases real-world problems are nonlinear.

TABLE 22.3 Comparison of the period of an oscillating body calculated from linear and nonlinear models.

Initial Displacement, θ_0	Period, s	
	Linear Model $(T = 2\pi\sqrt{l/g})$	Nonlinear Model [Numerical Solution of Eq. (22.36)]
$\pi/16$	1.5659	1.57
$\pi/4$	1.5659	1.63
$\pi/2$	1.5659	1.85

CASE STUDY 22.6   SPREADSHEET SOLUTION OF ORDINARY DIFFERENTIAL EQUATIONS

This case study uses the spreadsheet to solve ordinary differential equations. A spreadsheet to implement the method is contained on the supplementary disk associated with the text. Insert the disk, turn on the power switch, advance to the main menu, and activate the sixth selection. This causes Fig. 22.20 to appear on the screen.

The spreadsheet is designed to solve the following ordinary differential equation:

$$DY/DX = -Y + EXP(MX)$$

where M is a parameter. You can change the value of M in cell B3. You can also change the starting value of X and the step size in cells E3 and E4. The values of Y at START X for the Euler, Heun, and Ralston (second-order RK) methods can be changed in cells C9, D9, and E9. However, note that at X = 0, the value of Y is equal to $(2 + M)/(1 + M)$ and that the analytical solution can be written as

$$Y(X) = EXP(-X) + EXP(MX)/(1 + M)$$

Figure 22.20

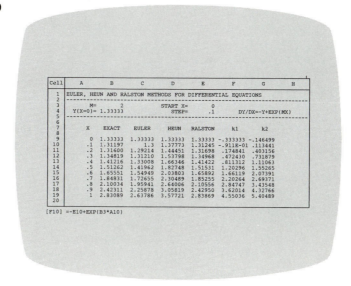

Now move the cursor to cell D10. The following equation is displayed on the bottom line of the spreadsheet:

$$D10 = D9 + (-D9 + EXP(B3*A9) - D10 + EXP(B3*A10))*E4/2$$

This equation states that the new value of Y (D10) using the Heun method is equal to the old value of Y (D9) plus the average value of the derivative times the step size (E4). The average value is based on the derivative evaluated at the beginning of the step (A9 and D9) and at the end of the step (A10 and D10). Note carefully that D10 appears on both sides of the equal sign. This creates a situation where the value of cell D10 changes each time the spreadsheet cell contents are updated using the F3 key. This is equivalent to iteration of the corrector equation employed by the Heun method. Now move the cursor to B3 to change the value of M to 3. Strike F3 two times and observe the iteration of the Heun method. Note that the values in column E for the Ralston method also change each time F3 is entered. Move the cursor to cell E11,

$$E11 = E10 + (F10/3 + 2G10/3)*E4$$

then to cell F10,

$$F10 = -E10 + EXP(B3*A10)$$

Note that E11 depends on F10 and F10 depends on E10. Therefore we see that E11 is computed on the basis of a number that subsequently changes (F10). Therefore correct values in the E, F, and G columns can only be computed one row at a time using the [F3] key. Note that both iteration of the Heun method and the line-by-line computation of the Ralston method occur each time [F3] is entered. Study all the terms in the spreadsheet to ensure that you completely understand all the methods.

Continue to strike [F3] until all the numbers in the spreadsheet remain fixed (Fig. 22.21). Compare the analytical solution with each of the three numerical methods.

Figure 22.21

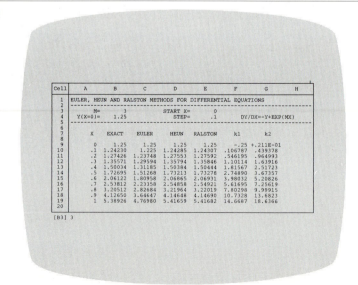

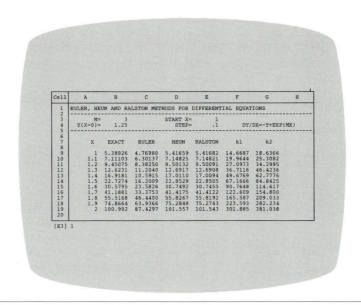

Cell	A	B	C	D	E	F	G	H
1	EULER, HEUN AND RALSTON METHODS FOR DIFFERENTIAL EQUATIONS							
2	---							
3		M=	3		START X=	1		
4	Y(X=0)=	1.25			STEP=	.1	DY/DX=-Y+EXP(MX)	
5	---							
6								
7	X	EXACT	EULER	HEUN	RALSTON	k1	k2	
8								
9	1	5.38926	4.76980	5.41659	5.41682	14.6687	18.6366	
10	1.1	7.11103	6.30137	7.14825	7.14821	19.9644	25.3082	
11	1.2	9.45075	8.38250	9.50132	9.50091	27.0973	34.2995	
12	1.3	12.6231	11.2040	12.6917	12.6908	36.7116	46.4236	
13	1.4	16.9181	10.5915	17.0110	17.0094	49.6769	62.7776	
14	1.5	22.7274	16.2009	22.8529	22.8505	67.1666	84.8425	
15	1.6	30.5795	23.5826	30.7492	30.7455	90.7648	114.617	
16	1.7	41.1881	33.3753	41.4175	41.4122	122.609	154.800	
17	1.8	55.5168	46.4400	55.8267	55.8192	165.587	209.033	
18	1.9	74.8664	63.9366	75.2848	75.2743	223.593	282.234	
19	2	100.992	87.4297	101.557	101.543	301.885	381.038	
20								

[E3] 1

Figure 22.22

Now let's continue the integration process over the range X = 1 to X = 2. This is accomplished by changing START X = 1 and transferring the contents of cells C19, D19, and E19 to cells C9, D9, and E9. Note that the value of Y at X = 0 depends only on M and, therefore does not change. Now strike [F3] several times to complete the computations. Note how closely the Heun and Ralston methods agree and how each is superior to Euler's method (Fig. 22.22).

Let's use the spreadsheet to study how the accuracy of the three methods varies as a function of step size while integrating the function from X = 0 to X = 1 with M = 4.

Enter appropriate values and strike [F3] until the calculations are complete. The results are shown in Table 22.4.

TABLE 22.4   Values of Y at X = 1 based on three numerical methods for various step sizes.

Exact	Euler	Heun	Ralston	Step
11.2875	1.0	18.9327	13.8236	1.0
11.2875	4.24452	13.8361	12.7748	0.5
11.2875	6.57223	11.7394	11.6186	0.2
11.2875	9.48461	11.4022	11.3767	0.1
11.2875	9.99422	11.3163	11.3105	0.05
11.2875	10.16340	11.2947	11.2933	0.025

Note how the accuracy of each method improves as the step size decreases while integrating to the same final value of X. The spreadsheet is a convenient tool to perform such numerical experiments because all the computations involved are available on one screen. On the other hand, note that the computational efficiency is poor because the whole spreadsheet is reviewed to perform update iterations on only a few cells.

PROBLEMS

General Engineering

22.1 Reproduce computations performed in Case Study 22.1 using your own software.

22.2 Perform the same computation as in Case Study 22.1, but with $k = \$60/\text{day}$.

22.3 Perform the same computation as in Case Study 22.1, but with a new equation for computer cost [replacing Eq. (22.2)],

Cost of individual computer ($)

$$= 1500\,(1\,+\,e^{-4.4\times10^{-5}N})$$

22.4 Repeat the falling parachutist problem (Example 1.2), but with the upward force due to drag as a second-order rate:

$$F_u = -cv^2$$

where $c = 24$ kg/m. Plot your results and compare with those of Example 1.1.

22.5 Population-growth dynamics are important in a variety of engineering planning studies. One of the simplest models of such growth incorporates the assumption that the rate of change of the population p is proportional to the existing population at any time t:

$$\frac{dp}{dt} = Gp \qquad (P22.5)$$

where G is a growth rate (per year). This model makes intuitive sense because the greater the population, the greater the number of potential parents.

At time $t = 0$ an island has a population of 10,000 people. If $G = 0.075$ per year, employ Euler's method to predict the population at $t = 20$ years, using a step size of 0.5 years. Plot p versus t on standard and semilog graph paper. Determine the slope of the line on the semilog plot. Discuss your results.

22.6 Although the model in Prob. 22.5 works adequately when population growth is unlimited, it breaks down when factors such as food shortages, pollution, and lack of space inhibit growth. In such cases, the growth rate itself can be thought of as being inversely proportional to population. One model of this relationship is

$$G = G'(p_{\max} - p) \qquad (P22.6)$$

where G' is a population-dependent growth rate (per people-year) and $p_{\max}$ is the maximum sustainable population. Thus when population is small ($p \ll p_{\max}$), the growth rate will be at a high, constant rate of $G'p_{\max}$. For such cases, growth is unlimited and Eq. (P22.6) is essentially identical to Eq. (P22.5). However, as population grows (that is, p approaches $p_{\max}$), G decreases until at $p = p_{\max}$ it is zero. Thus the model predicts that when the population reaches the maximum sustainable level, growth is nonexistent, and the system is at a steady state. Substituting Eq. (P22.6) into Eq. (P22.5) yields

$$\frac{dp}{dt} = G'(p_{\max} - p)p$$

For the same island studied in Prob. 22.5, employ Euler's method to predict the population at $t = 20$ years, using a step size of 0.5 years. Employ values of $G' = 10^{-5}$ per people-year and $p_{\max} = 20,000$ people. At time $t = 0$ the island has a population of 10,000 people. Plot p versus t and interpret the shape of the curve.

22.7 Suppose that, after falling for 10 s, the parachutist from Examples 1.1 and 1.2 pulls the rip cord. At this point, assume that the drag coefficient is instantaneously increased to a constant value of 50 kg/s. Compute the parachutist's velocity from $t = 10$ to 30 s with Euler's method. Employ all the parameters from Examples 1.1 and 1.2 and take the initial condition from the analytical solution. Plot v versus t for $t = 0$ to 30 s. Use the analytical solution for $t = 0$ to 10 s and the numerical solution obtained above for $t = 10$ to 30 s.

Chemical Engineering

22.8 Reproduce the computations performed in Case Study 22.2 using your own software.

22.9 Perform the first computation in Case Study 22.2, but for the case where $c_0 = 100$ mg/m, $Q = 10$ m^3/min, and $V = 50$ m^3.

22.10 Perform the second computation in Case Study 22.2, but for the system described in Prob. 10.9.

22.11 A mass balance for a chemical in a completely mixed reactor can be written as

$$V \frac{dc}{dt} = F - Qc - kVc^2$$

$$\text{Accumulation} = \frac{\text{feed}}{\text{rate}} - \text{outflow} - \text{reaction}$$

where V is volume (10 m^3), c is concentration, F is feed rate (200 g/min), Q is flow rate (1 m^3/min), and k is reaction rate (0.1 m^3/g/min). If $c(0) = 0$, solve the ODE until concentration reaches a stable level. Plot your results.

22.12 If $c_{in} = \bar{c}(1 - e^{-t})$, calculate the outflow concentration of a single, completely mixed reactor as a function of time. Use Euler's method to perform the computation. Employ values of $\bar{c} = 50$ mg/m^3, $Q = 5$ m^3/min, $V = 100$ m^3, and $c_0 = 10$ mg/m^3. Perform the computation from $t = 0$ to 50 min.

22.13 Aside from inflow and outflow, another way by which mass can enter or leave a reactor is by a chemical reaction. For example, if the chemical decays, the reaction can sometimes be characterized as a first-order reaction:

$$\text{Reaction} = -kVc$$

where k is a reaction rate (in minutes^{-1}), which can generally be interpreted as the fraction of the chemical that goes away per unit time. For example, $k = 0.1$ min^{-1} can be thought of as meaning that roughly 10 percent of the chemical in the reactor decays in a minute. The reaction can be substituted into the mass-balance equation [Eq. (22.8)] to give

$$V \frac{dc}{dt} = Qc_{in} - Qc - kVc$$

If $k = 0.2$ min^{-1}, $c_{in} = 70$ mg/min^3, $Q = 5$ m^3/min, and $V = 100$ m^3, what is the steady-state concentration of the reactor?

22.14 Repeat Prob. 22.13, but compute the transient concentration response if $c_0 = 10$ mg/m^3. Compute the response with Euler's method from $t = 0$ to 20 min.

Civil Engineering

22.15 Reproduce the computation performed in Case Study 22.3, using your own software.

22.16 Perform the steady-state computation in Case Study 22.3, but use a uniform load of 80 lb/ft and an $E = 2 \times 10^8$ lb/ft^2. Check your result by comparing it with the analytical solution.

22.17 Perform the steady-state computation in Case Study 22.3, but rather than using a constant wind force f employ a force that varies with height according to (recall Case Study 18.3)

$$f(z) = 200 \frac{z}{5 + z} e^{-2z/30}$$

Plot $y(z)$ versus z and compare the results with those for Case Study 22.3.

22.18 Determine the eigenvalues for the sailboat mast in Case Study 22.3 using nodes spaced at intervals of $L/4$ (see Fig. 22.12).

22.19 Employ the power method to determine the largest eigenvalue for the sailboat mast from Case Study 22.3. Use the segmentation scheme from Fig. 22.12b for your calculation.

22.20 Environmental and biomedical engineers must frequently predict the outcome of parasite-host or predator-prey relationships. A simple model of such interactions is provided by the following system of ODEs.

$$\frac{dH}{dt} = g_1 H - d_1 PH$$

and

$$\frac{dP}{dt} = -d_2 P + g_2 PH$$

where H and P are the numbers of hosts and parasites, respectively, and the d's and g's are death and growth rates, respectively, where the subscript 1 refers to the host and the 2 to the parasite. These relationships are called Lotka-Volterra equations. Notice that the deaths of the host and the growth of the parasite are dependent on both H and P. If $P(0) = 5$ and $H(0) = 20$ and if

$g_1 = 1$, $d_1 = 0.1$, $g_2 = 0.02$, and $d_2 = 0.5$, compute values of P and H from $t = 0$ to 10. Interpret your results.

Electrical Engineering

22.21 Reproduce the computations performed in Case Study 22.4 using your own software.

22.22 Perform the same computation as in the first part of Case Study 22.4, but with $R = 20\ \Omega$.

22.23 Solve the ODE in Case Study 6.4 using numerical techniques if $q = 0.1$ and $i = -3.281515$ at $t = 0$.

22.24 For a simple RL circuit, Kirchhoff's voltage law requires that (if Ohm's law holds)

$$L\frac{di}{dt} + Ri = 0$$

where i is current, L is inductance, and R is resistance. Solve for i, if $L = R = 1$ and $i(0) = 10^{-3}$. Solve this problem analytically and with a numerical method.

22.25 In contrast to Prob. 22.24, real resistors may not always obey Ohm's law. For example, the voltage drop may be nonlinear and the circuit dynamics described by a relationship such as

$$L\frac{di}{dt} + \left[\frac{-i}{I} + \left(\frac{i}{I}\right)^3\right]R = 0$$

where all other parameters are as defined in Prob. 22.24 and I is a known reference current equal to 1. Solve for i as a function of time under the same conditions as specified in Prob. 22.24.

22.26 Develop an eigenvalue problem for an LC network similar to the one in Fig. 22.16 but with only two loops. That is, omit the i_3 loop. Contrast your results with Eq. (21.4) and draw the network, illustrating how the currents oscillate in their primary modes.

Mechanical Engineering

22.27 Reproduce the computations performed in Case Study 22.5 using your own software.

22.28 Perform the same computation as in Case Study 22.5, but for a 3-ft-long pendulum.

22.29 Use a numerical method to duplicate the computation displayed in Fig. 6.11.

22.30 The rate of cooling a body can be expressed as

$$\frac{dT}{dt} = -k(T - T_a)$$

where T is the temperature of the body (in degrees Celsius), T_a is the temperature of the surrounding medium (also in degrees Celsius), and k is the proportionality constant (per minute). Thus, this equation specifies that the rate of cooling is proportional to the difference in temperature between the body and the surrounding medium. If a metal ball heated to 90°C is dropped into water that is held at a constant value of $T_a = 20°C$, use a numerical method to compute how long it takes the ball to cool to 30°C if $k = 0.1\ \text{min}^{-1}$.

22.31 The rate of heat flow (conduction) between two points on a cylinder heated at one end is given by

$$\frac{dQ}{dt} = \lambda A \frac{dT}{dx}$$

where λ is a constant, A is the cylinder's cross-sectional area, Q is heat flow, T is temperature, t is time, and x is the distance from the heated end. Because the equation involves two derivatives, we will simplify this equation by letting

$$\frac{dT}{dx} = \frac{100(L - x)(20 - t)}{100 - xt}$$

where L is the length of the rod. Combining the two equations gives

$$\frac{dQ}{dt} = \lambda A \frac{100(L - x)(20 - t)}{100 - xt}$$

Compute the heat flow for $t = 0$ to 15 s, if $\lambda = 0.3$ cal·cm/s·°C, $A = 10\ \text{cm}^2$, $L = 20$ cm, and $x = 2.5$ cm. The initial condition is that $Q = 0$ at $t = 0$. Plot your results.

Spreadsheet

22.32 Use an active spreadsheet that may be available to you (like the one on the Electronic TOOLKIT) to integrate the function

$$\frac{dy}{dx} = -y + \sin x$$

Use the Euler, Heun, and Ralston methods. Start from $x = 0$ with $y(x = 0) = 1$. Design the spreadsheet so that the upper value of x is equal to 10 times the step size. Try step sizes of 0.001, 0.01, 0.1, 1.0, 2.0, and 3.0. Discuss the accuracy of the methods. What do you observe about the number of iterations required for convergence of the Heun method?

22.33 Generate a plot of the values in Table 22.4 and discuss the results. Compare the plots with the theoretical error formulations.

22.34 Activate the spreadsheet associated with the text. Set M = 2 and perform an error analysis similar to Case Study 22.6.

22.35 In Case Study 22.6 values of k_1 and k_2 are provided for all calculations. What do k_1 and k_2 represent? Develop a plot of $|(k_1 - k_2)/k_1|$ versus the $|\%\text{ error}|$. Discuss the plot

and its significance in terms of what you know about the source of errors when performing numerical solutions of differential equations.

Miscellaneous

22.36 Read all the case studies in Chap. 22. On the basis of your reading and experience, make up your own case study for any one of the fields of engineering. This may involve modifying or reexpressing one of our case studies. However, it can also be totally original. As with our examples, it must be drawn from an engineering problem context and must demonstrate the use of numerical methods for solving ODEs. Write up your results using our case studies as models.

EPILOGUE: PART SIX

PT6.4 TRADE-OFFS

Table PT6.3 contains trade-offs associated with numerical methods for the solution of initial-value ordinary differential equations. The factors in this table must be evaluated by the engineer when selecting a method for each particular applied problem.

Simple self-starting techniques such as Euler's method can be used if the problem requirements involve a short range of integration. In this case, adequate accuracy may be obtained using small step sizes to avoid large truncation errors, and the round-off errors may be acceptable. Euler's method may also be appropriate for cases where the mathematical model has an inherently high level of uncertainty or has coefficients or forcing functions with significant errors as might arise during a measurement process. In this case, the accuracy of the model itself simply does not justify the effort involved to employ a more complicated numerical method. Finally, the simpler techniques may be best when the problem or simulation need only be performed a few times. In these applications, it is probably best to use a simple method that is easy to program and understand, despite the fact that the method may be computationally inefficient and relatively time-consuming to run on the computer.

If the range of integration of the problem is long enough to involve a large number of steps (say, > 1000), then it may be necessary and appropriate to use a more accurate technique than Euler's method. The fourth-order Runge-Kutta method and the fourth-order Adams method are popular and reliable for many engineering problems. In these cases, it may also be advisable to estimate the truncation error for each step as a guide to selecting the best step size. This can be accomplished with the adaptive RK or fourth-order Adams approaches. If the truncation errors are extremely small, it may be wise to increase the step size to save computer time. On the other hand, if the truncation error is large, the step size should be decreased to avoid accumulation of error. Milne's method should be avoided if significant stability problems are expected. The Runge-Kutta method is simple to program and convenient to use but may be less efficient than the multistep methods. However, the Runge-Kutta method is usually employed in any event to obtain starting values for the multistep methods.

If extremely accurate answers are required or if the function has large higher-order derivatives, Butcher's fifth-order Runge-Kutta method is sometimes desirable.

A large number of engineering problems may fall into an intermediate range of integration interval and accuracy requirement. In these cases, the second-order

TABLE PT6.3 Comparison of alternative methods for the numerical solution of ordinary differential equations. The comparisons are based on general experience and do not account for the behavior of special functions.

Method	Starting Values	Iterations Required	Global Error	Ease of Changing Step Size	Programming Effort	Comments
One step						
Euler's	1	No	$O(h)$	Easy	Easy	Good for quick estimates
Heun's	1	Yes	$O(h^2)$	Easy	Moderate	—
Improved polygon	1	No	$O(h^2)$	Easy	Moderate	—
Second-order Ralston	1	No	$O(h^2)$	Easy	Moderate	The second-order RK method that minimizes truncation error
Fourth-order RK	1	No	$O(h^4)$	Easy	Moderate	Widely used
Adaptive fourth-order RK or RK-Fehlberg	1	No	$O(h^5)$*	Easy	Moderate to difficult†	Error estimate allows step-size adjustment
Multistep						
Non-self-starting Heun	2	Yes	$O(h^3)$*	Difficult	Moderate to difficult†	Simple multistep method
Milne's	4	Yes	$O(h^5)$*	Difficult	Moderate to difficult†	Sometimes unstable
Fourth-order Adams	4	Yes	$O(h^5)$*	Difficult	Moderate to difficult†	

*Provided error estimate is used to modify the solution.
†With variable step size.

RK and the non-self-starting Heun methods are simple to use and are relatively efficient and accurate.

A variety of techniques are available for solving eigenvalue problems. For small systems or where only a few of the smallest or largest eigenvalues are required, simple approaches such as the polynomial and the power methods are available. For symmetric systems, Jacobi's, Given's, or Householder's method can be employed. Finally, the QR method represents a general approach for finding all the eigenvalues of symmetric and nonsymmetric matrices.

PT6.5 IMPORTANT RELATIONSHIPS AND FORMULAS

Table PT6.4 summarizes important information that was presented in Part Six. This table can be consulted to quickly access important relationships and formulas.

PT6.6 ADVANCED METHODS AND ADDITIONAL REFERENCES

Although we have reviewed a number of techniques for solving ordinary differential equations, there is additional information that is important in engineering

TABLE PT6.4 Summary of important information presented in Part Six.

Method	Formulation	Graphic Interpretation	Errors
Euler (First-order RK)	$y_{i+1} = y_i + hk_1$ $k_1 = f(x_i, y_i)$		Local error $\simeq O(h^2)$ Global error $\simeq O(h)$
Ralston's second-order RK	$y_{i+1} = y_i + h\left(\frac{1}{3}k_1 + \frac{2}{3}k_2\right)$ $k_1 = f(x_i, y_i)$ $k_2 = f\left(x_i + \frac{3}{4}h, y_i + \frac{3}{4}hk_1\right)$		Local error $\simeq O(h^3)$ Global error $\simeq O(h^2)$
Classic fourth-order RK	$y_{i+1} = y_i + h\left(\frac{1}{6}k_1 + \frac{1}{3}k_2 + \frac{1}{3}k_3 + \frac{1}{6}k_4\right)$ $k_1 = f(x_i, y_i)$ $k_2 = f\left(x_i + \frac{1}{2}h, y_i + \frac{1}{2}hk_1\right)$ $k_3 = f\left(x_i + \frac{1}{2}h, y_i + \frac{1}{2}hk_2\right)$ $k_4 = f(x_i + h, y_i + hk_3)$		Local error $\simeq O(h^5)$ Global error $\simeq O(h^4)$
Non-self-starting Heun	Predictor: (midpoint method) $y_{i+1}^0 = y_{i-1}^m + 2hf(x_i, y_i^m)$ Corrector: (trapezoidal rule) $y_{i+1}^j = y_i^m + h\dfrac{f(x_i, y_i^m) + f(x_{i+1}, y_{i+1}^{j-1})}{2}$	 	Predictor modifier: $E_p \simeq \frac{4}{5}\left(y_{i,u}^m - y_{i,u}^0\right)$ Corrector modifier: $E_c \simeq -\dfrac{y_{i+1,u}^m - y_{i+1,u}^0}{5}$
Fourth-order Adams	Predictor: (fourth Adams-Bashforth) $y_{i+1}^0 = y_i^m + h\left(\frac{55}{24}f_i^m - \frac{59}{24}f_{i-1}^m + \frac{37}{24}f_{i-2}^m - \frac{9}{24}f_{i-3}^m\right)$ Corrector: (fourth Adams-Moulton) $y_{i+1}^j = y_i^m + h\left(\frac{9}{24}f_{i+1}^{j-1} + \frac{19}{24}f_i^m - \frac{5}{24}f_{i-1}^m + \frac{1}{24}f_{i-2}^m\right)$	 	Predictor modifier: $E_p \simeq \frac{251}{270}\left(y_{i,u}^m - y_{i,u}^0\right)$ Corrector modifier: $E_c \simeq -\frac{19}{270}\left(y_{i+1,u}^m - y_{i+1,u}^0\right)$

practice[*]. The question of *stability* was introduced in Sec. 20.2.4. This topic has general relevance to all methods for solving ODEs. Further discussion of the topic can be pursued in Carnahan, Luther, and Wilkes (1969), Gear (1971), and Hildebrand (1974).

The issue of stability has special significance to the solution of *stiff equations*. These are equations with slowly and rapidly varying components. Although using a variable step size or higher-order methods can sometimes be helpful, special techniques are usually required for the adequate solution of stiff equations. You can consult Enright et al. (1975), Gear (1971), and Shampine and Gear (1979) for additional information regarding these techniques.

In Chap. 21, we introduced methods for solving *boundary-value* problems. Isaacson and Keller (1966), Keller (1968), Na (1979), and Scott and Watts (1976) can be consulted for additional information on standard boundary-value problems. Additional material on eigenvalues can be found in Ralston and Rabinowitz (1978), Wilkinson (1965), Fadeev and Fadeeva (1963), and Householder (1953).

In summary, the foregoing is intended to provide you with avenues for deeper exploration of the subject. Additionally, all the above references provide descriptions of the basic techniques covered in Part Six. We urge you to consult these alternative sources to broaden your understanding of numerical methods for the solution of differential equations.

---

[*]Books are referenced by the author here; a complete bibliography will be found at the back of this text.

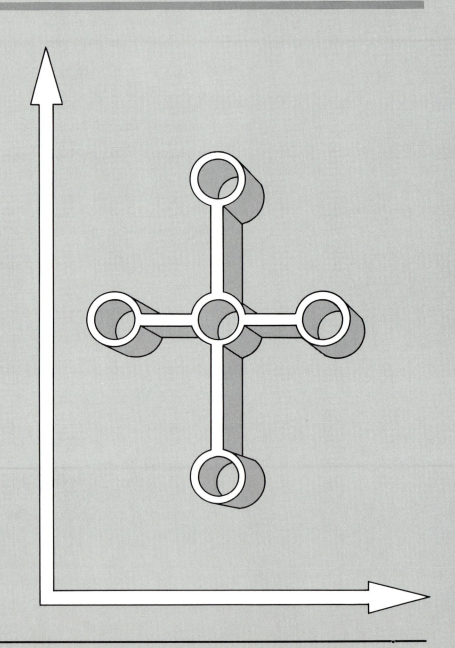

PART SEVEN

PARTIAL DIFFERENTIAL EQUATIONS

PT7.1 MOTIVATION

Given a function u that depends on both x and y, the partial derivative of u with respect to x at an arbitrary point (x, y) is defined as

$$\frac{\partial u}{\partial x} = \lim_{\Delta x \to 0} \frac{u(x + \Delta x, y) - u(x, y)}{\Delta x} \tag{PT7.1}$$

Similarly, the partial derivative with respect to y is defined as

$$\frac{\partial u}{\partial y} = \lim_{\Delta y \to 0} \frac{u(x, y + \Delta y) - u(x, y)}{\Delta y} \tag{PT7.2}$$

An equation involving partial derivatives of an unknown function of two or more independent variables is called a *partial differential equation* or *PDE*. For example,

$$\frac{\partial^2 u}{\partial x^2} + 2xy \frac{\partial^2 u}{\partial y^2} + u = 1 \tag{PT7.3}$$

$$\frac{\partial^3 u}{\partial x^2 \partial y} + x \frac{\partial^2 u}{\partial y^2} + 8u = 5y \tag{PT7.4}$$

$$\left(\frac{\partial^2 u}{\partial x^2} \right)^3 + 6 \frac{\partial^3 u}{\partial x \, \partial y^2} = x \tag{PT7.5}$$

$$\frac{\partial^2 u}{\partial x^2} + xu \frac{\partial u}{\partial y} = x \tag{PT7.6}$$

The *order* of a PDE is that of the highest-order partial derivative appearing in the equation. For example, Eqs. (PT7.3) and (PT7.4) are second- and third-order, respectively.

A partial differential equation is said to be *linear* if it is linear in the unknown function and all its derivatives, with coefficients depending only on the independent variables. For example, Eq. (PT7.3) is linear whereas Eqs. (PT7.5) and (PT7.6) are not.

Because of their widespread application in engineering, our treatment of PDEs will focus on linear, second-order equations. For two independent variables, such equations can be expressed in the following general form

$$A \frac{\partial^2 u}{\partial x^2} + B \frac{\partial^2 u}{\partial x \, \partial y} + C \frac{\partial^2 u}{\partial y^2} + D = 0 \tag{PT7.7}$$

TABLE PT7.1   **Categories into which linear, second-order partial differential equations in two variables can be classified.**

$B^2 - 4AC$	Category	Example
< 0	Elliptic	Laplace equation (steady-state with two spatial dimensions) $$\frac{\partial^2 T}{\partial x^2} + \frac{\partial^2 T}{\partial y^2} = 0$$
$= 0$	Parabolic	Heat conduction equation (time-variable with one spatial dimension) $$k\frac{\partial^2 T}{\partial x^2} = \frac{\partial T}{\partial t}$$
> 0	Hyperbolic	Wave equation (time-variable with one spatial dimension) $$\frac{\partial^2 y}{\partial x^2} = \frac{1}{c^2}\frac{\partial^2 y}{\partial t^2}$$

where A, B, and C are functions of x and y and D is a function of x, y, u, $\partial u/\partial x$, and $\partial u/\partial y$. Depending on the values of the coefficients of the second-derivative terms—A, B, C—Eq. (PT7.7) can be classified into one of three categories (Table PT7.1). This classification, which is based on the method of characteristics (for example, see Vichnevetsky, 1981 or Lapidus and Pinder, 1982), is useful because each category relates to specific and distinct engineering problem contexts that demand special solution techniques. It should be noted that for cases where A, B, and C depend on x and y, the equation may actually fall into a different category depending on the location in the domain for which the equation holds. For simplicity, we will limit the present discussion to PDEs that remain exclusively in one of the categories.

PT7.1.1   PDEs and Engineering Practice

Each of the categories of partial differential equations in Table PT7.1 conform to specific kinds of engineering problems. The initial sections of the following chapters will be devoted to deriving each type of equation for a particular engineering problem context. For the time being, we will discuss their general properties and applications and show how they can be employed in different physical contexts.

Elliptic equations are typically used to characterize *steady-state* systems. As in the *Laplace equation* in Table PT7.1, this is indicated by the absence of a time derivative. Thus, these equations are typically employed to determine the *steady-state* distribution of an unknown in two spatial dimensions.

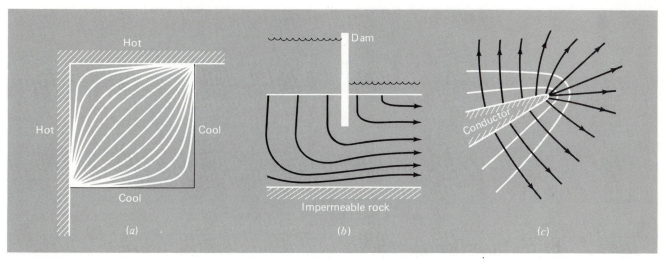

Figure PT7.1
Three steady-state distribution problems that can be characterized by elliptic PDEs. (a) Temperature distribution on a heated plate, (b) seepage of water under a dam, and (c) the electric field near the point of a conductor.

A simple example is the heated plate in Fig. PT7.1a. For this case, the boundaries of the plate are held at different temperatures. Because heat flows from regions of high to low temperature, the boundary conditions set up a *potential* that leads to heat flow from the hot to the cool boundaries. If sufficient time elapses, such a system will eventually reach the stable or steady-state distribution of temperature depicted in Fig. PT7.1a. The Laplace equation, along with appropriate boundary conditions, provides a means to determine this distribution. By analogy, the same approach can be employed to tackle other problems involving potentials, such as seepage of water under a dam (Fig. PT7.1b) or the distribution of an electric field (Fig. PT7.1c).

In contrast to the elliptic category, *parabolic equations* determine how an unknown varies in both space and time. This is manifested by the presence of both spatial and temporal derivatives in the *heat conduction equation* from Table PT7.1. Such cases are referred to as *propagation problems* because the solution "propagates" or changes in time.

A simple example is a long, thin rod that is insulated everywhere except at its end (Fig. PT7.2a). The insulation is employed to avoid complications due to heat loss along the rod's length. As was the case for the heated plate in Fig. PT7.1a, the ends of the rod are set at fixed temperatures. However, in contrast to Fig. PT7.1a, the rod's thinness allows us to assume that heat is distributed evenly over its cross section—that is, laterally. Consequently, lateral heat flow is not an issue and the problem reduces to studying the conduction of heat along the rod's longitudinal axis. Rather than focusing on the steady-state distribution in two

Figure PT7.2
(a) A long, thin rod that is insulated everywhere but at its end. The dynamics of the one-dimensional distribution of temperature along the rod's length can be described by a parabolic PDE. (b) Shows the solution consisting of distributions corresponding to the state of the rod at various times.

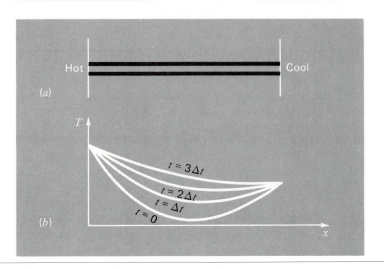

spatial dimensions, the problem shifts to determining how the one-dimensional spatial distribution changes as a function of time (Fig. PT7.2b). Thus, the solution consists of a series of spatial distributions corresponding to the state of the rod at various times. Using an analogy from photography, the elliptic case yields a portrait of a system's stable state whereas the parabolic case provides a motion picture of how it changes from one state to another. As with the other types of PDEs described herein, parabolic equations can be used to characterize a wide variety of other engineering problem contexts by analogy.

The final class of PDEs, the *hyperbolic* category, also deals with *propagation problems*. However, an important distinction manifested by the wave equation in Table PT7.1 is that the unknown is characterized by a second derivative with respect to time. As a consequence, the solution oscillates.

The vibrating string in Fig. PT7.3 is a simple physical model that can be described with the wave equation. The solution consists of a number of characteristic states with which the string oscillates. A variety of engineering systems such as vibrations of rods and beams, motion of fluid waves, and transmission of sound and electrical signals can be characterized by this model.

Figure PT7.3
A taut string vibrating at a low amplitude is a simple physical system that can be characterized by a hyperbolic PDE.

PT7.1.2 Precomputer Methods for Solving PDEs

Prior to the advent of digital computers, engineers relied on analytical or exact solutions of partial differential equations. Aside from the simplest cases, these solutions often required a great deal of effort and mathematical sophistication. In addition, many physical systems could not be solved directly but had to be simplified using linearizations, simple geometric representations, and other idealizations. Although these solutions are elegant and yield insight, they are limited with respect to how faithfully they represent real systems—especially those that are highly nonlinear and irregularly shaped.

PT7.2 ORIENTATION

Before proceeding to the numerical methods for solving partial differential equations, some orientation might be helpful. The following material is intended to provide you with an overview of the material discussed in Part Seven. In addition, we have formulated objectives to focus your studies in the subject area.

PT7.2.1 Scope and Preview

Figure PT7.4 provides an overview of Part Seven. Two broad categories of numerical methods will be discussed in the present part of this book. Finite-difference approaches, which are covered in Chaps. 23 and 24, are based on approximating the solution at a finite number of points. In contrast, finite-element methods, which are covered in Chap. 25, approximate the solution in pieces or "elements." Various parameters are adjusted until these approximations conform to the underlying differential equation in an optimal sense.

Chapter 23 is devoted to *finite-difference* solutions of *elliptic equations*. Before launching into the methods, we derive the Laplace equation for the physical problem context of the temperature distribution for a heated plate. Then, a standard solution approach, the *Liebmann method,* is described. We will illustrate how this approach is used to compute the distribution of the primary scalar variable, temperature, as well as a secondary vector variable, heat flux. The final section of the chapter deals with *boundary conditions*. This material includes procedures to handle different types of conditions as well as irregular boundaries.

In *Chapter 24* we turn to *finite-difference* solutions of *parabolic equations*. As with the discussion of elliptic equations, we first provide an introduction to a physical problem context, the heat-conduction equation for a one-dimensional rod. Then we introduce both explicit and implicit algorithms for solving this equation. This is followed by an efficient and reliable implicit method—the *Crank-Nicolson technique*. Finally, we describe a particularly effective approach for solving two-dimensional parabolic equations—the alternating-direction implicit, or *A.D.I., method*.

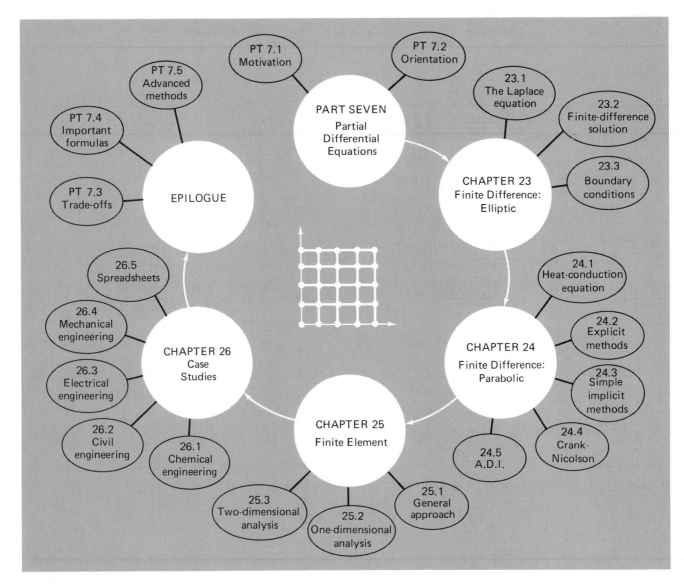

Figure PT7.4
Schematic representation of the organization of material in Part Seven: Partial Differential Equations.

Note that because they are somewhat beyond the scope of this book, we have chosen to omit hyperbolic equations. The epilogue of this part of the book contains references related to this type of PDE.

In *Chapter 25* we turn to the other major approach for solving PDEs—the *finite-element method*. Because it is so fundamentally different from the finite-difference approach, we devote the initial section of the chapter to a general overview. Then we show how the finite-element method is used to compute the steady-state temperature distribution of a heated rod. Finally, we provide an introduction to some of the issues involved in extending such an analysis to two-dimensional problem contexts.

Chapter 26 is devoted to case studies from all fields of engineering as well as to spreadsheet applications. Finally, a short review section is included at the end of Part Seven. This *epilogue* summarizes important information related to PDEs. This material includes a discussion of trade-offs that are relevant to their implementation in engineering practice. The epilogue also includes references for advanced topics.

PT7.2.2 Goals and Objectives

Study Objectives.   After completing Part Seven, you should have greatly enhanced your capability to confront and solve partial differential equations. Gen-

TABLE PT7.2   **Specific study objectives for Part Seven.**

1. Recognize the difference between elliptic, parabolic, and hyperbolic PDEs

2. Understand the fundamental difference between finite-difference and finite-element approaches

3. Recognize that the Liebmann method is equivalent to the Gauss-Seidel approach for solving simultaneous linear algebraic equations

4. Know how to determine secondary variables for two-dimensional field problems

5. Recognize the distinction between Dirichlet and derivative boundary conditions

6. Understand how to use weighting factors to incorporate irregular boundaries into a finite-difference scheme for PDEs

7. Know the difference between convergence and stability of parabolic PDEs

8. Understand the difference between explicit and implicit schemes for solving parabolic PDEs

9. Recognize how the stability criteria for explicit methods detract from their utility for solving parabolic PDEs

10. Know how to interpret computational molecules

11. Recognize how the A.D.I. approach achieves high efficiency in solving parabolic equations in two spatial dimensions

12. Understand the difference between the direct method and the method of weighted residuals for deriving element equations

13. Know how to implement Galerkin's method

14. Understand the benefits of integration by parts during the derivation of element equations; in particular, recognize the implications of lowering the highest derivative from a second to a first derivative

eral study goals should include mastering the techniques, having the capability to assess the reliability of the answers, and being able to choose the "best" method (or methods) for any particular problem. In addition to these general objectives, the specific study objectives in Table PT7.2 should be mastered.

Computer Objectives.   You have been provided with simple computer algorithms to implement the techniques discussed in Part Seven. All have utility as learning tools.

In addition, the supplementary software in the back of the book includes a section on spreadsheet applications of PDEs. This software is designed to accompany Case Study 26.5.

CHAPTER 23
Finite Difference: Elliptic Equations

Elliptic equations in engineering are typically used to characterize steady-state, boundary-value problems. Before demonstrating how they can be solved, we will illustrate how a simple case—the Laplace equation—is derived from a physical problem context.

23.1 THE LAPLACE EQUATION

As mentioned in the introduction to this part of the book, the Laplace equation can be used to model a variety of problems involving the potential of an unknown variable. Because of its simplicity and general relevance to most areas of engineering, we will use a heated plate as our fundamental context for deriving and solving this elliptic PDE. Case studies and homework problems will be employed to illustrate the applicability of the model to other engineering problem contexts.

Figure 23.1 shows an element on the face of a thin rectangular plate of thickness Δz. The plate is insulated everywhere but at its edges, where the temperature can be set

Figure 23.1
A thin plate of thickness Δz. An element is shown about which a heat balance is taken.

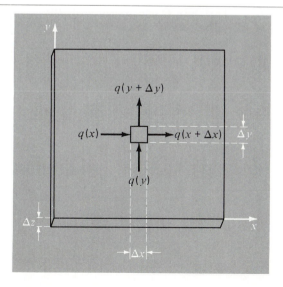

at a prescribed level. The insulation and the thinness of the plate mean that heat transfer is limited to the x and y dimensions. At steady state the flow of heat into the element over a unit time period Δt must equal the flow out as in

$$
\begin{aligned}
q(x)\,\Delta y\,\Delta z\,\Delta t \; + \; q(y)\Delta x\,\Delta z\,\Delta t \\
= q(x\,+\,\Delta x)\,\Delta y\,\Delta z\,\Delta t \; + \; q(y\,+\,\Delta y)\,\Delta x\,\Delta z\,\Delta t
\end{aligned} \tag{23.1}
$$

where $q(x)$ and $q(y)$ are the heat fluxes at x and y, respectively [cal/(cm^2·s)]. Dividing by Δz and Δt and collecting terms yields

$$
[q(x) - q(x\,+\,\Delta x)]\,\Delta y \; + \; [q(y) - q(y\,+\,\Delta y)]\,\Delta x = 0
$$

Multiplying the first term by $\Delta x/\Delta x$ and the second by $\Delta y/\Delta y$ gives

$$
\frac{q(x) - q(x\,+\,\Delta x)}{\Delta x}\,\Delta x\,\Delta y \; + \; \frac{q(y) - q(y\,+\,\Delta y)}{\Delta y}\,\Delta y\,\Delta x = 0 \tag{23.2}
$$

Dividing by $\Delta x/\Delta y$ and taking the limit results in

$$
-\frac{\partial q}{\partial x} - \frac{\partial q}{\partial y} = 0 \tag{23.3}
$$

where the partial derivatives result from the definitions in Eqs. (PT7.1) and (PT7.2).

Equation (23.3) is a partial differential equation that is an expression of the conservation of energy for the plate. However, unless heat fluxes are specified at the plate's edges, it cannot be solved. Because temperature boundary conditions are given, Eq. (23.3) must be reformulated in terms of temperature. The link between flux and temperature is provided by *Fourier's law of heat conduction* which can be represented as

$$
q_i = -k\rho C\,\frac{\partial T}{\partial i} \tag{23.4}
$$

where q_i is the heat flux in the direction of the i dimension [cal/(cm^2·s)], k is the coefficient of *thermal diffusivity* (cm^2/s), ρ is the density of the material (g/cm^3), C is the heat capacity of the material [cal/(g·°C)], and T is temperature (°C), which is defined as

$$
T = \frac{H}{\rho C V}
$$

where H is heat (cal) and V is volume (cm^3). Sometimes the term in front of the differential in Eq. (23.3) is treated as a single term

$$
k' = k\rho C \tag{23.5}
$$

where k' is referred to as the *coefficient of thermal conductivity* [cal/(s·cm·°C)]. In either case, both k and k' are parameters that reflect how well the material conducts heat.

Fourier's law is sometimes referred to as a *constitutive equation*. It is given this label because it provides a mechanism that defines the system's internal interactions.

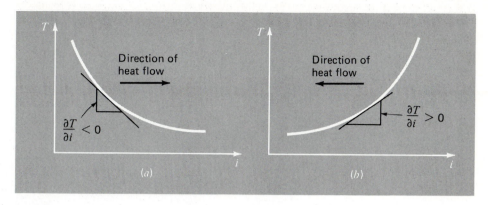

Figure 23.2
Graphical depiction of a temperature gradient. Because heat moves "downhill" from high to low temperature, the flow in (a) is from left to right in the positive i direction. However, due to the orientation of cartesian coordinates, the slope is negative for this case. Thus, a negative gradient leads to a positive flow. This is the origin of the minus sign in Fourier's law of heat conduction. The reverse case is depicted in (b) where the positive gradient leads to a negative heat flow from right to left.

Inspection of Eq. (23.4) indicates that Fourier's law specifies that heat flux perpendicular to the i axis is proportional to the *gradient* or slope of temperature in the i direction. The negative sign ensures that a positive flux in the direction of i results from a negative slope from high to low temperature (Fig. 23.2). Substituting Eq. (23.4) into Eq. (23.3) results in

$$\frac{\partial^2 T}{\partial x^2} + \frac{\partial^2 T}{\partial y^2} = 0 \tag{23.6}$$

which is the *Laplace equation*. Note that for the case where there are sources or sinks of heat within the two-dimensional domain, the equation can be represented as

$$\frac{\partial^2 T}{\partial x^2} + \frac{\partial^2 T}{\partial y^2} = f(x, y) \tag{23.7}$$

where $f(x, y)$ is a function describing the sources or sinks of heat. Equation (23.7) is referred to as the *Poisson equation*.

23.2 SOLUTION TECHNIQUE

The numerical solution of elliptic PDEs such as the Laplace equation proceeds in the reverse manner of the derivation of Eq. (23.6) from the preceding section. Recall that the derivation of Eq. (23.6) employed a balance around a discrete element to yield an algebraic difference equation characterizing heat flux for a plate. Taking the limit turned this difference equation into a differential equation [Eq. (23.3)].

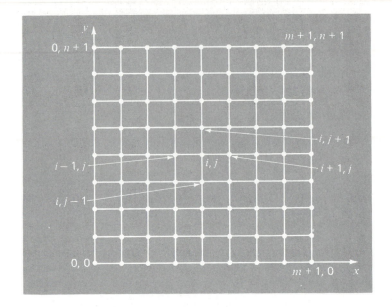

Figure 23.3
A grid used for the finite-difference solution of elliptic PDEs in two independent variables such as the Laplace equation.

For the numerical solution, finite-difference representations based on treating the plate as a grid of discrete points (Fig. 23.3) are substituted for the partial derivatives in Eq. (23.6). As described next, the PDE is transformed into an algebraic difference equation.

23.2.1   The Laplacian Difference Equation

Central differences based on the grid scheme from Fig. 23.3 are (recall Fig. 17.3)

$$\frac{\partial^2 T}{\partial x^2} = \frac{T_{i+1,j} - 2T_{i,j} + T_{i-1,j}}{\Delta x^2}$$

and

$$\frac{\partial^2 T}{\partial y^2} \simeq \frac{T_{i,j+1} - 2T_{i,j} + T_{i,j-1}}{\Delta y^2}$$

which have errors of $O[(\Delta x)^2]$ and $O[(\Delta y)^2]$, respectively. Substituting these expressions into Eq. (23.6) gives

$$\frac{T_{i+1,j} - 2T_{i,j} + T_{i-1,j}}{\Delta x^2} + \frac{T_{i,j+1} - 2T_{i,j} + T_{i,j-1}}{\Delta y^2} = 0$$

For the square grid in Fig. 23.3, $\Delta x = \Delta y$ and by collection of terms, the equation becomes

$$T_{i+1,j} + T_{i-1,j} + T_{i,j+1} + T_{i,j-1} - 4T_{i,j} = 0 \tag{23.8}$$

This relationship, which holds for all interior points on the plate, is referred to as the *Laplacian difference equation*.

In addition, boundary conditions along the edges of the plate must be specified to obtain a unique solution. The simplest case is where the temperature at the boundary

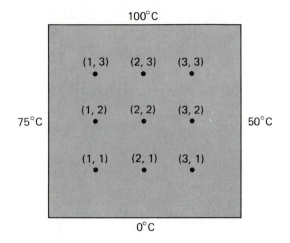

Figure 23.4
A heated plate where boundary temperatures are held at constant levels.. This case is called a Dirichlet boundary condition.

is set at a fixed value. This is called a *Dirichlet boundary condition*. Such is the case for Fig. 23.4, where the edges are held at constant temperatures $\overline{T}_{i,j}$. For the case illustrated in Fig. 23.4, a balance for node (1, 1) is, according to Eq. (23.8),

$$T_{2,1} + T_{0,1} + T_{1,2} + T_{1,0} - 4T_{1,1} = 0 \qquad (23.9)$$

However, $T_{0,1} = \overline{T}_{0,1} = 75$ and $T_{1,0} = \overline{T}_{1,0} = 0$, and, therefore, Eq. (23.9) can be expressed as

$$-4T_{1,1} + T_{1,2} + T_{2,1} = -75$$

Similar equations can be developed for the other interior points. The result is the following set of nine simultaneous equations with nine unknowns:

$$
\begin{aligned}
-4T_{1,1} + T_{2,1} \qquad\qquad + T_{1,2} \qquad\qquad\qquad\qquad\qquad\qquad &= -75 \\
T_{1,1} - 4T_{2,1} + T_{3,1} \qquad\qquad + T_{2,2} \qquad\qquad\qquad\qquad\qquad &= 0 \\
T_{2,1} - 4T_{3,1} \qquad\qquad\qquad + T_{3,2} \qquad\qquad\qquad &= -50 \\
T_{1,1} \qquad\qquad -4T_{1,2} + T_{2,2} \qquad\qquad + T_{1,3} \qquad\qquad &= -75 \\
T_{2,1} \qquad + T_{1,2} - 4T_{2,2} + T_{3,2} \qquad + T_{2,3} \qquad\qquad &= 0 \\
T_{3,1} \qquad\qquad + T_{2,2} - 4T_{3,2} \qquad\qquad + T_{3,3} &= -50 \\
T_{1,2} \qquad\qquad\qquad -4T_{1,3} + T_{2,3} \qquad &= -175 \\
T_{2,2} \qquad\qquad + T_{1,3} - 4T_{2,3} + T_{3,3} &= -100 \\
T_{3,2} \qquad\qquad\qquad + T_{2,3} - 4T_{3,3} &= -150
\end{aligned}
\qquad (23.10)
$$

23.2.2 The Liebmann Method

Most numerical solutions of the Laplace equation involve systems that are much larger than Eq. (23.10). For example, a 10-by-10 grid involves 100 linear algebraic equations. Solution techniques for these types of equations were discussed in Part Three. Because they are prone to round-off error, elimination methods are not usually employed for such systems. In addition, notice that there are a maximum of five unknown terms per line in Eq. (23.8). For larger-sized grids, this means that a significant number of the terms will be zero. When applied to such sparse systems, elimination methods waste great amounts of computer memory storing these zeros.

For the aforementioned reasons, elliptic PDEs are usually solved with approximate methods. The most commonly employed approach is Gauss-Seidel, which when applied to PDEs is also referred to as *Liebmann's method*. In this technique, Eq. (23.8) is expressed as

$$T_{i,j} = \frac{T_{i+1,j} + T_{i-1,j} + T_{i,j+1} + T_{i,j-1}}{4} \tag{23.11}$$

and solved iteratively for $j = 1$ to n and $i = 1$ to m. Because Eq. (23.8) is diagonally dominant, this procedure will eventually converge on a stable solution. *Overrelaxation* is often employed to accelerate the rate of convergence by applying the following formula after each iteration

$$T_{i,j}^{\text{new}} = \lambda T_{i,j}^{\text{new}} + (1 - \lambda)T_{i,j}^{\text{old}} \tag{23.12}$$

where $T_{i,j}^{\text{new}}$ and $T_{i,j}^{\text{old}}$ are the values of $T_{i,j}$ from the present and the previous iteration, respectively, and λ is a weighting factor which is set between 1 and 2.

As with the conventional Gauss-Seidel method, the iterations are repeated until the absolute values of all the percent relative errors $(\epsilon_a)_{i,j}$ fall below a prespecified stopping criterion ϵ_s. These percent relative errors are estimated by

$$|(\epsilon_a)_{i,j}| = \left| \frac{T_{i,j}^{\text{new}} - T_{i,j}^{\text{old}}}{T_{i,j}^{\text{new}}} \right| \times 100\% \tag{23.13}$$

EXAMPLE 23.1   Temperature of a Heated Plate with Fixed Boundary Conditions

Problem Statement: Use Liebmann's method (Gauss-Seidel) to solve for the temperature of the heated plate in Fig. 23.4. Employ overrelaxation with a value of 1.5 for the weighting factor and iterate to $\epsilon_s = 1\%$.

Solution: Equation (23.11) at $i = 1, j = 1$ is

$$T_{1,1} = \frac{0 + 75 + 0 + 0}{4} = 18.75$$

and applying overrelaxation yields

$$T_{1,1} = 1.5(18.75) + (1 - 1.5)0 = 28.125$$

For $i = 2, j = 1$,

$$T_{2,1} = \frac{0 + 28.125 + 0 + 0}{4} = 7.03125$$

and applying overrelaxation yields

$$T_{2,1} = 1.5(7.03125) + (1 - 1.5)0 = 10.546875$$

For $i = 3, j = 1$,

$$T_{3,1} = \frac{50 + 10.546875 + 0 + 0}{4} = 15.13671875$$

and applying overrelaxation yields

$$T_{3,1} = 1.5(15.13671875) + (1 - 1.5)0 = 22.70508$$

The computation is repeated for the other rows to give

$T_{1,2} = 38.67188$	$T_{2,2} = 18.45703$	$T_{3,2} = 34.18579$
$T_{1,3} = 80.12696$	$T_{2,3} = 74.469$	$T_{3,3} = 96.99554$

Because all the $T_{i,j}$'s are initially zero, all ϵ_a's for the first iteration will be 100%. For the second iteration the results are

$T_{1,1} = 32.51953$	$T_{2,1} = 22.35718$	$T_{3,1} = 28.60108$
$T_{1,2} = 57.95288$	$T_{2,2} = 61.63333$	$T_{3,2} = 71.86833$
$T_{1,3} = 75.21973$	$T_{2,3} = 87.95872$	$T_{3,3} = 67.68736$

The error for $T_{1,1}$ can be estimated as [Eq. (23.13)].

$$|(\epsilon_a)_{1,1}| = \left| \frac{32.51953 - 28.125}{32.51953} \right| \times 100\% = 13.5\%$$

Because this value is above the stopping criterion of 1%, the computation is continued. The ninth iteration gives the result

$T_{1,1} = 43.00061$	$T_{2,1} = 33.29755$	$T_{3,1} = 33.88506$
$T_{1,2} = 63.21152$	$T_{2,2} = 56.11238$	$T_{3,2} = 52.33999$
$T_{1,3} = 78.58718$	$T_{2,3} = 76.06402$	$T_{3,3} = 69.71050$

where the maximum error is 0.71%.

Figure 23.5 shows the results. As expected, a gradient is established as heat flows from high to low temperatures.

Figure 23.5
Temperature distribution for a heated plate subject to fixed boundary conditions.

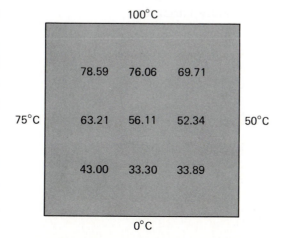

An analytical solution for Eq. (23.6) exists for the case where the upper boundary of the plate in Fig. 23.1 is set at $\bar{T}$ and the remaining faces are held at zero. The solution is (Holman, 1986)

$$T = \frac{2\bar{T}}{\pi} \sum_{n=1}^{\infty} \frac{(-1)^{n+1} + 1}{n} \sin\frac{n\pi x}{W} \frac{\sinh(n\pi y/W)}{\sinh(n\pi H/W)} \tag{23.13a}$$

where W and H are the width (x dimension) and height (y dimension), respectively, of the plate. This equation can be employed to evaluate the temperature distribution for each boundary condition specified in Fig. 23.4. Then, the total solution can be computed by superposition.

EXAMPLE 23.2   Comparison of Analytical and Numerical Solutions

Problem Statement: Use Eq. (23.13a) to generate an analytical solution for the heated plate in Fig. 23.4, and compare the results with those obtained numerically in Example 23.1. Also compare the analytical solution with numerical solutions employing step sizes that are half (7 interior nodes along each dimension) and quarter (15 interior nodes) the step sizes from Example 23.1 (3 interior nodes).

Solution: The results are summarized in Table 23.1. Notice that all the numerical results are reasonably close to the analytical solution and that the agreement improves as the mesh is made finer. In general, deceasing the step size by 50 percent reduces the relative errors by a factor of about 70 percent.

Obviously we do not usually have the luxury of an analytical solution. However, as in the previous example, they can prove useful for assessing the reliability of the numerical result for simple cases. Such assessments offer one way to test the approximate method and provide a measure of confidence for more complicated applications where analytical solutions are unavailable.

23.2.3 Secondary Variables

Because its distribution is described by the Laplace equation, temperature is considered to be the primary variable in the heated-plate problem. For this case, as well as for other problems involving PDEs, secondary variables may also be of interest. As a matter of fact, in certain engineering contexts, the secondary variable may actually be more important.

For the heated plate, a secondary variable is the rate of heat flux across the plate's surface. This quantity can be computed from Fourier's law of heat conduction. Central finite-difference approximations for the first derivatives (recall Fig. 17.3) can be substituted into Eq. (23.4) to give the following values for heat flux in the x and y dimensions.

$$q_x = -k' \frac{T_{i+1,j} - T_{i-1,j}}{2\,\Delta x} \tag{23.14}$$

TABLE 23.1 Comparison of exact analytical solution with approximate numerical solution for the heated plate depicted
in Fig. 23.4. Three numerical results are shown for
cases where the step size is progressively halved.
The numbers in parentheses represent the true percent
relative errors.

| Node | x/W and y/H | | | Exact |
	0.25	0.125	0.0625	
(1,1)	42.8572 (−0.622)	42.6710 (−0.185)	42.6166 (−0.057)	42.5924
(2,1)	33.2589 (−2.985)	32.6053 (−0.961)	32.3787 (−0.259)	32.2949
(3,1)	33.9286 (−1.290)	33.6184 (−0.364)	33.5276 (−0.093)	33.4965
(1,2)	63.1697 (0.541)	63.4020 (0.175)	63.4831 (0.048)	63.5132
(2,2)	56.2500 (0.000)	56.2500 (0.000)	56.2500 (0.000)	56.2500
(3,2)	52.4554 (−0.134)	52.4068 (−0.041)	52.3910 (−0.011)	52.3853
(1,3)	78.5714 (0.547)	78.8816 (0.154)	78.9723 (0.040)	79.0035
(2,3)	76.1161 (0.899)	76.5859 (0.287)	76.7470 (0.077)	76.8065
(3,3)	69.643 (0.371)	69.829 (0.105)	69.883 (0.027)	69.9021

and

$$q_y = -k' \frac{T_{i,j+1} - T_{i,j-1}}{2\,\Delta y} \tag{23.15}$$

The resultant heat flux can be computed from these two quantities by

$$q_n = \sqrt{q_x^2 + q_y^2} \tag{23.16}$$

where the direction of q_n is given by

$$\theta = \tan^{-1}\left(\frac{q_y}{q_x}\right) \tag{23.17}$$

for $q_x > 0$ and

$$\theta = \tan^{-1}\left(\frac{q_y}{q_x}\right) + 180 \tag{23.18}$$

for $q_x < 0$ where θ is in degrees. If $q_x = 0$, θ is 90°or 270° depending on whether q_y is positive or negative, respectively.

EXAMPLE 23.3   Flux Distribution for a Heated Plate

Problem Statement: Employ the results of Example 23.1 to determine the distribution of heat flux for the heated plate from Fig. 23.4. Assume that the plate is 40 × 40 cm and is made out of aluminum [$k' = 0.49$ cal/(s·cm·°C)].

Solution: For $i = j = 1$, Eq. (23.14) can be used to compute

$$q_x = -0.49 \frac{\text{cal}}{\text{s} \cdot \text{cm} \cdot \text{°C}} \frac{(33.29755 - 75)\text{°C}}{2(10 \text{ cm})} = 1.022 \frac{\text{cal}}{\text{cm}^2 \cdot \text{s}}$$

and [Eq. (23.15)]

$$q_y = -0.49 \frac{\text{cal}}{\text{s} \cdot \text{cm} \cdot \text{°C}} \frac{(63.21152 - 0)\text{°C}}{2(10 \text{ cm})} = -1.549 \frac{\text{cal}}{\text{cm}^2 \cdot \text{s}}$$

The resultant flux can be computed with Eq. (23.16):

$$q_n = \sqrt{(1.022)^2 + (-1.549)^2} = 1.856 \frac{\text{cal}}{\text{cm}^2 \cdot \text{s}}$$

and the angle of its trajectory by Eq. (23.17)

$$\theta = \tan^{-1}\left(\frac{-1.549}{1.022}\right) = -56.584°$$

Thus, at this point, the heat flux is directed down and to the right. Values at the other grid points can be computed; the results are displayed in Fig. 23.6.

Figure 23.6
Heat flux for a plate subject to fixed boundary temperatures. Note that the lengths of the arrows are proportional to the magnitude of the flux.

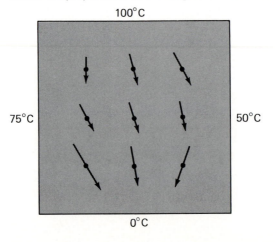

23.2.4 Computer Program for a Rectangular Plate

It is a simple proposition to prepare computer codes to predict the temperature and heat flux distributions for a plate. These programs should allow the user to specify a fixed temperature for each edge. The code could also be configured to permit the specification of a different temperature at each boundary node.

In Fig. 23.7, such a program is employed to perform the computations from Examples 23.1 and 23.3. Notice that as each iteration passes, the plate seems to be heating up until it converges on the steady-state distribution. This is the behavior that we might expect from a time-variable solution that is run to steady state. In fact, time-variable

Figure 23.7
Output from a computer program duplicating the computations from Examples 23.1 and 23.2.

```
RUN
horizontal segments = ? 4
horizontal length = ? 40
vertical segments = ? 4
vertical length = ? 40
coeff of thermal cond = ? .49
boundary temperatures:
     top = ? 100
    left = ? 75
bottom = ? 0
  right = ? 50

maximum iterations = ? 20

maximum error(%) = ? 1
relaxation factor = ? 1.5
- - - - - - - - - - - - - - - - - -
iteration =   0
- - - - - - - - - - - - - - - - - -
   0.0 100.0 100.0 100.0    0.0
  75.0   0.0   0.0   0.0   50.0
  75.0   0.0   0.0   0.0   50.0
  75.0   0.0   0.0   0.0   50.0
   0.0   0.0   0.0   0.0    0.0

- - - - - - - - - - - - - - - - - -
iteration =   1
- - - - - - - - - - - - - - - - - -
   0.0 100.0 100.0 100.0    0.0
  75.0  80.1  74.5  97.0   50.0
  75.0  38.7  18.5  34.2   50.0
  75.0  28.1  10.5  22.7   50.0
   0.0   0.0   0.0   0.0    0.0

- - - - - - - - - - - - - - - - - -
iteration =   2
- - - - - - - - - - - - - - - - - -
   0.0 100.0 100.0 100.0    0.0
  75.0  75.2  88.0  97.0   50.0
  75.0  58.0  61.6  71.9   50.0
  75.0  32.5  22.4  28.6   50.0
   0.0   0.0   0.0   0.0    0.0
```

```
- - - - - - - - - - - - - - - - - -
iteration =   3
- - - - - - - - - - - - - - - - - -
   0.0 100.0 100.0 100.0    0.0
  75.0  85.8  76.7  70.3   50.0
  75.0  66.2  68.3  51.0   50.0
  75.0  42.0  38.4  45.8   50.0
   0.0   0.0   0.0   0.0    0.0

- - - - - - - - - - - - - - - - - -
iteration =   4
- - - - - - - - - - - - - - - - - -
   0.0 100.0 100.0 100.0    0.0
  75.0  77.8  75.5  68.8   50.0
  75.0  70.2  55.4  51.8   50.0
  75.0  46.4  41.0  30.0   50.0
   0.0   0.0   0.0   0.0    0.0

- - - - - - - - - - - - - - - - - -
iteration =   5
- - - - - - - - - - - - - - - - - -
   0.0 100.0 100.0 100.0    0.0
  75.0  77.7  74.8  69.2   50.0
  75.0  60.5  53.6  51.5   50.0
  75.0  46.6  29.2  33.9   50.0
   0.0   0.0   0.0   0.0    0.0

- - - - - - - - - - - - - - - - - -
iteration =   6
- - - - - - - - - - - - - - - - - -
   0.0 100.0 100.0 100.0    0.0
  75.0  77.9  76.2  69.9   50.0
  75.0  61.5  55.9  52.4   50.0
  75.0  38.4  32.7  33.3   50.0
   0.0   0.0   0.0   0.0    0.0

- - - - - - - - - - - - - - - - - -
iteration =   7
- - - - - - - - - - - - - - - - - -
   0.0 100.0 100.0 100.0    0.0
  75.0  79.3  76.7  70.0   50.0
  75.0  64.1  57.0  53.0   50.0
  75.0  44.2  33.7  34.4   50.0
   0.0   0.0   0.0   0.0    0.0
```

(*Continues* **Figure 23.7**)

```
- - - - - - - - - - - - - - - - - -          - - - - - - - - - - - - - - - - - -
iteration =  8                               heat flux rates
- - - - - - - - - - - - - - - - - -          - - - - - - - - - - - - - - - - - -
    0.0 100.0 100.0 100.0   0.0               0.90  1.10  1.33
   75.0  78.4  75.9  69.4  50.0               0.99  1.08  0.89
   75.0  63.2  56.4  52.4  50.0               1.86  1.39  1.35
   75.0  42.7  33.4  34.0  50.0              - - - - - - - - - - - - - - - - - -
    0.0   0.0   0.0   0.0   0.0               trajectories(degrees)
                                             - - - - - - - - - - - - - - - - - -
- - - - - - - - - - - - - - - - - -           268.3  -78.6  -61.3
iteration =  9                                -62.0  -75.7  -80.3
- - - - - - - - - - - - - - - - - -           -56.6  -80.8  252.3
    0.0 100.0 100.0 100.0   0.0               Ok
   75.0  78.6  76.1  69.7  50.0
   75.0  63.2  56.1  52.3  50.0
   75.0  43.0  33.3  33.9  50.0
    0.0   0.0   0.0   0.0   0.0
maximum error =  .7116617
```

solution techniques (that is, those evolved for parabolic PDEs in two spatial dimensions) are the basis for alternative algorithms to solve elliptic equations. One of these methods—the A.D.I. technique—will be discussed in the next chapter.

23.3 BOUNDARY CONDITIONS

Because it is free of complicating factors, the rectangular plate with fixed boundary conditions has been an ideal context for showing how elliptic PDEs can be solved numerically. We will now elaborate on other issues that will expand our capabilities to address more realistic problems. These involve boundaries at which the derivative is specified and boundaries that are irregularly shaped.

23.3.1 Derivative Boundary Conditions

The fixed or Dirichlet boundary condition discussed to this point is but one of several types that are used with partial differential equations. A common alternative is the case where the derivative is given. This is commonly referred to as a *Neumann boundary condition*. For the heated-plate problem, this amounts to specifying the heat flux rather than the temperature at the boundary. One example is the situation where the edge is insulated. In this case, which is referred to as a *natural boundary condition,* the derivative is zero. This conclusion is drawn directly from Eq. (23.4) because insulating a boundary means that the heat flux (and consequently the gradient) must be zero. Another example would be where heat is lost across the edge by predictable mechanisms such as radiation and conduction.

Figure 23.8 depicts a node $(0, j)$ at the left edge of a heated plate. Applying Eq. (23.8) at the point gives

$$T_{1,j} + T_{-1,j} + T_{0,j+1} + T_{0,j-1} - 4T_{0,j} = 0 \tag{23.19}$$

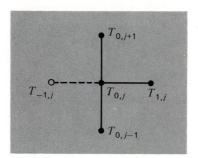

Figure 23.8
A boundary node $(0, j)$ on the left edge of a heated plate. To approximate the derivative normal to the edge (that is, the x derivative), an imaginary point $(-1, j)$ is located a distance Δx beyond the edge.

Notice that an imaginary point $(-1, j)$ lying outside the plate is required for this equation. Although this exterior fictitious point might seem to represent a problem, it actually serves as the vehicle for incorporating the derivative boundary condition into the problem. This is done by representing the first derivative in the x dimension at $(0, j)$ by the finite divided difference

$$\frac{\partial T}{\partial x} = \frac{T_{1,j} - T_{-1,j}}{2\Delta x}$$

which can be solved for

$$T_{-1,j} = T_{1,j} - 2\,\Delta x\,\frac{\partial T}{\partial x}$$

Now we have a relationship for $T_{-1,j}$ that actually includes the derivative. It can be substituted into Eq. (23.19) to give

$$2T_{1,j} - 2\,\Delta x\,\frac{\partial T}{\partial x} + T_{0,j+1} + T_{0,j-1} - 4T_{0,j} = 0 \tag{23.20}$$

Thus, we have incorporated the derivative into the balance.

Similar relationships can be developed for derivative boundary conditions at the other edges. The following example shows how this is done for the heated plate.

EXAMPLE 23.4 Heated Plate with an Insulated Edge

Problem Statement: Repeat the same problem as in Example 23.1 but with the lower edge insulated.

Solution: The general equation to characterize a derivative at the lower edge (that is, at $j = 0$) of a heated plate is

$$T_{i+1,0} + T_{i-1,0} + 2T_{i,1} - 2\,\Delta y\,\frac{\partial T}{\partial y} - 4T_{i,0} = 0$$

For an insulated edge, the derivative is zero and the equation becomes

$$T_{i+1,0} + T_{i-1,0} + 2T_{i,1} - 4T_{i,0} = 0$$

The simultaneous equations for temperature distribution on the plate in Fig. 23.4 with an insulated lower edge can be written in matrix form as

$$
\begin{bmatrix}
-4 & 1 & & 2 & & & & & & & & \\
1 & -4 & 1 & & 2 & & & & & & & \\
& 1 & -4 & & & 2 & & & & & & \\
1 & & & -4 & 1 & & 1 & & & & & \\
& 1 & & 1 & -4 & 1 & & 1 & & & & \\
& & 1 & & 1 & -4 & & & 1 & & & \\
& & & 1 & & & -4 & 1 & & 1 & & \\
& & & & 1 & & 1 & -4 & 1 & & 1 & \\
& & & & & 1 & & 1 & -4 & & & 1 \\
& & & & & & 1 & & & -4 & 1 & \\
& & & & & & & 1 & & 1 & -4 & 1 \\
& & & & & & & & 1 & & 1 & -4
\end{bmatrix}
\begin{Bmatrix}
T_{1,0} \\ T_{2,0} \\ T_{3,0} \\ T_{1,1} \\ T_{2,1} \\ T_{3,1} \\ T_{1,2} \\ T_{2,2} \\ T_{3,2} \\ T_{1,3} \\ T_{2,3} \\ T_{3,3}
\end{Bmatrix}
=
\begin{Bmatrix}
-75 \\ 0 \\ -50 \\ -75 \\ 0 \\ -50 \\ -75 \\ 0 \\ -50 \\ -175 \\ -100 \\ -150
\end{Bmatrix}
$$

Note that because of the derivative boundary condition, the matrix is increased to 12 $\times$ 12 in contrast to the 9 $\times$ 9 system in Eq. (23.10) to account for the three unknown temperatures along the plate's lower edge. These equations can be solved for

$$T_{1,0} = 71.91 \qquad T_{2,0} = 67.01 \qquad T_{3,0} = 59.54$$

$$T_{1,1} = 72.81 \qquad T_{2,1} = 68.31 \qquad T_{3,1} = 60.57$$

$$T_{1,2} = 76.01 \qquad T_{2,2} = 72.84 \qquad T_{3,2} = 64.42$$

$$T_{1,3} = 83.41 \qquad T_{2,3} = 82.63 \qquad T_{3,3} = 74.26$$

These results and computed fluxes (for the same parameters as in Example 23.3) are displayed in Fig. 23.9. Note that, because the lower edge is insulated, the plate's temperature is higher than for Fig. 23.5 where the edge temperature is fixed at zero. In addition, the heat flow (in contrast to Fig. 23.6) is now deflected to the right and moves parallel to the insulated wall.

Figure 23.9

Temperature and flux distribution for a heated plate subject to fixed boundary conditions except for an insulated lower edge.

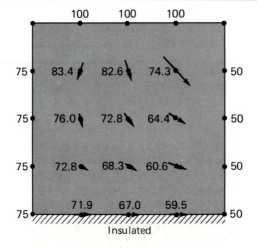

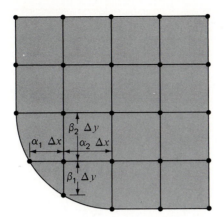

Figure 23.10
A grid for a heated plate with an irregularly shaped boundary. Note how weighting coefficients are used to account for the nonuniform spacing in the vicinity of the nonrectangular boundary.

23.3.2 Irregular Boundaries

Although the rectangular plate from Fig. 23.4 has served well to illustrate the fundamental aspects of solving elliptic PDEs, many engineering problems do not exhibit such an idealized geometry. For example, a great many systems have irregular boundaries.

Figure 23.10 is a system that can serve to illustrate how nonrectangular boundaries can be handled. As depicted, the plate's lower left boundary is circular, with radius h. Notice that we have affixed parameters—α_1, α_2, β_1, and β_2—to each of the lengths surrounding the node. Of course, for the plate depicted in Fig. 23.10, $\alpha_2 = \beta_2 = 1$. However, we will retain these parameters throughout the following derivation so that the resulting equation is generally applicable to any irregular boundary—not just one on the lower left-hand corner of a heated plate. The first derivatives in the x dimension can be approximated as

$$\left(\frac{\partial T}{\partial x}\right)_{i-1 \text{ to } i} = \frac{T_{i,j} - T_{i-1,j}}{\alpha_1 \Delta x} \tag{23.21}$$

and

$$\left(\frac{\partial T}{\partial x}\right)_{i \text{ to } i+1} = \frac{T_{i+1,j} - T_{i,j}}{\alpha_2 \Delta x} \tag{23.22}$$

The second derivatives can be developed from these first derivatives. For the x dimension, the second derivative is

$$\frac{\partial^2 T}{\partial x^2} = \frac{\partial}{\partial x}\left(\frac{\partial T}{\partial x}\right) \simeq \frac{\left(\dfrac{\partial T}{\partial x}\right)_{i \text{ to } i+1} - \left(\dfrac{\partial T}{\partial x}\right)_{i-1 \text{ to } i}}{\dfrac{\alpha_1 \Delta x + \alpha_2 \Delta x}{2}} \tag{23.23}$$

Substituting Eqs. (23.21) and (23.22) into (23.23) gives

$$\frac{\partial^2 T}{\partial x^2} \simeq 2 \frac{\left[\dfrac{T_{i-1,j} - T_{i,j}}{\alpha_1 \Delta x} + \dfrac{T_{i+1,j} - T_{i,j}}{\alpha_2 \Delta x}\right]}{\Delta x(\alpha_1 + \alpha_2)}$$

Collecting terms yields

$$\frac{\partial^2 T}{\partial x^2} \simeq \frac{2}{\Delta x^2} \left[\frac{T_{i-1,j} - T_{i,j}}{\alpha_1(\alpha_1 + \alpha_2)} + \frac{T_{i+1,j} - T_{i,j}}{\alpha_2(\alpha_1 + \alpha_2)} \right]$$

A similar equation can be developed in the y dimension:

$$\frac{\partial^2 T}{\partial y^2} \simeq \frac{2}{\Delta y^2} \left[\frac{T_{i,j-1} - T_{i,j}}{\beta_1(\beta_1 + \beta_2)} + \frac{T_{i,j+1} - T_{i,j}}{\beta_2(\beta_1 + \beta_2)} \right]$$

Substituting these equations in Eq. (23.6) yields

$$\frac{2}{\Delta x^2} \left[\frac{T_{i-1,j} - T_{i,j}}{\alpha_1(\alpha_1 + \alpha_2)} + \frac{T_{i+1,j} - T_{i,j}}{\alpha_2(\alpha_1 + \alpha_2)} \right]$$

$$+ \frac{2}{\Delta y^2} \left[\frac{T_{i,j-1} - T_{i,j}}{\beta_1(\beta_1 + \beta_2)} + \frac{T_{i,j+1} - T_{i,j}}{\beta_2(\beta_1 + \beta_2)} \right] = 0$$

(23.24)

As illustrated in the following example, Eq. (23.24) is applied to any node that lies adjacent to an irregular, Dirichlet-type boundary.

EXAMPLE 23.5   *Heated Plate with an Irregular Boundary*

Problem Statement: Repeat the same problem as in Example 23.1 but with the lower edge as depicted in Fig. 23.10.

Solution: For the case in Fig. 23.10, $\Delta x = \Delta y$, $\alpha_1 = \beta_1 = 0.732$, and $\alpha_2 = \beta_2 = 1$. Substituting these values into Eq. (23.24) yields the following balance for node (1, 1),

$$0.788675(T_{0,1} - T_{1,1}) + 0.57735(T_{2,1} - T_{1,1})$$

$$+ 0.788675(T_{1,0} - T_{1,1}) + 0.57735(T_{1,2} - T_{1,1}) = 0$$

Collecting terms, this equation can be expressed as

$$-2.73205T_{1,1} + 0.57735T_{2,1} + 0.57735T_{1,2} = -0.788675T_{0,1} - 0.788675T_{1,0}$$

The simultaneous equations for temperature distribution on the plate in Fig. 23.10 with a lower-edge boundary temperature of 75 can be written in matrix form as

$$\begin{bmatrix} -2.732 & 0.577 & & 0.577 & & & & & \\ 1 & -4 & 1 & & 1 & & & & \\ & 1 & -4 & & & 1 & & & \\ 1 & & & -4 & 1 & & 1 & & \\ & 1 & & 1 & -4 & 1 & & 1 & \\ & & 1 & & 1 & -4 & & & 1 \\ & & & 1 & & & -4 & 1 & \\ & & & & 1 & & 1 & -4 & 1 \\ & & & & & 1 & & 1 & -4 \end{bmatrix} \begin{Bmatrix} T_{1,1} \\ T_{2,1} \\ T_{3,1} \\ T_{1,2} \\ T_{2,2} \\ T_{3,2} \\ T_{1,3} \\ T_{2,3} \\ T_{3,3} \end{Bmatrix} = \begin{Bmatrix} -118.3 \\ -75 \\ -125 \\ -75 \\ 0 \\ -50 \\ -175 \\ -100 \\ -150 \end{Bmatrix}$$

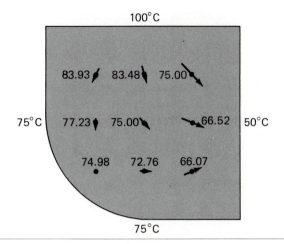

Figure 23.11
Temperature and flux distribution for a heated plate with a circular boundary.

These equations can be solved for

$$T_{1,1} = 74.98 \qquad T_{2,1} = 72.76 \qquad T_{3,1} = 66.07$$

$$T_{1,2} = 77.23 \qquad T_{2,2} = 75.00 \qquad T_{3,2} = 66.52$$

$$T_{1,3} = 83.93 \qquad T_{2,3} = 83.48 \qquad T_{3,3} = 75.00$$

These results along with the computed fluxes are displayed in Fig. 23.11. Note that the fluxes are computed in the same fashion as in Sec. 23.2.3 with the exception that $(\alpha_1 + \alpha_2)$ and $(\beta_1 + \beta_2)$ are substituted for the 2's in the denominators of Eqs. (23.14) and (23.15), respectively. Case Study 26.3 illustrates how this is done.

23.3.3 Incorporating Derivative Boundary Conditions and Irregular Boundaries into Software

Modifying a computer program to include derivative boundary conditions is a relatively straightforward task. This merely involves ensuring that additional equations are generated to characterize the boundary nodes at which the derivatives are specified. In addition, the code must be modified so that these equations incorporate the derivative in the fashion of Eq. (23.20). Problem 23.7 at the end of this chapter deals with this task.

Developing general software to characterize systems with irregular boundaries is a much more difficult proposition. For example, a fairly involved algorithm would be required to model the simple gasket depicted in Fig. 23.12. This would involve two major modifications. First, a scheme would have to be developed to conveniently input the configuration of the nodes and to identify which were at the boundary. Second, an algorithm would be required to generate the proper simultaneous equations on the basis of the input information. The net result is that general software for solving elliptic (and for that matter, all) PDEs is relatively complicated.

One method used to simplify such efforts is to require a very fine grid. For such cases, it is often assumed that the closest node serves as the boundary point. In this

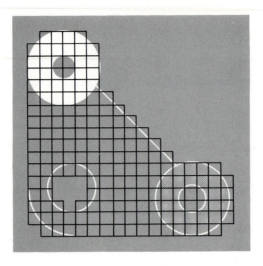

Figure 23.12
A finite difference grid superimposed on an irregularly
shaped gasket.

way, the analysis does not have to consider the weighting parameters from Sec. 23.3.2. Although this introduces some error, the use of a sufficiently fine mesh can make the resulting discrepancy negligible. However, this involves a trade-off due to the computational burden introduced by the increased number of simultaneous equations.

As a consequence of these considerations, numerical analysts have developed alternative approaches that differ radically from finite-difference methods. Although these *finite-element methods* are more conceptually difficult to comprehend, they can much more easily accommodate irregular boundaries. We will turn to these methods in Chap. 25. Before doing this, however, we will first describe finite-difference approaches for another category of PDEs—parabolic equations.

PROBLEMS

Hand Calculations

23.1 Use Liebmann's method to solve for the temperature of the square heated plate in Fig. 23.4 but with the upper boundary condition increased to 150°C and the right boundary decreased to 25°C. Use a relaxation factor of 1.2 and iterate to $\epsilon_s = 1\%$.

23.2 Compute the fluxes for Prob. 23.1 using the parameters from Example 23.3.

23.3 Repeat Example 23.1 but for the segmentation scheme depicted in Fig. P23.3.

23.4 Repeat Prob. 23.3 but for the case where the upper edge is insulated.

23.5 Repeat Examples 23.1 and 23.3 but for the case where the flux at the lower edge is directed downward with a value of 1 cal/(cm²·s).

23.6 Repeat Example 23.5 for the case where both the lower left and the upper right corners are rounded in the same fashion as the lower left corner of Fig. 23.10. Note that all boundary temperatures on upper and right sides are fixed at 100°C and all on lower and left sides are fixed at 75°C.

Computer-Related Problems

23.7 Develop a user-friendly computer program to implement Liebmann's method for a rectangular plate. Design the

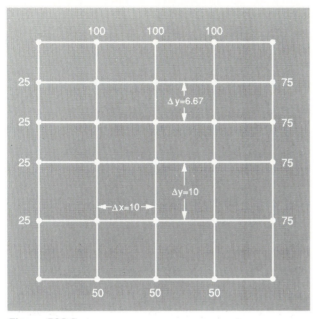

Figure P23.3

program so that it can compute both temperature and flux. Also, include the option of allowing either Dirichlet or derivative conditions at any of the boundaries. Test the program by duplicating the results of Examples 23.1, 23.3 and 23.4.

23.8 Employ the program from Prob. 23.7 to solve Probs. 23.1 and 23.2.

23.9 Employ the program from Prob. 23.7 to solve Probs. 23.3 through 23.5.

CHAPTER 24

Finite Difference: Parabolic Equations

The previous chapter dealt with steady-state PDEs. We now turn to the parabolic equations that are employed to characterize time-variable problems. In the latter part of this chapter, we will illustrate how this is done in two spatial dimensions for the heated plate. Before doing this, we will first show how the simpler one-dimensional case is approached.

24.1 THE HEAT CONDUCTION EQUATION

In a similar fashion to the derivation of the Laplace equation (Eq. 23.6), conservation of heat can be used to develop a heat balance for the differential element in the long, thin insulated rod shown in Fig. 24.1. However, rather than examine the steady-state case, the present balance also considers the amount of heat stored in the element over a unit time period Δt. Thus, the balance is in the form, inputs − outputs = storage, or

$$q(x) \, \Delta y \, \Delta z \, \Delta t - q(x + \Delta x) \, \Delta y \, \Delta z \, \Delta t = \Delta x \, \Delta y \, \Delta z \, \rho C \Delta T$$

Dividing by the volume of the element ($= \Delta x \, \Delta y \, \Delta z$) and Δt gives

$$\frac{q(x) - q(x + \Delta x)}{\Delta x} = \rho C \, \frac{\Delta T}{\Delta t}$$

Taking the limit yields

$$-\frac{\partial q}{\partial x} = \rho C \, \frac{\partial T}{\partial t}$$

Figure 24.1
A thin rod, insulated at all points except at its ends.

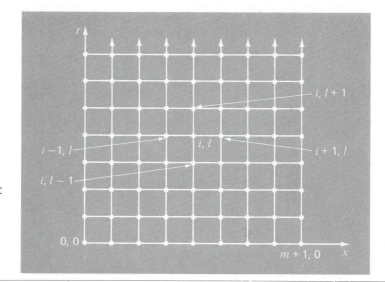

Figure 24.2
A grid used for the finite-difference solution of parabolic PDEs in two independent variables such as the heat-conduction equation. Note how, in contrast to Fig. 23.3, this grid is open-ended in the temporal dimension.

Substituting Fourier's law of heat conduction [Eq. (23.4)] results in

$$k \frac{\partial^2 T}{\partial x^2} = \frac{\partial T}{\partial t} \tag{24.1}$$

which is the *heat-conduction equation*.

Just as with elliptic PDEs, parabolic equations can be solved by substituting finite divided differences for the partial derivatives. However, in contrast to elliptic PDEs, we must now consider changes in time as well as in space. Whereas elliptic equations were bounded in all relevant dimensions, parabolic PDEs are temporally open-ended (Fig. 24.2). Because of their time-variable nature, solutions to these equations involve a number of new issues, notably stability. This, as well as other aspects of parabolic PDEs, will be examined in the following sections as we present two fundamental solution approaches—explicit and implicit schemes.

24.2 EXPLICIT METHODS

The heat-conduction equation requires approximations for the second derivative in space and the first derivative in time. The former is represented in the same fashion as for the Laplace equation by a centered finite divided difference:

$$\frac{\partial^2 T}{\partial x^2} = \frac{T_{i+1}^l - 2T_i^l + T_{i-1}^l}{\Delta x^2} \tag{24.2}$$

which has an error (recall Fig. 17.3) of $O[(\Delta x)^2]$. Notice the slight change in notation that superscripts are used to denote time. This is done so that a second subscript can be used to designate a second spatial dimension when the approach is expanded to the two-dimensional case.

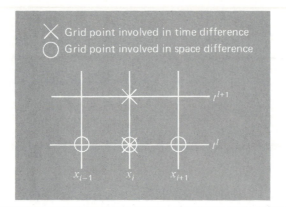

Figure 24.3
A computational molecule for
the explicit form.

A forward finite divided difference is used to approximate the time derivative

$$\frac{\partial T}{\partial t} = \frac{T_i^{l+1} - T_i^l}{\Delta t} \tag{24.3}$$

which has an error (recall Fig. 17.1) of $O(\Delta t)$.

Substituting Eqs. (24.2) and (24.3) into Eq. (24.1) yields

$$k \frac{T_{i+1}^l - 2T_i^l + T_{i-1}^l}{(\Delta x)^2} = \frac{T_i^{l+1} - T_i^l}{\Delta t} \tag{24.4}$$

which can be solved for

$$T_i^{l+1} = T_i^l + \lambda(T_{i+1}^l - 2T_i^l + T_{i-1}^l) \tag{24.5}$$

where $\lambda = k\,\Delta t/(\Delta x)^2$.

This equation can be written for all the interior nodes on the rod. It then provides an explicit means to compute values at each node for a future time based on the present values at the node and its neighbors. In this sense, it is very similar to Euler's method for solving systems of ODEs. That is, if we know the temperature distribution as a function of position at $t = 0$, we can compute the distribution at Δt based on Eq. (24.5).

A computational molecule for the explicit method is depicted in Fig. 24.3, showing the nodes that constitute the spatial and temporal approximations. This molecule can be contrasted with others in this chapter to illustrate the differences between approaches.

EXAMPLE 24.1   Explicit Solution of the One-Dimensional Heat-Conduction Equation

Problem Statement: Use the explicit method to solve for the temperature distribution of a long, thin rod with a length of 10 cm and the following values: $k' = 0.49$ cal/(s·cm·°C), $\Delta x = 2$ cm, and $\Delta t = 0.1$ s. At $t = 0$, the temperature of the rod is zero and the boundary conditions are fixed for all times at $T(0) = 100$°C and $T(10) = 50$°C. Note that the rod is aluminum with $C = 0.2174$ cal/(g·°C) and $\rho = 2.7$ g/cm^3. Therefore, $k = 0.49/(2.7 \cdot 0.2174) = 0.835$ cm^2/s and $\lambda = 0.835(0.1)/(2)^2 = 0.020875$.

Solution: Applying Eq. (24.5) gives the following value at $t = 0.1$ s for the node at $x = 2$ cm:

$$T_1^1 = 0 + 0.020875\,[0 - 2(0) + 100] = 2.0875$$

At the other interior points, $x = 4, 6$, and 8 cm, the results are

$$T_2^1 = 0 + 0.020875\,[0 - 2(0) + 0] = 0$$

$$T_3^1 = 0 + 0.020875\,[0 - 2(0) + 0] = 0$$

$$T_4^1 = 0 + 0.020875\,[50 - 2(0) + 0] = 1.0438$$

At $t = 0.2$ s, the values at the four interior nodes are computed as

$$T_1^2 = 2.0875 + 0.020875\,[0 - 2(2.0875) + 100] = 4.0878$$

$$T_2^2 = 0 + 0.020875\,[0 - 2(0) + 2.0875] = 0.043577$$

$$T_3^2 = 0 + 0.020875\,[1.0438 - 2(0) + 0] = 0.021788$$

$$T_4^2 = 1.0438 + 0.020875\,[50 - 2(1.0438) + 0] = 2.0439$$

The computation is continued, and the results at 3-s intervals are depicted in Fig. 24.4. The general rise in temperature with time indicates that the computation captures the diffusion of heat from the boundaries into the bar.

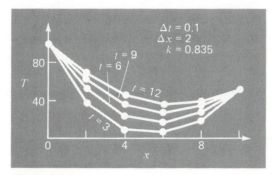

Figure 24.4
Temperature distribution in a long, thin rod as computed with the explicit method described in Sec. 24.2

24.2.1   Convergence and Stability

Convergence means that as Δx and Δt approach zero, the results of the finite-difference technique approach the true solution. *Stability* means that errors at any stage of the computation are not amplified but are attenuated as the computation progresses. It can

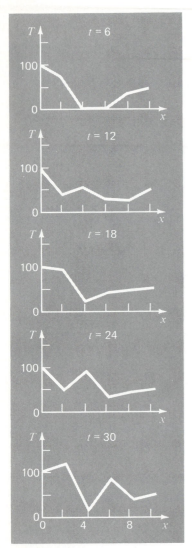

Figure 24.5
An illustration of instability. Solution of Example 24.1 but with $\lambda = 0.735$.

be shown (Carnahan et al., 1969) that the explicit method is both convergent and stable if $\lambda \leq 1/2$. Another way to formulate this criterion is

$$\Delta t \leq \frac{1}{2} \frac{\Delta x^2}{k} \tag{24.6}$$

In addition, it should be noted that setting $\lambda \leq 1/2$ could result in a solution in which errors do not grow but oscillate. Setting $\lambda \leq 1/4$ ensures that the solution will not oscillate. It is also known that setting $\lambda = 1/6$ tends to minimize truncation error (Carnahan et al., 1969).

Figure 24.5 is an example of instability caused by violating Eq. (24.6). This plot is for the same case as in Example 24.1 but with $\lambda = 0.735$ which is considerably greater than 0.5. As in Fig. 24.5, the solution undergoes progressively increasing oscillations. This situation will continue to deteriorate as the computation continues.

Although satisfaction of Eq. (24.6) will alleviate the instabilities of the sort manifested in Fig. 24.5, it also places a strong limitation on the explicit method. For example, suppose that Δx is halved to improve the approximation of the spatial second derivative. According to Eq. (24.6), the time step must be quartered to maintain convergence and stability. Thus, to perform comparable computations, the time steps must be increased by a factor of 4. Furthermore, the computation for each of these time steps will take twice as long because halving Δx doubles the total number of nodes for which equations must be written. Consequently, for the one-dimensional case, halving Δx results in an eightfold increase in the number of calculations. Thus, the computational burden may be large to attain acceptable accuracy. As will be described shortly, other techniques are available that do not suffer from such severe limitations.

24.2.2 Derivative Boundary Conditions

As was the case for elliptic PDEs (recall Sec. 23.3.1), derivative boundary conditions can be readily incorporated into parabolic equations. For a one-dimensional rod, this necessitates adding two equations to characterize the heat balance at the end nodes. For example, the node at the left end ($i = 0$) would be represented by

$$T_0^{l+1} = T_0^l + \lambda(T_1^l - 2T_0^l + T_{-1}^l)$$

Thus, an imaginary point is introduced at $i = -1$ (recall Fig. 23.8). However, as with the elliptic case, this point provides a vehicle for incorporating the derivative boundary condition into the analysis. Problem 24.1 at the end of the chapter deals with this exercise.

24.3 A SIMPLE IMPLICIT METHOD

As noted previously, explicit finite-difference formulations have problems related to stability. In addition, as depicted in Fig. 24.6, they exclude information that has a bearing on the solution. Implicit methods overcome both these difficulties at the expense of somewhat more complicated algorithms.

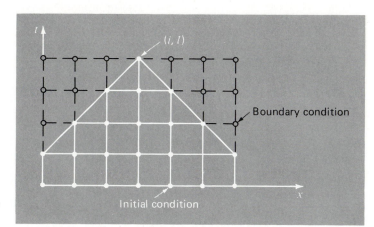

Figure 24.6
Representation of the effect of other nodes on the finite-difference approximation at node (i, l) using an explicit finite-difference scheme. The shaded nodes have an influence on (i, l) whereas the unshaded nodes, which in reality affect (i, l), are excluded.

The fundamental difference between explicit and implicit approximations is depicted in Fig. 24.7. For the explicit form, we approximate the spatial derivative at time level l (Fig. 24.7a). Recall that when we substituted this approximation into the partial differential equation, we obtained a difference equation (24.4) with a single unknown T_i^{l+1}. Thus, we can solve "explicitly" for this unknown as in Eq. (24.5).

In implicit methods, the spatial derivative is approximated at an advanced time level $l + 1$. For example, the second derivative would be approximated by (Fig. 24.7b)

$$\frac{\partial^2 T}{\partial x^2} \simeq \frac{T_{i+1}^{l+1} - 2T_i^{l+1} + T_{i-1}^{l+1}}{(\Delta x)^2} \tag{24.7}$$

which is second-order accurate. When this relationship is substituted into the original PDE, the resulting difference equation contains several unknowns. Thus, it cannot be solved explicitly by simple algebraic rearrangement as was done in going from Eq. (24.4) to (24.5). Instead, the entire system of equations must be solved simultaneously. This is possible because, along with the boundary conditions, the implicit formulations result in a set of linear algebraic equations with the same number of unknowns. Thus, the method reduces to the solution of a set of simultaneous equations at each point in time.

Figure 24.7
Computational molecules demonstrating the fundamental differences between (a) explicit and (b) implicit methods.

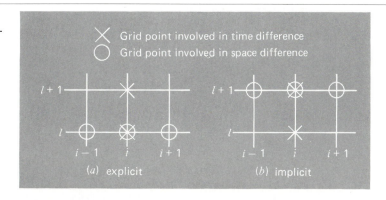

To illustrate how this is done, substitute Eqs. (24.3) and (24.7) into Eq. (24.1) to give

$$k\,\frac{T_{i+1}^{l+1} - 2T_i^{l+1} + T_{i-1}^{l+1}}{(\Delta x)^2} = \frac{T_i^{l+1} - T_i^l}{\Delta t}$$

which can be expressed as

$$-\lambda T_{i-1}^{l+1} + (1 + 2\lambda)T_i^{l+1} - \lambda T_{i+1}^{l+1} = T_i^l \tag{24.8}$$

where $\lambda = k\,\Delta t/(\Delta x)^2$. This equation applies to all but the first and the last interior nodes which must be modified to reflect the boundary conditions. For the case where the temperature levels at the ends of the rod are given, the boundary condition at the left end of the rod ($i = 0$) can be expressed as

$$T_0^{l+1} = f_0(t^{l+1}) \tag{24.9}$$

where $f_0(t^{l+1})$ is a function describing how the boundary temperature changes with time. Substituting Eq. (24.9) into Eq. (24.8) gives the difference equation for the first interior node ($i = 1$):

$$(1 + 2\lambda)T_1^{l+1} - \lambda T_2^{l+1} = T_1^l + \lambda f_0(t^{l+1}) \tag{24.10}$$

Similarly, for the last interior node ($i = m$),

$$-\lambda T_{m-1}^{l+1} + (1 + 2\lambda)T_m^{l+1} = T_m^l + \lambda f_{m+1}(t^{l+1}) \tag{24.11}$$

where $f_{m+1}(t^{l+1})$ describes the specified temperature changes at the right end of the rod ($i = m + 1$).

When Eqs. (24.8), (24.10), and (24.11) are written for all the interior nodes, the resulting set of m linear algebraic equations has m unknowns. In addition, the method has the added bonus that the system is tridiagonal. Thus, we can utilize the extremely efficient solution algorithms (recall Sec. 9.6) that are available for tridiagonal systems.

EXAMPLE 24.2   Simple Implicit Solution of the One-Dimensional Heat-Conduction Equation

Problem Statement: Use the simple implicit finite-difference approximation to solve the same problem as Example 24.1.

Solution: For the rod from Example 24.1, $\lambda = 0.020875$. Therefore, at $t = 0$, Eq. (24.10) can be written for the first interior node as

$$1.04175\,T_1^1 - 0.020875\,T_2^1 = 0 + 0.020875(100)$$

or

$$1.04175\,T_1^1 - 0.020875\,T_2^1 = 2.0875$$

In a similar fashion, Eqs. (24.8) and (24.11) can be applied to the other interior nodes. This leads to the following set of simultaneous equations:

$$\begin{bmatrix} 1.04175 & -0.020875 & & \\ -0.020875 & 1.04175 & -0.020875 & \\ & -0.020875 & 1.04175 & -0.020875 \\ & & -0.020875 & 1.04175 \end{bmatrix} \begin{Bmatrix} T_1^1 \\ T_2^1 \\ T_3^1 \\ T_4^1 \end{Bmatrix} = \begin{Bmatrix} 2.0875 \\ 0 \\ 0 \\ 1.04375 \end{Bmatrix}$$

which can be solved for the temperature at $t = 0.1$ s:

$$T_1^1 = 2.0047$$

$$T_2^1 = 0.0406$$

$$T_3^1 = 0.0209$$

$$T_4^1 = 1.0023$$

Notice how, in contrast to Example 24.1, all the points have changed from the initial condition during the first time step.

In order to solve for the temperatures at $t = 0.2$, the right-hand-side vector must be modified to account for the results of the first step as in

$$\begin{Bmatrix} 4.09215 \\ 0.04059 \\ 0.02090 \\ 2.04609 \end{Bmatrix}$$

The simultaneous equations can then be solved for the temperatures at $t = 0.2$ s:

$$T_1^2 = 3.9305$$

$$T_2^2 = 0.1190$$

$$T_3^2 = 0.0618$$

$$T_4^2 = 1.9653$$

Although the implicit method described is stable and convergent, it has the defect that the temporal difference approximation is first-order accurate whereas the spatial difference approximation is second-order accurate (Fig. 24.8). In the next section we present an alternative implicit method that remedies the situation.

Figure 24.8

A computational molecule for the simple implicit method.

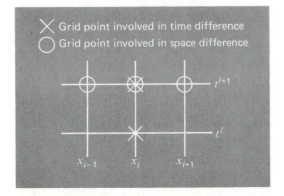

24.4   THE CRANK-NICOLSON METHOD

The *Crank-Nicolson method* provides an alternative implicit scheme that is second-order accurate in both space and time. To provide this accuracy, difference approximations are developed at the midpoint of the time increment (Fig. 24.9). To do this, the temporal first derivative can be approximated at $t^{l+1/2}$ by

$$\frac{\partial T}{\partial t} \simeq \frac{T_i^{l+1} - T_i^l}{\Delta t} \tag{24.12}$$

The second derivative in space can be determined at the midpoint by averaging the difference approximations at the beginning (t^l) and at the end (t^{l+1}) of the time increment

$$\frac{\partial^2 T}{\partial x^2} \simeq \frac{1}{2} \left[\frac{T_{i+1}^l - 2T_i^l + T_{i-1}^l}{(\Delta x)^2} + \frac{T_{i+1}^{l+1} - 2T_i^{l+1} + T_{i-1}^{l+1}}{(\Delta x)^2} \right] \tag{24.13}$$

Substituting Eqs. (24.12) and (24.13) into Eq. (24.1) and collecting terms gives

$$-\lambda T_{i-1}^{l+1} + 2(1 + \lambda)T_i^{l+1} - \lambda T_{i+1}^{l+1} = \lambda T_{i-1}^l + 2(1 - \lambda)T_i^l + \lambda T_{i+1}^l \tag{24.14}$$

where $\lambda = k\,\Delta t/(\Delta x)^2$. As was the case with the simple implicit approach, boundary conditions of $T_0^{l+1} = f_0(t^{l+1})$ and $T_{m+1}^{l+1} = f_{m+1}(t^{l+1})$ can be prescribed to derive versions of Eq. (24.14) for the first and the last interior nodes. For the first interior node

$$2(1 + \lambda)T_1^{l+1} - \lambda T_2^{l+1} = \lambda f_0(t^l) + 2(1 - \lambda)T_1^l + \lambda T_2^l + \lambda f_0(t^{l+1}) \tag{24.15}$$

and for the last interior node

$$-\lambda T_{m-1}^{l+1} + 2(1 + \lambda)T_m^{l+1} =$$
$$\lambda f_{m+1}(t^l) + 2(1 - \lambda)T_{m-1}^l + \lambda T^l + \lambda f_{m+1}(t^{l+1}) \tag{24.16}$$

Although Eqs. (24.14) through (24.16) are slightly more complicated than Eqs. (24.8), (24.10), and (24.11), they are also tridiagonal and, therefore, efficient to solve.

Figure 24.9
A computational molecule for the Crank-Nicolson method.

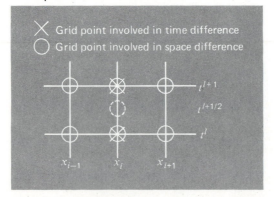

EXAMPLE 24.3 Crank-Nicolson Solution to the One-Dimensional Heat-Conduction Equation

Problem Statement: Use the Crank-Nicolson method to solve the same problem as in Examples 24.1 and 24.2.

Solution: Equations (24.14) through (24.16) can be employed to generate the following tridiagonal set of equations:

$$\begin{bmatrix} 2.01475 & -0.020875 & & \\ -0.020875 & 2.01475 & -0.020875 & \\ & -0.020875 & 2.01475 & -0.020875 \\ & & -0.020875 & 2.01475 \end{bmatrix} \begin{Bmatrix} T_1^1 \\ T_2^1 \\ T_3^1 \\ T_4^1 \end{Bmatrix} = \begin{Bmatrix} 4.175 \\ 0 \\ 0 \\ 2.0875 \end{Bmatrix}$$

which can be solved for the temperatures at $t = 0.1$ s:

$$T_1^1 = 2.0450$$

$$T_2^1 = 0.0210$$

$$T_3^1 = 0.0107$$

$$T_4^1 = 1.0225$$

In order to solve for the temperatures at $t = 0.2$ s, the right-hand-side vector must be changed to

$$\begin{Bmatrix} 8.1801 \\ 0.0841 \\ 0.0427 \\ 4.0901 \end{Bmatrix}$$

The simultaneous equations can then be solved for

$$T_1^2 = 4.0073$$

$$T_2^2 = 0.0826$$

$$T_3^2 = 0.0422$$

$$T_4^2 = 2.0036$$

24.4.1 Comparison of One-Dimensional Methods

Equation (24.1) can be solved analytically. For example, a solution is available for the case where the rod's temperature is initially at zero. At $t = 0$, the boundary condition at $x = L$ is instantaneously increased to a constant level of $\overline{T}$ while $T(0)$ is held at zero. For this case, the temperature can be computed by (Jenson and Jeffreys, 1977)

$$T = \overline{T} \left[\frac{x}{L} + \sum_{n=0}^{\infty} \frac{2}{n\pi} (-1)^n \sin\left(\frac{nx}{L}\right) \exp\left(\frac{-n^2\pi^2 kt}{L^2}\right) \right] \tag{24.17}$$

where L is the total length of the rod. This equation can be employed to compute the evolution of the temperature distribution for each boundary condition. Then, the total solution can be determined by superposition.

EXAMPLE 24.4   Comparison of Analytical and Numerical Solutions

Problem Statement: Compare the analytical solution from Eq. (24.17) with numerical results obtained with the explicit, simple implicit, and Crank-Nicolson techniques. Perform the comparison for the rod employed in Examples 24.1, 24.2, and 24.3.

Solution: Recall from the previous examples that $k = 0.835$ cm²/s, $L = 10$ cm, and $\Delta x = 2$ cm. For this case, Eq. (24.17) can be used to predict that the temperature at $x = 2$ cm and $t = 10$ s would equal 64.8018. Table 24.1 presents numerical predictions of $T(2, 10)$. Notice that a range of time steps are employed. These results indicate a number of properties of the numerical methods. First, it can be seen that the explicit method is unstable for high values of λ. This instability is not manifested by either implicit approach. Second, the Crank-Nicolson method converges more rapidly as λ is decreased and provides moderately accurate results even when λ is relatively high. These outcomes are as expected because Crank-Nicolson is second-order accurate with respect to both independent variables. Finally, notice that as λ decreases, the methods seem to be converging on a value of 64.73 that is different than the analytical result of 64.80. This should not be surprising because a fixed value of $\Delta x = 2$ is used to characterize the x dimension. If both Δx and Δt were decreased as λ was decreased (that is, more spatial segments were used), the numerical solution would more closely approach the analytical result.

TABLE 24.1   **Comparison of three methods of solving a parabolic PDE: the heated rod. The results shown are for temperature at $t = 10$ s at $x = 2$ cm, for the rod from Example 24.1 through 24.3. Note that the analytical solution is $T(2,10) = 64.8018$.**

Δt	λ	Explicit	Implicit	Crank-Nicolson
10	2.0875	208.75	53.01	79.77
5	1.04375	−9.13	58.49	64.79
2	0.4175	67.12	62.22	64.87
1	0.20875	65.91	63.49	64.77
0.5	0.104375	65.33	64.12	64.74
0.2	0.04175	64.97	64.49	64.73

As indicated by the previous example, the Crank-Nicolson method is the preferred numerical method for solving parabolic PDEs in one spatial dimension. Its advantages become even more pronounced for more complicated applications such as those involving unequally spaced meshes. Such nonuniform spacing is often advantageous where we have foreknowledge that the solution varies rapidly in local portions of the system. Further discussion of such applications and the Crank-Nicolson method in general can be found elsewhere (Ferziger, 1981; Lapidus and Pinder, 1982).

24.5 PARABOLIC EQUATIONS IN TWO SPATIAL DIMENSIONS

The heat-conduction equation can be applied to more than one spatial dimension. For two dimensions, its form is

$$\frac{\partial T}{\partial t} = k \left(\frac{\partial^2 T}{\partial x^2} + \frac{\partial^2 T}{\partial y^2} \right) \tag{24.18}$$

One application of this equation is to model the temperature distribution on the face of a heated plate. However, rather than characterizing its steady-state distribution, as was done in Chap. 23, Eq. (24.18) provides a means to compute the plate's temperature distribution as it changes in time.

24.5.1 Standard Explicit and Implicit Schemes

An explicit solution can be obtained by substituting finite-difference approximations of the form of Eqs. (24.2) and (24.3) into Eq. (24.18). However, as with the one-dimensional case, this approach is limited by a stringent stability criterion. For the two-dimensional case, the criterion is (Davis, 1984)

$$\Delta t \leq \frac{1}{8} \frac{(\Delta x)^2 + (\Delta y)^2}{k}$$

Thus, for a uniform grid ($\Delta x = \Delta y$), $\lambda = k \, \Delta t / (\Delta x)^2$ must be less than or equal to 1/4. Consequently, halving the step size results in a fourfold increase in the number of nodes and a 16-fold increase in computational effort.

As was the case with one-dimensional systems, implicit techniques offer alternatives that guarantee stability. However, the direct application of implicit methods such as the Crank-Nicolson technique leads to the solution of $m \times n$ simultaneous equations. Additionally, when written for two or three spatial dimensions, these equations lose the valuable property of being tridiagonal. Thus, matrix storage and computation time can become exorbitantly large. The method described in the next section offers one way around this dilemma.

24.5.2 The A.D.I. Scheme

The *alternating-direction implicit* or *A.D.I. scheme* provides a means for solving parabolic equations in two spatial dimensions using tridiagonal matrices. To do this, each time increment is executed in two steps (Fig. 24.10). For the first step, Eq. (24.18) is approximated by

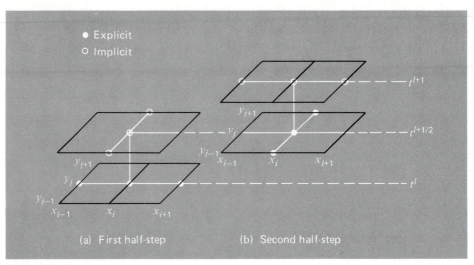

Figure 24.10
The two half-steps used in implementing the alternating-direction implicit scheme for solving parabolic equations in two spatial dimensions.

$$\frac{T_{i,j}^{l+1/2} - T_{i,j}^{l}}{\Delta t/2} = k\left[\frac{T_{i+1,j}^{l} - 2T_{i,j}^{l} + T_{i-1,j}^{l}}{(\Delta x)^2} + \frac{T_{i,j+1}^{l+1/2} - 2T_{i,j}^{l+1/2} + T_{i,j-1}^{l+1/2}}{(\Delta y)^2}\right] \quad (24.19a)$$

Thus, the approximation of $\partial^2 T/\partial x^2$ is written explicitly—that is, at the base point t^l where values of temperature are known. Consequently, only the three temperature terms in the approximation of $\partial^2 T/\partial y^2$ are unknown. For the case of a square grid ($\Delta y = \Delta x$), this equation can be expressed as

$$-\lambda T_{i,j-1}^{l+1/2} + 2(1 + \lambda)T_{i,j}^{l+1/2} - \lambda T_{i,j+1}^{l+1/2} =$$
$$\lambda T_{i-1,j}^{l} + 2(1 - \lambda)T_{i,j}^{l} + \lambda T_{i+1,j}^{l} \quad (24.19b)$$

which, when written for the system, results in a tridiagonal set of simultaneous equations.

For the second step from $t^{l+1/2}$ to t^{l+1}, Eq. (24.18) is approximated by

$$\frac{T_{i,j}^{l+1} - T_{i,j}^{l+1/2}}{\Delta t/2} = k\left[\frac{T_{i+1,j}^{l+1} - 2T_{i,j}^{l+1} + T_{i-1,j}^{l+1}}{(\Delta x)^2} + \frac{T_{i,j+1}^{l+1/2} - 2T_{i,j}^{l+1/2} + T_{i,j-1}^{l+1/2}}{(\Delta y)^2}\right] \quad (24.20)$$

In contrast to Eq. (24.19a), the approximation of $\partial^2 T/\partial x^2$ is now implicit. Thus, the bias introduced by Eq. (24.19a) will be partially corrected. For a square grid, Eq. (24.20) can be written as

$$-\lambda T_{i-1,j}^{l+1} + 2(1 + \lambda)T_{i,j}^{l+1} - \lambda T_{i+1,j}^{l+1} =$$
$$\lambda T_{i,j-1}^{l+1/2} + 2(1 - \lambda)T_{i,j}^{l+1/2} + \lambda T_{i,j+1}^{l+1/2} \quad (24.21)$$

Again, when written for a two-dimensional grid, the equation results in a tridiagonal system (Fig. 24.11). As in the following example, this leads to an efficient numerical solution.

EXAMPLE 24.5   A.D.I. method

Problem Statement: Use the A.D.I. method to solve for the temperature of the plate in Examples 23.1 through 23.3. At $t = 0$ assume that the temperature of the plate is zero and the boundary temperatures are instantaneously brought to the levels shown in Fig. 23.4. Employ a time step of 10 s. Recall from Example 24.1, that the coefficient of thermal diffusivity for aluminum is $k = 0.835$ cm²/s.

Solution: A value of $\Delta x = 10$ cm was employed to characterize the 40×40 cm plate from Examples 23.1 through 23.3. Therefore, $\lambda = 0.835(10)/(10)^2 = 0.0835$. For the first step to $t = 5$ (Fig. 24.11a), Eq. (24.19b) is applied to nodes (1,1), (1,2), and (1,3) to yield the following tridiagonal equations

$$\begin{bmatrix} 2.167 & -0.0835 & 0 \\ -0.0835 & 2.167 & -0.0835 \\ 0 & -0.0835 & 2.167 \end{bmatrix} \begin{Bmatrix} T_{1,1} \\ T_{1,2} \\ T_{1,3} \end{Bmatrix} = \begin{Bmatrix} 6.2625 \\ 6.2625 \\ 14.6125 \end{Bmatrix}$$

which can be solved for

$$T_{1,1} = 3.0160 \qquad T_{1,2} = 3.2708 \qquad T_{1,3} = 6.8692$$

In a similar fashion, tridiagonal equations can be developed and solved for

$$T_{2,1} = 0.1274 \qquad T_{2,2} = 0.2900 \qquad T_{2,3} = 4.1291$$

and

$$T_{3,1} = 2.0181 \qquad T_{3,2} = 2.2477 \qquad T_{3,3} = 6.0256$$

Figure 24.11
The A.D.I. method only results in tridiagonal equations, if it is applied along the dimension that is implicit. Thus, on the first step (a), it is applied along the y dimension and, on the second step (b), along the x dimension. These "alternating directions" are the root of the method's name.

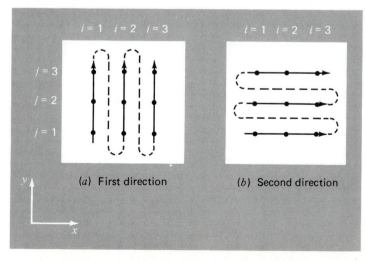

(a) First direction   (b) Second direction

For the second step to $t = 10$ (Fig. 24.11b), Eq. (24.21) is applied to nodes (1,1), (2,1), and (3,1) to yield

$$\begin{bmatrix} 2.167 & -0.0835 & 0 \\ -0.0835 & 2.167 & -0.0835 \\ 0 & -0.0835 & 2.167 \end{bmatrix} \begin{Bmatrix} T_{1,1} \\ T_{2,1} \\ T_{3,1} \end{Bmatrix} = \begin{Bmatrix} 12.0639 \\ 0.2577 \\ 8.0619 \end{Bmatrix}$$

which can be solved for

$$T_{1,1} = 5.5855 \qquad T_{2,1} = 0.4782 \qquad T_{3,1} = 3.7388$$

Tridiagonal equations for the other rows can be developed and solved for

$$T_{1,2} = 6.1683 \qquad T_{2,2} = 0.8238 \qquad T_{3,2} = 4.2359$$

and

$$T_{1,3} = 13.1120 \qquad T_{2,3} = 8.3207 \qquad T_{3,3} = 11.3606$$

The computation can be repeated, and the results for $t = 100$, 200, and 300 s are depicted in Fig. 24.12a through c. As expected, the temperature of the plate rises. After a sufficient time elapses, the temperature will approach the steady-state distribution of Fig. 23.5.

The A.D.I. method is but one of a group of techniques called *splitting methods*. Some of these represent efforts to circumvent shortcomings of A.D.I. For example, it is difficult to directly extend the A.D.I. method to three dimensions by using approximations at $l + 1/3$ and $l + 2/3$. Discussion of other splitting methods as well as more information on A.D.I. can be found elsewhere (Ferziger, 1981; Lapidus and Pinder, 1982).

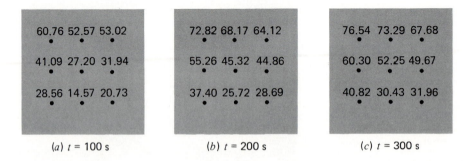

(a) $t = 100$ s   (b) $t = 200$ s   (c) $t = 300$ s

Figure 24.12
Solution for the heated plate from Example 24.4 at (a) $t = 100$ s, (b) $t = 200$ s, and (c) $t = 300$ s.

PROBLEMS

Hand Calculations

24.1 Repeat Example 24.1 but for the case where the rod is initially at 100 °C and the derivative at $x = 0$ is equal to 1 and at $x = 10$ is equal to 0. Interpret your results.

24.2 Repeat Example 24.1 but for a time step of $\Delta t = 0.05$ s. Compute results to $t = 0.2$ and compare with those in Example 24.1.

24.3 Repeat Example 24.2 but for the case where the derivative at $x = 10$ is equal to zero.

24.4 Repeat Example 24.3 but for $\Delta x = 2.5$ cm.

24.5 Repeat Example 24.5 but for the plate in Fig. P23.3.

24.6 The advection-diffusion equation is used to compute the distribution of concentration along the length of a rectangular chemical reactor (see Case Study 26.1),

$$\frac{\partial c}{\partial t} = D\,\frac{\partial^2 c}{\partial x^2} - U\,\frac{\partial c}{\partial x} - kc$$

where c is concentration (mg/m^3), $t =$ time (min), D is a diffusion coefficient (m^2/min), x is distance along the tank's longitudinal axis (m) where $x = 0$ at the tank's inlet, U is velocity in the x direction (m/min), and k is a reaction rate (min^{-1}) whereby the chemical decays to another form. Develop an explicit scheme to solve this equation numerically.

Computer-Related Problems

24.7 Develop a user-friendly computer program for the simple explicit method from Sec. 24.2. Test it by duplicating Example 24.1.

24.8 Modify the program in Prob. 24.7 so that it employs derivative boundary conditions.

24.9 Develop a user-friendly computer program to implement the simple implicit scheme from Sec. 24.3. Test it by duplicating Example 24.2.

24.10 Develop a user-friendly computer program to implement the Crank-Nicolson method from Sec. 24.4. Test it by duplicating Example 24.3.

24.11 Develop a user-friendly computer program for the A.D.I. method described in Sec. 24.5. Test it by duplicating Example 24.5.

CHAPTER 25
Finite-Element Method

To this juncture, we have employed *finite-difference* methods to solve partial differential equations. In these methods, the solution domain is divided into a grid of discrete points or nodes (Fig. 25.1*b*). The PDE is then written for each node and its derivatives replaced by finite divided differences. Although such *pointwise* approximation is conceptually easy to understand, it has a number of shortcomings. In particular, it becomes difficult to apply for systems with irregular geometry, unusual boundary conditions, or nonhomogeneous composition (recall Sec. 23.3.2).

The *finite-element* method provides an alternative that is better suited for such systems. In contrast to finite-difference techniques, the finite-element method divides the

Figure 25.1

(*a*) A gasket with irregular geometry and nonhomogeneous composition. (*b*) Such a system is very difficult to model with a finite-difference approach. This is due to the fact that complicated approximations are required at the boundaries of the system and at the boundaries between regions of differing composition. (*c*) A finite-element discretization is much better suited for such systems.

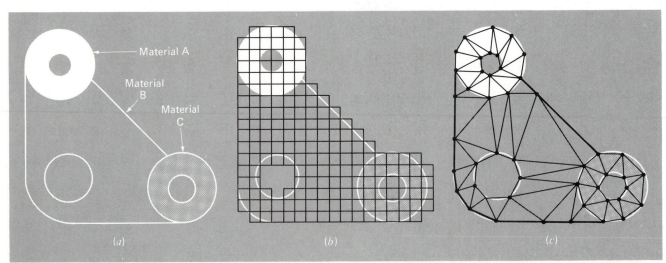

solution domain into simply shaped regions or "elements" (Fig. 25.1c). An approximate solution for the PDE can be developed for each of these elements. The total solution is then generated by linking together or "assembling" the individual solutions taking care to ensure continuity at the interelement boundaries. Thus, the PDE is satisfied in a *piecewise* fashion.

As in Fig. 25.1c, the use of elements, rather than a rectangular grid, provides a much better approximation for irregularly shaped systems. Further, values of the unknown can be generated continuously across the entire solution domain rather than at isolated points.

Because a comprehensive description is beyond the scope of this book, the present chapter provides a general introduction to the finite-element method. Our primary objective is to make you comfortable with the approach and cognizant of its capabilities. In this spirit, the following section is devoted to a general overview of the steps involved in a typical finite-element solution of a problem. This is followed by a simple example: a steady-state, one-dimensional heated rod. Although this example does not involve PDEs, it allows us to develop and demonstrate major aspects of the finite-element approach unencumbered by complicating factors. We can then discuss some issues involved in employing the finite-element method for PDEs.

25.1 THE GENERAL APPROACH

Although the particulars will vary, the implementation of the finite-element approach usually follows a standard step-by-step procedure. The following provides a brief overview of each of these steps. The application of these steps to engineering problem contexts will be developed in the subsequent sections.

25.1.1 Discretization

This step involves dividing the solution domain into finite elements. Figure 25.2 provides examples of elements employed in one, two, and three dimensions. The points of intersection of the lines that make up the sides of the elements are referred to as *nodes* and the sides themselves are called *nodal lines* or *planes*.

25.1.2 Element Equations

The next step is to develop equations to approximate the solution for each element. This involves two steps. First, we must choose an appropriate function with unknown coefficients that will be used to approximate the solution. Second, we evaluate the coefficients so that the function approximates the solution in an optimal fashion.

Choice of Approximation Functions.   Because they are easy to manipulate mathematically, polynomials are often employed for this purpose. For the one-dimensional case, the simplest alternative is a first-order polynomial or straight line,

$$u(x) = a_0 + a_1x \tag{25.1}$$

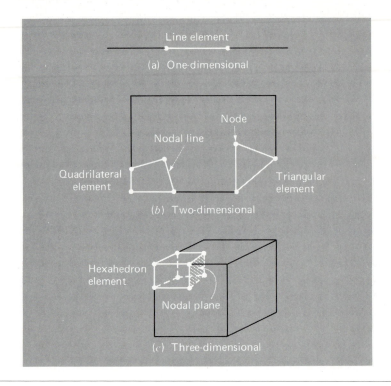

Figure 25.2
Examples of elements employed in one, two, and three dimensions.

where $u(x)$ is the dependent variable, a_0 and a_1 are constants, and x is the independent variable. This function must pass through the values of $u(x)$ at the end points of the element at x_1 and x_2. Therefore,

$$u_1 = a_0 + a_1 x_1$$

$$u_2 = a_0 + a_1 x_2$$

where $u_1 = u(x_1)$ and $u_2 = u(x_2)$. These equations can be solved using Cramer's rule for

$$a_0 = \frac{u_1 x_2 - u_2 x_1}{x_2 - x_1} \qquad\qquad a_1 = \frac{u_2 - u_1}{x_2 - x_1}$$

These results can then be substituted into Eq. (25.1) which, after collection of terms, can be written as

$$u = N_1 u_1 + N_2 u_2 \tag{25.2}$$

where

$$N_1 = \frac{x_2 - x}{x_2 - x_1} \tag{25.3}$$

and

$$N_2 = \frac{x - x_1}{x_2 - x_1} \tag{25.4}$$

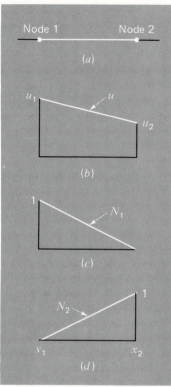

Figure 25.3
(*b*) A linear approximation or shape function for (*a*) a line element. The corresponding interpolation functions are shown in (*c*) and (*d*).

Equation (25.2) is called an *approximation* or *shape function* and N_1 and N_2 are called *interpolation functions*. Close inspection reveals that Eq. (25.2) is, in fact, the Lagrange first-order interpolating polynomial. It provides a means to predict intermediate values (that is, to interpolate) between given values u_1 and u_2 at the nodes.

Figure 25.3 shows the shape function along with the corresponding interpolation functions. Notice that the sum of the interpolation functions is always equal to one.

In addition, the fact that we are dealing with linear equations facilitates operations such as differentiation and integration. Such manipulations will be important in later sections. The derivative of Eq. (25.2) is

$$\frac{du}{dx} = \frac{dN_1}{dx} u_1 + \frac{dN_2}{dx} u_2 \tag{25.5}$$

According to Eqs. (25.3) and (25.4), the derivatives of the N's can be calculated as

$$\frac{dN_1}{dx} = -\frac{1}{x_2 - x_1} \qquad \frac{dN_2}{dx} = \frac{1}{x_2 - x_1} \tag{25.6}$$

and, therefore, the derivative of u is

$$\frac{du}{dx} = \frac{1}{x_2 - x_1} (-u_1 + u_2) \tag{25.7}$$

In other words, it is a divided difference representing the slope of the straight line connecting the nodes.

The integral can be expressed as

$$\int_{x_1}^{x_2} u \, dx = \int_{x_1}^{x_2} N_1 u_1 + N_2 u_2 \, dx$$

Each term on the right-hand side is merely the integral of a right triangle with base $x_2 - x_1$ and height u. That is,

$$\int_{x_1}^{x_2} Nu \, dx = \frac{1}{2} (x_2 - x_1) u$$

Thus, the entire integral is

$$\int_{x_1}^{x_2} u \, dx = \frac{(u_1 + u_2)}{2} (x_2 - x_1) \tag{25.8}$$

In other words, it is simply the trapezoidal rule.

Obtaining an Optimal Fit of the Function to the Solution. Once the interpolation function is chosen, the equation governing the behavior of the element must be developed. This equation represents a fit of the function to the solution of the underlying differential equation. Several methods are available for this purpose. Among the most common are the direct approach, the method of weighted residuals, and the variational approach. The outcome of all of these methods is analogous to curve fitting. However, instead of fitting functions to data, these methods specify relationships between the unknowns in Eq. (25.2) that satisfy the underlying PDE in an optimal fashion.

Mathematically, the resulting element equations will often consist of a set of linear algebraic equations that can be expressed in matrix form,

$$[k]\{u\} = \{F\} \tag{25.9}$$

where $[k]$ is an *element property* or *stiffness matrix*, $\{u\}$ is a column vector of unknowns at the nodes, and $\{F\}$ is a column vector reflecting the effect of any external influences applied at the nodes. Note that, in some cases, the equations can be nonlinear. However for the elementary examples described herein, and for many practical problems, the systems are linear.

25.1.3 Assembly

After the individual element equations are derived, they must be linked together or assembled to characterize the unified behavior of the entire system. The assembly process is governed by the concept of continuity. That is, the solutions for contiguous elements are matched so that the unknown values (and sometimes the derivatives) at their common nodes are equivalent. Thus, the total solution will be continuous.

When all the individual versions of Eq. (25.9) are finally assembled, the entire system is expressed in matrix form as

$$[K]\{u'\} = \{F'\} \tag{25.10}$$

where $[K]$ is the *assemblage property matrix* and $\{u'\}$ and $\{F'\}$ are column vectors for unknowns and external forces that are marked with primes to denote that they are an assemblage of the vectors $\{u\}$ and $\{F\}$ from the individual elements.

25.1.4 Boundary Conditions

Before Eq. (25.10) can be solved, it must be modified to account for the system's boundary conditions. These adjustments result in

$$[\overline{K}]\{u'\} = \{\overline{F}'\} \tag{25.11}$$

where the overbars signify that the boundary conditions have been incorporated.

25.1.5 Solution

Solutions of Eq. (25.11) can be obtained with techniques described previously in Part Three such as LU decomposition. In many cases, the elements can be configured so that the resulting equations are banded. Thus, the highly efficient solution schemes available for such systems can be employed.

25.1.6 Postprocessing

Upon obtaining a solution, it can be output in tabular form or displayed graphically. In addition, secondary variables can be determined and output.

Although the preceding steps are very general, they are common to most implementations of the finite-element approach. In the following section, we illustrate how they can be applied to obtain numerical results for a simple physical system—a heated rod.

25.2 FINITE-ELEMENT APPLICATION IN ONE DIMENSION

Figure 25.4 shows a system that can be modeled by a one-dimensional form of Poisson's equation

$$\frac{d^2T}{dx^2} = -f(x) \tag{25.12}$$

where $f(x)$ is a function defining a heat source along the rod and where the ends of the rod are held at fixed temperatures.

$$T(0,\ t) = T_1$$

and

$$T(L,t) = T_2$$

Notice that this is not a partial differential equation but rather is a boundary-value ODE. This simple model is used because it will allow us to introduce the finite-element approach without some of the complications involved in, for example, a two-dimensional PDE.

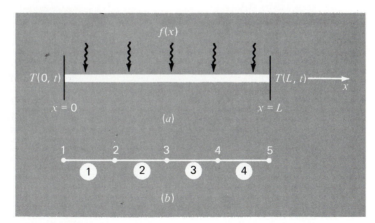

Figure 25.4
(a) A long, thin rod subject to fixed boundary conditions and a continuous heat source along its axis. (b) The finite-element representation consisting of four equal-length elements and five nodes.

EXAMPLE 25.1   Analytical Solution for a Heated Rod

Problem Statement: Solve Eq. (25.12) for a 10-cm rod with boundary conditions of $T(0,\ t) = 40$ and $T(10,\ t) = 200$ and a uniform heat source of $f(x) = 10$.

Solution: The equation to be solved is

$$\frac{d^2T}{dx^2} = -10$$

Assume a solution of the form

$$T = ax^2 + bx + c$$

which can be differentiated twice to give $T'' = 2a$. Substituting this result into the differential equation gives $a = -5$. The boundary conditions can be used to evaluate the remaining coefficients. For the first condition at $x = 0$,

$$40 = -5(0)^2 + b(0) + c$$

or $c = 40$. Similarly, for the second condition,

$$200 = -5(10)^2 + b(10) + 40$$

which can be solved for $b = 66$. Therefore, the final solution is

$$T = -5x^2 + 66x + 40$$

The results are plotted in Fig. 25.5.

Figure 25.5
The temperature distribution along a heated rod subject to a uniform heat source and held at fixed end temperatures.

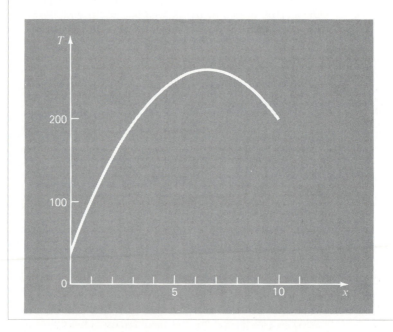

25.2.1 Discretization

A simple configuration to model the system is a series of equal-length elements (Fig. 25.4b). Thus, the system is treated as four equal-length elements and five nodes.

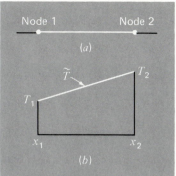

Figure 25.6
(a) An individual element. (b) The approximation function used to characterize the temperature distribution along the element.

25.2.2 Element Equations

An individual element is shown in Fig. 25.6a. The distribution of temperature for the element can be represented by the approximation function

$$\tilde{T} = N_1 T_1 + N_2 T_2 \tag{25.13}$$

where N_1 and N_2 are linear interpolation functions specified by Eqs. (25.3) and (25.4), respectively. Thus, as depicted in Fig. 25.6b, the approximation function amounts to a linear interpolation between the two nodal temperatures.

As noted in Sec. 25.1, there are a variety of approaches for developing the element equation. In the present section, we employ two of these. First, a *direct approach* will be used for the simple case where $f(x) = 0$. Then, because of its general applicability in engineering, we will devote most of the section to the *method of weighted residuals*.

The Direct Approach. For the case where $f(x) = 0$, a direct method can be employed to generate the element equations. The relationship between heat flux and temperature gradient can be represented by Fourier's law:

$$q = -k' \frac{dT}{dx}$$

where q is flux [cal/(cm^2·s)] and k' is the coefficient of thermal conductivity [cal/(s·cm·°C)]. If a linear approximation function is used to characterize the element's temperature, the heat flow into the element through node 1 can be represented by

$$q_1 = k' \frac{T_1 - T_2}{x_2 - x_1}$$

where q_1 is heat flux at node 1. Similarly for node 2,

$$q_2 = k' \frac{T_2 - T_1}{x_2 - x_1}$$

These two equations express the relationship of the element's internal temperature distribution (as reflected by the nodal temperatures) to the heat flux at its ends. As such, they constitute our desired element equations. They can be simplified further by recognizing that Fourier's law can be used to couch the end fluxes themselves in terms of the temperature gradients at the boundaries. That is,

$$q_1 = -k' \frac{dT(x_1)}{dx} \qquad q_2 = k' \frac{dT(x_2)}{dx}$$

which can be substituted into the element equations to give

$$\frac{1}{x_2 - x_1} \begin{bmatrix} 1 & -1 \\ -1 & 1 \end{bmatrix} \begin{Bmatrix} T_1 \\ T_2 \end{Bmatrix} = \begin{Bmatrix} -\dfrac{dT(x_1)}{dx} \\ \dfrac{dT(x_2)}{dx} \end{Bmatrix} \tag{25.14}$$

Notice that Eq. (25.14) has been cast in the format of Eq. (25.9). Thus, we have succeeded in generating a matrix equation that describes the behavior of a typical element in our system.

The direct approach has great intuitive appeal. Additionally, in areas such as mechanics, it can be employed to solve meaningful problems. However, in other contexts, it is often difficult or impossible to derive finite-element equations directly. Consequently, as described next, more general mathematical techniques are available.

The Method of Weighted Residuals. The differential equation (25.12) can be reexpressed as

$$0 = \frac{d^2T}{dx^2} + f(x)$$

The approximate solution [Eq. (25.13)] can be substituted into this equation. Because Eq. (25.13) is not the exact solution, the left side of the resulting equation will not be zero but will equal a residual.

$$R = \frac{d^2\tilde{T}}{dx^2} + f(x) \tag{25.15}$$

The method of weighted residuals (MWR) consists of finding a minimum for the residual according to the general formula

$$\int_D RW_i \, dD = 0 \qquad i = 1, 2, \ldots, m \tag{25.16}$$

where D is the solution domain and the W_i are linearly independent weighting functions.

At this point, there are a variety of choices that could be made for the weighting function (Box 25.1). The most common approach for the finite-element method is to employ the interpolation functions N_i as the weighting functions. When these are substituted into Eq. (25.16), the result is referred to as *Galerkin's method*.

$$\int_D RN_i \, dD = 0 \qquad i = 1, 2, \ldots, m$$

For our one-dimensional rod, Eq. (25.15) can be substituted into this formulation to give

$$\int_{x_1}^{x_2} \left[\frac{d^2\tilde{T}}{dx^2} + f(x) \right] N_i \, dx \qquad i = 1, 2$$

which can be reexpressed as

$$\int_{x_1}^{x_2} \frac{d^2\tilde{T}}{dx^2} N_i(x) \, dx = -\int_{x_1}^{x_2} f(x)N_i(x) \, dx \qquad i = 1, 2 \tag{25.17}$$

Box 25.1 *Alternative Residual Schemes for the MWR*

Several choices can be made for the weighting functions of Eq. (25.16). Each represents an alternative approach for the MWR.

In the *collocation* approach, we choose as many locations as there are unknown coefficients. Then the coefficients are adjusted until the residual vanishes at each of these locations. Consequently, the approximating function will yield perfect results at the chosen locations but will have a nonzero residual elsewhere. Thus, it is akin to the interpolation methods in Chap. 12. Note that collocation amounts to using the weighting function

$$W = \delta(x - x_i) \qquad \text{for } i = 1, 2, \ldots, n$$

where n is the number of unknown coefficients and $\delta(x - x_i)$ is the *Dirac delta function* which vanishes everywhere but at $x = x_i$, where it equals 1.

In the *subdomain method*, the interval is divided into as many segments or "subdomains" as there are unknown coefficients. Then the coefficients are adjusted until the average value of the residual is zero in each subdomain. Thus, for each subdomain, the weighting function is equal to 1 and Eq. (25.16) is

$$\int_{x_{i-1}}^{x_i} R \, dx = 0 \qquad \text{for } i = 1, 2, \ldots, n$$

where x_{i-1} and x_i are the bounds of the subdomain.

For the *least-squares* case, the coefficients are adjusted so as to minimize the integral of the square of the residual. Thus, the weighting functions are

$$W_i = \frac{\partial R}{\partial a_i}$$

which can be substituted into Eq. (25.16) to give

$$\int_D R \frac{\partial R}{\partial a_i} \, dD = 0 \qquad i = 1, 2, \ldots, m$$

or

$$\frac{\partial}{\partial a_i} \int_D R^2 \, dD = 0 \qquad i = 1, 2, \ldots, m$$

Comparison of the formulation with those of Chap. 11 shows that this is the continuous form of regression.

Galerkin's method employs the interpolation functions N_i as weighting functions. Recall that these functions always sum to 1 at any position in an element. For many problem contexts, Galerkin's method yields the same results as are obtained by variational methods. Consequently, it is the most commonly employed version of MWR used in finite-element analysis.

At this point, a number of mathematical manipulations will be applied to simplify and evaluate Eq. (25.17). Among the most important is the simplification of the left-hand side using *integration by parts*. Recall from calculus that this operation can be expressed generally as

$$\int_a^b u \, dv = uv \Big|_a^b - \int_a^b v \, du$$

If u and v are chosen properly, the new integral on the right-hand side will be easier to evaluate than the original one on the left-hand side. This can be done for the term on the left-hand side of Eq. (25.17) by choosing $N_i(x)$ as u and $(d^2\tilde{T}/dx^2)dx$ as dv to yield

$$\int_{x_1}^{x_2} N_i(x) \frac{d^2\tilde{T}}{dx^2} \, dx = N_i(x) \frac{d\tilde{T}}{dx} \Big|_{x_1}^{x_2} - \int_{x_1}^{x_2} \frac{d\tilde{T}}{dx} \frac{dN_i}{dx} \, dx \qquad i = 1, 2 \qquad (25.18)$$

Thus, we have taken the significant step of lowering the highest-order term in the formulation from a second to a first derivative.

Next, we can evaluate the individual terms that we have created in Eq. (25.18). For $i = 1$, the first term on the right-hand side of Eq. (25.18) can be evaluated as

$$N_1(x) \left.\frac{d\tilde{T}}{dx}\right|_{x_1}^{x_2} = N_1(x_2) \frac{d\tilde{T}(x_2)}{dx} - N_1(x_1) \frac{d\tilde{T}(x_1)}{dx}$$

However, recall from Fig. 25.3 that $N_1(x_2) = 0$ and $N_1(x_1) = 1$, and therefore

$$N_1(x) \left.\frac{d\tilde{T}}{dx}\right|_{x_1}^{x_2} = -\frac{d\tilde{T}(x_1)}{dx} \tag{25.19}$$

Similarly for $i = 2$,

$$N_2(x) \left.\frac{d\tilde{T}}{dx}\right|_{x_1}^{x_2} = \frac{d\tilde{T}(x_2)}{dx} \tag{25.20}$$

Thus, the first term on the right-hand side of Eq. (25.18) represents the natural boundary conditions at the ends of the elements.

Now before proceeding let us regroup by substituting our results back into the original equation. Substituting Eqs. (25.18) through (25.20) into Eq. (25.17) and rearranging gives for $i = 1$,

$$\int_{x_1}^{x_2} \frac{d\tilde{T}}{dx} \frac{dN_1}{dx} \, dx = -\frac{dT(x_1)}{dx} + \int_{x_1}^{x_2} f(x) \, N_1(x) \, dx \tag{25.21}$$

and for $i = 2$,

$$\int_{x_1}^{x_2} \frac{d\tilde{T}}{dx} \frac{dN_2}{dx} \, dx = \frac{dT(x_2)}{dx} + \int_{x_1}^{x_2} f(x) \, N_2(x) \, dx \tag{25.22}$$

Notice that the integration by parts has led to two important outcomes. First, it has incorporated the boundary conditions directly into the element equations. Second, it has lowered the highest-order evaluation from a second to a first derivative. This latter outcome yields the significant result that the approximation functions need to preserve continuity of value but not slope at the nodes.

Also notice that we can now begin to ascribe some physical significance to the individual terms we have derived. On the right-hand side of each equation, the first term represents one of the element's boundary conditions and the second is the effect of the system's forcing function—in the present case, the heat source $f(x)$. As will now become evident, the left-hand side embodies the internal mechanisms that govern the element's temperature distribution. That is, in terms of the finite-element method, the left-hand side will become the element property matrix.

To see this, let us concentrate on the terms on the left-hand side. For $i = 1$, the term is

$$\int_{x_1}^{x_2} \frac{d\tilde{T}}{dx} \frac{dN_1}{dx} \, dx \tag{25.23}$$

Recall from Sec. 25.1.2 that the linear nature of the shape function makes differentiation and integration simple. Substituting Eqs. (25.6) and (25.7) into Eq. (25.23) gives

$$\int_{x_1}^{x_2} \frac{T_1 - T_2}{(x_2 - x_1)^2} \, dx = \frac{1}{x_2 - x_1} (T_1 - T_2) \tag{25.24}$$

Similar substitutions for $i = 2$ [Eq. (25.22)] yield

$$\int_{x_1}^{x_2} \frac{-T_1 + T_2}{(x_2 - x_1)^2} \, dx = \frac{1}{x_2 - x_1} (-T_1 + T_2) \tag{25.25}$$

Comparison with Eq. (25.14) shows that these are similar to the relationships that were developed with the direct method using Fourier's law. This can be made even clearer by reexpressing Eqs. (25.24) and (25.25) in matrix form as

$$\frac{1}{x_2 - x_1} \begin{bmatrix} 1 & -1 \\ -1 & 1 \end{bmatrix} \begin{Bmatrix} T_1 \\ T_2 \end{Bmatrix}$$

Substituting this result into Eqs. (25.21) and (25.22) and expressing the result in matrix form gives the final version of the element equations

$$\underbrace{\frac{1}{x_2 - x_1} \begin{bmatrix} 1 & -1 \\ -1 & 1 \end{bmatrix} \{T\}}_{\text{Element stiffness matrix}} = \underbrace{\begin{Bmatrix} -\dfrac{dT(x_1)}{dx} \\ \dfrac{dT(x_2)}{dx} \end{Bmatrix}}_{\substack{\text{Boundary} \\ \text{condition}}} + \underbrace{\begin{Bmatrix} \displaystyle\int_{x_1}^{x_2} f(x) N_1(x) \, dx \\ \displaystyle\int_{x_1}^{x_2} f(x) N_2(x) \, dx \end{Bmatrix}}_{\text{External effects}}$$

$$\tag{25.26a}$$
$$\tag{25.26b}$$

Note that aside from the direct and the weighted residual methods, the element equations can also be derived using variational calculus (for example, see Allaire, 1985). For the present case, this approach yields equations that are identical to those derived above.

EXAMPLE 25.2   *Element Equation for a Heated Rod*

Problem Statement: Employ Eq. (25.26) to develop the element equations for a 10-cm rod with boundary conditions of $T(0, t) = 40$ and $T(10, t) = 200$ and a uniform heat source of $f(x) = 10$. Employ four equal-size elements of length $= 2.5$ cm.

Solution: The heat source term in Eq. (25.26a) can be evaluated by substituting Eq. (25.3) and integrating to give

$$\int_0^{2.5} 10 \frac{2.5 - x}{2.5} \, dx = 12.5$$

Similarly, Eq. (25.4) can be substituted into the heat source term of Eq. (25.26b), which can also be integrated to yield

$$\int_0^{2.5} 10 \, \frac{x - 0}{2.5} \, dx = 12.5$$

These results along with the other parameter values can be substituted into Eq. (25.26) to give

$$0.4T_1 - 0.4T_2 = -\frac{dT}{dx}(x_1) + 12.5$$

and

$$-0.4T_1 + 0.4T_2 = \frac{dT}{dx}(x_2) + 12.5$$

25.2.3 Assembly

Before the element equations are assembled, a global numbering scheme must be established to specify the system's topology or spatial layout. As in Table 25.1, this defines the connectivity of the element mesh. Because the present case is one-dimensional, the numbering scheme might seem so predictable that it is trivial. However, for two- and three-dimensional problems it offers the only means to specify which nodes belong to which elements.

TABLE 25.1 **The system topology for the finite-element segmentation scheme from Fig. 25.4b.**

	Node Numbers	
Element	**Local**	**Global**
1	1	1
	2	2
2	1	2
	2	3
3	1	3
	2	4
4	1	4
	2	5

Once the topology is specified, the element equation (25.26) can be written for each element using the global coordinates. Then they can be added one at a time to assemble the total system matrix (note that this process is explored further in Case Study 26.4). The process is depicted in Fig. 25.7.

1

FIGURE 25.7
The assembly of the equations for the total system

$$(a) \quad \begin{bmatrix} 0.4 & -0.4 & 0 & 0 & 0 \\ -0.4 & 0.4 & 0 & 0 & 0 \\ 0 & 0 & 0 & 0 & 0 \\ 0 & 0 & 0 & 0 & 0 \\ 0 & 0 & 0 & 0 & 0 \end{bmatrix} \begin{Bmatrix} T_1 \\ T_2 \\ 0 \\ 0 \\ 0 \end{Bmatrix} = \begin{Bmatrix} -dT(x_1)/dx + 12.5 \\ dT(x_2)/dx + 12.5 \\ 0 \\ 0 \\ 0 \end{Bmatrix}$$

$$(b) \quad \begin{bmatrix} 0.4 & -0.4 & 0 & 0 & 0 \\ -0.4 & 0.4 + 0.4 & -0.4 & 0 & 0 \\ 0 & -0.4 & 0.4 & 0 & 0 \\ 0 & 0 & 0 & 0 & 0 \\ 0 & 0 & 0 & 0 & 0 \end{bmatrix} \begin{Bmatrix} T_1 \\ T_2 \\ T_3 \\ 0 \\ 0 \end{Bmatrix} = \begin{Bmatrix} -dT(x_1)/dx + 12.5 \\ 12.5 + 12.5 \\ dT(x_3)/dx + 12.5 \\ 0 \\ 0 \end{Bmatrix}$$

$$(c) \quad \begin{bmatrix} 0.4 & -0.4 & 0 & 0 & 0 \\ -0.4 & 0.8 & -0.4 & 0 & 0 \\ 0 & -0.4 & 0.4 + 0.4 & -0.4 & 0 \\ 0 & 0 & -0.4 & 0.4 & 0 \\ 0 & 0 & 0 & 0 & 0 \end{bmatrix} \begin{Bmatrix} T_1 \\ T_2 \\ T_3 \\ T_4 \\ 0 \end{Bmatrix} = \begin{Bmatrix} -dT(x_1)/dx + 12.5 \\ 25 \\ 12.5 + 12.5 \\ dT(x_4)/dx + 12.5 \\ 0 \end{Bmatrix}$$

$$(d) \quad \begin{bmatrix} 0.4 & -0.4 & 0 & 0 & 0 \\ -0.4 & 0.8 & -0.4 & 0 & 0 \\ 0 & -0.4 & 0.8 & -0.4 & 0 \\ 0 & 0 & -0.4 & 0.4 + 0.4 & -0.4 \\ 0 & 0 & 0 & -0.4 & 0.4 \end{bmatrix} \begin{Bmatrix} T_1 \\ T_2 \\ T_3 \\ T_4 \\ T_5 \end{Bmatrix} = \begin{Bmatrix} -dT(x_1)/dx + 12.5 \\ 25 \\ 25 \\ 12.5 + 12.5 \\ dT(x_5)/dx + 12.5 \end{Bmatrix}$$

$$(e) \quad \begin{bmatrix} 0.4 & -0.4 & 0 & 0 & 0 \\ -0.4 & 0.8 & -0.4 & 0 & 0 \\ 0 & -0.4 & 0.8 & -0.4 & 0 \\ 0 & 0 & -0.4 & 0.8 & -0.4 \\ 0 & 0 & 0 & -0.4 & 0.4 \end{bmatrix} \begin{Bmatrix} T_1 \\ T_2 \\ T_3 \\ T_4 \\ T_5 \end{Bmatrix} = \begin{Bmatrix} -dT(x_1)/dx + 12.5 \\ 25 \\ 25 \\ 25 \\ dT(x_5)/dx + 12.5 \end{Bmatrix}$$

25.2.4 Boundary Conditions

Notice that as the equations are assembled, the internal boundary conditions cancel. Thus, the final result for $\{F\}$ in Fig. 25.7e has boundary conditions only for the first and the last nodes. Because T_1 and T_5 are given, these natural boundary conditions at the ends of the bar, $dT(x_1)/dx$ and $dT(x_5)/dx$, represent unknowns. Therefore, the equations can be reexpressed as

$$
\begin{aligned}
\frac{dT}{dx}(x_1) - 0.4T_2 &= -3.5 \\
0.8T_2 - 0.4T_3 &= 41 \\
-0.4T_2 + 0.8T_3 - 0.4T_4 &= 25 \\
-0.4T_3 + 0.8T_4 &= 105 \\
-0.4T_4 - \frac{dT}{dx}(x_5) &= -67.5
\end{aligned}
\tag{25.27}
$$

25.2.5 Solution

Equation (25.27) can be solved for

$$\frac{dT}{dx}(x_1) = 66 \qquad T_2 = 173.75 \qquad T_3 = 245$$

$$T_4 = 253.75 \qquad \frac{dT}{dx}(x_5) = -34$$

25.2.6 Postprocessing

The results can be displayed graphically. Figure 25.8 shows the finite-element results along with the exact solution. Notice that the finite-element calculation captures the overall trend of the exact solution and, in fact, provides an exact match at the nodes.

Figure 25.8
Results of applying the finite-element approach to a heated bar. The exact solution is also shown.

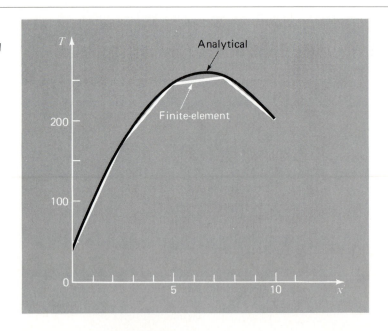

However, a discrepancy exists in the interior of each element due to the linear nature of the shape functions.

25.3   TWO-DIMENSIONAL PROBLEMS

Although the mathematical "bookkeeping" increases markedly, the extension of the finite-element approach to two dimensions is conceptually similar to the one-dimensional applications discussed to this point. It thus follows the same steps as were outlined in Sec. 25.1.

25.3.1   Discretization

A variety of simple elements such as triangles or quadrilaterals are usually employed for the finite-element mesh in two dimensions. In the present discussion, we will limit ourselves to triangular elements of the type depicted in Fig. 25.9.

25.3.2   Element Equations

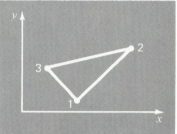

Figure 25.9
A triangular element.

Just as for the one-dimensional case, the next step is to develop an equation to approximate the solution for the element. For a triangular element, the simplest approach is the linear polynomial [compare with Eq. (25.1)]

$$u(x, y) = a_0 + a_{1,1}x + a_{1,2}y \tag{25.28}$$

where $u(x, y)$ is the dependent variable, the a's are coefficients, and x and y are the independent variables. This function must pass through the values of $u(x, y)$ at the triangle's nodes (x_1, y_1), (x_2, y_2), and (x_3, y_3). Therefore,

$$u_1 = a_0 + a_{1,1}x_1 + a_{1,2}y_1$$

$$u_2 = a_0 + a_{1,1}x_2 + a_{1,2}y_2$$

$$u_3 = a_0 + a_{1,1}x_3 + a_{1,2}y_3$$

or in matrix form

$$\begin{bmatrix} 1 & x_1 & y_1 \\ 1 & x_2 & y_2 \\ 1 & x_3 & y_3 \end{bmatrix} \begin{Bmatrix} a_0 \\ a_{1,1} \\ a_{1,2} \end{Bmatrix} = \begin{Bmatrix} u_1 \\ u_2 \\ u_3 \end{Bmatrix}$$

which can be solved for

$$a_0 = \frac{1}{2A_e} [u_1(x_2y_3 - x_3y_2) + u_2(x_3y_1 - x_1y_3) + u_3(x_1y_2 - x_2y_1)] \tag{25.29}$$

$$a_{1,1} = \frac{1}{2A_e} [u_1(y_2 - y_3) + u_2(y_3 - y_1) + u_3(y_1 - y_2)] \tag{25.30}$$

$$a_{1,2} = \frac{1}{2A_e} [u_1(x_3 - x_2) + u_2(x_1 - x_3) + u_3(x_2 - x_1)] \tag{25.31}$$

where A_e is the area of the triangular element

$$A_e = \frac{1}{2} [(x_2y_3 - x_3y_2) + (x_3y_1 - x_1y_3) + (x_1y_2 - x_2y_1)]$$

Equations (25.29) through (25.31) can be substituted into Eq. (25.28). After collection of terms, the result can be expressed as

$$u = N_1u_1 + N_2u_2 + N_3u_3 \tag{25.32}$$

where

$$N_1 = \frac{1}{2A_e} [(x_2y_3 - x_3y_2) + (y_2 - y_3)x + (x_3 - x_2)y]$$

$$N_2 = \frac{1}{2A_e} [(x_3y_1 - x_1y_3) + (y_3 - y_1)x + (x_1 - x_3)y]$$

$$N_3 = \frac{1}{2A_e} [(x_1y_2 - x_2y_1) + (y_1 - y_2)x + (x_2 - x_1)y]$$

As with the one-dimensional case, Eq. (25.32) provides a means to predict intermediate values for the element on the basis of the values at its nodes. Figure 25.10 shows the shape function along with the corresponding interpolation functions. Notice that the sum of the interpolation functions is always equal to 1.

As with the one-dimensional case, various methods are available for developing element equations based on the underlying PDE and the approximating functions. The resulting equations are considerably more complicated than Eq. (25.26). However, because the approximating functions are usually lower-order polynomials like Eq. (25.28), the terms of the final element matrix will consist of lower-order polynomials and constants.

25.3.3 Boundary Conditions and Assembly

The incorporation of boundary conditions and the assembly of the system matrix also become more complicated when the finite-element technique is applied to two- and three-dimensional problems. However, as with the derivation of the element matrix, the difficulty relates to the mechanics of the process rather than to conceptual complexity. For example, the establishment of the system topology which was trivial for the one-dimensional case becomes a matter of great importance in two and three dimensions. In particular, the choice of a numbering scheme will dictate the bandedness of the resulting system matrix and hence the efficiency with which it can be solved. Figure 25.11 shows a scheme that was developed for the heated plate formerly solved by finite-difference methods in Chap. 23.

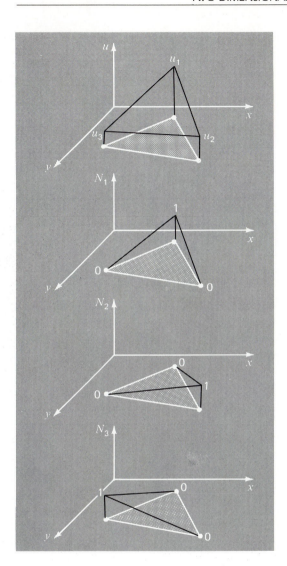

Figure 25.10
(*a*) A linear approximation
function for a triangular
element. The corresponding
interpolation functions are
shown in (*b*) through (*d*).

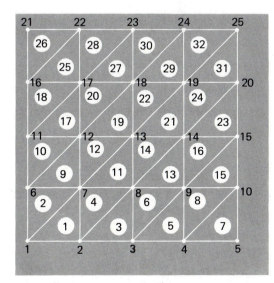

Figure 25.11
A numbering scheme for the
nodes and elements of a
finite-element approximation of
the heated plate that was
previously characterized by
finite differences in Chap. 24.

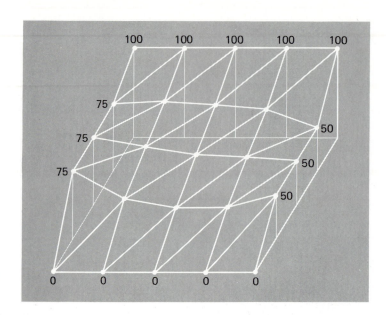

Figure 25.12
The temperature distribution of a heated plate as calculated with a finite-element method.

25.3.4 Solution and Postprocessing

Although the mechanics are complicated, the system matrix is merely a set of n simultaneous equations that can be used to solve for the values of the dependent variable at the n nodes. Figure 25.12 shows a solution that corresponds to the finite-difference solution from Fig. 23.9.

PROBLEMS

Hand Calculations

25.1 Repeat Example 25.1 but for $T(0, t) = 50$ and $T(10, t) = 100$ and a uniform heat source of 15.

25.2 Repeat Example 25.2 but for boundary conditions of $T(0, t) = 50$ and $T(10, t) = 100$ and a heat source of 15.

25.3 Apply the results of Prob. 25.2 to compute the temperature distribution for the entire rod using the finite-element approach.

25.4 Use Galerkin's method to develop an element equation for a steady-state version of the advection-diffusion equation described in Prob. 24.6. Express the final result in the format of Eq. (25.26) so that each term has a physical interpretation.

25.5 A version of the Poisson equation that occurs in mechanics is the following model for the vertical deflection of a bar

with a distributed load $P(x)$,

$$A_c E \frac{d^2 u}{dx^2} = P(x)$$

where A_c is the cross-sectional area, E is Young's modulus, u is deflection, and x is distance measured along the bar's length. If the bar is rigidly fixed ($u = 0$) at both ends, use the finite-element method to model its deflections for $A_e = 0.75$ ft^2, $E = 1.5 \times 10^8$ lb/ft^2, $L = 30$ ft, and $P(x) = 50$ lb/ft. Employ a $\Delta x = 6$ ft.

Computer-Related Problems

25.6 Develop a user-friendly program to model the steady-state distribution of temperature in a rod with a constant heat source using the finite-element method. Set up the program so that unequally spaced nodes may be used.

CHAPTER 26

Case Studies: Partial Differential Equations

The purpose of this chapter is to apply the methods from Part Seven to practical engineering problems. In *Case Study 26.1* a parabolic PDE is used to compute the time-variable distribution of a chemical along the longitudinal axes of a rectangular reactor. This example illustrates how the instability of a solution can be due to the nature of the PDE rather than to properties of the numerical method.

Case Studies 26.2 and 26.3 involve applications of the Poisson and Laplace equations to civil and electrical engineering problem contexts. Among other things this will allow you to see similarities as well as differences between field problems in these areas of engineering. In addition, they can be contrasted with the heated-plate problem that has served as our prototype system in this part of the book. *Case Study 26.2* deals with the deflection of a square plate whereas *Case Study 26.3* is devoted to computing the voltage distribution and charge flux for a two-dimensional surface with a curved edge.

Case Study 26.4 presents a finite-element analysis as applied to a series of springs. This application is closer in spirit to finite-element applications in mechanics and structures than was the temperature field problem used to illustrate the approach in Chap. 25.

Finally, *Case Study 26.5* shows how a spreadsheet can be employed to solve the Laplace equation. In particular, it illustrates how the two-dimensional layout of the spreadsheet is well-suited for modeling two-dimensional field problems.

CASE STUDY 26.1   ONE-DIMENSIONAL MASS BALANCE OF A REACTOR (CHEMICAL ENGINEERING)

Background: Chemical engineers make extensive use of idealized reactors in their design work. In Case Studies 10.2 and 22.2, we focused on single or coupled well-mixed reactors. These are examples of *lumped-parameter* systems (recall Sec. PT3.1.2).

Figure 26.1 depicts an elongated reactor with a single entry and exit point. This reactor can be characterized as a *distributed-parameter* system. If it is assumed that

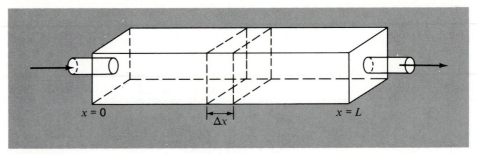

Figure 26.1
An elongated reactor with a single entry and exit point. A mass balance is developed around a finite segment along the tank's longitudinal axis in order to derive a differential equation for concentration.

the chemical being modeled is subject to first-order decay* and the tank is well-mixed vertically and laterally, a mass balance can be performed on a finite segment of length Δx as in

$$V \frac{\Delta c}{\Delta t} = \underbrace{Qc(x)}_{\text{Flow in}} - \underbrace{Q\left[c(x) \; + \; \frac{\partial c(x)}{\partial x} \Delta x\right]}_{\text{Flow out}} - \underbrace{DA_c \, \frac{\partial c(x)}{\partial x}}_{\text{Dispersion in}}$$

$$+ \underbrace{DA_c \left[\frac{\partial c(x)}{\partial x} \; + \; \frac{\partial}{\partial x} \frac{\partial c(x)}{\partial x} \Delta x\right]}_{\text{Dispersion out}} - \underbrace{kVc}_{\text{Decay reaction}}$$

(26.1)

where V = volume (m^3), Q = flow rate (m^3/h), c is concentration (moles/m^3), D is a dispersion coefficient (m^2/h), A_c is the tank's cross-sectional area (m^2), and k is the first-order decay coefficient (h^{-1}). Note that the dispersion terms are based on Fick's first law,

$$\text{Flux} = -D \, \frac{\partial c}{\partial x}$$

(26.2)

which is directly analogous to Fourier's law for heat conduction [recall Eq. (23.4)]. It specifies that turbulent mixing tends to move mass from regions of high to low concentration. The parameter D, therefore, reflects the magnitude of turbulent mixing.

If Δx and Δt are allowed to approach zero, Eq. (26.1) becomes

$$\frac{\partial c}{\partial t} = D \, \frac{\partial^2 c}{\partial x^2} - U \, \frac{\partial c}{\partial x} - kc$$

(26.3)

where $U = Q/A_c$ = the velocity of the water flowing through the tank. The mass balance for Fig. 26.1 is, therefore, now expressed as a parabolic partial differential equation. Equation (26.3) is sometimes referred to as the *advection-dispersion equation* with first-order reaction. At steady state, it is reduced to a second-order ODE,

*That is, the chemical decays at a rate that is linearly proportional to how much chemical is present.

$$0 = D \frac{d^2c}{dx^2} - U \frac{dc}{dx} - kc \qquad (26.4)$$

Prior to $t = 0$ the tank is filled with water that is devoid of the chemical. At $t = 0$, the chemical is injected into the reactor's inflow at a constant level of c_{in}. Thus, the following boundary conditions hold:

$$Qc_{in} = Qc_0 - DA_c \frac{dc_0}{dx}$$

and

$$c'(L, t) = 0$$

The second condition specifies that chemical leaves the reactor purely as a function of flow through the outlet pipe. That is, it is assumed that dispersion in the reactor does not affect the exit rate. Under these conditions, use numerical methods to solve Eq. (26.4) for the steady-state levels in the reactor. Note that this is an ODE boundary-value problem. Then solve Eq. (26.3) to characterize the transient response—that is, how the levels change in time as the system approaches the steady state. This application involves a PDE.

Solution: A steady-state solution can be developed by substituting centered finite differences for the first and the second derivatives in Eq. (26.4) to give

$$0 = D \frac{c_{i+1} - 2c_i + c_{i-1}}{\Delta x^2} - U \frac{c_{i+1} - c_{i-1}}{2\,\Delta x} - kc_i$$

Collecting terms gives

$$-\left(\frac{D}{U\,\Delta x} + \frac{1}{2} \right) c_{i-1} + \left(\frac{2D}{U\,\Delta x} + \frac{k\,\Delta x}{U} \right) c_i - \left(\frac{D}{U\,\Delta x} - \frac{1}{2} \right) c_{i+1} = 0 \qquad (26.5)$$

This equation can be written for each of the system's nodes. At the reactor's ends, this process introduces nodes that lie outside the system. For example, at the inlet node $(i = 0)$,

$$-\left(\frac{D}{U\,\Delta x} + \frac{1}{2} \right) c_{-1} + \left(\frac{2D}{U\,\Delta x} + \frac{k\,\Delta x}{U} \right) c_0 - \left(\frac{D}{U\,\Delta x} - \frac{1}{2} \right) c_1 = 0 \qquad (26.5a)$$

The c_{-1} can be removed by invoking the first boundary condition. At the inlet, the following mass balance must hold

$$Qc_{in} = Qc_0 - DA_c \frac{dc_0}{dx}$$

where c_0 is the concentration at $x = 0$. Thus, this boundary condition specifies that the amount of chemical carried into the tank by advection through the pipe must be equal to the amount carried away from the inlet by both advection and turbulent dispersion in the tank. A finite divided difference can be substituted for the derivative

$$Qc_{in} = Qc_0 - DA_c \frac{c_1 - c_{-1}}{2\,\Delta x}$$

which can be solved for

$$c_{-1} = c_1 + \frac{2\,\Delta x U}{D}\,c_{\text{in}} - \frac{2\,\Delta x U}{D}\,c_0$$

which can be substituted into Eq. (26.5a) to give

$$\left(\frac{2D}{U\,\Delta x} + \frac{k\,\Delta x}{U} + 2 + \frac{\Delta x U}{D}\right)c_0 - \left(\frac{2D}{U\,\Delta x}\right)c_1 = \left(2 + \frac{\Delta x U}{D}\right)c_{\text{in}} \qquad (26.6)$$

A similar exercise can be performed for the outlet where the original difference equation is

$$-\left(\frac{D}{U\,\Delta x} + \frac{1}{2}\right)c_{n-1} + \left(\frac{2D}{U\,\Delta x} + \frac{k\,\Delta x}{U}\right)c_n - \left(\frac{D}{U\,\Delta x} - \frac{1}{2}\right)c_{n+1} = 0 \qquad (26.7)$$

The boundary condition at the outlet is

$$Qc_n - DA_c\,\frac{dc_n}{dx} = Qc_n$$

As with the inlet, a divided difference can be used to approximate the derivative,

$$Qc_n - DA_c\,\frac{c_{n+1} - c_{n-1}}{2\,\Delta x} = Qc_n \qquad (26.8)$$

Inspection of this equation leads us to conclude that $c_{n+1} = c_{n-1}$. In other words, the slope at the outlet must be zero for Eq. (26.8) to hold. Substituting this result into Eq. (26.7) and simplifying gives

$$-\left(\frac{2D}{U\,\Delta x}\right)c_{n-1} + \left(\frac{2D}{U\,\Delta x} + \frac{k\,\Delta x}{U}\right)c_n = 0 \qquad (26.9)$$

Equations (26.5), (26.6), and (26.9) now form a system of n tridiagonal equations with n unknowns. For example, if $D = 2$, $U = 1$, $\Delta x = 2.5$, $k = 0.2$, and $c_{\text{in}} = 100$, the system is

$$\begin{bmatrix} 5.35 & -1.6 & & & \\ -1.3 & 2.1 & -0.3 & & \\ & -1.3 & 2.1 & -0.3 & \\ & & -1.3 & 2.1 & -0.3 \\ & & & -1.6 & 2.1 \end{bmatrix} \begin{Bmatrix} c_0 \\ c_1 \\ c_2 \\ c_3 \\ c_4 \end{Bmatrix} = \begin{Bmatrix} 325 \\ 0 \\ 0 \\ 0 \\ 0 \end{Bmatrix}$$

which can be solved for

$$c_0 = 76.44 \qquad c_1 = 52.47 \qquad c_2 = 36.06$$

$$c_3 = 25.05 \qquad c_4 = 19.09$$

These results are plotted in Fig. 26.2. As expected, the concentration decreases as the chemical flows through the tank due to the decay reaction. In addition to the above computation, Fig. 26.2 shows another case with $D = 4$. Notice how increasing the turbulent mixing tends to flatten the curve.

In contrast, if dispersion is decreased, the curve would become steeper as mixing became less important relative to advection and decay. It should be noted that if dispersion is decreased too much, the computation will become subject to numerical errors. This

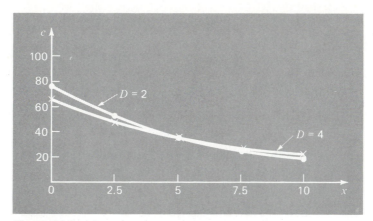

Figure 26.2
Concentration versus distance along the longitudinal axis of a rectangular reactor for a chemical that decays with first-order kinetics.

type of error is referred to as *static instability* to contrast it with the *dynamic instability* due to too large a time step during a dynamic computation. The criterion to avoid this static instability is

$$\frac{U \, \Delta x}{D} \leq 2$$

Thus, it occurs for cases where advection dominates over dispersion.

Aside from steady-state computations, numerical methods can be used to generate time-variable solutions of Eq. (26.3). Figure 26.3 shows results for $D = 2$, $U = 1$, $\Delta x = 2.5$, $k = 0.2$, and $c_{in} = 100$ where the concentration in the tank is 0 at time zero. As expected, the immediate impact is near the inlet. With time, the solution eventually approaches the steady-state level.

Figure 26.3
Concentration versus distance at different times during the build-up of chemical in a reactor.

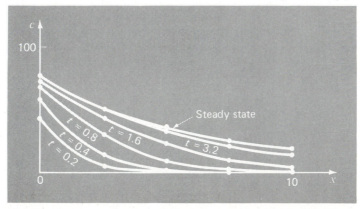

<u>CASE STUDY 26.2</u>   DEFLECTIONS OF A PLATE (CIVIL ENGINEERING)

Background: A square plate with simply supported edges is subject to an areal load q (Fig. 26.4). The deflection in the z dimension can be determined by solving the elliptic PDE (see Carnahan, Luther and Wilkes 1969)

$$\frac{\partial^2 z}{\partial x^4} + 2\frac{\partial^4 z}{\partial x^2 \, \partial y^2} + \frac{\partial^4 z}{\partial y^4} = \frac{q}{D} \tag{26.10}$$

subject to the boundary conditions that, at the edges, the deflection and slope normal to the boundary are zero. The parameter D is the flexural rigidity,

$$D = \frac{E\,\Delta z^3}{12(1-\sigma^2)} \tag{26.10a}$$

where E is the modulus of elasticity, Δz is the plate's thickness, and σ is Poisson's ratio.

If a new variable is defined as

$$u = \frac{\partial^2 z}{\partial x^2} + \frac{\partial^2 z}{\partial y^2}$$

Eq. (26.10) can be reexpressed as

$$\frac{\partial^2 u}{\partial x^2} + \frac{\partial^2 u}{\partial y^2} = \frac{q}{D} \tag{26.11}$$

Therefore, the problem reduces to successively solving two Poisson equations. First, Eq. (26.11) can be solved for u subject to the boundary condition that $u = 0$ at the edges. Then, the results can be employed in conjunction with

$$\frac{\partial^2 z}{\partial x^2} + \frac{\partial^2 z}{\partial y^2} = u \tag{26.12}$$

to solve for z subject to the condition that $z = 0$ at the edges.

Develop a computer program to determine the deflections for a square plate subject to a constant areal load. Test the program for a plate with 2-m-long edges, $q =$

Figure 26.4
A simply supported square plate subject to an area load.

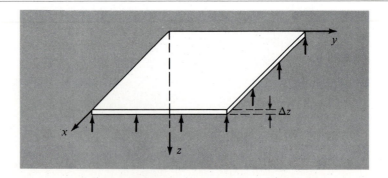

33.6 kN/m^2, $\sigma = 0.3$, $\Delta z = 10^{-2}$ m, and $E = 2 \times 10^{11}$ Pa. Employ $\Delta x = \Delta y = 0.5$ m for your test run.

Solution: Finite divided differences can be substituted into Eq. (26.11) to give

$$\frac{u_{i+1,j} - 2u_{i,j} + u_{i-1,j}}{\Delta x^2} + \frac{u_{i,j+1} - 2u_{i,j} + u_{i,j-1}}{\Delta y^2} = \frac{q}{D} \tag{26.13}$$

Equation (26.10a) can be used to compute $D = 1.832 \times 10^4$ N/m. This result along with the other system parameters can be substituted into Eq. (26.13) to give

$$u_{i+1,j} + u_{i-1,j} + u_{i,j+1} + u_{i,j-1} - 4u_{i,j} = 0.458$$

This equation can be written for all the nodes with the boundaries set at $u = 0$. The resulting equations are

$$
\begin{bmatrix}
-4 & 1 & & 1 & & & & & \\
1 & -4 & 1 & & 1 & & & & \\
& 1 & -4 & & & 1 & & & \\
1 & & & -4 & 1 & & 1 & & \\
& 1 & & 1 & -4 & 1 & & 1 & \\
& & 1 & & 1 & -4 & & & 1 \\
& & & 1 & & & -4 & 1 & \\
& & & & 1 & & 1 & -4 & 1 \\
& & & & & 1 & & 1 & -4
\end{bmatrix}
\begin{Bmatrix}
u_{1,1} \\ u_{2,1} \\ u_{3,1} \\ u_{1,2} \\ u_{2,2} \\ u_{3,2} \\ u_{1,3} \\ u_{2,3} \\ u_{3,3}
\end{Bmatrix}
=
\begin{Bmatrix}
0.458 \\ 0.458 \\ 0.458 \\ 0.458 \\ 0.458 \\ 0.458 \\ 0.458 \\ 0.458 \\ 0.458
\end{Bmatrix}
$$

which can be solved for

$$u_{1,1} = -0.315 \qquad u_{2,1} = -0.401 \qquad u_{3,1} = -0.315$$

$$u_{1,2} = -0.401 \qquad u_{2,2} = -0.515 \qquad u_{3,2} = -0.401$$

$$u_{1,3} = -0.315 \qquad u_{2,3} = -0.401 \qquad u_{3,3} = -0.315$$

These results in turn can be substituted into Eq. (26.12), which can be written in finite-difference form and solved for

$$z_{1,1} = 0.063 \qquad z_{2,1} = 0.086 \qquad z_{3,1} = 0.063$$

$$z_{1,2} = 0.086 \qquad z_{2,2} = 0.118 \qquad z_{3,2} = 0.086$$

$$z_{1,3} = 0.063 \qquad z_{2,3} = 0.086 \qquad z_{3,3} = 0.063$$

CASE STUDY 26.3   TWO-DIMENSIONAL ELECTROSTATIC FIELD PROBLEMS (ELECTRICAL ENGINEERING)

Background: Just as Fourier's law and the heat balance can be employed to characterize temperature distribution, analogous relationships are available to model field problems in other areas of engineering. For example, electrical engineers use a similar approach when modeling electrostatic fields.

Under a number of simplifying assumptions (see Paul and Nasar, 1987), an analog of Fourier's law can be represented in one-dimensional form as

$$D = -\epsilon \frac{dV}{dx}$$

where D is called the *electric flux density vector*, ϵ is the *permittivity* of the material, and V is the *electrostatic potential*.

Similarly, a Poisson equation for electrostatic fields can be represented in two dimensions as

$$\frac{\partial^2 V}{\partial x^2} + \frac{\partial^2 V}{\partial y^2} = -\frac{\rho_v}{\epsilon}$$

where ρ_v is the *volumetric charge density*.

Finally, for regions containing no free charge (that is $\rho_v = 0$), Eq. (26.14) reduces to a Laplace equation,

$$\frac{\partial^2 V}{\partial x^2} + \frac{\partial^2 V}{\partial y^2} = 0 \tag{26.14}$$

Employ numerical methods to solve Eq. (26.14) for the situation depicted in Fig. 26.5. Compute both the values for V and for D if $\epsilon = 2$.

Figure 26.5
(a) A two-dimensional system with a voltage of 1000 along the circular boundary and a voltage of 0 along the base. (b) The nodal numbering scheme.

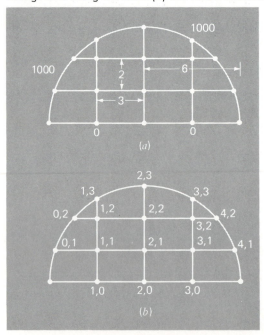

Solution: Using the approach outlined in Sec. 23.3.2, Eq. (23.24) can be written for node (1, 1) as

$$\frac{2}{\Delta x^2} \left[\frac{V_{1,1} - V_{0,1}}{\alpha_1(\alpha_1 + \alpha_2)} + \frac{V_{1,1} - V_{2,1}}{\alpha_2(\alpha_1 + \alpha_2)} \right] + \frac{2}{\Delta y^2} \left[\frac{V_{1,1} - V_{1,0}}{\beta_1(\beta_1 + \beta_2)} + \frac{V_{1,1} - V_{1,2}}{\beta_2(\beta_1 + \beta_2)} \right] = 0$$

According to the geometry depicted in Fig. 26.5, $\Delta x = 3$, $\Delta y = 2$, $\beta_1 = \beta_2 = \alpha_2 = 1$, and $\alpha_1 = 0.94281$. Substituting these values yields

$$0.12132V_{1,1} - 121.32 + 0.11438V_{1,1} - 0.11438V_{2,1} + 0.25V_{1,1}$$
$$+ 0.25V_{1,1} - 0.25V_{1,2} = 0$$

Collecting terms gives

$$0.73570V_{1,1} - 0.11438V_{2,1} - 0.25V_{1,2} = 121.32$$

A similar approach can be applied to the remaining interior nodes. The resulting simultaneous equations can be expressed in matrix form as

$$\begin{bmatrix} 0.73570 & -0.11438 & & -0.25000 & & \\ -0.11111 & 0.72222 & -0.11111 & & -0.25000 & \\ & -0.11438 & 0.73570 & & & -0.25000 \\ -0.31288 & & & 1.28888 & -0.14907 & \\ & -0.25000 & & -0.11111 & 0.72222 & -0.11111 \\ & & -0.31288 & & -0.14907 & 1.28888 \end{bmatrix} \begin{Bmatrix} V_{1,1} \\ V_{2,1} \\ V_{3,1} \\ V_{1,2} \\ V_{2,2} \\ V_{3,2} \end{Bmatrix} = \begin{Bmatrix} 121.32 \\ 0 \\ 121.32 \\ 826.92 \\ 250 \\ 826.92 \end{Bmatrix}$$

which can be solved for

$$V_{1,1} = 521.19 \qquad V_{2,1} = 421.85 \qquad V_{3,1} = 521.19$$
$$V_{1,2} = 855.47 \qquad V_{2,2} = 755.40 \qquad V_{3,2} = 855.47$$

These results are depicted in Fig. 26.6a.

In order to compute the flux (recall Sec. 23.2.3), Eqs. (23.14) and (23.15) must be modified to account for the irregular boundaries. For the present example, the modifications result in

$$D_x = -\epsilon \frac{V_{i+1,j} - V_{i-1,j}}{(\alpha_1 + \alpha_2)\,\Delta x}$$

and

$$D_y = -\epsilon \frac{V_{i,j+1} - V_{i,j-1}}{(\beta_1 + \beta_2)\,\Delta y}$$

For node (1, 1), these formulas can be used to compute the x and y components of the flux as

$$D_x = -2\,\frac{421.85 - 1000}{(0.94281 + 1)3} = 198.4$$

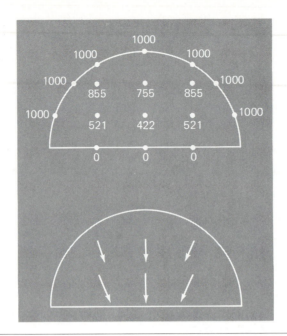

Figure 26.6
The results of solving the Laplace equation with correction factors for the irregular boundaries. (a) Potential and (b) flux.

and

$$D_y = -2 \frac{855.47 - 0}{(1+1)2} = -427.7$$

which in turn can be used to calculate the electric flux density vector

$$D = \sqrt{198.4^2 + (-427.7)^2} = 471.5$$

with a direction of

$$\theta = \tan^{-1}\left(\frac{-427.7}{198.4}\right) = -65.1°$$

The results for the other nodes are

Node	D_x	D_y	D	θ
2, 1	0.0	−377.7	377.7	− 90
3, 1	− 198.4	−427.7	471.5	245.1
1, 2	109.4	−299.6	281.9	− 69.1
2, 2	0.0	−289.1	289.1	− 90.1
3, 2	− 109.4	−299.6	318.6	249.9

The fluxes are displayed in Fig. 26.6b.

CASE STUDY 26.4   FINITE-ELEMENT SOLUTION OF A SERIES OF SPRINGS
(MECHANICAL ENGINEERING)

Figure 26.7 shows a series of interconnected springs. One end is fixed to a wall whereas the other is subject to a constant force F. Using the step-by-step procedure outlined in Chap. 25, a finite-element approach can be employed to determine the displacements of the springs.

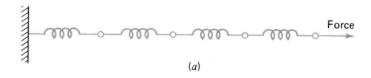

(a)

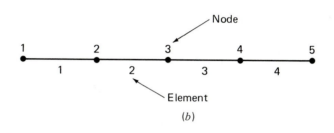

(b)

Figure 26.7
(a) A series of interconnected springs. One end is fixed to a wall whereas the other is subject to a constant force F. (b) The finite-element representation. Each spring represents an element. Therefore, the system consists of four elements and five nodes.

Discretization: The way to partition this system is obviously to treat each spring as an element. Thus, the system consists of four elements and five nodes (Fig. 26.7b).

Element Equations: Because this system is so simple, its element equations can be written directly without recourse to mathematical approximations. This is an example of the direct approach for deriving elements.

Figure 26.8 shows an individual element. The relationship between force F and displacement x can be represented mathematically by Hooke's law:

$$F = kx$$

Figure 26.8
A free-body diagram of a spring system.

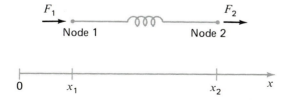

where k is the spring constant, which can be interpreted as the force required to cause a unit displacement. If a force F_1 is applied at node 1, the following force balance must hold:

$$F = k(x_1 - x_2)$$

where x_1 is the displacement of node 1 from its equilibrium position and x_2 is the displacement of node 2 from its equilibrium position. Thus, $x_2 - x_1$ represents how much the spring is elongated or compressed relative to equilibrium (Fig. 26.8). This equation can also be written as

$$F_1 = kx_1 - kx_2$$

For a stationary system, a force balance also necessitates that $F_1 = -F_2$ and, therefore,

$$F_2 = -kx_1 + kx_2$$

These two simultaneous equations specify the behavior of the element in response to prescribed forces. They can be written in matrix form as

$$\begin{bmatrix} k & -k \\ -k & k \end{bmatrix} \begin{Bmatrix} x_1 \\ x_2 \end{Bmatrix} = \begin{Bmatrix} F_1 \\ F_2 \end{Bmatrix}$$

or

$$[k]\{x\} = \{F\} \tag{26.15}$$

where the matrix $[k]$ is the *element property matrix*. For this case it is also referred to as the *element stiffness matrix*. Notice that Eq. (26.15) has been cast in the format of Eq. (25.9). Thus, we have succeeded in generating a matrix equation that describes the behavior of a typical element in our system.

Before proceeding to the next step—the assembly of the total solution—we will introduce some notation. The elements of $[k]$ and $\{F\}$ are conventionally superscripted and subscripted as in

$$\begin{bmatrix} k_{11}^{(e)} & -k_{12}^{(e)} \\ -k_{21}^{(e)} & k_{22}^{(e)} \end{bmatrix} \begin{Bmatrix} x_1 \\ x_2 \end{Bmatrix} = \begin{Bmatrix} F_1^{(e)} \\ F_2^{(e)} \end{Bmatrix}$$

where the superscript (e) designates that these are the element equations. The k's are also subscripted as k_{ij} to denote their location in the ith row and jth column of the matrix. For the present case, they can also be physically interpreted as representing the force required at node i to induce a unit displacement at node j.

Assembly: Before the element equations are assembled, all the elements and nodes must be numbered. This global numbering scheme specifies a system configuration or *topology*. That is, it documents which nodes belong to which element. Once the topology is specified, the equations for each element can be written with reference to the global coordinates.

The element equations can then be added one at a time to assemble the total system. The final result can be expressed in matrix form as [recall Eq. (25.10)]

$$[K]\{x'\} = \{F'\}$$

where

$$
[k] = \begin{bmatrix}
k_{11}^{(1)} & -k_{12}^{(1)} \\
-k_{21}^{(1)} & k_{22}^{(1)} + k_{11}^{(2)} & -k_{12}^{(2)} \\
& -k_{21}^{(2)} & k_{22}^{(2)} + k_{11}^{(3)} & -k_{12}^{(3)} \\
& & -k_{21}^{(3)} & k_{22}^{(3)} + k_{11}^{(4)} & -k_{12}^{(4)} \\
& & & -k_{21}^{(4)} & k_{22}^{(4)}
\end{bmatrix}
\tag{26.16}
$$

and

$$
\{F'\} = \begin{Bmatrix}
F_1^{(1)} \\
0 \\
0 \\
0 \\
F_2^{(4)}
\end{Bmatrix}
$$

and $\{x'\}$ and $\{F'\}$ are the expanded displacement and force vectors. Notice that as the equations were assembled, the internal forces cancel. Thus, the final result for $\{F'\}$ has zeros for all but the first and last nodes.

Before proceeding to the next step, we must comment on the structure of the assemblage property matrix [Eq. (26.16)]. Notice that the matrix is tridiagonal. This is a direct result of the particular global numbering scheme that was chosen (Table 26.1) prior to assemblage. Although it is not very important in the present context, the attainment of such a banded, sparse system can be a decided advantage for more complicated problem settings. This is due to the efficient schemes that are available for solving such systems.

Boundary Conditions: The present system is subject to a single boundary condition, $x_1 = 0$. Introduction of this condition and applying the global renumbering scheme reduces the system to (k's $= 1$)

$$
\begin{bmatrix}
2 & -1 \\
-1 & 2 & -1 \\
& -1 & 2 & -1 \\
& & -1 & 1
\end{bmatrix}
\begin{Bmatrix}
x_2 \\
x_3 \\
x_4 \\
x_5
\end{Bmatrix}
=
\begin{Bmatrix}
0 \\
0 \\
0 \\
F
\end{Bmatrix}
$$

The system is now in the form of Eq. (25.11) and is ready to be solved.

Although reduction of the equations is certainly a valid approach for incorporating boundary conditions, it is usually preferable to leave the number of equations intact when performing the solution on the computer. Some schemes for accomplishing this are outlined in Box 26.1. Whatever the method, once the boundary conditions are incorporated, we can proceed to the next step—the solution.

Solution: Using one of the approaches from Part Three, such as the efficient tridiagonal solution technique delineated in Chap. 9, the system can be solved for (with all k's $= 1$ and $F = 1$)

$$x_2 = 1 \qquad x_3 = 2 \qquad x_4 = 3 \qquad x_5 = 4$$

Box 26.1   Computer-Oriented Methods for Incorporating Boundary Conditions

---

Several methods have been developed to incorporate boundary conditions without reducing the dimensions of $[K]$. One of the most straightforward (and consequently among the easiest to program) was suggested by Payne and Irons (1963). In this approach, the diagonal term of $[K]$ associated with the boundary nodal variable is multiplied by a relatively large number. At the same time, the corresponding element of $\{F\}$ is replaced by the value of the boundary condition multiplied by the same large number. In essence, this approach artificially forces the

"unknown" nodal variable to take on the given boundary value.

An alternative approach has been proposed by Felippa and Clough (1970). If i is the subscript of the prescribed boundary variable, element k_{ii} is set equal to unity and the remaining terms in the ith row and ith column of $[K]$ are set to zero. The ith term of $\{F'\}$ is replaced by the value of the boundary condition. Each of the remaining values of $\{F'\}$ are modified by subtracting the boundary condition multiplied times the appropriate column term from the original $[K]$ matrix.

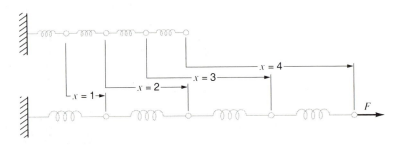

Figure 26.9
(a) The original spring system. (b) The system after the application of a constant force. The displacements are indicated in the space between the two systems.

Postprocessing: The results can now be displayed graphically. As in Fig. 26.9, the results are as expected. Each spring is elongated a unit displacement.

CASE STUDY 26.5   SPREADSHEET SOLUTION OF THE LAPLACE EQUATION

Background: This case study uses the spreadsheet to solve the Laplace equation. A spreadsheet to implement the method is contained on the supplementary disk associated with the text.

Solution: Activate your computer and type NUMMET following the A› prompt. Advance to the main menu, and select the Laplace equation example. This causes Fig. 26.10 to appear on the screen.

The domain of the solution is the rectangular area shown. The values that appear outside the rectangular area are boundary conditions. These boundary conditions can be changed by moving the cursor to the appropriate location and entering desired values. All other elements of the spreadsheet are inactive. Study the equations in all other cells to ensure yourself that you understand the method completely.

Move the cursor to cell A6 and enter a value of 100. Do the same thing for cells A8, A10, A12, A14, and A16. Next strike [F3] to perform one iteration. Note how the

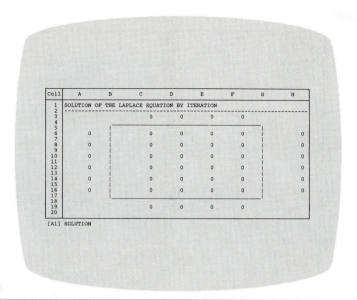

Figure 26.10

new boundary values begin to propagate across the rectangle. Strike [F3] several more times. Observe how the values in cells C6 and C16 become closer and closer, eventually producing a symmetric solution as expected. The final solution is shown in Fig. 26.11.

The spreadsheet solution is convenient and has the advantage of providing insight into the behavior of the relaxation procedure. The main disadvantage is that the spreadsheet must be reprogrammed if the geometry of the problem changes.

Figure 26.11

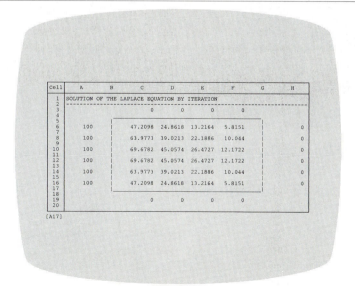

PROBLEMS

Chemical Engineering

26.1 Reproduce the computation in Case Study 26.1 using your own software.

26.2 Perform the same computation as in Case Study 26.1 but use $\Delta x = 1.25$.

26.3 Develop a finite-element solution for the steady-state system of Case Study 26.1.

26.4 Compute mass fluxes for the steady-state solution of Case Study 26.1 using Fick's first law.

26.5 Compute the steady-state distribution of concentration for the tank shown in Fig. P26.5. The PDE governing this system is

$$D\left(\frac{\partial^2 c}{\partial x^2} + \frac{\partial^2 c}{\partial y^2}\right) - kc = 0$$

and the boundary conditions are as shown. Employ a value of 0.5 for D and 0.2 for k.

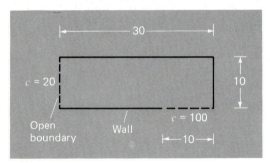

Figure P26.5

Civil Engineering

26.6 Reproduce the computation in Case Study 26.2 using your own software.

26.7 Perform the same computation as in Case Study 26.2 but use $\Delta x = 0.4$ and $\Delta y = 0.5$ m.

26.8 The flow through porous media can be described by the Laplace equation

$$\frac{\partial^2 h}{\partial x^2} + \frac{\partial^2 h}{\partial y^2} = 0$$

where h is head. Use numerical methods to determine the distribution of head for the system shown in Fig. P26.8.

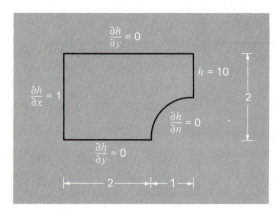

Figure P26.8

26.9 The velocity of water flow through the porous media can be related to head by D'Arcy's law

$$q_n = -K \frac{dh}{dn}$$

where K is the hydraulic conductivity and q_n is discharge velocity in the n direction. If $K = 5 \times 10^{-4}$ cm/s, compute the water velocities for Prob. 26.8.

Electrical Engineering

26.10 Reproduce the computation in Case Study 26.3 using your own software.

26.11 Perform the same computation as in Case Study 26.3 but for the system depicted in Fig. P26.11.

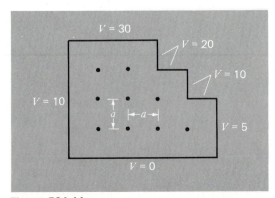

Figure P26.11

26.12 Perform the same computation as in Case Study 26.3 but for the system depicted in Fig. P26.12.

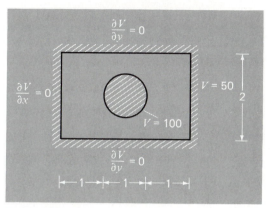

Figure P26.12

Mechanical Engineering

26.13 Reproduce the computation in Case Study 26.4 using your own software.

26.14 Perform the same computation as in Case Study 26.4 but change the spring constants to

Spring	1	2	3	4
k	0.5	2	0.75	1.5

Also change the force to 2.

26.15 Perform the same computation as in Case Study 26.4 but use five springs with

Spring	1	2	3	4	5
k	0.2	0.4	1.5	0.7	0.9

Spreadsheet

26.16 Use an active spreadsheet that may be available to you to solve the Laplace equation for a square plate with nine internal node points. Solve the equation when the right-hand boundary has a value of 200 and all other boundary values are zero. Repeat the solution with the left-hand boundary at 200 with all other values zero. Discuss the numerical and physical significance of the results.

26.17 Activate the spreadsheet used in Case Study 26.5, input the left-hand boundary, and iterate until convergence is obtained. Now move the cursor to cells H10 and H12 and input values of 10,000 and strike [F3]. Note which elements of the rectangle change during the first iteration. Now strike [F3] several times, noting the propagation of the right-hand boundary over the other cells of the rectangle.

26.18 Repeat Prob. 26.17 using boundary values of your own selection.

Miscellaneous

26.19 Read all the case studies in Chap. 26. On the basis of your reading and experience, make up your own case study for any one of the fields of engineering. This may involve modifying or reexpressing one of our case studies. However, it can also be totally original. As with our examples, it must be drawn from an engineering problem context and must demonstrate the use of numerical methods for solving PDEs. Write up your results using our case studies as models.

EPILOGUE: PART SEVEN

PT7.3   TRADE-OFFS

The primary trade-offs associated with numerical methods for the solution of partial differential equations involve choosing between *finite-difference* and *finite-element* approaches. The *finite-difference* methods are conceptually easier to understand. In addition, they are easy to program for systems that can be approximated with uniform grids. However, they are difficult to apply to systems with complicated geometries.

Finite-difference approaches can be divided into categories depending on the type of PDE that is being solved. *Elliptic* PDEs can be approximated by a set of linear algebraic equations. Consequently, the *Liebmann Method* (which, in fact is Gauss-Seidel) can be employed to obtain a solution iteratively.

One-dimensional parabolic PDEs can be solved in two fundamentally different ways: explicit or implicit approaches. The *explicit method* steps out in time in a fashion that is similar to Euler's technique for solving ODEs. It has the advantage that it is simple to program but has the shortcoming of a very stringent stability criterion. In contrast, stable implicit methods are available. These typically involve solving simultaneous tridiagonal algebraic equations at each time step. One of these approaches, the *Crank-Nicolson method,* is both accurate and stable for moderate time steps and, therefore, is widely used for one-dimensional parabolic problems.

Two-dimensional parabolic PDEs can also be modeled explicitly. However, their stability constraints are even more severe than for the one-dimensional case. Special implicit approaches, which are generally referred to as splitting methods, have been developed to circumvent this shortcoming. These approaches are both efficient and stable. One of the most common is the *A.D.I.*, or alternating direction implicit, method.

All the above *finite-difference* approaches become unwieldy when applied to problems involving nonuniform grids and heterogeneous conditions. Finite-element methods are available that handle such systems in a superior fashion.

Although the *finite-element* method is based on some fairly straightforward ideas, the mechanics of generating a good finite-element code for two- and three-dimensional problems is not a trivial exercise. In addition, it can be computationally expensive for large problems. However, it is vastly superior to finite-difference approaches for systems involving complicated shapes. Consequently, its expense and conceptual "overhead" are often justified because of the detail of the final solution.

PT7.4 IMPORTANT RELATIONSHIPS AND FORMULAS

Table PT7.3 summarizes important information that was presented regarding the finite-difference methods in Part Seven. This table can be consulted to quickly access important relationships and formulas.

TABLE PT7.3 Summary of finite-difference methods.

	Computational molecule	Equation
Elliptic PDEs Liebmann's method		$T_{i,j} = \dfrac{T_{i+1,j} + T_{i-1,j} + T_{i,j+1} + T_{i,j-1}}{4}$
Parabolic PDEs (one-dimensional) Explicit method		$T_i^{l+1} = T_i^l + \lambda(T_{i+1}^l - 2T_i^l + T_{i-1}^l)$
Implicit method		$-\lambda T_{i-1}^{l+1} + (1 + 2\lambda) T_i^{l+1} - \lambda T_{i+1}^{l+1} = T_i^l$
Crank-Nicolson method		$-\lambda T_{i-1}^{l+1} + 2(1 + \lambda) T_i^{l+1} - \lambda T_{i+1}^{l+1}$ $= \lambda T_{i-1}^l + 2(1 - \lambda) T_i^l + \lambda T_{i+1}^l$

PT7.5 ADVANCED METHODS AND ADDITIONAL REFERENCES

Carnahan, Luther, and Wilkes (1969), Rice (1983), Ferziger (1981), and Lapidus and Pinder (1982) provide useful surveys of methods and software for solving PDEs. You can also consult Ames (1977), Gladwell and Wait (1979), Vichnevetsky (1981, 1982), and Zienkiewicz (1971) for more in-depth treatments. Additional information on the finite-element method can be found in Allaire (1985), Huebner and Thornton (1982), Stasa (1985), and Baker (1983). Aside from elliptic and hyperbolic PDEs, numerical methods are also available to solve hyperbolic equations. Nice introductions and summaries of some of these methods can be found in Lapidus and Pinder (1982), Ferziger (1981) and Forsythe and Wasow (1960).

Appendix A: Supplementary Software

This textbook contains a computer software supplement. The software will operate on IBM-PC or compatible computers that have a capacity of 256K or greater. The software consists of seven examples based on the spreadsheet from the ELECTRONIC TOOLKIT. These spreadsheets illustrate or compare important numerical methods discussed in various sections of the text. These spreadsheets are discussed in detail in the Case Study sections in the text. Several homework problems are also based on the software supplement.

In order to use the software, activate your computer and then insert the disk and type the following after the prompt:

A > NUMMET

A title screen will appear on the monitor screen as shown in Fig. 6.12a. The Return key advances the screen to the main menu as shown in Fig. 6.12b. Various selections from the main menu are made using the [↑] and [↓] keys or the numeric keys.

A completed spreadsheet example appears on the screen following the selection of any item from the main menu. Most of the cells of the spreadsheet examples contain equations and related information to implement various numerical methods. The contents of these cells cannot be modified. Other cells contain information such as equation parameters, starting conditions, or initial guesses. You can change the contents of these cells. The exact location of active and inactive cells varies for each example. The active cells are shown in color or are highlighted on monochrome monitors. A detailed explanation of each spreadsheet on the software supplement is contained within the Case Study sections of the text.

The contents of each cell of the spreadsheet can be reviewed on the bottom line of the screen by using the [↑], [↓], [←], and [→] keys. This information will help you understand the numerical method more completely. Similarly, you can use the above keys to proceed to the input cells of the spreadsheet. The F3 key is used to update the value of all cells in the spreadsheet after the appropriate input information has been entered. The F2 key allows access to an insert type edit mode. This allows you to easily correct errors while inputting information to a cell. The F10 key is used as a toggle to display or delete key functions on the bottom of the screen. The End key is used to return to the main menu.

It is often convenient to produce a hard copy of your spreadsheet analyses. This can be accomplished by using the upper case [PrtSc] key, provided your computer is connected to an appropriate printer.

Appendix B: The Fourier Series

The Fourier series can be expressed in a number of different formats. Two equivalent trigonometric expressions are

$$f(t) = a_0 + \sum_{k=1}^{\infty} [a_k \cos (k\omega_0 t) + b_k \sin (k\omega_0 t)]$$

or

$$f(t) = a_0 + \sum_{k=1}^{\infty} [c_k \cos (k\omega_0 t + \theta_k)]$$

where the coefficients are related by (see Fig. B.1)

$$c_k = \sqrt{a_k^2 + b_k^2}$$

and

$$\theta_k = -\tan^{-1}\left(\frac{b_k}{a_k}\right)$$

Figure B.1
Relationships between rectangular and polar forms of the Fourier series coefficients.

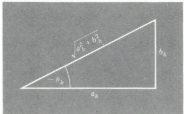

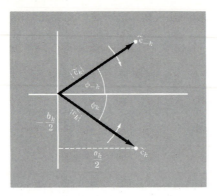

Figure B.2
Relationships between complex exponential and real coefficients of the Fourier series.

In addition to the trigonometric formats, the series can also be expressed in terms of the exponential function as

$$f(t) = \tilde{c}_0 + \sum_{k=1}^{\infty} [\tilde{c}_k e^{ik\omega_0 t} + \tilde{c}_{-k} e^{-ik\omega_0 t}] \tag{B.1}$$

where (see Fig. B.2)

$$\tilde{c}_0 = a_0$$

$$\tilde{c}_k = \frac{1}{2}(a_k - ib_k) = |\tilde{c}_k| e^{i\phi}k$$

$$\tilde{c}_{-k} = \frac{1}{2}(a_k + ib_k) = |\tilde{c}_k| e^{-i\phi}k$$

where $|\tilde{c}_0| = a_0$ and

$$|\tilde{c}_k| = \frac{1}{2}\sqrt{a_k^2 + b_k^2} = \frac{c_k}{2}$$

and

$$\phi_k = \tan^{-1}\left(\frac{-b_k}{a_k}\right)$$

Note that the tilde signifies that the coefficient is a complex number.

Each term in Eq. (B.1) can be visualized as a rotating phasor (the arrows in Fig. B.2). Terms with a positive subscript rotate in a counterclockwise direction whereas those with a negative subscript rotate clockwise. The coefficients $\tilde{c}_k$ and $\tilde{c}_{-k}$ specify the position of the phasor at $t = 0$. The infinite summation of the spinning phasors, which are allowed to rotate at $t = 0$, is then equal to $f(t)$.

Appendix C: ELECTRONIC TOOLKIT Integrated Computational Software for Engineers and Scientists

The ELECTRONIC TOOLKIT is a general problem-solving software package for engineers, scientists, and mathematicians. It is an integrated and improved version of two previous software packages called NUMERICOMP and ENGINCOMP. The TOOLKIT was conceived and designed by Raymond P. Canale from the University of Michigan and Steven C. Chapra from the University of Colorado. The algorithms to interpret functions, check errors, input data, and generate plots were programmed by Theodore A.D. Slawecki of EnginComp Software, Inc. Quality assurance testing was performed by Joseph Campbell and Eric Van Steenlandt from the University of Michigan.

INTRODUCTION

The ELECTRONIC TOOLKIT has six numerical methods and three problem-solving programs. These problems have utility in a wide variety of engineering, science, and mathematics applications. The TOOLKIT is designed to be used on IBM PC and compatible personal computers that have at least 256K memory and a color graphics display card.

The utility of the TOOLKIT is greatly enhanced when it is possible to obtain hard copies of the tables and graphs. However, some special characters used by the TOOLKIT may not be displayed correctly unless your printer can accommodate the extended ASCII character set and graphics. If your printer does not have these capabilities, your output may be blank or contain inappropriate characters.

Text and graphics screens are transferred to printers using the [Shift] and [PrtSc] keys. However, this requires the use of the IBM system program called GRAPHICS which is located on your DOS 2.0 or higher disk. You must execute this program prior to running the TOOLKIT with the command A>GRAPHICS to transfer graphics screens to a printer.

This appendix will take you through an example problem for each of the nine programs on the TOOLKIT. These examples illustrate the features and limitations of each program and allow you to solve other problems.

TOOLKIT MAIN MENU

Insert your disk into your computer and type TOOLKIT following the A>prompt. After the time required for the program to load, Screen 1 should appear on your screen or monitor. This screen has information about the title and source of the software. After loading the TOOLKIT program, the disk should be removed to avoid accidental erasures.

Now enter [Return] to display the TOOLKIT Main Menu, as shown in Screen 2. This menu has a selection of nine programs and options to read data, to save data, or to exit. Note that the first six programs on the TOOLKIT are contained within a common box. This signifies that these programs share a common data base. New data can be created or modified within any of the six programs. Changes in the data made in any one program are automatically transferred to the other programs. Newly created data in any program destroys the old data in all the other programs. Data can be transferred from the spreadsheet to the other five integrated programs. The reverse process is not possible. The Trapezoidal Rule, Graphics, and Bisection programs share the same function of X.

The data and equations associated with the six integrated programs, as well as the other three programs, can be saved or read from a disk. Up to 10 different data sets can be stored on a single disk. Any number of disks can be used to accumulate a library of TOOLKIT data files.

SPREADSHEET

The Spreadsheet program is activated by entering selection [1] from the TOOLKIT Main Menu (Screen 2). Selection [1] is addressed using either the numeric or up and down arrow keys. Note how you can use these keys to scroll up and down the menu. (Appendix I contains a description of the function of all the other special keys used in the TOOLKIT). Screen 3 shows the Spreadsheet menu. This menu allows you to create a new spreadsheet, modify the current spreadsheet, or exit to the Main Menu. A previously saved spreadsheet may be read from a disk using selection [10] from the Main Menu. To do this you return the Main menu and select [10] to read data. The table that appears will have the names of all of the data files you have saved on your disk. After you select the appropriate file name, hit [End]. This loads the data from the disk and returns you to the Main Menu. You can now return to the Spreadsheet program, and the stored information will be available to you using selection [2].

Let us create a new spreadsheet by use of selection [1]. Screen 4 shows the appearance of a TOOLKIT spreadsheet. The spreadsheet consists of an

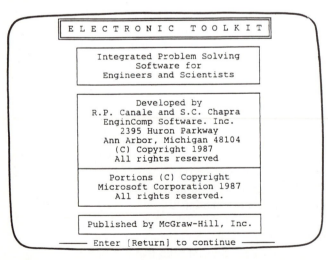

Screen 1

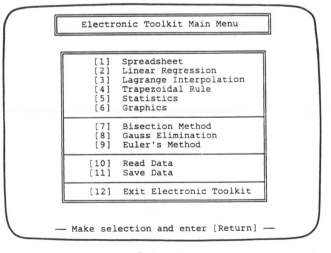

Screen 2

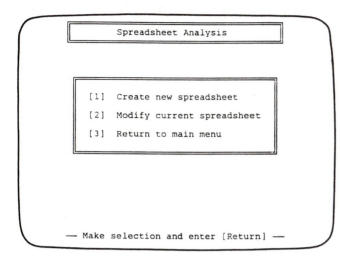

Screen 3

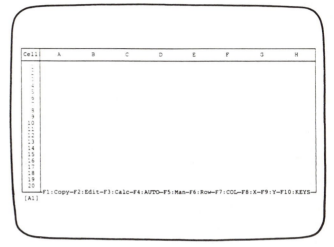

Screen 4

interconnected network of computational elements called *cells.* These cells can contain either text, numeric constants, or mathematical equations. The mathematical equations may be defined in terms of the names of cells on the spreadsheet. Thus, the cells of the spreadsheet can be linked by mathematical equations. The advantage of spreadsheet computations is this ability to define relationships among cells by equations. Thus, changes in any cell entry affect the values of all coupled cells. The computation strategy of the spreadsheet involves scanning the entire network and updating values in all the cells.

Note that the spreadsheet has 8 columns (A through H) and 20 rows. These define 160 separate cells or locations on the spreadsheet. It is not possible to modify this display or the length and width of the TOOLKIT spreadsheet. The location of the cursor defines the active cell. The cursor is moved using the arrow keys. The location of the cursor is shown in the brackets on the extreme left of the last line on the screen. The letter designates the column, and the number defines the row. The rest of the bottom line of the screen shows the contents of the active cell. The bottom of Screen 4 also shows the function of special keys available with the spreadsheet. This will be described in detail below.

The Spreadsheet program distinguishes between text and numeric data by the first character of the input. The program assumes a cell entry is numeric if the first character is a number. Thus, a cell entry such as "1st"

is treated as a number with a value of 1. The program disregards the nonnumeric "st" part that follows. The program treats the cell entry as text if the first character is nonnumeric except for an equal sign (=). The cell is also defined as text if the first character is a single quote ('). If the first character is an equal sign, the program processes the cell entry as an equation. If a space is entered before the equal sign, the program assumes the entry is text. Cells that contain text or that have not been defined have numeric values of zero.

The Spreadsheet program allows you to use up to 71 characters of text to describe a single cell. These characters will be shown on the bottom of the screen, but only the first 9 characters will be displayed within the spreadsheet cell. Constants are displayed with a maximum of 9 characters. Very small and very large numbers may be automatically converted into exponential form.

The complexity and length of spreadsheet formulas is limited. The actual allowable number of characters or mathematical operations varies depending on the form of the equation. Extremely long equations can be broken into parts if necessary to accommodate the Spreadsheet program requirements. An error signal is given if an equation is entered that is either too long or complex.

The equations in cells use standard mathematical operators (such as $+,-,*,/,$ and $^\wedge$) and the special mathematical functions listed in Appendices II and III. In particular, note that the sum function, SUM (X.Y), is convenient for many spreadsheet computations. This function adds the cell contents over the specified range. For example the operation SUM (A1.C5) adds the cell contents of A1 through C5. Cells with alphabetic information or blanks have numerical values of zero as shown in Screen 5.

The [F1] key allows you to copy the contents of one cell to other cells. This key displays a request on the right-hand side of the last line of the screen for information about where the current cell contents will be copied. The cell contents can be copied to a single cell or to a block of cells. This will result in the loss of information previously entered into a cell. A block of cells is defined by specifying the location of the top left and botton right cells. The [F1] key can be used to copy text, numbers, or equations from one cell to another cell or to a range of cells. Text and numbers are copied exactly as in the original cell. Equations are copied in relative manner. This means that the terms of new equations are defined relative to the position of the terms in the original equation. For example in Screen 6, E3 = SUM (A1.C5). When this equation is copied over the range F3 to H3, cell F3 will contain SUM (B1.D5), G3 will contain SUM (C1.E5), and H3 will contain SUM (D1.F5). Copy commands that result in equations that range beyond the limits of the spreadsheet give an error message.

Mistakes in cell entries can be corrected using the [Backspace] key. However, this is sometimes inconvenient because this key erases characters. Therefore, the Spreadsheet program has insert-type editing capabilities. This means new characters are inserted rather than copied over the old characters. The edit mode is activated automatically when a syntax error is detected or manually using the [F2] key. The [Del] and arrow keys can be used to modify cell characters in the edit mode. The [Return] key is used to leave the edit mode and save the changes. The [Esc] key allows you to leave the edit mode without saving the changes. This restores the original contents of the cell.

You should be alert to the fact that spreadsheet computations with complex interrelationships among cells are often difficult to interpret. Sometimes, a second pass through the columns and rows of the spreadsheet will change the values of some cells. The [F3] key is used to update all cell entries. You can choose between automatic updating by the spreadsheet or manual updating using the [F4] and [F5] keys. It is convenient to perform updates by either rows or by columns depending on the layout of the computations. [F6] updates the cells by rows, while [F7] updates by columns.

The [F8] and [F9] keys are used to transfer numbers contained in the spreadsheet to other parts of the TOOLKIT program. The Spreadsheet, Linear Regression , Lagrange Interpolation, Trapezoidal Rule, and Graphics programs on the TOOLKIT share a common data base consisting of paired X and Y values. The Statistics program shares only the X values. These data can be input from any one of the above programs. However, the [F8] and [F9] keys can also be used to define values for the common data base from the spreadsheet. [F8] tells the program that we wish to define X data values. After hitting [F8], the message "Copy X-values from:" appears. You can then supply an individual or range of cell locations. Screen 7 shows a sample using C10 through C17. [F9] allows similar assignment of Y data information contained in E10 through E17. The [F10] key is a switch that turns the special key function reference line on or off.

The [End] key is used to exit the spreadsheet and return to the menu. If you would like to modify the contents of the spreadsheet, use selection [2] because selection [1] destroys the previous spreadsheet. Thus, if you have only a few modifications to make, use of selection [2] is preferred.

You may also wish to save the information on the spreadsheet on a disk for later use. To do this return to the Main Menu by selection [3]. After selecting [11] from the Main Menu, the Save Data table is displayed (Screen 8). When this is done, be sure you have a properly formatted disk (not the TOOLKIT disk) in the A drive . The program allows you to store up to 10 different files and the names of any existing files on the disk will appear in the table. Move the cursor to an open position, add the name of the new file and press [End]. This automatically saves the file and returns you to the Main Menu. You can save a new file under an old file name, but the old file will be erased to make room for the new one. Files can be deleted using the [Del] key. Files saved in the above manner may be read in a similar fashion using selection [10] from the Main Menu.

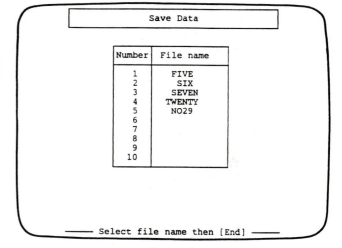

Screen 5

Cell	A	B	C	D	E	F	G	H
1								
2		3	1	2	1	radis	2	
3			1		5	sine	4	
4		5	3	5	3			
5		6	pie		1		arc	
6		7	6	8	1			
7			7		cost		2	
8		9	8	8	1	whole		
9								
12		99						

[B12] =SUM(B1.G10)

Screen 7

EXAMPLE PROBLEM   f(X) = X + 3*sin(X)

DATA	RESULTS
0	0
2	4.72789
4	1.72959
7	8.97096
8	10.9680
10	8.36793
12.5	12.3010
14	16.9718

[C10] 0   Copy X-values from: C10.C17

Screen 6

Cell	A	B	C	D	E	F	G	H
1	1	1	1					
2	1	1	1					
3	1	1	1		15	10	20	25
4	1	1	1					
5	1	1	1					

[E3] =SUM(A1.C5)   Copy current cell to: F3.H3

Screen 8

Save Data

Number	File name
1	FIVE
2	SIX
3	SEVEN
4	TWENTY
5	NO29
6	
7	
8	
9	
10	

Select file name then [End]

LINEAR REGRESSION

The second program on the TOOLKIT is Linear Regression. This program can be used to fit a straight trend line to data using the least-squares criterion for best fit. The menu for Linear Regression is activated by selection [2] from the TOOLKIT Main Menu (Screen 2). The Linear Regression menu, shown in Screen 9, allows you to enter or modify data, perform regression, or plot the data and the calculated regression line. Previously saved data may be read from a disk using selection [10] from the Main Menu.

New data points can be entered by selection [1]. However, do not use this selection now because this will result in the loss of the data we generated in the spreadsheet. Rather use selection [2] to display a table that contains the data we have already created (use Screen 10). Note that the table can accommodate up to 16 pages of data, with each page containing 10 points, or a total of 160 points. Values for X and Y are entered according to the location of the shaded horizontal cursor bar. This cursor can be moved to any location using the arrow keys. The [Ins] key can be used to insert a line of data between existing lines. Likewise, the [Del] key deletes a line of data from the table. A new page is advanced using the [PgDn] key. Old pages can be reviewed using the [PgUp] key. The values in the data tables (and other places in the TOOLKIT) may be input in either standard or scientific form using E or D format. The maximum acceptable value is 1E18. A data input or modification session is terminated by the [End] key. This returns you to the Linear Regression menu.

We will now perform regression on these data using selection [3]. The results appear in Screen 11. The screen shows the calculated equation of the regression line, as well as the coefficient of determination, the correlation coefficient, and the standard error of estimate. These results characterize how well the regression line approximates the trend of the data. The program also allows you to input a value for X and interpolate (or extrapolate) for a value of Y based on the calculated regression equation. This is convenient for estimating values of Y based on the trend line. You can get back to the menu by entering [Return] after interpolating as many times as you like.

Once you are back in the Linear Regression menu, you can plot the line and data. Plots are developed using selection [4]. The program prompts you for the dimensions of the plot using the wide horizontal cursor. Plot parameters cannot be input using scientific notation. Screen 12 shows plot parameters that define the limits of X and Y. An error message is given if you enter inappropriate values for the minimum and maximum values of either X or Y, or if you fail to enter all values. Default values for the plot parameters may be shown on the screen. These default values may be the extreme values of X and Y from the data or the values used for a previous plot. These default values will help you modify the design of a plot by reminding you of the old values. Any or all of the plot parameters can be changed by entering new values using the up and down arrow keys. Screen 13 shows the regression line and data points from the example problem we started in the spreadsheet. Individual data points may not be distinguishable on the plot if they are close together compared to the scale on the X and Y axes. A plot parameter input session is completed using the [End] key. If you wish to change the appearance of the plot, simply enter [Return] to bring you back to the Linear Regression menu. Select [4] again and enter appropriate plot parameters. Note that convenient even divisions on the plot are obtained if the X range is a multiple of 5 and the Y range is a multiple of 6. After entering [Return], you are back in the Linear Regression menu where you can enter new data, modify your old data, or return to the TOOLKIT Main Menu.

LAGRANGE INTERPOLATION

The next program on the TOOLKIT is Lagrange Interpolation. This program can be used to fit data with up to an eighth-order polynomial. The polynomial can be used to compute intermediate values of Y between given values of X. The menu for Lagrange Interpolation is displayed by selection [3] from the TOOLKIT Main Menu (Screen 2). The Lagrange Interpolation menu is shown in Screen 14. This menu allows you to enter or modify data, perform interpolation, plot the data and the calculated interpolation polynomial, or check the accuracy of the computations. Previously saved data may be read from a disk using selection [10] from the Main Menu.

```
┌─────────────────────────────────────┐
│  ┌───────────────────────────────┐  │
│  │     Linear Regression         │  │
│  │     For Curve Fitting         │  │
│  └───────────────────────────────┘  │
│                                      │
│  ┌───────────────────────────────┐  │
│  │  [1]  Enter new data          │  │
│  │                               │  │
│  │  [2]  Modify old data         │  │
│  │                               │  │
│  │  [3]  Perform regression      │  │
│  │                               │  │
│  │  [4]  Plot data and function  │  │
│  │                               │  │
│  │  [5]  Return to main menu     │  │
│  └───────────────────────────────┘  │
│                                      │
│ — Make selection and enter [Return] —│
└─────────────────────────────────────┘
```

Screen 9

```
┌─────────────────────────────────────┐
│  ┌───────────────────────────────┐  │
│  │ Calculated Regression Parameters │ │
│  └───────────────────────────────┘  │
│                                      │
│          Regression equation         │
│  ┌───────────────────────────────┐  │
│  │ Y =  .4106481 +  1.056559X    │  │
│  └───────────────────────────────┘  │
│                                      │
│        Regression coefficients       │
│  ┌──────────────────────────┬─────┐ │
│  │ Number of data points    │  8  │ │
│  │ Coefficient of determination│0.86123│
│  │ Correlation coefficient  │0.92802│
│  │ Standard error estimate  │2.2706│ │
│  └──────────────────────────┴─────┘ │
│                                      │
│             Interpolate              │
│  ┌──────────────┬──────────────┐    │
│  │      X       │      Y       │    │
│  ├──────────────┼──────────────┤    │
│  │              │              │    │
│  └──────────────┴──────────────┘    │
│                                      │
│ ——— Enter X value or [Return] ———   │
└─────────────────────────────────────┘
```

Screen 11

```
┌─────────────────────────────────────┐
│  ┌───────────────────────────────┐  │
│  │       Modify old data         │  │
│  └───────────────────────────────┘  │
│                                      │
│  ┌────┬────────────┬────────────┐   │
│  │    │  X value   │  Y value   │   │
│  ├────┼────────────┼────────────┤   │
│  │ 1  │     0      │     0      │   │
│  │ 2  │     2      │  4.727893  │   │
│  │ 3  │     4      │  1.729593  │   │
│  │ 4  │     7      │  8.97096   │   │
│  │ 5  │     8      │ 10.96807   │   │
│  │ 6  │    10      │  8.367937  │   │
│  │ 7  │   12.5     │ 12.30103   │   │
│  │ 8  │    14      │ 16.97182   │   │
│  │ 9  │            │            │   │
│  │ 10 │            │            │   │
│  └────┴────────────┴────────────┘   │
│                                      │
│  ———— Enter data then [End] ————    │
└─────────────────────────────────────┘
```

Screen 10

```
┌─────────────────────────────────────┐
│  ┌───────────────────────────────┐  │
│  │     Plot data and function    │  │
│  └───────────────────────────────┘  │
│                                      │
│  ┌──────────────┬──────────────┐    │
│  │  Parameter   │    Value     │    │
│  ├──────────────┼──────────────┤    │
│  │  Minimum X   │      0       │    │
│  ├──────────────┼──────────────┤    │
│  │  Maximum X   │     14       │    │
│  ├──────────────┼──────────────┤    │
│  │  Minimum Y   │      0       │    │
│  ├──────────────┼──────────────┤    │
│  │  Maximum Y   │  16.97182    │    │
│  └──────────────┴──────────────┘    │
│                                      │
│ ——— Enter parameters then [End] ——— │
└─────────────────────────────────────┘
```

Screen 12

New data points are entered directly by selection [1]. However, this will destroy the old data. Instead, we will continue to use the data that we generated in the Spreadsheet, which remains as our common data base until it is modified.

Lagrange Interpolation is implemented through selection [3], as shown in Screen 15. First, you must input the order of the polynomial. Any order up to a maximum of 8 is acceptable if enough data are provided. The order of the polynomial must be less than, or equal to, the number of data points minus one. Next, you input the number of the first point to be matched by the polynomial. This is necessary because the program allows you to fit a polynomial to various subsets of your input data. For example, if you input a total of five data points, you could fit a second-order polynomial to the first three points; the second, third, and fourth points; or the last three points. Thus, for this case, the number of the first point to be matched by the polynomial could be one, two, or three. The fourth point cannot be used because only two points would remain and three points are required to define the polynomial. The program provides appropriate default values and ranges to help you. The program will give you an error message if you attempt to interpolate using data that have the same value of X. The program will not attempt to determine an interpolating polynomial if only one data point is entered. After you have input an acceptable order and number of first point, the program immediately executes the Lagrange Interpolation algorithm. Screen 16 shows the calculated coefficients of the interpolating polynomial. Also note that you can perform interpolation computations in a manner similar to that used in the Linear Regression program. After interpolating as many times as you like, you can get back to the Lagrange menu by entering [Return].

You can compute additional interpolating polynomials through selection [3]. This facilitates comparison of results using different combinations of orders and first points. The accuracy of the polynomial can be checked by entering selection [5]. This selection causes the program to use the computed polynomial to calculate values for Y at values of X where data have been input. A table is then displayed where calculated and input Y values are compared. If the values are identical, you can usually be assured that the computation

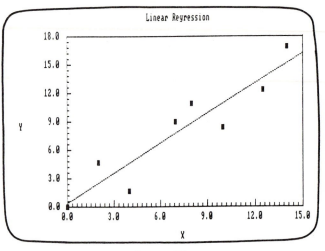

Screen 13

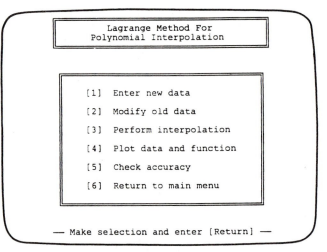

Screen 14

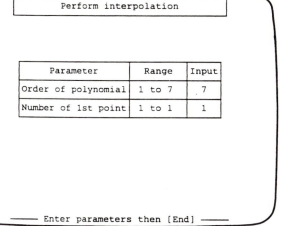

Screen 15

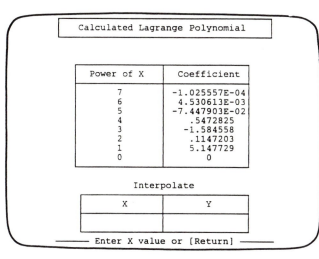

Screen 16

proceeded correctly. An example is shown in Screen 17. It is important that you test the accuracy of your results because high-order polynomials are sometimes subject to round-off errors.

Once you return to the Lagrange Interpolation menu, you can enter new data, modify old data, or plot the polynomial and data. Plots are developed using selection [4]. They are constructed in a manner exactly the same as that used in the Linear Regression program. Screen 18 shows a seventh-order polynomial and corresponding data. After entering [Return], you are back in the Lagrange Interpolation menu.

TRAPEZOIDAL RULE

The Trapezoidal Rule can be used to integrate a known function of X or tabular data. The menu for the Trapezoidal Rule is activated by selection [4] from the TOOLKIT Main Menu (Screen 2). The Trapezoidal Rule menu is shown in Screen 19. This menu allows you to enter and integrate a function of X, integrate tabular data, or plot the integral as the area under a curve. Previously saved functions and data may be read from a disk using selection [10] from the Main Menu.

The first task is to enter a new function using selection [1]. If you try to integrate or plot before entering a function, the program responds with a descriptive error message. Screen 20 shows the appearance of the screen after you have entered an appropriate function. As with all the functions in· the TOOLKIT except Euler's Method, X is the only variable allowed. If you want to abort the function-entry step, the [Esc] key will return you to the Trapezoidal Rule menu. Also note that the function entered here will automatically be transformed to the Graphics and Bisection programs.

The program checks for syntax errors when you input a function. Messages are displayed at the bottom of the screen if syntax errors are detected. The program also displays the error-free part of your function at the bottom of the screen. This gives you a clue regarding the character that prompted the error message. Screen 21 shows an error made entering f(X). You can make immediate corrections by moving the cursor with the arrow keys or by continuing to input characters with

Check accuracy

X value	Calculated Y value	Input Y value
0	0	0
2	4.727893	4.727893
4	1.729593	1.729593
7	8.97096	8.97096
8	10.96807	10.96807
10	8.367937	8.367937
12.5	12.30103	12.30103
14	16.97182	16.97182

———— Enter [Return] to continue ————

Screen 17

Trapezoidal Rule
For Numerical Integration

[1] Enter function

[2] Integrate function

[3] Integrate new data

[4] Integrate existing data

[5] Plot integral as area

[6] Return to main menu

— Make selection and enter [Return] —

Screen 19

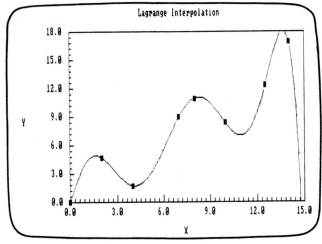

Screen 18

Enter function

f(X) = X^2/10

———— Enter f(X) then [Return] ————

Screen 20

appropriate modifications. Note that new characters are inserted rather than copied over the old characters. You can also use the [Backspace] and [Del] keys to erase characters. The [Home] and then the [Crtl][End] keys can be used to delete the whole function.

Several computational errors can occur during the execution of the Trapezoidal Rule. A list of syntax and execution errors is given in Appendix IV. Screen 22 shows the appearance of the screen when an execution error is encountered. If these errors occur, you should check your data and input parameters to determine what went wrong. Note that the program gives you some clue by providing the value of X when the error was encountered. Similar errors might occur in other programs on the TOOLKIT.

The function is integrated using selection [2] from the Trapezoidal Rule menu. The program requests values for the lower and upper limits of integration and the number of segments. The upper limit must exceed the lower limit, otherwise an error message is displayed. The maximum number of segments is 9999. The program computes and displays the integral in the box at the bottom of Screen 23. The Trapezoidal Rule menu is activated again by entering [Return].

Discreet data can also be integrated by using the TOOLKIT. Selection [3] can be used to enter new data, but this destroys any old data. We will continue the example from the previous problem by integrating the data

from the spreadsheet using selection [4] from the menu (Screen 24). Various values from within the table could be changed using the arrow keys and the [PgUp], [PgDn], [Ins], and [Del] keys. However the data must **always** appear in ascending order of X. When all the data are input to the table, the [End] key results in an automatic calculation of the integral, which is displayed in the box at the bottom of the screen.

Plots that portray the integral as the area under a curve defined by either the function of X or data are constructed with selection [5] from the Trapezoidal Rule menu. The plot dimensions are defined in a manner exactly the same as that used in the Linear Regression program except that the range of integration is used for default values for X. Screens 25 and 26 show plots generated by the program for both continuous functions of X and tabular data.

Note that the segments are shown on the plots. The function of X is shown for the entire plot range. The shaded portion under the curve represents the integral of the function over the limits of integration. This plot allows you to estimate the adequacy of the Trapezoidal Rule for various functions and step sizes.

You can return to the Trapezoidal Rule menu by entering [Return]. You can now enter a new function or data, modify data, or return to the TOOLKIT Main Menu to gain access to other programs.

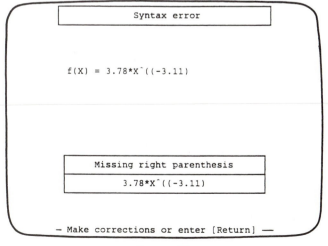

Screen 21

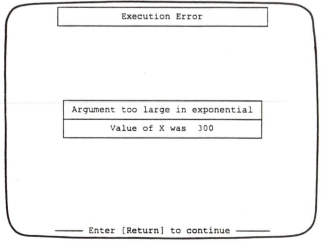

Screen 22

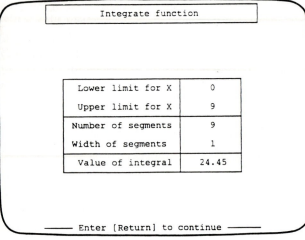

Integrate function	
Lower limit for X	0
Upper limit for X	9
Number of segments	9
Width of segments	1
Value of integral	24.45

Enter [Return] to continue

Screen 23

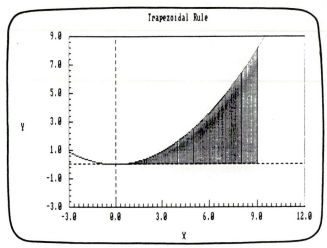

Screen 25

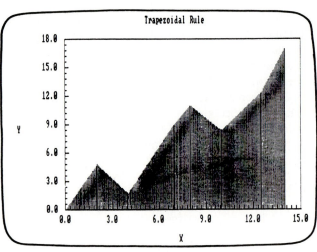

	X value	Y value
1	0	0
2	2	4.727893
3	4	1.729593
4	7	8.97096
5	8	10.96807
6	10	8.367937
7	12.5	12.30103
8	14	16.97182
9		
10		

Value of integral	104.3326

Enter [Return] to continue

Screen 24

Screen 26

STATISTICS

The Statistics program computes statistical characteristics of data and plots histograms. The menu for Statistics is activated by selection [5] from the TOOLKIT Main Menu (Screen 2). The Statistics Main Menu is shown in Screen 27. This menu allows you to analyze new data, modify old data, or generate a histogram. Previously saved data may be read from a disk using selection [10] from the Main Menu.

New data are entered by selection [1]. This action will result in the loss of the data we generated in the Spreadsheet. Let us proceed in this manner. Note that any time old data are in danger of being lost, the TOOLKIT provides a warning message. Screen 28 shows a display of two tables following the selection [1]. The left-hand table accepts the values for X that you supply. The right-hand table displays computations of various statistical measures of the average and distribution of the data. As you input, insert, or delete data from the left-hand table, the components of the right-hand table are automatically updated. Sometimes illegal computations may arise, such as the determination of the geometric mean of a negative number or the variance of one X value. In these cases the program gives an "undefined" message rather than the usual numerical result.

Screen 29 shows example data and calculations. A data input session is normally terminated by the [End] key. This returns you to the Statistics menu. The [Esc] key can also be used at any time to either abort a data entry or to retreat to a previous menu. You now can modify the data using selection [2].

A histogram is generated by using selection [3] from the Statistics menu. Screen 30 shows a table with convenient data values. Default values for X and Y are generated from the data in Screen 29. Convenient plotting divisions are obtained on the X axis if you select a range for X that is an even multiple of 5. The Y values represent the frequency, or number, of values that occur in a given interval of X. This interval of X is called the *Bar width*. *Bar start* represents the value of X where the bars begin. Bars are constructed between *Bar start* and the maximum value of X. If a value of Y occurs for an X that falls exactly on the division between two bars, it is counted as a member of the higher, or right-hand bar. Screen 31 shows the histogram that results from the data in Screen 30.

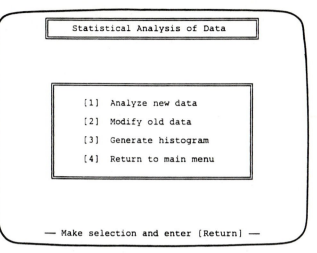

Screen 27

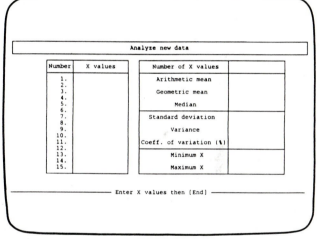

Screen 28

Analyze new data

Number	X values
1.	6
2.	5
3.	6
4.	7
5.	6
6.	4
7.	8
8.	6
9.	5
10.	7
11.	3
12.	8
13.	2
14.	9
15.	6

Number of X values	15
Arithmetic mean	5.866667
Geometric mean	5.511905
Median	6
Standard deviation	1.884776
Variance	3.552381
Coeff. of variation (%)	32.12687
Minimum X	2
Maximum X	9

———— Enter X values then [End] ————

Screen 29

Generate histogram

Parameter	Value
Minimum X	0
Maximum X	10
Minimum Y	0
Maximum Y	6
Bar width	1
Bar start	.5

———— Enter parameters then [End] ————

Screen 30

You can use various values of the plot parameters in Screen 30 to construct histograms that display the distribution of your data in different ways. Screens 32 and 33 show two more histograms that describe the same data. You will get an error message at the bottom of the screen if your maximum value for X does not exceed the minimum value. An error message will also be given if your *Bar width* is not a positive value.

You get back to the Statistics menu by entering [Return] or [Esc]. Here you can enter new data, modify old data, modify the histogram, or return to the TOOLKIT Main Menu to gain access to other programs.

GRAPHICS

The Graphics program on the TOOLKIT is different from the IBM system program called GRAPHICS, although they have the same name. The TOOLKIT Graphics program can be used to plot a function of X or data on either linear, semilog, or log-log scales.

The Graphics program on the TOOLKIT is activated by entering selection [6] from the TOOLKIT Main Menu (Screen 2). The Graphics program menu is shown in Screen 34. This menu allows you to enter a function, enter new data, modify old data, or plot the function and data. Previously saved functions and data may be read from a disk using selection [10] from the Main Menu.

The function you previously input from the Trapezoidal Rule is shown following selection [1]. This function can be erased using the [Home] and then the [Ctrl][End] keys. A new function is entered in a manner exactly the same as in the Trapezoidal Rule (as shown in Screen 35). When the complete function is entered, press [Return].

New data points are entered in a manner exactly the same as in the Linear Regression method by using selection [2]. This action will result in the loss of the data we generated in the Statistics program. New data are shown in Screen 36.

Let us now proceed with selection [4] to plot the new function and data found on Screens 35 and 36. The program prompts you for the type of plot, the dimensions of the plot, and the plot labels in Screen 37. First, note that the wide horizontal cursor is used to select among linear

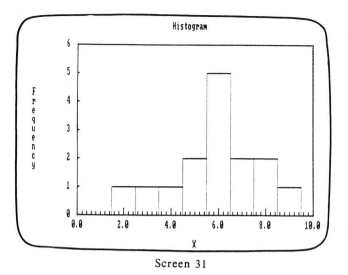

Screen 31

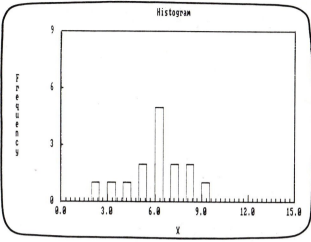

Screen 33

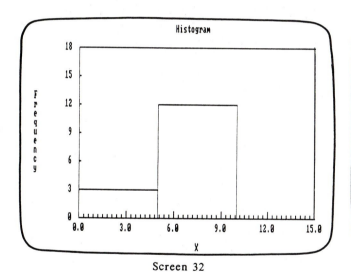

Screen 32

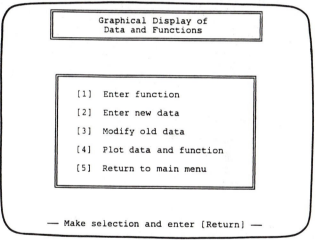

Screen 34

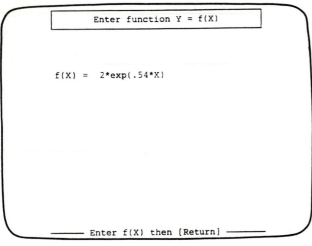

```
┌─────────────────────────────────────┐
│      ┌─────────────────────────┐     │
│      │  Enter function Y = f(X) │     │
│      └─────────────────────────┘     │
│                                      │
│                                      │
│   f(X)  =  2*exp(.54*X)              │
│                                      │
│                                      │
│                                      │
│                                      │
│       ──── Enter f(X) then [Return] ────  │
└─────────────────────────────────────┘
```

Screen 35

```
┌─────────────────────────────────────┐
│      ┌─────────────────────────┐     │
│      │     Enter new data       │     │
│      └─────────────────────────┘     │
│   ┌──────┬──────────┬──────────┐    │
│   │      │ X value  │ Y value  │    │
│   ├──────┼──────────┼──────────┤    │
│   │  1   │    .5    │   2.1    │    │
│   │  2   │   1.2    │   3.9    │    │
│   │  3   │    2     │   6.2    │    │
│   │  4   │    3     │   9.9    │    │
│   │  5   │    4     │  17.7    │    │
│   │  6   │   4.5    │  22.2    │    │
│   │  7   │   3.5    │  12.1    │    │
│   │  8   │          │          │    │
│   │  9   │          │          │    │
│   │ 10   │          │          │    │
│   └──────┴──────────┴──────────┘    │
│       ──── Enter data then [End] ────  │
└─────────────────────────────────────┘
```

Screen 36

semilog, and log-log plots. The cursor is moved horizontally by using the left and right arrow keys. You enter [Return] when the cursor is located on the type of plot you wish to generate. Next, you use a similar procedure to plot either the points alone, the function alone, or both. Now the Graphics program constructs the bottom of Screen 37 and requests information on the plot dimensions, labels for the axes, and a title. Screen 37 shows plot parameters that define the limits of X and Y for a linear plot of both the data and the function. Screen 38 shows a similar table for a semilog plot. In this case the vertical scale is determined by the logarithm of the minimum value of Y and the number of log cycles. A maximum of six logarithmic cycles can be plotted. Default values for the plot parameters may be shown on the screen. These default values may be the minimum and maximum values of X and Y from the data, or the values used for a previous plot.

A plot parameter input session is completed using the [End] key. An error message is given if you enter inappropriate values for the minimum amd maximum values of either X or Y, if you give a negative or zero value for the number of log cycles, or if you fail to enter all values. After acceptable plotting parameters are entered, plots are automatically constructed as shown in Screens 39 and 40. These plots correspond to the data on Screens 37 and 38.

If you wish to modify the appearance of a plot, simply enter [Return] or [Esc] to bring you back to the Graphics menu. Select [4] again and change the appropiate plot parameters. Screen 41 shows incorrect parameters for a log-log plot and an associated error message. Screens 42 and 43 show corrected parameter values and a plot of the function on Screen 35 after syntax error corrections are made. You can get back to the Graphics menu by entering [Return]. This menu allows you to modify the function, data, or plot. You can also return to the TOOLKIT Main Menu to gain access to other programs.

BISECTION METHOD

Let us say that you are given the function $f(X)=0=X^2-X-6$. You would like to use the Bisection Method program to determine its roots. The program for

Plot data and function		
Linear	Semi-log	Log-log
Points	Function	Both
Parameter		Value
Minimum X		0
Maximum X		5
Minimum Y		0
Maximum Y		24
X axis label	X	
Y axis label	Y	
Plot Title	EXAMPLE	

—— Enter parameters then [End] ——

Screen 37

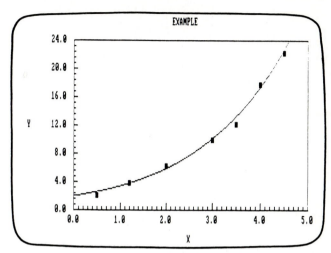

Screen 39

Plot data and function		
Linear	Semi-log	Log-log
Points	Function	Both
Parameter		Value
Minimum X		0
Maximum X		5
Log (min Y)		0
Y cycles		2
X axis label	X	
Y axis label	Y	
Plot Title	EXAMPLE	

—— Enter parameters then [End] ——

Screen 38

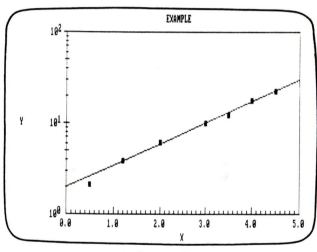

Screen 40

Plot data and function

Linear	Semi-log	Log-log
Points	Function	Both

Parameter	Value
Minimum X	0
Maximum X	1
Log (min Y)	0
Y cycles	8
X axis label	X
Y axis label	Y
Plot Title	EXAMPLE

——— Cycles must be from 1 to 6 ———

Screen 41

Plot data and function

Linear	Semi-log	Log-log
Points	Function	Both

Parameter	Value
Minimum X	0
Maximum X	1
Log (min Y)	0
Y cycles	3
X axis label	X
Y axis label	Y
Plot Title	EXAMPLE

——— Enter parameters then [End] ———

Screen 42

the Bisection Method on the TOOLKIT is activated by entering selection [7] from the Main Menu (Screen 2). The Bisection Method menu is shown in Screen 44. This menu allows you to enter a function, solve for roots, or plot a function. A previously saved function may be read from a disk using selection [10] from the Main Menu.

The first task is to enter the function using selection [1]. However, you must first delete the previous function using the [Home] and then the [Ctrl][End] keys. When the function has been properly input into the program, you can get back to the Bisection Method menu by entering [Return]. Now you can either plot the function to help you visually estimate the roots or proceed directly to solve for a root.

A plot-parameter input session is initiated using selection [3] and is performed exactly as in the Linear Regression program. After proper plot parameters are entered, a plot is automatically constructed as shown in Screen 45. If you wish to change the appearance of the plot, simply enter [Return] to bring you back to the Bisection Method menu. Select [3] again and change the appropriate plot parameters. The plot displayed in Screen 45 shows that the function has two roots where f(X) crosses the X axis. These values are approximately -2 and 3. These observations can be used to establish upper and lower bounds for the root as required to implement the Bisection Method.

Roots of f(X) are computed by entering selection [2] from the Bisection Method menu. Default values for the maximum iterations and error are provided by the program. Modifications of these defaults, along with upper and lower bounds for the root, are shown in Screen 46. The program will give you an error message if any of these inputs are not defined. The program also responds with an appropriate error message if you enter a zero value for the iterations or error. The maximum number of iterations is 9999. Another error message appears if your bounds for the root are equal or if the lower bound exceeds the upper bound. You are also notified with a message if you select upper and lower bounds that do not bracket a root or if you bracket an even number of roots.

After receiving proper inputs values and [End], the program proceeds immediately to calculate the root in the selected interval. The results for our problem are displayed in Screen 47. Note that both the actual number

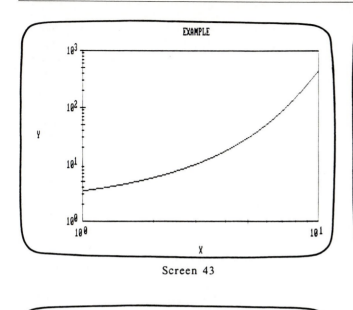

Screen 43

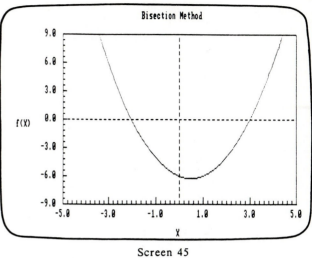

Screen 45

```
┌─────────────────────────────────────┐
│  ┌───────────────────────────────┐  │
│  │    Bisection Method for       │  │
│  │    Roots of Equations         │  │
│  └───────────────────────────────┘  │
│                                     │
│  ┌───────────────────────────────┐  │
│  │                               │  │
│  │   [1]   Enter function        │  │
│  │                               │  │
│  │   [2]   Solve for root        │  │
│  │                               │  │
│  │   [3]   Plot function         │  │
│  │                               │  │
│  │   [4]   Return to main menu   │  │
│  │                               │  │
│  └───────────────────────────────┘  │
│                                     │
│ — Make selection and enter [Return] — │
└─────────────────────────────────────┘
```

Screen 44

```
┌───────────────────────────────────────┐
│   ┌─────────────────────────────────┐ │
│   │         Solve for root          │ │
│   └─────────────────────────────────┘ │
│                                        │
│   ┌─────────────────────────┬───────┐ │
│   │  Maximum iterations     │  25   │ │
│   │  Actual iterations      │       │ │
│   ├─────────────────────────┼───────┤ │
│   │  Maximum error (%)      │  .1   │ │
│   │  Actual error (%)       │       │ │
│   ├─────────────────────────┼───────┤ │
│   │  Lower bound for root   │   0   │ │
│   │  Upper bound for root   │  10   │ │
│   ├─────────────────────────┼───────┤ │
│   │  Value of root          │       │ │
│   │  f(X) at root           │       │ │
│   └─────────────────────────┴───────┘ │
│                                        │
│  ──── Enter parameters then [End] ──── │
└───────────────────────────────────────┘
```

Screen 46

of iterations and the actual error are shown. The computations terminate when either the iteration or error criterion is met or the value of the function evaluated at the estimated root is zero. For our case, the error criterion is met first. The value of the root is displayed with a highlight. Finally, the value of f(X) at the estimated value of the root is shown as a check. This result should be close to zero. Once a root has been determined, we can display the solution process by selection [3] from the Bisection Method menu. After entering the proper plot parameters, the upper and lower bounds are illustrated on the screen as two bars. An arrow is used to represent the estimated root. Press any key to continue the iteration process. The final root is shown with an arrow and vertical line as shown in Screen 48.

After entering [Return], you are back in the Bisection Method menu, where you can recompute the root of f(X) with different iteration or error criterion or upper and lower bounds. Alternatively, you can change the function or return to the TOOLKIT Main Menu to gain access to other programs.

GAUSS ELIMINATION

The eighth program on the TOOLKIT is Gauss Elimination. It is not linked to any of the other programs. This program can be used to solve from 1 to a maximum of 15 simultaneous linear algebraic equations. The menu for Gauss Elimination is displayed by selection [8] from the TOOLKIT Main Menu (Screen 2). This menu (Screen 49) allows you to enter the coefficients for a new system of equations, modify a previously entered system, solve the system of equations, or check the accuracy of the calculated solutions. A previously saved system may be read from a disk using selection [10] from the Main Menu.

A new system is entered by selection [1]. If you try to modify an old system, solve the system or check the accuracy before a system has been entered, an appropriate error message is displayed. Let us say that you want to solve the following system of equations:

$$2X_1 + X_2 - 6X_3 = 10$$
$$X_1 + X_2 - X_3 = 5$$
$$-3X_1 + 5X_2 - 4X_3 = -8$$

The solution of this system is given by values of X_1, X_2, and X_3 that simultaneously satisfy the above equations.

The number of equations is input as shown in Screen 50. If you attempt to input more than 15 equations, an error message is displayed. After you enter the number of equations to be solved, the program proceeds immediately to Screen 51 for input of the system coefficients. The program automatically defines and displays the indices of the system coefficients in the left-hand column of the screen. The right-hand column contains space for you to input coefficient values. The coefficients of the first equation are entered by typing values for a(1,1), a(1,2), a(1,3), and c(1). Note that the up and down arrow keys can be used to move the cursor to any location and conveniently modify table entries. The [PgDn] key advances the page and sets up the screen for input of the second equation. Note that the indices in the left-hand column are automatically changed. The third equation is entered in a similar fashion. Any coefficient in the system can be conveniently changed by using the [PgDn], [PgUp], and arrow keys. A coefficient input session is completed by entering [End], which returns you to the Gauss Elimination menu. An error message is displayed if you fail to input values for all the system coefficients.

Let us solve for the unknowns with selection [3]. The solution algorithm begins immediately, and after a brief period for computation, the solutions appear on the screen as shown in Screen 52. An appropriate message is given if your system does not have a unique solution. The accuracy of the computed results can be checked by entering [Return] to get back to the menu and then choosing selection [4]. Screen 53 shows the results of the accuracy check. The accuracy of the solutions is evaluated by calculating the left-hand side of each equation with the recently computed values of X. This calculated value is then compared to the input value on the right-hand side of each equation. This capability of the TOOLKIT is provided because the solutions of some systems are subject to considerable round-off error. Systems of this type are called *ill-conditioned*. The computed values of X are usually satisfactory if the calculated values in the left-hand column are close to the known values in the right-hand column.

After entering [Return], you are back in the Gauss Elimination menu. You can now enter a new system,

Solve for root	
Maximum iterations	25
Actual iterations	12
Maximum error (%)	.1
Actual error (%)	8.1367E-02
Lower bound for root	0
Upper bound for root	10
Value of root	3.000488
f(X) at root	2.4414E-03

—— Enter [Return] to continue ——

Screen 47

Gauss Elimination Method for Systems of Linear Equations		

[1]   Enter new system

[2]   Modify old system

[3]   Solve system

[4]   Check accuracy

[5]   Return to main menu

— Make selection and enter [Return] —

Screen 49

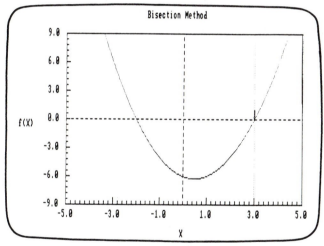

Bisection Method

Screen 48

Enter new system		

Parameter	Range	Input
Number of equations	1 to 15	3

—— Enter parameters then [Return] ——

Screen 50

```
┌──────────────────────────────────────┐
│        ┌──────────────────────┐       │
│        │    Enter new system   │       │
│        └──────────────────────┘       │
│                                        │
│    ┌─────────────┬──────────────┐     │
│    │ Coefficient │    Value      │     │
│    ├─────────────┼──────────────┤     │
│    │  a[ 1, 1]   │       2       │     │
│    │  a[ 1, 2]   │       1       │     │
│    │  a[ 1, 3]   │      -6       │     │
│    │  c[ 1]      │      10       │     │
│    └─────────────┴──────────────┘     │
│                                        │
│   ── Enter coefficients then [End] ──  │
└──────────────────────────────────────┘
```

Screen 51

```
┌──────────────────────────────────────┐
│        ┌──────────────────────┐       │
│        │     Check accuracy     │       │
│        └──────────────────────┘       │
│                                        │
│  ┌──────────┬────────────┬──────────┐ │
│  │ Equation │ Calculated │  Input   │ │
│  │  Number  │ Right Side │Right side│ │
│  ├──────────┼────────────┼──────────┤ │
│  │    1     │     10     │    10    │ │
│  │    2     │      5     │     5    │ │
│  │    3     │     -8     │    -8    │ │
│  └──────────┴────────────┴──────────┘ │
│                                        │
│    ── Enter [Return] to continue ──    │
└──────────────────────────────────────┘
```

Screen 53

```
┌──────────────────────────────────────┐
│        ┌──────────────────────┐       │
│        │       Solutions        │       │
│        └──────────────────────┘       │
│                                        │
│      ┌──────────┬──────────────┐      │
│      │ Variable │    Value      │      │
│      ├──────────┼──────────────┤      │
│      │   X[1]   │   4.102564    │      │
│      │   X[2]   │    .7179486   │      │
│      │   X[3]   │   -.1794872   │      │
│      └──────────┴──────────────┘      │
│                                        │
│    ── Enter [Return] to continue ──    │
└──────────────────────────────────────┘
```

Screen 52

```
┌──────────────────────────────────────┐
│        ┌──────────────────────┐       │
│        │   Euler's Method For   │       │
│        │  Differential Equations│       │
│        └──────────────────────┘       │
│                                        │
│   ┌────────────────────────────────┐  │
│   │ [1]  Enter differential equation│  │
│   │                                 │  │
│   │ [2]  Enter integration conditions│ │
│   │                                 │  │
│   │ [3]  Display solution as graph  │  │
│   │                                 │  │
│   │ [4]  Display solution as table  │  │
│   │                                 │  │
│   │ [5]  Return to main menu        │  │
│   └────────────────────────────────┘  │
│                                        │
│  ── Make selection and enter [Return] ──│
└──────────────────────────────────────┘
```

Screen 54

modify the old system, or return to the TOOLKIT Main Menu to gain access to other programs.

EULER'S METHOD

The ninth program on the TOOLKIT is Euler's Method. It is not linked to any of the other programs. This program can be used to obtain the solution of a first-order ordinary differential equation of the form dY/dX = f(X,Y).

The menu for Euler's Method (Screen 54) is displayed by selection [9] from the TOOLKIT Main Menu (Screen 2). This menu allows you to enter the right-hand side of the differential equation, enter integration conditions, and display the solution as a graph or a table. A previously saved function may be read from a disk using selection [10] from the Main Menu. The right-hand side of the differential equation is entered by selection [1]. If you attempt to solve or plot before having entered an equation, an appropriate error message is displayed. Note that the program assumes that Y is the dependent variable and that X is the independent variable. These cannot be modified, therefore you may have to change your variable names to accommodate this convention. Screen 55 shows an example for dY/dX = -Y + SIN (X).

Next, return to the Euler's Method menu by entering [Return] and use selection [2] to define the integration conditions. The first and second integration conditions are the initial and final values of X, which define the range of the integration. The third integration condition is the initial value of Y, which is the starting point for the dependent variable. The fourth and fifth integration conditions are the integration step size and the output interval. The last condition is a convenient option because you may not want to display all your calculations if your step size is small. The output interval must be a multiple of the step size. For example, if you are working with an integration range of 0 to 1 with a step size of 0.001, you might elect an output interval of 0.1 to limit the amount of calculations displayed. The program will give you an error message if the initial and final values of X are equal or if the initial value exceeds the final value. An error message is also displayed if the step size or output interval is zero

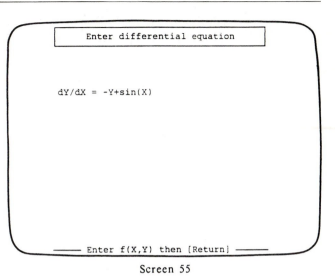

Screen 55

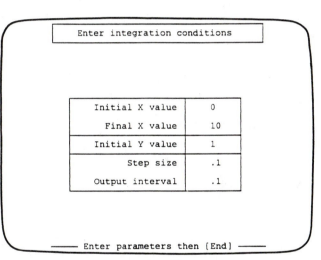

Screen 56

or greater than the range of the integration. Finally, an error message is displayed if the output interval is not a multiple of the step size. For example, if you are integrating from X = 0 to X = 10 with a step size of 2, an output interval of 3 is inappropriate. Suitable integration conditions for this example are shown in Screen 56.

After returning to Euler's Method menu, you can elect to compute and display the solution in either graphic or tabular form. The graphic form more conveniently provides an overall picture of the solution, whereas the tabular form is appropriate when exact numerical results are important. Graphic solutions are displayed by selection [3] with results as shown in Screen 57. The dimensions for the plot are established in a manner identical to that used in the Linear Regression program except that range of integration is used for the default values of X.

Numerical output is displayed using selection [4] from the Euler's Method menu. Some numerical results are shown in Screen 58. This numerical output is equivalent to the graphic output shown in Screen 57.

After entering [Return], you are back in Euler's Method menu. You can enter a new differential equation, modify the integration conditions, display the solution in graphic or tabular form, or return to the TOOLKIT Main Menu to gain access to other programs.

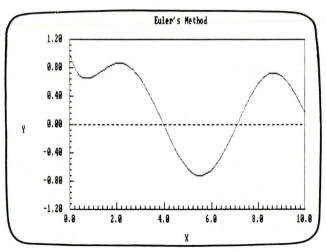

Screen 57

X value	Y value
0.0	1
0.1	.9
0.2	.8199834
0.3	.757852
0.4	.7116188
0.5	.6793987
0.6	.6594014
0.7	.6499255
0.8	.6493547
0.9	.6561549
1.0	.6688721
1.1	.686132
1.2	.7066395
1.3	.7291795
1.4	.7526174

Solutions

—— Enter [Return] to continue ——

Screen 58

APPENDIX 1
Functions of Special Keys for the TOOLKIT

Key	Function
[Esc]	Aborts data entry or retreats to previous menu
[ ]	Deletes a character using backspace
[ ]	Moves the cursor up for table entries or for menu selection
[ ]	Moves the cursor down for table entries or for menu selection
[ ]	Moves the cursor to the left for table entries or editing
[ ]	Moves the cursor to the right for table entries or editing
[End]	Completes a data-entry session
[Home]	Moves the cursor to the top of a table or menu or to the beginning of a function
[Del]	Deletes a character, table entry, or file
[Return]	Completes a data entry or advances the screen or cursor
[PgUp]	Moves data or coefficient screens backward one page
[PgDn]	Moves data or coefficient screens forward one page
[Shift][PrtSc]	Copies screen contents on printer
[Ctrl] [End]	Deletes all characters to the right of cursor
[Home], [Ctrl][End]	Deletes entire function
[F1]	Copies the contents of a spreadsheet cell to another cell or group of cells
[F2]	Activates edit mode for spreadsheet cell entries
[F3]	Scans spreadsheet and performs update calculations for all cell values
[F4]	Establishes automatic update mode for spreadsheet
[F5]	Establishes manual update mode for spreadsheet
[F6]	Updates spreadsheet cells by rows
[F7]	Updates spreadsheet cells by column
[F8]	Defines X data for integrated data set
[F9]	Defines Y data for integrated data set
[F10]	Toggle to display function of special keys

APPENDIX II
Available TOOLKIT Functions

Function	Equivalent
Sign of argument	SGN (X)
Integer part of argument	INT (X)
Absolute value of argument	ABS (X)
Square root	SQR (X)
Random number	RND (X)
Trigonometric sine (radian argument)	SIN (X)
Trigonometric cosine (radian argument)	COS (X)
Trigonometric tangent (radian argument)	TAN (X)
Trigonometric arc-tangent (radian values)	ATN (X)
Exponential	EXP (X)
Natural Logarithm (base e)	LOG (X)
Sums spreadsheet cells X to Y	SUM (X.Y)

APPENDIX III
Derived TOOLKIT Functions

Function	Equivalent
Secant	SEC (X)=1/COS(X)
Cosecant	CSC(X)=1/SIN(X)
Cotangent	COT(X)=1/TAN(X)
Inverse sine	ARCSIN(X)=ATN(X/SQR(-X*X+1))
Inverse cosine	ARCCOS(X)=ATN(X/SQR(-X*X+1))+1.5708
Inverse secant	ARCSEC(X)=ATN(SQR(X*X-1))+(SGN(X)-1*1.5708
Inverse cosecant	ARCCSC(X)=ATN(1/SQR(X*X-1))+(SGN(X)-1)*1.5708
Inverse cotangent	ARCCOT(X)=-ATN(X)+1.5708
Hyperbolic sine	SINH(X)=(EXP(X)-EXP(-X))/2
Hyperbolic cosine	COSH(X)=(EXP(X)+EXP(-X))/2
Hyperbolic tangent	TANH(X)=(EXP(X)-EXP(-X))/(EXP(X)+EXP(-X))
Hyperbolic secant	SECH(X)=2/(EXP(X)+EXP(-X))
Hyperbolic cosecant	CSCH(X)=2/(EXP(X)-EXP(-X))
Hyperbolic cotangent	COTH(X)=(EXP(X)+(EXP(-X))/-EXP(X)-EXP(-X))
Inverse hyperbolic sine	ARCSINH(X)=LOG(X+SQR(X*X+1))
Inverse hyperbolic cosine	ARCCOSH(X)=LOG(X+SQR(X*X-1))
Inverse hyperbolic tangent	ARCTANH(X)=LOG(1+X)/(1-X))/2
Inverse hyperbolic secant	ARCSECH(X)=LOG((SQR(-X*X+1)+1)/X)
Inverse hyperbolic cosecant	ARCCSCH(X)=LOG((SGN(X)*SQR(X*X+1)+1)/X)
Inverse hyperbolic cotangent	ARCCOTH(X)=LOG((X+1)/(X-1))/2
Common logarithm (base 10)	LOG10(X)=LOG(X)/ 2.302528

APPENDIX IV
TOOLKIT Error Messages

Syntax Error Messages	Execution Error Messages
Missing operand	Division by zero
Missing operator	Argument too large for modulo
Missing right parentheses	Square root of a negative number
Missing left parentheses	Logarithm of a negative number
Illegal character	Argument too large in exponential
	Formula too complex
	Illegal exponentiation
	Overflow

BIBLIOGRAPHY

Al-Khafaji, A. W., and J. R. Tooley, *Numerical Methods in Engineering Practice*, Holt, Rinehart and Winston, New York, 1986.

Allaire, P. E., *Basics of the Finite Element Method*, William C. Brown, Dubuque, IA, 1985.

Ames, W. F., *Numerical Methods for Partial Differential Equations*, Academic Press, New York, 1977.

Ang, A. H-S., and W. H. Tang, *Probability Concepts in Engineering Planning and Design, Vol. 1: Basic Principles*, Wiley, New York, 1975.

Atkinson, K. E., *An Introduction to Numerical Analysis*, Wiley, 1978.

Atkinson, L. V., and P. J. Harley, *An Introduction to Numerical Methods with Pascal*, Addison-Wesley, Reading, MA, 1983.

Baker, A. J., *Finite Element Computational Fluid Mechanics*, McGraw-Hill, New York, 1983.

Bathe, K.-J., and E. L. Wilson, *Numerical Methods in Finite Element Analysis*, Prentice-Hall, Englewood Cliffs, NJ, 1976.

Booth, G. W. and T. L. Peterson, "Nonlinear Estimation," I. B. M. Share Program Pa. No. 687 WLNL1 (1958).

Branscomb, L. M., "Electronics and Computers: An Overview," *Science*, **215**:755 (1982).

Brigham, E. O., *The Fast Fourier Transform*, Prentice-Hall, Englewood Cliffs, NJ, 1974.

Butcher, J. C., "On Runge-Kutta Processes of Higher Order," *J. Austral. Math. Soc.*, **4**:179 (1964).

Carnahan, B., H. A. Luther, and J. O. Wilkes, *Applied Numerical Methods*, Wiley, New York, 1969.

Chapra, S. C., and R. P. Canale, *Introduction to Computing for Engineers*, McGraw-Hill, New York, 1986.

Cheney, W., and D. Kincaid, *Numerical Mathematics and Computing*, 2d ed., Brooks/Cole, Monterey, CA, 1985.

Chirlian, P. M., *Basic Network Theory*, McGraw-Hill, New York, 1969.

Cooley, J. W., P. A. W. Lewis and P. D. Welch, "Historical Notes on the Fast Fourier Transform," *IEEE Trans. Audio Electroacoust.*, AU-15(2): 76–79 (1977).

Dantzig, G. B., *Linear Programming and Extensions*, Princeton University Press, Princeton, NJ, 1963.

Davis, P. J., and P. Rabinowitz, *Methods of Numerical Integration*, Academic Press, New York, 1975.

Dijkstra, E. W., "Go To Statement Considered Harmful," *Commun. ACM*, **11**(3):147–148 (1968).

Draper, N. R., and H. Smith, *Applied Regression Analysis*, 2d ed., Wiley, New York, 1981.

Enright, W. H., T. E. Hull, and B. Lindberg, "Comparing Numerical Methods for Stiff Systems of ODE's," *BIT*, **15**:10(1975).

Fadeev, D. K. and V. N. Fadeeva, *Computational Methods of Linear Algebra*, Freeman, San Francisco, 1963.

Ferziger, J. H., *Numerical Methods for Engineering Application*, Wiley, New York, 1981.

Forsythe, G. E., M. A. Malcolm, and C. B. Moler, *Computer Methods for Mathematical Computation*, Prentice-Hall, Englewood Cliffs, NJ, 1977.

Forsythe, G. E., and W. R. Wasow, *Finite-Difference Methods for Partial Differential Equations*, Wiley, New York, 1960.

Gabel, R. A., and R. A. Roberts, *Signals and Linear Systems*, Wiley, New York, 1987.

Gear, C. W., *Numerical Initial-Value Problems in Ordinary Differential Equations*, Prentice-Hall, Englewood Cliffs, NJ, 1971.

Gerald, C. F., and P. O. Wheatley, *Applied Numerical Analysis*, 3d ed., Addison-Wesley, Reading, MA, 1984.

Gladwell, J., and R. Wait, *A Survey of Numerical Methods of Partial Differential Equations*, Oxford University Press, New York, 1979.

Guest, P. G., *Numerical Methods of Curve Fitting*, Cambridge University Press, New York, 1961.

Hamming, R. W., *Numerical Methods for Scientists and Engineers*, 2d ed., McGraw-Hill, New York, 1973.

Hartley, H. O., "The Modified Gauss-Newton Method for Fitting Non-linear Regression Functions by Least Squares," *Technometrics* 3: 269–280 (1961).

Hayt, W. H., and J. E. Kemmerly, *Engineering Circuit Analysis*, McGraw-Hill, New York, 1986.

Heideman, M. T., D. H. Johnson and C. S. Burrus, "Gauss and the History of the Fast Fourier Transform," *IEEE ASSP Mag.*, 1(4): 14–21 (1984).

Henrici, P. H., *Elements of Numerical Analysis*, Wiley, New York, 1964.

Hildebrand, F. B., *Introduction to Numerical Analysis*, 2d ed., McGraw-Hill, New York, 1974.

Hornbeck, R. W., *Numerical Methods*, Quantum, New York, 1975.

Householder, A. S., *The Theory of Matrices in Numerical Analysis*, Blaisdell, New York, 1964.

Huebner, K. H., and E. A. Thornton, *The Finite Element Method for Engineers*, Wiley, New York, 1982.

Hull, T. E., and A. L. Creemer, "The Efficiency of Predictor-Corrector Procedures," *J. Assoc. Comput. Mach.*, **10**:291 (1963).

Isaacson, E., and H. B. Keller, *Analysis of Numerical Methods*, Wiley, New York, 1966.

Jacobs, D. (ed.), *The State of the Art in Numerical Analysis*, Academic Press, London, 1977.

James, M. L., G. M. Smith, and J. C. Wolford, *Applied Numerical Methods for Digital Computations with FORTRAN and CSMP*, 3d ed., Harper & Row, New York, 1985.

Keller, H. B., *Numerical Methods for Two-Point Boundary-Value Problems*, Wiley, New York, 1968.

Lapidus, L., and G. F. Pinder, *Numerical Solution of Partial Differential Equations in Science and Engineering*, Wiley, New York, 1981.

Lapidus, L., and J. H. Seinfield, *Numerical Solution of Ordinary Differential Equations*, Academic Press, New York, 1971.

Lapin, L. L., *Probability and Statistics for Modern Engineering*, Brooks/Cole, Monterey, CA, 1983.

Lawson, C. L., and R. J. Hanson, *Solving Least Squares Problems*, Prentice-Hall, Englewood Cliffs, NJ, 1974.

Luenberger, D. G., *Introduction to Linear and Nonlinear Programming*, Addison-Wesley, Reading, MA, 1973.

Lyness, J. M., "Notes on the Adaptive Simpson Quadrature Routine," *J. Assoc. Comput. Mach.*, **16**:483 (1969).

Malcolm, M. A., and R. B. Simpson, "Local Versus Global Strategies for Adaptive Quadrature," *ACM Trans. Math. Software*, **1**:129 (1975).

Maron, M. J., *Numerical Analysis, A Practical Approach*, Macmillan, New York, 1982.

Muller, D. E., "A Method for Solving Algebraic Equations Using a Digital Computer," *Math. Tables Aids Comput.*, **10**:205 (1956).

Na, T. Y., *Computational Methods in Engineering Boundary Value Problems*, Academic Press, New York, 1979.

Noyce, R. N., "Microelectronics," *Sci. Am.* **237**:62 (1977).

Oppenheim, A. V., and R. Schafer, *Digital Signal Processing*, Prentice-Hall, Englewood Cliffs, NJ, 1975.

Ortega, J., and W. Rheinboldt, *Iterative Solution of Nonlinear Equations in Several Variables*, Academic Press, New York, 1970.

Prenter, P. M., *Splices and Variational Methods*, Wiley, New York, 1975.

Press, W. H., B. P. Flanner, S. A. Teukolsky, and W. T. Vetterling, *Numerical Recipes: The Art of Scientific Computing*, Cambridge University Press, Cambridge, 1986.

Rabinowitz, P., "Applications of Linear Programming to Numerical Analysis," *SIAM Rev.*, **10**:121–159 (1968).

Ralston, A., "Runge-Kutta Methods with Minimum Error Bounds," *Match. Comp.*, **16**:431 (1962).

————, and P. Rabinowitz, *A First Course in Numerical Analysis*, 2d ed., McGraw-Hill, New York, 1978.

Ramirez, R. W., *The FFT, Fundamentals and Concepts*, Prentice-Hall, Englewood Cliffs, N. J., 1985.

Rice, J. R., *Numerical Methods, Software and Analysis*, McGraw-Hill, New York, 1983.

Ruckdeschel, F. R., *BASIC Scientific Subroutine*, Vol. 2, Byte/McGraw-Hill, Peterborough, NH, 1981.

Scarborough, J. B., *Numerical Mathematical Analysis*, 6th ed., Johns Hopkins Press, Baltimore, MD, 1966.

Scott, M. R., and H. A. Watts, "A Systematized Collection of Codes for Solving Two-Point Boundary-Value Problems," in *Numerical Methods for Differential Equations*, L. Lapidus and W. E. Schiesser, eds., Academic Press, New York, 1976.

Shampine, L. F., and R. C. Allen, Jr., *Numerical Computing: An Introduction*, Saunders, Philadelphia, 1973.

————, and C. W. Gear, "A User's View of Solving Stiff Ordinary Differential Equations," *SIAM Review*, **21**:1 (1979).

Simmons, E. F., *Calculus with Analytical Geometry*, McGraw-Hill, New York, 1985.

Stark, P. A., *Introduction to Numerical Methods*, Macmillan, New York, 1970.

Stasa, F. L., *Applied Finite Element Analysis for Engineers*, Holt, Rinehart and Winston, New York, 1985.

Stewart, G. W., *Introduction to Matrix Computations*, Academic Press, New York, 1973.

Swokowski, E. W., *Calculus with Analytical Geometry*, 2d ed., Prindle, Weber and Schmidt, Boston, 1979.

Taylor, J. R., *An Introduction to Error Analysis*, University Science Books, Mill Valley, CA, 1982.

Tewarson, R. P., *Sparse Matrices*, Academic Press, New York, 1973.

Thomas, G. B., Jr., and R. L. Finney, *Calculus and Analytical Geometry*, 5th ed., Addison-Wesley, Reading, MA, 1979.

Van Valkenburg, M. E., *Network Analysis*, Prentice-Hall, Englewood Cliffs, NJ, 1974.

Varga, R., *Matrix Iterative Analysis*, Prentice-Hall, Englewood Cliffs, NJ, 1962.

Vichnevetsky, R., *Computer Methods for Partial Differential Equations, Vol. 1: Elliptical Equations and the Finite Element Method*, Prentice-Hall, Englewood Cliffs, NJ, 1981.

————, *Computer Methods for Partial Differential Equations, Vol. 2: Initial Value Problems*, Prentice-Hall, Englewood Cliffs, NJ, 1982.

Wilkinson, J. H., and C. Reinsch, *Linear Algebra: Handbook for Automatic Computation*, Vol. II, Springer-Verlag, Berlin, 1971.

————, *The Algebraic Eigenvalue Problem*, Oxford University Press, Fair Lawn, NJ, 1965.

Wold, S., "Spline Functions in Data Analysis," *Technometrics*, 16(1): 1–11 (1974).

Yakowitz, S., and F. Szidarovsky, *An Introduction to Numerical Computation*, Macmillan, New York, 1986.

Young, D. M., *Iterative Solution of Large Linear Systems*, Academic Press, New York, 1971.

Zienkiewicz, O. C., *The Finite Element Method in Engineering Science*, McGraw-Hill, London, 1971.

Index